HANDBOOK OF ADHESIVE TECHNOLOGY

HANDBOOK OF ADHESIVE TECHNOLOGY

edited by

A. Pizzi
University of the Witwatersrand
Johannesburg, South Africa

K. L. Mittal
Skill Dynamics, an IBM Company
Thornwood, New York

MARCEL DEKKER, INC.   NEW YORK · BASEL · HONG KONG

Library of Congress Cataloging-in-Publication Data

Handbook of adhesive technology / edited by A. Pizzi, K. L. Mittal.
     p. cm.
  Includes bibliographical references and index.
  ISBN 0-8247-8974-1 (alk. paper)
  1. Adhesives. I. Pizzi, A. (Antonio). II. Mittal, K. L.
TP968.H347 1994
668'.3--dc20

94-4800
CIP

The publisher offers discounts on this book when ordered in bulk quantities. For more information, write to Special Sales/Professional Marketing at the address below.

This book is printed on acid-free paper.

Marcel Dekker, Inc.
270 Madison Avenue, New York, New York 10016

Current printing (last digit):
10 9 8 7 6 5 4 3 2

PRINTED IN THE UNITED STATES OF AMERICA

Preface

Bonding different materials together by means of an adhesive may appear to most people as a mundane occurrence. In reality a great deal of technology backs the apparently simple action of bonding. Thus, a complex and advanced technology, or series of technologies, has arisen to deal with adhesives and their applications in many fields. The diversity of substrates and the continuous introduction of new processes and materials has ensured that the field of adhesives technology is one of the more swiftly expanding manufacturing endeavors. Some excellent handbooks on adhesives already exist although there are very few indeed. However, the expansion and diversity of this field has by necessity limited the number of technologies and relevant aspects described in such volumes. This volume is no exception to such a trend.

The editors and authors do not pretend that overlaps with other similar works do not exist since basic background is often necessary to understand more advanced concepts. This volume however covers some aspects of technology that are not described in other volumes of this type. It also often looks at already reported technologies from a very different angle. It is hoped that such a volume will help to fill some of the technological gaps between the existing literature and industrial reality.

The volume is divided into four main sections, the first being an introductory overview. The remaining three sections are concerned with (1) fundamental aspects, (2) adhesive classes, and (3) some fields in which application of adhesives is very extensive. All the contributors are known specialists in their fields who practice their specialties on a daily basis. Their chapters are the results of considerable knowledge and experience in their particular niches.

It is a pleasant duty for the editors and authors, on completing a volume of this nature, to acknowledge the help willingly given by friends, colleagues, their companies, and their institutions. Without their help and encouragement most of the chapters presented would not have seen the light of day. Last, but definitely not least, our thanks go to Marcel Dekker, Inc. and its staff for originating this book, for their help and encouragement, and for prompting us to finish it.

A. Pizzi
K. L. Mittal

Contents

v

III. Adhesive Classes

IV. Applications of Adhesives

Contributors

Naim Akmal* Department of Chemistry, University of Cincinnati, Cincinnati, Ohio

Andrew G. Bachmann Dymax Corporation, Torrington, Connecticut

Monika Bauer Branch Lab Teltow, Fraunhofer-Institute of Applied Materials Research, Teltow, Germany

Melissa G. D. Baumann Wood Adhesives Science and Technology, Forest Products Laboratory, USDA–Forest Service, Madison, Wisconsin

P. R. Borgmeier College of Engineering, University of Utah, Salt Lake City, Utah

Anthony H. Conner Wood Adhesives Science and Technology, Forest Products Laboratory, USDA–Forest Service, Madison, Wisconsin

Eckhard H. Cordes Quality Assurance Technology, Mercedes-Benz AG, Bremen, Germany

Paul Cranley Coatings, Adhesives, and Sealants Laboratory, The Dow Chemical Company, Freeport, Texas

Dennis J. Damico Product Development and Synthesis, Lord Industrial Adhesives, Erie, Pennsylvania

D. W. Dahringer Materials and Technology Integration Research, AT&T Bell Laboratories, Murray Hill, New Jersey

Guy D. Davis Department of Surface Sciences, Martin Marietta Laboratories, Baltimore, Maryland

Sadhan K. De Rubber Technology Centre, Indian Institute of Technology, Kharagpur, India

---

Current affiliation: Teledyne Analytical Instruments, City of Industry, California

K. L. DeVries   College of Engineering, University of Utah, Salt Lake City, Utah

Neal J. Earhart   Ciba Additives, CIBA-GEIGY Corporation, Ardsley, New York

Ken Geddes   International Sales Department, Crown Berger Limited, Darwen, Lancashire, England

T. M. Goulding   Consultant, Johannesburg, South Africa

David N.-S. Hon   Department of Forest Resources, Clemson University, Clemson, South Carolina

Douglas Horsey   Ciba Additives, CIBA-GEIGY Corporation, Ardsley, New York

Brian K. Irons*   Columbia Research Laboratories, Madison, Wisconsin

John Johnston   Applications Engineering, tesa tape, inc., Charlotte, North Carolina

Fred A. Keimel   Adhesives & Sealants Consultants, Berkeley Heights, New Jersey

Jerome M. Klosowski   Sealants Science and Technology, Dow Corning Corporation, Midland, Michigan

Gerrit Knobloch   Additives Division, CIBA-GEIGY Corporation, Basel, Switzerland

Om S. Kolluri   HIMONT Plasma Science, Foster City, California

Alan L. Lambuth   Timber and Wood Products Division, Boise Cascade Corporation, Boise, Idaho

Dennis G. Lay   Coatings, Adhesives, and Sealants Laboratory, The Dow Chemical Company, Freeport, Texas

Loren D. Lower   Sealants Science and Technology, Dow Corning Corporation, Midland, Michigan

Alan M. Lyons   Materials and Technology Integration Research, AT&T Bell Laboratories, Murray Hill, New Jersey

Eliakim Mizrahi   Orthodontics/Dental Research Institute, University of the Witwatersrand, Johannesburg, South Africa

M. Nardin   Centre de Recherches sur la Physico-Chimie des Surfaces Solides, CNRS, Mulhouse, France

Eberhard W. Neuse   Department of Chemistry, University of the Witwatersrand, Johannesburg, South Africa

Ambu Patel   PEF Additives, CIBA-GEIGY Corporation, Ardsley, New York

Charles L. Pearson   Consultant, Swift Adhesives Division, Reichhold Chemicals, Inc., Downers Grove, Illinois

A. Pizzi†   Department of Chemistry, University of the Witwatersrand, Johannesburg, South Africa

---

Current affiliations:
*School of Pharmacy, University of Wisconsin, Madison, Wisconsin
†Faculté des Sciences, Université de Nancy, Epinal, France

William G. Repensek Department of Sales, Thiem Automotive Division, National Starch and Chemical Company, Oak Creek, Wisconsin

Richard D. Rich New Business Development, Loctite Corporation, Newington, Connecticut

Bryan H. River Wood Adhesives Science and Technology, Forest Products Laboratory, USDA–Forest Service, Madison, Wisconsin

Joseph R. Robinson School of Pharmacy, University of Wisconsin, Madison, Wisconsin

Jürgen Schneider Branch Lab Teltow, Fraunhofer-Institute of Applied Materials Research, Teltow, Germany

J. Schultz Centre de Recherches sur la Physico-Chimie des Surfaces Solides, CNRS, Mulhouse, France

Terry Sellers, Jr. Mississippi Forest Products Laboratory, Mississippi State University, Mississippi State, Mississippi

D. Kent Shaffer Department of Surface Sciences, Martin Marietta Laboratories, Baltimore, Maryland

A. M. Usmani Firestone, Carmel, Indiana

W. J. van Ooij* Armco Research & Technology, Middletown, Ohio

Peter Walker SCT, Atomic Weapons Establishment Plc, Aldermaston, Berkshire, England

Current affiliation: Department of Materials Science and Engineering, University of Cincinnati, Cincinnati, Ohio

I
Review Topics

1

Historical Development of Adhesives and Adhesive Bonding

Fred A. Keimel *Adhesives & Sealants Consultants, Berkeley Heights, New Jersey*

I. INTRODUCTION

The history of adhesives and sealants is closely related to the history of humankind. Some of what are thought of as relatively "new" uses of adhesives have their origins in ancient times, and although most of these materials have been subject to vast changes, others have been changed very little over time. As new materials are developed, a review of the history of uses can lead one to see where they might be applied to improve old applications, and sometimes to satisfy requirements of entirely new applications.

II. EARLY HISTORY OF ADHESIVES AND SEALANTS

"Insects, fish and birds know the art of producing mucous body fluids suitable for gluing. The load-carrying capacity of the hardened glue, as exemplified by egg-fastening and nest-building, is comparable to that of modern structural adhesives" [1, p. 1]. As humankind evolved, inquisitive persons observed and thought about insect and bird building and repair of nests with mud and clay. They encountered spider webs and naturally occurring "sticky" plant and asphaltic materials that entrapped insects, birds, and small mammals.

Unlike species that use an inherited instinct to perform a single task, human beings adopted the techniques of many species. They observed the natural phenomenon of sticky substances, then gathered and used these materials in locations away from their origins, exemplified today by the recently discovered Stone Age natives of South America's Amazon region and those in the interior of Borneo and New Guinea.

As rains fell, and then drying set in, many sticky materials regained their sticky properties, and some of the leaves used by ancient peoples to wipe sticky residues from their hands retained small quantities of water. Observing this, the first crude waterproof containers were manufactured using what we now call *pressure-sensitive adhesives*.

Our early ancestors used mud, clay, snow, and other natural materials to keep vermin, wind, and inclement weather out of their dens, warrens, caves, and other habitations. Today we use materials called *sealants* to perform similar functions in the construction and maintenance of modern buildings.

Straw and other vegetable material found its way into the muds and clays and reinforced them, forming the first crude *composites*. These materials later developed into bricks, which were in turn joined with the same or other materials used as mortars.

As human beings developed tools and weapons, sharp stones had to be fastened to handles to make axes and spears. Some of these were bound with vines, fibers, pieces of animal skin, or tendons or other body parts, and some had natural self-adhering properties to supplement the use of knots. To enhance the joining process, observing users soon smeared on sticky materials found locally.

When some natural materials fell on rocks heated by the sun, they softened and became sticky, and later hardened in the cool of the night. Observers made use of these natural *phase-change materials* as they chanced upon them. When lightning started fires, some materials melted and then cooled in interesting shapes. Observers, using the fires to harden their sharpened stick weapons, put out the fires by rubbing their sticks on the ground, and some contacted and melted resins, which when cooled, again hardened. Thus was born the technology we now call *hot-melt adhesives*.

Some of the other materials used by early human beings as adhesives are now called beeswax, rosin, rubber, shellac, sulfur, tar, and vegetable gums. Later, as people developed bows and arrows, it was found that feathers fastened to an arrow shaft helped to stabilize the arrow's flight. The same sticky or heat-softened materials soon supplemented the use of natural fibers to attach the feathers.

If Noah really did build an ark, the seams had to be sealed to keep out the water. And early human beings must have floated their possessions across bodies of water in bark or leaf containers with the seams sealed with sticky, waterproof materials.

Prehistoric peoples made pottery, and contrary to the Bible admonition in Jeremiah 19:11, ''as one breaks a potter's vessel, so that it can never be mended,'' they often used rosin to reassemble broken vessels to retain food buried with the dead, as we know from remnants found in archaeological digs.

Bituminous cements were used to fasten ivory eyeballs in statues in 6000-year-old Babylonian temples, and combinations of egg whites and lime were used by the Goths 2000 years ago to fasten Roman coins to wood, bonds that remain intact today [2, p. ix]. ''Bitumen was supposedly the mortar for the Tower of Babble; beeswax and pine tar were used in caulking Roman vessels that dominated the Mediterranean Sea'' [3, p. 62]. ''Plant gums and mucilage have been known and in use since very early times, reference being made to them in the Bible; they seem to have been of commercial value for several thousand years, especially in India, Asia, Africa, Australia, and China'' [4, p. vii].

In historic times the Egyptians used crude animal and casein glues to laminate wood for bows and furniture, including wood veneers, many of which have endured to modern times in that dry climate. To make these products it is likely that they were familiar with the production of *bonded abrasives* in the form of sand bonded to papyrus or cloth with animal glue. They developed starch pastes for use in bonding papyrus to textiles and to bond leather, and a plaster of calcined gypsum identical to today's plaster of Paris. Later the Greeks used slaked lime as a mortar, and both the Greeks and Romans mixed the lime with volcanic ash and sand to create a material still known as *pozzolanic cement*. This was used in the construction of the Roman Pantheon and Colosseum. Thus was born the rude beginnings of the art and science we now call adhesive bonding technology.

III. MODERN ADHESIVES AND SEALANTS

From the earliest days, the materials that we later called cements, glues, gums, mucilage, mortars, resins, pastes, and finally, adhesives and sealants, were used interchangeably. Only in modern times have we attempted to differentiate between adhesives and sealants. For the most part it has been a vain attempt, as many so-called adhesives also serve as sealants, and all sealants have adhesive properties. Some polyurethane and silicone sealants have strength properties similar to those of structural adhesives. Only seals, which have no adhesive properties (gaskets, O-rings, stuffing boxes, etc.), have been excluded from the technical definitions, but even here, seals and sealants are often combined in the literature and in use, as they often perform in similar applications. Mixtures of glycerin and litharge, alone and with additives, were used for many years [5, p. 358] as both an adhesive and a sealant, and are still used in the repair and restoration of older aquariums.

In his book *The Technology of Adhesives* [6], John Delmonte tells us that the first commercial glue plant was founded in Holland in 1690, that casein glues appear to have been manufactured in Germany and Switzerland in the early nineteenth century, and that the first U.S. patent (number 183,024) on a casein glue was issued in 1876. He mentions that starch adhesives were used on postage stamps when they were first issued in 1840, and that the first U.S. patent (number 61,991) on a dextrin adhesive was issued in 1867.

Before the advent of synthetic resin adhesives, semisynthetic cellulosic materials were developed, but when they were first dissolved in solvents and used as an adhesive is not clear from the literature. "Historically, the first thermoplastic synthetic adhesive (only partly synthetic) was the cellulose ester cellulose nitrate, often called nitrocellulose, and it is still one of the most important. Later, other esters such as the acetate were developed, as well as certain mixed esters" [1, p. 295].

Inorganic sodium silicate adhesives had minor commercial use in 1867, but it was not until 1900 that their use as a glue became of commercial importance as a replacement for starch in the production of corrugated and solid fiber paperboard [5, p. 279]. Very fine silicate frit mixed with phosphoric acid was used as a dental cement [5, p. 376] before the turn of the century. Magnesium chloride inorganic cements were used at least as far back as 1876 in hospital kitchen floors, as they provide resistance to greases and oils [5, pp. 355–356].

There is little agreement in the literature about the dates when various adhesives and sealants were first developed or used in a specific application. This is due to simultaneous developments in many parts of the world and the fact that references in the literature are almost exclusively from the more developed countries. Table 1 shows Delmonte's [6, p.4] viewpoint on the times of adhesive developments, up to the year of publication of his work. In the accompanying text he notes that "The developments are tabulated according to their first public disclosure, whether by patent or citation in technical literature."

Some experts trace the roots of the first modern adhesives technology to 1839, when Charles Goodyear discovered that a mixture of rubber and sulfur changed from a plastic to an elastic state when heated. In 1843 this process was termed *vulcanization* by Thomas Hancock, who is believed to have used his hard rubber (Ebonite) for bonding to metals, possibly discovering its effectiveness when trying to remove the mixture from metal containers used in its preparation. As it also bonded to natural rubber during vulcanization, it was used for many years as the only practical means of joining metal to rubber— but it had serious limitations as a thermoplastic [7, pp. 1–3].

Table 1 Chronological Developments of Adhesives in the United States

Year	Material
1814	Glue from animal bones (patent)
1872	Domestic manufacture of fish glues (isinglass)
1874	First U.S. fish glue patent
1875	Laminating of thin wood veneers attains commercial importance
1909	Vegetable adhesives from cassava flour (F. G. Perkins)
1912	Phenolic resin to plywood (Baekeland-Thurlow)
1915	Blood albumen in adhesives for wood (Haskelite Co.)
1917	Casein glues for aircraft construction
1920–1930	Developments in cellulose ester adhesives and alkyd resin adhesives
1927	Cyclized rubber in adhesives (Fischer-Goodrich Co.)
1928	Chloroprene adhesives (McDonald–B. B. Chemical Co.)
1928–1930	Soybean adhesives (I. F. Laucks Co.)
1930	Urea–formaldehyde resin adhesives
1930–1935	Specialty pressure-sensitive tapes: rubber base (Drew–Minnesota Mining & Mfg. Co.)
1935	Phenolic resin adhesive films (Resinous Products & Chemical Co.)
1939	Poly(vinyl acetate) adhesives (Carbide & Carbon Chemicals Co.)
1940	Chlorinated rubber adhesives
1941	Melamine–formaldehyde resin adhesives (American Cyanamid Corp.) and Redux by de Bruyne (Aero Research Ltd.)
1942	Cycleweld metal adhesives (Saunders-Chrysler Co.)
1943	Resorcinol–formaldehyde adhesives (Penn. Coal Products Co.)
1944	Metlbond adhesives (Havens, Consolidated Vultee-Aircraft Corp.)
1945	Furane resin adhesives (Delmonte, Plastics Inst.) and Pliobond (Goodyear Tire & Rubber Co.)

Source: Ref. 6.

The rubber cement used in early rubber-to-metal bonding was a simple dispersion of rubber sheeting in benzene and later toluene or other solvent. It was brushed on the metal and dried prior to contact with the bulk rubber to be bonded to the metal by vulcanization in a heated press. In 1862, Charles Sanderson, in a British patent (number 3288), specified that metal be brass plated by electrodeposition to obtain a strong bond to rubber [7, p. 3]. In 1911 the process was used in the production of rubber rolls, but was not used as a general commercial process until the 1920–1930 period.

Efforts to bond rubber to metal without the use of metal plating led to what is believed to be the first research efforts in *surface preparation* prior to adhesive bonding. Strong and durable bonds of rubber to metal were necessary for rubber shock mounts for automobiles in the late 1920s, but they were limited to proprietary formulations used on specific metals. In 1927 solvent-based thermoplastic rubber cements for metal-to-rubber bonding were prepared from rubber "cyclized" by treatment with sulfuric or other strong acids. With these rubber cements strong bonds could be made to either vulcanized or unvulcanized rubber.

Thermosetting solvent-based rubber cements for rubber-to-metal bonding, based on halogenated rubber compounds, first became available between World Wars I and II, but like much of the rubber-to-metal bonding technology, most of the work was proprietary and only glimpses of the technology involved can be found in the patent literature. The first use of natural rubber–based "tacky" adhesives on a backing is credited to Henry Day, who was issued a U.S. patent (number 3,965) in 1845. James Corbin of Minnesota Mining & Manufacturing Co. (now 3M Company), in a 1952 paper, "Practical Applications of Pressure-Sensitive Adhesives" [8, p. 139], states that 1925 is generally considered to be the birth date of the pressure-sensitive tape industry. He mentions that prior to that time, both cloth-backed surgical tapes and cloth-backed friction tape for use by electricians were in limited use. Both were apparently tried as masking tapes for the new two-toned automobiles, but failed to resist paint penetration and to strip clean. A crepe-paper backing, impregnated with animal glue and glycerin and coated with a pressure-sensitive adhesive (PSA), was developed in 1925.

Synthetic rubber, a dimethylbutadiene, was developed as a substitute for natural rubber in Germany during World War I and saw limited use as an adhesive. In the early 1930s, neoprene rubber (then called Duprene) became available to adhesive manufacturers in the United States, and shortly thereafter in Great Britain. Today, neoprene rubber adhesives are available as both thermoplastic and cross-linking systems in both solvent and emulsion formulations. Neoprene rubber is the major base resin for *contact adhesives*. A limited amount of neoprene rubber is also used in sealants.

It was not until the commercialization of synthetic plastics resins in the 1930s that an almost unlimited variety of base materials became available for compounding into adhesives and sealants. Most of the thermoplastic resins were soluble in organic solvents and were used as solvent adhesives for molded plastic articles of the same base composition and sometimes for other materials. Poly(vinyl chloride) (PVC), a thermoplastic developed in 1927, is used today in solvent formulations to bond PVC articles such as coated fabrics, films, foams, and pipe. In the early 1930s, phenolics came into importance as adhesive resins. Before that time they were used as coating varnishes [9, p. 239]. "About 1931 development of the use of a new phenolic resin for plywoods and veneers began" [9, p. 239].

Poly(vinyl acetate) was used as a solvent-based adhesive in the 1930s, and later as a hot melt, but was not of commercial importance until its introduction in the 1940s as an emulsion adhesive used mainly to bond paper and wood. Today, in emulsion form as a white glue, it is the most widely used thermoplastic adhesive worldwide. Vinyl acetate–ethylene (VAE) emulsion adhesives, with over 55% vinyl acetate content, were developed in the early 1950s but did not become of commercial importance in the United States until the mid-1960s.

Acrylic adhesives first appeared about 1937; "the acrylic resins may be considered as belonging to the vinyl family" [1, p. 305]. Today, acrylic adhesives appear in many forms: as both pressure-sensitive and non-pressure sensitive formulations in organic solvent and emulsion forms; as monomer and polymer cements; as anaerobics; as cyanoacrylates; as so-called reactive or "honeymoon" two-part systems; and as radiation curing formulations. "Commercial production of acrylic polymers began in the late 1920s, but it was not until 1958 that the first acrylic sealant was developed" [10, p. 226]. "The solvent-based acrylic sealants were first introduced to the construction industry in about 1960" [11, p. 121].

Urea–formaldehyde adhesives were patented in 1920 but were first commercialized around 1937. During World War II, starch was modified with urea resins to make both waterproof adhesives and impregnants for paper, which led in the 1940s to phenolic-impregnated paper for the first durable honeycomb core for lightweight rigid honeycomb panels.

Prior to World War II only in Germany was bonding to synthetic rubber being done. Polyisocyanate adhesives for rubber-to-metal bonding were developed under Otto Bayer in Germany during World War II. During the war there was widespread bonding of synthetic rubbers to metals in other countries, but documentation is almost nonexistent. It was only with the development of high-strength *toughened* phenolic thermosetting adhesives during World War II for metal-to-metal bonding that high-strength bonding of vulcanized rubber to metal became practical. Today, both vulcanized and unvulcanized rubber may be bonded to most materials of commercial importance, with a variety of room- or elevated-temperature setting- or curing-type adhesives.

During World War II, synthetic rubber and resin-modified phenolics were used to bond aluminum sheets (available only in $\frac{1}{16}$-in. thickness at that time) into billets from which airplane propellers were carved, thus replacing laminated wood, which often shattered on impact with a bullet. Similar adhesives were used to bond rubber to metal in a variety of vibration-damping applications. "The most successful widely known product of the new technology was the automotive bonded brake lining first introduced in 1947, and now regarded as a symbol of quality and integrity" [12, p. 490].

In a book entitled *Adhesives* [2] published in 1943, only six of 150 pages are devoted to synthetic adhesives, and many of these are combined with animal glue and other natural adhesives. There are chapters entitled "Flour Pastes and Starch Adhesives," "Dextrin Adhesives," "Casein Adhesives," "Vegetable Glues," "Animal Glues," "Sodium Silicate Adhesives," "Rubber Dispersions and Solutions as Adhesives," "Rosin and Its Derivatives," "Wax Adhesives," "Putties," and other chapters on adhesives from natural raw materials. In one chapter, "Miscellaneous Adhesives," there is a single formulation where a synthetic, poly(vinyl alcohol), is combined with starch. There is one chapter, "Gums and Resins (Natural and Synthetic)," with no mention of any synthetic material, and a single small chapter, "Adhesives Derived from Synthetic Material," where phenol–formaldehyde, urea–formaldehyde, and acrylic resins are mentioned, which suggests that they can be blended with animal glues to produce strong, waterproof adhesives. Also mentioned are poly(vinyl acetate), used alone or combined with ethyl cellulose. There is no mention of the rubber-modified phenolic adhesives developed during World War II, possibly because such formulations were classified as "secret."

One interesting omission in the book *Adhesives* is the use of poly(vinyl butyral) as the adhesive in safety glass. In 1936, Carbide and Carbon Chemicals Corporation first describes the use of poly(vinyl butyral) for laminating "high-test" safety glass [13, p. 165]. But in this book, poly(vinyl acetate), used as an adhesive for cellulose nitrate or cellulose acetate film, is mentioned as one laminating material for safety glass. This omission was particularly evident to the author of the present article, as poly(vinyl butyral) was a major product of my employer, E.I. DuPont, at their Plastics Division in North Arlington, New Jersey, in 1941. It had two major uses, as a safety glass laminating adhesive and as a box-toe softener for leather shoes.

To see just how far progress in adhesives and sealants extended during World War II, one has only to compare the book *Adhesives* with a book completed three years later, in December 1946. *The Technology of Adhesives* [6] had 516 pages, over 4000 index entries, and 1900 references. It covers in great detail the history, chemistry, theoretical background, testing, and technology of adhesives. It "seems" to have been written decades after the other volume. The term "pressure-sensitive adhesives," not found in the first volume, has 13 index entries, and similarly, "hot melts" has six index entries. Resorcinol–formaldehyde for wood bonding, introduced commercially in 1943, is covered in detail in the second volume, and an entire chapter, "Cementing of Organic Plastics," covers both thermoplastic and thermosetting materials, whereas the other volume mentions neither.

Again, this was of particular interest to the author, as in 1941 I helped with the formulation of a number of the solvent cements for acrylics used in the fabrication and repair of transparent acrylic aircraft enclosures. These adhesives, by the way, are still being sold by a number of vendors and are widely used by sign, incubator, and other fabricators of acrylic plastics. ASTM Committee D-20 on Plastics was organized in the United States in 1937–1938, and adhesives were a regular topic of discussion. From this committee came the nucleus of members who organized ASTM Committee D-14 on Adhesives in 1944.

Silicone adhesives were introduced commercially in 1944 [5, p. 213]. "In 1960 the silicone sealants were introduced to the construction industry" [11, p. 86]. Silicones are useful at both high and low temperatures and are available today as solvent-based moisture-curing adhesives, one-part moisture-curing adhesives and sealants, two-part curing adhesives and sealants, and pressure-sensitive adhesives.

According to one author, epoxy–phenolic adhesives for high-temperature applications were developed during World War II at Forest Product Laboratories in Madison, Wisconsin, and nitrile phenolic adhesives shortly after World War II [9, pp. 153, 156]. A patent for epoxy resins was applied for in Germany in 1934, "and the inventor disclosed that it could be hardened with equivalent amounts of amines, diamines, or polyamines and that it showed strong adhesion" [14, p. 8]. Epoxy resins are believed to have been commercialized in the United States first by the former Jones Dabney Company sometime after 1942.

Polyurethanes had their commercial beginning with the work of Otto Bayer in Germany in 1937. "In addition, American patent literature revealed that in the early 1940s much study was directed toward the use of diisocyanates as adhesive assistants, particularly in adhering elastomers to metals and fibers" [15, p. 4]. ". . .The following working definition of polyurethanes may be derived—they are polymers produced by addition reactions between polyisocyanates (difunctional or higher) and hydroxyl-rich compounds (at least two hydroxyl groups per molecule) such as glycols polyesters, polyethers, etc." [15, p. 3]. Today, polyurethane adhesives are available as solvent-based moisture-adhesives, thermoplastic hot melts, thermosetting systems, and emulsions.

During World War II, from 1939 to 1945, under the pressure of wartime shortages and the development of new and improved weapons of war, great progress was made in adhesives and adhesive bonding. However, due to wartime secrecy, much that went on has never been formally published. The homopolymer polyisobutylene was used in pressure-sensitive adhesives (PSAs) in 1939 as a replacement for natural rubber PSAs. Today, butyl rubber, the copolymer, has minor use in adhesives but is widely used in sealants. Polyvinyl acetals [poly(vinyl formal) and poly(vinyl butyral)] were used as flexibilizers for phenolic resins to make tough metal-bonding adhesives.

Styrene–butadiene rubber (SBR) adhesives, used to replace natural rubber adhesives, saw limited use during World War II, but commercialization took place during the 1950s. Today, in terms of monetary value, SBR adhesives are the most important adhesives in the United States. Their use in sealants is minor.

When glass-fiber reinforcements were used in organic resins in the 1940s, they lost much of their strength during prolonged exposure in water. In 1947 silanes were found to be effective primers or "coupling agents." "Silane monomers may be used in integral blends of fillers and liquid resins in the preparation of composites. The modified polymer 'adhesive' in this case is termed a matrix resin'' [16, p. 4]. In a chapter entitled ''The Chemistry of Tackifying Terpene Resins,'' we learn that terpene resins were first produced for adhesive applications in the early 1950s, first for pressure-sensitive adhesives and were then combined with wax in early synthetic resin hot melts [17, pp. 396–397].

Anaerobic materials were discovered in the 1940s but were not commercialized until the early 1950s as a new form of acrylic adhesives, termed ''anaerobics'' by their inventor, Vernon Krieble, then a professor at Trinity College in Hartford, Connecticut. Their first use was as ''threadlocking'' sealants, to lock nuts on threaded fasteners as a replacement for metal lock washers, and to lock threaded fasteners in tapped holes in metal parts. They were the first products termed ''sealants'' to have a viscosity lower than that of water. Today, such anaerobic adhesives and sealants are used in almost all mechanical equipment that is subject to vibration.

Polysulfide rubber was first produced in 1929, and the liquid polymers were used in sealants and as flexibilizers for epoxy adhesives around 1950. ''In 1952 the polysulfide sealant was introduced to the construction industry'' [11, p. 74]. ''In the 1950s the first butyl rubber caulks appeared in the construction market'' [11, p. 108] and ''latex caulks'' [vinyl acrylic and poly(vinyl acetate)] appeared sometime after 1956.

Polyester resins (alkyds) were commercialized for coatings use in 1926, and unsaturated polyesters were used as thermoset fiberglass composite matrix resins in the 1940s, but the early resins made poor adhesives. When flexibilized resins appeared in the 1950s, they were used as adhesives. Today, unsaturated polyesters are widely used as adhesives for thermoset plastics bonding, and even for metal bonding in most countries, but are seldom used as adhesives in the United States, where the more expensive epoxy adhesives are used in similar applications. The saturated polyesters, used as thermoplastic hot-melt adhesives, seem to have appeared in the literature first in the 1954–1957 period.

Polyethylene seems to have been mentioned first for use in a hot-melt adhesive in a 1954 patent application. Patent 2,894,925 was issued in 1959 [18, p. 62]. Today, polyethylene is the most important of the hot-melt adhesives in terms of tonnage, and is second, after ethylene–vinyl acetate (EVA), in dollar value in the United States. EVA (containing less than 55% vinyl acetate) adhesives, developed in the late 1960s, ''wet'' more substrates, had better low-temperature properties, and was compatible with more formulating ingredients—but all at a higher price.

By the early 1960s, the raw materials used in adhesive formulations were so numerous that in the first handbook on the subject, the editor said: ''It would be a virtual impossibility for any single volume to list all the ingredients which might conceivably be employed in an adhesive compound. Such a list would encompass practically every known chemical compound currently available in the United States'' [19, p. 11].

It was only in the late 1950s and early 1960s that raw material suppliers established marketing programs that specifically targeted the adhesives industry. Before that time, a person formulating adhesives of more than a single chemical type had to have an extensive knowledge of the product lines of hundreds of supplier firms. For this reason, almost all formulators had backgrounds in the coatings or rubber industries. Many of today's adhesive manufacturing firms reflect the earlier period by combining adhesives and sealants with the coatings or rubber areas of their businesses. Almost all synthetic resins used in adhesive formulations were used previously in coatings or rubber technology. The few people in the adhesives technical areas that were not from the coatings or rubber areas were mechanical engineers, who could evaluate the physical properties of the compounds developed by the chemists and the strength and durability of bonded assemblies.

From the chemists has come the *classification* of adhesives and sealants by chemical type, and from the mechanical engineers the classification as either "structural" or "nonstructural." Neither is a "pure" system, since many adhesives and sealants have more than a single chemical base resin, and many "structural"-based resin systems are used in nonstructural applications. In the chapter "Structural Adhesives" we are told that the term *structural adhesive* came into general use in the 1960–1970 period, but to this day all definitions are inadequate [9, Chap. 7]. "Adhesive manufacturers and their advertising departments now miss no opportunity to use, or abuse, the word. Companies which formerly sold urethane, acrylic, or anaerobic adhesives now call their products 'structural urethane,' 'structural acrylic,' or 'structural anaerobic' adhesives. Recently, this usage has further escalated, and these products are now called 'second generation' or 'third generation' structural adhesives" [9, pp. 133–134].

α-Cyanoacrylates were discovered in 1949, but "the adhesive properties of α-cyano-acrylates were first recognized during the investigation of a series of 1,1-disubstituted ethylenes, in the laboratories of Eastman Kodak" [20, p. 1]. Cyanoacrylate adhesives were first offered commercially by Eastman Kodak, their developer, in 1958 [9, p. 305].

Nylon epoxy adhesives were developed in the early 1960s. These extremely "tough" adhesives were used to laminate helicopter rotor blades and in honeycomb core-to-skin bonding [9, p. 157].

Urethane sealants were first used in in-plant assembly applications. "In 1960 the two-component urethane sealants were introduced to the construction industry. . . . The properties of the urethanes, in general, are intermediate between the polysulfides and the silicones. . . . The two names, 'urethane' and 'polyurethane,' are both used when referring to this class of sealants" [11, p. 93].

Thermoplastic rubber block copolymers, with completely new adhesive performance, were developed in 1965 [21]. The first commercial product was Shell Chemical's Kraton 101, of styrene–polybutadiene–styrene composition. This development led to the carboxy-terminated nitrile (CTBN) rubber modifiers used to flexibilize epoxy and other brittle resin adhesives in the late 1960s. Today, the thermoplastic rubber block copolymer adhesives are used in hot melt–, solvent- and water-based adhesives, and as hot melt– and solvent-based sealants. Major applications are as pressure-sensitive adhesives, construction adhesives and sealants, and general assembly adhesives.

Polymercaptan sealants were commercialized in 1969. "The polymercaptans are a new group of sealants just entering the sealant market. . . . The polymercaptans, with respect to their properties, are intermediate between the polysulfides and the urethanes" [11, p. 102].

Some adhesive-based resins are also used as additives, modifiers, or curing agents in other adhesive formulations. For example, starch is used in urea and phenol–formaldehyde adhesives as an extender. Poly(vinyl alcohol) is often combined with poly(vinyl acetate) to control solubility in warm or cold water in paper adhesives of the type used in schools. Polyamide higher-molecular-weight resins were first used commercially as solid hot-melt adhesives for leather shoe bonding in 1953, while the lower-molecular-weight liquid resins were first used in the mid-1950s as curing agents for epoxy resin adhesives. Today, many base resins are combined at the molecular level by the raw material suppliers. This is the case with acrylics, which have been combined with many other polymers to provide a large number of specialty resins for specific customers and applications.

In recent years a technique has been developed in which large quantities of adhesive resins called *compatibilizers* are used between layers of noncompatible extruded films, primarily in packaging applications. With this technology, up to seven layers of film (foil, paper, or plastic), up to three of which may be adhesive, may be combined to offer properties unlike that of any single-film material. The adhesive layer(s) may also contribute special properties to the multilayer composite film, in addition to acting as the compatibilizer. It is interesting to note that a particular plastic resin with adhesive properties may be used alone as an extruded plastic film, or in a multilayer composite film, where no compatibilizer is required, in which case it should not be counted in the adhesive statistics. This example is just one of hundreds where an intermediate or final user makes a decision to use a product sold as an adhesive, or a product sold as a nonadhesive as an adhesive, which make difficult the job of an analyst compiling industry statistics. Other common examples are the use of products labeled as coatings, encapsulants, dipping or potting compounds, modified concrete, paints, solvents, tar, thermoplastics or thermoset resins in many forms, varnishes, wheat and other flour, and so on, as adhesives. Conversely, products labeled as adhesives are often used as coatings or for other applications.

Today, even in the most developed countries, natural adhesives dominate the market because they are less expensive than synthetic-based materials, and they perform the intended function. Natural rubber is still the most widely used base material in pressure-sensitive adhesives. The first such modern uses were "flypaper" to trap flying insects, and medical bandages and tapes. Because of restrictions on the use of pesticides in many countries, both natural rubber and "sticky" synthetic materials have returned full circle to one of their original uses in trapping rodents and other small mammals. Natural rubber solvent solution adhesives are widely used throughout the world as general-purpose adhesives.

It is important to note that many adhesive technologists including brazing, soldering, and welding of metals as adhesive bonding [1, Chap. 10]. "Welding, brazing, soldering and gluing have flow processes as a common denominator. . . . Soldering is a true adhesive bonding method, as the flow process is restricted to the metallic adhesive" [2, p. 2]. This is not as unreasonable as might be thought, as a close study of the subjects show that there is much in common with other hot-melt joining of materials. It is just that the temperatures are often higher with the metals. However, indium and other low-temperature-melting metal alloys are often used interchangeably and at temperatures comparable to those of thermoplastic synthetic-resin hot-melt adhesives in joining metal to themselves and to other materials. FEP Teflon (copolymer of tetrafluoroethylene and hexafluoropropylene) and other high-temperature-melting thermoplastics are used today as hot-melt adhesives at temperatures equivalent to or exceeding those used for ordinary metal solders.

Inorganic adhesives and cements are also often classified differently by various experts. For example, sodium silicate is always classified as an adhesive when used in smaller quantities for bonding in electrical and electronic applications and for bonding paper and corrugated paperboard. But when used in larger quantities in furnace construction, they may be grouped with the portland and other hydraulic cements used in construction. Again, this makes it difficult to compile industry statistics. In the United States, the Department of Commerce has a Standard Industrial Classification (SIC) system for statistical purposes in which hydraulic cements are listed with stone, clay, and glass products as SIC 3241. Under chemicals and allied products are listed adhesives and sealants as SIC 2891 (including silicates) and dental adhesives under SIC 2844. One other class of materials, not usually considered as adhesives, that are used in large quantities

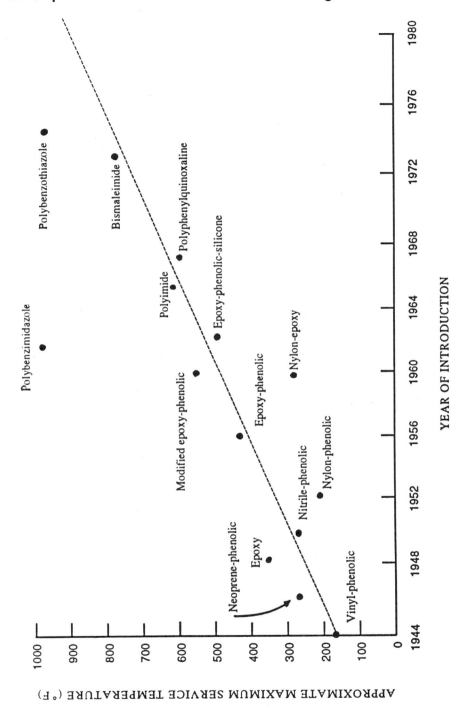

Figure 1 Sequence of development of heat-resistant adhesives. (Adapted from Ref. 23.)

worldwide for bonding thermoplastics are organic solvents. When they contain dissolved polymers, all adhesive technologists consider them as plastic cements. But when used alone, they are usually left out of the literature and the statistics. The best modern reference to their use as adhesives is Chapter 8 in the *Adhesives Technology Handbook* [22].

The history of the modern adhesives and sealants industry is closely tied to the development of the aircraft and aerospace industries. From the earliest flights to the most modern aerospace equipment, light weight has been one of the most vital considerations. Adhesive bonding was an ideal joining method for the early wood and textile aircraft, and today is the most important joining method for aluminum, titanium, and other metals in advanced military air- and spacecraft and some advanced commercial airplanes. Except for a very few very high-temperature brazed panels, all honeycomb panels are adhesive bonded.

Figure 1 shows the approximate maximum service temperatures for adhesives and the approximate year of introduction [23]. The maximum service temperatures of the highest-temperature adhesives are not indicative of usefulness for prolonged exposure at these temperatures but show that systems are available for certain applications. These adhesives tend to be brittle rather than tough and are usually much more difficult to apply than are the lower-temperature systems. Other newer adhesives are usually considered experimental rather than production systems.

One of the more interesting uses of modern adhesives and sealants is by museums in the repair and restoration of antiquities. Nitrocellulose-based adhesives are widely used in such applications, as are epoxies and polyurethanes. In the United States, the Guggenheim Museum has made exhaustive, expensive, and highly scientific evaluations of the effects on durability of such repairs on irreplaceable artifacts from the past. Thus adhesives and sealants have come full circle back to some of their original uses.

REFERENCES

1. R. Houwink and G. Salomon, eds. *Adhesion and Adhesives*, Vol. 1, 2nd ed., Elsevier, New York, 1965.
2. Felix Braude, *Adhesives*, Chemical Publishing Co., Brooklyn, N.Y., 1943.
3. Norbert M. Bikales, ed., *Adhesion and Bonding*, Wiley-Interscience, New York, 1971.
4. F. Smith, *The Chemistry of Plant Gums and Mucilage*, Reinhold, New York, 1959.
5. N. A. De Bruyne and R. Houwink, eds., *Adhesion and Adhesives*, Elsevier, London, 1951.
6. John Delmonte, *The Technology of Adhesives*, Reinhold, New York, 1947.
7. S. Buchan, *Rubber to Metal Bonding*, Lockwood, London, 1959.
8. *Adhesion and Adhesives: Fundamentals and Practice*, Society of Chemical Industry, London, and Wiley, New York, 1954.
9. Gerald L. Schneberger, ed., *Adhesives in Manufacturing*, Marcel Dekker, New York, 1983.
10. Adolfas Damusis, ed., *Sealants*, Reinhold, New York, 1967.
11. John Philip Cook, *Construction Sealants and Adhesives*, Wiley-Interscience, New York, 1970.
12. F. A. Keimel, in *Kirk-Othmer Encyclopedia of Chemical Technology*, Vol. 1, 3rd ed., Wiley, New York, 1978.
13. *ASTM Symposium on Plastics*, Feb. 22–23, 1944.
14. Robert L. Patrick, *Treatise on Adhesion and Adhesives*, Vol. 2, *Materials*, Marcel Dekker, New York, 1969.
15. Bernard A. Dombrow, *Polyurethanes*, 2nd ed., Reinhold, New York, 1965.
16. Edwin P. Plueddemann, *Silane Coupling Agents*, Plenum Press, New York, 1982.
17. Lieng-Huang Lee, ed., *Adhesion Science and Technology*, Plenum Press, New York, 1975.

18. Robert S. Willard, *Adhesive Patents 1955–1963*, Padric Publishing, Mountainside, N.J., 1964.

19. E. Patrick McGuire, ed., *Adhesives Raw Materials Handbook 1964*, Padric Publishing, Mountainside, N.J., 1964.

20. Henry Lee, ed., *Cyanoacrylate Resins: The Instant Adhesives*, Pasadena Technology Press, Pasadena, Calif., 1981.

21. J. T. Harlan, Jr., U.S. patent 3,239,478 (1965).

22. Arthur H. Landrock, *Adhesives Technology Handbook*, Noyes Publications, Park Ridge, N.J., 1985.

23. H. Schwartz, Structural adhesives development, in *Treatise on Adhesion and Adhesives*, Vol. 4, *Structural Adhesives* (National Materials Advisory Board, National Research Council, ed.), Marcel Dekker, New York, 1976.

II
Fundamental Aspects

2

Theories and Mechanisms of Adhesion

J. Schultz and M. Nardin *Centre de Recherches sur la Physico-Chimie des Surfaces Solides, CNRS, Mulhouse, France*

There are agents in nature able to make the particles of joints stick together by very strong attraction and it is the business of experimental philosophy to find them out. *Sir Isaac Newton*

I. INTRODUCTION

The adhesion phenomenon is relevant to many scientific and technological areas and has become in recent years a very important field of study. The main application of adhesion is bonding by adhesives, this technique replacing, at least partially, more classical mechanical attachment techniques such as bolting or riveting. It is considered to be competitive primarily because it allows us to save weight, to ensure a better stress distribution, and offers better aesthetics since the glue line is practically invisible. Applications of bonding by adhesives can be found in many industries, particularly in such advanced technical domains as the aeronautical and space industry, automobile manufacture, and electronics. Adhesives have also been introduced in such areas as dentistry and surgery.

Adhesives joints are not, however, the only applications of adhesion. Adhesion is involved whenever solids are brought into contact, as in coatings, paints, and varnishes; multilayered sandwiches; polymer blends; filled polymers; and composite materials. Since the final performance of these multicomponent materials depends significantly on the quality of the interface that is formed between the solids, it is understandable that a better knowledge of the adhesion phenomenon is required for practical applications.

Adhesion began to create real interest in scientific circles only about 50 years ago. At that time adhesion became a scientific subject in its own right but is still a subject in which empiricism and technology are slightly in advance of science, although the gap between theory and practice has been shortened considerably. In fact, the term *adhesion* covers a wide variety of concepts and ideas, depending on whether the subject is broached from a molecular, microscopic, or macroscopic point of view or whether one talks about formation of the interface or failure of the formed system. The term *adhesion* is therefore ambiguous, meaning both the establishment of interfacial bonds and the mechanical load required to break an assembly. As a matter of fact, one of the main difficulties in the study of adhesion mechanisms lies in the fact that the subject is at the boundary of several scientific fields, including macromolecular science, physical chemistry of surfaces and interfaces, materials science, mechanics and micromechanics of fracture, and rheology. Consequently, the study of adhesion uses various concepts, depending very much on one's field of expertise, and therefore treatment of the phenomena observed can be considerably different. This variety of approaches is emphasized by the fact that many theoretical models of adhesion have been proposed, which together are both complementary and contradictory:

1. Mechanical interlocking
2. Electronic theory
3. Theory of boundary layers and interphases
4. Adsorption (thermodynamic) theory
5. Diffusion theory
6. Chemical bonding theory.

Among these models, one usually distinguishes rather arbitrarily between mechanical and specific adhesion, the latter being based on the various types of bonds (electrostatic, secondary, chemical) that can develop between two solids. Actually, each of these theories is valid to some extent, depending on the nature of the solids in contact and the conditions of formation of the bonded system. Therefore, they do not negate each other and their respective importance depends largely on the system chosen.

II. MECHANISMS OF ADHESION

A. Mechanical Interlocking

The mechanical interlocking model, proposed by MacBain and Hopkins in 1925 [1], conceives of mechanical keying, or interlocking, of the adhesive into the cavities, pores, and asperities of the solid surface to be the major factor in determining adhesive strength. One of the most consistent examples illustrating the contribution of mechanical anchoring was given many years ago by Borroff and Wake [2], who have measured the adhesion between rubber and textile fabrics. These authors have clearly proved that penetration of the protruding fiber ends into the rubber was the most important parameter in such adhesive joints. However, the possibility of establishing good adhesion between smooth surfaces leads to the conclusion that the theory of mechanical keying cannot be considered to be universal. To overcome this difficulty, following the approach suggested primarily by Gent and Schultz [3,4], Wake [5] has proposed that the effects of both mechanical interlocking and thermodynamic interfacial interactions could be taken into account as multiplying factors for estimating the joint strength G:

$$G = (\text{constant}) \times (\text{mechanical keying component}) \times (\text{interfacial interactions component})$$

Therefore, according to the foregoing equation, a high level of adhesion should be achieved by improving both the surface morphology and physicochemical surface properties of substrate and adhesive. However, in most cases, the enhancement of adhesion by mechanical keying can be attributed simply to the increase in interfacial area due to surface roughness, insofar as the wetting conditions are fulfilled to permit penetration of the adhesive into pores and cavities.

Work by Packham and co-workers [6–9] has further stressed the notable role played by the surface texture of substrates in determining the magnitude of the adhesive strength. In particular, they have found [6] that high values of peel strength of polyethylene on metallic substrates were measured when a rough and fibrous type of oxide surface was formed on the substrate. More recently, Ward et al. [10–12] have emphasized the improvement in adhesion, measured by means of a pull-out test, between plasma-treated polyethylene fibers and epoxy resin. In that case, long-time plasma treatments create a pronounced pitted structure on the polyethylene surface, which can easily be filled by the epoxy resin by means of good wetting.

One of the most important criticisms of the mechanical interlocking theory, as suggested in different studies [9,13,14], is that improved adhesion does not necessarily result from a mechanical keying mechanism but that the surface roughness can increase the energy dissipated viscoelastically or plastically around the crack tip and in the bulk of the materials during joint failure. Effectively, it is now well known that this energy loss is often the major component of adhesive strength.

B. Electronic Theory

The electronic theory of adhesion was proposed primarily by Deryaguin and co-workers [15–19] in 1948. These authors have suggested that an electron transfer mechanism between the substrate and the adhesive, having different electronic band structures, can occur to equalize the Fermi levels. This phenomenon could induce the formation of a double electrical layer at the interface, and Deryaguin et al. have proposed that the resulting electrostatic forces can contribute significantly to the adhesive strength. Therefore, the adhesive–substrate junction can be analyzed as a capacitor. During interfacial failure of this system, separation of the two plates of the capacitor leads to an increasing potential difference until a discharge occurs. Consequently, it is considered that adhesive strength results from the attractive electrostatic forces across the electrical double layer. The energy of separation of the interface G_e is therefore related to the discharge potential V_e as follows:

$$G_e = \frac{h\epsilon_d}{8\pi} \left(\frac{\partial V_e}{\partial h} \right)^2 \tag{1}$$

where h is the discharge distance and ϵ_d the dielectric constant. Moreover, according to such an approach, adhesion could vary with the pressure of the gas in which the measurement is performed. Hence Deryaguin et al. have measured, by means of a peel test, the work of adhesion at various polymer–substrate interfaces, such as poly(vinyl chloride)–glass and natural rubber–glass or steel systems, in argon and air environments at various gas pressures. A significant variation in peel energy versus gas pressure was indeed evidenced and very good agreement between the theoretical values, calculated from Eq. (1), and the measured values of G_e was obtained whatever the nature of the gas used. However, several other analyses [5,20] have not confirmed these results and seem to indicate that the good agreement obtained previously was rather casual. According to Deryaguin's approach, the adhesion depends on the magnitude of the potential barrier at the substrate–adhesive interface. Although this potential barrier does exist in many cases (see, e.g., [21,22]), no clear correlation between electronic interfacial parameters and work of adhesion is usually found. Moreover, for systems constituted of glass substrate coated with a vacuum-deposited layer of gold, silver, or copper, von Harrach and Chapman [23] have shown that the electrostatic contribution to peel strength, estimated from the measurement of charge densities, can always be considered as negligible. Furthermore, as already mentioned, the energy dissipated viscoelastically or plastically during fracture experiments plays a major role on the measured adhesive strength, but it is not included conceptually in the electronic theory of adhesion. Finally, it could be concluded that the electrical phenomena often observed during failure processes are the consequence rather than the cause of high bond strength.

C. Theory of Weak Boundary Layers: Concept of Interphase

It is now well known that alterations and modifications of the adhesive and/or adherend can be found in the vicinity of the interface leading to the formation of an interfacial zone exhibiting properties (or properties gradient) that differ from those of the bulk materials. The first approach to this problem is due to Bikerman [24], who stated that the cohesive strength of a weak boundary layer (WBL) can always be considered as the main factor in determining the level of adhesion, even when the failure appears to be interfacial. According to this assumption, the adhesion energy G is always equal to the cohesive energy G_c(WBL) of the weaker interfacial layer. This theory is based primarily on probability considerations showing that the fracture should never propagate only along the adhesive–substrate interface for pure statistical reasons and that cohesive failure within the weaker material near the interface is a more favorable event. Therefore, Bikerman has proposed several types of WBLs, such as those resulting from the presence at the interface of impurities or short polymer chains.

Two main criticisms against the WBL argument can be invoked. First, there is much experimental evidence which shows clearly that purely interfacial failure does occur for many different systems. Second, although the failure is cohesive in the vicinity of the interface in at least one of the materials in contact, this cannot necessarily be attributed to the existence of a WBL. According to several authors [25,26], the stress distribution in the materials and the stress concentration near the crack tip certainly imply that the failure must propagate very close to the interface, but not at the interface.

However, the creation of interfacial layers has received much attention in the last years and has led to the concept of ''thick interface'' or ''interphase,'' widely used in adhesion science [27]. Such interphases are formed whatever the nature of both adhesive and substrate, their thickness being between the molecular level (a few angstroms or nanometers) and the microscopic scale (a few micrometers or more). Many physical, physicochemical, and chemical phenomena are responsible for the formation of such interphases, as shown from examples taken from our own recent work [28]:

1. The orientation of chemical groups or the overconcentration of chain ends to minimize the free energy of the interface [29]
2. Migration toward the interface of additives or low-molecular-weight fraction [30]
3. The growth of a transcrystalline structure, for example, when the substrate acts as a nucleating agent [31]
4. Formation of a pseudoglassy zone resulting from a reduction in chain mobility through strong interactions with the substrate [32]
5. Modification of the thermodynamics and/or kinetics of the polymerization or cross-linking reaction at the interface through preferential adsorption of reaction species or catalytic effects [33,34]

It is clear that the presence of such interphases can strongly alter the strength of multicomponent materials and that the properties of these layers must not be ignored in the analysis of adhesion measurements data. A complete understanding of adhesion, allowing performance prediction, must take into account potential formation of these boundary layers.

D. Adsorption (or Thermodynamic) Theory

The thermodynamic model of adhesion, generally attributed to Sharpe and Schonhorn [35], is certainly the most widely used approach in adhesion science at present. This theory is based on the belief that the adhesive will adhere to the substrate because of interatomic and intermolecular forces established at the interface, provided that an intimate contact is achieved. The most common interfacial forces result from van der Waals and Lewis acid–base interactions, as described below. The magnitude of these forces can generally be related to fundamental thermodynamic quantities, such as surface free energies of both adhesive and adherend. Generally, the formation of an assembly goes through a liquid–solid contact step, and therefore criteria of good adhesion become essentially criteria of good wetting, although this is a necessary but not sufficient condition.

In the first part of this section, wetting criteria as well as surface and interface free energies are defined quantitatively. The estimation of a reversible work of adhesion W from the surface properties of materials in contact is therefore considered. Next, various models relating the measured adhesion strength G to the free energy of adhesion W are examined.

1. Wetting Criteria, Surface and Interface Free Energies, and Work of Adhesion

In a solid–liquid system, wetting equilibrium may be defined from the profile of a sessile drop on a planar solid surface. Young's equation [36], relating the surface tension γ of materials at the three-phase contact point to the equilibrium contact angle θ, is written as

$$\gamma_{SV} = \gamma_{SL} + \gamma_{LV} \cos \theta \tag{2}$$

The subscripts S, L, and V refer, respectively, to solid, liquid, and vapor phases, and a combination of two of these subscripts corresponds to the given interface (e.g., SV corresponds to a solid–vapor interface). The term γ_{SV} represents the surface free energy of the substrate after equilibrium adsorption of vapor from the liquid and is sometimes lower than the surface free energy γ_S of the solid in vacuum. This decrease is defined as the spreading pressure π ($\pi = \gamma_S - \gamma_{SV}$) of the vapor onto the solid surface. In most cases, in particular when dealing with polymer materials, π could be neglected and, to a first approximation, γ_S is used in place of γ_{SV} in wetting analyses. When the contact angle has a finite value ($\theta > 0°$), the liquid does not spread onto the solid surface. On the contrary, when $\theta = 0°$, the liquid totally wets the solid and spreads over the surface spontaneously. Hence a condition for spontaneous wetting to occur is

$$\gamma_S \geq \gamma_{SL} + \gamma_{LV} \tag{3}$$

or

$$S = \gamma_S - \gamma_{SL} - \gamma_{LV} \geq 0 \tag{4}$$

the quantity S being called the spreading coefficient. Consequently, Eq. (4) constitutes a wetting criterion. It is worth noting that geometrical aspects or processing conditions, such as surface roughness of the solid and applied external pressure, are able to restrict the applicability of this criterion.

However, a more fundamental approach leading to the definition of other wetting criteria is based on analysis of the nature of forces involved at the interface and allows calculation of the free energy of interactions between two materials to be made. For low-surface-energy solids such as polymers, many authors have estimated the thermodynamic surface free energy from contact-angle measurements. The first approach was an empirical one developed by Zisman and co-workers [37–39]. They established that a linear relationship often exists between the cosine of the contact angle, cos θ, of several liquids and their surface tension, γ_{LV}. Zisman introduced the concept of critical surface tension, γ_c, which corresponds to the value of the surface energy of an actual or hypothetical liquid that will just spread on the solid surface, giving a zero contact angle. However, there is no general agreement about the meaning of γ_c and Zisman himself has always emphasized that γ_c is not the surface free energy of the solid but only a closely related empirical parameter.

For solid–liquid systems, taking into account Dupré's relationship [40], the adhesion energy W_{SL} is defined as

$$W_{SL} = \gamma_S + \gamma_{LV} - \gamma_{SL} = \gamma_{LV}(1 + \cos\theta) \tag{5}$$

in agreement with Eq. (2) and neglecting the spreading pressure. Fowkes [41] has proposed that the surface free energy γ of a given entity can be represented by the sum of the contributions of different types of interactions. Schultz et al. [42] have suggested that γ may be expressed by only two terms: a dispersive component (London's interactions) and a polar component (superscripts D and P, respectively), as follows:

$$\gamma = \gamma^D + \gamma^P \tag{6}$$

The last term on the right-hand side of this equation corresponds to all the nondispersion forces, including Debye and Keesom interactions, as well as hydrogen bonding. Fowkes [43] has also considered that the dispersive part of these interactions between solids 1 and 2 can be well quantified as twice the geometric mean of the dispersive component of the surface energy of both entities. Therefore, in the case of interactions involving only dispersion forces, the adhesion energy W_{12} is given by

$$W_{12} = 2(\gamma_1^D\gamma_2^D)^{1/2} \tag{7}$$

By analogy with the work of Fowkes, Owens and Wendt [44] and then Kaelble and Uy [45] have suggested that the nondispersive part of interactions between materials can be expressed as the geometric mean of the nondispersive components of their surface energy, although there is no theoretical reason to represent all the nondispersive interactions by this type of expression. Hence the work of adhesion W_{12} becomes

$$W_{12} = 2(\gamma_1^D\gamma_2^D)^{1/2} + 2(\gamma_1^P\gamma_2^P)^{1/2} \tag{8}$$

For solid–liquid equilibrium, a direct relationship between the contact angle θ of the drop of a liquid on a solid surface and the surface properties of both products is obtained from Eqs. (5) and (8). By contact-angle measurements of droplets of different liquids of known surface properties, the components γ_S^D and γ_S^P of the surface energy of the substrate can then be determined.

More recently, it has been shown, in particular by Fowkes and co-workers [46–49], that electron acceptor and donor interactions, according to the generalized Lewis acid–base concept, could be a major type of interfacial force between the adhesive and the substrate. This approach is able to take into account hydrogen bonds, which are often involved in adhesive joints. Moreover, Fowkes and Mostafa [47] have suggested that the contribution of the polar (dipole–dipole) interactions to the thermodynamic work of adhesion could generally be neglected compared to both dispersive and acid–base contributions. They have also considered that the acid-base component W^{ab} of the adhesion energy can be related to the variation of enthalpy, $-\Delta H^{ab}$, corresponding to the establishment of acid–base interactions at the interface, as follows:

$$W^{ab} = f(-\Delta H^{ab})n^{ab} \tag{9}$$

where f is a factor that converts enthalpy into free energy and is taken equal to unity, and n^{ab} is the number of acid–base bonds per unit interfacial area, close to about 6 μmol/m^2. Therefore, from Eqs. (7) and (9), the total work of adhesion W_{12} becomes

$$W_{12} = 2(\gamma_1^D\gamma_2^D)^{1/2} + f(-\Delta H^{ab})n^{ab} \tag{10}$$

The experimental values of the variation of enthalpy $(-\Delta H^{ab})$ can be estimated from the work of Drago and co-workers [50,51], who proposed the following relationship:

$$-\Delta H^{ab} = C^A C^B + E^A E^B \tag{11}$$

where C^A and E^A are two quantities that characterize the acidic material at the interface, and similarly, C^B and E^B characterize the basic material. The validity of Eq. (11) was clearly evidenced for polymer adsorption on various substrates [49]. Another estimation of $(-\Delta H^{ab})$ can be carried out from the semiempirical approach defined by Gutmann [52], who has proposed that each material may be characterized by two constants: an electron acceptor number AN and an electron donor number DN. For solid surfaces, similar numbers, K_A and K_D, respectively, have been defined and measured by inverse gas chromatography [53–55]. In this approach, the enthalpy $(-\Delta H^{ab})$ of formation of acid–base interactions at the interface between two solids 1 and 2 is now given by [52,53]

$$-\Delta H^{ab} = K_{A1}K_{D2} + K_{A2}K_{D1} \tag{12}$$

This expression was applied successfully by Schultz et al. [55] to describe fiber–matrix adhesion in the field of composite materials.

Finally, it must be mentioned that acid–base interactions can also be analyzed in terms of Pearson's hard–soft acid–base (HSAB) principle [56,57]. At present, the application of this concept to solid–solid interactions and thus to adhesion is under investigation.

2. Models Relating the Adhesion Strength G to the Adhesion Energy W

Although described in Section II.F, these models also apply to other types of interfacial interactions. One of the most important models in adhesion science, usually called the *rheological model* or *model of multiplying factors*, was proposed primarily by Gent and Schultz [3,4] and then reexamined using a fracture mechanics approach by Andrews and Kinloch [58] and Maugis [59]. In this model, the peel adhesion strength is simply equal to the product of W by a loss function Φ, which corresponds to the energy irreversibly dissipated in viscoelastic or plastic deformations in the bulk materials and at the crack tip and depends on both peel rate v and temperature T:

$$G = W\Phi(v,T) \tag{13}$$

As already mentioned, the value of Φ is usually far higher than that of W, and the energy dissipated can then be considered as the major contribution to the adhesion strength G. In the case of assemblies involving elastomers, it has been clearly shown in various studies [3,4,58,60–62] that the viscoelastic losses during peel experiments, and consequently, the function Φ, follow a time–temperature equivalence law such as that of Williams et al. [63].

It is more convenient to use the intrinsic fracture energy G_0 of the interface in place of W in Eq. (13), as follows:

$$G = G_0\Phi(v,T) \tag{14}$$

Effectively, when viscoelastic losses are negligible (i.e., when performing experiments at very low peel rate or high temperature), $\Phi \to 1$ and G must tend toward W. However, the resulting threshold value G_0 is generally 100 to 1000 times higher than the thermodynamic work of adhesion, W.

From a famous fracture analysis of weakly cross-linked rubbers called the *trumpet model*, de Gennes has derived [64] an expression similar to equation (14) when the crack propagation rate v is sufficiently high. He distinguished three different regions along the trumpet starting from the crack tip: a hard, a viscous, and finally, a soft zone. The length of the hard region is equal to $v\tau$, where τ is the relaxation time, and then the viscous region extends to a distance $\lambda v\tau$. Factor λ is the ratio of the high-frequency elastic modulus to the zero-frequency elastic modulus of the material, and obviously represents the viscoelastic behavior of the rubber. Hence according to this approach, it is shown that the total adhesive work is given by the following expression, similar to Eq. (14):

$$G \sim G_0\lambda \tag{15}$$

where G_0 is the intrinsic fracture energy for low velocities (i.e., when the polymer near the crack behaves as a soft material).

Carré and Schultz [65] have reexamined the significance of G_0 on cross-linked elastomer–aluminum assemblies and proposed that it can be expressed as

$$G_0 = Wg(M_c) \tag{16}$$

where g is a function of molecular weight M_c between cross-link nodes and corresponds to a molecular dissipation. Such an approach is based on Lake and Thomas's argument [66], which states that to break a chemical bond somewhere in a chain, all bonds in the chain must be stressed close to their ultimate strength. More recently, de Gennes [67] has proposed further analysis of this problem. He postulates that the main energy dissipation near the interface could be due to the extraction of short segments of chains in the junction zone during crack opening, this phenomenon being called the *suction process*. From a volume balance and a stress analysis, the following expression of the intrinsic fracture energy G_0 is obtained for low fracture velocity:

$$G_0 = \sigma_c a^2 v L \tag{17}$$

where σ_c is a threshold stress that can be considered as a material constant to a first approximation, and a^2, v, and L are, respectively, the cross-sectional area, the number per unit interfacial area, and the extended length of chain segments sucked out during the crack propagation. At present, no experimental verification of this approach has yet been published. Obviously, this analysis holds only for values of L less than L_e (i.e., the critical length at which physical entanglements between macromolecular chains just occur). The case where $L > L_e$ implies at least a disentanglement process, but above all, a process of chain scission, which is analyzed below.

Concerning the adhesion phenomena occurring at the fiber–matrix interface in composite materials, Nardin and Schultz [68] have recently proposed that the shear strength τ_i

of the interface, measured by means of a fragmentation test on single fiber composites, is related directly to the free energy of adhesion W, calculated from Eq. (10), according to

$$G \sim \kappa \tau_i = \left(\frac{E_m}{E_f} \right)^{1/2} W \tag{18}$$

In this expression, κ is a constant equal to about 0.5 nm, corresponding to a mean intermolecular distance when only physical interactions (dispersive and acid–base interactions) are involved; E_m and E_f are the elastic moduli of the matrix and the fiber, respectively. This model is equivalent to this of Gent and Schultz [3,4] for a cylindrical geometry and in the case of pure elastic stress transfer between both materials. It is very well verified experimentally for various fiber–matrix systems. The influence of the formation of interfacial layers exhibiting mechanical behavior completely different from that of the bulk matrix has also been examined [31].

Finally, it is worth examining the analyses concerning tack, in other words, the instantaneous adhesion when a substrate and an adhesive are put in contact for a short time t (of the order of 1 s) under a given pressure. This tack phenomenon is of great importance for processing involving hot-melt or pressure-sensitive adhesives. First, it has clearly been shown [69] that the viscoelastic characteristics of the adhesive, in particular its viscous modulus, plays a major role on the separation energy. Recently, de Gennes [70] has suggested that the measured tack could be related to both the free adhesion energy W and the rheological properties of the bulk adhesive, as follows:

$$G_{\text{tack}} \sim \begin{cases} W \dfrac{\mu_\infty}{\mu_0} & \text{for weakly cross-linked elastomers} \tag{19} \\[2em] W \dfrac{t}{\tau} & \text{for uncross-linked elastomers} \tag{19'} \end{cases}$$

where μ_∞ and μ_0 are the high-frequency and zero-frequency moduli of the adhesive, respectively, and τ is the reptation time of the macromolecular chains (see the next section). The latter equation holds for time t much larger than this reptation time. The experimental verification of this approach is under investigation.

E. Diffusion Theory

The diffusion theory of adhesion is based on the assumption that the adhesion strength of polymers to themselves (autohesion) or to each other is due to mutual diffusion (interdiffusion) of macromolecules across the interface, thus creating an interphase. Such a mechanism, mainly supported by Voyutskii [71], implies that the macromolecular chains or chain segments are sufficiently mobile and mutually soluble. This is of great importance for many adhesion problems, such as healing and welding processes. Therefore, if interdiffusion phenomena are involved, the joint strength should depend on different factors, such as contact time, temperature, nature and molecular weight of polymers, and so on. Actually, such dependences are experimentally observed for many polymer–polymer junctions. Vasenin [72] has developed, from Fick's first law, a quantitative model for the diffusion theory that correlates the amount of material w diffusing in a given x direction across a plane of unit area to the concentration gradient $\partial c / \partial x$ and the time t:

$$\partial w = -D_f \, \partial t \, \frac{\partial c}{\partial x} \tag{20}$$

where D_f is the diffusion coefficient. To estimate the depth of penetration of the molecules that interdiffused into the junction region during the time of contact t_c, Vasenin assumed that the variation of the diffusion coefficient with time is of the form $D_d t_c^{-\beta}$, where D_d is a constant characterizing the mobility of the polymer chains and β is on the order of 0.5. Therefore, it is possible to deduce the depth of penetration l_p as well as the number N_c of chains crossing the interface, which are given by

$$l_p \sim k(\pi D_d t_c^{1/2})^{1/2} \tag{21}$$

$$N_c = \left(\frac{2N\rho}{M} \right)^{2/3} \tag{22}$$

where k is a constant, N is Avogadro's number, and ρ and M are, respectively, the density and the molecular weight of the polymer. Finally, Vasenin assumed that the measured peel energy G was proportional to both the depth of penetration and the number of chains crossing the interface between the adhesive and the substrate. From Eqs. (21) and (22), G becomes

$$G \sim K \left(\frac{2N\rho}{M} \right)^{2/3} D_d^{1/2} t_c^{1/4} \tag{23}$$

where K is a constant that depends on molecular characteristics of the polymers in contact. Experimental results and theoretical predictions from Eq. (23) were found [72] in very good agreement in the case of junctions between polyisobutylenes of different molecular weights. In particular, the dependence of G on $t_c^{1/4}$ and $M^{-2/3}$ was clearly evidenced.

One important criticism of the model proposed by Vasenin is that the energy dissipated viscoelastically or plastically during peel measurements does not appear in Eq. (23). Nevertheless, in his work, the values of coefficients K and D_d are not theoretically quantified but determined only by fitting. Therefore, it can be assumed that the contribution of hysteretic losses to the peel energy is implicitly included in these constants.

In fact, the major scientific aspect of interdiffusion phenomena is concerned with the dynamics of polymer chains in the interfacial region. Recently, the fundamental understanding of the molecular dynamics of entangled polymers has advanced significantly due to the theoretical approach proposed by de Gennes [73], extended later by Doi and Edwards [74] and Greassley [75]. This new approach stems from the idea that the chains cannot pass through each other in a concentrated polymer solution, a melt, or a solid polymer. Therefore, a chain with a random coil conformation is trapped in an environment of fixed obstacles. This constraint confines each chain inside a tube. De Gennes has analyzed the motion, limited mainly to effective one-dimensional diffusion along a given path, of a polymer chain subjected to such a confinement. He described this type of motion as wormlike and gave it the name *reptation*. The reptation relaxation time τ associated with the movement of the center of gravity of the entire chain through the polymer was found varying with the molecular weight M as M^3. Moreover, the diffusion coefficient D, which defines the diffusion of the center of mass of the chain, takes the form $D \sim M^{-2}$.

One of the most important and useful applications of the reptation concept concerns crack healing, which is primarily the result of the diffusion of macromolecules across the interface. Healing process was studied particularly by Kausch and co-workers [76]. The problem of healing is to correlate the macroscopic strength measurements to the microscopic description of motion. The difference between self-diffusion phenomena in the bulk polymer and healing is that the polymer chains in the former case move over

totally miscible with PPO, it was reasonably expected that the copolymer organizes at the interface, due to the fact that each block dissolves in the respective homopolymer. The molecular weight of these blocks is always superior to the critical molecular weight M_e, for which entanglements of chains occur in the homopolymers. Experimentally, Brown employed partially or fully deuterated copolymers in order to be able to determine the deuterium on the fracture surface after separation by secondary-ion mass spectrometry (SIMS) and forward-recoil spectroscopy (FRES) [85]. A scission of the copolymer chains near the junction point of both blocks is observed, indicating that the diblock copolymers are well organized at the interface, whatever their molecular weights, with their junction accurately located at the PMMA–PPO interface. Moreover, Brown has proposed [86] a molecular interpretation of the toughness of glassy polymers, which can also be applied to the failure of interfaces between immiscible polymers. This approach stems from the idea that the cross-tie fibrils, which exist between primary fibrils in all crazes, can transfer mechanical stress between the broken and unbroken fibrils and thus strongly affect the failure mechanics of a craze. It is based on a simple model of crack tip stress concentration. Finally, assuming that all the effectively entangled chains in the material are drawn into the fibril, the fracture energy G of a polymer is found directly related to the square of both the areal density v of entangled chains and the force f required to break a polymer chain:

$$G \sim v^2 f^2 \frac{D}{S} \tag{25}$$

where D is the fibril diameter and S is the stress at the craze–bulk interface, which is assumed to be constant. Brown has considered [86] that diblock copolymer-coupled interfaces between PMMA and PPO are ideal experimental systems for testing the validity of his model. Indeed, a linear dependence of the interfacial fracture energy G on the diblock copolymer surface density v, in logarithmic scales, is observed for copolymers of different molecular weights. A slope of 1.9 ± 0.2 was found for the master straight line in good agreement with Eq. (25). Nevertheless, it is worth noting that Brown's results involving chain scission at the interface and leading to a dependence of G on v^2 are in contradiction with both previous examples, where linear relationships between G and v are established.

III. CONCLUDING REMARKS

Adhesion is a very complex field beyond the reach of any single model or theory. Given the number of phenomena involved in adhesion, the variety of materials to be bonded, and the diversity of bonding conditions, the search for a unique, universal theory capable of explaining all the experimental facts is useless. In practice, several adhesion mechanisms can be involved simultaneously. However, it is generally assumed that the adsorption or thermodynamic theory defines the main mechanism exhibiting the widest applicability. It describes the achievement of intimate contact and the development of physical forces at the interface. This is a necessary step for interlocking, interdiffusion, and chemical bonding mechanisms to occur subsequently, further increasing the adhesive strength.

Finally, one can consider that the measured adhesive strength of an assembly could be expressed as a function of three terms relating, respectively, to (1) the interfacial molecular interactions, (2) the mechanical and rheological properties of bulk materials, and (3) the characteristics of the interphase. The first two terms have received a great deal of attention during recent decades, as a result of studies in physical chemistry of surfaces and fracture mechanics. The third term constitutes the real challenge for a real and complete understanding of adhesion.

REFERENCES

1. J. W. McBain and D. G. Hopkins, *J. Phys. Chem.* 29: 88 (1925).
2. E. M. Borroff and W. C. Wake, *Trans. Inst. Rubber Ind.* 25: 190, 199, 210 (1949).
3. A. N. Gent and J. Schultz, *Proc. 162nd ACS Meeting 31*(2): 113 (1971).
4. A. N. Gent and J. Schultz, *J. Adhesion 3*: 281 (1972).
5. W. C. Wake, *Adhesion and the Formulation of Adhesives*, Applied Science Publishers, London, 1982.
6. J. R. Evans and D. E. Packham, in *Adhesion*, Vol. 1 (K. W. Allen, ed.), Applied Science Publishers, London, 1977, p. 297.
7. D. E. Packham, in *Developments in Adhesives*, Vol. 2 (A. J. Kinloch, ed.), Applied Science Publishers, London, 1981, p. 315.
8. D. E. Packham, in *Adhesion Aspects of Polymeric Coatings* (K. L. Mittal, ed.), Plenum Press, New York, 1983, p. 19.
9. P. J. Hine, S. El Muddarris, and D. E. Packham, *J. Adhesion 17*: 207 (1984).
10. N. H. Ladizesky and I. M. Ward, *J. Mater. Sci. 18*: 533 (1983).
11. M. Nardin and I. M. Ward, *Mater. Sci. Technol. 3*: 814 (1987).
12. N. H. Ladizesky and I. M. Ward, *J. Mater. Sci. 24*: 3763 (1989).
13. H. E. Bair, S. Matsuoka, R. G. Vadimsky, and T. T. Wang, *J. Adhesion 3*: 89 (1971).
14. T. T. Wang and H. N. Vazirani, *J. Adhesion 4*: 353 (1972).
15. B. V. Deryaguin and N. A. Krotova, *Dokl. Akad. Nauk SSSR 61*: 843 (1948).
16. B. V. Deryaguin, *Research 8*: 70 (1955).
17. B. V. Deryaguin, N. A. Krotova, V. V. Karassev, Y. M. Kirillova, and I. N. Aleinikova, *Proc. 2nd International Congress on Surface Activity III*, Butterworth, London, 1957, p. 417.
18. B. V. Deryaguin and V. P. Smilga, in *Adhesion: Fundamentals and Practice*, Maclaren, Ministry of Technology, UK, ed., London, 1969, p. 152.
19. B. V. Deryaguin, N. A. Krotova, and V. P. Smilga, *Adhesion of Solids*, Studies in Soviet Science, Plenum Press, New York, 1978.
20. C. L. Weidner, *Adhesives Age 6*(7): 30 (1963).
21. J. Krupp and W. Schnabel, *J. Adhesion 5*: 296 (1973).
22. L. Lavielle, J. L. Prévot, and J. Schultz, *Angew. Makromol. Chem. 169*: 159 (1989).
23. G. von Harrach and B. N. Chapman, *Thin Solid Films 13*: 157 (1972).
24. J. J. Bikerman, *The Science of Adhesive Joints*, Academic Press, New York, 1961.
25. W. D. Bascom, C. O. Timmons, and R. L. Jones, *J. Mater. Sci. 19*: 1037 (1975).
26. R. J. Good, *J. Adhesion 4*: 133 (1972).
27. L. H. Sharpe, *Proc. 162nd ACS Meeting 31*(2): 201 (1971).
28. J. Schultz, *J. Adhesion 37*: 73 (1992).
29. J. Schultz, A. Carré, and C. Mazeau, *Intern. J. Adhesion Adhesives 4*: 163 (1984).
30. J. Schultz, L. Lavielle, A. Carré, and P. Comien, *J. Mater. Sci. 24*: 4363 (1989).
31. M. Nardin, E. M. Asloun, F. Muller, and J. Schultz, *Polymer Adv. Technol. 2*: 161 (1991).
32. M. Nardin, A. El Maliki, and J. Schultz, *J. Adhesion 40*:93 (1993).
33. J. Schultz and A. Carré, *J. Appl. Polymer Sci. Appl. Polymer Symp. 39*: 103 (1984).
34. V. Péchereaux, J. Schultz, X. Duteurtre, and J. M. Gombert, *Proc. 2nd International Conference on Adhesion and Adhesives*, Adhecom '89 1: 103 (1989).
35. L. H. Sharpe and H. Schonhorn, *Chem. Eng. News 15*: 67 (1963).
36. T. Young, *Phil. Trans. Roy. Soc. 95*: 65 (1805).
37. H. W. Fox and W. A. Zisman, *J. Colloid Sci. 5*, 514 (1950); 7: 109, 428 (1952).
38. E. G. Shafrin and W. A. Zisman, *J. Am. Ceram. Soc. 50*: 478 (1967).
39. W. A. Zisman, in *Advances in Chemistry Series*, Vol. 43 (R. F. Gould, ed.), American Chemical Society, Washington, 1964, p. 1.

40. A. Dupré, in *Théorie Mécanique de la Chaleur*, Gauthier-Villars, Paris, 1869, p. 369.
41. F. M. Fowkes, *J. Phys. Chem. 67*: 2538 (1963).
42. J. Schultz, K. Tsutsumi, and J. B. Donnet, *J. Colloid Interface Sci. 59*: 277 (1977).
43. F. M. Fowkes, *Ind. Eng. Chem. 56*: 40 (1964).
44. D. K. Owens and R. C. Wendt, *J. Appl. Polymer Sci. 13*: 1740 (1969).
45. D. H. Kaelble and K. C. Uy, *J. Adhesion 2*: 50 (1970).
46. F. M. Fowkes and S. Maruchi, *Org. Coatings Plastics Chem. 37*: 605 (1977).
47. F. M. Fowkes and M. A. Mostafa, *Ind. Eng. Chem. Prod. Res. Dev. 17*: 3 (1978).
48. F. M. Fowkes, *Rubber Chem. Technol. 57*: 328 (1984).
49. F. M. Fowkes, *J. Adhesion Sci. Technol. 1*: 7 (1987).
50. R. S. Drago, G. C. Vogel, and T. E. Needham, *J. Am. Chem. Soc. 93*: 6014 (1971).
51. R. S. Drago, L. B. Parr, and C. S. Chamberlain, *J. Am. Chem. Soc. 99*: 3203 (1977).
52. V. Gutmann, *The Donor–Acceptor Approach to Molecular Interactions*, Plenum Press, New York, 1978.
53. C. Saint-Flour and E. Papirer, *Ind. Eng. Chem. Prod. Res. Dev. 21*: 337, 666 (1982).
54. E. Papirer, H. Balard, and A. Vidal, *European Polymer J. 24*: 783 (1988).
55. J. Schultz, L. Lavielle, and C. Martin, *J. Adhesion 23*: 45 (1987).
56. R. G. Pearson, *J. Am. Chem. Soc. 85*: 3533 (1963).
57. R. G. Pearson, *J. Chem. Ed. 64*: 561 (1987).
58. E. H. Andrews and A. J. Kinloch, *Proc. Roy. Soc. London A332*: 385, 401 (1973).
59. D. Maugis, *J. Mater. Sci. 20*: 3041 (1985).
60. A. N. Gent and R. P. Petrich, *Proc. Roy. Soc. London A310*: 433 (1969).
61. G. R. Hamed, in *Treatise on Adhesion and Adhesives*, Vol. 6 (R. L. Patrick, ed.), Marcel Dekker, New York, 1989, p. 33.
62. M. F. Vallat and J. Schultz, *Proc. International Rubber Conference, Paris, II*(16): 1 (1982).
63. M. L. Williams, R. F. Landel, and J. D. Ferry, *J. Am. Chem. Soc. 77*: 3701 (1955).
64. P. G. de Gennes, *C.R. Acad. Sci. Paris Sér. II 307*: 1949 (1988).
65. A. Carré, and J. Schultz, *J. Adhesion 17*: 135 (1984).
66. G. J. Lake and A. G. Thomas, *Proc. Roy. Soc. London A300*: 103 (1967).
67. P. G. de Gennes, *J. Phys. Paris 50*: 2551 (1989).
68. M. Nardin and J. Schultz, *C.R. Acad. Sci. Paris Sér. II 311*: 613 (1990).
69. Mun Fu Tse, *J. Adhesion Sci. Technol. 3*: 551 (1989).
70. P. G. de Gennes, *C.R. Acad. Sci. Paris Sér. II 312*: 1415 (1991).
71. S. S. Voyutskii, *Autohesion and Adhesion of High Polymers*, Wiley-Interscience, New York, 1963.
72. R. M. Vasenin, in *Adhesion: Fundamentals and Practice*, Maclaren, [idem ref 18] London, 1969, p. 29.
73. P. G. de Gennes, *J. Chem. Phys. 55*: 572 (1971).
74. M. Doi and S. F. Edwards, *J. Chem. Phys. Faraday Trans. 74*: 1789, 1802, 1818 (1978).
75. W. W. Graessley, *Adv. Polymer Sci. 47*: 76 (1982).
76. K. Jud, H. H. Kausch, and J. G. Williams, *J. Mater. Sci. 16*: 204 (1981).
77. P. G. de Gennes, *C.R. Acad. Sci. Paris Sér. B 291*: 219 (1980); *292*: 1505 (1981).
78. S. Prager and M. Tirrell, *J. Chem. Phys. 75*: 5194 (1981).
79. Y. M. Kim and R. P. Wool, *Macromolecules 16*: 1115 (1983).
80. S. Buchan and W. D. Rae, *Trans. Inst. Rubber Ind. 20*: 205 (1946).
81. E. P. Plueddemann, *Silane Coupling Agents*, Plenum Press, New York, 1982.
82. A. N. Gent and A. Ahagon, *J. Polymer Sci. Polymer Phys. Ed. 13*: 1285 (1975).
83. P. Delescluse, J. Schultz, and M. E. R. Shanahan, in *Adhesion 8* (K. W. Allen, ed.), Elsevier, London, 1984, p. 79.
84. H. R. Brown, *Macromolecules 22*: 2859 (1989).
85. H. R. Brown, V. R. Deline and P. F. Green, *Nature 341*: 221 (1989).
86. H. R. Brown, *Macromolecules 24*: 2752 (1991).

3

Application of Plasma Technology for Improved Adhesion of Materials

Om S. Kolluri   *HIMONT Plasma Science, Foster City, California*

I. INTRODUCTION

Adhesion, whether the bonding of polymers or the adhesion of coatings to polymer surfaces, is a recurring and difficult problem for all industries that use these materials as key components in their products. Designers must often select specially formulated and expensive polymeric materials to ensure satisfactory adhesion (albeit even these materials often require surface preparation). In some cases, entire design concepts must be abandoned due to the prohibitive cost of the required polymer or the failure of crucial bonds.

Historically, surface treatments to improve adhesion of coatings to plastics consisted of mechanical abrasion, solvent wiping, solvent swell that was followed by acid or caustic etching, flame treatment, or corona surface treatment. Each of these treatments has limitations, thus providing a strong driving force for the development of alternative surface preparation methods. Many of the common methods mentioned are accompanied by safety and environmental risks, increased risk of part damage, and expensive pollution and disposal problems.

Mechanical abrasion or sand blasting is operator sensitive, dirty, difficult to do on small parts, and often does not reach hidden areas of complex-part geometries. Although more effective than solvent-based methods, acid etching can easily result in overtreated and damaged parts in addition to serious hazard and disposal problems. For example, other than plasma treatment, the most effective method for improving the bonding of materials to fluoropolymers has been to etch the surface with a material commonly referred to as *sodium etch*. The process consists of a brief immersion of the component to be bonded in a solution of sodium naphthalene in tetrahydrofuran or other suitable solvent. Although sodium etch is quite effective in treating fluoropolymers, concerns with operator safety and the problems of disposal have caused many users to seek alternatives.

Flame and corona, although useful in oxidizing the surfaces of plastics, have limited utility in many applications. In addition, the transitory nature of these modifications prevents their widespread use in many applications. Corona treatment is limited both to the materials that are responsive to this method of surface preparation and the part configuration itself. Complex shapes cannot easily be treated, as the treatment quality is a function of the distance of the part from the electrode. Thus small-diameter holes and surfaces that are difficult to access in complex parts prove particularly troublesome when treating these parts with corona discharge. The result is an uneven surface treatment at best. Since corona discharge surface treatments are typically conducted in ambient air, the process is subject to change from day to day as the environment changes in the location where corona treatment is being conducted.

For many industrial applications of plastics that are dependent on adhesive bonding, cold gas plasma surface treatment has rapidly become the preferred industrial process. Plasma surface treatment, which is conducted in a vacuum environment, affords an opportunity to minimize or eliminate the barriers to adhesion through three distinct effects: (1) removal of surface contaminants and weakly bound polymer layers, (2) enhancement of wettability through incorporation of functional or polar groups that facilitate spontaneous spreading of the adhesive or matrix resin, and (3) formation of functional groups on the surface that permit covalent bonding between the substrate and the adhesive or matrix resin. Since plasma treatment is a process of surface modification, the bulk properties of the material are retained. The nature of the process also allows precise control of the process parameters and ensures repeatability of the process in industrial applications. Finally, several studies have demonstrated that these surface modifications can be achieved with minimum impact on the environment.

II. PLASMA PROCESSING EQUIPMENT

While most, if not all, plasma equipment consists of similar components, the design of the reactor chamber, the distribution of power, the excitation frequency, and the gas dynamics can all be critical parameters influencing the efficiency and properties of plasma reactions. An extensive amount of work has been published that shows a direct correlation between excitation frequency and plasma reactivity. Manufacturers of plasma equipment employing radio-frequency (RF) excitation use either low frequencies (i.e., less than 400 kHz) or the higher frequencies at 13.56 or 27.12 MHz as specified by the Federal Communications Commission. For applications involving the treatment of plastics, 13.56 MHz is the preferred frequency. Also important is whether the material being treated is in a primary or a secondary plasma. Older equipment using large cylindrical barrels typically comprises secondary plasma systems (Fig. 1). The plasma is created either between closely spaced, paired electrodes that may function as shelves or in the annulus between the vessel's outer wall and a ring electrode, when employed. Treatment of materials placed within the working volume depends on the diffusion of active species created in the primary plasma (i.e., within the RF field). Diffusion of these active species is very dependent on pressure; the higher the pressure, the shorter the mean free path. The mean free path is the distance that active species can probably travel before undergoing collisions that deactivate radicals or neutralize ions. Therefore, when using a secondary plasma, the concentration of active species varies either across the diameter of a barrel system or between electrode pairs, as the case may be. Thus, by the physical laws of nature, the treatment within the working volume of a secondary plasma system cannot be uniform. By contrast, when working within the RF field, or primary plasma, the gas is constantly being excited. Thus polymeric articles being treated are immersed in a constant concentration of active species. Further, since diffusion is not a mechanistic limitation; significantly higher operating pressures may be used. This allows higher process gas flow rates, assuring that off-gassing species from the polymer are sufficiently diluted, providing the full benefits of the desired process gas. In addition, the primary plasma is rich in

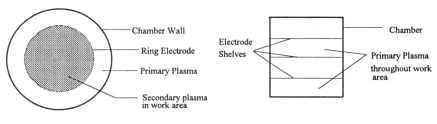

Barrel Type Reactor   Primary Plasma Reactor

Figure 1 Typical reactor configurations.

ultraviolet radiation, which is often an important initiation step in polymer reactions. Since ultraviolet (UV) radiation is line of sight, uniform treatment of multiple parts can only be obtained when working within the primary plasma. Otherwise, any part in the shadow of another will receive different radiation, and therefore the effectiveness of the treatment is expected to vary.

The types of reactors used for the deposition of plasma polymers have been varied. Glass and/or quartz reactors or aluminum chambers with metal parallel-plate electrodes seem to predominate in the literature, although several investigators have used inductively or capacitively coupled systems with external electrodes. High rates of deposition are found in the glow area, with the rate of deposition decreasing as we move farther away from the glow discharge region. Consequently, primary plasma systems that use a 13.56-MHz RF source are favored. The RF excitation used by various equipment manufacturers can be as low as 2 to 4 kHz or can be the more typical 13.56 MHz (high frequency). Microwave plasma systems have also been used for the deposition of plasma polymers. Previous studies have shown that the densities of films deposited by low-frequency systems are significantly lower than those of films deposited by either the high-frequency or microwave plasma systems. The choice of equipment used for plasma polymerization and deposition is thus dictated by the rate of deposition desired, the film properties that can be obtained by the various systems, and practical considerations such as the size of the parts to be treated and processing rates that are feasible in any given system.

III. PLASMA TECHNOLOGY

A. Plasma Processes

There are many definitions of the term *plasma*, according to the various disciplines with which it is involved. It has often been referred to as the fourth state of matter; the generation of plasma being analogous to the transitions that occur when energy is supplied to a material, causing solids to melt and liquids to become gases. Sufficient additional energy supplied to a gas creates a plasma. In the case of cold gas plasma, typical of that used in this work, the process is excitation of a gas at reduced pressure by RF energy. Typically, a plasma is composed of a large concentration of highly excited atomic, molecular, ionic, and radical species. While on an atomic scale, plasma generation cannot be construed as a room-temperature equilibrium process, as the bulk of the material remains near room temperature. The plasma contains free electrons as well as other metastable particles, which upon collision with the surfaces of polymers placed in the plasma environment break covalent chemical bonds, thus creating free radicals on the polymer surface [1]. The free radicals will then undergo additional reactions, depending on the gases present in the plasma or subsequent exposure to gases in the atmosphere. The result is that these gas–radical reactions form a surface that is potentially very different from that of the starting bulk polymer. Since the process is conducted in a reactor under very controlled conditions, the end result is very reproducible.

Plasma processing is not one process but a "field of opportunities" that can be classified into three overlapping categories: (1) plasma activation, (2) plasma-induced grafting, and (3) plasma polymerization. Plasma activation is the alteration of surface characteristics by the substitution of chemical groups or moieties for groups normally present on the polymer chain being modified. The assumed mechanism is free-radical creation and coupling of these free radicals with active species from the plasma environment. Depending on the process gas selected, a large variety of chemical groups can be incorporated into the surface. These groups may be hydroxyl, carbonyl, carboxylic, amino, or peroxyl groups. Most important, the insertion or substitution of these groups in the polymer chain is under the control of the operation. In this manner, the surface energies and the surface chemical reactivity of plastics can be altered completely without affecting their bulk properties.

Plasma-induced grafting offers another method by which plastic surfaces can be modified. If a noble gas is employed to generate a plasma, a multitude of free radicals are created along the polymer backbone. If after the plasma is extinguished but prior to the introduction of air, an unsaturated monomer such as ally alcohol is introduced into the reaction chamber, it will add to the free radical, yielding a grafted polymer. The range of functional and reactive sites that can be incorporated onto a surface is increased significantly with this technique. This process differs from activation in that instead of functional modification of the surface polymer chains, material is added on to the polymer backbone.

The third category of plasma processes, plasma deposition, utilizes gases or vapors that fractionate and undergo polymerization under the influence of RF energy. For example, methane, CH_4, under the influence of plasma will deposit as a polyhydrocarbon that has a density approaching 1.6 g/cm^3. Any material that can be introduced into the process chamber is a potential candidate as a feed material for plasma polymerization. The properties of materials polymerized in this manner are very different from polymers obtained from these materials via conventional polymerization methods. These properties include a high degree of cross-linking and the ability to form pinhole-free films that adhere tenaciously to various substrates.

B. Factors Influencing Adhesion

The strength of an adhesive joint is influenced by several factors [2–4]. Removal of contaminants and process aids provides a means for the adhesive to interlock with the substrate surface rather than with a boundary layer that is merely resting on the surface. Increasing the surface energy of the substrate above the surface tension of the adhesive makes it possible for the adhesive to wet the entire surface of the polymer substrate. The increase in the apparent surface area of contact serves to increase the strength of the adhesive bond. Figure 2 illustrates this process.

Ablation of the surface layers of the exposed polymer can result in a micro-roughened surface that increases the area of contact between the adhesive and the substrate. Finally, modification of the surface chemistry in a manner that facilitates covalent bonding

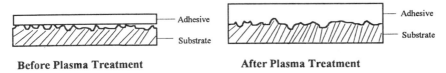

Before Plasma Treatment   **After Plasma Treatment**

Figure 2   Effect of plasma treatment on surface wetting.

between the adhesive and the substrate surface further enhances adhesion strength. These changes are accomplished by competing molecular reactions that take place on the surface of a polymer substrate in a plasma.

1.   *Ablation:* removal by evaporation of surface material. Ablation is the key process by which contaminants are removed from the surface of materials placed in a plasma. As the molecular weight of the contaminants is reduced due to chain scission, they become volatile enough to be removed by the vacuum system. Ablation of the surface layers of the polymer can also take place in a plasma and occur through a similar mechanism. If the substrate consists of a blend or alloy of materials that react differently in a plasma, differential ablation of these components can be used to create a micro-roughened surface.

2.   *Activation:* act of substituting atoms in the polymer molecule with chemical groups from the plasma. The surface energy of the polymer placed in a plasma can be increased very rapidly by plasma-induced oxidation, nitration, hydrolization, or amination. The higher surface energy of the polymer surface increases its *wettability*, which describes the ability of a liquid to spread over and penetrate the surface. The increase in apparent bonded surface area that results serves to increase the strength of the bond. The process of activation can also be used to substitute surface polymer groups with those that facilitate covalent bonding between the polymer substrate and the adhesive.

C.   Plasma Activation and Reactions on the Surface

The use of plasma surface treatment to improve adhesion is well known [5–18,44] and several literature sources provide an in-depth discussion of the nature of gas plasmas and their chemistries [1,5,7,10–12]. Although any gas can be ionized using RF excitation, gases such as O_2, N_2, He, Ar, NH_3, N_2O, CO_2, CF_4, and air or some combination of these gases are generally used for surface treatment.

One of the more common plasma processes used to enhance the adhesion of polymers is surface treatment in an oxygen plasma. An oxygen plasma is aggressive in its reactivity and forms numerous components. Within an oxygen plasma O^+, O^-, O_2^+, O_2^-, O, O_3, ionized ozone, metastably excited O_2, and free electrons are generally observed. The ionization of oxygen into the various species found in an oxygen plasma can be represented by the following reaction scheme:

$$O_2 + e \rightarrow O_2^- \rightarrow O + O^-$$

$$e + O \rightarrow O^+ + 2e$$

$$e + O_2 \rightarrow O_2^+ + 2e$$

$$e + 2O_2 \rightarrow O_2^- + O_2$$

$$O_2^- + O \rightarrow O_3 + e$$

These reactions represent a small sampling of those that occur in an oxygen glow discharge.

As the components formed during the ionization recombine, they release energy and photons, emitting a faint blue glow and much UV radiation. The photons in the UV region have enough energy to break the carbon–carbon and carbon–hydrogen bonds in the materials on the surface that are exposed to the plasma. In the case of contaminants, the net effect appears to be degradative, such that lower-molecular-weight materials are created. These lower-molecular-weight materials are subsequently removed by the vacuum. In this manner the surface that has been exposed to a plasma is cleaned. Lower-molecular-weight polymer fractions that comprise the weak boundary layers on the surface are also removed in this manner. Several reports documenting the efficacy of plasma surface cleaning have been published [1,13,15,19,20].

Once the contaminants have been removed, the virgin polymer surface is exposed to the plasma environment. The electrons, ions, and free radicals in the plasma act on this exposed polymer, creating free radicals in the molecular chains on the surface [1,21,24]. The free radicals that are created on the polymer surface by this process can then react with the various molecular and active species present in the plasma environment. In a low-pressure oxygen plasma, the following oxidation reaction scheme has been suggested:

$$RH + O^{\cdot} \rightarrow R^{\cdot} + {}^{\cdot}OH$$

$$R^{\cdot} + O_2 \rightarrow RO_2{}^{\cdot}$$

$$RO_2{}^{\cdot} + R'H \rightarrow RO_2H + R''$$

$$RO_2{}^{\cdot} + R'' \rightarrow RO_2R'$$

Here the RO_2H and RO_2R' indicate the formation of acids and esters. Not indicated in this reaction scheme are the possible formation of alcohols, ethers, peroxides, and hydroperoxides.

Thus in addition to the reactions resulting from the bombardment of the surface by photons, ions, and neutral particles, all of the active species in the plasma react with the polymer surface. The by-products, consisting of CO_2, H_2O, and low-molecular-weight hydrocarbons are readily removed by the vacuum system. The use of co-reactants can serve to modify the surface chemistry obtained with a single gas chemistry or to accelerate the reaction kinetics. For example, in an oxygen plasma, breaking of the carbon–carbon and carbon–hydrogen bonds is the rate-limiting step. When tetrafluoromethane is introduced as a coreactant, the O_2/CF_4 plasma yields excited forms of O, OF, CO, CF_3, CO_2, and F. Since fluorine or fluorine-containing species are more effective in breaking the carbon–carbon and carbon–hydrogen bonds, the reaction rate is accelerated. The permanent nature of these changes on the polymer surface have been confirmed by spectroscopic analyses and documented in several studies [25–28]. The use of other gases permits incorporation of other functional groups on the polymer surface. Examples include the use of ammonia, nitrogen, and oxides of nitrogen plasmas that are used to incorporate nitrogen in the surface and create nitrogen-based functional groups such as primary and secondary amines [29,30].

One result of such surface modification of the polymer surface is an increase in the surface energy of the polymer and an attendant improvement in surface wetting. As stated earlier, adequate wetting of the surface by the adhesive contributes to the improvement in bond strength by increasing the apparent area of contact over which the load is distributed. Published studies suggest that this improvement in wetting contributes directly to the observed improvement in the strength of the adhesive bond [33–35]. Another factor that contributes to improved adhesion is an increase in surface area of the polymer surface through microroughening. This occurs through the process of ablation of the polymer surface through exposure to a plasma. This is particularly the case when the plasma is highly reactive, as in the case when oxygen is used as one of the gas components that is being ionized. The nature of the gas being ionized to create the plasma is not the only factor that determines the extent of ablative etching. The nature of the polymer that is exposed to the plasma also plays a key role. Studies have shown that etching through ablation of surface polymer layers does occur in the case of polymers such as polyethylene, PET, and nylon 66 [11,36], whereas polyaramid materials such as Kevlar appear to be resistant to microroughening through ablation of the polymer chains [37].

Evidence has been presented in several studies which indicate that the strength of the adhesive bond is dependent on the particular functional group that has been created on the surface of the polymer. In some cases a direct correlation is drawn relating the nature of the chemical groups on the surface, the nature of the adhesive used, and the observed improvement in adhesion [11,35,38]. In other cases, the improvements are related to effects of hydrogen bonding and specific surface chemical interactions that do not necessarily result in covalent bonding between the polymer surface and the adhesive [39]. It is not unlikely that these conclusions are left to the reader to infer from the adhesion data presented along with the data describing the nature of the surface chemistry as determined by XPS analysis [40,41].

As these examples illustrate, selection of the process gas determines how the plasma will alter the polymer. Very aggressive plasmas can be created from relatively benign gases. Oxidation by fluorine free radicals that are generated when tetrafluoromethane is included as one of the gases is as effective as oxidation by the strongest mineral acid solution. The primary difference is that the by-products of the plasma process do not require special handling since the active species recombine to their original stable and nonreactive form outside the RF field. In all cases, profound changes in the chemical nature of the polymer surface are implemented, changes that are permanent in nature. The stability of these surface changes is a function of the materials themselves and the storage conditions used [42]. For instance, plasticizers that can migrate to the surface or contaminants in the storage area that can be attracted to these high-energy surfaces will negate the effects of the chemical changes that have been created on the surface of these materials. Contact-angle measurements and ESCA analyses of plasma-treated surfaces have confirmed the permanent and long-lasting nature of plasma surface modification of polymers. For example, plasma-modified FEP was shown to retain its surface chemical characteristics over an 18-month observation period [43]. Similar phenomena have been observed by other investigators for other materials, such as polyethylene and polystyrene (unpublished data, HIMONT Plasma Science Applications Laboratory). These changes ultimately lead to significant improvements in adhesion strength, as the data in Table 1 suggest.

D. Plasma-Induced Grafting

As effective as these surface modification processes might be, they present limitations in terms of the extent to which the surfaces of polymers can be modified. Plasma-induced grafting offers another method by which chemical functional groups can be incorporated. In this process, free radicals are generated on the surface of a polymer through the use of an inert gas plasma. Because of the nonreactive nature of the inert gas plasma, surface chemical modification of the polymer does not occur. If the polymer surface that has been

Table 1   Lap Shear Strength of Untreated and Plasma-Treated Surfaces

Material	Plasma chemistry	Adhesive	Bond strength (psi)	Failure mode	Ref.
Vectra	Control	Epoxy	939	Adhesive	8
A625	Oxygen plasma	Scotchweld	1598	Cohesive	
	Ammonia plasma	2216 (3M)	1240	Cohesive	
Noryl	Control	Epoxy	617	Adhesive	8
731	Oxygen plasma	Scotchweld	1485	Adhesive	
	Ammonia plasma	2216 (3M)	1799	Cohesive	
Ultem	Control	Epoxy	186	Adhesive	8
1000	Oxygen plasma	Scotchweld	1939	Cohesive	
	Ammonia plasma	2216 (3M)	2056	Cohesive	
Rynite	Control	Epoxy	683	Adhesive	16
530/935	Plasma treatment A		5875	Cohesive	
	Plasma treatment B		6067	Cohesive	
Tefzel	Control	Epoxy	10		12
	Ammonia plasma	Uniset	202		
	O_2/SF_6 plasma	D276	293		

treated in this fashion is exposed to vapors of unsaturated monomers, these monomers then get attached to the surface of the polymer. A variety of vinyl monomers are available and the possibilities for incorporating many different chemical functional groups are endless. Unlike surface modification, this is a two-step process that adds a degree of complexity. Few studies have appeared in the open literature, with the majority of such processes being used in proprietary applications [45,46]. This process is mentioned here as an option that is available to the surface engineer.

E.   Plasma Film Deposition

In comparison to the processes described above, plasma polymerization offers an entirely new avenue for adhesion improvement when bonding different materials. For example, films deposited from a methane plasma have been shown to improve dramatically the adhesion properties of many materials when tested in both the dry and wet state [45]. The process of plasma film deposition is often called plasma polymerization, although the process that takes place is not polymerization in the classical sense. Gases in plasma may undergo polymerization, usually through a free-radical initiation process. When a gas is ionized by RF energy, the resulting plasma contains free electrons as well as other metastable particles. When the process gas mixture used consists wholly or in part of hydrocarbon gases, the hydrocarbon molecule is fractured into free-radical fragments. These free-radical fragments become the sites at which the polymerization process is initiated. As the molecular weight of the plasma polymerized product increases, it is deposited on to the substrate placed within the plasma chamber. Since the fragmentation of the feed gas in the plasma generates free-radical species for initiating the polymerization process, gases such as methane (CH_4), which have zero functionality, can be used to form plasma polymers. In addition to methane, plasma polymers have been formed from other hydrocarbon gases, such as ethylene or propylene, fluorocarbon monomers such as tetrafluoroethylene, and organosilicon compounds such as hexamethyldisiloxane (HMDSO) or vinyltrimethylsilane (VTMS). Due to the complex nature of the fragmentation process, the resulting polymer structure is unlike any that can be deduced from conventional polymerization mechanisms [47].

The physics of plasma polymerization processes has been described in depth else-where in sufficient detail for the interested reader [45,48]. The conditions used during glow discharge polymerization determine not only the structure of the resulting film but also the rate at which these films are deposited onto the target substrate materials [45,49,50]. The degree to which the monomer is fragmented is dependent on the amount of energy supplied per unit weight of monomer that is allowed to flow through the reactor. When sufficient energy is supplied to break all the bonds of the monomer molecule, the recombination or polymerization process becomes atomic in nature. In addition, the structure of the plasma polymers can be varied by changing reaction conditions, including the use of comonomers or the introduction of oxygen, nitrogen, or ammonia into the reaction chamber during the polymerization process. These studies have developed a correlation between the power input, type of monomer used, and monomer flow rate to the density and the type of active species in the plasma. These factors, in turn, determine the rate of deposition and the film structure [50,52]. Table 2 shows typical deposition rates for some common plasma-polymerized films.

While plasma of ammonia, mixtures of hydrogen and nitrogen, and oxides of nitrogen have been used to incorporate nitrogen atoms in the surface layers of the polymer [29,30], the level of nitrogen incorporation has been less than 10 at % [30]. In contrast, films deposited from allyl amine have shown to contain up to 25 at % nitrogen as measured by spectroscopic methods [53]. Despite this high nitrogen content, however, the authors report a lower than expected concentration of amino groups. Other studies have shown concentrations of up to 2 molecules/nm^2 of reactive amine groups on the surface of films deposited from allyl amine onto FEP substrates. These surface concentrations were determined by derivatization of the amine groups with fluoroscein isothiocyanate and subsequent detection of the fluoroscein chromophore by optical spectroscopic methods [54]. Since ESCA analysis does not always allow precise determination of functional sites, the earlier data may reflect limitations of the analytical methods used.

In a similar vein, hydroxyl and carboxylic acid functionalities can be incorporated by plasma-polymerizing acrylic acid [55] or ally alcohol [53]. Another technique commonly employed to incorporate specific atomic species is the use of co-reactants along with the primary monomer. In one such example, ammonia or acrylonitrile was used as the co-reactant during the deposition of films from a methane plasma [56]. Two additional

Table 2 Deposition Rates for Various Compounds

Compound	Deposition rate, $D(\text{Å} \times 10^8 \text{ g/cm}^2 \cdot \text{min})$	D/D_0
Hexamethyldisiloxane	$233 = D_0$	1.00
Acrylic acid	28	0.12
Styrene	173	0.74
Tetramethyldisiloxane	191	0.82
Divinyltetramethyldisiloxane	641	2.75
Ethylene	42	0.18
Benzene	110	0.47

Source: Ref. 45.

techniques that are available to the surface engineer interested in modifying plasma-deposited films are plasma surface modification of the deposited film in a second process step and wet chemical reaction methods. As an example, carbonyls formed during the plasma deposition of films from N-vinylpyrrolidone were reacted with lithium aluminum hydride and sodium borohydride to convert these carbonyls to hydroxyl groups [57]. It should be noted that the use of plasma-deposited films for adhesion enhancement is not limited to polymeric substrates. Such films have also been deposited onto inorganic materials such as mica [55] and metal substrates such as aluminum and steel in an effort to improve adhesion of these materials to polymers [45].

IV. CONCLUDING REMARKS

One hundred years ago, Sir William Crooke stated: ''Investigation of the fourth state of matter—plasma—will be one of the most challenging and exciting fields of human endeavor.'' Plasma technology today is gaining increasing acceptance and recognition as an important industrial process for the surface modification of materials. Plasma processing is not one process but a ''field of opportunities'' that is poised to open up the development of complex new materials and products. By using this technology, the product designer is unlimited in the choice of materials at his or her disposal. That the surfaces of materials can be modified easily and effectively for a variety of end uses has become an established fact. The synthesis and deposition of ultrathin films with unique properties offers a means for cost-effective surface engineering in the quest for improved functionality of existing products and the development of products that were out of the realm of possibility just a few years ago.

REFERENCES

1. H. V. Boenig, *Plasma Science and Technology*, Cornell University Press, Ithaca, N.Y., 1982.
2. E. Rantz, *Adhesives Age* (May 1987).
3. W. Prane, *Adhesives Age* (June 1989).
4. E. M. Petrie, *Adhesives Age* (May 1989).
5. P. W. Rose and E. Liston, *Proc. SPE 43rd ANTEC*, 1985.
6. M. Londshien and W. Michaeli, *ANTEC*, Montreal, Quebec, Canada, 1991.
7. S. L. Kaplan and P. W. Rose, *SPE Tech. Papers 34* (Apr. 1988).
8. S. L. Kaplan and P. W. Rose, *Plastics Eng. 44*(5) (1988).
9. S. L. Kaplan, P. W. Rose, H. X. Nguyen, and H. W. Chang, *SAMPE Quart. 19*(4) (1988).
10. H. X. Nguyen, G. Riahi, G. Wood, and A. Poursartip, *33rd International SAMPE Symposium*, Anaheim, Calif., 1988.
11. O. S. Kolluri, S. L. Kaplan, and P. W. Rose, *SPE Adv. Polymer Composites '88 RETEC*, 1988.
12. G. P. Hansen, R. A. Rushing, R. W. Warren, S. L. Kaplan, and O. S. Kolluri, Achieving optimum bond strength with plasma treatment, *Technical Paper AD89-537*, Society of Manufacturing Engineers, Dearborn, Mich., 1989.
13. L. C. Jackson, *Adhesives Age* (Sept. 1978).
14. W. Yang and N. Sung, *Polymer. Mater. Sci. Eng. 62* (1990).
15. R. L. Bersin, *Adhesives Age* (Mar. 1972).
16. S. Sangiuolo and W. E. Hansen, *International Coil Winding Association Technical Conference*, Rosemont, Ill., 1990.
17. H. Shonhorn, F. W. Ryan, and R. H. Hansen, *J. Adhesion 2* (Apr. 1970).

18. H. Schonhorn and R. H. Hansen, *Appl. Polymer Sci. 11* (1967).

19. W. E. Hansen and M. Hozbor, Gas plasma cleaning for electrical product manufacture, *International Coil Winding Association Conference,* Boston, MA, 1991.

20. O. S. Kolluri, Surface cleaning with plasma: an environmentally safe alternative, *HIMONT Plasma Science Technical Note,* HIMONT, Foster City, Calif., 1992.

21. N. Morosoff, B. Crist, M. Bumgarner, T. Hsu, and H. K. Yasuda, *Symposium on Plasma Chemistry of Polymers* (K. C. Shen and Mitchell, eds.), Marcel Dekker, New York, 1976.

22. D. J. Conley, S. J. Rzad, M. C. Burrell, and J. Chera, *Polymer. Mater. Sci. Eng. 62* (1990).

23. R. D. Cormia and O. S. Kolluri, *Res. Develop.* (July 1990).

24. G. W. Pitt, J. E. Lakenan, D. M. Fogg, and B. A. Strong, *SAMPE Quart.* (Oct. 1991).

25. L. J. Gerenser, *Polymer. Mater. Sci. Eng. 62* (1990).

26. T. G. Vargo, J. A. Gardella, and L. Salvati, *J. Polymer Sci. A Polymer Chem. 27* (1989).

27. D. J. D. Moyer and J. P. Wightman, *Surface Interface Anal. 14* (1989).

28. J. E. Klemberg-Sapieha, L. Martinu, E. Sacher, and M. R. Wertheimer, *ANTEC*, Montreal, Quebec, Canada, 1991.

29. J. R. Hollahan, B. B. Stafford, R. D. Falb, and S. T. Payne, *J. Appl. Polymer Sci. 13* (1969).

30. N. Inagaki, S. Tasaka, J. Ohkubo, and H. Kawai, *Polymer. Mater. Sci. Eng. 62* (1990).

31. B. Tung, D. M. Polynski, and S. T. Terney, *SAMPE Quart. 19*(3) (1988).

32. J. E. Klemberg-Sapieha, O. M. Kittel, L. Martinu, and M. R. Wertheimer, *J. Vacuum Sci. Technol. A9*(6) (1991).

33. E. Liston, *Polymer. Mater. Sci. Eng. 62* (1990).

34. S. P. Wesson and R. E. Allred, *Proc. ACS Division of Polymeric Materials: Science and Engineering,* 1988, p. 58.

35. S. Mujin, H. Baorong, W. Yisheng, T. Ying, H. Weiqiu, and D. Youxian, *Composite Sci. Technol. 34* (1989).

36. T. Yasuda, T. Okuno, M. Miyama, and H. K. Yasuda, *Polymer. Mater. Sci. Eng. 62* (1990).

37. R. E. Allred, E. W. Merrill, and D. K. Roylance, *Proc. ACS Symposium on Composites,* Seattle, Wash., 1983.

38. G. Dagli and N. H. Sung, *Polymer. Mater. Sci. Eng.* (1987).

39. I. Iyengar and D. E. Erickson, *J. Appl. Polymer Sci. 2* (1967).

40. S. Nowak, H. P. Haerri, and L. Schlapbach, *Polymer. Mater. Sci. Eng. 62* (1990).

41. G. M. Porta, D. F. Foust, M. C. Burrell, and B. R. Karas, *ANTEC*, 1991.

42. H. S. Munro and D. I. McBriar, *Polymer. Mater. Sci. Eng. 56* (1987).

43. C. B. Hu and D. D. Solomon, *Polymer Prepri. 28*(2) (1981).

44. J. Osterndorf, R. Rosty, and M. J. Bodnar, *SAMPE J. 25*(4) (1989).

45. H. Yasuda, *Plasma Polymerization,* Academic Press, New York, 1985.

46. B. Das, *SAMPE J. 28*(2) (1992).

47. P. W. Rose, O. S. Kolluri, and R. D. Cormia, *34th International SAMPE Symposium,* Reno, Nev., May 1989.

48. D. C. Schram, G. M. W. Kroesen, and J. J. Buelens, *Polymer. Mater. Sci. Eng. 62* (1990).

49. R. d'Agostino, F. Fracassi, and F. Illuzi, *Polymer. Mater. Sci. Eng. 62* (1990).

50. G. Smolinsky and M. J. Vasile, *Symposium on Plasma Chemistry of Polymers* (K. C. Shen and Mitchell, eds.), Marcel Dekker, New York, 1976.

51. O. S. Kolluri, S. L. Kaplan, and D. A. Frazier, *4th International Symposium on Surface Modification Technologies,* Minerals, Metals and Materials Society, Paris, Nov. 1991.

52. K. Yanagihara, M. Kimura, K. Numata, and M. Niinomi, *Polymer. Mater. Sci. Eng. 56* (1987).

53. W. R. Gombotz and A. S. Hoffman, *Polymer. Mater. Sci. Eng. 56* (1987).

54. H. J. Griesser and R. C. Chatelier, *Polymer. Mater. Sci. Eng. 62* (1990).

55. D. L. Cho, P. M. Claesson, C. G. Gölander, and K. S. Johansson, *Polymer. Mater. Sci. Eng. 62* (1990).

56. R. A. Engelman and H. K. Yasuda, *Polymer. Mater. Sci. Eng. 62* (1990).

57. R. E. Marchant, D. Yu, X. Li, and M. J. Danilich, *Polymer. Mater. Sci. Eng. 62* (1990).

4

Silane and Other Adhesion Promoters in Adhesive Technology

Peter Walker *Atomic Weapons Establishment Plc, Aldermaston, Berkshire, England*

I. INTRODUCTION

It is axiomatic that the paramount property of an adhesive is adhesion to the substrate, but adhesion is also critical in the performance of surface coatings, which must adhere to protect and in obtaining optimum mechanical properties in particulate- and fiber-filled composites. Many surface preparation techniques have been employed to achieve high initial adhesion; these range from removal of surface contamination (solvent and vapor degreasing) to changes in substrate profile (grit blasting) to chemical modification (phosphating of steel, anodic treatments of aluminum). None of these methods solve the most critical problem in adhesion technology: that of the damaging effect of water on organic/inorganic bonds. Hydrolytic stability is essential in many technologies.

It has been shown by Walker that many types of organic coating lose up to 85% of their initial adhesion under water-soaked conditions [1], that adhesives show a marked loss of bond strength in water [2,3], and that glass-fiber-reinforced composites are readily degraded [4]. To improve the initial bond strength between adhesives and substrate, adhesion promoters may be used. These function by improving substrate wetting or by secondary bonding by van der Waals forces, dipole–dipole interactions, hydrogen bonding, or acid–base reactions. Relatively weak forces in the range 5 to 8 kcal/mol are involved. If hydrolytic stability of the bond is to be achieved, use of a coupling agent that is capable of forming primary chemical bonds, 50 to 250 kcal/mol [5], is required. The important distinction here is that in the nature of the bond formed, only coupling agents form primary bonds and can therefore be expected to produce water-resistant bonding. A coupling agent is therefore defined as a compound capable of chemical reaction with both the polymer and the substrate, although there is some evidence that reaction with the polymer is not necessarily a prerequisite. It should be noted that a coupling agent can function as an adhesion promoter; the reverse is not true. Current views are that only a limited range of organometallic compounds are true coupling agents.

II. EVIDENCE FOR COUPLING ACTIVITY

Interfacial bonding studies have attracted the attention of many workers using a variety of spectroscopic techniques. Evidence for coupling activity has been demonstrated in the cases shown in Table 1.

Table 1 Evidence for Coupling Activity

Bond	Coupling agent/substrate	Method[a]	Ref.
Si—O—Al	Silane/Al_2O_3	FTIR	6
Cr—O—Al	Volan/aluminum	ESCA	7
Zr—O—Al	Zirconate/aluminum		8
Si—O—Fe	Silane/iron	SIMS	9
Si—O—Si	Silane/SiO_2	NMR	10
Si—O—Pb	Silane/lead oxide	FTIR	11
Ti—O—Si	Titanate/SiO_2	FTIR	6

[a]FITR, Fourier transform infrared; ESCA, electron spectroscopy for chemical analysis; SIMS, secondary-ion mass spectrometry; NMR, nuclear magnetic resonance.

III. MECHANISM OF ADHESION PROMOTION

Mechanistic theories of adhesion promotion have been described in detail by Rosen [11] with sole reference to silanes, but it is likely that many of the proposed theories apply equally well to coupling agents in general. The proposed mechanisms are described below.

A. Chemical Bond Theory

As applied to silanes, the theory postulates that trialkoxysilane groups chemically bond to silanols on the mineral substrate surface by reaction of the hydrolyzed alkoxy group forming interfacial bonds of 50 to 100 kcal/mol [12] to 50 to 250 kcal/mol [13]. The organofunctional groups of the silane bond chemically to the polymer molecules. Both reactions were considered to be essential if true coupling is to be achieved. Although originally postulated for silanes, it is equally applicable to other adhesion promoters, including titanates and zirconates. It has been pointed out that although covalent bonds may be formed between polymer and mineral surfaces, some covalent oxane bonds are easily hydrolyzed [12,14], and examples of hydrolytically stable bonding has been achieved in the absence of chemical reaction with the polymer [12,15].

B. Deformable Layer Theory

This theory postulates that the interface zone is plastic, allowing stresses between the polymer and mineral surfaces to self-relieve without bond rupture. Internal stresses are thus reduced. It has been suggested that in the case of a silane, the film is too thin to allow this [11]. However, it is possible that the presence of the coupling agent might cause preferential adsorption. Credence for this view is given by the number of workers who report that the amount of adhesion promoter used is critical and that excessive usage may result in adhesion failure [16,17].

C. Surface Wettability Theory

Erickson and Plueddemann suggest that particularly in filled systems, complete wetting of the mineral surface will improve adhesion by physical adsorption that would exceed the cohesive strength of the polymer. However, it is difficult to see how physical adsorption provides bond reinforcement when the polymer is in competition with water and possibly other weakly bonded surface layers, and where chemical bonding is also present [12].

D. Restrained Layer Theory

In effect, this theory postulates a chemical reaction between promoter, polymer, and mineral substrate as in the chemical bond theory but also suggests that the presence of a region of intermediate modulus between polymer and substrate which transfers stress from the high modulus surface to the relatively low modulus polymer. Adhesive technology has long recognized this principle in specially formulated primers for use when bonding rubbery polymers to metals.

E. Reversible Hydrolytic Bond Theory

Best regarded as a combination theory, it postulates the chemical reactions between coupling agent, substrate, and polymer of the chemical bond theory together with the rigid interface of the restrained layer theory and the plastic interface of the deformable layer theory. It allows for stress relaxation by the reversible breaking of stressed bonds without loss of adhesion in the presence of water. It also argues that when Si-O or Ti-O substrate bonds are broken by the intrusion of water, they may re-form with some recovery in adhesion. It is likely that hydrogen bonding is a particularly important aspect of this theory, especially in the case of silanes [18]. Recovery of adhesion between urethane and epoxide coatings and metal substrates on drying out after water immersion has been demonstrated by Walker [19–21]. It is now generally accepted that some silane coupling agents do not need to react with the polymer chemically to provide enhanced initial and wet adhesion [16,22].

F. Oxide Reinforcement

This theory postulates that a primary mechanism by which silane coupling agents improve initial and wet adhesion is by reaction with the oxide surface on a metal to increase the cohesive strength of the oxide [23] and certainly, in the case of aluminum oxide, increase the wet strength of the oxide by inhibiting hydration of the oxide [24]. This has the effect of causing any failure to occur in the new weakest layer (i.e., the adhesive or coating). This may also explain, in part, the differences in bond strength achieved with different coupling agents and different metals, as it may be the nature of the oxide film and the degree of reinforcement that varies rather than any intrinsic property of a particular metal–adhesion promoter combination.

It seems unlikely that any single theory can explain the mechanism of adhesion promotion for such diverse systems as particulate- and fiber-filled composites, surface coatings, and adhesives applied to the complete range of metallic and other mineral substrates. Plueddemann opines that all theories of adhesion describe factors that are involved in bonding through silane coupling agents [16], and this view is likely to apply generally to the entire field of adhesion promotion.

G. Other Mechanisms

In addition to the stated theories of adhesion promotion, there are other mechanisms that may be both operative and important and are of general application.

1. Interpolymer Networks/Chain Tangling

It has been suggested that in the case of silane coupling agents, interdiffusion of siloxanol segments with polymer molecules may be a factor in bonding thermoplastic matrices. Interpenetrating polymer networks (IPNs) need not necessarily involve cross-linking of the silane or other coupling agent and the polymer matrix. Plueddemann expresses the view that to establish a strong interpenetrating boundary layer involves a tricky interplay of mechanical and chemical interreaction at the interface [16]. A similar mechanism has been suggested for titanates [17].

2. Acid–Base Reactions

A comprehensive account of acid–base reactions is covered in detail in volumes 4 (No. 4), 5, and 8 (1990) and volume 5 (No. 1) (1991) of the *Journal of Adhesion Science and Technology*, and a detailed account is beyond the scope of this chapter. Since different metal oxides have different isoelectric points in water and may therefore be regarded as acidic or basic, addition of material having acidic properties to adhesives to be used on basic substrates, or basic materials for use on acidic substrates, may improve adhesion [25]. Work with epoxide and polyurethane coatings of similar composition to adhesives have shown them to be basic in nature; the pH of water-soaked surfaces from which they were stripped is known to be 8 to 10. The aminosilanes APES and AAMS are strongly basic and when applied to oxide surfaces having isoelectric points in the range 9.1 (Al^{3+}) to 12.0 (Fe^{2+}) may be expected to produce a basic surface. Neither silane could therefore be expected to enhance the adhesion of a basic polymer by an acid–base reaction, although both have been shown to improve the initial and wet adhesion of epoxides and urethanes. This is not to argue that acid–base reactions are unimportant in adhesion promotion technology.

IV. METHODS OF USE IN ADHESIVE TECHNOLOGY

In general, adhesion promoters may be used as pretreatments or as additives. In the former case the promoter is used either as a solution in a suitable solvent or solvent mixture or as a formulated primer [26]; in the latter case they may be incorporated into the adhesive in a self-bonding concept [25]. The technique of filler treatment can be regarded as representing both approaches, although there is no evidence of this being used in filled adhesive systems. There are advantages and disadvantages inherent in both approaches: the pretreatment method allows a specific adhesion promoter to be used on a specific substrate to obtain optimum adhesion but has the disadvantage of introducing a process that is not under the control of the manufacturer. In theory the self-bonding additive concept is almost universally desirable, but in practice there are several critical parameters that need to be recognized, including potential polymer reactions, depletion by water, and shelf life. In adhesive technology as opposed to surface coating technology, the additive approach may not be as effective [27,28].

V. SILANES

Silanes of the general structure R-Si(OR')$_3$, where R is an organofunctional group and R' a hydrolyzable group, constitute the most technologically important group of adhesion promoters in use today and have a solid background of associated theory. Silane molecules are bifunctional, containing polar silanol groups and organofunctional groups capable of reaction with polymers. The range of silanes available commercially is large and continually expanding. Typical silane adhesion promoters are shown in Table 2. Other silanes that are attracting increasing interest are the fluoralkyl-functional silanes, the chemistry and uses of which have been reviewed extensively by Owen and Williams [29], cationic methacrylate, and cationic styryl silanes [30].

Table 2 Typical Silane Adhesion Promoters Commercially Available

Chemical description	Structure	Functional group	
		With polymer	With substrate
3-Chloropropyltrimethyl oxysilane	$ClCH_2CH_2CH_2Si(OCH_3)_3$	Chloro	Methoxy
Vinyltriethoxysilane	$CH_2{=}CHSi(OC_2H_5)_3$	Vinyl	Ethoxy
γ-Methylacryloxypropyl trimethoxysilane	$CH_2{=}C{-}C\text{-}OCH_2CH_2CH_2Si(OCH_3)_3$ with CH_3 and O groups	Methacryloxy	Methoxy
γ-Glycidoxypropyl trimethoxysilane	$CH_2CHCH_2OCH_2CH_2CH_2Si(OCH_3)_3$ O	Aliphatic epoxide	Methoxy
γ-Mercaptopropyl trimethyoxysilane	$HSCH_2CH_2CH_2Si(OCH_3)_3$	Mercapto	Methoxy
γ-Aminopropyltriethoxy silane	$NH_2CH_2CH_2CH_2Si(OC_2H_5)_3$	Amino	Ethoxy
N-β-(Aminoethyl) aminopropyl trimethoxysilane	$NH_2CH_2CH_2NHCH_2CH_2CH_2Si(OCH_3)_3$	Aminodiamino	Methoxy

A. Silane Coupling Reactions

The reactions of interest in silane coupling are summarized below.

(a) Hydrolysis of the silane group:

$$R{-}SiX_3 + 3H_2O \xrightarrow{\text{pH or catalyst}} R{-}Si(OH)_3 + 3HX \qquad (1)$$

where HX is usually an alcohol.

(b) Hydrogen bonding to the surface:

$$R{-}Si(OH)_3 + HO{-}Si \longrightarrow R{-}Si{-}O \cdots O{-}Si \qquad (2)$$

(c) Condensation with the surface:

$$R-Si(OH)_3 + HO-Si \longrightarrow R-\overset{\displaystyle |}{\underset{\displaystyle |}{Si}}-O-Si + H_2O \tag{3}$$

(d) Polymerization:

$$2nR-Si(OH)_3 \longrightarrow HO\left[\underset{OH}{\overset{R}{\underset{|}{\overset{|}{Si}}}}-O-\underset{OH}{\overset{R}{\underset{|}{\overset{|}{Si}}}}-O\right]_n + 2nH_2O \tag{4}$$

(e) Reaction with the polymer:

$$C\overset{\diagup\quad\diagdown}{\underset{O}{\quad}}C + RNH_2 \longrightarrow HO-C-C-NHR$$

$$\longrightarrow HO-C-C-\overset{R}{\underset{|}{N}}-C-C-OH \tag{5}$$

In this case reaction of the primary amine group in an aminosilane with an epoxide group.

Hydrolysis, Eq. (1), may take place on the surface by reaction with surface water or in solution prior to application, occurring rapidly in neutral or slightly acidic water solutions and only slowly in hydrocarbon solvents [31]. Aminosilanes are autocatalytic and do not depend entirely on hydrolysis for aqueous solubility [32]. Polymerization, Eq. (4), may occur not only on the surface as the silane triols, $RSi(OH)_3$, condense to form oligomeric siloxanes as in Eq. (3) via disiloxanols and trisiloxanols but also in solution before application. The speed at which this occurs and the oligomers became insoluble depends on silane concentration, solution pH, the presence of soluble catalytic salts [16], and the type of silane [33]. The pH factor is particularly important in silane technology.

The hydrolyzed silanol group will react with inorganic surface hydroxyl groups to form hydrogen bonds, Eq. (2), followed by condensation to form oxane bonds, Eq. (3). It should be noted that both hydrogen and oxane bond formation is reversible. Equation (5), reaction with the polymer, is a typical example of many possible reactions, depending on the functional groups on the silane and the polymer. In work with surface coatings, Walker [34] has postulated a variety of possible chemical reactions between silanes and the functional groups present in epoxide and polyurethane compositions.

B. Nature of Silane Films on Metals and Glass

Bascom [35] employed a variety of techniques in his study of the structure of silane films deposited on glass and metal substrates and concluded that vinyl-, amino-, and chloro-functional silanes films were deposited as polysiloxanes, some of which could easily be removed from the surface by organic solvents or water. Contact-angle measurements on the remaining strongly retained material indicated it to be of an open polymeric structure since it was easily penetrated by the wetting liquids. That the critical surface tension of a silane film is not an important factor in adhesion promotion by silanes is indicated by many measured values below the minimum Y_c value of about 35 dyn/cm for polyesters and 43 dyn/cm for epoxides for optimum wetting to occur [16]. It is suggested that on glass the performance of reactive silanes parallel reactivity rather than polarity (as described by the solubility parameters) of the organofunctional groups [36]. Plueddemann concludes that reactivity of the silane in copolymerization is much more significant than polarity or wettability [16].

Films deposited from nonpolar solvents are relatively thick (>1000 Å) and resistant to desorption; films from polar solvents are generally thinner (< 100 Å) and easily disrupted by polar solvents. An adsorbed silane film can consist of different strata: a silane interface with covalent bonding [10], a relatively cross-linked intermediate layer, and a superimposed layer of relatively un-cross-linked material. In practice, adsorbed films on both glass and metals are discontinuous and consist of discrete islands or agglomerates, called the button-down theory [37].

The molecular structure of silane films has been shown to depend to a great extent on the pH of the solution from which it was deposited. Using modified infrared spectroscopy to examine films of γ-aminopropyltrimethoxysilane (APS) adsorbed on iron and aluminum surfaces, Boerio and Williams [38] demonstrated that the nature of the film was highly pH dependent. When deposited from solutions of pH below 9.5, the films were of the structure indicated by the expected interaction with oxides with the amine functional groups uppermost and available for reaction with the polymer. When the solution pH was greater than 9.5, the film structure was reversed, suggesting a reaction between the amino groups and the surface. In this case the organofunctional groups were not available for reaction with the polymer. Further, this upside-down structure resulted in less hydrolytically stable bonds.

In an investigation of epoxide joints on iron and titanium using γ-APS as a primer, Boerio [39] concluded that although the film structures formed by γ-APS adsorbed onto the two metals were very similar, the performance of the films as adhesion promoters was very different. He concluded that the performance was determined by the orientation of the APS molecules at the oxide surface rather than by the overall structure of the film. The orientation was determined by the isoelectric point of the oxide and the pH at which the films were adsorbed onto the oxide [39,40]. A comprehensive account of the structure of APS silane films is provided by Ishido and co-workers [41].

C. Performance of Silanes in Adhesive Technology

Boerio and co-workers [32,39] showed that the average shear stress of epoxide/titanium lapshear joints primed with γ-APS at either pH 10.4 or 8.0 showed almost no decrease in strength after water immersion at 60°C for 60 days; unprimed controls lost 75% of their original strength. Epoxide/iron joints primed with APS at pH 8.0 retained 75% of their original strength after 60 days. Kaul and co-workers [24,42] investigated the strength of epoxide/aluminum single lapshear joints primed with γ-APS and showed that the use of γ-APS resulted in a lower dry bond strength than the unprimed control, and thicker films produced even lower bond strengths. The strength retention of unprimed joints after water immersion at 55°C was approximately 80% of the original, whereas the joints primed and standard dried (1 h at 25°C under vacuum) maintained only 50% of the dry strength. Joints primed and dehydrated (10 days at 110°C under vacuum) before bonding showed more than 95% retention.

In a particularly useful paper, Gledhill and co-workers [43] investigated the effects of silane type, solution pH, solution age, and drying on the bond strength of a bisphenol A diglycidyl ether–based epoxide in a butt joint configuration. Several silanes were investigated. Unprimed joints showed a fall in bond strength from 37 to 5.8 MPa after 1500 h of immersion in water at 60°C. Joints treated with an aqueous 1% solution of γ-glycidoxypropyltrimethoxysilane (GPMS) aged for periods up to 24 days before application showed a retained bond strength of between 17.5 and 34.4 after the same immersion time, with a peak retention between 30 and 90 min of aging. In a similar experiment using a solution of 95 parts of ethanol and 4 parts of water there was no solution age dependency, the recorded joint strengths were lower, and there was no evidence of increased water resistance. Attempts to accelerate the drying of the aqueous γ-GPMS film resulted in a marked reduction in bond strength. In a study of bismaleimide adhesives the same authors showed that the use of an aqueous solution of γ-APS increased the bond strength of the unprimed joint from 9.7 MPa to 23 MPa, but was highly dependent on solution pH.

Kerr and Walker [28] investigated the bond strength of a two-pack polyamide-cured adhesive and a diphenylmethane diisocyanate–cured polyester adhesive on mild steel, stainless steel, and aluminum in a butt tensile configuration using a range of silanes as pretreatment primers and additives. It was shown that not all silanes were effective adhesion promoters on all substrates. The most effective silanes were γ-mercaptopropyltrimethoxysilane (MPS) on stainless and mild steel and N-β-(aminoethyl)-γ-aminopropyltrimethoxysilane (AAMS) on degreased aluminum and stainless steel, with the urethane, where a 20% improvement in bond strength was achieved. AAMS was the most effective on degreased aluminum with the epoxide. On grit-blasted substrates, considerably higher bond strengths were achieved. In the comparative trials all the silanes were found to be more effective when used as pretreatments rather than as additives, a finding directly opposed to that found in the case of surface coatings, where the opposite was true. It was considered that this was a function of viscosity and curing time. After exposure to 100% relative humidity for periods up to 2 years, stainless steel specimens coated with MPS and γ-GPMS showed an equilibrium bond strength retention more than double that of the uncoated controls. On glass the retention values were four to five times greater.

Hong and Boerio described a particularly interesting practical use of silanes in obtaining good adhesion to mineral oil–contaminated steel substrates [44]. They showed that the addition of 5 wt % of γ-GPMS to amidoamine–cured epoxide adhesives [Epon 828 and V115 (both Shell Chemical Co.)] to oil-contaminated mild steel lapshear specimens increased the initial bond strength from 968 psi to 1556 psi. More surprisingly, the specimens immersed in boiling water for 12 h increased to 1681 psi, whereas the nonsilane control decreased to 665 psi.

VI. ZIRCONATES

Organometallic compounds based on zirconium are actively being promoted as adhesion promoters and are claimed to function as coupling agents. Zirconium compounds appear to have widespread potential for use in the polymer industries since they exist in both water and organic solvent–soluble forms. The aqueous chemistry is dominated by hydrolysis, depending on zirconium and hydrogen ion concentration and the nature and concentration of anions present. Depending on the ligand present the polymeric species in solution can be cationic, anionic, or neutral [17]. Simplified structural representations of polymeric zirconium species are shown in Fig. 1.

Cationic: zirconium oxychloride

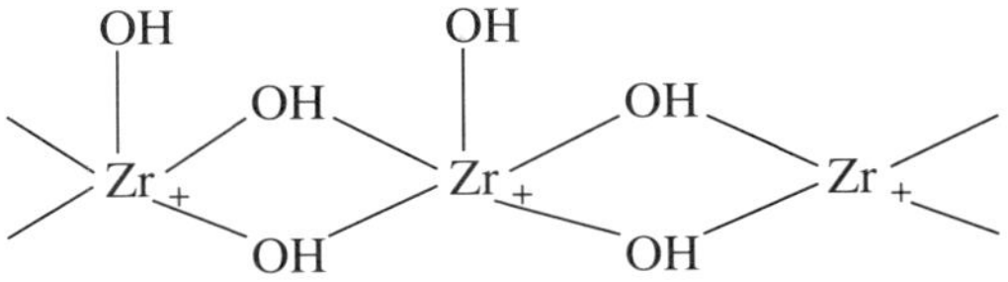

Anionic: ammonium zirconium carbonate

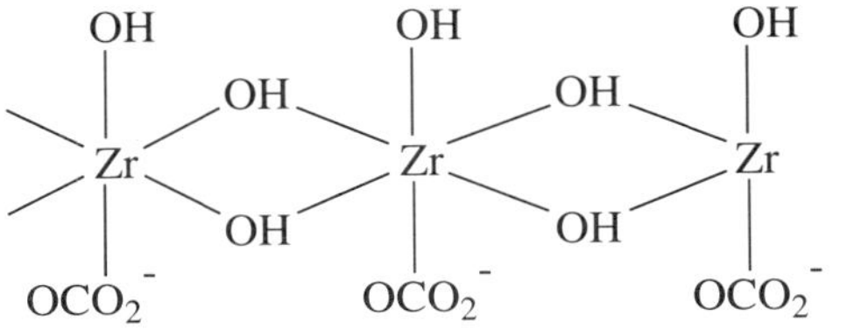

Neutral: zirconium acetate

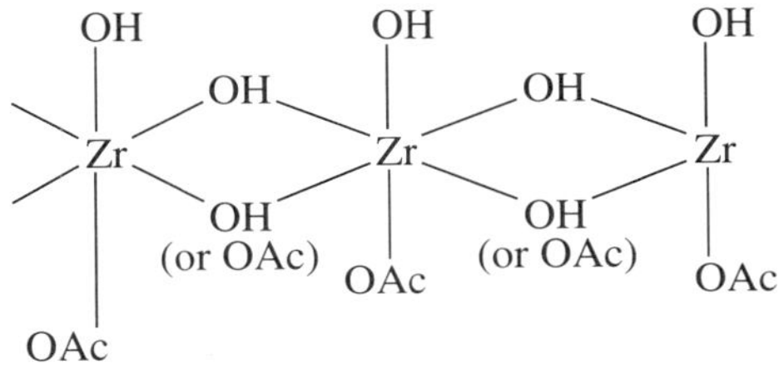

Figure 1 Structural representations of polymeric zirconium species.

Solvent soluble compounds include zirconium acetylacetonate, zirconium methacrylate, and the family of neoalkoxyl zirconates. Some commercially available zirconates are shown in Table 3. Wang [8] has described the synthesis of a soluble linear Schiff base zirconium–based coordination polymer ($N,N'N''N'''$-tetrasalicylidene-3,3′-diaminobenzidene) zirconium, and other hybrid copolymers, and has demonstrated improved adhesion on glass and aluminum substrates for poly(methyl methacrylate), polyethylene, and polypropylene when used as hot-melt compounds.

Studies on a reactive PVA copolymer (commercial vinyl alcohol), vinyl acetate copolymers stabilized by *N*-methylol acrylamide, and unreactive PVA homopolymer emulsions have shown that the resistance of these materials to cold- and boiling-water immersion tests can be improved considerably by the addition of zirconium oxychloride, zirconium hydroxychloride, and zirconium nitrate. These improvements in water resistance are considered to occur in the former case by interaction between a polynuclear zirconium species and functional groups on the polymer and in the latter by reaction with the colloidal stabiliser. In these cases the zirconium compounds are functioning by an insolubilization process rather than as an adhesion promoter per se, but the end result is an increase in wet adhesion [45].

Other zirconium-containing coupling agents are the zircoaluminates, which are described in the technical literature [46] as inorganic polymer backbone materials of low molecular mass containing specific atom ratios of zirconium and aluminum with two organic ligands: one for overall molecular stability and the second to confer organo func-

Table 3 Typical Zirconate Coupling Agents

Chemical description	Structure
Neoalkoxytrisneodecanoyl zirconate	$RO\text{-}Zr\left[O\text{-}\overset{\overset{O}{\|}}{C}\text{-}C_9H_{19}\right]_3$
Neoalkoxytris(dodecanoyl)benzene sulfonyl zirconate	$RO\text{-}Zr\left[O\text{-}\overset{\overset{O}{\|}}{\underset{\underset{O}{\|}}{S}}\text{-}\langle\text{--}\rangle\text{-}C_{12}H_{25}\right]_3$
Neoalkoxytris(ethylenediaminoethyl) zirconate	$RO\text{-}Zr(O\text{-}C_2H_4\text{-}NH\text{-}C_2H_4\text{-}NH_2)_3$
Neoalkoxytris(*m*-aminophenyl) zirconate	$RO\text{-}Zr(O\text{-}C_6H_4\text{-}NH_2)_3$
Zirconium propionate	(structure)

$$X = OH \quad or \quad O_2CC_2H_5$$

tionality. The commercial range includes amino, carboxy, and methacryloxy-functional compounds. It is claimed that the addition of these materials to adhesives and surface coatings improves their wet adhesion materially [47].

VII. TITANATES

A comprehensive patented range of titanates is marketed by Kenrich Petrochemicals, Inc. under the trade name Ken-React, and most of the published data on titanates emanates from this source. It is claimed that a typical titanate coupling agent provides six functions [17], although only three may be considered relevant to their use as adhesion promoters: the reaction of the alkoxy group of the titanate with free protons on the mineral surface to form an organic monomolecular layer on the substrate; transesterification resulting in cross-linking with carboxyl and hydroxyl groups in the polymer; and possibly chain entanglement. Titanate coupling agents are unique in that their reaction with free protons on the substrate surface results in a monomolecular layer on the mineral surface whether it be filler particle or metallic substrate. The reaction proceeds according to the equation

$$R'O\text{=}Ti(OR)_3 + M\text{=}OH \rightarrow MOTi(OR)_3 + R'OH \tag{6}$$

where $R'O$ is a hydrolyzable moiety. Cassidy and Yager [48] speculate that ester linkages

Table 4 Typical Titanate Coupling Agents

Chemical description	Structure	Type
Isopropyl tri(N-ethylaminoethylamino) titanate	CH_3 | $CH_3\text{-}CH\text{-}O\text{-}Ti(\text{-}O\text{-}C_2H_4NHC_2H_4NH_2)_3$	Monoalkoxy
Isopropyl triisostearoyl titanate	CH_3 | $CH_3\text{-}CH\text{-}O\text{-}Ti(\text{-}O\text{-}CCH_{17}H_{35})_3$	Monoalkoxy
Titanium di(dioctylpyrophosphate)oxy acetate	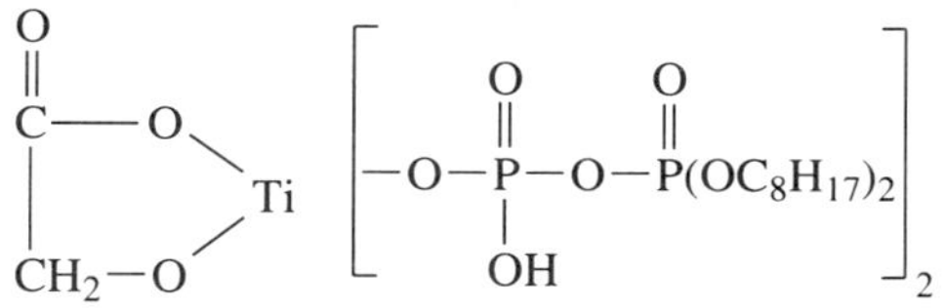 (see structure below)	Monoalkoxy
Tetraisopropyl di(dioctylphosphito) titanate	$(CH_3{-}CH{-}O{-})_4Ti\left[(H{-}\overset{O}{\overset{\|}{P}}(OC_8H_{17})_2\right]$	Coordinate
Neoalkoxytri[p-N-(β-aminoethyl)amino phenyl] titanate	$RO{-}Ti(OC_6H_4NHC_2H_4NH_2)_3$	Neoalkoxy

are hydrolyzed and coordination or condensation occurs between the resulting hydroxyl groups and substrate surface groups. Calvert and co-workers [49] infer the presence of strong bonds between isopropoxytitanium tristearate and SiO_2 and Al_2O_3 by the failure to remove the coupling agent by an extended hot-water treatment.

XPS studies by Yang and co-workers [50] showed that aluminum and steel surfaces treated with di(dioctyl)pyrophosphate were covered with the titanate coupling agent, and in the case of steel, the octyl groups of the titanate molecule were uppermost, confirming the view that titanates modify hydrophilic metal oxide surfaces with a hydrophobic organic layer. The possibility has been raised that acidic surface sites on glass may catalyze condensation with surface silanols when chelate titanium acetyl acetonate is used [8]. The range of chemical types include monoalkoxy, chelate, coordinate, neoalkoxy, and cycloheteroatom. A very few of the wide range of commercial titanate coupling agents available are shown in Table 4.

Although there are many references to the improvement in adhesion of surface coatings obtained by the use of titanates, numerical data are sparse [51] and few are available on their use in conjunction with adhesives. The literature contains conflicting evidence on the value of titanate adhesion promoters, and in an investigation of eight titanates tested with an acrylic resin only two titanates performed better than the nontitanate control. It has been claimed that alkyl titanates are effective coupling agents for polyethylene [52–54]. Using isopropyl triisostearoyl titanate as a primer for polyethylene/

Al$_2$O$_3$ joints, Sung and co-workers suggested that it was unlikely that this particular titanate functioned as an adhesion promoter in this system, notwithstanding the observation that heating the titanate above 70°C in vacuo resulted in a significant increase in peel strength [55].

Calvert and co-workers [49] have demonstrated the presence of isopropyl isostearate and isopropyl laurate in commercial isopropyl triisostearoyl titanate and conclude that this is the reason why the commercial product does not function as an adhesion promoter; treatment at 70°C in vacuo removes these fatty acid esters. It has been suggested to the author that failures to obtain improvements in adhesion using titanates can be remedied by isolating the pure compound (B. Nordenheim, private communication, 1988). It is possible that improvements in adhesion are more likely when commercial titanates are used as additives rather than as pretreatment primers.

It is only fair to say that the trade literature [17] is emphatic that it is critical to use the correct amount of titanate coupling agent. The use of excessive amounts is probably the most significant factor in application failure tests. It is strongly recommended that selected titanates should be examined in a range of concentrations from 0.1 to 2.0% by mass in a filled system and even lower for unfilled systems. Excess titanate will result in unreacted alkoxy groups on the surface and in a loss of adhesion of the polymer. This could lead to the mistaken conclusion that a particular titanate was unsuitable or even harmful.

In general, titanates with the more polar organic moieties such as isopropyl tri(*N*-ethylenediamio)ethyl titanate and neoalkoxy tri[-*P*-*N*-(β-aminoethyl)amino phenyl] titanate are recommended for adhesion promotion to polar substrates. Titanates with relatively nonpolar moieties, such as aliphatic carboxy titanates and isopropyl tri(dioctylphosphato)titanate, will adhere better to the nonpolar substrates.

VIII. CHROMIUM-CONTAINING PROMOTERS

Adhesion promoters containing chromium fall into two main classes: inorganic and organic complexes. Examples of the former are the chromate conversion coatings used extensively in the aerospace industries for pretreatment of aluminum and its alloys. Although acting as anticorrosion coatings in their own right, they improve paint adhesion substantially [56]. Chromium conversion coatings that may be of chromium phosphate (amorphous, accelerated, or nonaccelerated) may be applied by brush dip or spray. The crystalline chromium phosphate type is normally restricted to steel surfaces. In a simulated sterilization test using epoxide can coatings on an aluminum surface, Paramanov and co-workers showed that the use of a chromate conversion coating was essential for satisfactory adhesion [57].

Examples of the organic type are the coordination complexes of trivalent chromium chloride with carboxylic acids (Volan manufactured by Du Pont). The methacrylate–chrome complex is well known in fiberglass technology. In water solution the chromium chlorides hydrolyze to form basic salts that form oligomeric salts through olation of hydroxyl groups on the adjacent chromium molecules. Hydroxyl groups also bond to silanol groups on the glass surface via hydrogen bonding and possibly covalent oxane bonds. The organic acid group develops a fairly stable bond to chromium by being coordinated to adjacent chromium atoms [8]. Other chromium complexes of functional carboxylic acids have been proposed [58]. Following is a typical structure:

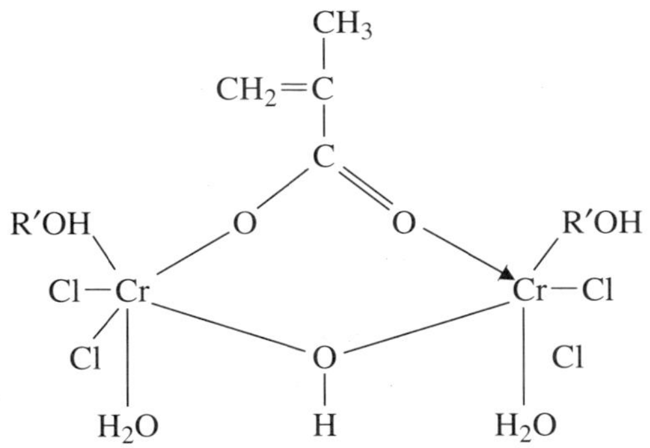

$$
\begin{array}{c}
CH_3 \\
| \\
CH_2{=}C \\
| \\
C \\
O \diagdown \quad \diagup O \\
\end{array}
$$

R'OH⟍   R'OH⟋
Cl—Cr⟍   ⟋Cr—Cl
Cl | O   Cl
H_2O   H   H_2O

A trivalent chromium fumarato-coordination compound, Volan 82, has been claimed to be an effective adhesion promoter for polyethylene coatings on aluminum. The toxicity of chromium compounds must place a question mark against their continued use as adhesion promoters.

IX. OTHER ADHESION PROMOTERS

In addition to the organometallic adhesion promoters, a large number of inorganic, organometallic, and organic compounds have been investigated, usually in specific adhesives and coatings or on selected substrates. A comprehensive account of coupling agents in use prior to 1969 is provided by Cassidy and Yager [48]. Although much of the reported information on other adhesion promoters concerns their use with surface coatings or filled systems, it is likely that many would also be suitable for use with adhesives and have therefore been included in the interests of completeness.

A. Phosphorus-Containing Compounds

Tritolyl phosphate (TTP) has been examined as a pretreatment for E-glass in epoxide laminates and thermoplastic adhesives for bonding poly(vinyl chloride) to aluminum, steel to zinc, and acrylonitrile–butadiene styrene to aluminum [59]. Mono- and diphosphate esters have been claimed to be suitable adhesion-promoting primers for acrylic adhesives on metal [60,61] unsaturated acid phosphates have been suggested as primers for use on metals to be bonded with free radical–initiated adhesives [42] and thiophosphates esters for adhesives to be used on plastics, ceramics, and metals [62].

B. Amines

Hydroxybenzamines of the general formula have been claimed to improve the adhesion of a wide range of coatings to zinc and cadmium and other metallic substrates when used as either pretreatment primers or additives [63]. Ethylenically unsaturated hydroxy-functional amines have been claimed to improve the adhesion of water-based systems [64], and amines have been examined as adhesion promoters for aromatic isocyanate–cured adhesives on glass and other substrates [65]. Primary aliphatic amines are claimed to improve the bondability of polyolifines [66] and a oxyethylated polyethylene polyamine claimed for polymer-to-polymer bonding [67].

4-Ethylpiperidine has shown promise as an adhesion promoter [68], aminoethyl-

$$HO \longrightarrow \underset{R'}{\overset{}{\bigcirc}} - CH_2 - N \overset{R}{\underset{H}{\diagdown}}$$

piperidine has been shown to be beneficial in steel–epoxide systems [48], and primers based on piperidine derivatives improve the adhesion of adhesives to aluminum and stainless steel [69]. Pyridene derivatives such as 2-(2-methylethyl)pyridene are claimed to be effective primers for crystalline and nonpolar polymers to be bonded with cyanoacrylate adhesives [70].

C. Organic Resins

A wide variety of organic resins have been claimed to act as adhesion promoters on many different substrates. Mahajan and Ghatge have reported that the use of a liquid epoxide resin (epoxide equivalent 260) materially improved the initial adhesion of polysulfide sealants to anodized aluminum alloy [71]. Abietate-terminated polysulfide polymers, epoxide-terminated polysulfide polymers, and abietate-terminated polyesters have also been claimed to improve the adhesion of polysulfide sealants [72]. Oxygenated fluorocarbon primers are claimed to improve the bonding of polyacetal and polyamide–imide substrates [73]. A primer based on a ethylene–vinyl mercaptoacetate copolymer has been shown to improve the adhesion of epoxide adhesives to steel [74]. Primers containing diorganopolysiloxanes are claimed to improve the adhesion to silicone elastomers [75], and aromatic polyether resins with aminophosphonic acid groups are stated to improve the adhesive bonding of steel, galvanized iron, and aluminum [76].

D. Miscellaneous Promoters

DeNicola and Bell report the use of bibenzoylmethane and 1-(O-hydroxyphenyl)-3-phenyl-1,3-propanediol as wet adhesion promoters for epoxide resin adhesives on low-carbon mild steel [77]. Metal chelating O-hydroxybenzlamine compounds are stated to produce adhesion-promoting films on metals [78], and improved adhesion to titanium is claimed for metal alkoxide primers. Oxazolidines containing trialkoxy or triaryloxysilyl groups are claimed to be adhesion promoters on metallic substrates [79]. Improved adhesion of epoxide resin adhesives to copper substrates can be achieved by pretreatment in a weak solution of a benzoheterocyclic(thiol) compound [80] and benzotriazole and derivatives have been claimed to improve the bond between vinyl polymers and steel [81] and the adhesion of photosensitive compounds to polymeric substrates [82]. Pesetski and Aleksandrova describe the use of dicarboxylic acids as primers for polyamide films on copper [83].

X. EFFECTS OTHER THAN ADHESION PROMOTION

It should be noted that side effects are possible when using adhesion promoters/coupling agents as additives in adhesives and coatings, usually beneficial but not invariably so. Beneficial effects from the use of titanates include deagglomeration, improved wetting, improved corrosion resistance, increased hydrophobicity, electrical conductivity in conducting systems, and acid resistance. The overall rheology of filled systems may be changed by both titanates and silanes requiring products to be reformulated. Silanes, particularly amino silanes, may function as curing agents or accelerators in epoxide and urethane adhesives, thereby reducing the pot life of a mixed system. These side effects should be considered in the selection or rejection of an adhesion promoter even if adhesion is the primary concern.

XI. CONCLUSIONS

In attempting to cover such a wide and diverse topic as adhesion promoters in a short paper, the author is aware of many gaps in the information presented and the omission of many aspects of theoretical and practical interest, it is hoped that the references quoted will fill many of these gaps. It has been demonstrated that adhesion promoters/coupling agents have a major role to play in the development of adhesive technology, but only if due regard is paid to the importance of matching the promoter to the substrate and the polymer, pH effects, solution age, dosage, and film thickness. Plueddemann concludes that the performance of coupling agents may depend as much on the physical properties resulting from the method of application as on the chemistry involved.

REFERENCES

1. P. Walker, *Offic. Digest 37*: 1561 (1965).
2. D. J. Falconer, N. MacDonald, and P. Walker, *Chem. Ind.* 1230 (July 1964).
3. R. A. Gledhill, S. J. Shaw, and D. A. Tod, *Intern. J. Adhesion Adhesives 10*: 192 (1990).
4. J. Bjorksten and L. L. Yaeger, *Mod. Plastics 29*: 124 (1952).
5. D. C. Bradley, D. C. Hancock, and W. Wardlaw, *J. Chem. Soc.*, 2773 (1952).
6. S. Naviroj, S. R. Culler, J. L. Koenig, and H. Ishida, *J. Colloid Interface Sci.* 97: 308 (1984).
7. Q. Yang and Q. Zhou, in *Coordination Compounds in Adhesive Chemistry* (L. Lee, ed.), Plenum Press, New York, 1984, p. 799.
8. B. Wang, Ph.D. dissertation University of Massachusetts, Feb. 1989.
9. M. Gettings and A. J. Kinlock, *J. Mater. Sci. 12*: 2511 (1977).
10. H. Ishida and J. L. Koenig, *J. Colloid Interface Sci. 64*: 555 (1978).
11. M. R. Rosen, *J. Coating Technol. 50*: 644 (1978).
12. P. W. Erickson and E. P. Plueddemann, *Composite Materials*, Vol. 6, Academic Press, New York, 1974, Chap. 6.
13. W. Southeng, *Polymer Interface and Adhesion*, Marcel Dekker, New York, 1982.
14. W. D. Bascom, *Composite Materials*, Vol. 6, Academic Press, New York, 1974, Chap. 2.
15. P. Walker, *J. Oil Colour Chemists' Assoc. 66*: 415 (1982).
16. E. P. Plueddemann, *Silane Coupling Agents*, Plenum Press, New York, 1982.
17. *Technical Bulletin KR-1084-2*, Kenrick Petrochemicals, Inc., Bayonne, 1987.
18. E. P. Plueddemann, *J. Paint Technol. 42*: 600 (1970).
19. P. Walker, *J. Oil Colour Chemists' Assoc. 65*: 436 (1982).
20. P. Walker, *J. Oil Colour Chemists' Assoc. 66*: 188 (1983).
21. P. Walker, *J. Oil Colour Chemists' Assoc. 67*: 108 (1984).
22. P. Walker, *J. Oil Colour Chemists' Assoc. 67*: 126 (1984).
23. P. Walker, *J. Adhesion Sci. Technol. 5*: 279 (1991).
24. A. Kaul and N. H. Sung, *Polymer Eng. Sci. 26*: 768 (1980).
25. J. C. Bolger and A. S. Michaels, in *Interface Conversion for Polymer Coatings* (P. Weiss and G. D. Cheeves, eds.), American Elsevier, New York, 1969.
26. E. P. Plueddemann, *Prog. Org. Coatings 11*: 297 (1983).

27. P. Walker, in *Surface Coatings*, Vol. 1 (A. D. Wilson, J. W. Nicholson, and H. J. Prosser, eds.), Elsevier, New York, 1987, Chap. 6.
28. C. Kerr and P. Walker, in *Adhesion*, Vol. 11 (K. W. Allen, ed.), Elsevier, New York, 1987, Chap. 2.
29. M. J. Owen and D. E. Williams, *J. Adhesion Sci. Technol. 5*: 307 (1991).
30. E. P. Plueddemann, *J. Adhesion Sci. Technol. 5*: 261 (1991).
31. E. P. Plueddemann, *Soc. Plast. Ind. 19A* (1969).
32. *Silane Coupling Agents*, Dow Corning Corp., Midland, 1981.
33. *Silane Coupling Agents*, Dow Corning Corp., Midland, 1970.
34. P. Walker, *J. Coatings Technol. 52*: 49 (1980).
35. W. D. Bascom, *Macromolecules 5*: 792 (1972).
36. E. P. Plueddemann, *J. Paint Technol. 40*: 516 (1968).
37. E. P. Plueddemann, *J. Adhesion Sci. Technol. 5*: 261 (1991).
38. F. J. Boerio and J. W. Williams, *Application of Surface Science*, Vol. 7, North-Holland, Amsterdam, 1981.
39. F. J. Beorio and R. G. Dillingham, in *Proc. International Symposium on Adhesive Joints: Formation Characterisation and Testing* (K. L. Mittal, ed.), Plenum Press, New York, 1982.
40. F. J. Boerio, *Polymer Prepr. Am. Chem. Soc. Div. Polymer Chem. 24*: 204 (1983).
41. H. Ishida, C. Chang, and J. L. Koenig, *Polymer 23*: 251 (1982).
42. A. Kaul, N. H. Sung, I. J. Chin, and C. S. P. Sung, *Polymer Prepr. 26*: 113 (1985).
43. R. A. Gledhill, S. J. Shaw, and D. A. Tod, *Intern. J. Adhesion Adhesives 10*: 192 (1990).
44. S. G. Hong and F. J. Boerio, *J. Adhesion 32*: 67 (1990).
45. Anon., *Data Sheets 114C* and *114D*, Nov. 1983, and *Data Sheet 157* Nov. 1984, Magnesium Elektron, Twickenham, Surrey, England.
46. B. L. Cohen, *High Solids Coatings 9*: 2 (1984).
47. F. M. Young and L. L. Rouch, *Am. Chem. Soc. Div. Org. Coatings Plast. Chem. Prepr. 27*: 110 (1967).
48. P. E. Cassidy and B. J. Yager, *Tracor Document TTO-AV-7362-U*, NASA Contract NA58-24073, 1971.
49. P. D. Calvert, R. R. Lalanandham, and D. R. M. Walton, in *Interfacial Coupling by Alkoxytitanium and Zirconium Tricarboxylates* (K. L. Mittal, ed.), Plenum Press, New York, 1983.
50. C. Q. Yang, J. F. Moulder, and W. Fateley, *J. Adhesion Sci. Technol. 2*: 11 (1988).
51. Cleveland Society for Coatings Technology, *J. Coating Technol. 51*: 38 (1979).
52. Japanese patent 81,999,266.
53. S. J. Monte and G. Sugerman, in *Adhesion Aspects of Polymeric Coatings* (K. L. Mittal, ed.), Plenum Press, New York, 1983.
54. C. L. Gray, Jr., W. L. MacCarthy, and T. F. McLaughline, *Mod. Packaging 34*: 143 (1961).
55. N. H. Sung, A. Kaul, S. Ni, C. S. P. Sing, and I. J. Chin, *Proc. 35th Annual Technical Conference of the Reinforced Plastics/Composites Institute*, 1980.
56. P. Walker, *Proc. American Electroplaters and Surface Finishers Society, 73rd Annual Technical Conference*, Chicago, Paper R5, Session R, July 1987.
57. V. A. Paramonov, R. K. Italova, I. M. Katser, and V. A. Litvinenko, *Prot. Metals 19*: 343 (1983).
58. J. A. Robertson and J. W. Treibilock, *Tappi 58*: 106 (1975).
59. L. Deutsch, M. E. Shrader, I. Lerner, and F. J. D'oria, *Proc. Annual Technical Conference, 22nd SPI Reinforced Plastics Division*, Washington, D.C., 1967.
60. European patent application 334,492.
61. Japanese unexamined patent 01/297,482.
62. Japanese unexamined patent 01/132,677.
63. U.S. patent 4,357,181.
64. U.S. patent 3,356,655.

65. F. Liang and P. Dreyfuss, *Papers 4, 8, 18*, American Chemical Society Division of ORPL, Cleveland, 1983.
66. European patent application 2,199,240.
67. Russian patent 1,388,410.
68. S. M. Wilhelm, Y. Tanizawa, C. Y. Lu, and N. Hackerman, *Corrosion Sci. 22*: 791 (1982).
69. Japanese unexamined patent 62/250,076.
70. Japanese unexamined patent 62/195,071.
71. S. S. Majahn and N. D. Ghatge, *Intern. J. Polymer. Mater. 10*: 141 (1983).
72. U.S. patent 3,838,078.
73. Japanese unexamined patent 62/125,092.
74. R. G. Schmidt and J. P. Bell, *J. Adhesion 27*: 135 (1989).
75. European patent application 336,854.
76. U.S. patent application 4,781,984.
77. European patent application 084,111.
78. European patent application 276,079.
79. European patent application 253,776.
80. U.S. patent 4,428,987.
81. European patent application 233,987.
82. European patent application 247,549.
83. S. S. Pesetski and O. N. Aleksandrova, *Colloid J. USSR 52*: 251 (1990).

5

Testing of Adhesives

K. L. DeVries and P. R. Borgmeier   *University of Utah, Salt Lake City, Utah*

I. INTRODUCTION

The molecular mechanisms by which materials can adhere to one another have not been determined unambiguously. To date, no one has been able to predict reliably the strength of an adhesive joint based purely on the properties of the adhesives and adherends. Rather, to determine the strength of a joint, one must resort to testing. As will hopefully be made clear in this chapter, testing is not always as straightforward as it might appear superficially. Testing is, however, an important aspect of adhesive science and technology. Tests are conducted for a wide variety of purposes. There are a large number of different standard adhesive tests available to the technologist, engineer, or scientist, depending on his or her goals. It is essential not only that the proper test be selected (or designed) and that care be exercised in conducting the test, but also that the results be interpreted properly. The latter aspect often requires considerable background and insight.

II. STANDARD TESTS

In the United States, the American Society for Testing and Materials (ASTM) is the organization that has assumed the responsibility for the "development of standards on characterization and performance of materials, product systems and services, and the promotion of related knowledges." The testing specifications of other organizations (such as the military in their "Mil-Specs") often parallel those specified by ASTM. ASTM operates as a source of voluntary consensus standards. Most other countries (or groups of countries) have similar organizations, and there is considerable interaction and interchange between these groups from the various countries. For example, many of the standards adopted by European and Asian groups find their basis in ASTM, and vice versa. This is accomplished through coordinating committees from the various countries that meet frequently. While the standards of most of these countries could be cited in this chapter, the authors will rely almost exclusively on ASTM since these are the standards with which they are most familiar. The interested reader can usually find comparable standards in his or her own country. ASTM publishes an Annual Book of Standards that updates the test methods and other details on a yearly basis. The changes are seldom revolutionary but rather, evolutionary. New standards do appear, however, and all standards are reviewed periodically, at which time they may be eliminated or updated. The responsibility for developing new standards, revising existing standards, and approving standards for publication falls on ASTM volunteer committees. Of the more than 30,000 members of ASTM, nearly two-thirds serve on such committees. This is no small task. The 1990 Annual Book of Standards was composed of 68 volumes, divided among 16 sections. It is not uncommon for these volumes to be more than 500 pages long and some have nearly twice this many pages. The D-14 committee has the primary responsibility for adhesives. Volume 15.06 is the ASTM publication that covers most adhesive standards. Including indexes and the like, it has some 485 pages covering 116 standards.

To this point we have used the term *standard* in a very general sense. ASTM has a hierarchy of types of communications of this general nature, which are defined as follows:

1. *Classifications* are a systematic arrangement or division of materials, products, systems, or services into groups based on similar characteristics (origin, properties, composition, etc.) in which the instructions or options do not recommend specific courses of action.
2. *Guides* provide a series of options or instructions but do not recommend a specific course of action. The purpose here is to offer guidance based on a consensus but not to establish fixed procedures.
3. *Practices* outline definitive procedures for conducting specific operations or functions that do not produce specific test results (comparative test methods).
4. *Specifications* are a precise statement of a set of requirements to be satisfied by a material, system, service, and so on, and the procedures to be used to determine if the requirements are satisfied.
5. *Terminology* is a document that helps standardize the terminology, their definitions, descriptions, symbols, abbreviations, acronyms, and so on. The relevant example here is D-907, Standard Terminology of Adhesives, which was originally published in 1947 and most recently approved by D-14 in 1990. This provides definitions for several hundred terms in common use in adhesive science and technology. Despite this very worthy effort, the use of many terms is somewhat ambiguous and a reader/ researcher must seek to determine a word's exact meaning by looking at the context in which it is used.
6. *Test methods* are definitive procedures for identification, measurement, and/or evaluation of qualities, characteristics, or properties of materials, products, systems, or services. These will, in general, produce test results.

III. SOME SELECTED STANDARDS

Tests are performed for a wide variety of different reasons. Many tests have as a goal to compare different materials, procedures, products, and so on. For such comparisons to be meaningful, it is important that some type of standard procedure be used to obtain the information that will be used for comparison. This is, of course, one of the most important reasons for having standards. The goal is to separate, as much as possible, the results obtained from differences due to the laboratory or operator. In principle, one should be able to compare results from one operator (say, in Europe) with those in another place (say, in the United States).

Furthermore, all tests are not conducted with equal rigor. Quantitative testing is generally expensive and time consuming. Testing organizations have at times, therefore, developed qualitative tests that might be used to eliminate obvious unlikely candidates quickly. ASTM-D-3808, Spot Adhesion Test, is a good example of such a test. It is well established both theoretically and empirically that all adhesive–substrate pairs are not equally compatible. An adhesive that might tightly adhere to one substrate may form very weak bonds with other substrates with varying degrees between these extremes. To prepare and test standard quantitative test geometries (some of which will be described subsequently) for a large number of candidate adhesives for a given substrate could be prohibitively expensive. ASTM D-3808 suggests the following alternatives:

1. Prepare candidates substrates using techniques similar to those expected in service.
2. Mix a quantity of the candidate adhesive according to the procedures specified by the adhesive manufacturer.
3. Small spots of the adhesive (typically 5 to 10 mm in diameter) are placed on the substrates using application and curing techniques comparable to those expected in service or specified by the manufacturer.
4. The operator then uses a thin stainless steel spatula (or similar probe) to pry or lift the spot from the substrate. The operator then uses his or her senses to access ease of separating the spot from the substrate.

Based on this operation, a decision is made as to whether this system is worthy of further quantitative analysis. Some quantitative tests are the subject of the next section.

A. Tests for Adhesive Joint Strength

A relatively large number of tests have been proposed and formalized for evaluating the strength of an adhesive. Although there are some important exceptions (some covered in the following sections), the majority of the long-standing adhesive joint strength tests fall in three categories: tensile, shear, and peel.

1. Tensile Tests

Given the choice, a designer seldom uses adhesives in a direct tensile loading mode. The primary reason for this is probably related to the fact that by overlapping, scarfing, and so on, the contact area of the adhesive may be markedly increased over that of a simple butt joint. The finger joints used to produce longer pieces of lumber or in other wood construction are a familiar example of such techniques. Figure 1 clearly demonstrates the difference in bond area between a butt joint and one of the types of finger joint assemblies described in ASTM D-4688. It is noted that this geometry also results in a change in the stress state. The butt joint appears to be dominated by tensile stresses (superficially at

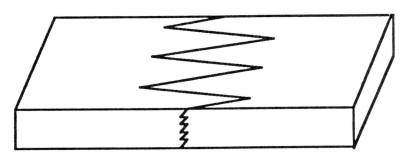

Figure 1 Typical finger joint.

least). Differences in elastic–viscous–plastic properties between the adhesive and adher-ends induce shear stresses along the bond line even for the case of the pure butt joint, but the dominate "direct" stresses are tensile. (This is perhaps an opportune point to note that as explained in more detail later, the stress distribution in joints is almost never as simple as it might first appear.) Clearly, for the tensile-loaded fingerjoint, shear stresses as well as tensile stresses are applied directly to the bond line. Another reason why tensile joints are avoided in design is because it has been observed empirically that many adhesives exhibit lower strength and high sensitivity to alignment when exposed to butt-type tensile stresses.

However, a variety of standard tensile tests are available. Sample geometries for some of the more common of these are shown schematically in Fig. 2. Figure 2a is a schematic representation of the configuration from ASTM D-897. It is used (with slight geometric alterations) for various adherends, ranging from wood to metal. It is often called "pi tensile test" because the diameter is generally chosen to yield a cross-sectional area of 1 square inch. Detailed specifications for U-shaped grips machined to slip over the collar on the bonded spool-shaped specimen, to help maintain alignment during loading, are given in the same standard. Our experience has been, however, that even with such grips, reasonable care in sample manufacture, and acceptable testing machine alignment, it is difficult to apply a really centric load. As a consequence, such experiments often exhibit quite large data scatter [1,2]. The specimens for ASTM D-897 are relatively costly to manufacture. To reduce this cost, ASTM D-2094 specifies the use of simpler bar and rod specimens for butt tensile testing. The D-2094 half specimens show in Fig. 2b are

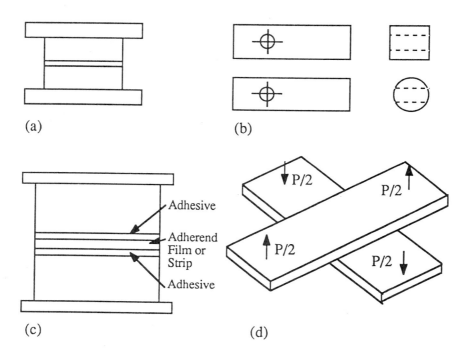

(a)

(b)

Adhesive

Adherend
Film or
Strip

Adhesive

(c)

(d)

P/2

P/2

P/2

P/2

Figure 2 Tensile adhesion test geometries: (a) pi tensile test ASTM D-897; (b) bar and rod tensile test, ASTM D-2094; (c) sandwich tensile test, ASTM D-257; (d) cross-lap tension test, ASTM D-1344.

loaded by pins through the holes. The testing specification also describes a fixture to assist in sample alignment, but this still remains a problem.

The specimens of both ASTM D-897 and D-2094 can be adapted for use with materials that cannot readily be manufactured into conventional specimen shapes, by the approach shown schematically in Fig. 2c. Such materials as plates of glass or ceramics and thin polymer films can be sandwiched between the spool-like sections made of materials that can be more readily machined such as metals. Obviously, to obtain meaningful results from this test, the strength of the adhesive bond to the spool segments must be greater than that to the thin film or sheet.

ASTM D-1344 describes a cross-lap specimen of the type shown in Fig. 2d for determining tensile properties of adhesive bonds. Wood, glass, sandwich, and honeycomb materials have been tested as samples in this general configuration. Even under the best of circumstances, one would not anticipate the stress distribution in such a case to be very uniform. The exact stress distribution is highly dependent on the relative flexibilities of both the cross beams and the adhesive. Certainly, caution must be exercised when comparing tensile strength from this test with data obtained from other tensile tests. Probably for these reasons, this test is scheduled by ASTM for discontinuation.

The results from these "tensile" tests are normally reported as the force at failure divided by the cross-sectional area. Such average stress information can be misleading. The importance of alignment has already been discussed. Even when alignment is "perfect" and the bonds are of uniform thickness over the complete bond area, the maximum stresses in the bond line can differ markedly from the average stress [3,4]. The distribution of stresses along the bond line is a strong function of the adhesive joint geometry. This is demonstrated in Fig. 3, which shows both the normal and shear stress distribution as a function of position for specimens of the general shape shown in Fig. 2a and b. These calculations assume an elastic adhesive that is much less rigid than the steel adherends and with a Poisson ratio of 0.49. The various curves in Fig. 3 represent differing diameter/ adhesive thickness ratios, shown as the parameters listed for each curve. Note that both the normal stresses and the induced shear stress vary dramatically over the bonded surface. As described later, an understanding of stress and strain distribution in the adhesives and adherends is useful in fracture mechanics analysis. The average stress results commonly reported from standard tensile tests must be used with great caution in attempts at predicting the strength of different joints, even where the joints may be superficially similar. References 2 and 4 use numerical and experimental techniques to explore the effect of adhesive thickness on the strength of butt joints. The analysis and associated experimental results in these references illustrate that the strength of the joint and the locus of the point from which failure initiates are highly dependent on the adhesive thickness/diameter ratio.

2. Shear Tests

Some of the most commonly used adhesive joint strength tests fall under the general category of lap shear tests. Such samples are relatively easy to construct and closely resemble the geometry of many practical joints. The stress distribution for lap joints is far from uniform, but again the test results are commonly reported as load at failure divided by the area of overlap. The maximum stress generally differs markedly from this average value. More important, perhaps, failure of the joint is, in all likelihood, more closely related to the value of the induced cleavage stresses at the bond termini than to the shear stresses. Figure 4 shows several of the commonly used lap shear specimen geometries, (i.e., those corresponding to ASTM D-1002, D-3165, and D-3528).

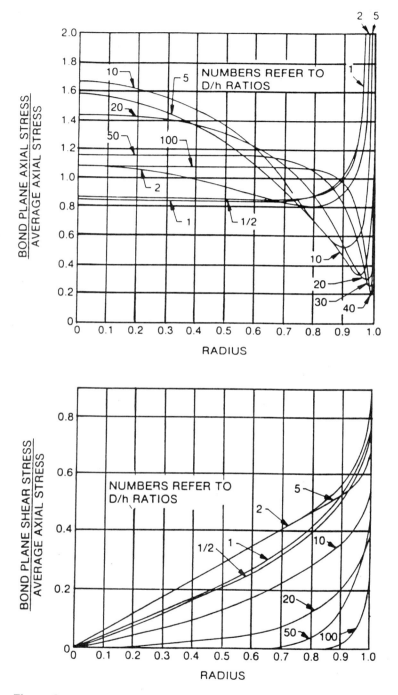

Figure 3 Axial and shear stress distributions in butt joints, $v = 0.49$. *D/h* is diameter/thickness ratio.

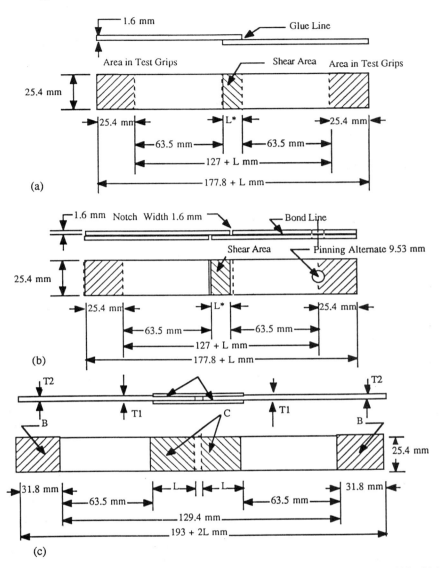

Figure 4 Standard lap shear geometries: (a) simple lap joint test, ASTM D-1002; (b) laminated lap shear joint test, ASTM D-3165; (c) double lap joint test ASTM D-3528.

Because of flow of adhesive out the sides and edges as well as other difficulties in preparing and aligning the parts when preparing individual lap specimens, samples are frequently prepared from two relatively large sheets and the specimens are "saw" cut from the resulting laminated sheet. This is illustrated in Figs. 4b and 5 and described in ASTM D-3165. It has long been recognized that lap shear specimens such as those represented in Fig. 4a and b must distort, so that the forces applied to the sample fall on the same line of action. This induces cleavage stresses in the adhesive near the bond termini. Double lap shear specimens described in ASTM D-3528 are proposed as a means of alleviating this problem (see Fig. 4c). However, based on our computer analysis and

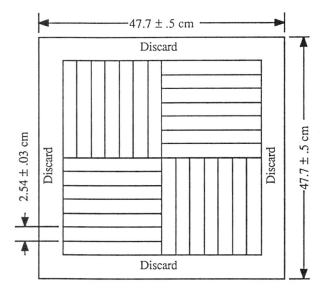

Figure 5 Large panel used to manufacture ASTM D-3165 specimens. The panel is cut into strips for testing.

experimental studies, we feel that failure of double lap joints is still dominated by cleavage stresses [5].

As was the case for tensile specimens, the stress distribution in lap joints is intricately related to the details of the specimen geometry [2,6–16]. Such factors as amount of overlap, adhesive thickness, adherend thickness, relative stiffness (moduli) of the adhesive and adherend, and other factors critically influence the stress distribution. The maximum stresses occur near the bond termini. Figure 6 shows the results of an elastic finite element analysis for the stresses in this region for a specimen of ASTM D-1002 geometry. Note that the tensile stresses are consistently higher than the shear. This fact, coupled with the fact that adhesives have typically been shown to be more susceptible to cleavage than shear failure, is consistent with the observation that cleavage rather than shear stresses usually dominate lap joint fracture.

A reason that one should be very cautious in attempting to use average shear stress criteria to predict failure of lap joints is illustrated in Fig. 7. This figure shows the force required to cause failure in ASTM D-1002 lap joint specimens using steel adherends of differing thicknesses. Note that for a given applied load, the "average shear stress" for all adherend thickness would be the same (i.e., the applied force divided by the area of overlap). If there was a one-to-one correspondence between this stress and failure, one would expect failure load to be independent of the adherend thickness. It is obvious from this figure that this is not the case. Similar results were obtained by Guess et al. [8]. The trends shown in Fig. 7 can be quantitatively explained using numerical methods and fracture mechanics [2,8]. Fracture mechanics is discussed later in the chapter.

A further indication of the popularity of lap shear tests is the number and variety of such tests that have been standardized as illustrated in the following examples. ASTM D-905, Standard Test Method for Strength Properties of Adhesive Bonds in Shear by Compression Loading, describes test geometry, a shearing tool, and procedures for testing

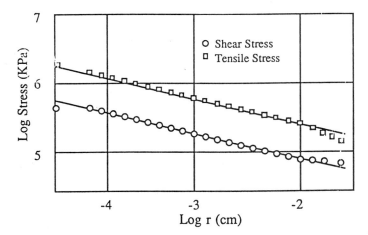

Figure 6 Log stress plot for the lap shear test.

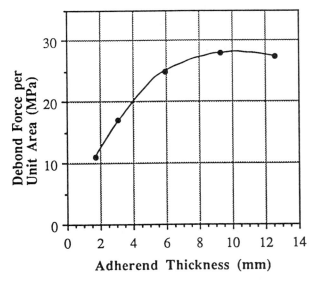

Figure 7 Force required to cause failure versus adherend thickness for ASTM D-1002 specimens using adherends of differing thicknesses.

wood and similar materials. ASTM D-906, Standard Test Method for Strength Properties of Adhesive in Plywood Type Construction in Shear by Tension Loading, provides specifications for specimen shape and dimensions, grips and jaws, and testing procedures for testing plywood materials. ASTM D-2339 is a very similar test for plywood construction materials. ASTM D-3930 describes several adhesives tests, including a block shear test for wood-based home construction materials. D-3931 is similar to D-905, but for testing gap-filling adhesives, ASTM D-3163 and D-3164 are similar to D-1002, except for use with plastics and plastic to metal instead to metal to metal. ASTM D-3983, Standard Test Method for Measuring Strength and Shear Modulus of Non Rigid Adhesives by a Thick Adherend Tensile Lap Specimen, describes a dual-transducer slip gauge, specimen geometry, and methods for determination of adhesive joint strength and modulus. ASTM D-4027 has a similar purpose but uses a more complex "modified-rail test" apparatus. ASTM D-4501 describes a shear tool and holding block arrangement for use in testing the force required to remove a square block of material bonded to a larger plate. D-4501 describes a shear test in which the specimen is a block bonded to a larger plate. D-4562 describes a shear test in which the specimen is a pin bonded inside a collar. The test uses a press to force the pin through the collar, which rests on a support cylinder. The test results in this case are the force required to initiate failure divided by the bonded area between the pin and collar. ASTM E-229 also uses a pin-in-collar type of specimen except that here torsional loading causes the failure. D-4896, Standard Guide for Use of Adhesive-Bonded Single Lap-Joint Specimen Test Results, is intended to give insight into the interpretation of the results from all the lap shear tests. At other points in ASTM Vol. 15.06 and in this chapter, reference is made to other standard tests that use "shear-type specimens" to explore moisture, other environmental, fatigue, and creep effects in adhesive joints.

3. Peel Tests

Another common type of tests is the peel test. Figure 8 shows four common types of peel specimens. One can understand the test described in ASTM D-1876 by examining Fig. 8a. In this adhesive peel resistance test, often called the T-peel test, two thin 2024-T3 aluminum or other sheets typically 152 mm wide by 305 mm long are bonded over an area 152 mm wide by 229 mm long. The samples are then usually sheared or sawed into strips 25 mm wide by 305 mm long (at times, the sample is tested as a single piece). The 76-mm-long unbonded regions are bent at right angles, as shown in Fig. 8a to act as tabs for pulling with standard tensile testing grips in a tensile testing machine.

In a related test, one of the adhering sheets is either much stiffer than the other or is firmly attached to a rigid support. Various jigs have been constructed to hold the stiffer segment at a fixed angle to the horizontal and, by using rollers or other means, allow it to "float" so as to maintain the peel point at a relatively fixed location between the grips and at a specific peel angle, as show schematically in Fig. 8b (see, e.g., ASTM D-3167).

ASTM D-1781 describes the climbing drum peel test that incorporates light, hollow drums in spool form. The sample to be peeled is attached on one end to the central (smaller) part of spool. The other end of the sample is affixed to the clamp attached to the top of the crosshead of the tensile testing machine, as illustrated in Fig. 8c. Flexible straps are wrapped around the larger-diameter part of the spool and attached to the other crosshead of a loading machine. Upon loading, the flexible straps unwind from the drum as the peel specimen is wound around it and the drum travels up (hence the name "climbing drum peel test"), thereby peeling the adhesive from its substrate.

One of the simplest peel tests to conduct is the 180° peel test described in ASTM 903, Standard Test Method for Peel or Stripping Strength of Adhesive Bonds. In this test, one adherend is much more flexible than the other, so that upon gripping and pulling the two unbonded ends, the sample assumes the configuration shown in Fig. 8d.

Clearly, peel strength is not an inherent fundamental property of an adhesive. The value of the force required to initiate or sustain peel is not only a function of the adhesive type but also depends on the particular test method, rate of loading, nature, thickness of the adherend(s), and other factors [17]. Regardless, the peel test has proven to be a useful test for a variety of purposes.

The authors have been impressed with the interesting studies at the University of Akron [18,19], where Gent and his associates have used a peel test to measure the "work of adhesion." The work of adhesion is essentially synonymous with adhesive fracture energy, discussed in Section III.C. Gary Hamed has published an article reviewing some of the work at Akron [17]. This paper describes how peel tests have been used to, among other things, (1) verify the usefulness of WLF time–temperature superposition, (2) investigate dependence of the adhesive fracture energy on bond thickness, (3) study adherend thickness effects on adhesive strength, and (4) examine the effects of peel angle. J. R. Huntsberger has also written an interesting discussion on the interpretation of peel test results [20]. Others who have analyzed the stresses, energy dissipation, slip-stick phenomena, and other aspects of peel adhesion include Kaelble [21,22], Igarashi [23], Gardon [24], Dahlquist [25], Bikerman [26], and Wake [27].

Another type of test, somewhat related to both tensile and peel, are the cleavage tests, such as described ASTM D-1062 and D-3807. This test uses a specimen that resembles the compact tension specimen used for fracture of metals except that there is an adhesive bond line down the sample center. The stresses for such a geometry are nonuniform, but typical test results are given as force per width (i.e., failure-loaded divided bond width). ASTM D-3807 uses a specimen fabricated by bonding two narrow, long rectangular beams together to form a split double cantilever beam. The force required to initiate separation between the beams is measured and reported as average load per unit width of beam. ASTM 3433, 3762, and 5041 also make use of cleavage specimens. These are discussed in Section III.C because they are commonly used for determination of adhesive fracture energy. D-1184, Standard Test Method for Flexure Strength of Adhesive Bonded Laminated Assemblies, makes use of standard beam theory to calculate the interlaminar shear strength in laminated beams loaded to failure.

B. Environmental and Related Considerations

Adhesives are often used in applications where they are exposed to continuous or intermittent loads over long periods. It is difficult to duplicate such conditions in the laboratory. Neither is adhesive testing and/or observation under actual service conditions a very feasible alternative. The designer is not usually able or willing to await the results of years of testing before using the adhesive, and to tie up testing equipment and space for such long periods would be prohibitively expensive. There are, however, companies, universities, and other industry groups that have loading racks or other test systems where samples are exposed to dead weight or other loadings while exposed to "natural-weathering" conditions. It is advantageous, however, to have these backed up with (and an attempt made to relate them to) accelerated tests. These accelerated tests are generally experiments in which extreme conditions are used to increase the rate of degradation and deterioration of the adhesive joint. Although it is seldom possible to establish a one-to-one correlation between the rate of deterioration in the accelerated test and actual weather-aging conditions, it is hoped that the short-term tests will, at the very least, provide a relative ranking of adhesive–adherend pairs, surface preparation, bonding conditions, and so on, and/or provide some insight into relative expected lifetimes. As with all tests, the tester/designer should use all of his or her knowledge, common sense, and insight in interpreting the data.

Some accelerated tests are surprisingly simple and intended to give only highly qualitative information, while others have been formulated into standard tests intended to

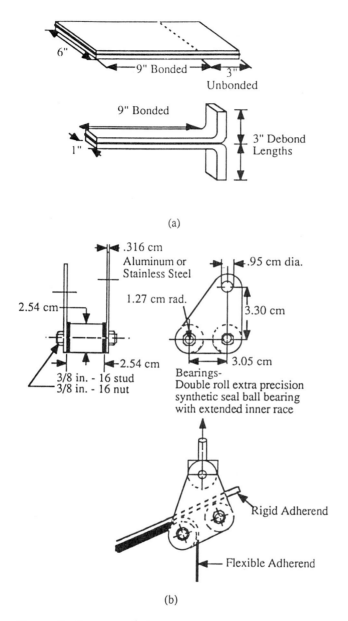

(a)

(b)

Figure 8 Some standard peel test geometries: (a) T-peel test specimen, ASTM D-1876; (b) typical testing jig used in ASTM D-1876; (c) climbing drum peel test, ASTM D-1781; (d) 180° peel test, ASTM 903.

yield more quantitative results. Since heat and moisture, to which adhesive joints are commonly exposed, are environmental factors known to greatly influence adhesive durability, most accelerated tests involve these two agents.

As an example of a simple qualitative test we would like to cite a test devised by the late E. Plueddemann of Dow Corning [28]. Plueddemann was perhaps the world's fore-

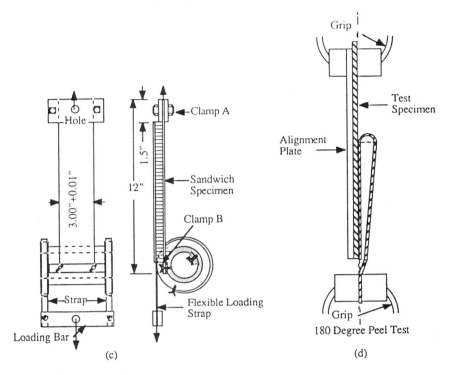

Figure 8 (continued)

most researcher in the area of silane coupling agents, bifunctional compounds with one end of the molecule designed to react with oxygen or similar molecules on the substrate (e.g., an oxide layer on a metal) and the other end designed to react with the polymer in the adhesive [29–31]. In this way, a covalent "bridge" is developed between the adherend and the adhesive. One of the main goals of these treatments is to reduce moisture deterioration of the bond line. Accordingly, Plueddemann had need for a test to access quickly this aspect of the wide variety of silanes produced and differing substrates. He devised the following simple test for this purpose.

In his test, a thin film of adhesive on a glass microscope slide or a metal coupon is cured and soaked in hot water until the film can be loosened with a razor blade. There is usually a sharp transition between samples that exhibited cohesive failure in the polymer and those which exhibited more of an interfacial failure. Since the diffusion of water into the interface is very rapid in this test, the time to failure is dependent only on interfacial properties and may differ dramatically between unmodified epoxy bonds and epoxy bonds primed with an appropriate silane coupling agent. The time to debond in the hot water for various silane primers differed by several thousandfold when used with a given epoxy. In parallel tests, a thick film of epoxy adhesive on nonsilaned aluminum coupon showed about the same degree of failure after 2 h in 70°C water as a silaned joint exhibited after more than 150 days (3600 h) under the same conditions.

The authors have, several times, heard Plueddemann express the opinion that he would be willing to guarantee that an adherend–silane–adhesive system that could withstand a few months of exposure to the conditions of his accelerated test would last many decades under normal outside exposure conditions. He was quick to point out that this guarantee does not cover other types of deteriorations of the adherends or adhesive (e. g., corrosion or polymer degradation) and that because of his age and health, he would not be around to honor the guarantee. Nevertheless, he was very convinced (and convincing) that his test was an ''acid test'' much more severe than most practical adhesive joints would ever experience in their lifetime.

Perhaps, the best known test of this type is the Boeing wedge test, a form of which is standardized in ASTM D-3762. Figure 9 shows this type of specimen and a typical plot of results reported by McMillan and his associates at Boeing [32,33]. Theidman et al. [34] have also used the wedge to investigate coupling agents.

Since adhesives have long been used in the wood/lumber business, where outdoor exposure is inevitable, many of the standard accelerated tests were originally developed for these materials. Such tests are increasingly finding uses for other materials. The most common accelerated aging tests are:

1. ASTM D-1101, Standard Test Method for Integrity of Glue Joints in Structural Laminated Wood Products for Exterior Use. Two methods are outlined in this standard for using an autoclave vessel to expose the joint alternately to water at vacuum pressure (ca. 635 mmHg) and low temperature with a high-pressure stage (ca. 520 kPs), followed by a high-temperature drying stage (ca. 65°C circulated dry air). After the prescribed number of cycles (typically one or two), the samples are visually inspected for signs of delamination.

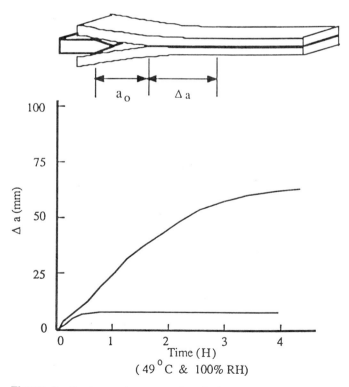

Figure 9   Boeing wedge test and typical results.

2. ASTM D-1183, Standard Test Methods for Resistance of Adhesives to Cyclic Laboratory Aging Conditions. This standard describes several different test procedures in which the joints of interest are subjected to cycles made up of stages at different relative humidities and temperatures, high-temperature drying cycles, and/ or immersed in water for specified periods. The joints are then evaluated by standard strength tests (lap joint, tensile, or other) to ascertain the extent of degradation in strength.

3. ASTM D-2559, Specifications for Adhesives for Structural Laminated Wood Products for Use Under Exterior (Wet Use) Exposure Conditions. Like ASTM D-1101, this test makes use of an autoclave-vacuum chamber to impregnate specimens with water followed by drying in a hot (65°) air-circulating oven. A more quantitative measure of degradation is obtained in this test by measuring lap shear compressive strength and measuring deformation as well as visual evaluation to determine the extent of delamination.

4. ASTM D-3434, Standard Test Method for Multiple-Cycle Accelerated Aging Test (Automatic Boil Test) for Exterior Wet Use Wood Adhesives. This standard describes the construction of apparatus to expose adhesive joints automatically to alternate boil/dry cycles. A typical cycle is composed of (a) submerging the specimen for 10 min in boiling water, (b) drying for 4 min with 23°C air circulating at 1.75 m/s, and (c) exposing the specimen for 57 min to 107°C air circulating at 1.75 m/s. At a prescribed number of cycles, 10 specimens are withdrawn and their tensile shear strength measured and compared to tests on samples that have not been exposed to the accelerated testing conditions.

5. ASTM D-3632, Standard Practice for Accelerated Aging of Adhesive Joints by the Oxygen-Pressure Method. This test is intended to explore degradation in elastomer based and other adhesives that may be susceptible to oxygen degradation. The practice involves subjecting specimens to controlled aging environments for specified times and then measuring physical properties (shear or tensile strength or other). The controlled environment consists of elevated temperature (70°C) and high pressure (2MPa) oxygen.

6. ASTM D-4502, Standard Test Method for Heat and Moisture Resistance of Wood–Adhesive Joints. Rather than using an expensive autoclave as in D-1101 and D-2559, moisture aging in this test is accomplished in *moist aging jars*. The samples are exposed to prescribed temperature–humidity cycles in the jars heated in ovens. The strength of the aged samples is measured by standard methods and compared to similar virgin samples.

There are tests that have been developed for use with solid polymers specimens that find some use with adhesives. Ultraviolet (UV) radiation (e.g., as present in sunlight) is known often to have detrimental effects on polymers. Accordingly, a popular accelerated weathering (aging) test, ASTM G-53, "Standard Practice for Operating Light- and Water-Exposure Apparatus (Fluorescent UV-Condensation Type) for Exposure of Nonmetallic Materials," describes use of a "weatherometer" that incorporates UV radiation moisture and heat. These commercially available devices consist of a cabinet in which samples are mounted on aluminum panels, which in turn are stacked edgewise on a sloped rack along either side of the cabinet. These samples are then alternately exposed to two stages in periodic cycles: a condensation stage followed by a UV-drying stage. The first stage is accomplished by heating water in a partially covered tank below the specimens in the bottom of the cabinet. The specimens are maintained at a constant temperature (typically

30 to 50°C) which is lower than the water temperature. This results in moisture condensation on the specimens. This stage might last for 1 to 4 h as selected by the operator. This is followed by the UV-drying stage. An array of special fluorescent light tubes are situated along each side of the cabinet parallel to rack-mounted samples. The operator selects a temperature higher than the condensation temperature (typically 40 to 70°C) that the system automatically maintains in the cabinet for a fixed period (usually 1 to 40 h) while the specimens are exposed to the UV-radiation. The weatherometer is equipped with a timer, a float-controlled water supply for the tank, and other controls, so that it can continuously cycle through this two-stage cycle for months or even years with minimum operator care. Samples are removed periodically and their strength measured by standard techniques for comparison with virgin samples and aged and virgin samples of other materials. Many adherends are opaque to UV light, and hence one might question the use of this weatherometer to explore weathering aging of adhesives used with such adherends. However, even here, having such a commercial automated device might be very useful. Both the condensation and radiation-heat curing stages are very analogous to environmental exposures experienced by practical joints. This, along with the automated and ''standard'' nature of the equipment, often makes the technique quite attractive. More important, there are problems where the UV part of the aging may be critically important. An example would be the bonding of a thin transparent cover film to a thin sheet containing the reflective elements in sheeting used to make reflecting road signs. This device would have obvious advantages in such cases and, indeed, has become the standard for use in evaluating weather durability in that industry.

It is recognized that most accelerated tests do not duplicate or even closely approximate actual service conditions. As a case in point, most joints will, in all probability, never be exposed to boiling water. It is hoped, however, that resistance to boiling for a few hours or days may provide some valid evidence (or at least insight) into the durability of a laminated part after years of exposure to high ambient humidity and temperature. While such accelerated tests are never perfect, they may be the only alternative to observing a part in actual service for decades. The authors feel this philosophy is well stated in Section 1.1 of ASTM D-1183: ''It is recognized that no accelerated procedure for degrading materials correlates perfectly with actual service conditions, and that no single or small group of laboratory test conditions will simulate all actual service conditions. Consequently, care must be exercised in the interpretation and use of data obtained in this test.'' ASTM D-3434 includes the following statement about its significance: ''The test method assumes that boil/dry cycling is an adequate and useful accelerated aging technique.'' Evaluation of long-term durability of adhesives in wood joints under severe service conditions, including extended exterior exposure, is a complex field, and no entirely reliable short-term test is known to ensure that a new type of adhesive system will resist satisfactorily all of the chemical, moisture, microorganism, and solve. ffects that such severe service may involve. Except for the effects of microorganisms and similar biological influences, this test method has proven very useful for comparison purposes to distinguish between adhesive systems of different degrees of durability to the usual temperature, moisture, and cyclic moisture conditions. It has proven very useful to distinguish between bondlines, made with adhesives of proven chemical and biological durability, that when properly used in production resist the mechanical and moisture effects that such joints must withstand in severe servi e over extended periods of exposure. It does not, however, in itself, assure that new types of adhesives will always withstand actual exterior or other severe service.

Other environmental related tests include ASTM D-904, Standard Practice for Exposure of Adhesive Specimens to Artificial (Carbon-Arc Type) and Natural Light; ASTM D-1828, Standard Practice for Atmospheric Exposure of Adhesive-Bonded Joints and Structures; ASTM D-1879, Standard Practice for Exposure of Adhesive Specimens to High Energy Radiation; and ASTM D-3310, Standard Test Methods for Determining Corrosivity of Adhesive Materials.

In practical joints, adhesives are not always loaded statically or loaded for short periods of time. To help evaluate the performance of stressed adhesive joints as a function of time, tests have been developed to determine response of adhesive joints to creep and cyclic loading. ASTM D-1780, Standard Practice for Conducting Creep Tests of Metal-to-Metal Adhesives, makes use of a deadload weight-lever loading frame to measure creep of lap shear specimens. ASTM D-2793, Standard Test Method for Creep of Adhesives in Shear by Compression Loading (Metal-to-Metal), describes the constructions and procedures for use of creep test apparatus in which the sustained loading is maintained by springs. ASTM D-2294 is similar to D-2293, except that here the spring-loaded apparatus loads the lap specimen in tension. ASTM D-4680, Standard Test Method for Creep and Time to Failure of Adhesives in Static Shear by Compression Loading (Wood-to-Wood), describes the construction of a spring-loaded apparatus and testing procedures for a creep apparatus for use with relatively large wood specimens. ASTM D-2918 and D-2919 describe tests to measure the durability of adhesive joints in peel and lap shear, respectively. The tests and recommended fixtures are intended to hold specimens under sustained loadings while exposed to environments such as moisture, air, vapors, water, or other environments.

ASTM D-3166, Standard Test Method for Fatigue Properties of Adhesives in Shear by Tension Loading (Metal-to-Metal), provides procedures for testing and measurement of the fatigue strength of lap specimens. It makes use of conventional tensile testing machines capable of applying cyclic axial loads. Researchers have also made beneficial use of the concepts of fracture mechanics to evaluate the fatigue crack growth rate per cycle, da/dn, as a function of stress intensity factor. For this purpose, Mostovoy and Ripling, [35] for example, have used fracture mechanics specimens similar to those described in ASTM D-3433. Fracture mechanics is discussed at the end of this chapter as well as in many books.

Adhesive joints are frequently exposed to sudden dynamics loads, and hence a knowledge of how adhesives react to impact loading is important for some applications. ASTM D-950, Impact Strength of Adhesive Bonds, describes sample configuration and testing apparatus for measuring the impact strength of adhesive bonds. The method is generally analogous to the Izod test method used for impact studies on a single material.

ASTM D-2295 describes apparatus that utilizes tubular quartz lamps to investigate failure of adhesive joint samples at high temperatures, and ASTM D-2257 outlines procedures for testing samples at low temperatures (-268 to $-55°C$). ASTM also provides specific standards to investigate failure-related properties of adhesives that are less directly related to mechanical strength. Such properties include resistance to growth and attack by bacteria, fungi, mold, or yeast (D-4300 and D-4783), chewing by rodents (D-1383), eating by insects (D-1382), resistance to chemical reagents (D-896), and so on. It is enough to make the adhesive designer or researcher paranoid. Not only are stresses, temperatures, moisture, and age working against him or her, but now it appears that microorganisms and the animal kingdom want to take their toll on any adhesively bonded structure.

Although the primary purpose of this chapter is to discuss mechanical testing and strength of adhesive joints, the reader should be aware that ASTM covers a wide variety of tests to measure other properties. ASTM, for example, includes standard tests to measure the viscosity of uncured adhesives, density of liquid adhesive components, non-volatile content of adhesives, filler content, extent of water absorption, stress cracking of plastics by liquid adhesives, odor, heat stability of hot-melt adhesives, ash content, and similar properties or features of adhesives.

Of particular interest to the adhesive technologist are surface treatments. ASTM has adopted standard practices for treating surfaces to better adhesives. ASTM D-2093, Standard Practice for Preparation of Plastics Prior to Adhesive Bonding, describes physical, chemical, and cleaning treatments for use on a wide variety of polymer adherends. D-2651, Standard Practice for Preparation of Metal Surfaces for Adhesive Bonding, describes techniques, cleaning solutions and methods, etchants or other chemical treatment, and so on, for metal adherends, including aluminum alloys, steel, stainless steel, titanium alloys, copper alloys, and magnesium alloys. ASTM D-2675 is concerned with the analysis and control of etchant effectiveness for aluminum alloys. ASTM D-3933 provides a standard practice for phosphoric acid anodizing of aluminum surfaces to enhance adhesion.

C. Fracture Mechanics Techniques

As noted earlier, the most commonly used standards for determination of adhesive joint properties and characteristics suggest reporting the results in terms of average stress at failure. If average stress criteria were generally valid, one would anticipate that a doubling of the bond area should result in a proportionate increase in joint strength. Recent studies in our laboratory demonstrate that such criteria may lead to erroneous predictions [36]. This study involved leaving half of the overlap area unbonded. All of the samples had the same amount of overlap. Part of the samples (the control) were bonded over the complete overlap region. Part of the overlap area near one of the bond termini without any adhesive and the other part had a 50% unbonded region centered in the overlap region. In neither case was the reduction in strength commensurate with the reduction in bond area. That is, for the end debonds, the reduction in load at failure averaged less than 25% and for the center debonds it was reduced by less than 10%. This is consistent with standards such as ASTM D-3165, Strength Properties of Adhesives in Shear by Tension Loading of Laminated Assemblies, which address the fact that average lap shear strength is dependent on different bond areas of the joint and that the adhesive can respond differently to small bond areas compared to large bond areas.

To determine failure loads for different geometries, other methods are required. This is due to the fact that the stress state in the bond region is complex and cannot be approximated by the average shear stress. A method that is gaining in popularity for addressing this problem is the use of fracture mechanics. With the use of modern computers, the stress state and displacement even in complex adhesive bonds can be determined with good accuracy. A fracture mechanic approach uses these stresses combined with the displacements and strains and conservation of energy principles to predict failure conditions. At least three ASTM standards are based on fracture mechanics concepts (ASTM D-3433, D-3762, and D-5041). The premise behind fracture mechanics is accounting for changes in energies associated with the applied load, the test sample, and the creation of new surface or area. Since fracture mechanics techniques are well documented in the literature (see, e.g., Ref. 37 and 38), only a brief review is presented here.

The methods suggested here are formulated using methods similar to those postulated by Griffith [39]. Griffith hypothesized that all real "elastic" bodies have inherent cracks in them. He hypothesized that a quantity of energy to make the most critical of these cracks grow would need to come from the strain energy in the body and work applied by loads. Conservation of energy dictates that a crack can grow only when the strain energy released as the crack grows is sufficient to account for the energy required to create the new "fracture" surface. In Griffith's original work, he considered only perfectly elastic systems. He conducted his confirmation experiments on glass, which behaves as a nearly ideal elastic material. In this case the *fracture energy* is very closely associated with the *chemical surface energy*. Indeed, Griffith was able to establish such a correlation by measuring the surface tension of glass melts and extrapolating back to room temperature. Most engineering solids are not purely elastic and the energy required to make a crack grow involves much more than just the chemical surface energy. In fact, the energy dissipated by other means often dominates the process and may be several orders of magnitude higher than the chemical surface energy term. As explained in modern texts on fracture mechanics, this has not prevented the use of fracture mechanics for analysis of fracture in other quasi-elastic systems, where typically the other dissipation mechanisms are lumped into the fracture energy term G_c (discussed later). Techniques are also being developed to use the basic approach for systems that experience extensive viscoelastic and plastic deformation. These approaches are more complex. Here we confine our attention to quasi-elastic cases. Significant research and development work has gone into methods of increasing the dissipative processes required for a crack to grow, thereby, "toughening" the materials. As a consequence, another name for the fracture energy is *fracture toughness* G_c.

While Griffith's earlier fracture mechanics work was for cohesive fracture, the extension to adhesive systems is logical and straightforward. In the latter case, one must account for the strain energy in the various parts of the system, including adherends and adhesive, as well as recognizing that failure can proceed though the adhesive, the adherend, or the interphase region between them. Here too, energy methods can be useful in determining where the crack might grow since one might anticipate that it would use the path requiring the least energy.

In principle, the fracture mechanics approach is straightforward. First, one makes a stress-strain analysis, including calculation of the strain energy for the system with the assumed initial crack and the applied loads. Next, one performs an energy balance as the crack proceeds to grow in size. If the energy released from the stress field plus the work of external forces is equal to that required to create new surface, the crack grows. A difficulty in performing these analyses is that for many practical systems, the stress state is very complex and often not amenable to analytical solutions. With the use of modern computers and computational codes, this is becoming progressively less of a problem. Finite element codes are refined to the point that very accurate stress–strain results can be obtained, even for very complex geometries (such as adhesive joints). Furthermore, it is possible to incorporate complex material behavior into the codes. The latter capability has not reached its full potential to date, at least in part, because the large strain deformation properties of the materials involved, at the crack tip, are not well defined.

A relatively simple example may help in understanding the application of fracture mechanics. First, let's assume that we have two beams bonded together over part of their length, as shown in Fig. 10a. Here we will make two additional convenient (but not essential) assumptions: (1) the adhesive bond line is sufficiently thin that when bonded as

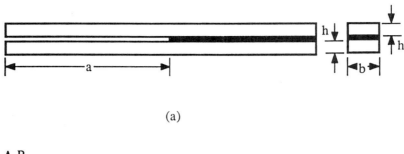

(a)

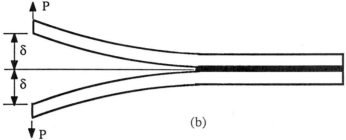

(b)

Figure 10 Double cantilever beam specimen showing nomenclature and specimen deflection.

shown in Fig. 10b, the energy stored in the adhesive is negligible compared to that in the unbonded length of the beams a; and (2) the length a is sufficiently long that when loaded, the energy due to bending dominates (i.e., the shear energy and other energy stored in the part of the beam to the right of a is negligible).

The end deflection of the unbonded segment of each beam due to a load P is found in mechanics of material texts to be [40]

$$\delta = \frac{Pa^3}{3EI} \tag{1}$$

where a is the unbonded length, E the modulus of elasticity for the material from which beams are manufactured, and I the moment of inertia for a rectangular cross section. The work done by the forces P in deforming this double cantilever is therefore

$$W = \frac{1}{2}P(2\,\delta) \tag{2}$$

which for a conservative elastic system equals the energy U, where

$$U = P\delta = \frac{P^2 a^3}{3EI} \tag{3}$$

If we assume Griffith-type fracture behavior, the energy released as the crack grows (a increases) must go into the formation of new fracture surface. The fracture energy G_c (sometimes called energy release rate) is therefore

$$G_c = \frac{\Delta U}{\Delta A} \tag{4}$$

Substituting Eq. (3) into (4) and taking the limit leads to

$$G_c = \frac{\partial U}{\partial A} = \frac{1}{b}\frac{\partial U}{\partial a} = \frac{P^2 a^2}{EIb} = \frac{12P^2 a^2}{Eh^3 b^2} \tag{5}$$

where

$$\partial A \approx 2a(b) \tag{6}$$

We thus see that such beam systems could be used to determine G_c. Indeed, ASTM D-3433 makes use of such beams for this measurement. It should be noted that the equation given in Section 11.1.1 of D-3433 has a similar form:

$$G_c = \frac{4P^2(3a^2 + h^2)}{Eb^2 h^3} \tag{7}$$

This equation can be used for smaller values of a since it includes effects of shear and other factors that, as noted above, were neglected in our simple analysis. Inspection of Eq. (7) shows that as a becomes large compared to h, this equation reduces to that derived above. By tapering the beams, ASTM shows how a related geometry can be used for G_c determination where the data analysis is independent of the crack length. The advantage here is that one need not monitor crack length. An advantage of double cantilever beam specimens is that measurement can be made for several different values of a for a given beam, making it possible to get several data points from a single sample. Conversely, if G_c is known, say from another test configuration, fracture mechanics can be used to determine the load required to initiate fracture.

The case described above is the exception rather than the rule. Most practical problems do not have stress and displacement fields easily obtained by analytical methods. Numerical methods can be applied to obtain the necessary stress and displacements needed to calculate the energy release rates. The basic consideration for calculating energy release rates is accounting for the difference in energy before and after crack extension dA. This condition can be approximated for a discrete system by using finite element methods (FEMs), that is, by using a computer to calculate the strain energy change as a crack area is incrementally increased.

Two different methods of calculating the energy release rate numerically are outlined briefly. The first, the *compliance method*, is readily applied using FEM [37]. This method requires only two computer runs for each energy release rate desired. The first computer run is used to calculate the total relative displacement, u^a, of the sample for the crack length a, under a constant force F^a. In the second computer run, a finite element node is released, extending the crack to a length of $a + \Delta a$. The second run allows calculation of the total relative displacement, $u^{a+\Delta a}$, of the sample for a crack length $a + \Delta a$ under the same constant load F^a. Using the discrete form of the energy release rate for a linear elastic body leads to

$$G = \frac{F^a(u^{a+\Delta a} - u^a)}{2\Delta A} \tag{8}$$

The discrete form of the energy release rate under constant displacement assumptions for the same crack length can easily be shown to be

$$G = \frac{F^a u^a \left(\dfrac{1 - u^a}{u^a + \Delta^a}\right)}{2\Delta A} \tag{9}$$

A third computer run can be used to extend the crack by releasing the next node in front of the crack tip. This increases the crack length to $a + 2\Delta a$. Letting $u^{a+\Delta a}$ equal the new relative displacement and u^a equal the total relative displacement from run 2, the energy release rate for the new crack length can be computed using Eq. (8) or (9) with the addition of only one computer run. This process can be repeated, each time determining a new energy release rate per single computer run, thereby producing a curve of energy release rate versus crack length. The crack length for which the energy release rate is equal to G_c is a critical crack length where failure should ensue.

A second method for calculating energy release rate is the *crack closure integral* (CCI) *method* [37,41–43]. This method uses FEM techniques to calculate the energy that is needed to close a crack extension. It is assumed that for an elastic system this energy is equivalent to the energy that is needed to create the crack's new surface, giving an alternative method of calculating the energy release rate from that just discussed.

The CCI method is also readily accomplished using FEM. The energy release rate calculated as outlined in the following is for a constant stress crack extension assumption. In general, this method requires four computer runs for each energy release rate calculated. The method can be shortened to three computer runs if verifying that the crack was closed is not necessary. In this case, only the second, third and fourth steps explained below need to be performed.

The first computer run is used to calculate the displacements (or locations) of the finite element nodes near the crack tip for crack length a under constant load F^a. Let u^{BX1} and u^{BY1} represent the displacements of the first bottom node in the X and Y directions, respectively, from run 1. Let u^{TX1} and u^{TY1} represent the displacements of the first top node in the X and Y directions, respectively, from run 1. Since we have not yet released this first node for run 1, the top and bottom nodes are coincidental and should have the same displacement. This initial displacement indicates the point to which the nodes will have to be moved when the crack is closed in runs 3 and 4.

In the second computer run, a node is released extending the crack length to a length of $a + \Delta a$. This allows calculation of the new nodal positions of the same nodes that were released from run 1 while under the constant load F^a. Let u^{BX2} and u^{BY2} represent the displacements of the bottom node in the X and Y directions, respectively, from run 2. Let u^{TX2} and u^{TY2} represent the displacements of the top node in the X and Y directions, respectively, from run 2. The distance between the top node and bottom node in the X and Y directions indicates how far the nodes will need to be moved in each direction to close the crack. Figure 11 shows computer runs 1 and 2 in a pictorial schematic and summary of the two steps.

In the third run, vertical unit forces are applied to the released nodes in the direction needed to bring them back to their original vertical positions. The nodal displacements of these released nodes of the sample for a crack length of $a + \Delta a$ under the same constant load F^a are noted. Let u^{BY3} represent the displacement of the bottom node in the Y direction from run 3. Let u^{TY3} represent the displacements of the top node in the Y direction from run 3. The difference between the node displacements from runs 2 and 3 indicates how far the vertical unit force moved each node toward its original position.

The fourth run is the same as run 3 except that the forces are now horizontal. Figure 12 shows computer runs 3 and 4 in a pictorial schematic and a summary of the two steps.

Runs 3 and 4 provide a measure of how much each node is moved back toward each original position by unit forces. Since the material is modeled as linear elastic, the force required to move each node back to its original position can be determined simply by

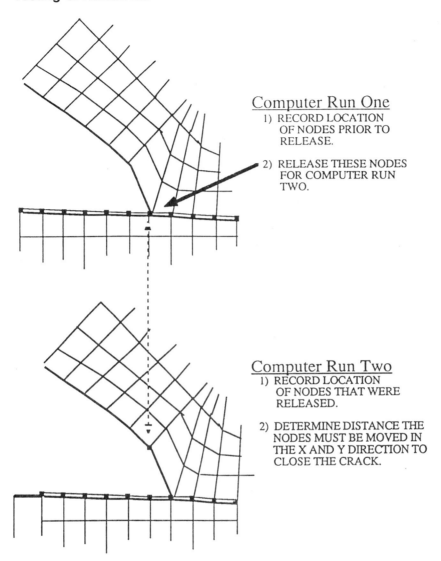

Computer Run One
1) RECORD LOCATION
   OF NODES PRIOR TO
   RELEASE.

2) RELEASE THESE NODES
   FOR COMPUTER RUN
   TWO.

Computer Run Two
1) RECORD LOCATION
   OF NODES THAT WERE
   RELEASED.

2) DETERMINE DISTANCE THE
   NODES MUST BE MOVED IN
   THE X AND Y DIRECTION TO
   CLOSE THE CRACK.

Figure 11 Pictorial of the first and second steps in the CCI method of determining energy release rates.

increasing these forces in proportion to the total displacements required to bring the nodes back to their original position. We will call these forces F^X and F^Y. The energies E^X and E^Y required to close the crack in the X and Y directions can now be calculated by determining the work of these forces on the crack closure displacements since for an elastic system, the work is stored as strain energy.

Based on this analysis, the energy release rates can be partitioned into the energy release rates associated with crack opening displacement (G_I) and shear displacement normal to the crack surface (G_{II}):

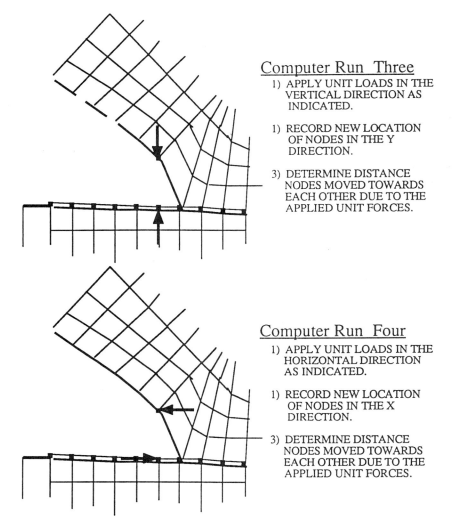

Figure 12 Pictorial of the third and fourth steps in the CCI method of determining energy release rates.

$$G_{\mathrm{I}} = \frac{\Delta E_1^X - \Delta E_2^X}{\Delta A}$$

$$G_{\mathrm{II}} = \frac{\Delta E_1^Y - \Delta E_2^Y}{\Delta A} \tag{10}$$

The total energy release rate can be determined from

$$G = G_{\mathrm{I}} + G_{\mathrm{II}} \tag{11}$$

In Eqs. (10) and (11), G represents the total energy release rate under constant-stress crack extension assumptions, G_{I} represents the portion of G in the mode I direction, and G_{II} represents the portion of G in the mode II direction. A couple of concluding comments to this section might be in order. Each of these methods of solving for the fracture energy has advantages. The compliance method is simpler and less expensive to perform. The crack closure method, on the other hand, facilitates calculation of crack growth analysis under different loading (i.e., crack opening or shear). There is evidence that cracks in materials behavior differently under the various loadings. As a consequence, a given crack growth may depend on the relative amounts of mode I and mode II stresses at its tip. It is also advantageous to have the two methods of determining G. Since they provide a systematic check on each other, that might be used to verify for code, programming, or other errors.

Computer-based energy release rates might be a very useful design tool. An example might be used to illustrate such use. Assume that nondestructive evaluation techniques are used to locate and determine the size and shape of a debond region in the interphase region of an adhesive bond in a structure. An example of such a bond might be between a rocket motor case and a solid propellant grain. A reasonable question would be: Is this "flawed" region "critical" in that it might cause failure under prescribed service load conditions? For many structures (e.g., the rocket case noted) this might literally be a million dollar question. A reliable answer could mean the difference between safely using the structure or needing to repair or discard it. Trying to answer the question experimentally would generally be very expensive. An alternative approach might make use of fracture mechanics. One might use FEM to model the structure, including the debond region that has been identified. Using the techniques outlined above, the energy release rate can be calculated for assumed increases in the size of debond. If the critical fracture toughness (G_c) is known (perhaps determined from independent tests with standard specimens), it can be compared with the value calculated. If the energy release rate calculated for the debond in the structure is equal to G_c, the debond is apt to grow. If it is larger than G_c, the crack should accelerate. Values of energy release rate significantly lower than G_c would indicate a noncritical region. Other types of irregularities, cohesive or adhesive, might be treated in a similar manner. Obviously, such an approach requires not only adequate FEM techniques and computers but also nondestructive means to identify and quantitatively measure potential critical regions.

IV. CONCLUSIONS

There are a great many properties that affect adhesive quality. One of the more important is the strength of joints formed with the adhesive and adherends. We have discussed a number of standard tests commonly used to "measure" adhesive joint strength. These methods are valuable and serve many purposes. In this chapter, however, we point out that the use of the "standard" results from these tests to predict the strength of other joints that differ in even seemingly minor details is questionable. A more reliable comparison between adhesives and joints might be to compare more fundamental properties, such as moduli, adhesive fracture toughness, and so on. This is a basic premise of fracture mechanics. With improved computation facilities and codes, these methods show promise for using results from standard tests to predict the performance of other (perhaps more complex) practical adhesive joints. As such, they should become very powerful design tools.

REFERENCES

1   G. P. Anderson, S. Chandapeta, and K. L. DeVries, in *Adhesively Bonded Joints: Testing, Analysis, and Design*, ASTM STP-981 (W. S. Johnson, ed.), American Society for Testing and Materials, Philadelphia, 1988.
2   G. P. Anderson and K. L. DeVries, in *Adhesion and Adhesives*, Vol. 6 (R. Patrick, K. L. DeVries, and G. P. Anderson, eds.), Marcel Dekker, New York, 1988.

3 G. P. Anderson, K. L. DeVries, and G. Sharon, in *Delamination and Debonding of Materials*, ASTMP STP-876 (W. S. Johnson, ed), American Society for Testing and Materials, Philadelphia, 1985.

4 K. C. Grammol, M.S. thesis Mechanical Engineering Department, University of Utah, 1984.

5 J. K. Strozier, K. J. Ninow, K. L. DeVries, and G. P. Anderson, in *Adhesion Science Review* (H. F. Brinson, J. P. Wightman, and T. C. Ward, eds.), Commonwealth Press, Radford, VA., 1987, p. 125.

6 O. Volkersen, *Luftfahrt Forsch. 15*: 41 (1938).

7 M. Goland and E. Reissner, *J. Appl. Mech. 11*: 17 (1944).

8 T. R. Guess, R. E. Allred, and F. P. Gerstle, *J. Testing Eval., 5*(2): 84 (1977).

9 A. J. Kinloch, *Adhesion and Adhesives Science and Technology*, Chapman & Hall, New York, 1987.

10 G. R. Wolley and D. R. Carver, *J. Aircr. 8*(19): 817 (1971).

11 R. D. Adams and N. A. Peppiatt, *J. Strain Anal. 9*(3): 185 (1974).

12 S. Amijima, A. Yoshida, and T. Fujii, in *Proc. 2nd International Conference on Composite Materials*, Toronto, Ontario, Canada, Metallurgical Society of AIME, New York, 1978.

13 J. K. Sen and R. M. Jones, *AIAA J. 18*(10): 1237 (1980).

14 J. K. Sen and R. M. Jones, *AIAA J. 18*(11): 1376 (1980).

15 W. N. Sharpe and T. J. Muha, in *Proc. Army Symposium on Solid Mechanics*, AMMRC-MS-75-8, Army Materials and Mechanics Research Center, Watertown, Mass., 1974, p. 23.

16 J. L. Hart-Smith, *Douglas Aircraft Report 6922*, Long Beach, California, 1980.

17 G. R. Hamed, in *Adhesion and Adhesives*, Vol. 6, (R. L. Patrick, K. L. DeVries, and G. P. Anderson, eds.), Marcel Dekker, New York, 1988.

18 A. N. Gent and G. R. Hamed, *J. Appl. Polymer Sci., 21*: 2817 (1977).

19 A. N. Gent and G. R. Hamed, *Polymer Eng. Sci., 17*: 462 (1977).

20 J. R. Huntsberger, in *Adhesion and Adhesives*, Vol. 6 (R. L. Patrick, K. L. DeVries, and G. P. Anderson, eds.), Marcel Dekker, New York, 1988.

21 D. H. Kaelble, *Physical Chemistry of Adhesion*, Wiley-Interscience, New York, 1971.

22 D. H. Kaelble and C. L. Ho, *Trans. Soc. Rheol. 18*: 219 (1974).

23 T. Igarashi, in *Adhesive Joints, Formation, Characteristics and Testing* (K. L. Mittal, ed.), Plenum Press, New York, 1984.

24 J. L. Gardon, *J. Appl. Polymer Sci. 7*: 643 (1963).

25 C. A. Dahlquist, in *Treatise on Adhesion and Adhesives*, Vol. 2 (R. L. Patrick, ed.), Marcel Dekker, New York, 1969.

26 J. J. Bikerman, *The Science of Adhesive Joints*, Academic Press, New York, 1968.

27 W. C. Wake, *Adhesion and the Formulation of Adhesives*, Applied Science Publishers, London, 1982.

28 E. P. Plueddemann, *J. Adhesion Sci. Technol. 2*: 179 (1988).

29 E. P. Plueddemann, *Silane Coupling Agents*, Plenum Press, New York, 1982.

30 F. J. Boerio, in *Adhesion and Adhesives*, Vol. 6 (R. L. Patrick, K. L. DeVries, and G. P. Anderson, eds.), Marcel Dekker, New York, 1988.

31 S. Naviroj, J. L. Koenig, and H. Ishida, *J. Adhesion 18*: 93 (1985).

32 J. C. McMillian, in *Bonded Joints and Preparation for Bonding: AGARD Lecture Series 102*, Technical Editing and Reproduction, London, 1979, Chap. 7.

33 J. C. McMillian, *Developments in Adhesives*, 2nd ed., Applied Science Publishers, London, 1981, p. 243.

34 W. Theidman, F. C. Tolan, P. J. Pearce, and C. E. M. Morris, *J. Adhesion, 22*: 197 (1987).

35 S. Mostovoy and E. J. Ripling, *J. Adhesion Sci. Technol. 913*: 513 (1975).

36 P. R. Borgmeier, K. L. DeVries, J. K. Strozier, and G. P. Anderson, in *Mechanics of Plastics and Plastic Composites* (V. K. Stokes, ed.), American Society of Mechanical Engineers, New York, 1989.

37 D. Broek, *Elementary Engineering Fracture Mechanics*, 4th ed., Kluwer, Boston, 1986.

38 D. K. Felbeck and A. G. Atkins, *Strength and Fracture of Engineering Solids*, Prentice-Hall, Englewood Cliffs, N.J., 1984.

39 A. A. Griffith, *Phil. Trans. Roy. Soc. London, A221*: 163–198 (1970).

40 F. B. Beer and E. R. Johnston, Jr., *Mechanics of Materials*, McGraw-Hill, New York, 1981.

41 G. R. Irwin, in *Adhesion and Adhesives*, Vol. 1 (R. L. Patrick, K. L. DeVries, and G. P. Anderson, eds.), Marcel Dekker, New York, 1966.

42 G. P. Anderson, S. J. Bennett, and K. L. DeVries, *Analysis and Testing of Adhesive Bonds*, Academic Press, New York, 1981.

43 E. F. Rybicki, D. W. Schmuesser, and J. Fox, *J. Composite Mater. II*: 470 (1977).

6

Physical Testing of Pressure-Sensitive Adhesive Systems

John Johnston   *tesa tape, inc., Charlotte, North Carolina*

I. INTRODUCTION

It is assumed in almost all cases that any pressure-sensitive adhesive to be tested has already been applied to a flexible carrier, which is either in tape form or which can be cut into tapes for testing, virtually all test methods making this assumption. If this is not the case, it would be necessary to coat the adhesive onto a suitable flexible carrier, usually 25-μm polyester film, which may need to be suitably preprimed, the prime coat used depending on the adhesive type. The coat weight chosen should be that used for the practical application of that adhesive, or if this adhesive is still under development, a series of coat weights can be run to determine which provides optimum performance. An exponential relationship will be found between coat weight and resulting adhesion.

The need for the physical testing of a pressure-sensitive adhesive can vary considerably, such as the determination that a given adhesive will perform satisfactorily for its intended use, that it meets a specific standard, that uniformity exists within a given population or between populations, and that it could be used to compare one system to similar systems—all of which demand that any test method must be accurate and reproducible. The thermoplastic nature of pressure-sensitive systems can make this objective very difficult to achieve without a full understanding of their behavior and without observing a number of precautions.

Recognizing that the behavior of a pressure-sensitive system varies according to temperature and to rate, and that a pressure-sensitive adhesive deforms easily under pressure, which affects the degree of contact, the conditions under which any pressure-sensitive adhesive is both prepared for testing and then tested must be rigidly controlled. Otherwise, considerably different values will be found for each uncontrolled evaluation of the same adhesive system. This also applies to the geometry of the test, where the bending of a flexible adhesive carrier plays a key part, as in peel adhesion testing, the relative stiffness of the backing altering the intended angle of the test.

The nature of the bond that is formed with a specific adhesive depends both on the adhesive design and on the nature of the surface to which it is adhered. This applies not only to its material of construction, but also whether it is porous or nonporous, the degree of surface roughness, and from this, the contact that can be obtained, 100% contact during testing being a rarity. So now we must add to our protocol for standardization of testing, standardized test surfaces in both material and surface roughness, recognizing that non-porous and porous substrates will behave quite differently, keying into the micro-irregularities of a porous surface being one means of attachment.

The internationally recognized standard material chosen for a nonporous test surface is stainless steel [1,7,11,13,14] although there are differing opinions as to which grade or surface roughness. In the United States, a standardized cardboard, Standard Reference Material 1810 from the National Bureau of Standards, is used for a porous test surface, other testing authorities again having their own choice. Recognizing that any result obtained using a standardized test surface will be applicable to that surface only, accurate data for alternate surfaces will require a retest with that new surface.

The application of any stress that a pressure-sensitive adhesive encounters in practical use ranges from a very rapid rate, as in unwinding a pressure-sensitive tape on an automatic application machine, to a continuous slow stress as would be the case with a packaging tape in use or with a double-sided tape used as a mounting tape. In the former case, the viscous component of the adhesive has little or no time to respond, and under excessively rapid rates, separation may even be in the form of a brittle fracture. This behavior would be duplicated if the adhesive were stressed at extremely low temperature—the time-temperature relationship [15]. In the second case of a continuous low stress, there is ample time for viscous response and resultant molecular disentangle-ment, and failure may well be cohesive, again with a corresponding behavior if tested at a high temperature. For the most part, both in use and in testing, there is simultaneous elastic and plastic response to stress by pressure-sensitive adhesives, each to a greater or lesser extent, depending on the rate and/or temperature.

Regarding test panels, it is common practice to reuse standard test panels over and over again, and the need for a contamination-free surface at each test is essential to obtaining reliable data. While it appears that a specific pressure-sensitive adhesive leaves no residue on removal, and the panel may appear perfectly clean, a simple dusting, as is done for fingerprint identification, will show that there is indeed an adhesive residue remaining after each test, and further, this micro-trace can be quite difficult to remove. There are standard procedures as to the correct solvent for test panel cleaning [2], but with the variety of pressure-sensitive systems now available, the correct method is the one that removes the last-applied adhesive system most satisfactorily and leaves the test panel clean and dry. It may well involve the need for two solvents, one to remove the adhesive, another to remove the last traces of the first solvent. Because of this inherent adhesive micro-contamination, even following the standard cleaning procedures, there are those who prefer to restrict the use of a series of test panels to a specific adhesive system, to standardize their test surfaces further and to prevent any cross-contamination, and others who use microscope slides as their standardized surface, the slides being discarded after each test. Other possible sources of contamination of both the panel and the sample under test are perspiration from the fingers, dust, and moisture, all of which can act as a weak boundary layer between the adhesive and the test panel. Although care in handling and panel cleaning can correct the perspiration and dust, the humidity of the air controls the latter, an testing must be in a controlled-humidity environment, normally 50% relative humidity, and at a temperature of 23°C (73.5°F) [3], to restrict the testing to a known condition.

If all of the foregoing is appreciated, not only will the observed behavior of a pressure-sensitive adhesive during testing be better understood, but it will also be possible either to adapt existing test methods or to devise others that will be more appropriate to the practical application of a specific adhesive.

Finally, it should be recognized that when stress is applied to a pressure-sensitive adhesive, it is one of tensile, one of shear, or a combination of each. In the specific end use of given pressure-sensitive adhesive system, a clear understanding should exist as to what type of stresses can be encountered, to ensure that the test methods applied bear a relationship to use. Various standard test methods can now be considered.

II. ADHESION

Although slight differences occur here and there, the universally recognized method to determine how well a pressure-sensitive system adheres is to apply the adhesive, which has been precoated onto a flexible carrier, to a rigid standardized test panel under very controlled conditions and environment, using a roller of standard weight, dimensions, and construction, in order to apply a controlled application pressure, then finding out what force is required to separate the two, by peeling the flexible carrier at a controlled rate, at either 180° [9] or 90° [12], (Fig. 1a and b), the rate of both roll-down and peel usually being at 12 in. (300 mm) per minute, with various dwell times between roll-down and testing, from immediate to 20 min. The 90° test is usually chosen when the thickness of the backing prevents a 180° peel, but these two angles of peel tell a different story, the 90° peel being one of tensile, while the 180° peel is a combination of tensile and shear [16]. Also, it is to be noted that while both the 90° and the 180° tests are being peeled at the same tester rate, the peel front of the 90° test is moving at twice the rate of 180° test, so the adhesion results of the two angles of peel are not truly comparable.

To maintain a constant 90° during peel, a special jig is necessary that moves the test panel at the same rate as the tape is being stripped (Fig. 2a). An alternative method [9] is to apply the tape to a free-rolling drum, which automatically maintains 90° by rotating during the stripping process (Fig. 2b).

The work done in peeling is divided between that necessary to separate the adhesive from the test surface and that required to bend the backing, and the latter value can be a significant component [17]. Further, the basic assumption is that the geometry of the peel process is a constant. With more rigid backings this is not the case, the backing both creating a restrictive action to the intended angle of peel and also requiring a considerable amount of energy to bend. The resulting adhesion with the same coat weight adhesive using various backings over a range of backing thickness can be seen in Fig. 3. There are two ways that the effect of the backing can be investigated and/or overcome. One is to use the same backing in a series of thicknesses, each coated with the adhesive under evaluation at a constant desired coat weight. By plotting the adhesion values obtained, then extrapolating, an adhesion value can be obtained that eliminates the effect of the backing. An alternative method is to attach the adhesive-coated strip, adhesive side exposed, to a test panel, using a high-adhesion double-sided tape, then applying to this a strip of 25 μm polyester film very slightly wider than the test tape, following the same standard preparation and test procedures as if the polyester strip were the tape under test and the exposed adhesive surface were the test panel (Fig. 1c). This would then enable tapes of varying stiffness backings to be compared without the backings interfering with the test geometry. This method is also applicable to irregularly shaped pressure-sensitive coated pieces, the recorded adhesion values obtained being correlated with the changes in dimensions of the piece and brought to that of a standard width by calculation (Fig. 1d).

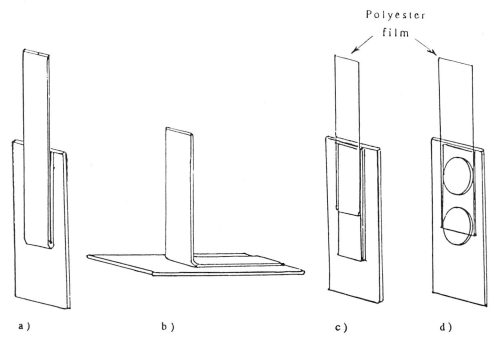

Figure 1 Peel adhesion testing: (a) 180° peel; (b) 90° peel; (c) reverse 180° peel; (d) adhesion of irregular shapes.

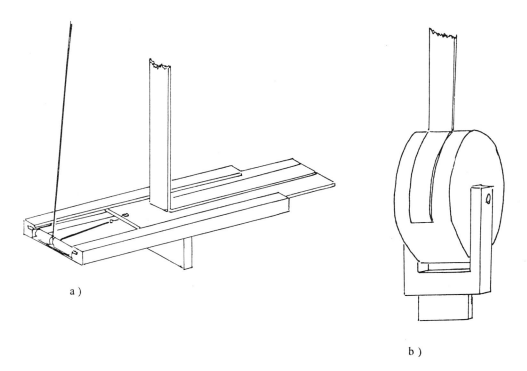

Figure 2 90° peel adhesion testing: (a) controlled 90° peel adhesion; (b) drum peel.

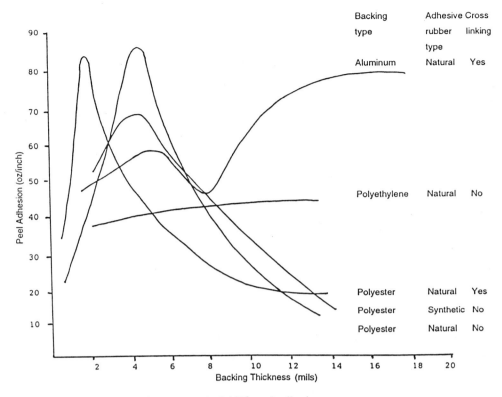

The following table appears within the figure:

Backing type	Adhesive rubber type	Cross linking
Aluminum	Natural	Yes
Polyethylene	Natural	No
Polyester	Natural	Yes
Polyester	Synthetic	No
Polyester	Natural	No

Figure 3 Effect of backing on nominal 180° peel adhesion.

Adhesion testing of double-sided tapes is achieved by applying 25μm polyester film of suitable width to one side of the tape, then carrying out a standard adhesion test on the other, recognizing that the polyester film will add to the value obtained as work done in peeling.

A. Tack

Tack is an indication of how quickly an adhesive can wet out, and so come into intimate contact with, a particular surface, and eventually reach its optimum adhesion. In effect, what we attempting to measure is the rate of change of adhesion with time. This rate of change is exponential in character, can be manipulated by the degree of applied pressure, and much of the event can occur in fractions of a second. The methods of measuring tack are many [18], with considerably different techniques and geometries, often being unique adhesion test methods, but they can be divided into three broad categories: those using a rolling ball, those using a modification of the peel test, and those using some form of probe.

There are three major methods of tack test using a rolling ball method, each with its own variations, to suit a specific need, all involving rolling a small ball down a ramp and over the adhesive surface. The Douglas method (Fig. 4a), useful for adhesives with a high tack value, uses a 12-in. (300-mm)-radius ramp and a 1⅛-in.-diameter steel ball, but differing sizes and types of ball are also used, depending on individual preference, one method using a table tennis ball, to create the minimum of pressure. The tack value is

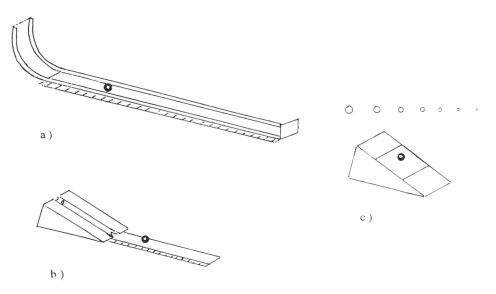

a)

c)

b)

Figure 4   Rolling ball tack testing: (a) Douglas; (b) PSTC; (c) Dow.

proportional to the reciprocal of the distance rolled, which also implies that the farther the ball rolls, the less reliable the method becomes, useful to 15 in. or so of ball travel. It has its problems, such as the contact area of the adhesive on the ball differing with the softness of the adhesive, the tendency of the tape to climb up the receding side of the ball, and the fact that after one revolution of the ball, the surface is now adhesive contaminated. However, it is a useful comparative test for similar adhesive systems. PSTC Test Method No. 6 for Rolling Ball Tack (Fig. 4b) uses a 6-in. (150-mm) ramp at 21.5° with a slight curve at the exit, and a $^7/_{16}$-in.-diameter steel ball. It is useful for adhesives with a lower tack value, with the same relationship and problems as the Douglas, useful for only the first few inches of travel. The Dow method (Fig. 4c) uses a 30° ramp, the adhesive-coated material being secured face up, to the ramp, with 3 in. of adhesive exposed. Then a series of steel balls each differing in diameter from the next by $^1/_{32}$ in., are rolled individually in decreasing order, over the 3-in. (75-mm) section of adhesive on the ramp, until one ball is stopped in the 3-in. (75-mm) section. The tack value is taken as the numerator of the diameter of that ball. In all cases, the need for controlled conditions, and thorough cleaning and drying of the balls used, is essential.

The commonest modified peel test, known as the quick-stick, or Chang test [19], uses the same equipment as for 90° peel (Fig. 2a), but the tape is laid on the test panel without any applied external pressure, then peeled immediately at 90°. The flaws in this method are that the adhesive continues to wet out the test panel with time, as it remains in contact, the precise time from application to testing being variable, but of much greater significance is the fact that a zone of pressure follows the peeling action, governed somewhat by the flexibility of the backing, so that the tape effectively applies itself to the panel with pressure immediately following the peel zone [20] (Fig. 5a). In effect, it becomes a 90° adhesion test. One other well-known method, known as the loop tack method [21] (Fig. 5b), makes a loop of a 1-in.-wide sample of the tape, adhesive exposed; then this exposed adhesive is touched to a horizontal test panel with minimal pressure, to bring a definite

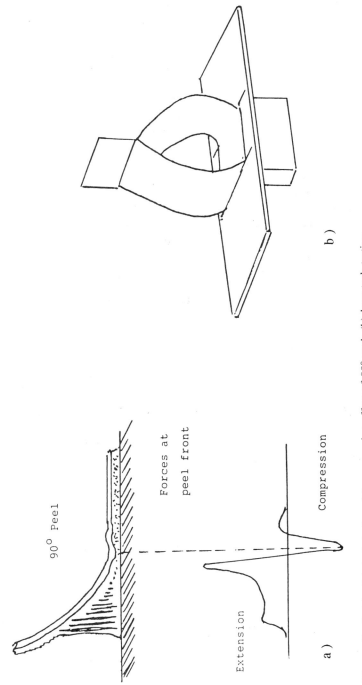

Figure 5 Quick-stick tack testing: (a) compressive effect of 90° peel; (b) loop tack testing.

area into contact. The force required to remove the loop, which is tested as for a 90° peel, is recorded as the tack value. This is essentially two Quick-Stick tests, back to back, and therefore has the same flaws.

There have been various methods of probe tack testing, which have included using a ball bearing with supporting ring [22] or a large hemispherical probe [23], but one method now dominates, which is the Hammond or Polyken probe tack tester [19] (Fig. 6a), which has the latitude of a variable pressure, a variable dwell time, a variable rate of removal of the probe, probes of various construction, and even the option of a heated probe. Bringing the probe and adhesive surface together to ensure uniform contact can be difficult, even when using a probe of small diameter, the standard being a flat surface stainless steel probe of 0.5-cm^2 area, as there is a tendency to trap a minute air bubble. A slightly domed probe can overcome this [25]. Too low an application pressure will give nonuniform contact; too great a pressure causes the value obtained to plateau out, as will too great a dwell time before separation of the probe. The standard test conditions are 100-g/cm^2 loading, with a 1-s dwell time and 1-cm/s removal, but this can be adjusted to suit certain adhesive systems. The separation mechanism is in actuality a very low-angle circular peel (Fig. 6b), at very high speed, as has been shown by high-speed photography [26]. Although the probe tack method has not yet been fully adopted, when understood, it gives the most meaningful data.

Variations in test results for tack from test to test for a given sample are common, possibly because such a small area of the adhesive surface is under examination at each test, and for this reason, whatever test method is used for tack testing, a number of checks are necessary to obtain a statistically meaningful result.

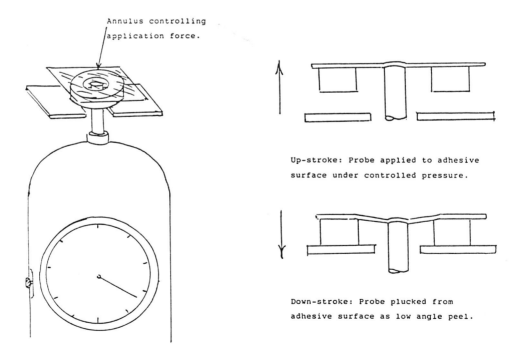

Annulus controlling application force.

Up-stroke: Probe applied to adhesive surface under controlled pressure.

Down-stroke: Probe plucked from adhesive surface as low angle peel.

Figure 6   Kendall probe tack testing.

B. Shear

The universally accepted method for shear is a static load test [27], where a known surface area of the adhesive-coated product is applied under controlled conditions to a standard test surface, usually stainless steel or a standard cardboard, or even the product's own backing. Then this is secured vertically, plus a 2° tiltback, to prevent any possible low-angle peel. A fixed load is then applied (Fig. 7a), and the time taken for failure is recorded or any slip that occurs in a given time measured. A trip mechanism can be set up so that the falling weight stops a stopclock to denote the failure time. This test can be carried out at elevated temperature, and variations of the basic concept exist, such as mounting the test surface horizontally [28] with a small radius of curvature at the edge of the test panel to redirect the tail of the tape under test to the vertical, in order to hang a suitable load (Fig. 7b). A further variation, used in the electrical industry to measure adhesive thermoset characteristics, is a high-temperature shear test [10], where a controlled area overlap bond is formed, usually 0.5 in. (12 mm) by 0.5 in. (12 mm), either face to face or face to backing. The adhesive is then given a thermoset cycle, followed by a high-temperature shear test (Fig. 7c).

While a static load test is the commonly accepted procedure, in practice many variations in areas and weights are used, to compensate for the varying qualities of adhesive evaluated, so that the test results will fall into a similar time frame, so it becomes difficult to compare different adhesive systems from accumulated data. It has the disadvantage of giving variable results for the same adhesive system and is essentially a pass/fail test, as many products remain in place at the end of the test period. A more reliable method is to convert the method to a dynamic one, but allowing a sufficiently slow rate to allow for any molecular disentanglement that may occur. This can be carried out on a modern tensile tester at a rate in the area of 0.01 in. (0.25 mm)/min, with a suitably compensated chart speed. A sample length of 0.25 in. (6 mm) has been found to be sufficient, the width used being proportional to the result obtained. This method generates results of greater reproducibility. The data obtained by any method using nonporous test panels are not translatable to porous substrates. Experience has shown that the shear

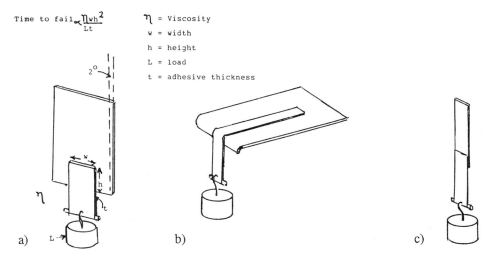

Time to fail $\propto \dfrac{\eta w h^2}{Lt}$

η = Viscosity
w = width
h = height
L = load
t = adhesive thickness

Figure 7 Shear testing: (a) 178° shear; (b) horizontal shear; (c) overlap shear.

properties of pressure sensitives to porous and nonporous substrates can be quite different, and each must be judged on its own merits.

C. Adhesion to Release Liner

The test methods in existence to evaluate the adhesion of pressure sensitives to release liners are modified adhesion tests, such as the 180° or 90° peel test, with the liner adhered to a test panel, or a T-peel, where the sample is freely suspended. The values obtained by the latter method are considerably affected by the stiffness of the liner, which alters the angle of peel.

One method occasionally used to evaluate the relationship of silicone release liners or other materials of very low surface energy to a pressure sensitive is to bring a coated pressure sensitive into contact with the liner under pressure to carry out an adhesion test. This is of little value as the adhesive is unable to wet out and so come into intimate contact with the release liner. The adhesive must first be coated onto the liner, dried, or cooled in the case of a hot melt, and then a carrier laminated to it, as is standard practice for transfer coating. As low values of adhesion can be expected, the force required to bend the backing may dominate, so a thin flexible backing should be used, to maintain a constant peel angle, 25-μm polyester being satisfactory. Then a standard 180° peel test can be carried out with the release liner secured to a test panel.

III. AGING TESTS

The usual function of an aging test should be to challenge the product artificially in an accelerated manner in one or more environments that can be expected, so that either after prolonged storage or after long periods of use, its behavior can be predicted.

The popular accelerated aging temperature used to predict natural aging is 65°C (150°F), with the adhesive-coated products either in roll or sheet form independently supported in a controlled-humidity environment such that they occupy less than one-fourth of the available space to ensure ample circulating air. Typical commercial products are exposed to 80% humidity using a saturated reagent-grade ammonium sulfate bath [29], while for electrical products this is taken up to 90% using a bath of 37% by weight solution of glycerin [30]. The time period to predict behavior after 1 year of natural aging varies from 4 to 7 days at 65°C (150°F), depending on the authority, the change in physical characteristics before and after high-temperature aging being compared. A reduction in physical characteristics of less than 10% after the aging test is normally considered satisfactory performance.

The chemical makeup of many pressure sensitives makes them prone to oxidation, and although antioxidants can compensate for this, it is still necessary to evaluate whether the antioxidant is functioning satisfactorily. Any test to evaluate resistance to oxidation becomes a comparative test in change of performance, usually of adhesion and tack, the severity of the test depending on individual requirements for the finished product. One simple but effective test is to expose the adhesive to 120°C (250°F) in a forced-air oven, for various periods of time, evaluating samples before and after heat exposure, a 1-h exposure being a good starting point. A more severe test is the use of an oxygen bomb, at lower temperatures, with high-pressure oxygen. Again, a good starting point is an overnight (16-h) test at 38°C (100°F) and 300 psi.

Exposure to ultraviolet (UV) light can have similar degrading effects, either from sunlight or even exposure to fluorescent lighting. The need for stability to UV light will depend on the adhesive end use, many adhesives never encountering such conditions. As with the oxidation test, the adhesive should be exposed to a controlled source of UV light, typically around 300 to 400 nm in wavelength, for prolonged periods, with the samples under test mounted on a slowly rotating turntable to ensure uniform exposure, the adhesion and tack before and after exposure being compared [31]. Commercial UV light sources are available. The source should be far enough away from the samples, usually around 18 to 30 in. (450 to 750 mm) or so, to eliminate a possible secondary exposure to excessive heat from the light source, certainly no more than 50°C (120°F). The UV intensity at the test surface should be determined with a suitable UV-light meter, adjusting the UV source location if necessary to ensure the same UV exposure for each test cycle. One proposed intensity is 2250 $\mu W/cm^2$ at 12 in. (300 mm) from the source with the samples at 18 in. (450 mm). As the UV source will deteriorate with time, it will be necessary to replace it periodically. The time of exposure is dependent on the marketing objectives for that product, but one can expect the test period to be as prolonged as 180 h.

An alternative UV exposure test method is the use of a commercial weatherometer [32] in which carbon arcs are used as the UV source, the samples being mounted on a carousel rotating around the source. The carbon arcs need changing frequently, to provide the necessary exposure time, again the time of exposure depending on individual preference. The weatherometer has the added advantage of being able to provide, in addition, a water spray at chosen intervals, to simulate outdoor exposure.

Degradation by heat, resulting in breakdown of the various polymer structures, is another cause of loss of properties as can occur during the processing of hot melt systems or in adhesive systems intended for high-temperature applications. Here the effect will not be seen so much in a change of tack or adhesion (these characteristics may even be improving) but in deterioration of the ability of the adhesive to resist shear forces, which can be dramatic. The actual change in molecular weight and its distribution can be determined by gel permeation chromatography [33]. The lower-molecular-weight polymer systems as used in hot melt pressure sensitives can be handled with conventional equipment and technology, but the much higher molecular weight of other natural and synthetic rubber-based systems will require more sophisticated equipment, including the use of heated columns.

IV. LOW-TEMPERATURE TESTING

Low-temperature testing can be used to determine whether the adhesive can be applied effectively at that low temperature, or whether it functions satisfactorily when in use at that temperature. Both require the use of a cold box with access to manipulate the samples and test equipment, the test equipment being either totally or partially enclosed within the box. The sample to be tested and all materials needed to set the test up must be in the cold box at the chosen temperature for at least 4 h before testing begins. Test panels will need to be washed with acetone to remove any condensate, then dried thoroughly with a lint-free wipe just prior to application of the adhesive. Testing can then proceed as normally, using whatever test methods are felt to be appropriate, usually including a peel adhesion test. In a functional test that involves adhesive separation from a test surface, if the adhesion value obtained is sufficiently high to reflect a satisfactory bond, any adhesive transfer due to brittle fracture that occurs under these conditions is usually discounted.

V. HIGH-TEMPERATURE TESTING

Here the adhesive is applied under standard conditions to whatever test surface is felt appropriate. The test, usually adhesion or shear, is then carried out in a high-temperature environment, after allowing sufficient time for the assembly to reach the test temperature before beginning the test. Visual examination for signs of cohesive failure is a necessary part of high-temperature adhesion testing.

One method used to evaluate the upper working temperature of block copolymer systems is to measure the shear adhesion failure temperature (SAFT), a useful method to discover exactly what has been gained in upper working temperature limits, when resins are added to the styrene domain of polystyrene end-block systems, to increase the effective glass transition temperature. Typically, the test is set up as a standard shear test, either to the standard stainless steel panel or to polyester, with an area of 1 by 1 in. (25 by 25 mm) and a load of 1 kg. The setup is placed in an oven which can be controlled accurately so that the temperature is increased by 2.0°C (3.6°F)/min, the temperature at which failure occurs being recorded as the shear adhesion failure temperature.

The above tests discussed above are general in nature. Specific products may call for specific tests aimed at the needs of the market.

VI. PACKAGING

Packaging in industry is universal; the need in most cases is the ability for a pressure-sensitive adhesive that will hold well to a cardboard surface under stress. A simple introductory test would be to determine whether the adhesive wets out the surface well enough to delaminate the cardboard. Although this is a subjective test, it can be used very satisfactorily to rate the cardboard keying ability of various systems. This test can be performed both initially and after allowing the adhesive to remain in contact at ambient conditions for a prolonged period, usually overnight. An estimate is made of the percentage fiber tear. A cardboard surface is not random, but the cellulose fibers lie in a definite direction, and at an angle, predetermined by the papermaking machine. Because of this, any cardboard used should be examined initially to determine which way the fibers lie. This is discovered by stripping and adhesive tape from the cardboard surface in each of the four directions. The cross direction of the card will show little or no fiber tear in either direction, while fiber tear should occur to some degree at right angles to this, denoting the machine direction of the cardboard. Because the fibers lie at an angle, the degree of fiber tear in the machine direction can be different depending on the direction of stripping. Testing should be carried out in the machine direction, stripping both ways, and recording the estimate fiber tear in each direction.

There are various box closure tests in use, some proprietary, all of which naturally call for the standard cardboard. If the cardboard used in the final package is available, the tests should be repeated with this specific material, as many are coated or treated, and the behavior on such surfaces can be quite different from those obtained with the standard cardboard.

On conventional packaging tape test is an inverted 90° peel test [30], using either standard cardboard or cardboard of choice as test surface. The time taken to strip a 1-in. (25-mm) length of a 1-in. (25-mm) wide sample using a 200-g load is measured (Fig. 8a). By using a window of silicone-coated release paper to define the 1-in. (25-mm) length,

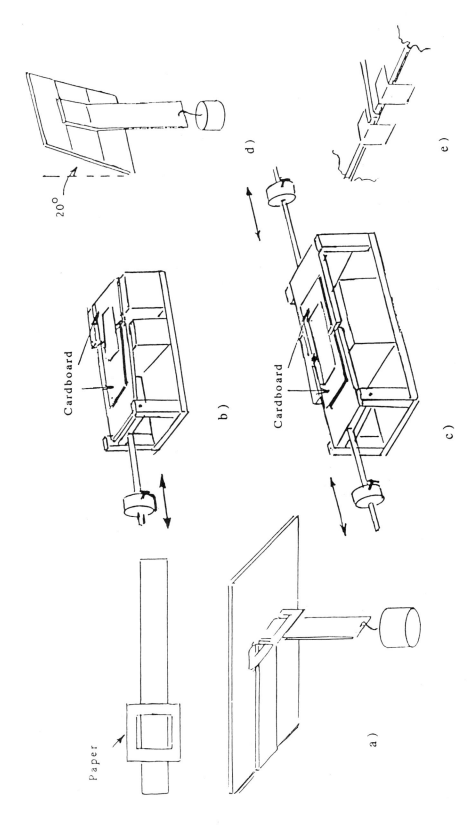

Figure 8 Cardboard box testing: (a) 90° inverted peel; (b) single-arm closure test; (c) double-arm closure test; (d) 20° peel; (e) L-clip.

Paper

Cardboard

Cardboard

20°

a)

b)

c)

d)

e)

only that section of cardboard is contacted, and the sample falls away at the completion of the test.

By using a piece of equipment arranged to create a synthetic box closure, a stress can then be applied to a 2-in. (50-mm) length of packaging tape placed equally across two cardboard surfaces, one or both of which can be variously loaded, and can pivot, as for a conventional box closure (Fig. 8b and c). In the former case, the stress is one of shear to the tape on the moving surface, and low-angle peel to the tape on the fixed surface. In the latter case, they are both shear. The time taken to fail is recorded. Both testing techniques are used commercially. There are also those who prefer a static 20° peel from cardboard in place of either of these methods, measuring the time taken to strip a given length (Fig. 8d).

Then there are the pragmatists who use a new box, then apply 2-in. (50-mm) lengths of packaging tape as L-clips to hold the box flaps in place, 1 in. (25 mm) on the flap and 1 in. (25 mm) on the body of the box, allowing the spring action of the new cardboard to create the stress (Fig. 8e). The boxes are then aged for 1 week under various conditions, finally rating the success or failures. A variation of this is the use of boxes overstuffed with latex foam to create the stress. A further commercial method is to fill the boxes suitably, as would be expected in practice, then arrange them on a vibration table such that the flaps are vertical rather than horizontal, and carry out a standard container shipping test.

A. The Appliance Industry

While packaging needs remain a major concern of the appliance industry, one other cause for concern with pressure sensitives is the potential for staining on painted surfaces. A pressure-sensitive adhesive system may come into contact with a painted surface, either temporarily, during the manufacturing process, or for prolonged periods of storage, following which, on removal of the adhesive tape, slight discoloration may be observed. This may be apparent immediately or may not become apparent until some time later.

One suitable test to evaluate staining [31] is to use freshly painted test panels using the paint under consideration, to which are applied strips of the tape being considered in precise locations, plus a control known to be satisfactory, marking the reverse side of the panels as to the exact location and type of each sample under evaluation. Sufficient areas of the panels are left exposed to provide a good comparison later. These panels are then aged. For evaluation of in-process adhesive systems, the temperatures that the system can encounter during processing are used, at double the exposure time. For long-term storage evaluation, 1 week at 65°C (150°F) is usually sufficient. The adhesive tapes are then removed, and any adhesive residue is cleaned off with a suitable solvent, ensuring that the solvent chosen does not also affect the paint. This can be a low-boiling-point aliphatic hydrocarbon, but isopropyl alcohol may be adequate. The panels are then examined carefully under both daylight and artificial light conditions for signs of color change. A Macbeth light box can be a useful source of light. For latent staining [32] using these stripped and clearly identified panels, one half of the previous location of each adhesive tape is obscured with small metal plates or other suitable covering. The panels are then exposed to UV light for several hours, 4 h being typical, then examined for any latent staining. The UV setup can be as outlined previously for aging, taking care to prevent any temperature buildup from the heat of the light source.

VII. HIGH-TEMPERATURE MASKING TAPES

This testing is in addition to the high-temperature testing described previously. High-temperature masking tape can be applied to a variety of surfaces—plain, primed, or painted metal, glass, rubber, or chrome finished—and testing must include them all. The simplest test is to apply the tape to the appropriate surface, subject it to heat for the time period expected in practical use, then strip the tape from the surface at various rates: both immediately, while the panel is still hot, and after 1 h, when the panel has been allowed to return to ambient temperature, visually examining the surface for any signs of cohesive failure.

However, masking tapes must conform to curves, and the adhesive must hold at high temperatures under these stressed states. Also, it is common practice to use paper or plastic drapes to cover other areas that may accidentally receive the applied paint. These drapes, or "aprons," are set into place by the same masking tape and quite often adhered to the backing of masking tape already applied directly to the surface to be painted. Using a high-velocity, high-temperature oven, samples of the masking tape under evaluation should be set in the oven at the evaluation temperature in a manner duplicating this use: that is, in curves of various radii and applied to its own backing in similar curves with apron paper attached, in both slight and excessive curves and applied to its own backing, these test panels being mounted vertically such that they are under the effect of the high-velocity hot air. After the test cycle, the samples are examined carefully for signs of lifting. All of the previous tests can be duplicated to include applying the various paints that the masking tape may encounter in use, then investigating any effect that the paint has had on the adhesive.

VIII. ELECTRICAL TAPES

Almost all electrical tapes have thermosetting adhesive characteristics. A suitable test to measure satisfactory thermosetting has been described in section II.B. Many electrical tapes are used to wrap coils of various diameters, and there may be a tendency for the end of the tape to lift away after application, known as *flagging*. This can occur either after the coil is taped and held in storage awaiting the next process, or during a thermoset cycle, in preparation for a varnish dip, or in the varnish dip process itself. Any flagging test must cover all eventualities. How much challenge the tape receives will depend on the diameter of the coils wrapped: the smaller the diameter, the greater the stress and therefore the greater the tendency to flag. One suitable test is to use a 0.5-in. (12-mm)-diameter mandrel and wrap the tape around the mandrel with an exact 0.5-in. (12-mm) overlap. These should be set aside and examined 1 week later for any tendency to flag. While the storage conditions for this 1-week period would be expected to be ambient, temperatures as high as 30°C (85°F) and humidities of 80 to 90% can be expected in a coil manufacturer's working environment and so must be brought into the evaluation study. The test assemblies, or others set up simultaneously, are then subjected to the manufacturer's recommended cure cycle for that adhesive, or if unknown, 1 h at 150°C (300°F), and examined again for flagging. The cured assemblies are then dipped for 20 min in the commercial varnish that the adhesive would normally encounter and reexamined for flagging. When sufficient experience has been gained, including comparative studies, this may be modified to a 5-min dip in the solvent system used in the varnish.

In the case of an electrical tape used as a harness wrap, the flagging test can be modified [33] by using a 0.125-in. (3 mm)-diameter rod, spiral wrapping the rod with the tape under test, ensuring a 50% overlap, then examining for any tendency to flag after an aging cycle. Electrical tapes may come into contact with current-carrying fine bare copper wires, a potential condition for electrolytic corrosion. While the tendency for electrolytic corrosion can be, and is, estimated indirectly from the reciprocal of the insulation resistance of an adhesive tape, insulation resistance is determined using the total cross section of the tape, whereas electrolytic corrosion is related to the adhesive only. A more reliable estimate can be made directly, by attaching two 32-gauge fine bare copper wires 0.25 in. (6 mm) apart along the adhesive surface of a 6-in. (150-mm) length of tape using the standard roll-down application method. Care should be taken to avoid kinks in the wire. With 250 V DC applied across the wires, this assembly is placed in an enclosure that provides a saturated atmosphere and then aged to 50°C (120°F) for 20 h. After removing the applied voltage and allowing to cool, the wires are carefully removed and the tensile test performed on each wire. The direct electrolytic corrosion is reported as the ratio of the tensile strengths of the two wires, no corrosion giving a factor of 1.

A much simpler version of this test, which can be used as a screening test, is to apply the adhesive tape to a freshly polished soft copper panel using a 100-grit aluminum oxide polishing wheel or 150-grit paper, then heat-aging this at 100°C (212°F) for 72 h. After allowing the tape to cool, it is removed and the panel inspected for any indication of staining.

Electrical tapes can also encounter various solvents as used in varnishes, or oils as used in transformers, and here a test can be devised that is specific to the end use, by immersing a test panel on which the adhesive tape under consideration has been applied, as for an adhesion test, in the liquid in question, the time being that which would be encountered in practice. After removal from the liquid, the excess liquid is removed carefully, using an absorbent, and a standard adhesion test is carried out, the result being compared to its original value. In most cases, some adhesive edge transfer is permissible. There are other tests related to electrical properties, such as dielectric strength and insulation resistance, but these are more related to the backing and thus will not be discussed here.

IX. SPECIAL TESTS

There may be a tendency, with time, for a pressure-sensitive adhesive tape to curl away slowly from the surface to which it is applied, usually due to a prolonged slight stress, coupled with stress relaxation within the adhesive. Conventional testing may not indicate this weakness. One way to study this is to use a modified form of the 90° inverted peel test shown in Fig. 8a. The surface under evaluation should be marked into equal units, as for example, six 1-in. (25-mm) divisions. The adhesive tape under evaluation can be applied along this marked area, as for a conventional adhesion test, together with a control product, known to adhere satisfactorily. Then in place of the usual 200-g loading, a much lighter load is applied, sufficient to apply a very slight stress. This will be in the area of 10 or 20 g, depending on the adhesive. Any adhesive system that has a tendency to release slowly from a surface will gradually peel away, sometimes within minutes, while a satisfactory product will hold indefinitely.

This can be modified to provide more meaningful data, by setting up with the peel equipment, as shown in Fig. 2a, and running as a 90° peel test, but with the test run at

0.01 in. (25 mm)/min, the recorder being adjusted appropriately. An alternative method that has provided interesting data is to begin this test at 0.01 in. (25 mm)/min, then stop the tensile tester under stress but allow the recorder to continue. Stress relaxation will be shown by the rate of drop of the peel value obtained when the tester was stopped.

The degree of initial contact that can be obtained from a given pressure-sensitive adhesive-coated material depends not only on the surface to which it is applied, but also on the roughness of the adhesive surface itself, which may come from prolonged contact with its own irregular backing, such as a glass cloth, or embossed or heavily creped paper or plastic facing or carrier. One simple method to evaluate surface contact is to apply the adhesive-coated product to a piece of plate glass as for a standard peel test, then examine the degree of contact area as seen by the degree of light reflection difference at the adhesive/glass interface, observing from the reverse side of the glass. An additional method is to use carbon paper. By applying the adhesive-coated product to the carbon paper, with a standard roll-down, then carefully removing it, the degree of ink removed and the pattern seen is a measure of the potential contact. The carbon paper may need reinforcement to prevent tearing on removal of the tape.

A. BUTT TENSILE TEST

The work done in testing pressure-sensitive adhesive in the 0.01-in. (25-mm)/min area has yielded an interesting new research test method, the Butt Tensile Test. The standard 90° peel test is one of tensile, but there is a continuous change in the adhesive under examination as the peel front recedes. By rearranging the geometry so that a limited fixed area of adhesive is under a continuous tensile stress, a new and informative test can be performed on the adhesive. For this test a tensile tester with the ability to record accurately values covering the range 2.0 lb (1000 g) is necessary, test results by this method typically falling into this range. First, the adhesive-coated tape under evaluation is prepared for testing by securing it, adhesive out, to a microscope slide in such a manner that a narrow strip of exposed adhesive at the end of the slide can be evaluated (Fig. 9a). The opposite end of the microscope slide is wrapped to give the tester jaws better grip. This assembly is then mounted in the upper jaws of the tensile tester. With the tester set on the 2 lb (1000 g) scale and the return speed of the tester set at a suitably low value, starting at 5 in. (125mm)/min, the narrow strip of exposed adhesive at the end of the microscope slide is then brought into contact with a polished horizontal test surface that has been mounted in the lower jaw of the tester, adjustments being made to the slide if necessary, to ensure that the two surfaces are parallel (Fig. 9b).

Slight pressure is then brought to bear to ensure intimate contact. One way that this can be achieved is the use of a tester fitted with an extension/compression strain gauge, setting the zero of the recorder at a position 5% from the zero of the full-scale deflection, followed by recalibration of the scale. The lower jaws are slowly raised until sufficient compression is applied to the bond to return the recorder pen to zero. The drive is then stopped and the application force held for 10 s. The lower jaw is then lowered slowly until the applied stress returns to zero. The setup is now allowed to relax for 2 min to relieve any stress within the adhesive. A tensile test is now carried out on the bond at 0.01 in. (0.25 mm)/min with the recorder set at 1.0 in. (25 mm)/min. Typical curves that can be generated are shown in Fig. 10. The shape of the curves obtained are independent of width, which only affects the amplitude, so compensation can be made for the sample

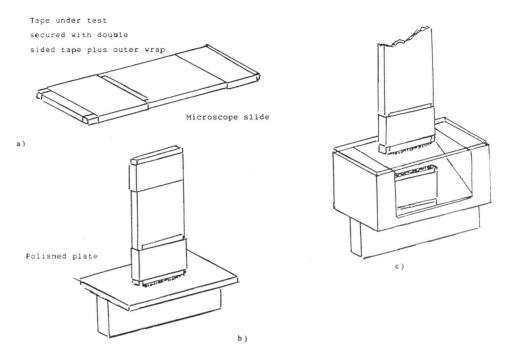

Figure 9 Butt tensile test: (a) sample preparation; (b) tensile tester setup; (c) 45° mirror setup.

width available, by calculation, as for adhesion testing. As the fibers are delicate, the equipment should be made as vibration free as possible.

Another variation in the test surface is to use a second microscope slide and using a suitable setup, mount a 45° mirror below. Then it can be ensured visually that adequate contact is being achieved (Fig. 9c), and further, the mechanism of failure can be studied. Initially, this method was used successfully to ensure the feasibility of the test method.

A description of these curves and a possible interpretation follow. The very slow speed of testing allows for cavitation, molecular disentanglement, and the formation of adhesive fibers, which are then drawn out until adhesive failure at the polished surface takes place. The length of these fibers is many times the original thickness of the adhesive. One can make the analogy that the adhesive consists of a broad spectrum of polymer structures. At one end of this spectrum, fibers are formed that require a considerable force to elongate, but the degree of elongation is low. At the other end of the spectrum, fibers are formed with very high elongation, which it takes little effort to extend. The former are those that contribute to adhesion while the latter contribute to tack. If the adhesive polymer structure is chemically cross-linked or has pseudo-cross-links, as in a block copolymer, the fiber extension process is restricted and orientation occurs, resulting in the formation of a secondary peak. The ratio of the primary and secondary peaks is a measure of the cure level for that adhesive. If a series of tests are carried out on an adhesive at various stages in its cure cycle, not only will the secondary peak gradually build, but the tail of the curve is also gradually lost, no doubt due to the lower-molecular-weight components, the most mobile, being taken up in the curing process (Fig. 10).

Adhesive systems showing high adhesion levels by conventional methods naturally show a high value of the primary peak, the area under the curve denoting the work done in

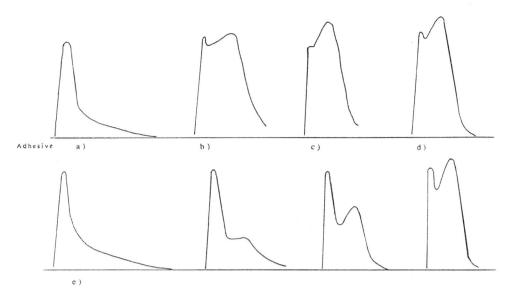

Figure 10 Typical butt tensile test graphs: (a) undercured, (b) cross-linked, (c) overcured, and (d) block copolymer adhesives; (e) stages in the cross-linking of a pressure sensitive adhesive.

separation. Un-cross-linked systems show no secondary peak (Fig. 10a). Adhesives with a high degree of cure show a high-value secondary peak (Fig. 10b and c) the pseudo-cross-linking of block copolymers showing the highest, and depending on adhesive design, often higher than the primary peak (Fig. 10d).

This technique can therefore provide various pieces of information in a single test. It can be used to evaluate the degree of cure of a given system once a standard has been determined and evaluated (Fig. 10e). The tail is a reflection of the low-molecular-weight component; adhesives with a low-molecular-weight plasticizer will show an exceptionally long tail. Such systems in practical use would be expected to show a tendency to lift after application. The method can thus highlight the cause of failure and identify poor design. Other potential uses for this test method will present themselves as the user becomes more familiar with its operation.

X. CONCLUSIONS

In conclusion, anyone testing pressure-sensitive adhesive systems should not only be fully acquainted with the fundamental characteristics of pressure-sensitive adhesives and the pitfalls in testing to ensure that the data gathered in whatever test method being used are meaningful, but he or she should also develop the skill to manipulate these test methods to meet specific needs.

REFERENCES

1. *Test Methods for Pressure Sensitive Tapes*, 9th ed., Pressure Sensitive Tape Council, Chicago, Ill., (1989).
2. Ref. 1, Appendix C.
3. Ref. 1, Appendix A.

4. Ref. 1, PSTC 14.
5. Ref. 1, PSTC 21.
6. Ref. 1, PSTC 22.
7. *Standard Test Methods for Pressure Sensitive Adhesive Tapes for Electrical and Electronic Applications*, D1000-88, American Society for Testing and Materials, Philadelphia, 1988.
8. Ref. 7, Sections 70–74.
9. Ref. 7. Sections 43–47.
10. Ref. 7, Sections 54–58.
11. *Peel Adhesion of Pressure Sensitive Tape at 180°*; D3330-90, American Society for Testing and Materials, Philadelphia, 1990.
12. Finat, *Pressure Sensitive Laminates, Suppliers and Users Technical Manual*, 1980.
13. *Test Methods Manual*, Association des Fabricants Européens de Rubans Auto-Adhésifs, 1990.
14. *Specification for Pressure Sensitive Tapes for Electrical Purposes*, Part 2, Methods of Test, Publication 454-2, International Electrotechnical Commission.
15. J. O. Hendricks and C. A. Dalquist, in *Adhesion and Adhesives*, Vol. 2, R. Houwink and G. Salomon (eds.), Elsevier, Amsterdam, 1967.
16. D. H. Kaelble, *Adhesives Age* (May 1960).
17. J. Johnston, *Adhesives Age 11*(4): 29 (1968).
18. J. Johnston, Tack, *Proc. Pressure Sensitive Tape Council Technical Seminar*, 1983, pp. 126–146.
19. F. S. Chang, *Rubber Chem. Technol. 20*: 847 (1957).
20. D. H. Kaelble, *Trans. Soc. Rheo. 9*(2): 135 (1965).
21. Test Procedures and Conversion Tables, V, Loop tack test, *Publication 6512*, Gelva, Monsanto, p. 20.
22. K. Kamagala, T. Sailo and M. Toyama, *J. Adhesion 20*: 281 (Oct. 1970).
23. R. Bates *J. Appl. Polymer Sci. 20*: 2941 (1976).
24. F. H. Hammond, Jr., *Symposium on Recent Developments in Adhesive Science*, Vol. 20, ASTM Special Publication American Society for Testing and Materials, Philadelphia, 1963, pp. 123–134.
25. *Standard Test Method for Pressure Sensitive Tack of Adhesives Using an Inverted Probe Method*, D2979-82, Note 1, American Society for Testing and Materials, Philadelphia, 1982.
26. C. A. Dalquist in *Adhesion: Fundamentals and Practice*, Ministry of Technology, Maclaren, London, 1969, pp. 143–151.
27. *Standard Test Method for Holding Power of Pressure Sensitive Tapes*, D3654-88, American Society for Testing and Materials, Philadelphia, 1988.
28. *Tapes, Pressure Sensitive and Gummed, Methods of Inspection, Sampling and Testing*, Federal Test Method Standard 147C.
29. *Recommended Practice for Maintaining Constant Relative Humidity by Means of Aqueous Solutions*, E104-85, Method C, Ammonium sulfate solution.
30. Ref. 29, Method A, Aqueous glycerin solution.
31. *Tape, Pressure Sensitive Film, Office Use*, Federal Specification LT 90E, Section 4.3.4.4.
32. *Standard Practice for Operating Light-Exposure Apparatus (Carbon Arc Type) with and Without Water for Exposure of Non-metallic Materials*, G23-81, American Society for Testing and Materials, Philadelphia, 1981.
33. L. H. Tung, *J. Appl. Polymer Sci. 24*: 953 (1979).

7

Durability of Adhesive Joints

Guy D. Davis and D. Kent Shaffer   *Martin Marietta Laboratories, Baltimore, Maryland*

I. INTRODUCTION

Although obtaining high initial bond strength is relatively easy, obtaining good bond durability in aggressive environments is comparatively more difficult. The most important factor leading to bond degradation is moisture. Moisture is pervasive over much of the world and is responsible for the vast majority of bond failures, either in the field during service or in the laboratory during research and development.

The importance of adhesive bond durability will vary depending on the particular application and environment. One of the most critical cases is military and civilian aircraft, especially those that operate in tropical coastal or marine locations. At the other extreme, less critical applications include those subject to low stresses and protected from harsh environments, such as interior furniture in temperate climates [1] or even wrapped gift packages. Increased durability generally requires additional initial cost through more expensive materials or processing. Consequently, it makes little sense for a bond to have a significantly longer lifetime than the item or system of which it is a part. In this chapter we concentrate on applications where durability is a critical issue, an emphasis natural to the topic and to our involvement in the aerospace industry.

The rate of bond degradation depends on a number of variables that can be grouped into three categories: environment, material, and stress. The environment is dominated by temperature and moisture. It can also include the concentration of aggressive ions, such as chlorides, and the presence of fuels, deicers, and other fluids. The material grouping is all-inclusive and includes the adherend, the adhesive, and the interphase between them. Finally, the stresses to which the bond is subject either during or after exposure also influence its lifetime or residual strength.

Each of these factors will be discussed in more detail. We also review the means to enhance durability. Entire books can and have been written on durability; in this short chapter, we can only touch on the subject. For more details, the reader is referred to the many reviews in the literature [1–11].

II. ENVIRONMENT

As already mentioned, moisture is the bane of most adhesive bonds. It is nearly impossible to keep water from a bond [12]: It can readily diffuse through the adhesive or the adherend, if it is permeable, as a composite might be. Moisture can also wick or travel along the interface and it can migrate via capillary action through cracks and crazes in the adhesive. Once moisture is present, it can attack the bond by [12]:

1. Reversibly altering the adhesive (e.g., plasticization)
2. Swelling the adhesive and inducing concomitant stresses
3. Disrupting secondary bonds across the adherend–adhesive interface
4. Irreversibly altering the adhesive (e.g., hydrolysis, cracking, or crazing)
5. Hydrating or corroding the adherend surface

The first few of these processes are reversible to one extent or another. Provided that bond degradation has not proceeded too far, if the joint is dried out (which may be a long process), the bond can regain some of its lost strength [10,13]. There appears to be a critical water concentration below which either no weakening occurs [1,14] or whatever weakening that does occur is reversible [13,15]. This critical water concentration is dependent on the materials used in the joint and is likely to be dependent on the temperature and stress as well. At higher moisture levels, some strength may be recovered upon drying, but at a certain point, the failure becomes near catastrophic and is beyond recovery.

Upon moisture penetration, the locus of failure almost always switches from cohesive within the adhesive to at or near the interface. Because metal oxide surfaces are polar, they attract water molecules, which can disrupt any dispersive (van der Waals) bonds across the interface. This disruption can be seen thermodynamically by the work of adhesion in an inert medium, W_A, which can be represented as [1]

$$W_A = \gamma_a + \gamma_s - \gamma_{as} \tag{1}$$

where γ_a and γ_s are the interfacial free energies of the adhesive and substrate, respectively, and γ_{as} is the surface free energy. In the presence of a liquid such as water, the work of adhesion, W_{Al}, becomes

$$W_{Al} = \gamma_{al} + \gamma_{sl} - \gamma_{as} \tag{2}$$

where γ_{al} and γ_{sl} are now the interfacial free energies of the adhesive–liquid and substrate–liquid interfaces, respectively. In an inert environment, the work of adhesion for a bonded system will be positive, indicating a stable interface, whereas in the presence of water, the work of adhesion may become negative, indicating an unstable interface that may dissociate. Table 1 shows, in fact, that moisture will displace epoxy adhesives from iron (steel), aluminum, and silicon substrates and promote disbonds. In contrast, although moisture weakens epoxy–carbon fiber bonds, these remain thermodynamically stable. Industrial experience with both metal and composite joints confirms these predictions [1].

The data presented in Table 1 illustrate the potential disastrous results when relying solely on dispersive bonds across the interface between an epoxy adhesive and metals or ceramics. To illustrate this danger we have produced demonstration specimens that exhibit good initial strength but fall apart under their own weight when a drop of water is placed at the crack tip.

Table 1 Work of Adhesion for Various Interfaces

Interface	Work of adhesion (mJ/m^2)		Evidence of interfacial debonding after immersion
	In inert medium	In water	
Epoxy–ferric oxide	291	− 255	Yes
Epoxy–alumina	232	− 137	Yes
Epoxy–silica	178	− 57	Yes
Epoxy–carbon fiber	88–90	22–44	No

Source: Ref. 1.

III. MATERIALS

The adherend, adhesive, and interphase, among them, determine bond durability. For example, the simple disruption of the dispersive forces already described indicates that joints made with composite adherends will be inherently more stable than those made with metallic adherends. To increase durability, most adherends undergo surface treatments designed to alter the surface chemistry or morphology to promote primary covalent chemical bonds and/or physical bonds (mechanical keying or interlocking) to maximize, supplement, or replace secondary dispersive bonds. These treatments are discussed in another chapter and elsewhere [1,3,11,16–18]. One intent of each treatment is to provide interfacial bonding that is resistant to moisture intrusion.

Formation of durable chemical bonds is an obvious means to stabilize the interface and has been demonstrated for phenolic–alumina joints [19] and for silane coupling agents [20,21]. However, for most structural joints using epoxy adhesives, mechanical interlocking on a microscopic scale is needed between the adhesive–primer and adherend for good durability. In these cases, even if moisture disrupts interfacial chemical bonds, a crack cannot follow the convoluted interface between the polymer and oxide and the joint remains intact unless this interface or the polymer itself is destroyed.

The scale of the microscopic surface roughness is important to assure good mechanical interlocking and good durability. Although all roughness serves to increase the effective surface area of the adherend and therefore to increase the number of primary and secondary bonds with the adhesive–primer, surfaces with features on the order of tens of nanometers exhibit superior performance to those with features on the order of micrometers. Several factors contribute to this difference in performance. The larger-scale features are fewer in number and generally are smoother (even on a relative scale), so that interlocking is less effective. Depending on the particular treatment used, there may also be loosely bound detritus that prevent bonding to an integral adherend [22]. In addition, the larger-scale roughness frequently allows trapped air and surface contaminants to remain at the bottoms of troughs and pores [22,23]. These unbonded regions limit joint performance by reducing both chemical and physical bonds and serving as stress concentrators. By contrast, smaller-scale microroughness generates strong capillary forces as the primer wets the surface, drawing the polymer into all the ''nooks and crannies'' of the oxide and displacing trapped air and some contaminants to form a microcomposite interphase [23]. Indeed, cross-sectional micrographs show complete filling of the micropores [4,6,11,16,22,24].

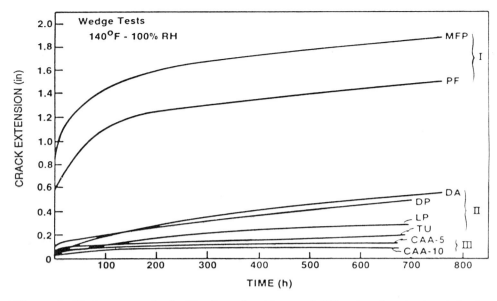

Figure 1 Wedge test results for Ti adherends with several different surface treatments having differing degrees and scales of roughness. Specimens were exposed to 100% relative humidity at 60°C. (From Ref. 25.)

This dependence on the degree and scale of roughness is illustrated in Fig. 1, which shows wedge test results of titanium bonds with several surface preparations. Because the titanium surface is stable under these conditions (see below), differences in the joint performance can be attributed solely to differences in the polymer-to-oxide bonds and correlate very well with adherend roughness. The poorest-performing group of pretreatments [class I: phosphate fluoride (PF) and modified phosphate fluoride (MPF)] produced relatively smooth surfaces. The intermediate group [class II: Dapcotreat (DA), dry Pasa Jell (DP), liquid Pasa Jell (LP), and Turco (TU)] exhibited macrorough surfaces with no microroughness. They had significant improvements in durability over the smooth adherends, but not as good as class III pretreatments [chromic acid anodization (CAA), 5 and 10 V], which provided a very evolved microroughness. Subsequent class III tests using sodium hydroxide anodization (SHA) and plasma spray provide further evidence to this correlation [26–28]. Both of these give very good durability performance and exhibit high levels of microroughness.

A. Adherends

The adherend often establishes ultimate joint durability. The morphology of its surface determines the degree of physical bonding (mechanical interlocking) with the polymer, and its chemistry, in part, determines the degree and type of chemical bonding. Furthermore, the stability of the adherend and its surface determines the ultimate limit of durability. Once the adherend becomes degraded, the bond line is reversibly damaged and the joint fails.

Each material exhibits its own form of degradation and conditions under which the degradation occurs. For aluminum adherends, moisture causes hydration of the surface

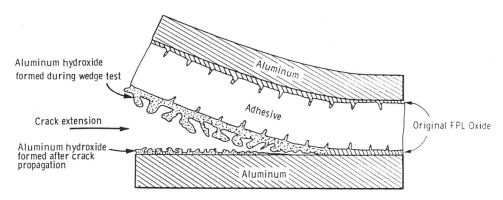

Figure 2 Schematic diagram of an aluminum adherend where the surface hydrates to cause bond failure.

[i.e., the Al_2O_3 that is formed during the surface treatment is transformed into the oxyhydroxide $AlOOH$ (boehmite)]. The transformation to boehmite results in an expansion of the interphase (the volume occupied by the boehmite is larger than that originally occupied by the Al_2O_3). This expansion and the corresponding change in surface morphology induce high stresses at the bond line. These stresses, coupled with the poor mechanical strength of the boehmite, cause crack propagation near the hydroxide–metal interface, as shown schematically in Fig. 2 [4].

Bond durability can be improved significantly with surface treatments that stabilize the oxide against hydration. One such treatment is phosphoric acid anodization (PAA) [24], which provides an oxide coating that is inherently hydration resistant. Its stability is due to a layer of phosphate incorporated into the outer Al_2O_3 surface during anodization; only when this phosphate layer goes into solution does the underlying Al_2O_3 hydrate to $AlOOH$ [29]. Another means of providing a hydration-resistant surface is its treatment with a hydration inhibitor [30]. Figure 3 shows wedge tests results for a Forest Product Laboratory (FPL) bond [31], an FPL bond pretreated with nitrilotrismethylenephosphonic (NTMP) acid [32], and a PAA bond. The monolayer coverage of NTMP stabilizes the surface against hydration and provides wedge test bond performance similar to that of PAA-treated adherends.

Steel adherends are also subject to corrosion in moist environments. Unfortunately, no general etch or anodization treatment has been developed that provides superior bond durability [11,16,33]. In part, this is due to the lack of a coherent, adherent stable oxide–iron oxides do not protect the underlying substrate from the environment. Equally important, the different steel metallurgies react to chemical treatments differently—a procedure that may give good results for one steel may give very poor results for another, similar steel.

In contrast to aluminum and titanium structural bonds where performance can be optimized for most aerospace applications, steel bonds are often designed to minimize cost as long as certain performance standards are met [34]. If feasible, many manufacturers prefer to select adhesives or primers that provide adequate strength and durability with untreated steel rather than to prepare the surface for bonding.

The most common surface treatments are grit blasting or other mechanical abrasion processes that clean the surface and provide a more chemically reactive oxide. Although

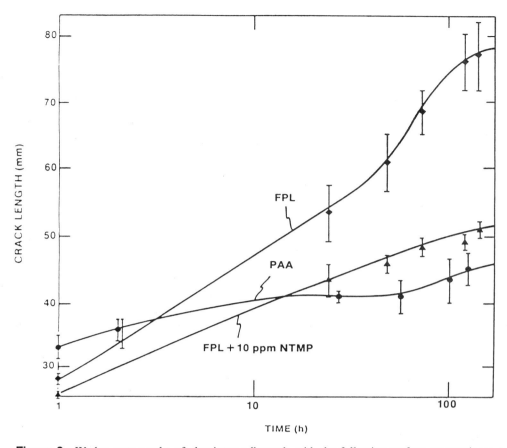

Figure 3 Wedge tests results of aluminum adherends with the following surface preparations: FPL, PAA, and FPL, followed by an NTMP treatment.

the result is not as good as that of the common aluminum and titanium treatments, performance is frequently adequate. Improvements to grit blasting, based on either performance or cost, have been reported for individual steels [35–41]; however, rankings of different treatments commonly vary from researcher to researcher because of different steels or exposure/test conditions.

Deposited coatings often provide better bond durability than that of native surface treatments. For example, optimized conversion coatings can provide a microscopically rough surface that is resistant to corrosion [42–46]. They serve to stabilize the surface from degradation and to form physical bonds with the adhesive–primer. Again, differences in the adherend metallurgy can cause differences in the coating morphology and chemistry. Nonetheless, such conversion coatings and other similar deposited coatings provide the best durability for steel bonds.

Joints made with steel and other metals can also be subject to cathodic disbondment if they are immersed in an electrolyte and subjected to a cathodic potential, such as that created when the adherend is in electrical contact with a more electrochemically active metal [47]. Although corrosion of the adherend is suppressed via cathodic protection, the rate of bond failure is increased. After an induction period that depends on the imposed cathodic potential and temperature [47], interphasal debonding occurs [48,49]; such disbondment does not occur in the absence of a cathodic potential. Several mechanisms have been proposed to explain this phenomenon: hydrogen evolution at the steel substrate [47], degradation and weakening of the polymer in the high-pH environment generated at the interface [48–51], osmotic pressure resulting from lead chlorides formed at the interface by dehydrohalogenation of the polymer [49], and breaking of secondary and primary bonds at the interface [1].

Bond failure can also occur if the surface is anodic relative to another joint component. An example would be clad aluminum adherends where a thin layer of pure aluminum overlays the base alloy. Such a surface layer is designed to be more corrosion resistant than the alloy but to act as a sacrificial anode should corrosion occur. Although this approach works well for corrosion protection of the substrate material, it can be a disaster for bonded material if the adherend surface–interface corrodes [1,52]. As a result, U.S. companies tend to use unclad aluminum for bonding and provide other means of corrosion protection, such as painting. On the other hand, European companies commonly use clad adherends, but with a thicker oxide (chromic acid anodization) [6,16,53–55] that provides bondline corrosion protection.

In contrast to aluminum and steel, titanium adherends are stable under conditions of moderately elevated temperatures and humidity. Although moisture has been shown to accelerate the crystallization of the amorphous oxide of titanium adherends anodized in CAA to anatase [56], the crystallization, along with the resulting morphology change, is very slow relative to the changes observed with aluminum and steel. As shown by the wedge test results presented in Fig. 1, the adherend surfaces underwent no change in morphology or crystallinity. Failure of the CAA specimens remained within the adhesive, with the physical bonds provided by the microscopically rough oxide remaining intact [56]. Identical results were also observed for wedge tests performed in boiling water (i.e., crack propagation for CAA adherends was entirely within the adhesive) [57].

At elevated temperatures where titanium alloys would be the adherend of choice, a different failure mechanism becomes important. Because the solubility of oxygen increases, the oxygen in a CAA or other oxide dissolves into the metal, leaving voids or microcracks at the metal/oxide interface. Consequently, stresses are concentrated at small areas at the interface and the joint fails at low stress levels [57]. Such phenomena have been observed for adherends exposed to 600°C for as little as 1 h or 300°C for 710 h prior to bonding and for bonds using a high-temperature adhesive cured at 400°C.

To prevent this failure mode, thick oxides, such as those grown by CAA, must be avoided in high-temperature applications. One surface treatment that is compatible with high temperatures is a plasma-sprayed metallic titanium coating [57]. By properly controlling the deposition parameters, a fractal-like microrough coating is obtained that provides physical bonding with the adhesive. Only a thin native oxide is present, and this is apparently insufficient to cause the type of failure described above. Plasma-sprayed adherends have been heated to 450°C for 165 h prior to bonding and tensile testing and have been bonded with a 400°C-curing adhesive and wedge-tested at 230°C for 1000 h [16,57]. In both cases, failure occurred within the adhesive, indicating a stable interphase.

Another means by which temperature can influence bond durability is through stresses that develop when different parts of a joint have different coefficients of thermal expansion (CTEs). This consideration is especially important when different classes of materials are being bonded together. Typical CTE values for selected materials are given in Table 2. Polymer CTEs usually are 10 to 100 times those of other materials. Stresses begin to develop across the interphase once the adhesive cures to a solid (rubbery) state and the joint begins to cool [58]. As long as the adhesive is above the glass transition temperature (T_g), it will generally be compliant enough to relax and accommodate these interphasal stresses. However, once T_g is reached, the adhesive is less compliant and stresses begin to build up. Thus the thermal stresses in a joint will depend on the CTE differences between the substrate(s), adhesive, and any overlying layers or films and on the degree of cooling below T_g. One way to minimize these stresses is to blend low- and high-CTE polymers to match the CTE of the substrate—a procedure most relevant to polymeric substrates [59]. Another is to incorporate mineral fillers into the adhesive to reduce its CTE [58]. In cases of mismatched adherends (e.g., composite to aluminum), a near room-temperature-curing adhesive may be the best solution [60,61].

B. Adhesives/Primers

The effects of water and temperature on the adhesive itself are also of utmost importance to the durability of bonded structures. In the presence of moisture, the adhesive can be affected in a number of ways, depending on its chemistry and how rapidly the water permeates through and causes significant property changes [38,62–64]. The potential efficacy of moisture penetration on the locus of failure of bonded joints has been discussed in the preceding section. As expected, elevated temperature conditions tend to degrade joint strength at a faster rate.

Of primary importance in moist environments is the plasticization, or softening, of the adhesive, a process that depresses T_g and lowers the modulus and strength of the elastomer [65–67]. Plasticization of the adhesive may also allow disengagement from a microrough adherend surface to reduce physical bonding and thus reduce joint strength and durability [30]. On the other hand, it may allow stress relaxation or crack blunting and improve durability [68].

Brewis et al. [66] studied the effects of moisture and temperature on the properties of aluminum–epoxy joints by measuring changes in the mechanical strength properties of the

Table 2 Coefficients of Thermal Expansion

Material	CTE (10^{-6} K^{-1})
Epoxy (above T_g)	190
Epoxy (below T_g)	68
Poly(methyl methacrylate) (above T_g)	530
Poly(methyl methacrylate) (below T_g)	260
Low-density polyethylene	100
High-density polyethylene	130
Polystyrene	700
Aluminum	29
Steel	11
Titanium	9
Soda glass	8.5
Wood (along grain)	3–5
Wood (across grain)	35–60

Source: Ref. 58.

soaked adhesive. The T_g values of the wet adhesive and relative strengths of wet and dry joints were evaluated for up to 2500 h. They concluded that the joint weakening effect of water was due to plasticization of the adhesive, which, in turn, was dependent on the rate of water diffusion within the adhesive.

The softening behavior has also been observed with FM-1000 aluminum single lap joints exposed at 100% relative humidity at 50°C for 1000 h [67]. As shown in Fig. 4, wet and dry joints exhibited similar strength–temperature relationships, with the former being shifted to a lower temperature by 30 to 50°C, a quantity close to the water-induced depression of T_g. Hence in this case, the T_g depression acts as a shift factor that defines the strength–temperature relationship between the dry and wet adhesive so that at a given temperature, a wet joint exhibits lower strength than a dry one.

Water entering a joint can also cause swelling, which tends to introduce stresses to weaken the bonded system. Weitsman [69] has shown that normal stresses resulting from swelling (3%) of an epoxide adhesive are manifested at the edges of the joint; however, after an initial rise, the stress concentration decreases with time, suggesting that they do not contribute to long-term structural weakening.

As discussed earlier, chemical bonds between the adhesive and adherend help to stabilize the interface and increase joint durability. Aluminum joints formed with phenolic adhesives generally exhibit better durability than those with epoxy adhesives [1,68,70]. This is partly attributable to strongly interacting phenolic and aliphatic hydroxyl groups that form stable primary chemical bonds across the interface [19,71,72]. Nonetheless, epoxy adhesives are more widely used due to their greater toughness and lower temperatures and pressures required during cure.

Silanes and other coupling agents can be applied to various substrates or incorporated into an adhesive–primer to serve as hybrid chemical bridges to increase the bonding between organic adhesive and inorganic adherend surfaces [73–76]. Such bonding increases the initial bond strength and also stabilizes the interface to increase the durability of the resulting joint. Silanes can be represented by the formula $YRSiX_3$, where X is a

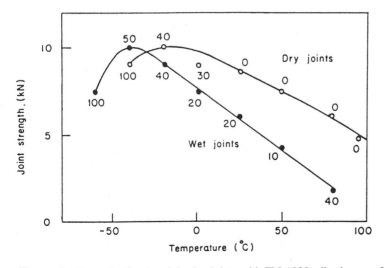

Figure 4 Strength of wet and dry lap joints with FM-1000 adhesive as a function of temperature: ○, dry joints; ●, joints preconditioned for 1000 h at 50°C and 100% relative humidity. (From Ref. 67.)

Table 3 Silane Coupling Agents

Adhesive type	Amino				Epoxy		Mercapto	Methacryloxy	Vinyl	
	A-1100	A-1110	A-1120	A-1130	A-186	A-187	A-189	A-174	A-151	A-172
Acrylic	X	X	X	X	A	X	A	A	—	—
Butyl	A	—	—	—	—	A	—	X	—	—
Epoxy	X	X	X	X	A	A	A	—	—	—
Neoprene	A	—	—	—	—	—	X	—	—	—
Nitrocellulose	X	A	A	A	—	—	—	—	—	—
Nitrile	A	—	—	—	—	—	X	—	—	—
Phenolic	A	A	—	—	A	A	X	—	—	—
Polyester	A	A	—	—	—	A	—	X	A	A
Polysulfide	A	—	—	—	—	X	X	—	—	—
Polyurethane	A	X	A	A	—	X	X	—	—	—
Vinyl	A	A	X	X	—	—	—	—	—	—
Hot melts	—	—	—	—	A	A	—	X	X	X

[a]X, generally effective; A, alternate.
Source: Refs. 76 and 77.

hydrolyzable group, such as alkoxy; Y is a functional organic group, such as amino, methacryloxy, or epoxy; and R is a small aliphatic linkage [76]. Generally applied from aqueous solutions at a pH of 4, the silane hydrolyzes and condenses upon drying to leave an oligomeric polysiloxane film on the metal oxide substrate with a stable covalent M—O—Si—R—Y structure [74]. The R group is chosen to react with the polymeric matrix to complete the coupling. Several coupling agents appropriate for different adhesives are listed in Table 3 [76,77].

Silane-based primers have been shown to be effective in increasing the environmental resistance of joints prepared for aluminum [78] and titanium [79] alloys. Plueddeman has shown that the resulting interphase can be designed for maximum water resistance by employing hydrophobic resins and coupling agents and by providing a high degree of cross-linking [80].

Corrosion-inhibiting adhesive primers are commonly applied onto bonding surfaces soon after the surface treatment [75]. Their primary function is to wet the adherend and penetrate the "nooks and crannies" to form both chemical and physical bonds. They also perform other functions essential for durable bonds: creation of a stable surface, prevention of contamination or mechanical damage of surfaces that have been chemically etched or anodized, and corrosion inhibition to the bonded and nonbonded areas of the assembly. Primer systems are normally pigmented with chromates, such as calcium, strontium, or zinc, which provide corrosion inhibition. The mechanism of inhibition by chromates in the presence of moisture involves the passivation of the aluminum surface and the prevention of cathodic evolution of hydrogen by the reduction of hexavalent chromium to the trivalent state [81]. Figures 5 and 6 illustrate the improvement in joint performance using a corrosion preventive primer (BR-127) [11,16].

IV. STRESS

The stresses that a joint experiences during environmental exposure also influence its durability (i.e., it exhibits either decreased lifetime or decreased residual strength) [5,7,10,82,83]. As with moisture, there may be a critical stress level below which failure

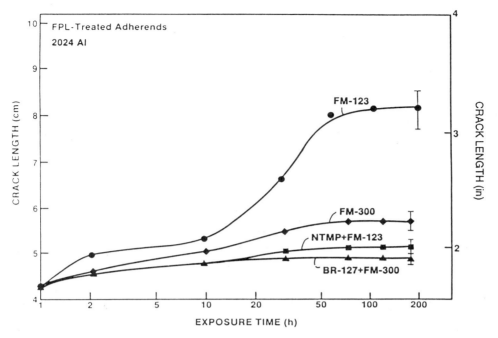

Figure 5 Wedge test results for FPL-etched aluminum adherends: with FM-123 (moisture-wicking) adhesive, with FM-300 (moisture-resistant) adhesive, with an NTMP treatment and FM-123, and with BR-127 primer and FM-300.

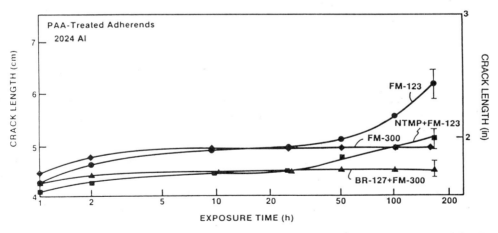

Figure 6 Wedge test results for PAA aluminum adherends: with FM-123 (moisture-wicking) adhesive, with FM-300 (moisture-resistant) adhesive, with an NTMP treatment and FM-123, and with BR-127 primer and FM-300.

does not occur [84,85] or is not accelerated (depending on the moisture level). The type of stress is also important. For example, cyclic stresses degrade the bond more rapidly than constant stresses [5,7].

The stresses on a joint make primary and secondary chemical bonds, both within the polymer itself and across the polymer–oxide interface, more susceptible to environmental attack by lowering the activation energy for bond breaking [1,12,68]. The stresses can also increase the rate of transport of moisture in the adhesive, possibly via crazing or the formation of microcracks [1,12,68,86]. Joints subjected to thermal "spikes" or cycling, such as those present in high-speed military aircraft, are particularly vulnerable to this type of aging. Thus weight gains in bonded composite systems that encountered one to four spikes (0°C–150°C–0°C) per day were proportional to the total number of spikes [87,88], suggesting that water was entering microcracks formed during thermal cycling. Additionally, studies of the chemical hydrolysis of epoxides by water (80°C) indicated that although unstressed samples were unaffected for up to 3 months, stressed systems induced hydrolysis of ester groups within days [89,90].

The stress at the crack tip of a wedge test specimen, together with the presence of moisture at the tip, serve to make this test specimen more severe than soaked lap shear specimens or similar types and a better evaluation of relative durability. In fact, Boeing has correlated the results of wedge tests from actual aircraft components with their in-service durability [5,91,92]. Wedge test specimens fabricated from components that had exhibited service disbonds showed significant crack growth during the first hour of exposure, whereas those fabricated from good components showed no crack growth during this time period. In contrast, lap shear specimens and porta shear specimens all demonstrated high bond strengths regardless of the service conditions.

V. MEANS TO IMPROVE

Although the degradation of a bond is likely to be inevitable, there are means by which to slow the process, some of which were discussed above. For convenience of the discussion, these methods can be classified as environment, materials, and design related.

Changing the environment to which a bond is exposed is probably the most effective means of ensuring good durability; bonded structures are not likely to degrade at moderate temperatures and in low humidity. Unfortunately, this usually is not a feasible option. However, it may be possible to protect the bond from its external environment, at least for a period of time. Ways to design a joint to do this are discussed below.

Material selection and preparation are perhaps more feasible options than changing the environment. Although not an economical solution, substitution of titanium for aluminum would solve many moisture-related problems. Selection of more water-resistant adhesives and/or corrosion-resistant primers is more common and was illustrated in Figs. 5 and 6 [11,16]. Here selection of a water-resistant adhesive (FM-300) decreased the final crack length in the wedge test by 1 to 2 cm for both FPL and PAA surfaces, compared to a water-wicking adhesive (FM-123). The use of a chromate-containing (corrosion-resistant) primer (BR-127) further decreased the final crack length (Section III. B).

Surface preparation is another commonly used means to increase durability. We have already seen that the durability of microrough surfaces is superior to smooth surfaces or to surfaces with only larger-scale roughness (Fig. 1) and that hydration-resistant aluminum surfaces provide further improvements (see Section III. A). Figures 5 and 6 also illustrated this enhancement. For a given adhesive, PAA surfaces, which are more hydration resistant, show less crack growth than FPL surfaces, and for both FPL and PAA surfaces, treatment with a hydration inhibitor (NTMP) gives superior durability to untreated surfaces. This improvement of PAA surfaces over FPL surfaces has also been demonstrated in the field, most notably in Vietnam, where FPL-treated joints suffered a large number of disbonds; PAA-treated joints were significantly more reliable [10].

Proper design of a joint or structure is also necessary to maximize durability. Although moisture cannot be prevented from reaching a bond line, it can be slowed or reduced in quantity. One way is to prevent pooling or other accumulation of water by designing the geometry to promote runoff or including adequate drain holes. Maintenance is then required to ensure that the holes do not become plugged. Sealants are also used to slow moisture ingress from joint edges and seams [93]. Again, proper application is necessary to prevent moisture accumulation and to ensure the absence of an easy path to the interface. One category of sealants, water-displacing corrosion inhibitors (WDCIs), can even creep under existing water films, displacing the moisture and eliminating the corrosive environment [93].

Another approach to improve durability involves overdesigning the bond so that the actual stresses experienced are a small fraction of the stresses that the joint is capable of withstanding. Stresses are thereby reduced to below any critical level, and the load can be carried even if moisture creates a disbond over a portion of the joint. Of course, this approach may not be feasible from a cost or weight standpoint. Alternatively, the bond can be designed so that moisture has a long diffusion path to reach a critical area—the same general principle by which sealants work.

VI. SUMMARY

Long-term durability is one of the most important properties of many adhesive bonds. Although it can be difficult to achieve in aggressive environments, modern materials and processes have proven successful in increasing durability. Moisture is the cause of most environmentally induced bond failures. It can weaken or disrupt secondary (dispersion-force) bonds across the adhesive–adherend interface, especially those involving high-energy surfaces such as metals; as a result, the joint may need to rely solely on primary (covalent or ionic) or physical (mechanical interlocking) bonds. Subsequently, more severe degradation can occur with hydration or corrosion of the adherend surface. At this point the joint will fail regardless of the type of bonding at the interface.

Most means of improving durability involve slowing the degradation mechanisms or providing additional bonding schemes (e.g., primary and/or physical bonds) that are less susceptible to degradation. Surface preparations that provide physical bonds and a hydration-resistant surface are typical examples. The use of coupling agents or phenolic-based adhesives (with aluminum adherends) are other examples where stable chemical bonds are formed at the interphase and slow bond degradation.

ACKNOWLEDGMENTS

We would like to gratefully acknowledge many valuable discussions with J. S. Ahearn, C. O. Arah, H. M. Clearfield, D. K. McNamara, T. K. Shah, and J. D. Venables. This work was funded in part by ONR under contract N00014-C-92-004.

REFERENCES

1. A. J. Kinloch, *Adhesion and Adhesives: Science and Technology*, Chapman & Hall, London, 1987).
2. A. J. Kinloch, *J. Mater. Sci. 17*:617 (1982).
3. A. J. Kinloch, ed., *Durability of Structural Adhesives*, Applied Science, London, 1983.
4. J. D. Venables, *J. Mater. Sci. 19*:2431 (1984).

5. J. A. Marceau and E. W. Thrall, in *Adhesive Bonding of Aluminum Alloys* (E. W. Thrall and R. W. Shannon, eds.), Marcel Dekker, New York, 1985, p. 177.

6. W. Brockman, O. D. Hennemann, H. Kollek, and C. Matz, *Intern. J. Adhesion Adhesives* 6:115 (1986).

7. L. R. Pitrone and S. R. Brown, in *Adhesively Bonded Joints: Testing, Analysis, and Design*, ASTM STP 981 (S. Johnson, ed.), American Society for Testing and Materials, Philadelphia, 1988, p. 289.

8. H. F. Brinson, chm., *Engineered Materials Handbook*, Vol. 3; *Adhesives and Sealants*, ASM International, Metals Park, Ohio, 1990.

9. W. Brockmann, in *The Science and Technology of Adhesive Bonding*, (L. H. Sharpe and S. E. Wentworth, eds.), Gordon and Breach, New York, 1990, p. 53.

10. J. D. Minford, in *Adhesive Bonding* (L.-H. Lee, ed.), Plenum Press, New York, 1991, p. 239.

11. H. M. Clearfield, D. K. McNamara, and G. D. Davis, in *Adhesive Bonding* (L.-H. Lee, ed.), Plenum Press, New York, 1991, p. 203.

12. J. Comyn, in *Durability of Structural Adhesives* (A. J. Kinloch, ed.), Applied Science, London, 1983, p. 85.

13. N.-H. Sung, in *Engineered Materials Handbook*, Vol. 3; *Adhesives and Sealants*, (H. F. Brinson, chm.), ASM International, Metals Park, Ohio, 1990, p. 622.

14. R. A. Gledhill and A. J. Kinloch, *J. Adhesion* 8:11 (1976).

15. A. J. Kinloch, *J. Adhesion* 10:193 (1979).

16. H. M. Clearfield, D. K. McNamara, and G. D. Davis, in *Engineered Materials Handbook*, Vol. 3, *Adhesives and Sealants* (H. F. Brinson, chm.), ASM International, Metals Park, Ohio, 1990, p. 259.

17. E. W. Thrall and R. W. Shannon, eds., *Adhesive Bonding of Aluminum Alloys*, Marcel Dekker, New York, 1985.

18. R. F. Wegman, *Surface Preparation Techniques for Adhesive Bonding*, Noyes Publications, Park Ridge, N.J., 1989.

19. B. F. Lewis, W. M. Bowser, J. L. Horn, Jr., T. Luu, and W. H. Weinberg, *J. Vacuum Sci. Technol.* 11:262 (1974).

20. M. Gettings and A. J. Kinloch, *J. Mater. Sci.* 12:2511 (1977).

21. J. D. Miller and H. Ishida, in *Fundamentals of Adhesion* (L. H. Lee, ed.), Plenum Press, New York, 1991, p. 291.

22. J. A. Bishopp, E. K. Sim, G. E. Thompson, and G. C. Wood, *J. Adhesion* 26:237 (1988).

23. G. D. Davis, *Surface Interface Anal.* 17:439 (1991).

24. J. A. Marceau, in *Adhesive Bonding of Aluminum Alloys*, (E. W. Thrall and R. W. Shannon, eds.), Marcel Dekker, New York, 1985, p. 51.

25. S. R. Brown, in *Proc. 27th National SAMPE Symposium*, SAMPE, Azusa, Calif., 1982, p. 363.

26. A. C. Kennedy, R. Kohler, and P. Poole, *Intern. J. Adhesion Adhesives* 3:133 (1983).

27. J. A. Filbey, J. P. Wightman, and D. J. Polgar, *J. Adhesion* 20:283 (1987).

28. D. K. Shaffer, H. M. Clearfield, C. P. Blankenship, Jr., and J. S. Ahearn, in *Proc. 19th SAMPE Technical Conference*, SAMPE, Azusa, Calif., 1987, p. 291.

29. G. D. Davis, T. S. Sun, J. S. Ahearn, and J. D. Venables, *J. Mater. Sci.* 17:1807 (1982).

30. G. D. Davis, J. S. Ahearn, L. J. Matienzo, and J. D. Venables, *J. Mater. Sci.* 20:975 (1985).

31. H. W. Eichner and W. E. Schowalter, *Report 1813*, Forest Products Laboratory, Madison, Wis., 1950.

32. J. D. Venables, M. E. Tadros, and B. M. Ditchek, U.S. patent 4,308,079 (1983).

33. D. K. McNamara and J. S. Ahearn, *Intern. Mater. Rev.* 32:292 (1987).

34. W. Brockmann, in *Durability of Structural Adhesives* (A. J. Kinloch, ed.), Applied Science, London, 1983, p. 281.

35. W. J. Russell, R. Rosty, S. Whalen, J. Zideck, and M. J. Bodnar, Preliminary study of

adhesive bond durability on 4340 steel substrates, *Technical Report ARSCD-TR-81020*, ARRADCOM, Dover, N.J., Apr. 1981.

36. A. T. Devine, Adhesive bonded steel: bond durability as related to selected surface treatments, *Technical Report ARLCD-TR-77027*, ARRADCOM, Dover, N.J., Dec. 1977.
37. D. Trawinski, *SAMPE Quart. 16*:1 (1984).
38. J. D. Minford, in *Durability of Structural Adhesives* (A. J. Kinloch, ed.), Applied Science, London, 1983, p. 135.
39. H. L. David, *Metal Deform. 19*:27 (1973).
40. A. V. Pocius, C. J. Almer, R. D. Wald, T. H. Wilson, and B. E. Davidian, *SAMPE J. 20*:11 (1984).
41. T. Smith, *J. Adhesion 17*:1 (1984).
42. D. L. Trawinski, D. K. McNamara, and J. D. Venables, *SAMPE Quart. 15*:6 (1984).
43. D. Trawinski, *Advancing Technology in Materials and Processes*, SAMPE, Azusa, Calif., 1985, p. 1065.
44. C. Bishof, A. Bauer, R. Kapeele, and W. Possant, *Intern. J. Adhesion Adhesives 5*:97 (1985).
45. H. E. Chandler, *Metals Progr. 121*:38 (1982).
46. E. Janssen, *Mater. Sci. Eng. 42*:309 (1980).
47. A. Stevenson, in *Engineered Materials Handbook*, Vol. 3, *Adhesives and Sealants*, (H. F. Brinson, chm.), ASM International, Metals Park, Ohio, 1990, p. 628.
48. J. F. Watts, *Surface Interface Anal. 12*:497 (1988).
49. F. J. Boerio, S. J. Hudak, M. A. Miller, and S. G. Hong, *J. Adhesion 23*:99 (1987).
50. J. S. Hammond, J. W. Holubka, J. W. DeVries, and R. A. Dickie, *Corrosion Sci. 21*:239 (1981).
51. J. F. Watts and J. E. Castle, *J. Mater. Sci. 18*:2987 (1983).
52. R. J. Reil, *SAMPE J. 7*, 16 (1971).
53. *MIL-A-8625C*, U.S. Military Specification.
54. *Process Specification 4352*, rev. J, Bell Helicopter, Textron, Inc., June 1980.
55. *Process Specification TH6.7851*, Fokker VFW B.V., The Netherlands, Aug. 1978.
56. M. Natan and J. D. Venables, *J. Adhesion 15*:125 (1983).
57. H. M. Clearfield, D. K. Shaffer, S. L. VanDoren, and J. S. Ahearn, *J. Adhesion 29*:81 (1989).
58. J. Comyn, *Engineered Materials Handbook*, Vol. 3, *Adhesives and Sealants* (H. F. Brinson, chm.), ASM International, Metals Park, Ohio, 1990, p. 616.
59. H. K. Charles, in *Engineered Materials Handbook*, Vol. 3, *Adhesives and Sealants* (H. F. Brinson, chm.), ASM International, Metals Park, Ohio, 1990, p. 579.
60. C. O. Arah, D. K. McNamara, J. S. Ahearn, A. Berrier, and G. D. Davis, in *Proc. 5th International Symposium on Structural Adhesive Bonding*, Dover, N.J., 1987, p. 440.
61. C. O. Arah, D. K. McNamara, H. M. Hand, and M. F. Mecklenburg, *J. Adhesion Sci. Technol. 3*:261 (1989).
62. R. A. Jurf and J. R. Vinson, *J. Mater. Sci. 20*:2979 (1985).
63. C. Kerr, N. C. MacDonald, and S. J. Orman, *J. Appl. Chem. 17*:62 (1967).
64. W. D. Bascom, *J. Adhesion 2*:161 (1970).
65. D. M. Brewis, J. Comyn, R. J. A. Shalash, and J. L. Tegg, *Polymer 21*:357 (1980).
66. D. M. Brewis, J. Comyn, and R. J. A. Shalash, *Intern. J. Adhesion Adhesives 2*:215 (1982).
67. R. J. A. Shalash, Ph.D. thesis, Leicester Polytechnic, 1980.
68. A. J. Kinloch, in *Durability of Structural Adhesives* (A. J. Kinloch, ed.), Applied Science, London, 1983, p. 1.
69. Y. Weitsman, *J. Composite Mater. 11*:378 (1977).
70. W. Brockmann, in *Engineered Materials Handbook*, Vol. 3, *Adhesives and Sealants* (H. F. Brinson, chm.), ASM International, Metals Park, Ohio, 1990, p. 663.
71. A. Knop and L. Pilato, *Phenolic Resins: Chemistry, Applications and Performance—Future Directions*, Springer-Verlag, New York, 1985, Chap. 11.

72. F. L. Tobiason, in *Handbook of Adhesives*, 3rd ed. (I. Skeist, ed.), Van Nostrand Reinhold, New York, 1990, Chap. 17.

73. E. P. Plueddemann, in *Industrial Adhesion Problems* (D. M. Brewis and D. Briggs, eds.), Orbital Press, Oxford, 1985, p. 148.

74. E. P. Plueddemann, *Silane Coupling Agents*, Plenum Press, New York, 1982.

75. W. D. Bascom, in *Engineered Materials Handbook*, Vol. 3, *Adhesives and Sealants* (H. F. Brinson, chm.), ASM International, Metals Park, Ohio, 1990, p. 254.

76. W. O. Buckley and K. J. Schroeder, in *Engineered Materials Handbook*, Vol. 3, *Adhesives and Sealants* (H. F. Brinson, chm.), ASM International, Metals Park, Ohio, 1990, p. 175.

77. *Organofunctional Silanes—A Profile*, technical brochure, Union Carbide Corporation, 1983.

78. R. L. Patrick, J. A. Brown, N. M. Cameron, and W. G. Gehman, *Appl. Polymer Symp. 16*:87 (1981).

79. F. J. Boerio and R. G. Dillingham, in *Adhesive Joints* (K. L. Mittal, ed.), Plenum Press, New York, 1984, p. 541.

80. E. P. Plueddemann, in *Fundamentals of Adhesion* (L.-H. Lee, ed.), Plenum Press, New York, 1991, p. 279.

81. D. B. Boises, B. J. Northan, and W. P. McDonald, Corrosion inhibitors in primers for aluminum, *Paper 31*, National Association of Corrosion Engineers, Houston, 1969.

82. J. L. Cotter, in *Developments in Adhesives*, Vol. 1 (W. C. Wake, ed.), Applied Science, London, 1977, p. 1.

83. G. F. Carter, *Adhesives Age 10*:32 (1967).

84. B. W. Cherry and K. W. Thompson, *Adhesion*, Vol. 1 (K. W. Allen, ed.), Applied Science, London, 1977, p. 251.

85. E. J. Ripling, S. Mostovoy, and C. F. Bersch, *J. Adhesion 3*:145 (1971).

86. G. Good, in *Engineered Materials Handbook*, Vol. 3, *Adhesives and Sealants* (H. F. Brinson, chm.), ASM International, Metals Park, Ohio, 1990, p. 651.

87. C. E. Browning, *Polymer Eng. Sci. 18*:16 (1978).

88. C. E. Browning, in *Proc. 22nd National SAMPE Symposium*, SAMPE, Azusa, Calif., 1977, p. 365.

89. M. K. Antoon, J. L. Koenig, and T. Serafini, *J. Polymer Sci. Phys. Ed. 19*:1567 (1981).

90. M. K. Antoon and J. L. Koenig, *J. Polymer Sci. Phys. Ed. 19*:197 (1981).

91. J. A. Marceau, Y. Moji, and J. C. McMillan, *Adhesives Age 20* (Oct. 1977).

92. M. H. Kuperman and R. E. Horton, in *Engineered Materials Handbook*, Vol. 3, *Adhesives and Sealants* (H. F. Brinson, chm.), ASM International, Metals Park, Ohio, 1990, p. 801.

93. T. J. Reinhart, in *Engineered Materials Handbook*, Vol. 3, *Adhesives and Sealants* (H. F. Brinson, chm.), ASM International, Metals Park, Ohio, 1990, p. 637.

8
Analysis of Adhesives

David N.-S. Hon *Clemson University, Clemson, South Carolina*

I. INTRODUCTION

An adhesive is any substance, inorganic or organic, natural or synthetic, that is capable of bonding substances together by surface attachment. The bonding power of an adhesive depends heavily on its molecular weight or the size of the molecules. Under a proper bonding process, the adhesive with higher molecular weight provides a stronger bond. Hence adhesive is a high-molecular-weight substance or a polymer. Polymers are composed of repeating units (i.e., monomers) that are linked together into long chains that can be linear, branched, or cross-linked. If a polymer contains two different types of monomers, it is a copolymer. A linear polymer is a thermoplastic. At elevated temperatures it melts and flows as a liquid. The irregular, cross-linked three-dimensional polymer network is a thermoset. Once the polymerization is completed, the polymer cannot be softened or melted. It is hard, infusible, and insoluble. Hence a thermoset adhesives is the most durable but is also difficult to characterize as compared to the thermoplastic.

Analysis or characterization is an essential step in working with adhesives. As a rule, such efforts are directed toward a specific purpose that may focus on structural determination, curing reaction, material design at a molecular level, process control, or failure analysis. In this chapter we provide a general review of several physical methods frequently used for analysis of adhesives. In view of the prolific literature on the subject, it is not intended to give a comprehensive treatment of the theory and experimental aspects. The examples chosen for this review are illustrative and not exhaustive.

II. TECHNIQUES FOR ANALYSIS AND CHARACTERIZATION

Perhaps the most apt definition of analysis of adhesives is their qualitative and quantitative characterization. Since the early days of adhesive development, the elementary chemical analysis of adhesives has provided valuable information on the structure and purity of materials. The use of such techniques for analysis, however, has been decreasing over the last 30 years. This is due to the complexity of the polymeric structures of adhesives as well as the development of many powerful and sophisticated instrumentations. More recently, advances in computer technology have been combined with analytical instruments to give speed, resolution, simplicity and minimal sample requirements that were unimaginable two decades ago.

Because of the statistical nature of the polymerization process, most polymers are composed of mixtures of molecules having a range of molecular weights. A complete description of the molecular weight distribution of a polymer is important to understand its physical, rheological, and mechanical properties. Hence a separation technique, gel permeation chromatography (GPC), which has proven to be useful for determination of molecular weight and molecular weight distribution is discussed. The use of spectroscopic methods for the characterization of adhesive systems has provided important molecular-level descriptions of these systems. Thus Fourier transform infrared (FTIR) spectroscopy and nuclear magnetic resonance (NMR) are described. Many adhesives are network polymers that are insoluble and as a result are not as easily characterized by the conventional methods, including GPC, which require dissolution. For these adhesives, thermal techniques have been used popularly to study the chemical kinetics of curing reactions, curing behavior, and degradation reactions and the transition of molecules. Thus three thermal analysis techniques—differential thermal analysis (DTA), differential scanning calorimetry (DSC), and dynamic mechanical thermal analyzer (DMTA)—are described.

A. Average Molecular Weight and Molecular Weight Distribution

The molecular weight of an adhesive is of prime importance in its preparation, application, and performance. The useful bonding and mechanical properties of an adhesive are heavily dependent on its molecular weight. Normally, bonding power does not begin to develop in adhesives until a minimum molecular weight above 5000 is achieved. Above that size, there is a rapid increase in the mechanical performance of adhesives as their molecular weights increase.

The molecular weights of adhesives can be determined by chemical or physical methods of functional group analysis or by measurement of the colligative properties, light scattering, ultracentrifugation, or dilute solution viscosity [1]. With the exception of some types of end-group analysis, all molecular weight methods require dissolution of the adhesive.

Unlike low-molecular-weight compounds where the molecular weight in the sample is uniform, polymer samples are usually polydisperse. They are composed of polymer chains of varying length and hence exhibit a distribution of molecular weights. Therefore, in expressing polymer molecular weights, various average expressions are used. These averages are defined in terms of the molecular weight M_i and the number of moles n_i or the weight w_i of the component molecules by the following equations:

Number-average molecular weight:

$$\overline{M}_n = \frac{\Sigma n_i M_i}{\Sigma n_i} = \frac{\Sigma w_i}{\Sigma w_i / M_i} \tag{1}$$

Weight-average molecular weight:

$$\overline{M}_w = \frac{\Sigma n_i M_i^2}{\Sigma n_i M_i} = \frac{\Sigma w_i M_i}{\Sigma w_i} \tag{2}$$

Z-average molecular weight:

$$\overline{M}_z = \frac{\Sigma n_i M_i^3}{\Sigma n_i M_i^2} = \frac{\Sigma w_i M_i^2}{\Sigma w_i M_i} \tag{3}$$

(Z + 1)-average molecular weight:

$$\overline{M}_{z+1} = \frac{n_i M_i^4}{\Sigma n_i M_i^3} = \frac{\Sigma w_i M_i^3}{\Sigma w_i M_i^2} \tag{4}$$

Viscosity-average molecular weight:

$$\overline{M}_v = \left[\frac{\Sigma n_i M_i^{1+a}}{\Sigma n_i M_i} \right]^{1/a} = \left[\frac{\Sigma w_i M_i^a}{\Sigma w_i} \right]^{1/a} \tag{5}$$

A typical molecular weight distribution (MWD) is depicted in Fig. 1. The M_n, M_v, M_w, and M_z are labeled.

It is clear from these equations that for M_w and M_z the molecules of greater mass contribute more to the average than do less massive molecules. Conversely, M_n is very sensitive to the presence of low-molecular-weight tails. In the study of polymerization kinetics, M_n is a critical parameter. The glass transition temperature (T_g) of an adhesive is also particularly sensitive to low-molecular-weight species. Knowledge of M_n is also necessary for evaluating the level of functionalization in adhesives with specific end groups. In studying rheology, various average molecular weights and the shape and breadth of the distribution are important. Polydispersity ratios such as M_w/M_n and M_z/M_w can give some insight into the latter problem. Adhesion, toughness, tensile strength, brittleness, and environmental stress-crack resistance are a few of the many properties affected by MWD.

In this section, gel permeation chromatography (GPC) is introduced for characterization of MW and MWD. Several valuable texts are available [2,3]. GPC is a liquid column

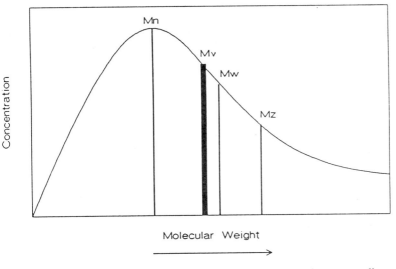

Figure 1 Typical polymer molecular weight distribution and corresponding average molecular weights.

chromatographic technique in which an adhesive solution is introduced onto a column packed with a rigid porous gel and is carried through the column by a solvent or solvents. Ideally, size separation is achieved by differential pore permeation. Under the influence of the solvent stream passing down the column, the smaller molecules in the adhesive go into and out of more pores in the packing than do the larger molecules. Hence larger molecules are eluted from the column earlier than smaller ones and are detected by means of some suitable instruments. The separation is based on the hydrodynamic volume of a polymer molecule. This hydrodynamic volume is converted to a molecular weight or equivalent molecular weight compared to the hydrodynamic volume of a calibrated polymer by means of a calibration curve (see Fig. 2). Specific molecular weights in GPC can be determined only from a calibration curve. Calibration requires chromatographing several samples of the specific polymer type that have narrow molecular weight distributions and known molecular weights covering the entire range of interest. The peak retention

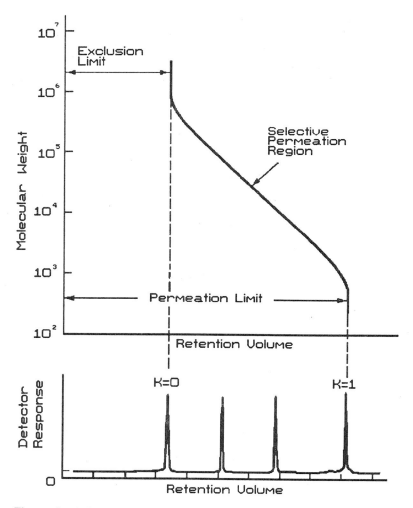

Figure 2   Calibration curve for a gel permeation chromatographic column.

volumes are then plotted graphically against the known molecular weight average. The molecular weight average of the unknown is determined from the calibration plot and the peak retention volume and is in the units of the calibration curve, M_w, M_n, or M_v. Totally excluded molecules elute in the column have a zero distribution coefficient ($K = O$). For small molecules that can enter all the pores of packing, $K = 1$. Thus intermediate-sized molecules are eluted between these two limits. Figure 2 is a typical GPC chromatogram. Notice that the first peak to be eluted corresponds to the largest molecular weight species in the sample. It should be noted that GPC requires the polymer sample to be dissolved in a solvent. For this reason, cross-linked adhesives cannot be analyzed by the GPC technique.

The essential components of the instrumentation (Fig. 3) are a solvent reservoir, a solvent delivery system (pump), sample injection system, packed columns, detector(s), and a data processing system. The heart of the instrumentation is the fractionation column where the separation takes place. The most common packing material used has been a semirigid cross-linked polystyrene gel. The concentration of the polymer molecules eluting from GPC columns is monitored continuously by a detector. The most widely used detector in GPC is the differential refractometer, which measures the difference in refractive index between solvent and solute. Other detectors commonly used for GPC are a functional group detector and an ultraviolet detector. A chromatogram obtained on a phenol–formaldehyde prepolymer adhesive is shown in Fig. 4. The various fractions of the polymer can be estimated from a calibration curve. Determination of the molecular weight distribution and kinetic investigations of formaldehyde-containing phenolics have been reported [4–7]. Resorcinol–formaldehyde condensate have been characterized [8]. The use of N,N-dimethylformamide (DMF) as a solvent with Poragel/Styragel columns allows analysis of phenol–formaldehyde and melamine–formaldehyde condensates [9]. Epoxy resins have been characterized by GPC under various conditions depending on

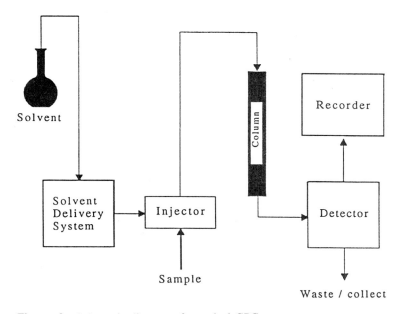

Figure 3 Schematic diagram of a typical GPC apparatus.

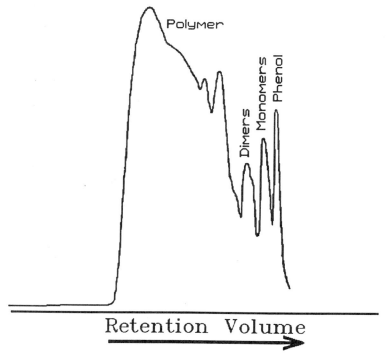

Figure 4 GPC chromatogram of a phenol–formaldehyde prepolymer adhesive.

their molecular weights. Typically, tetrahydrofuran was used as the solvent and ultraviolet absorption as the detection mode. For low-molecular-weight resins, <1000, the uses of Biorad SX-2 and combinations of Styragel and Biorad SX-2 for higher-molecular-weight materials have been reported [10].

B. Infrared Spectroscopy

Infrared (IR) spectroscopy is the fastest and cheapest of the spectroscopic techniques used by organic and polymer chemists. It is the measurement of the absorption of IR frequencies by organic compounds placed in the path of the beam of light. The samples can be solids, liquids, or gases and can be measured in solution or as neat liquid mulled with potassium bromide (KBr) or mineral oil. Recent development in attenuated total reflection and diffuse reflectance techniques have made the analysis of solid adhesives possible.

IR radiation is electromagnetic radiation in the wavelength range that is adjacent to and of less energy than visible radiation. The IR region starts at a wavelength of about 0.7 μm and ends at a wavelength of about 500 μm. Many chemists refer to the radiation in the vibrational IR region in terms of a unit called wavenumbers (ν), which is the number of waves of the radiation per centimeter (cm^{-1}).

While molecules absorb radiation, some parts of the molecule (i.e., the component atoms or groups of atoms) vibrate at the same frequency as the incident radiant energy. After absorbing radiation, the molecules vibrate at an increased amplitude. When molecular vibrations result in a change in the bond dipole moment, as a consequence of change in the electron distribution in the bond, it is possible to stimulate transitions between energy levels by interaction with electron magnetic radiation of the appropriate frequency. In effect, when the vibrating dipole is in phase with the electric vector of the incident radiation, the vibrations are enhanced and there is transfer of energy from the incident radiation to the molecule. It is the detection of this energy absorption that constitutes IR spectroscopy. The complex motions of the atoms in a molecule due to the twisting, bending, rotating, and vibrating actions produce an absorption spectrum that is characteristic of the functional groups comprising the molecule and of the overall molecular configuration as well. Thus IR spectroscopy readily distinguishes between aliphatic and aromatic compounds. Table 1 gives characteristic infrared absorption bands for some functional groups. The total structure of an unknown may not be readily identified from the infrared spectrum, but perhaps the type or class of compound can be deduced. Once the key functional groups have been established as present (or, equally important, as absent), the unknown spectrum is compared with spectra of known compounds. Several collections of spectra are available [12,13].

IR instrumentation is divided into dispersive and nondispersive types. Before the advent of the Fourier transform infrared (FTIR) spectrophotometer, a dispersive instrument that depends on gratings and prisms to disperse the infrared radiation geometrically was necessary for IR spectroscopy. Most dispersive spectrometers are double-beam instruments. The dispersed IR radiation is passed over a slit system, by means of a scanning device, and thus the frequency range falling on the detector is isolated. The data indicate the amount of energy transmitted through a sample as a function of frequency, and as a result, an IR spectrum can be obtained. However, the sensitivity of the technique is relatively low, because a large percentage of the available energy from the source of radiation does not fall on the open slits and is lost to the technique. Fortunately, the energy limitation can be minimized by using interferometers of the Michelson type rather than the conventional prism and grating instruments. This technique is called FTIR spectroscopy [14].

The Michelson interferometer is shown schematically in Fig. 5. It consists of two mutually perpendicular plane mirrors, one of which can move at a constant rate along the axis and one of which is stationary. Between the fixed mirror and the movable mirror is a beam splitter where a beam of radiation from an external source can be partially reflected to the fixed mirror and partially transmitted to the movable mirror. After each beam is reflected back to the beam splitter, it is again partially reflected and partially transmitted. Thus a portion of the beams that traveled in the path to both the fixed and movable mirrors reaches the detector. If the two path lengths are the same, no phase difference between the beams occurs, and they combine constructively for all frequencies present in the original beam. For different path lengths, the amplitude of the recombined signals depends on the frequency and the distance the mirror moved. For example, low frequencies interfere destructively (they have a phase shift of 180°) for relatively large movements of the mirror, whereas high frequencies require relatively small movements for this condition to occur. The resulting interferogram contains information on the intensity of each frequency in the spectrum. These data can be calculated by a mathematical operation known as the Fourier transform to yield the IR spectrum.

FTIR spectroscopy offers dramatic improvement in the signal-to-noise ratio over dispersive infrared spectroscopy due to the multiplexing (or Fellgett) advantage and the throughput (or Jacquinot) gain. The multiplexing advantage arises because all of the resolution elements are observed all the time. The large sampling area and the absence of narrow slits in the interferometer produces the throughput gain. The accuracy of frequen-

Table 1 Simplified Correlation Chart

	Type of vibration		Frequency (cm^{-1})	Wavelength (μm)	Intensity[a]
C–H	Alkanes	(stretch)	3000–2850	3.33–3.51	s
	–CH$_3$	(bend)	1450 and 1375	6.90 and 7.27	m
	–CH$_2$–	(bend)	1465	6.83	m
	Alkenes	(stretch)	3100–3000	3.23–3.33	m
		(out-of-plane bend)	1000–650	10.0–15.3	s
	Aromatics	(stretch)	3150–3050	3.17–3.28	s
		(out-of-plane bend)	900–690	11.1–14.5	s
	Alkyne	(stretch)	ca. 3300	ca. 3.03	s
	Aldehyde		2900–2800	3.45–3.57	w
			2800–2700	3.57–3.70	w
C–C	Alkane	Not interpretatively useful			
C=C	Alkene		1680–1600	5.95–6.25	m-w
	Aromatic		1600 and 1475	6.25 and 6.78	m-w
C≡C	Alkyne		2250–2100	4.44–4.76	m-w
C=C	Alkyne		1740–1720	5.75–5.81	s
	Ketone		1725–1705	5.80–5.87	s
	Carboxylic acid		1725–1700	5.80–5.88	s
	Ester		1750–1730	5.71–5.78	s
	Amide		1670–1640	6.00–6.10	s
	Anhydride		1810 and 1760	5.52 and 5.68	s
	Acid chloride		1800	5.56	s
C–O	Alcohols, ethers, esters, carboxylic acids, anhydrides		1300–1000	7.69–10.0	s
O–H	Alcohols, phenols				
	Free		3650–3600	2.74–2.78	m
	H-bonded		3500–3200	2.86–3.13	m
	Carboxylic acids		3400–2400	2.94–4.17	m
N–H	Primary and secondary amines and amides (stretch)		3500–3100	2.86–3.23	m
		(bend)	1640–1550	6.10–6.45	m-s
C–N	Amines		1350–1000	7.4–10.0	m-s
C=N	Imines and oximes		1690–1640	5.92–6.10	w-s
C≡N	Nitriles		2260–2240	4.42–4.46	m
X=C=Y	Allenes, ketenes, isocyanates, isothiocyanates		2270–1950	4.40–5.13	m-s
N=O	Nitro (R–NO$_2$)		1550 and 1350	6.45 and 7.40	s
S–H	Mercaptans		2550	3.92	w
S=O	Sulfoxides		1050	9.52	s
	Sulfones, sulfonyl chlorides, sulfates, sulfonamides		1375–1300 and 1200–1140	7.27–7.69 and 8.33–8.77	s
C–X	Fluoride		1400–1000	7.14–10.0	s
	Chloride		800–600	12.5–16.7	s
	Bromide, iodide		<667	>15.0	s

[a]s, Strong; m, moderate; w, weak.
Source: Ref. 11.

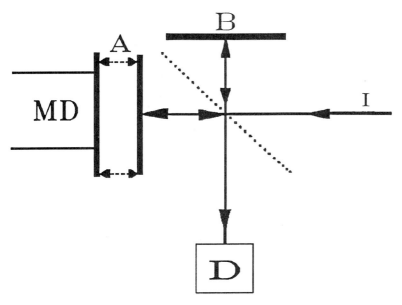

Figure 5 Michelson interferometer. I, unmodulated incident beam; A, moving mirror; B, stationary mirror; D, detector; MD, mirror drive.

cy determination (Connes' advantage) in FTIR, which is made possible by the use of the He–Ne laser interferometer to reference the position of the moving mirror, is another advantage. In addition to the advantages inherent in the interferometric acquisition of the data, the computer, which is an integral part of the system, offers many possibilities for manipulation of the stored data (such as spectral subtraction or addition) to improve the capabilities for rapid and accurate quantitative measurements. Detailed discussions of the principles of FTIR have been given by Bell [14] and Griffiths [15]. The application of FTIR has been popularly used for characterization of adhesives. Sojka et al. [16] used FTIR to study the reaction of phenol with hexamethylenetetramine at 100°C. Pearce et al. [17] analyzed cured novolak and resol adhesives. The curing kinetics and mechanism, degradation processes and chemical reactions of epoxies with coupling agents have also been studied by FTIR [18]. The completeness of the cross-linking reaction can be shown by FTIR spectra such as that shown in Fig. 6. Spectra are shown for a mixture of epoxy resin and curing agent immediately after being mixed and after being cured at 160°C for 2.5 h [19].

When a beam of radiation encounters an interface between two media, approaching it from the side of higher refractive index, total reflection occurs if the angle of incidence is greater than some critical angle, the value of which is given by

$$\alpha = \mathrm{Sin}^{-1} \frac{n_2}{n_1} \qquad (6)$$

when n_1 and n_2 are the two indices of refraction, with $n_1 > n_2$. Not so generally realized, although predicted by electromagnetic theory, is the fact that in total reflection, some portions of the energy of the radiation actually cross the boundary and return. If the less

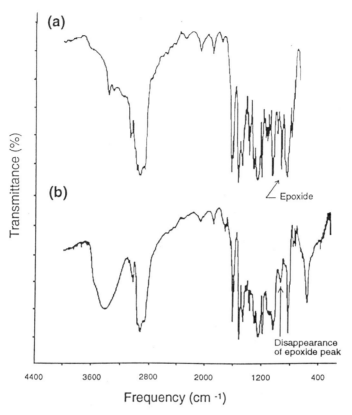

Figure 6 FTIR spectra of epoxy polymer before and after standard curing schedule: (a) freshly mixed adhesive; (b) after being cured at 160°C for 2.5 h.

dense medium absorbs at the wavelength of the radiation, the reflected beam will contain less energy than the incident, and a wavelength scan will produce an absorption spectrum. This principle has been found useful in the IR. The distance to which the radiation appears to penetrate in internal reflection depends on the wavelength but is on the order of 5 μm or less in the mid-IR region. This phenomenon is generally known as attenuated total reflection (ATR) [20]. The ATR method enables a reflection spectrum to be obtained which is superficially very similar to an absorption spectrum. To obtain measurable absorption spectra, it is a normal practice to use multiple reflection prisms (Fig. 7), and these are available as standard spectrometer accessories. A common material for the prisms is thallium bromoiodide (KRS-5). A trapezoidal prism having an angle of incidence of 45° giving 25 reflections is useful for adhesive materials. The sample is clamped securely to provide good optical contact with the prism and, where possible, the sample should be placed on both surfaces of the prism to give optimum sensitivity. ATR has been found most useful with opaque materials that must be observed in the solid state. Applications include studies of polymeric materials, adsorbed surface films, paints, and adhesives.

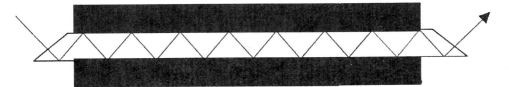

Figure 7 Multiple internal reflection setup.

C. Nuclear Magnetic Resonance Spectroscopy

Nuclear magnetic resonance (NMR) spectroscopy is a method of great interest and importance for the study of adhesives. The reasons are that (1) individual chemical groups in adhesives often give signals that can be resolved, (2) the NMR signals are sensitive to environment, and (3) the theory is well understood and the relationship between spectral parameters and the information of interest (such as concentration or structure) is relatively straightforward.

NMR involves the interaction of radio waves and the spinning nuclei of the combined atoms in a molecule. The nuclei of certain isotopes, such as ^1H, ^{19}F, ^{31}P, ^{13}C, ^{29}Si, and others, have an intrinsic spinning motion around their axes, which generates a magnetic moment along the axis of spin. (For the vast majority of polymeric materials, only ^1H exists in high concentration and has been the subject of the majority of applications of NMR to date. However, important developments in the technique have enabled other important constituent elements in polymers in which magnetic nuclei only exist naturally in low abundance to be investigated.) The simultaneous application of a strong external magnetic field H_o and the radiation from a second and weaker radio-frequency source H_1 (applied perpendicular to H_o) to the nuclei results in transitions between energy states of the nuclear spin. NMR phenomenon occurs when these nuclei undergo transition from one alignment in the applied field to an opposite one. This process is illustrated in Fig. 8 for a hydrogen nucleus.

The energy absorption is a quantized process, and the energy absorbed must equal the energy difference between the two states involved.

$$\Delta E_{absorbed} = (E_{-1/2} \text{ state} - E_{+1/2} \text{ state}) = h\nu \tag{7}$$

In practice, this energy difference is a function of the strength of the applied magnetic field, H_o. The relationship between these energy levels and the frequency ν of absorbed radiation can be calculated as follows:

$$E = -M \left(\frac{\mu}{I} \frac{h}{2\pi} \right) B_o \tag{8}$$

where M is the magnetic quantum number, μ the nuclear magnetic spin, B_o the applied magnetic field, and I the spin angular momentum.

At a given radio frequency, all protons absorb at the same effective field strength, but they absorb at different applied field strengths. It is this applied field strength that is measured and against which the absorption is plotted. The NMR spectrum consists of a set of resonances (or spectral lines) corresponding to the different types of hydrogen atoms in the sample. There are three basic measurements that can be obtained from a set of

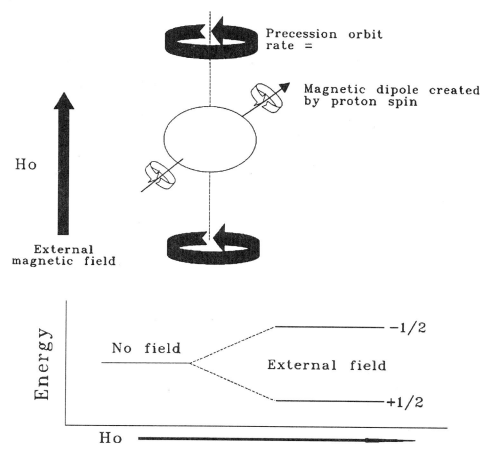

Figure 8 (a) Precession of a proton's magnetic dipole in a magnetic field; (b) splitting of energy levels for a hydrogen nucleus in a magnetic field.

resonances: (1) the number of signals, which is related to the number of protons presented in the molecule; (2) the area under the resonance, which is proportional to the amount of species present in the sample; (3) the position of the resonance or chemical shift, which is indicative of the identity of the species; and (4) the linewidth of the resonance, which is related to the molecular environment of the particular ^{1}H. The fact that the resonance area is proportional to the concentration of the species is the basis of quantitative NMR. By taking the ratios of different resonances corresponding to different species, the composition of multicomponent systems can be obtained.

In a given molecule, protons with the same environment absorb at the same applied field strength; protons with different environments absorb at different field strengths. A set of protons with the same environment are considered to be equivalent; the number of signals in the NMR spectrum shows how many sets of equivalent protons a molecule contains. The position of the signals in the spectrum indicates the types of proton (primary, secondary, tertiary, aromatic, benzylic, acetylenic, vinylic, etc.) in the molecule. These protons of different kinds have different electronic environments, which determine the number and location of the signals generated. When a molecule is placed in a magnetic field, circulation of electrons about the proton itself generates a field that acts against or reinforces the applied field. In each situation, the proton is to be shielded or deshielded. Shielding and deshielding thus shifts the absorption upfield and downfield, respectively. For example, the proton attached to the carbonyl carbon of acetaldehyde (CH^3—CHO) is more deshielded, due to the electron-withdrawing properties of the carbonyl oxygen, than the protons of the methyl group of an alkane (R—CH_3), which are surrounded by a higher electron density since there is no electron-withdrawing group. The carbonyl proton absorbs downfield from the methyl protons. The reference point from which chemical shifts are measured is not the signal from a naked proton, but the signal from a reference compound, usually tetramethylsilane, $(CH_3)_4Si$ (TMS).

The position of the absorption relative to TMS is called the chemical shift; its designation is δ and its units are ppm. Thus

$$\text{chemical shift (ppm)} = \frac{\text{chemical shift from TMS (Hz)}}{\text{spectrometer frequency (Hz)}} \times 10^6 \qquad (9)$$

For protons there is an alternative scale that expresses the chemical shifts in ppm on a τ (tau) scale, so that

$$\tau = 10 - \delta \qquad (10)$$

Most chemical shifts have δ values between 0 and 15. A small δ value represents a small downfield shift, and vice versa. A simplified correlation chart for proton chemical shift value is shown in Fig. 9.

The real power of NMR derives from its ability to define complete sequences of groups or arrangements of atoms in the molecule. The absorption band multiplicities (splitting patterns) give the spatial positions of the nuclei. These splitting patterns arise through reciprocal magnetic interaction between spinning nuclei in a molecular system facilitated by the strongly magnetic binding electrons of the molecule in the intervening bonds. This coupling, called spin-spin coupling or splitting, causes mutual splitting of the otherwise sharp resonance lines into multiplets. The strength of the spin-spin coupling or coupling constant, denoted by J, is given by the spacings between the individual lines of the multiples. The number of splittings of a multiplet adjacent to n equivalent spins is given by

$$s = 2nI + 1 \qquad (11)$$

where s is the number of lines and I is the spin of the nucleus causing the splitting. Since ^1H has a spin of ½, this reduces to $n + 1$ for proton spectra. The intensities of the multiplets also have a predictable ratio and turn out to be related to the coefficients of the binomial expansion $(a + b)^n$. These are given by Pascal's triangle, where each coefficient is the sum of the two terms diagonally above it:

```
            1
          1   1
        1   2   1
      1   3   3   1
    1   4   6   4   1
  1   5  10  10   5   1
1   6  15  20  15   6   1
     and so forth
```

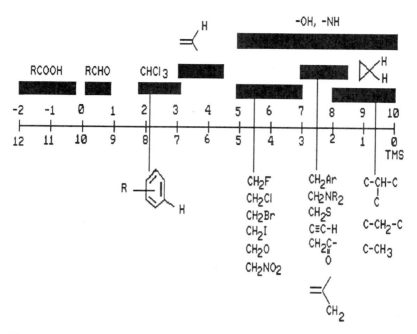

Figure 9 Simplified correlation chart for proton chemical shift values.

For simple spectra, then, we can predict the number of splittings and their intensities from the multiplicity and intensity rules given above.

An NMR instrument normally has a strong magnet with a homogeneous field, a radio-frequency transmitter and receiver, and a computer to store, compile, and integrate the signals. A sample holder positions the sample relative to the magnetic field so that the sample will be exposed continuously to a homogeneous magnetic field. The sample holder may also have a variable temperature control. NMR instruments are built with different magnetic field strengths. They are listed according to the radio frequencies required for the proton to resonate. They can have magnetic fields that require protons to absorb from 60 to 600 MHz to resonate.

A few words about sample preparation. The typical NMR is obtained from a sample in a 5-mm thin-walled glass tube containing about 0.4 mL of sample. Sample concentrations for routine work can be as low as 0.01 M, but concentrations greater than 0.2 M are preferred for good signal-to-noise ratio. Liquid samples are seldom run as neat liquids, since their greater viscosity will lead to broader lines. Instead, the liquids and solids are dissolved in a suitable solvent that does not show any peaks in the region of interest. Common solvents include CCl_4, $CDCl_3$, D_2O, acetone-d_6, and DMSO-d_6.

Many studies have been done for polymer system with respect to monomer composition and the average stereochemical configuration present along polymeric chains [21,22]. Both solid-state and conventional solution NMR techniques provide information on molecular motion, chain flexibility, and in some cases, crystallinity and network formation due to chain entanglement or cross-links [23–25]. The use of NMR spectroscopy for solid polymers has been reviewed by McBrierty [26–28], who has covered molecular motion studies in addition to the structural characterization of these systems in great detail. Jelinski [29] addressed the subject of chemical information and problem solving for

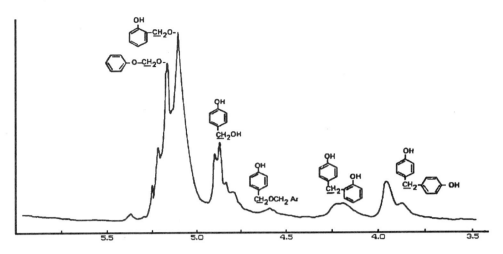

Figure 10 Nuclear magnetic resonance spectrum of a phenol–formaldehyde resol resin.

both solution and solid-state polymer studies. In addition to elucidating chemical structure, NMR can also be used for a particular facet of a structure, such as chain length or moles of a branched polymer, and in the study of polymer motion by relaxation measurements. Kinetic studies of curing reactions at temperatures in the range -150 to $+200°C$ are another application. Useful information can also be obtained from complex mixtures, such as the total methylene linkage of a phenol–formaldehyde adhesive. NMR has also been used to distinguish the structure of transient molecules involved in resin formation [30], species involved in HMTA cure, and final structure of phenolic oligomers [31]. Figure 10 is the NMR spectrum of a phenol–formaldehyde resol in acetone-d_6 solution. The resonances of the protons in various structures are identified.

D. Thermal Analysis

Thermal analysis includes a group of techniques in which specific physical properties of a material are measured as a function of temperature. The techniques include the measurement of temperatures at which changes may occur, the measurement of the energy absorbed (endothermic transition) or evolved (exothermic transition) during a phase transition or a chemical reaction, and the assessment of physical changes resulting from changes in temperature. Hence thermal analysis has provided important contributions in the characterization of adhesives, and a great deal has been written on this subject. It yields a "fingerprint" that may uniquely characterize the adhesive and assess its thermal stability. Thermal analysis data may also permit the evaluation of the kinetic parameters for the chemical changes that may have taken place during the heating process. For insoluble network adhesives, thermal techniques have been used to establish the degree and rate of cure, to study the chemical kinetics of curing reactions and the curing behavior itself, and to study degradation reactions.

Differential thermal analysis (DTA) and differential scanning calorimetry (DSC) are techniques that monitor either the heat evolution or absorption for any reactions that are occurring in a sample. They are discussed below.

1. Differential Thermal Analysis

DTA is a technique by which phase transitions or chemical reactions can be followed through observation of heat absorbed or evolved. It is especially suited to studies of structural changes within a solid adhesive at elevated temperatures. The temperature difference between a sample and an inert reference material is monitored while both are subjected to a linearly increasing environmental temperature. Figure 11 illustrates the principle. Two small crucibles, placed in suitably shaped cavities in a metal block, contain the sample (S) and the reference (R). The two junctions of a thermocouple are inserted into the crucibles so as to give directly the temperature difference between them. A separate thermocouple is placed in cavity B to measure the temperature of the block. The entire assembly is then heated under the control of a linear programmer. With constant heating, any transition or thermally induced reaction in the sample will be recorded as a peak or dip in an otherwise straight line. An endothermic process will cause the thermocouple junction in the sample to lag behind the junction in the reference material, and hence develop a voltage, whereas an exothermic event will produce a voltage of opposite sign. It is customary to plot exotherms upward and endotherms downward.

Conventional DTA can give good qualitative data about temperatures and signs of transitions, but it is difficult to obtain quantitative information about the sample or the heat of transition. Figure 12 shows a typical DTA thermogram of a linear high-pressure polyethylene blend [32]. This polymer, upon heating, undergoes three phase changes from its high-pressure form (115°C) to cocrystalline form (124°C) to a linear form (134°C). The 115°C peak was associated with the high-pressure polyethylene, whereas the 134°C peak was shown to be proportional to the linear content of the system. Recently, it was shown that there is reasonable correlation between the gelation time and the tempera-

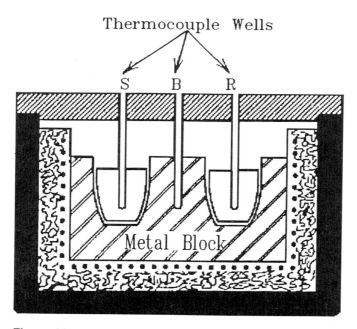

Figure 11 Equipment for DTA.

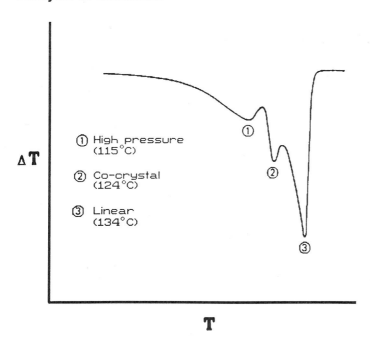

Figure 12 Typical DTA thermogram of a linear-high-pressure polyethylene blend.

ture corresponding to the peak of the isotherm on curves for an epoxy resin system [33]. The DTA method has yielded accurate kinetic results for a phenol–formaldehyde resin system, in addition to providing an insight to the actual chemistry of the curing process [34].

2. Differential Scanning Calorimetry

Quantitative results can be obtained by converting the sample compartment of a DTA apparatus into a differential calorimeter. The instrument, DSC, is built based on this principle. In this setup, the sample and reference are heated directly with separate heating coils as shown in Fig. 13. A heating coil makes the temperature of the reference material increase at a constant rate. A second heating coil is placed in the sample. The sample and the reference are kept at equal temperature. When a phase change or weight loss occurs, the sample and reference temperature become slightly different, which generates a current in the thermocouple system measuring the temperature difference between the two cells. The current activates a relay, causing extra power to be directed to the cell at the lower temperature. In this manner the temperatures of the reference and sample cells are kept virtually equal throughout. The quantity of electrical energy used in heating the sample and the reference is measured accurately and continuously. In turn, the electrical energy is an exact measure of the number of calories used in heating the cells.

The resultant thermogram is similar to a DTA trace but more accurate and reliable. Endothermic changes are recorded as heat input into the sample, and exothermic changes as heat input into the reference. The area of the peaks is an exact measure of heat input involved. Differences in heat capacity or thermal conductivity do not affect the results. From the data, accurate quantitative analytical results can be obtained.

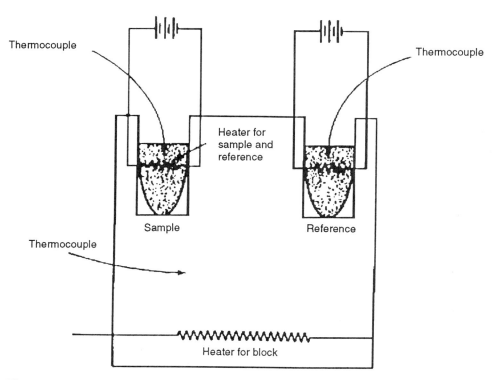

Figure 13   Equipment for DSC.

DSC gives much information about molecular ordering, including the glass transition temperature T_g, melting temperature (T_m), heat of fusion, and entropy of fusion (Fig. 14). The melting behavior in DSC permits a determination of the extent of crystallinity. DSC is the more frequently used technique in the measurement of heats of reactions for the types of polymerization under discussion because it does give a quantitative measure of the heat and the rate of the curing reaction. Fava [35] described kinetic measurements involving the heat of reaction for measuring the extent of cure in an epoxy resin. The three methods for obtaining isothermal cure curves using the DSC technique are isothermal operation, analysis of thermograms with different scan rates, and scans on partly cured resins. From DSC curves, the state of cure can be monitored and the kinetic parameters of cure can be determined. Using this technique, kinetic studies have been made of polymerization of vinyl acetate and phenol–formaldehyde, and curing of epoxy resins. A kinetic study of isothermal cure of epoxy resin has been carried out [36,37]. Kinetic parameters associated with the cross-linking process of formaldehyde–phenol and formaldehyde–melamine copolymers have been obtained from exotherms of a single DSC temperature scan [38]. The other major application of DSC is the measurement of glass transition temperature [39,40]. In the absence of endothermic or exothermic reactions, the DSC heat flow output is proportional to the sample heat capacity, and the T_g may be determined from the characteristic discontinuity in heat capacity. The T_g of a cross-linked polymer in general shows an increase with increasing degree of cross-linking, and thus provides a useful index of the degree of cure. The T_g is dependent on the chain flexibility and the free volume associated with the chemical structure as well as the overall cross-link density.

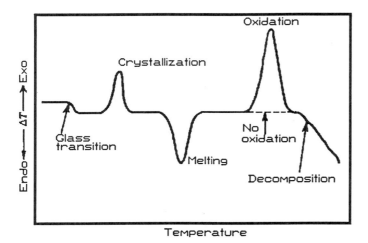

Figure 14 Schematic DSC curve.

3. Dynamic Mechanical Thermal Analyzer

Many adhesives exhibit time-dependent, reversible viscoelastic properties in deformation. Hence a viscoelastic material can be characterized by measuring its elastic modulus as a function of temperature. The modulus depends both on the method and the time of measurement. Dynamic mechanical tests are characterized by application of a small stress in a time-varying periodic or sinusoidal fashion. For viscoelastic materials when a sinusoidal deformation is applied, the stress is not in phase with displacement. A complex tensile modulus ($E*$) or shear modulus ($G*$) can be obtained:

$$E* = E' + iE'' \qquad (\text{or} \quad G* = G' + iG'') \tag{12}$$

where E' (or G') is the in-phase component (or real or storage part of dynamic modulus) and E'' (or G'') is the out-phase component imaginary (or imaginary or loss part of dynamic modulus). The mechanical loss factor corresponding to the damping of the material can be defined by

$$\tan \delta = \frac{E''}{E'} \qquad (\text{or} \quad \frac{G''}{G'}) \tag{13}$$

Tan δ describes the significance of viscous dissipation in a material and is often the parameter chosen to relate dynamic data to molecular or structural motion in plastics materials. Dynamic modulus for polymers is strongly dependent on temperature and frequency in the transition zones. It is interesting to study E' (or G') and tan δ by changing frequency at constant temperature and changing temperature at constant frequency. Dynamic mechanical spectroscopy by studying $E*$ (or $G*$)—versus temperature and frequency—can give information about relaxation processes (i.e., main chain relaxation from glass to rubber associated with the glass transition process and secondary transitions related to movements of side chains or to motions of small parts of the main chain). The elastic modulus–temperature curve for a typical amorphous polymer is given in Fig. 15 and shows five different regions: (1) the glassy region where the modulus of most amorphous polymers is on the order of 10^{11} dyn/cm^2; (2) the transition region, where the modulus changes rapidly with temperature; (3) the rubbery plateau region, which is

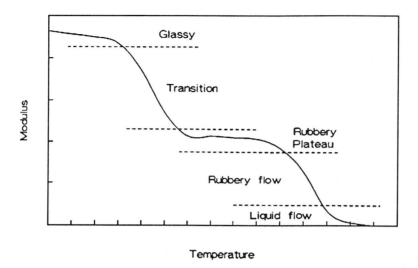

Figure 15   Modulus–temperature curve for an amorphous adhesive.

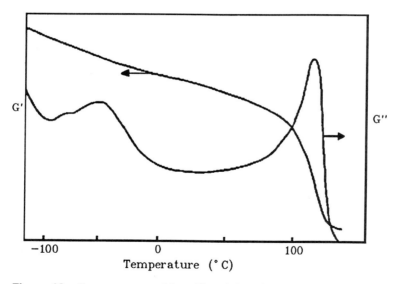

Figure 16   Shear storage modulus, G', and shear loss modulus, G'', as a function of temperature for a carbon fiber-reinforced epoxy adhesive cured at 120°C.

apparent only when polymers form a loose network by cross-linking or by pigment re-enforcement; (4) the rubbery flow region; and (5) the liquid flow region. The modulus in these various regions of viscoelastic behavior is affected differently when the molecular weight of the polymer is changed. For adhesive applications, the curing of adhesives can be examined by DMTA. The modulus increases with cure. The tan δ value usually rises at first, then falls rapidly as the cross-link density becomes higher and progressively inhibits conformation changes. Dynamic mechanical properties of epoxy resins cured with various kinds of hardeners have been evaluated by Kamon and Furukawa [41]. The data in Fig. 16 show T_g relaxation spectra in a carbon fiber–reinforced epoxy resin; the position and area of the loss peaks in composites indicate the degree and direction of reinforcement, in addition to the degree of cure in the matrix of polymer.

III. CONCLUSIONS

The examples cited in this chapter clearly establish that physical analysis techniques, such as GPC, FTIR, NMR, DSC, DGA, and DMTA, are broadly applicable in adhesive research, product development, manufacturing, and quality control or assurance programs. It is frequently desirable to use several different techniques to study a specific adhesive to obtain detailed information for a particular property. In some cases it may be necessary to conduct analysis or characterization with two or more techniques simultaneously.

REFERENCES

1. A. R. Cooper, ed., *Determination of Molecular Weight*, Wiley, New York, 1989.
2. W. W. Yau, J. J. Kirkland, and D. D. Bly, *Modern Size-Exclusion Liquid Chromatography: Practice of Gel Permeation and Gel Filtration Chromatography*, Wiley, New York, 1979.
3. S. A. Borman, *Anal. Chem. 55*: 384A (1983).
4. J. W. Aldersley, V. M. R. Bertram, G. R. Harper, and B. P. Stark, *Brit. Polymer J. 1*: 101 (1969).
5. J. W. Aldersley and P. Hope, *Angew. Makromol. Chem. 24*: 137 (1972).
6. E. R. Wagner and R. J. Greff, *J. Polymer Sci. A1 9*: 2193 (1971).
7. M. Duval, B. Bloch, and S. Kohn, *J. Appl. Polymer Sci. 16*: 1585 (1972).
8. P. Hope, R. Anderson, and A. S. Bloss, *Brit. Polymer J. 5*: 67 (1973).
9. B. Feurer and A. Gourdenne, *Polymer Prepr. 15(2)*: 279 (1974).
10. H. Batzer and S. A. Zahir, *J. Appl. Polymer Sci. 19*: 585 (1975).
11. D. L. Pavia, G. M. Lampman, and G. S. Kriz, Jr., *Introduction to Spectroscopy: A Guide for Students of Organic Chemistry*, Saunders College, Philadelphia, 1979, pp. 13–80.
12. W. W. Simons, ed., *The Sadtler Handbook of Infrared Spectra*, Sadtler Research Laboratory, Philadelphia, 1978.
13. D. O. Hammel, *Atlas of Polymer and Plastics Analysis*, 2nd ed., Hanser, Munich, 1978–1981.
14. R. J. Bell, *Introductory Fourier Transform Spectroscopy*, Academic Press, New York, 1982.
15. P. Griffiths, in *Transform Techniques in Chemistry* (P. Griffiths, ed.), Plenum Press, New York, 1978.
16. S. A. Sojka, R. A. Wolfe, and G. S. Guenther, *Macromolecules 14*: 1539 (1981).
17. Y. Zaks, J. Jo, D. Raucher, and E. M. Pearce, *J. Appl. Polymer Sci. 27*: 913 (1982).
18. E. Mertzel and J. L. Koenig, *Adv. Polymer Sci. 75*: 74 (1986).
19. E. M. Yorkgitis, N. S. Eiss, Jr., C. Tran, G. L. Wilkes, and J. E. McGrath, *Adv. Polymer Sci. 72*: 79 (1985).
20. N. J. Harrick, *Internal Reflection Spectroscopy*, Wiley, New York, 1967.
21. F. A. Bovey, *High Resolution NMR of Macromolecules*, Academic Press, New York, 1972.
22. J. J. Ivin, *Pure Appl. Chem. 55*: 1529 (1983).
23. V. J. McBrierty and D. C. Douglas, *J. Polymer Sci. Macromol. Rev. 16*: 195 (1981).
24. B. C. Gerstein, *Anal. Chem. 55*: 781A, 899A (1983).
25. J. R. Havens and J. L. Koenig, *Appl. Spectr. 37*: 226 (1983).
26. V. J. McBrierty, *Polymer 15*: 503 (1974).
27. V. J. McBrierty and D. C. Douglas, *Phys. Rep. 63*: 61 (1980).

28. V. J. McBrierty, *Magnetic Reson. Rev. 8*: 165 (1983).
29. L. W. Jelinski, *Chem. Eng. News 62*(45): 26 (1984).
30. P. W. Kopf and E. R. Wagner, *J. Polymer Sci. Polymer Chem. Ed. 11*: 939 (1973).
31. J. C. Woodbrey, H. P. Higginbottom, and H. M. Culbertson, *J. Polymer Sci. A 3*: 1079 (1965).
32. B. H. Clampett, *Anal. Chem. 35*: 1834 (1963).
33. A. G. Ulukhanov, V. A. Kapitskii, M. S. Akutin, and L. D. Skokova, *Plastmassy 9*: 59 (1981).
34. Z. Katovic, *J. Appl. Polymer Sci. 11*: 95 (1967).
35. R. A. Fava, *Polymer 9*: 137 (1968).
36. J. Sickfield and B. Heinze, *J. Thermal Anal. 6*: 689 (1974).
37. S. Sourour and M. R. Kamal, *Thermochim. Acta 14*: 41 (1976).
38. R. Kay and A. R. Westwood, *European Polymer J. 11*: 25 (1975).
39. M. J. Richardson and N. G. Savill, *Polymer 16*: 753 (1975).
40. J. H. Flynn, *Thermochim. Acta 8*: 69 (1974).
41. T. Kamon and H. Furukawa, *Adv. Polymer Sci. 80*: 174 (1986).

9
Fracture of Adhesive-Bonded Wood Joints

Bryan H. River   *Forest Products Laboratory, USDA–Forest Service, Madison, Wisconsin*

I. INTRODUCTION

Adhesives are arguably the most important fastening system used with forest products. Large volumes of adhesives are used successfully in wood-, particle-, and fiber-based industries. In fact, large and important industries such as panel products would not exist without adhesives. However, the sometimes unpredictable and misunderstood behavior of wood–adhesive joints, particularly fracture, is a major constraint to improving the performance of existing products and the development of new wood–adhesive marriages. In this chapter I briefly examine current understanding of fracture mechanisms in wood–adhesive joints. The discussion is limited to joints bonded with those adhesives having sufficient strength and rigidity to cause fracture in the wood adherends. Primarily, these are the rigid, thermosetting adhesives such as phenol and urea–formaldehyde, nonrigid epoxy and thermosetting poly(vinyl acetate) adhesives, and some thermoplastic types such as poly(vinyl acetate).

The fracture of strong wood–adhesive joints (e.g., in the catastrophic rupture of a large laminated beam) may be viewed (and heard) as a macrocracking process. It may also be viewed microscopically and heard by acoustic emission technology in flakeboard as a microcracking process resulting from shrinkage. These examples are not meant to imply that beams do not fail by microprocesses or that particleboard does not fail by macroprocesses. Both of these examples of fracture begin with the microscopic initiation of a crack at some flaw in the material or, in this case, the bonded joint.

The initial flaw can be a discontinuity, such as a void, or an abrupt change in material properties. By nature, wood contains innumerable discontinuities, such as the cell cavity and transition zones between cell wall layers. An adhesive may contain air bubbles or fillers with properties different from the resin. A rough wood surface may not be completely wetted by the adhesive, leaving voids at the interface. The adhesive and wood also have different mechanical properties. When a joint or bonded material is subjected to some force, the resultant stress is heightened or concentrated around the discontinuities far above the average stress in the joint or material. Fracture results when the stress at a discontinuity reaches the ultimate stress or strength of either the adhesive, the adherend, or the interface.

The stress conditions around a cracklike discontinuity can be described by the stress intensity factor (K), which is a function of the applied load, the size of the cracklike flaw, and the material. Fracture occurs when the stress intensity factor reaches a critical level, called the critical stress intensity factor or fracture toughness (K_c). Fracture mechanics relates the applied stress at which a material fractures to the critical stress intensity factor and the critical flaw size of the material:

$$\sigma_a = \frac{K_c}{\pi a}$$

where σ_a is the applied stress at fracture, K_c the critical stress intensity factor, and a the crack length or flaw size. The stress intensity factor has been found useful for describing the fracture behavior of many materials. However, the stress intensity surrounding discontinuities in adhesive joints is extremely difficult to define because of the dissimilar materials combined in the joint. Therefore, the sensitivity of adhesive joints to stress and discontinuities is usually measured and described in terms of the energy required to initiate a crack or the energy released in forming a new crack surface (G_c) (Fig. 1). K_c, the critical stress intensity factor, and G_c are related through the elastic properties of the material:

$$G_c = \frac{K_c^2}{E} (1 - \nu^2)$$

where E is the tensile modulus of isotropic adherends and ν is Poisson's ratio of adherends.

The crack initiation energy can readily be determined from a mode I cleavage test using a double cantilever beam specimen [1,2]. This method has been applied to wood–adhesive joints by many researchers [3–17]. The test requires the beam compliance (C), load at crack initiation (P_c), crack length at initiation (a), specimen thickness (t), and change in beam compliance $(dC/da = $ change in displacement of load points/change in crack length) (Fig. 1b):

$$G_c = \frac{P_c^2}{2t} \frac{dC}{da}$$

Anderson and others [18] have reviewed these and other fracture test methods for evaluating adhesive bonds.

The load or force that causes stress around a discontinuity may be applied externally to a material or structure. More often, in bonded joints the force arises from differential shrinkage and swelling of the bonded members or particles. In adhesive joints and most wood products, forces tending to cleave the joint (mode I loading) are of primary importance. Sliding shear (mode II) and torsional shearing (mode III) forces are less important. However, most wood joints experience a combination of mode I with either mode II or mode III shear.

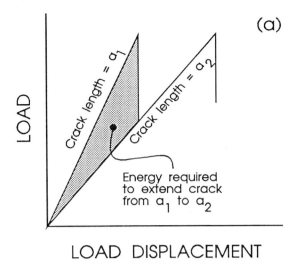

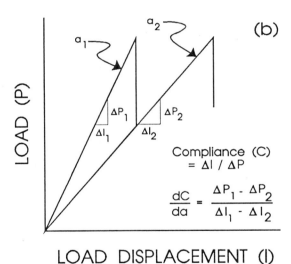

Figure 1 (a) Energy expended in forming new crack surface (shaded area), determined as the difference in the area under the load–displacement curve before crack extension (a_1) minus the area of the load–displacement curve after crack extension (a_2); (b) calibration procedure for determining the change in beam compliance with change in crack length (dC/da).

Whereas the fracture of bonded joints is due to the presence of a critical discontinuity in a field of stress, the stress intensity or crack initiation energy of bonded joints and materials at which fracture occurs is also a function of the properties of the wood and the adhesive, the environment at a given time, changes in the environment, and external forces on the joint or bonded material. These relationships are explored in the following sections.

II. FRACTURE BEHAVIOR

A. Wood

Since wood fracture usually dominates the performance of well-made joints, it is worth-while before focusing on the bonded joint and the influence of the adhesive to examine how wood itself fractures. At the molecular level, Porter [19] found that wood fractures in the amorphous, water-accessible regions of the cell wall rather than in the crystalline regions. These regions are also most susceptible to change as a result of varying temperature, moisture content, and chemicals. At the microscopic level, wood fractures in different locations depending on the type of cell, direction of load, temperature, moisture content, speed of test, grain angle, wood pH, and aging.

Anatomical features such as the S1, S2, and S3 layers of the cell wall (Fig. 2) are especially important in the fracture of wood and wood–adhesive joints. There are three general types of fracture at the microscopic level [20]: transwall, intrawall, and intercellu-lar (Fig. 3). Transwall cracks may be parallel to the longitudinal cell axis (Fig. 3a) or transverse (Fig. 3d), but in either case the cell lumen is exposed. Transwall fractures are common in thin-walled cells such as softwood earlywood tracheids, hardwood vessels, and parenchyma cells. Longitudinal transwall fracture of thick-walled latewood cells is unusual. When such fracture occurs, it is extremely fibrous and is called fine-fiber failure [21]. Transverse transwall fracture (Fig. 3d) is rare in thick-walled cells (such as hard-wood fibers and softwood latewood tracheids) as a result of their great tensile strength parallel to the cell axis. Such fracture does occur in compression wood of softwoods and at the tips of splinters in tough wood. These thick-wall cells are more likely to produce a diagonal combined shear and tension transwall fracture following the helical angle of the S2 layer microfibrils (not pictured). This is the manner in which a crack grows across the grain in tough wood. Intrawall fracture (Fig. 3b) is also very common in thick-walled cells. An intrawall crack travels within the cell wall, leaving the cell lumen intact.

Intrawall fracture initiates at the discontinuities between the layers of the secondary wall (Fig. 2). The cell wall consists of microfibrils of cellulose helically wound around the

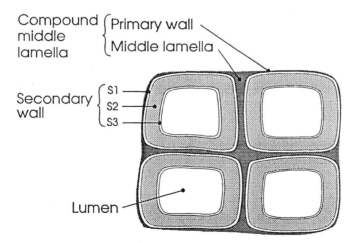

Figure 2 Transverse cross section of four wood cells showing the compound middle lamella joining them, the three layers (S1, S2, and S3) of the secondary wall, and the cell lumen.

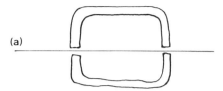

(a)

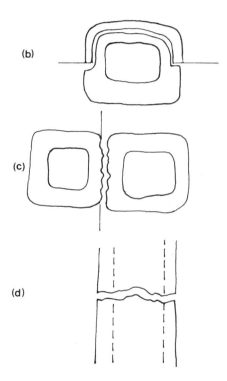

(b)

(c)

(d)

Figure 3 Schematic diagrams of fracture modes in wood: (a) longitudinal transwall; (b) intrawall; (c) intercellular; (d) transverse transwall.

longitudinal cell axis. The cell wall layers are differentiated by the angles of the microfibrils in each layer. The microfibrils in the outermost (S1) and innermost (S3) layers are wound at a large angle around the longitudinal axis of the cell. The microfibrils in the S2 layer sandwiched between the S1 and S3 layers are wound at a small angle around the longitudinal cell axis. The transition between these layers is often gradual, yet it still presents a material discontinuity. Mark [22,23] clearly pinpointed the S1–S2 interphase as the site of crack initiation in the fracture of solid wood. Intercellular cracks (Fig. 3c) travel in the compound middle lamella (CML), leaving the secondary wall and cell lumen intact.

Investigators have shown the preferential fracture of wood at various cell wall interfaces, depending on the temperature at fracture. Woodward [24] found fracture predominately in the S1 layer in the range from 20 to 77°C. At the lower end of the scale, the crack path jumped back and forth across the middle lamella from the S1 layer of one cell to the S1 layer of a contiguous cell. At the higher temperature, the crack tended to stay within the S1 of a given cell from one end to the other. Furthermore, fractures of the lignin-rich CML are rare at normal temperatures, but they are likely to occur under hot, wet conditions [20].

On a larger scale, the type of fracture varies with the density of the tissue through which a crack is growing. Fracture in the longitudinal–tangential (LT) plane is dominated by longitudinal transwall fracture of the first-formed earlywood cells. A mixture of transwall and intrawall fracture is common in the longitudinal–radial (LR) and planes intermediate to the LR and LT planes as a result of alternating high- and low-density bands of the earlywood and latewood cells. Fracture patterns similar to those described for wood have been observed in solid wood joints and in wood particles bonded with droplets of adhesive [25,26].

B. Adhesive

The fracture toughness of wood in terms of crack initiation energy ranges from 50 to 1000 J/m^2, whereas the crack initiation energies for typical thermosetting polymers are in the range 100 to 300 J/m^2 [27]. It seems interesting that wood joints bonded with conventional thermosetting adhesive also have fracture toughness values of about 100 to 300 J/m^2 (Table 1). Much higher values are possible if the adhesive is toughened by the addition of fillers or plasticizers.

Plasticizers used to reduce the modulus of thermosetting adhesives to match more closely the wood moduli perpendicular to the grain have a marked effect on the fracture

Table 1 Fracture Toughness for Wood–Adhesive Joints

Mode	Adherend	Adhesive[a]	Fracture toughness K_{Ic} (kPa m$^{1/2}$)	Fracture toughness G_{Ic} (J/m^2)	Ref.
Cleavage	Beech	PVA	—	1206	11
		PVA/phenol	—	390	
		PF	—	170	
		RF/filler	—	390	
		EP/P	—	200	
		Ep/60P[b]	—	1180	
		EP	—	200–340	13
		EP/20P	—	280–460	
		EP/40P	—	460–790	
		EP/60P[b]	—	450–1070	
	Douglas-fir	UF/filler	—	250	6
		EPI	—	900	
		ISO	—	300	
		PRF	—	800	
		PF	—	200	
		PF/PVA	—	700	
	Aspen	PRF	255	—	28

Table 1 (*Continued*)

Mode	Adherend	Adhesive[a]	K_{Ic} (kPa m$^{1/2}$)	G_{Ic} (J/m^2)	Ref.
	Unknown	UF/filler	—	530	29
		UF/PVA	—	640	
		PF/PVA	—	640	
		UF/MF/filler	—	700	
		PRF	—	870	
	Yellow poplar	Casein	380	—	30
		EP	430	—	
		PRF	470	—	
		PVA	680	—	
	Spruce	PVA	310	—	
	Douglas-fir	PVA	550	—	
	Southern pine	PVA	560	—	
	Walnut	PVA	600	—	
	Ash	PVA	680	—	
	Maple	PVA	790	—	
	Western redcedar	PRF	280	—	31
		Solid wood	180	—	
	Southern pine	PRF	520	—	
		Solid wood	430	—	
	Hard maple	PRF	690	—	
		Solid wood	490	—	
	Douglas-fir	Solid wood	410	—	32
	Southern pine				33
	Earlywood	PRF	520	—	
	Latewood	PRF	400	—	
	Southern pine	Solid wood	494	—	34
	Douglas-fir FB[c]				
	560 kg/m^2	PF	88	—	35
	800 kg/m^2	PF	350	—	
	Douglas-fir LVL[c]	PF	360	—	
	Douglas-fir	PRF	290	—	36
	Southern pine	PRF	480	—	
Sliding shear	Kaba	PVA	—	1280	37
	Southern pine	PRF	1670	—	36
	Douglas-fir	PRF	1830	—	
	Southern pine	Solid wood	1980	—	34
Torsion shear	Radiata pine	PRF	—	480	38

[a]Adhesive abbreviations are as follows: EP, amine-cured epoxy; EP/20P, EP/40P, and EP/60P, amine-cured epoxy with 20, 40, and 60 parts polysulfide flexibilizer; EPI, emulsion–polymer isocyanate; ISO, isocyanate; MF, melamine–formaldehyde; MUF, melamine/urea–formaldehyde; PF, phenol–formaldehyde; PF/PVA, phenol–formaldehyde flexibilized with poly(vinyl acetate); PVA, poly(vinyl acetate); PRF, phenol/resorcinol–formaldehyde; RF, resorcinol–formaldehyde; UF, urea–formaldehyde; UF/filler, UF with wheat flour; UF/MF/filler, UF/MF copolymer with wheat flour.

[b]Thick layer.

[c]FB, flakeboard; LVL, laminated veneer lumber.

toughness. The addition of 20 parts of poly(vinyl acetate) to phenol–formaldehyde (PF/PVA) adhesive increased mode I fracture toughness by 340%, from 200 J/m^2 to almost 700 J/m^2 (Table 1) [6]. Less rigid thermosetting adhesives, such as emulsion polymer/isocyanate (EPI), produced joints with toughness as high as 900 to 1000 J/m^2 (Table 1) [6]. In this case the toughness varied with the amount of isocyanate cross-linking agent. Toughness first increased as the amount of isocyanate was increased from 0 to about 6 parts per 100 parts of emulsion polymer, but then decreased with further additions (not shown in Table 1). When Takatani and Sasaki [13] added polysulfide rubber flexibilizer (P) to epoxy resin (EP) adhesive, the fracture toughness of bonded joints increased from about 200 J/m^2 to 300 J/m^2 (Table 1). The toughest joints were those made with thick adhesive layers, in which case the crack initiation energies rose as high as 900 to 1200 J/m^2 (Table 1). Many other studies showed that flexible or semirigid adhesives produce joints having higher short-term strength and fracture toughness compared to rigid adhesives [13,31,39,40]. Takatani and others [11] observed that flexible adhesive improves the fracture toughness of joints made with rigid adherends such as spruce, beech, and oak; however, rigid adhesive improves the toughness of joints made with flexible adherends such as balsa.

Very high fracture toughness values for wood–adhesive joints can be attributed to a combination of adhesive plastic deformation and reduction of microcracking of the wood around the crack tip. A flexible adhesive layer, especially a thick layer, distributes the concentrated stress over a larger area (volume) and lowers the level of the peak stress (Fig. 4). This apparently inhibits microcracking in the adjacent wood. Reduction of microcracking is indicated by the lower percentages of wood failure and lower counts of acoustic emission [41] per unit of new fracture surface in joints made with nonrigid adhesives compared to rigid thermosetting adhesives.

The fracture surface of a conventional urea–formaldehyde adhesive (Fig. 5) shows distinctive smooth brittle fracture surfaces formed when the adhesive layers cracked as a result of shrinkage stress that developed during cure [42]. A moderately toughened urea–formaldehyde bonded joint (Fig. 6) shows three distinct types of fracture surfaces arising under differing conditions: (A) cure shrinkage, (B) vacuum-pressure soak-dry (VPSD) treatment, and external loading to fracture (C). The rough surfaces are contrasted to the smooth cure-shrinkage crack surfaces. The crack caused by cyclic VPSD treatment (B) shows signs of plastic deformation. However, the plastic deformation does not have any directional properties. It appears to have occurred when the adhesive was in a weakened state, such as might occur from the absorption of water. Crack surface (C) occurred during testing when the material was dry and strong. Initially, it propagated at a high rate from the adhesive's interface with the lower adherend toward the upper interface. As the crack slowed, the adhesive deformed plastically, leaving striations in the upper corner. The fracture surface (C) suggests strength and toughness. In contrast, a phenol–formaldehyde adhesive layer (Fig. 7) shows extreme plastic deformation and directionality. Both these traits suggest a tough, strong adhesive layer. In contrast to the blocky fracture surface that resulted from an adhesive layer that was precracked by cure shrinkage (Fig. 8), the phenol–formaldehyde adhesive remained uncracked until externally loaded to failure (Fig. 9). In addition to the plastic deformation seen at high magnification (Fig. 7), there are no preexisting cracks in the adhesive layer. During testing to failure, when the primary crack jumps across the adhesive layer from one interphase to the opposite, the cracked adhesive surface is most often sloped (arrow).

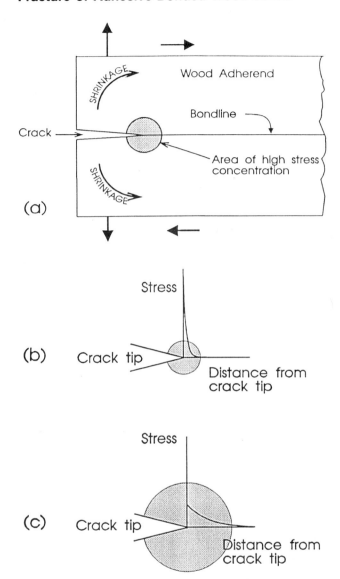

Figure 4 (a) Fracture process zone (area of stress concentration) surrounding the area or volume of the bondline immediately ahead of the crack tip when the joint is subjected to cleavage, shear, or shrinkage forces; (b) small process zone and high stress concentration with rigid adherend and adhesive; (c) large process zone and low stress concentration with flexible adhesive and adherend.

A rigid brittle thermosetting adhesive such as the unmodified urea–formaldehyde shown in Figs. 5 and 8 does not have the ability to arrest a growing crack, as evidenced by extensive brittle fracturing even before testing. A modified, toughened thermosetting adhesive does have this ability. Figure 10 shows an arrested crack in a fillet of amine-modified urea–formaldehyde adhesive in southern pine particleboard subjected to 10

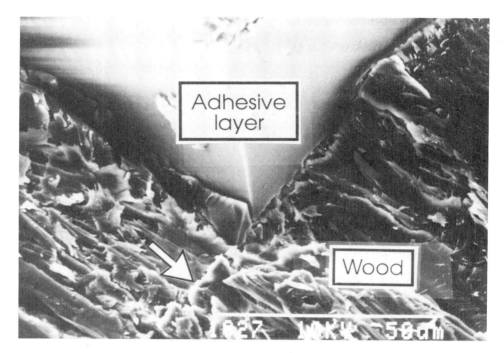

Figure 5 Smooth (glassy) fracture surface of a brittle urea–formaldehyde adhesive layer fractured by stress developed in the adhesive layer as it cured. Note the tensile rupture of the cells at the wood surface (arrow) caused by the cure-shrinkage crack in the adhesive.

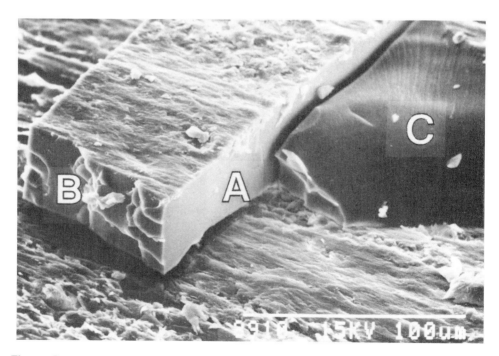

Figure 6 Fracture surfaces of an amine-modified urea–formaldehyde adhesive showing three distinct types of fracture surface: (A) cure-shrinkage crack surface; (B) vacuum-pressure soak-dry crack surface; (C) crack surface created during loading to failure.

Figure 7 Fracture surface of a phenol–formaldehyde adhesive showing striations (arrow) indicative of plastic deformation, yielding, and toughness.

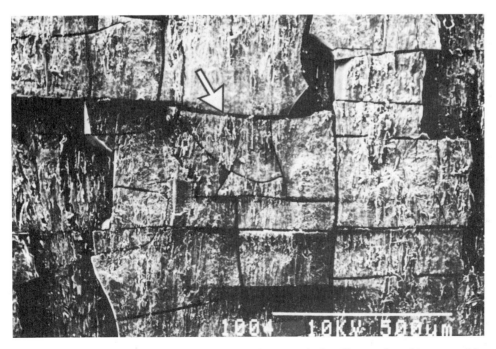

Figure 8 Fracture surface pattern produced by cleavage of a bondline weakened by precracking (arrow) of the adhesive layer as it shrinks during cure (unmodified UF adhesive).

Figure 9 Overview of fracture surface of phenol–formaldehyde bonded joint showing exposed adhesive layer (A) without preexisting cure-shrinkage cracks and surface of lower adherend (B). The sloped test fracture surfaces (arrow) characteristic of phenol–formaldehyde and toughened urea–formaldehyde adhesive layers show where the crack jumped from one interphase to the opposite as the crack traveled in the fiber direction.

VPSD cycles. The rounded crack tip shows plastic deformation and blunting. A new sharp notch can be seen forming at the root of the blunt crack tip. The ability to blunt cracks in the adhesive layer or in the fillet of adhesive between particles or flakes tends to force fracture in the wood, as illustrated in Fig. 11 and discussed by River and others [42].

C. Joints

1. Crack Initiation

Fracture of wood and bonded joints and materials begins at a geometric or material discontinuity where displacement of the adherends (due to external or internal stress) creates the greatest stress concentration and where either the adherend or the adhesive is the weakest. Examples of geometric discontinuities in adhesive-bonded wood joints are the square-cut ends of overlapped adherends, voids at the tips of fingers in fingerjoints, voids in reconstituted boards, voids in the adhesive layer, and even the square-cut ends of individual fibers. Examples of material discontinuities are the juncture of adherends of different density, the interface between adhesive and adherends of differing moduli, earlywood and latewood bands of widely different density, and the transition zone between the low fibril angle S1 and high fibril angle S2 layers of the cell wall. When adhesive bonds near this zone are sheared, the microfibrils in the S1 layer appear to undergo a rolling-shear failure [43]. Adhesive penetration of the cell wall was shown to

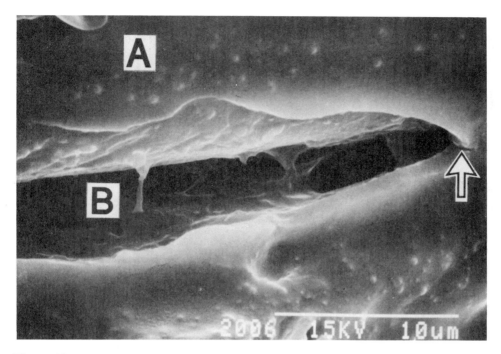

Figure 10 Fillet of adhesive (A) of amine-toughened urea–formaldehyde adhesive in Southern pine flakeboard showing an arrested crack (B) after 10 vacuum-pressure soak-dry cycles. Note the plastic deformation and blunting at the rounded crack tip and the beginning of new crack growth in the sharp notch at the end of the blunt crack tip (arrow).

affect fracture positively in the vicinity of the S2–S3 interphase [44]. An epoxy adhesive applied soon after mixing was of sufficiently low molecular weight to penetrate the cell wall from the lumen. Subsequently, when the adhesive layer was stripped from the wood surface, fracture occurred in the S2 layer. The same adhesive applied some hours after mixing was higher in viscosity (and thus molecular weight) and did not penetrate the cell wall as deeply. In this case, fracture occurred in the S3 layer and S2–S3 interphase.

The idea of an intrinsic or inherent flaw size in wood was explored by Schniewind and Lyon [32] who found the intrinsic flaw to be 3 mm. The same idea was applied to wood-based panels by Ilcewicz and Wilson [46] and to solid-wood joints by Kyokong and others [28]. Ilcewicz and Wilson used a modified fracture model based on Eringen's nonlocal theory [46] to determine the fracture toughness of flakeboard in tension perpendicular to the panel. According to their model, the critical-stress intensity factor of the flakeboard is a function of the intrinsic flaw size (which they determined to be 8.6 mm), the intrinsic strength of the board (determined to be 4.5 MPa), and the "characteristic dimension." The characteristic dimension in the original model for the fracture behavior of metal is the atomic distance of the metal. Ilcewicz and Wilson [45] substituted the flake thickness for the atomic distance in their modified model for flakeboard. They found the critical stress intensity factor (K_{Ic}) of the flakeboard was indeed a function of the characteristic dimension as well as the resin content of the board. Furthermore, the effect of flake thickness decreased as the resin content in the board increased from 5% to 11%.

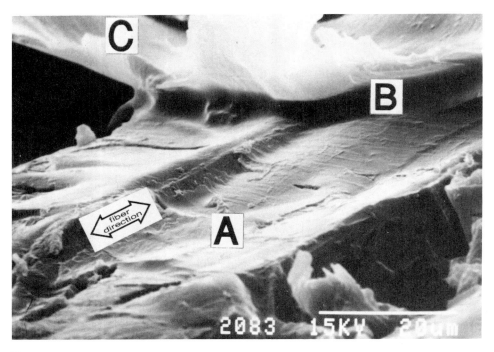

Figure 11 Fracture surface of Southern pine flakeboard showing unfractured fillet of phenol–formaldehyde adhesive: (A) original surface of wood flake; (B) adhesive fillet; (C) fragment of S1 layer of secondary wall from second wood flake.

Based on this relationship, the authors predicted that K_{Ic} would become independent of resin content at about 17% and at this point the dependency of K_{Ic} would shift from the flake thickness to some anatomical substructure, independent of resin content, such as the average lumen diameter of the cells in the flakes. Similar relationships of fracture toughness to board density, resin content, and particle size were reported by Niemz and Schadlich [47]. It seems clear that the geometric discontinuities in reconstituted materials can be minimized by using lower-modulus, more conformable woods such as aspen rather than oak, thinner flakes or strands, higher compaction ratios, and higher resin content.

Research by Kyokong and others [28] lent credibility to Ilcewicz and Wilson's hypothesis. They applied Eringen's nonlocal theory to solid poplar (*Populus tremuloides*) joints bonded with resorcinol adhesive, substituting the average vessel lumen diameter of aspen (100 μm) as the characteristic dimension. They were able to show that the nonlocal theory using this dimension correlated very closely with the fracture toughness of the joints as determined by classic (local) theory.

In solid wood members, considerable effort is devoted to minimizing geometric discontinuities through the use of scarf and finger joints instead of butt and lap joints. Scarf joints of sufficiently low slope can achieve 85 to 90% of the strength of solid wood [48]. Scarf joints effectively minimize material discontinuities between earlywood and latewood as well as geometric discontinuity. However, uniform-density wood, such as white pine, is likely to produce more efficient scarf joints than wood with growth rings of widely varying density, such as southern pine. However, 100% efficiency is unlikely to be obtained in any case because of mismatched wood and adhesive properties. Finger joints are less efficient because the flat portion of each finger tip represents a small butt joint and geometric as well as material discontinuity. These can be effectively minimized by cutting sharp rather than blunt tips [49]. Tool wear presents a practical limitation to tip sharpness in machined finger joints, especially in higher-density woods or woods with high-density latewood bands.

Impression finger joints take tip sharpness to the extreme and would seem to approach a well-made scarf joint in freedom from geometric discontinuity. Impression joints are formed by pressing a heated die with knife-edged serrated surfaces into the end-grain surfaces to be joined. This process eliminates damage caused by cutting and has the advantage of producing essentially a side-grain surface for gluing. But because of the maximum compressibility of the wood at the fingertip by the die, impression joints are limited to woods with density less than about 0.5 [50]. Even though many structural woods are lower in density, they possess latewood bands of much higher density. However, elimination of the geometric discontinuity by the impression process densifies the fingertips but not the valleys; this results in a material discontinuity and thus stress concentration. Fracture typically occurs across the roots of the fingers as a result of these closely spaced stress concentrations.

2. Crack Growth Stability

Once it initiates, a crack may propagate in one of several ways. It may fracture completely and catastrophically as glass (unstable) (Fig. 12a), it may fracture in several moderate increments of growth with intermediate arrest points (stable/unstable or stick/slip) (Fig. 12c), or it may fracture by tearing or continuous small increments (stable) (Fig. 12b and d). The preferred joint is that which requires a high crack initiation energy and produces stable crack growth (Fig. 12b). Such a joint requires a great amount of energy for complete failure to occur.

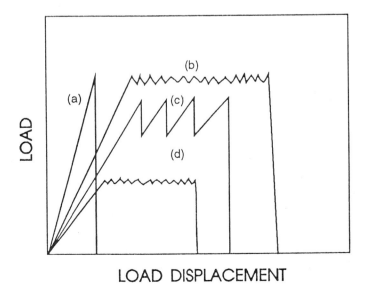

Figure 12 Characteristic crack growth behavior and fracture toughness of wood adhesive joints: (a) strong/unstable; (b) strong/stable; (c) strong/moderately unstable; (d) weak/stable.

Polymers, including adhesives, exhibit these behaviors. As such, an adhesive influences the fracture behavior of the joint in several ways. If the adhesive is formulated, applied, or cured improperly, its cohesive strength and toughness may be lower than that of the wood; if the adhesive does not properly wet or penetrate the wood, the adhesion strength may be lower than the cohesive strength of the wood. Under these circumstances, the crack will travel preferentially in the weaker adhesive layer or joint interface. The crack initiation energy will be low, and crack growth will be stable with little difference between the crack initiation and arrest energies. Such behavior would be expected from a starved or filtered joint or one between inactivated wood surfaces. Similar behavior would be expected from a joint made with adherends whose surfaces have been damaged by crushing during machining or by chemical degradation. In this case the crack travels in the wood or wood interphase but does not deviate far from the plane of the wood surface. The fracture surface produced by this type of crack growth behavior is often termed *shallow wood failure*. Poor adhesion also produces this type of crack growth behavior and shallow wood failure.

Stable/unstable or stick/slip crack growth occurs when the adhesive is properly formulated, applied, and cured and the grain direction is purposely directed toward the bondline. The adhesive is stronger than the wood and tough to moderately tough. The weak planes in the wood force the crack toward the bondline. Under these conditions, the crack will travel in the wood near the interphase, or in the interphase, and occasionally cross the adhesive layer [42]. The crack initiation energy will be moderately high to high, depending on the plasticity and strength of the adhesive and the species of wood. As the joint is loaded, some energy will be stored in elastic deformation of the adhesive and the adherends, and some will be consumed in plastic deformation and microcracking in the wood surrounding the crack tip. Once crack growth begins, the crack tip will advance and consume the stored energy at a high rate. The crack will arrest when the stored energy level drops below a certain level. Arrest will likely not occur in the adhesive or the interphase but in a region of lower wood density where a large amount of energy is rapidly consumed. The crack tip may remain fixed in this region or grow slowly by microcracking ahead of the primary crack tip, as additional energy is stored in the adhesive and the adherends. The later scenario seems more probable in wood joints, considering the weak interphase between the S1 and S2 layers. Rapid propagation occurs when the stored energy again reaches the critical level. This type of joint is created specifically for testing the fracture toughness of bonded wood joints. It also occurs at random in real joints as a result of the natural grain variation and variation in the plane of cut. The fracture surface resulting from this type of joint and crack growth behavior is typically interphaseal or shallow-wood, although somewhat deeper wood failure may occur in low-density regions. It is also typified by occasional to frequent crossings of the adhesive layer from one interphase to the other. These crossings are due to variations of the strength and modulus of the adherends on either side of the adhesive layer. Crossings are facilitated by a precracked (cure-shrinkage cracked) adhesive layer (Fig. 8) or hindered by a continuous tough adhesive layer (Fig. 9).

Stable crack growth also occurs when the adhesive is strong and tough and has established good adhesion, the wood surface is sound, and the grain angle is parallel or away from the bondline. Under these conditions, the crack deviates into the wood according to the mechanisms described by Wang and others [51] and Knauss [52] and remains there. The fracture toughness of the joint is essentially determined by the toughness of the wood. The fracture surface does not necessarily follow the plane of the bondline. It is more likely to follow the grain angle, producing what is often termed *deep wood failure*. In this type of fracture, the crack advances by continuous transwall cracking of the thin-walled cells and intrawall or diagonal transwall cracking of the thick-walled cells. Stable crack growth will also occur when the adhesive establishes good adhesion but is weaker than the wood, as for example with elastomeric and some thermoplastic adhesives (100% adhesive failure). These adhesives are too weak to store sufficient energy in the adhesive or the adherends to support rapid crack propagation. Instead, the adhesive tears slowly when it reaches its ultimate tensile stress. In testing, this rate is controlled by the rate of crosshead movement. There is essentially no difference between the crack initiation and crack arrest energies.

The brittleness index [53] is a normalization of the energy released during a period of rapid crack growth with respect to the energy stored in the joint just at the onset of crack growth. It provides a quantitative measure of the behavior described here.

$$I = \frac{G_c - G_a}{G_c}$$

where I is the brittleness index, G_c the energy required to initiate crack growth, and G_a the energy remaining at crack arrest. According to this measurement, an ideally brittle (Fig. 12a) (unstable) material that fails suddenly and completely will have an index (I) value of 1 and an ideally plastic (stable) material that fails by continuous tearing (no difference between initiation and arrest energies) will have an I value of 0. Practically, most wood–adhesive joints fail by a combination of unstable and stable crack growth and have I values between 1 and 0. Joints tending toward stable crack growth (Fig. 12b and d) will have I values in the range 0.8 to 0.95. Joints tending toward more unstable behavior (Fig. 12c) will have I values in the range 0.4 to 0.8.

The stability of crack growth behavior was also quantified by acoustic emissions per unit of new surface area formed by crack propagation [41]. Large bursts of acoustic emissions (AEs) occurred during unstable crack growth. Fewer but continuous AEs occurred during stable crack growth. A joint producing 100% bursts of AE during fracture would correspond to a brittleness index of 1 and be perfectly brittle. A joint producing 100% continuous AE would correspond to a brittleness index of 0 and be perfectly plastic. The ratio of burst to continuous AE provides a measure of the stability or instability of the crack growth. The burst to continuous AE ratio of bonded wood joints fell in the range of 0.5 (quite stable) to about 5 (quite brittle). The maximum crack initiation energy tended to be associated with a ratio of about 1.3 times as many burst AEs as continuous AEs.

In the Suzuki and Schniewind study, the fracture toughness of joints made with a variety of conventional, modified conventional, and nonconventional wood adhesives was linearly associated with the AEs per unit area of new crack surface formed during testing. The maximum fracture energy and AEs as a function of various modifications such as filler content, fortifier, and adhesive layer thickness were clearly evident from the relationship. Two different relationships between fracture toughness and AEs were also evident. Nonrigid adhesives, including epoxy, poly(vinyl acetate) (PVA), emulsion polymer isocyanate (EPI), and PVA-modified phenol–formaldehyde (PF), were approximately 2.8 times tougher at a given AE count per unit of area as were conventional rigid thermosetting adhesives, including urea–formaldehyde (UF), PF, phenol–resorcinol formaldehyde (PRF), and isocyanate. The less rigid adhesives absorb or dissipate more energy before cracking than do the rigid adhesives. Viewed another way, for a given level of energy expended to create a new fracture surface, a joint bonded with a rigid adhesive produced approximately 20 times as many acoustic events (microcrack formations) as did a joint bonded with a nonrigid adhesive. Obviously, some of these additional events are in the adhesive, but most are in the wood, given a high percentage of wood failure. In a sense, the less rigid adhesive protects the weak S1–S2 interface by reducing the stress concentration at the crack tip.

3. Crack Deflection

A natural crack initiated in the center of the adhesive layer in a symmetric joint between symmetric isotropic (metal) adherends will tend to propagate through the center of the adhesive layer. However, in wood joints, there is a strong tendency for the crack to travel in the wood near the joint. This condition should be expected in joints made with the lower-density species or in species with the low-density earlywood such as the Southern pines (*Pinus* spp.). However, wood failure is not uncommon in joints made with high-density species even when there is a starter crack in the adhesive layer before testing. There seem to be some rational explanations for this behavior.

First, a crack will deviate toward one or the other adherend if it is softer (lower in modulus) than the adhesive [51]. This is a common condition in wood joints bonded with rigid thermosetting adhesives. The tension modulus of the wood perpendicular to the grain is typically in the range 400 to 1200 MPa [54], while the tensile modulus of adhesives used with wood will be in the range 1200 to 4700 MPa at the same moisture level [55–57].

Second, shear forces that develop in the vicinity of the crack tip direct it toward one or the other adherend [52]. Shear forces arise in a cleavage specimen from unequal moduli of the two adherends and the adhesive. Unequal moduli of two wood adherends is virtually certain as a result of the variable morphology and density of any two pieces of wood. When load is applied to the cracked joint, this inequality induces shear stress around the crack tip and thereby directs it toward one adherend or the other.

Once the crack enters the wood as a result of these mechanisms, it will travel preferentially along the weak radial–longitudinal (RL) and tangential–longitudinal (TL) planes. Unless these planes again intersect the bondline, the adhesive will not be likely to fracture beyond that point. If the fiber direction in both adherends is oriented toward the bondline (this is done purposely in some fracture toughness test methods), the crack will be forced to remain close to the adhesive layer. In this case the local density and modulus of the two adherends seems to determine on which side of the adhesive layer the fracture occurs. Since these properties vary continually, it is not unusual for the crack tip to jump repeatedly from one adherend, across the adhesive layer, to the opposite adherend according to the mechanism of Wang and others [51] and Knauss [52]. Given a locale with earlywood on one side of the adhesive layer and latewood on the other side, the crack may not travel preferentially on the earlywood side. Pervasive adhesive penetration of the earlywood may raise the density and modulus to the extent that latewood on the opposite adherend is more amenable to crack growth.

4. Adhesive Layer Thickness

Shear strength studies of joints bonded with rigid thermosetting adhesives over many years has resulted in the prescription that the best joints are those with an adhesive layer in the thickness range 0.05 to 0.15 mm. Ebewele and others [3], for example, found an optimal thickness between 0.07 and 0.08 mm (Fig. 13). Other research based on fracture mechanics [13,28,30] has helped to define this relationship, although not its cause. Apparently, below some minimum thickness, a joint is adhesive starved and the inter-phase is rife with voids. Above the optimum thickness, stress concentrations are height-

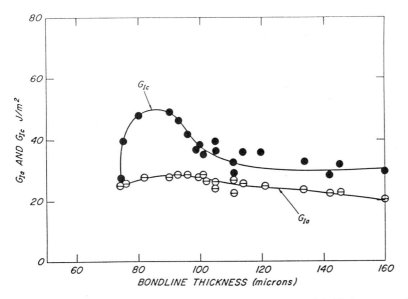

Figure 13 Effect of bondline thickness on the cleavage crack initiation energy (G_{Ic}) and cleavage crack arrest energy (G_{Ia}) of hard maple specimens bonded with rigid thermosetting PRF adhesive. (From Ref. 3.)

ened by cure-shrinkage stresses in the adhesive layer. The narrow optimal thickness range disappears if the adhesive modulus is greatly reduced. In the study by Takatani and Sasaki [13], an epoxy adhesive was flexibilized by the addition of 20, 40, and 60 parts of polysulfide. These additions decreased the adhesive modulus from 2200 MPa to 1600, 670, and 160 MPa, respectively. The last two moduli are in the range of the tensile modulus of wood perpendicular to the grain used to test fracture toughness (beech, MOE = 590 MPa). Joints of the nonflexibilized adhesive had a slight optimum at 0.3 mm thickness; however, there was actually little difference in toughness (G_{Ic} = 220 J/m^2) over the entire range of adhesive layer thickness from 0.1 to 1.5 mm. The addition of 20 parts of polysulfide removed the optimum at 0.3 mm thickness and increased toughness to 330 J/m^2. The big change came with the addition of 40 to 60 parts of polysulfide. Although these additions failed to increase toughness of joints with the thin adhesive layer, toughness increased dramatically with each increment of adhesive thickness. In these joints the crack initiation energy increased from about 330 J/m^2 to 1100 J/m^2 in specimens with adhesive layers 1.5 mm thick.

It is probable that very high fracture toughness values in wood joints bonded with thicker, lower-modulus adhesive layers may be due to the enhancement of an existing energy-dissipating mechanism such as microcracking of the wood as well as the adhesive.

5. Grain Angle

Structural joints are purposely not designed with the fibers intersecting the plane of the bondline as a result of the weakness of this design. However, this relationship can be of great importance in the delamination of structural joints, where there are unavoidable local grain deviations, such as around knots. Furniture and picture frames often contain mitered joints in which the fibers intersect. However, the joints are often pinned or doweled for added strength.

Generally, fracture toughness increases with increasing grain angle, although there is usually a minimum toughness between about 10 and 30° (Fig. 14) [3,16,30]. The shape of the relationship varies from almost flat to very steep as a function of species and the type of adhesive. There is usually a minimum in the range 15 to 30° above the bondline and a maximum at 90°. Both the stress intensity factor (K_{Ic}) and the strain-energy release rate (G_{Ic}) follow the same trend [30]. The basis for the shape of this relationship is unclear. Ruedy and Johnson speculate that it is due to variation of adhesive penetration and stored energy at the crack tip with grain angle, while Ebewele and others [3] think the relationship is caused by variation of the Cook–Gordon ''weak-interface'' crack-stopping mechanism with grain angle [58]. White [33] attributed the increased toughness at large grain angle to increased penetration and reinforcement of the interphase. The depth of penetration of loblolly pine (*Pinus taeda*) by a resorcinol adhesive increased nine times and the fracture toughness doubled as the grain angle was increased from 0 to 45°.

A grain angle effect has also been reported in the shear strength of bonded joints. When the grain direction runs with the applied force (Fig. 15a), the principal stress across the grain direction is in compression (closing mode). When the grain direction runs against the applied force (Fig. 15b), the principal stress across the grain is in tension (opening mode). The shear strength of the joint is highest when the grain direction runs

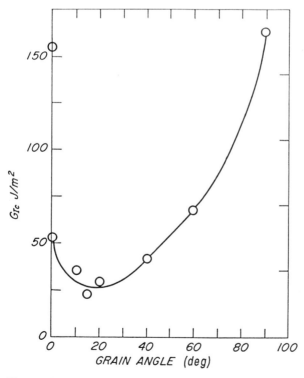

Figure 14 Effect of adherend grain angle to the bondline on the fracture toughness of bonded wood joints in cleavage. (From Ref. 3.)

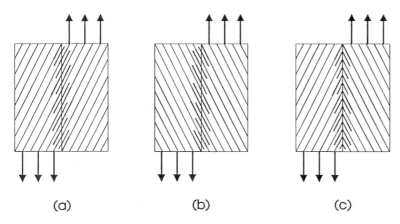

(a)   (b)   (c)

Figure 15 Schematic of joints with nonplanar grain orientations: (a) both adherends in compression (closing) mode perpendicular to the grain; (b) both adherends in tension (opening) mode perpendicular to the grain; (c) one adherend in each mode. Grain direction indicated by fine lines. Adhesive penetration indicated by heavy lines.

with the applied force. This effect is due to the disparity between the tensile and compression strengths of the wood and to adhesive penetration. When the grain direction is not parallel to the surface, the adhesive is very likely to penetrate the wood deeply. When such a joint is loaded in the opening mode, the strength of the cell wall, particularly the S1 layer and the S1–S2 interphase, is still the limiting factor. Adhesive penetration adds little to the tensile strength of wood perpendicular to the grain. However, in the closing mode, plugs of adhesive in the cell lumens increase the compression strength across the grain. They also distribute the shear force away from the plane of the joint. This was evident in the results of Furuno and others [43], who used tensile single lap joints for their experiments. Joints loaded in the closing mode were 25% stronger than parallel-grain joints and 45% stronger than joints loaded in the opening mode. Fracture of a joint with one adherend in the opening mode and one in the closing mode (Fig. 15c) occurred at low strength as a result of the opening mode. Swietliczny [59] conducted a similar experiment using compression block shear specimens and found the opposite results. The difference can be attributed to the method of loading, particularly the support of the specimen during loading, which inhibited opening or tensile fracture.

6. Moisture

The moisture content of both the wood and the adhesive affect the fracture behavior of adhesive bonded joints. Wood joints are especially sensitive to moisture effects as a result of the porosity and permeability of wood, which allows ready access by water to both the interior of the wood member and the adhesive layer. Irle and Bolton [57] showed that the superior durability of wood-based panels bonded with an alkaline PF adhesive compared to panels bonded with a UF adhesive was due to the ability of the phenolic adhesive to absorb and be plasticized by water. In the plasticized state, the phenolic adhesive is able to reduce stress concentrations that otherwise fracture the wood or the adhesive in urea-bonded panels.

Another important effect of moisture is due to a change in the moisture level, or content, of the wood member in a dynamic service environment. In thick members, changes in moisture content and the moisture-dependent dimension in the center fall behind changes that occur at the surface of the member. The difference in dimension creates stress in the member and bonded joints in the member. Adhesive bonds also restrain the swelling and shrinking of bonded members with different swell/shrink coefficients resulting from grain direction, growth-ring angle, or species. Moisture gradients and differential swelling or shrinking of the adherends are common cause of fracture of joints or materials. In this regard the size of the bonded members and the mechanical properties of the adhesive and the adherends have important roles in determining the magnitude of the stresses (and stress concentrations) that arise from moisture changes. The most severe stresses arise as both the adhesive and the wood dry because of the attendant differential increases in the adherend and adhesive moduli.

Simply changing the growth-ring orientation in adjoining laminate can alter the possibility of fracture in the vicinity of the joint caused by a change in moisture content of the laminated member. Laufenberg [60] studied the effects of growth-ring orientation in parallel Douglas-fir laminates. By finite element analysis, he showed that maximum stresses occurred at the edge of the laminate when one lamina had flat grain and the other vertical grain. He also found that a difference of growth-ring angles of only 15° was likely to produce splits or delamination as a result of moisture content cycling.

Nestic and Milner [61] also examined the effects of growth-ring orientation and found vast differences, particularly in the peak tensile stresses perpendicular to the grain, that depended on the difference between growth-ring orientation of adjoining laminae. The authors also found that the closer the pith was to a bondline, the greater the stress in the bondline induced by a moisture content change in the wood.

When the laminae are cross-laminated, the stresses are even more severe. Adherends thicker than roughly 5 mm will create sufficient stress to fracture the wood when bonded in a cross-laminated configuration. The most severe stresses arise as both the wood and the adhesive dry out, with an accompanying increase in strength and modulus. However, the stresses imposed by differential swelling of wood members are also severe in the case of an adhesive that is overly plasticized and weakened at high moisture contents. The effects of wood and adhesive properties and the environment on fracture behavior are complex. The effects of internal stress generated by wood on adhesives with varying sensitivity to moisture have been described [62]. Gillespie [62] compared the effects of medium-density, high-swelling maple (*Acer saccharum*) to low-density, low-swelling pine (*Pinus strobus*) using the same adhesives. The joints of maple bonded with moisture-sensitive adhesives (PVA, catalyzed PVA, and casein) were destroyed or suffered severe and irrecoverable loss of strength from soaking. Similar joints of pine recovered all or most of their original strength upon redrying.

Internal stress may detract significantly from the apparent strength of a joint even if it is insufficient to fracture the joint. For example, if the internal tensile stress in a joint is equal to one-half the ultimate stress or strength of the weakest material, the available tensile strength of the joint is lowered by 50%.

7. Geometry

External loads are imposed on a bonded joint or structure by the dead weight of the structure and its contents, accumulated snow, wind, and people. The average stresses in a joint created by these loads can be calculated from structural analysis, but the maximum stress at joint edges is more difficult to determine. These stresses have been examined in some detail. Discussions by Walsh and others [63] and Glos and Horstmann [40] of the effects of external forces on the stresses in bonded wood lap joints are notable. Walsh and others applied plastic failure and fracture mechanics criteria to study the effects of the ratio of lap length (L) to primary member thickness (T) on the average failure stress of double-lap joints. The authors conclude that the plastic failure criterion (uniform stress) governs failure of the joint only at a very low L/T ratio (<1). The fracture mechanics criterion governs to an L/T of 8. The authors developed the following conservative empirical design formula:

$$\frac{\sigma_{nom}}{\tau_y} = (2T)^{-0.5} \left[1 + \frac{1}{8} \left(\frac{L}{T} \right) \right]$$

for $L/T = 2$ to 8 and $t/T = 0.5$ to 2, where σ_{nom} is the allowable axial stress in the adherends, τ_y the design shear stress in the joint, and t the lap adherend thickness.

Glos and Horstmann [40] systematically studied the effects of various joint design factors on the fracture of double-lap joints. These factors included (1) grain angle between two side-grain to side-grain members, (2) length of overlap, (3) shape of bonded area, (4) wood density, (5) type of adhesive, and (6) end distance. (End distance is defined as the distance to which the unloaded portion of the bonded members extends beyond the joint.) The authors found that all factors had a strong effect on joint fracture except the shape of the bonded area. Most interesting was the finding that creating a finite end distance increased the strength of joints with lap length/member thickness (L/T) ratios of 3 or less, but decreased the strength for L/T ratios above 3. The explanation given is that in short joints (those governed by the plastic failure criteria), the critical stresses perpendicular to the grain at the end of the overlap are spread over a larger area than in joints in which the adherend ends abruptly at the end of the joint. In long joints (those governed by fracture mechanics criteria), creating a finite end distance increases the sharpness of the notch from 90° compared to 0°. The increase in notch sharpness increases the stress concentration at a given load and thus reduces the average stress in the joint at which fracture occurs.

III. FRACTURE-BASED DESIGN

A. Structural Joints

There are no standard design methods for adhesive-bonded wood joints, let alone design methods based on fracture mechanics. This is obviously due in part to the complexity of the fracture behavior of wood joints and materials. The lack of adequate design methods has obviously been a hindrance in furthering the use of adhesives in structural assembly joints for wood structures. However, studies have demonstrated the power of fracture mechanics for developing generalized methods for predicting the behavior of adhesively bonded joints and materials. Conventional strength tests have not been able to predict such behavior.

Komatsu and others [64] found that the strength of double-lap joints was proportional to the bond area for relatively short overlaps. However, for long overlaps, stress concentrations and fracture mechanics controlled the strength of the joint. The authors developed the following fracture-based design equation:

$$\sigma_{max} = \sqrt{G_c E_x S}$$

where σ_{max} is the shear strength of the joint, G_c the critical strain energy release rate, E_x the elastic tensile modulus of wood adherends along the grain, and S the geometrical joint parameter.

Wernersson and Gustafsson [65] developed a nonlinear fracture mechanics relationship based on pure shear for predicting the performance of lap joints of varying geometry and adherend properties based on the adhesive brittleness ratio:

$$\frac{\tau_f^2}{G_f}$$

where τ_f is the ultimate shear stress of the bondline obtained from uniform stress test method, and G_f is the total fracture energy of the bondline. Wernersson [66] used this brittleness ratio to show how the failure of different types of joints is controlled by various criteria. Joints with a low brittleness ratio exhibit ductile behavior with uniform plastic deformation along the bondline. The joint strength is proportional to the local bond strength. Joints with a high brittleness ratio exhibit brittle behavior, with strength independent of the local bond strength. Joint strength is governed by fracture energy. The strength of joints with an intermediate brittleness ratio is affected by the local strength but also by the fracture energy and the shape of the stress–strain curve of the materials.

Based on his analysis, Wernersson proposed that the optimal adhesive properties, in terms of producing the strongest joint, are not necessarily those that produce the highest wood failure. However, Wernersson also acknowledges that this conclusion does not take into account the effects of time, temperature, or moisture. When long-term effects are considered, it is still too early to reject the long-standing requirements for high wood failure and maximum allowable cyclic delamination as indicators of the probable permanence of structural joints.

Komatsu [38] also applied fracture mechanics to the design of bonded cross-lapped knee joints that experience a torsion shear loading. Specimens were tested with the angle of the knee at 90°, 120°, and 150°. The crack initiation energies for the three angles were, respectively, 480, 600, and 1100 J/m². Failures at 90° and 120° were largely brittle (tension perpendicular to the grain) and had a greater correspondence to the lower torsional shear fracture toughness values than failure at 150°. The 150° joints showed a fairly uniform distribution of five different types of fracture. Some of the difference in toughness and type of fracture is no doubt due to the greater proportion of sliding shear forces in the 150° joint. Overall, Komatsu concluded that the fracture mechanics analysis gave a better prediction of strength than a method based on elastic torsional theory.

B. Wood-Based Panels

Lei and Wilson [35,67] developed a model for the fracture toughness (K'_{Ic}) of flakeboards bonded with phenol–formaldehyde resin adhesive. The model is based on the initial crack length (a) in the specimen, the average size of the inherent flaw (Ω) in the solid wood, the expected increase in crack length resulting from nonbonded interflake cracks and voids (Δa), and the K_{Ic} value of the wood used to make the flakes.

$$K'_{Ic} = K_{Ic} e^{(\Omega - 1/\mu)} \frac{a^{1/2} Y(a/W)}{(a + \delta a)^{0.5} Y[(a + \Delta a)/W]}$$

Other factors are the compliance (Y) for the specimen geometry, the board width (W), and the average length of the cracks or voids in the flakeboard ($1/\mu$). The experiment revealed that the K_{Ic} value of the flakeboard was the same as the K_{Ic} value of solid wood when the average crack length was equal to 2.5 mm, the same value as determined by Schniewind and Lyons [32] for solid Douglas-fir. Another study [35] showed that the solid wood density and the compaction ratio of the flakeboard also affect the average crack length and thus the fracture toughness of the flakeboard. The lengths of nonbonded voids and interflake crack decreased as the compaction ratio increased up to a board density of 780 kg/m^3. Mechanical damage to the flakes at a higher compaction ratio lowered fracture toughness.

IV. SUMMARY AND CONCLUSIONS

Understanding the causes and mechanisms of fracture in adhesive-bonded joints and materials is important to improving their performance, developing products based on new combinations of materials and adhesives, predicting the performance of new materials, and developing design methods for structural joints. In this chapter I have briefly discussed some aspects of wood and adhesive fracture, the influence of wood and adhesive properties upon joint fracture, the effects of environment and joint geometry on fracture, and the attempts to develop design methods for bonded joints and materials based on fracture mechanics.

Microstructure, in particular the discontinuities in the walls of thick-walled cells, is a controlling factor in the fracture of well-made joints bonded with rigid, thermosetting adhesives. The properties of the adhesive, however, play a major role in ameliorating the weaknesses of thin-walled cells and the discontinuities in thick-walled cells. Good wetting and chemical adhesion are important to bond performance, but they are not in themselves sufficient for maximum fracture toughness of bonded wood joints and materials. Hard, brittle adhesives, especially those that do not effectively penetrate the wood cell cavities and the cell wall, promote transwall cracking of thin-walled cells and microcracking and intrawall fracture of thick-walled cells. Less rigid adhesives that penetrate the cell lumens and cell wall distribute stress and inhibit microcracking in the wood.

The best adhesive for improved fracture toughness (1) does not develop shrinkage stresses during cure, (2) has a modulus close to that of wood perpendicular to the grain, (3) has a modulus that changes in parallel with the wood modulus as moisture content changes, (4) penetrates small-lumen, thick-walled cells but does not overpenetrate large-lumen thin-walled cells, and (5) can infiltrate the cell wall to reinforce the weak interphase between cell-wall layers.

The behavior of bonded joints and materials can be predicted successfully on the basis of material properties through applying the principles of fracture mechanics. However, much research is still required to achieve a method that is generally applicable to all adhesives, species, and joint geometries or material constructions. One field of particular importance and complexity revolves around the important effects of time, moisture, and temperature, and their interactions. At present, without extensive and long-term testing, there is no way to predict or evaluate the trade-offs between high short-term fracture toughness in joints or materials bonded with semirigid adhesives and reduced stress-rupture resistance of these adhesives under conditions of elevated moisture, temperature, or prolonged loading. An understanding of these relationships and the development of a model to predict the effects of trade-offs could lead to a new generation of wood-based materials and efficient adhesive-bonded wood structures.

REFERENCES

1. S. Mostovoy, P. B. Crosley, and E. J. Ripling, *J. Mater.* 2(3): 661 (1967).
2. E. J. Ripling, S. Mostovoy, and H. T. Corten, *J. Adhesion* 3(2): 107 (1971).
3. R. O. Ebewele, B. H. River, and J. A. Koutsky, *Wood Fiber* 11(3): 197 (1979).
4. R. O. Ebewele, B. H. River, and J. A. Koutsky, *Wood Fiber* 12(1): 40 (1980).
5. R. O. Ebewele, B. H. River, and J. A. Koutsky, *J. Adhesion* 14: 189 (1982).
6. M. Suzuki and A. P. Schniewind, *Mokuzai Gakkaishi* 30(1): 60 (1984).
7. C. T. Scott, B. H. River, and J. A. Koutsky, *J. Testing Evaluation* 20(4): 259 (1992).
8. B. H. River and E. A. Okkonen, *J. Testing Evaluation* 21(1): 21 (1993).
9. H. Sasaki and P. F. Walsh, *J. Soc. Mater. Sci. (Japan)* 26(284): 453 (1977).
10. M. Takatani, R. Hamada, and H. Sasaki, *Mokuzai Gakkaishi* 30(2): 124 (1984).
11. M. Takatani, R. Hamada, and H. Sasaki, *Mokuzai Gakkaishi* 30(2): 130 (1984).
12. M. Takatani, R. Hamada, S. Hayashi, and H. Sasaki, *Mokuzai Gakkaishi* 31(8): 657 (1985).
13. M. Takatani and H. Sasaki, *Wood Res.* 66: 30 (1980).
14. K. Komatsu, H. Sasaki, and T. Maku, *Wood Res.* 59/60(3): 80 (1976).
15. T. Iwasaki and M. Suzuki, *Mokuzai Gakkaishi* 35(8): 696 (1989).
16. J. S. Mijovic and J. A. Koutsky, *Wood Sci.* 11(3): 164 (1979).
17. B. H. River, C. T. Scott, and J. A. Koutsky, *Forest Prod. J.* 39(11/12): 23 (1989).
18. G. P. Anderson, S. J. Bennett, and K. L. DeVries, *Analysis and Testing of Adhesive Bonds*, Academic Press, New York, 1977.
19. A. W. Porter, *Forest Prod. J.* 14(8): 325 (1964).
20. Z. Koran, *Svensk Papperstid.* 71(17): 567 (1968).
21. R. W. Meyer, *Microstruct. Sci.* 4: 121 (1976).
22. R. E. Mark, *Cell Wall Mechanics of Tracheids*, Yale University Press, New Haven, Conn., 1967, pp. 65–66.
23. R. E. Mark, in *Adhesion in Cellulosic and Wood-Based Composites* (J. F. Oliver, ed.), Plenum Press, New York, 1981.
24. C. Woodward, *Wood Sci.* 13(2): 83 (1980).
25. J. B. Wilson and R. L. Krahmer, *Forest Prod. J.* 26(11): 42 (1976).
26. Z. Koran and R. C. Vasishth, *Wood Fiber* 3(4): 202 (1972).
27. A. J. Kinloch and R. J. Young, *Fracture Behavior of Polymers*, Applied Science, London, 1983.
28. B. Kyokong, F. J. Keenan, and S. J. Boyd, *Wood Fiber Sci.* 18(4): 499 (1986).
29. M. Suzuki and S. Iwakiri, *Mokuzai Gakkaishi* 32(4): 242 (1986).
30. T. C. Ruedy and J. A. Johnson, in *Proc. First International Conference on Wood Fracture*, Banff, Alberta, Canada, 1978, Forintek Canada Corp. Western Forest Products Laboratory, Vancouver, British Columbia, Canada, 1979.
31. M. S. White and D. W. Green, *Wood Sci.* 12(3): 149 (1980).
32. A. P. Schniewind and D. E. Lyon, *Wood Fiber Sci.* 7(1): 45 (1973).
33. M. S. White, *Wood Sci.* 10(1): 6 (1977).
34. D. E. Kretschmann, D. W. Green, and V. Malinauskas, in *Proc. 1991 International Timber Engineering Conference*, Vol. 3, TRADA, London, 1991, pp. 391–398.
35. Y.-K. Lei and J. B. Wilson, *Wood Sci.* 12(3): 154 (1980).
36. F. W. Smith and D. T. Penney, *Wood Sci.* 12(4): 227–235 (1980).
37. M. Suzuki, *Mokuzai Gakkaishi* 32(12): 972–977 (1986).
38. K. Komatsu, Application of fracture mechanics to the strength of cross-lapped glued timber joints, *FRI Bulletin 61*, Forest Research Institute, New Zealand Forest Service, Rotorua, New Zealand, 1984.
39. H. Wernersson and P. J. Gustafsson, in *Proc. International Conference on Timber Engineering*, Seattle, Wash., Forest Products Research Society, Madison, Wis., 1988.
40. P. Glos and H. Horstmann, in *Proc. CIB-W18A Meeting 22*, Vol. I, Paper 22-7-8, Interna-

tional Council for Building Research Studies and Documentation, Working Commission W18A: Timber Structures, Berlin, 1989.

41. M. Suzuki and A. P. Schniewind, *Wood Sci. Technol. 21*: 121 (1987).
42. B. H. River, R. O. Ebewele, and G. E. Myers, manuscript submitted for publication in *Holz Roh- Werkstoff* (1993).
43. T. Furuno, H. Saiki, T. Goto, and H. Harada, *Mokuzai Gakkaishi 29*(1): 43–53 (1983).
44. H. Saiki, *Mokuzai Gakkaishi 30*(1): 88–92 (1984).
45. L. B. Ilcewicz and J. B. Wilson, *Wood Sci. 14*(2): 65–72 (1981).
46. A. C. Eringen, *Continuum Physics*, Vol. 4, *Polar and Nonlocal Field Theories*, Academic Press, New York, 1976.
47. P. Niemz and S. Schadlich, *Holz Roh- Werkstoff 50*: 389–391 (1991).
48. T. M. Wardle, Glued scarf and finger joints for structural timber, *Information Bulletin E/IB/13*, Timber Research and Development Association, Hughenden Valley, High Wycombe, Bucks, England, 1967.
49. M. L. Selbo, *Forest Prod. J. 13*(9): 390–400 (1963).
50. M. D. Strickler, *Forest Prod. J. 17*(10): 23–28 (1976).
51. S. S. Wang, J. F. Mandell, and F. J. McGarry. Effects of crack elevation in TDCB adhesive fracture test, *Research Report R 76-3*, Department of Materials Science, School of Engineering, MIT, Cambridge, Mass., 1976.
52. W. G. Knauss, *J. Composite Mater. 5*: 176–192 (1971).
53. R. O. Ebewele, B. H. River, and J. A. Koutsky, *J. Appl. Polymer Sci. 31*: 2275–2302 (1986).
54. Forest Products Laboratory, *Wood Handbook: Wood as an Engineering Material*, Agriculture Handbook 72, U.S. Department of Agriculture, Washington, D.C., 1987.
55. W. Clad, *Holz Roh- Werkstoff 23*(2); 58–67 (1965).
56. W. T. Simpson and V. R. Soper, Stress–strain behavior of films of four adhesives used with wood, *Research Note FPL-0198*, USDA Forest Service, Forest Products Laboratory, Madison, Wis., 1968.
57. M. A. Irle and A. J. Bolton, *Holzforschung 42*: 53–58 (1988).
58. J. Cook and J. E. Gordon, *Proc. Roy. Soc. A 282*: 508–520 (1964).
59. V. M. Swietliczny, *Holztechnologie 21*(2): 83–87 (1980).
60. T. L. Laufenberg, *Forest Prod. J. 32*(5): 42–48 (1982).
61. R. Nestic and H. R. Milner, *J. Inst. Wood Sci. 12*(4): 225–231 (1991).
62. R. H. Gillespie, Effect of internal stresses on bond strength of wood joints, *NTIS PB-258 832/5ST*, prepared for Department of Housing and Urban Development, U.S. Department of Agriculture Forest Service, Forest Products Laboratory, Madison, Wis., 1976.
63. P. F. Walsh, R. H. Leicester, and A. Ryan, *Forest Prod. J. 23*(5): 30–33 (1973).
64. K. Komatsu, H. Sasaki, and T. Maku, *Wood Res. 61*: 11–24 (1976).
65. H. Wernersson and P. J. Gustafsson, The complete stress-slip curve of wood-adhesives in pure shear, *Report TVSM-7039*, Lund Institute of Technology, Lund, Sweden, 1987.
66. H. Wernersson, in *Proc. 1991 International Timber Engineering Conference*, London, 1991.
67. Y.-K. Lei and J. B. Wilson, *Wood Sci. 13*(3): 151–156 (1981).

10
Spectroscopic Techniques in Adhesive Bonding

W. J. van Ooij* *Armco Research & Technology, Middletown, Ohio*

I. INTRODUCTION

This chapter reviews the many different spectroscopic techniques that can be used to study the composition and performance of adhesively bonded materials. Such a review was deemed necessary because many new analytical techniques have been developed lately and a large number of interesting applications have been published recently. The most widely used spectroscopic techniques are x-ray photoelectron spectroscopy (XPS, also called ESCA), Auger electron spectroscopy (AES), secondary ion mass spectrometry (SIMS), and infrared spectroscopies (FTIR). The distinction should be made here between *microscopic* and *spectroscopic* techniques for surface and interface characterization. Both types of techniques are invariably applied for the development and failure analysis of adhesives. Examples of microscopic techniques are optical microscopy, scanning electron microscopy (SEM), transmission electron microscopy (TEM), scanning transmission microscopy (STEM), scanning tunneling microscopy (STM), and atomic force microscopy (AFM). In general, spectroscopic techniques provide qualitative analysis of the chemistry of a surface or an interface, although in certain cases quantitative analysis is possible. Microscopic techniques are primarily employed to study surface and interface morphology of adhesives and adherends. When these techniques are applied to failure surfaces, important information on failure modes can be obtained.

This review is limited to spectroscopic techniques. Excellent reviews already exist on the use of microscopic methods in adhesive bonding technology [1,2].

In general, the applications of spectroscopic techniques in the study of various aspects of adhesive bonding that have been reported are the following:

Surface characterization of materials prior to bonding; cleanliness, surface contamination, oxide thickness, and so forth

Modification of surfaces to improve bondability; examples are modification of polymer surfaces by plasma or corona treatments, anodization or other treatments of metals, plasma film depositions, deposition of functional silanes or other coupling agents to metal surfaces, and so forth

---

Current affiliation: University of Cincinnati, Cincinnati, Ohio.

Analysis of interfaces between materials, especially for failure analysis after aging and
   testing of adhesively bonded systems
Study of the cure of adhesives

The outline of this chapter is as follows. The spectroscopic techniques that can be
used for surface or interface characterization of adhesively bonded materials are listed in
Table 1. The most popular techniques are then discussed briefly in terms of the type of
information they provide and where they can be applied. Their limitations are also
described briefly. Since just a handful of techniques are used on a regular basis, notably
XPS,* AES, SIMS, FTIR, and Raman spectroscopy, only these techniques will be
discussed in detail. Recent and ongoing instrumental developments are described and
specific applications of each of these techniques are presented and discussed. Finally, a
bibliography containing many references to textbooks and important articles is given.

II. OVERVIEW OF SPECTROSCOPIC TECHNIQUES

Table 1 lists a wide range of spectroscopic techniques with details on the type of
information that these techniques can provide, their sampling depth, their sensitivity, and
their major limitations. Some key references are provided for each technique. The
techniques are classified in five major categories: ion spectroscopies, electron spectroscopies, x-ray spectroscopies, vibrational spectroscopies, and miscellaneous techniques. This distinction is, of course, arbitrary and is based on the type of signal that is
recorded in each technique. It would be equally justifiable to classify the techniques on the
basis of the primary beam or excitation source or even on sampling depth [3]. However, in
each group a distinction has been made between those techniques that are well known and
currently widely used and techniques that are either variations of the major techniques or
still in a developmental stage. The techniques that are closely related or that are variations
of the same main technique are grouped together.

Another way of comparing techniques is to group them according to the combination
of excitation (i.e., signal in) and response (i.e., signal out). This is frequently done in the
literature [4]. This has been done here for a number of the techniques listed in Table 1.
The results are shown in Table 2, which illustrates that a technique has been developed or
proposed for almost all combinations of ions, electrons, and photons.

Table 1 indicates that the sampling depth for the various techniques varies from 1
monolayer to several millimeters. In general, the ion-based techniques, for instance SIMS
and ISS, have the lowest sampling depths because the mean free path of ions in solids is
not more than one or two monolayers. The penetration and escape depths for photons are
much higher, and therefore, the techniques that are based on the detection of electromagnetic radiation, such as FTIR and XRF, give information on microns in metals and even
millimeters in organics. This does not mean, however, that these techniques cannot detect
monolayers. In suitable samples, both FTIR techniques and XRF can detect monolayers
with high sensitivity, but it is difficult to restrict the signal acquisition to the monolayer
only because of the larger sampling depths.

Methods based on electron detection have intermediate sampling depths. The sam-
pled thickness in techniques such as AES and XPS is of the order of 50 Å. Since the
escape depth of an electron is dependent on its energy, the sampling depth in XPS and

---

*See Table 1 for a definition of all acronyms used in the text.

Table 1 Spectroscopic Techniques for Use in Adhesive Bonding Studies

Technique	Acronym	Sampling Depth	Information	Sensitivity	Principle	Limitation	Ref.
A. Ion spectroscopies							
Secondary ion mass spectrometry	SIMS	5 Å	Mass spectrum; mapping; SNMS also depth profile	All elements	Ion beam excitation	UHV; qualitative; matrix effects	7–9
Static SIMS	SSIMS			«1 monolayer			6
Time-of-flight SIMS	TOFSIMS						28
Surface analysis by laser ionization	SALI						12
Fast atom bombardment SIMS	FABSIMS						6
Secondary neutrals mass spectroscopy	SNMS						8
Laser microprobe mass analysis/ spectrometry	LAMMA, LAMMS	1 μm	Mass spectrum of small area	All elements	Laser excitation	Reproducibility; damage	105
Laser ionization mass analysis	LIMA						
Laser ionization mass spectrometry	LIMS						
Ion scattering spectroscopy	ISS	2 Å	Elements; heavier than primary ion	<1 monolayer	Ion beam scattering	Spectral resolution	106
Low-energy ion scattering spectroscopy	LEIS(S)						107
Medium-energy ion scattering spectroscopy	MEIS(S)						
High-energy ion scattering spectroscopy	HEIS(S)						
Rutherford backscattering spectroscopy	RBS	> 1 μm	Depth profile; nondestr.	High	Scattering; MeV ions	Equipment cost few elem.	101–104
Neutral scattering spectrometry	NSS						108
Nuclear reaction analysis	NRA	1 μm					111
Electron-stimulated desorption spectroscopy	ESD	5 Å	Mass spectrum; small spot	<1 monolayer	e⁻ beam excitation	Poor sensitivity; e⁻ damage	109
Electron-induced ion desorption	EIID						
Electron-stimulated desorption ion angular distribution	ESDIAD						110

Table 1 *(Continued)*

Technique	Acronym	Sampling Depth	Information	Sensitivity	Principle	Limitation	Ref.
Single photon ionization	SPI	as SIMS	Mostly as SIMS; more quantitative	Lower than SIMS	Postionization of sputtered neutrals	Lower sensitivity than SIMS	12
Surface analysis by resonance ionization of sputtered atoms	SARISA						
Resonantly-enhanced multiphoton ionization	REMPI						
Non-resonant multiphoton ionization spectroscopy	NRMPI						
Hydrogen forward scattering spectroscopy	HFS	1–5 μm	H detection		H^+, $4He^+$ beams	H only	103
Forward recoil scattering spectroscopy	FRS						133
B. Electron spectroscopies							
Auger electron spectroscopy	AES	10–50 Å; to 1 μm in depth profiling; mapping	Elements; >He; depth profiles	0.1 at ~%	e^- beam excitation; e^- detection	UHV; conductors only; limited chemical info	14, 18
Scanning auger microscopy	SAM						
X-ray induced auger electron spectroscopy	XAES						
Ion-induced auger electron spectroscopy	IIAES						8
Ion neutralization spectroscopy	INS						112
Appearance potential spectroscopy	APS						113, 131
X-ray photoelectron spectroscopy	XPS	As in AES; UPS lower	Elements; >H; binding states	0.1 at ~%	$h\nu$ excitation; e^- detection	UHV; mapping limited; low sensitivity	14, 114
Electron spectrocopy for chemical analysis	ESCA						
UV photoelectron spectroscopy	UPS						
Angular-resolved UV photoelectron spectroscopy	ARUPS, ARPES						115

Technique	Abbrev.	Size	Detects	Sensitivity	Excitation/detection	Notes	Ref.
Inverse photoemission spectroscopy	IPES						17
Bremsstrahlung isochromat spectroscopy	BIS						

C. X-ray spectroscopies

Technique	Abbrev.	Size	Detects	Sensitivity	Excitation/detection	Notes	Ref.
Electron energy loss spectroscopy	EELS	50 Å	All elements	0.1 at – %	e⁻ excitation in SEM/TEM	Low sensitivity	17
Scanning low energy electron energy loss microscopy	SLEELM						116
Ionization low spectroscopy	ILS						118, 132
Energy-dispersive x-ray analysis	EDXA	1 μm	elements; >B	0.01 at – %	e⁻ beam; hν detection	No chemical info; not surface sensitive	119
Wavelength dispersive x-ray analysis	WDXA						
Electron probe microanalysis	EPMA	PIXE less					120
Particle-induced x-ray emission	PIXE						
Extended x-ray fine structure spectroscopy	EXAFS	varies	Chemical states	Low	Oscill. in X-ray spectra	Slow; synchroton	121, 122
Surface extended x-ray fine structure spectroscopy	SEXAFS						
Near-edge x-ray absorption fine structure	NEXAFS						
X-ray absorption near edge structure	XANES						
X-ray fluorescence	XRF	10 μm	Elements	0.001 at – %	hν excitation and detection	No chemical info; no mapping	—
X-ray diffraction	XRD	10 μm	Crystal structure	Low	X-rays	Crystalline only	—

D. Vibrational spectroscopies

Technique	Abbrev.	Size	Detects	Sensitivity	Excitation/detection	Notes	Ref.
Fourier transform infrared spectroscopy	FITR	10 μm	Molecules, functional groups	Low	Excitation of bonds by hν	Mainly qualitative	38, 40, 124
Diffuse-reflectance infrared Fourier transform spectr.	DRIFT						

Table 1 Continued

Technique	Acronym	Sampling Depth	Information	Sensitivity	Principle	Limitation	Ref.
Attenuated total reflection spectroscopy	ATR						
Reflection-absorption infrared spectroscopy	RAIR						
Multiple reflection absorption infrared spectroscopy	MRAIR						
Grazing incidence reflection (spectroscopy)	GIR(S)						
Multiple reflection infrared spectroscopy	MRS						
Multiple internal reflection (spectroscopy)	MIR						
External reflection spectroscopy	ERS						
Surface reflectance infrared spectroscopy	SRIRS						
Photoacoustic spectroscopy	PAS						
Emission spectroscopy	EMS						
Photothermal beam deflection spectroscopy	PBDS						
Internal reflection spectroscopy	IRS						
Raman spectroscopy	RS	10 μm; 50 Å in SERS	Bonds and molecules	Low	Scattered photons	Low sensitivity; qualitative	35, 62, 70
Laser raman spectroscopy	LRS						
Fourier transform raman spectroscopy	FTRS						
Hadamer transform raman spectroscopy	HTRS						
Surface-enhanced raman spectroscopy	SERS						
Resonance raman spectroscopy	RRS						

Technique	Abbrev.	Depth/Sensitivity	Information		Principle	Limitation	Ref.
High-resolution electron energy loss spectroscopy	HREELS	50 Å	Molecular vibrations		e^- excitation	Low resolution	37
Inelastic electron tunneling spectroscopy	IETS	1 monolayer	Molecular vibrations	Low	Excitation by voltage	Sample preparation	125
Ellipsometry	—	—	Film thickness	—	Polaried light	Sample transparent	88, 91
Bombardment-induced light emission	BLE						126, 127
Glow-discharge optical spectroscopy	GDOS	> 10 μm	Depth profile of elements	High	Sputtering by Ar ions	Quantitative	99, 100
E. Other techniques							
Low-energy electron diffraction	LEED	50 Å	Crystalline surface structure			Limited applicability	128, 129, 130
Elastic low-energy electron diffraction	ELEED						
Inelastic low-energy electron diffraction	ILEED						
Reflection high-energy electron diffraction	RHEED						
Mössbauer spectroscopy	MS	High	Chemical environment of atom (e.g. Fe)	Low	Absorption of γ-rays by nucleus	Limited no. of elements	134
Nuclear magnetic resonance	NMR	Bulk samples	Chemical state and free spins	High	Resonance in magnetic fields	No surface info	—
Electron spin resonance	ESR						
Surface composition analysis by neutral and ion impact radiation	SCANIIR						8
Electrochemical impedance spectroscopy	EIS		Impedance of coated metal			Modeling	95, 96

185

Table 2   Principles of Some Spectroscopic Techniques

Primary beam	Signal				
	Ions	Electrons	Photons	Neutrals	Vibrations
Ions	SIMS ISS RBS GDMS HFS	IAES INS	PIXE GDOS SCANIIR	SNMS SARISA	BLE GDOS
Electrons	ESD	AES ILS EELS LEED SLEELM	EDXA WDXA XES EPMA		HREELS
Neutrals	FAB			NSS	BLE
Photons	LAMMA SALI	XPS XAES UPS IPES	(S)EXAFS XANES XRF Ellipsometry		FTIR SERS LRS
Voltage	FIM	STM FEM			IETS

AES is not the same for all elements detected in the sample. Further, by varying the angle between the sample surface and analyzer, the sampling depth can be varied, resulting in a nondestructive quantitative concentration depth profile in the range 5–50 Å. This feature is especially useful in XPS.

The type of information provided by the techniques listed in the tables also varies greatly. Many spectroscopic techniques give qualitative and/or quantitative elemental composition. The vibrational techniques, however, generally provide information on the molecular structure. SIMS, especially in the static mode (SSIMS or TOFSIMS), can yield information on molecular structures and even orientation of monolayers [5–10]. This is particularly useful for the study of the adsorption of coupling agents on metals or to determine the effects of plasma treatments on polymer surfaces [11]. TOFSIMS instruments also have capabilities for determining the two-dimensional distributions of elements or molecular species at the surface, similar to the capabilities (for elements only) offered by AES and EDXA or WDXA.

The major technique for determining a depth profile of elemental compositions is AES. A newer technique for this purpose is SNMS, which has a better interface resolution (due to a lower sampling depth) than AES and a better sensitivity for many elements than AES [10,12,13]. In SNMS the neutrals emitted in the SIMS process are ionized and then mass-analyzed. The emission of neutrals is much less matrix-dependent than the emission of positive or negative ions detected in regular SIMS. Depth profiling can also be done in regular SIMS (the so-called dynamic SIMS version), but this technique then requires extensive calibration of sputtering rates and elemental sensitivities. Depth profiling in both AES and SIMS techniques is done by sputtering, usually by means of a beam of Ar^+ ions.

The limitations of some of the more popular techniques are also given in Table 1. For many techniques, especially the more surface-sensitive ones (ion- and electron-based methods), an UHV environment is required. This requirement, of course, increases the cost of the equipment, but it also reduces the flexibility and applicability of the technique. Other limitations of certain techniques as indicated in the table are the difficult or complex sample preparation procedures, low sensitivity (long acquisition times), and poor resolution or element selectivity (e.g., ISS). Another limitation of some of the more sophisticated techniques is that they are not commercially available. To carry out certain techniques, it may be necessary to modify commercial instruments.

III. PRINCIPLES OF SELECTED SPECTROSCOPIC TECHNIQUES AND APPLICATIONS

Despite the enormous number of spectroscopic techniques that have been described and developed, only a limited number are commercially available and are actually used in the study and development of adhesive bonding materials. These techniques will be described in more detail in this section.

The techniques highlighted here are XPS, AES, SIMS, various forms of FTIR, Raman spectroscopies, and HREELS. This selection is based on their relative ease of application and interpretation, their commercial availability, and the unique capabilities that each technique possesses for the study of an aspect of adhesive bonding. These capabilities are also highly complementary. The applications discussed are chosen to illustrate the applications in three major areas described earlier: surface characterization, modification of metal or polymer surfaces, and analysis of interfaces.

A. Electron and Ion Spectroscopies

1. Principles

The three important techniques in this category that are discussed are XPS (also frequently called ESCA), AES, and SIMS. The basic principles of these techniques are discussed only superficially. Recent literature on these three techniques with examples of applications to materials science problems is abundant [3–14]. The surface analysis technique ion scattering spectroscopy (ISS), frequently discussed along with XPS, AES, and SIMS, is not considered in this chapter. Excellent recent reviews of this technique are available [15,16].

(a) XPS and AES. The principles of XPS are summarized in Fig. 1. Essentially a solid surface is ionized by low-energy x-rays, e.g., AlKα of 1486.6 eV. The photoelectrons emitted by the surface are collected and their energy distribution analyzed. The characteristic peaks observed in the photoelectron spectrum represent the various electron orbitals with binding energies lower than the exciting x-ray energy and are therefore specific for the elements present in the surface layers. The electrons in the energy region of interest have a mean free path in solids in the range 5–20 Å. The sampling depth is roughly three times this length, i.e., 15–60 Å. The technique can detect all elements with the exception of hydrogen, and an important feature of XPS is that the photoelectron peaks can shift somewhat. These shifts are dependent on the chemical state of the element. The technique, therefore, can give a quantitative analysis of all elements > H in the outermost 60 Å of the sample and provides some information on the functional groups

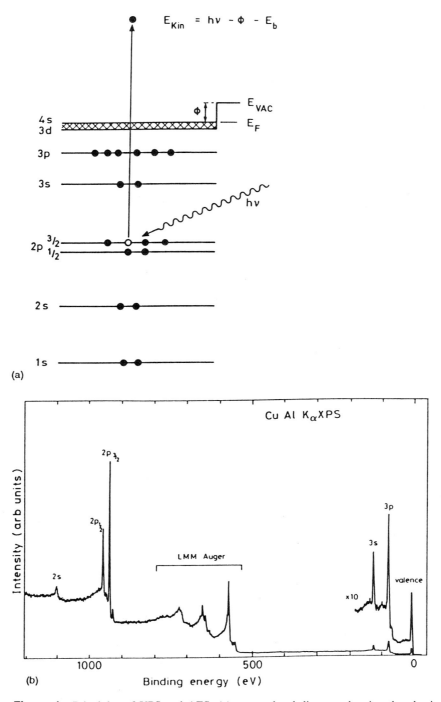

Figure 1 Principles of XPS and AES; (a) energy level diagram showing the physical basis of XPS; (b) XPS spectrum of a clean copper surface; (c) energy level diagram showing the physical basis of AES; (d) AES spectrum of a clean copper surface in the direct (top) and differentiated form (bottom). (From Ref. 17.)

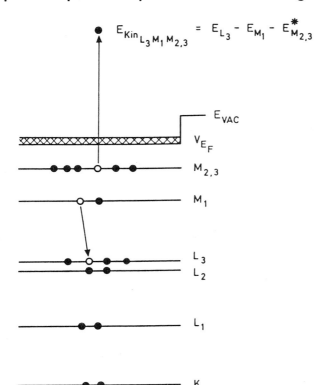

$$E_{Kin_{L_3 M_1 M_{2,3}}} = E_{L_3} - E_{M_1} - E^*_{M_{2,3}}$$

(c)

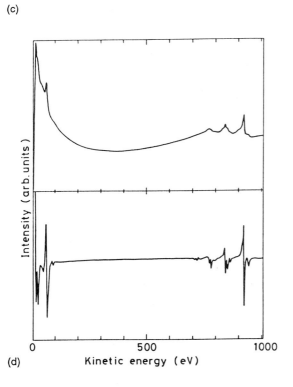

(d)

or oxidation states of these elements. Some molecular structure information can be derived from a detailed analysis of the valence band of the material, which, however, has very low intensity. A more detailed analysis of the valence band structure is possible in UPS.

XPS has been in use since the early 1970s. Currently there are five manufacturers of commercial instruments at a cost between \$300,000 and \$1 million. The strengths of XPS are its ease of operation, its quantification, and its applicability to a wide range of materials and sample forms (powders, wires, foils, chunks, etc.); in addition, both conductors and insulators can be analyzed. Sample preparation is minimal. Since the sampling depth is several dozens of monolayers, the technique is not so sensitive to surface contamination as the ion-based techniques SIMS and ISS. The major limitations of the technique are its rather poor spatial resolution (although recent improvements have been made, pushing the resolution on some commercial instruments to about 10 μm), the rather low detection limits (about 0.1 at.-% for many elements, requiring many hours of data acquisition), and the rather limited chemical information obtainable on organic materials, which is nowhere near that of, for instance, NMR and FTIR.

AES differs from XPS in that the surface is ionized by a finely focused electron beam of 5–30 kV (Fig. 1). The secondary electrons have no specific information content in AES, but the Auger electrons, which are emitted shortly after the secondary electrons, and which involve transitions between different orbitals, are recorded as a function of their kinetic energy. Such Auger electrons are also emitted after ionization by x-rays in XPS and consequently the same Auger transitions lines are also observed in XPS. In XPS, this aspect of the technique is sometimes referred to as XAES (Table 1). Because of a very high background, AES spectra are conveniently represented as differentiated spectra. The peak-to-peak heights (more accurately, peak areas in the nondifferentiated spectrum) are proportional to the number of atoms in the probed sample volume.

Since the electron energy ranges are approximately the same in XPS and AES, the surface sensitivities and sampling depths in these two techniques are very similar. In both techniques, quantification is performed, in a first approximation, by dividing the areas under the peaks (or the peak heights) by the appropriate sensitivity factor for the elements, followed by normalizing to 100%. Sensitivity factors are usually provided by the equipment manufacturer and they have been determined experimentally, although they can be calculated using ionization cross-sections (in XPS) and backscattering factors (AES). The use of these standard sets of sensitivity factors enables atomic concentrations to be determined with an accuracy of about 1–5% [14,17]. Since the electrons detected for the various elements in a sample differ in energy, the depths from which their signals originate, and hence for which their concentrations are calculated, are not the same for all elements detected. This applies to both XPS and AES.

The strength of AES lies in its small spot analysis capability. Modern instruments equipped with field emission electron guns have a spot size of about 100 Å. The lateral resolution for mapping elemental distributions is therefore less than 0.1 μm, i.e., considerably better than in EDX. The reason for this is that in AES the signal stems primarily from the surface layers; hence the broadening of the primary beam, as occurs in EDXA, does not affect the resolution in AES very much. Another capability of AES is to provide elemental concentration depth profiles using a simultaneous sputtering process by energetic ions (usually Ar^+). Depth profiling can be automated under computer control. In XPS depth profiling can also be done but only intermittently, and another problem is that electrons are collected from a much wider area than in AES and, therefore, edge effects due to nonhomogeneous sputtering rates across the width of the argon beams are more likely to occur.

There are several limitations in AES. The most important one is that spectra cannot be collected (or can be collected with great difficulty) from insulating materials. Charge neutralization procedures that can be applied routinely do not exist. As opposed to surface charging in SIMS (see below), the charge in AES is of a negative sign. This limits the application of AES to metals, semiconductors, or thin films (e.g., oxides). Other limitations are the electron beam damage that easily occurs with certain materials, especially organic films, and the rather limited chemical information that can be extracted from AES spectra [18]. In principle, chemical shifts occur as in XPS, but they are more complicated because several orbitals are involved in each Auger transition and in most commercial instruments the energy resolution is not good enough to resolve such shifts. By the same token, peak overlap occurs in certain cases, especially in the energy range where many transition elements have their major AES lines, such as the range Mn–Zn (600–1000 eV). As an example, Mn in steel is very difficult to resolve and materials containing Fe, Ni, and Co would require sophisticated peak subtraction software to analyze. Other limitations are that quantification is less reliable than in XPS as a result of electron backscattering phenomena, which can be estimated but not with high accuracy. Further, the ionization probability by the primary electron beam depends on the energy of these primary electrons. In the derivative mode, errors can be introduced because peak shapes change with chemical state of the elements. This problem can be resolved, however, by determining the peak areas before spectra differentiation and by using sensitivity factors derived for the nondifferentiated peaks. Similar to XPS, the detection limits in AES are not very low, i.e., for most elements of the order of 0.1 at.-% or worse.

Current developments in AES are mainly in the areas of improved electron guns with higher brightness and smaller spot sizes, and multichannel detectors with improved sensitivities. Improvements of energy resolution will enable chemical states to be studied in more detail, leading to better analysis of complex mixtures with partly overlapping peaks.

(b) SIMS. The major ion beam technique that is currently going through a period of rapid development is SIMS [4–9]. There are several variations of the technique (Table 1), but the principle common to all is that a solid surface is bombarded by energetic ions. Ions that are commonly used are Ar^+, O_2^+, O^-, Cs^+, Ga^+, and others. Their energy can be in the range 5–30 keV. The impact of these ions results in the emission of secondary ions, neutral atoms, molecular fragments, and electrons. A mass spectrometer collects the ions (positive or negative) in the form of a mass spectrum. The major variations of the SIMS technique are the following:

Static SIMS. In this form of SIMS, the total primary ion dose is so low (i.e., around 10^{12} ions/cm^2, or even less) that in the course of the experiment (1–5 min) all primary ions impinge on a fresh surface; the result is that the mass spectrum does not change with time and represents a fragmentation pattern that can be taken as a fingerprint of the material [19,20]. Both small and large ions (up to the molecular ion, or oligomers, if present) are emitted [21,23]. The sampling depth of this technique is not more than 1–2 monolayers ($\approx$5 Å). SSIMS is unique in that it detects all elements (including hydrogen) and at the same time provides molecular information on the outermost surface layers of the sample. For example, it can easily distinguish polyethylene from polypropylene (as shown in Fig. 2), detects inorganic contaminants in these polymers, and also indicates surface oxidation from the presence of O-containing ions. The latter aspect is useful for studying surface

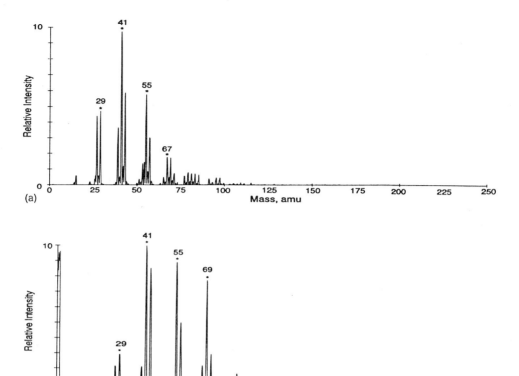

Figure 2 Positive quadrupole SIMS spectrum of surface of polyethylene (a) and polypropylene (b). (From Ref. 20.)

modifications of polymers by plasma or corona techniques [11,24]. For elemental detection, the technique is very sensitive, especially for alkali metals (ppm level). Although the technique is primarily used in a qualitative mode, quantitative correlations between peak ratios and elemental concentrations in XPS of the same sample have been demonstrated [25,26].

Dynamic SIMS.   In this version of SIMS, the total ion dose is much higher than in static SIMS, up to a factor of 10^4. Therefore, sputtering is now the dominant process and an elemental depth profile is obtained [27]. Organic structural information is no longer contained in the spectra, because organics decompose to the elements and CH_x fragments. Therefore, concentration depth profiling of organic materials, with retention of some of the molecular structure, is impossible. However, some information can be obtained if one of the constituents is specifically isotopically labeled, for instance by deuterium. As in AES, the profiling process can be automated. As a result of the lower sampling depth in SIMS, the depth resolution is better than in AES. On the other hand, the quantification of elemental concentrations is more complicated in SIMS because ion emission is strongly matrix-dependent.

Imaging SIMS. In this variation, the ion beam is rastered across a surface and a two-dimensional distribution of elements or organic materials is obtained. If this version is combined with the static SIMS mode, mapping of each peak observed in the spectrum can be performed, so even in mixtures of many organic compounds each component can be individually mapped as long as the component has at least one specific peak in the spectrum. For metals, oxides, and semiconductors, the mapping capabilities are similar to that of AES, although the sensitivity for many elements in SIMS is much higher. An example of mapping of an organic compound is shown in Fig. 3.

Secondary Neutrals Mass Spectrometry. This version of SIMS detects specifically the neutral molecules or atoms that are emitted in the SIMS process [12,13]. The postionization of these neutrals is performed by low-energy electrons or by lasers used in a resonant or nonresonant mode. Since in SIMS about 99% of the emitted species are neutral particles, the postionization process increases the sensitivity of certain elements. The species that had originally been emitted as ions are, of course, detected also, but they constitute only a small fraction of the total signal. The importance of SNMS is that matrix effects are largely eliminated. In general, the emission of a species (i.e., charged and uncharged) does not depend on the chemical state or the matrix, but on the sputtering

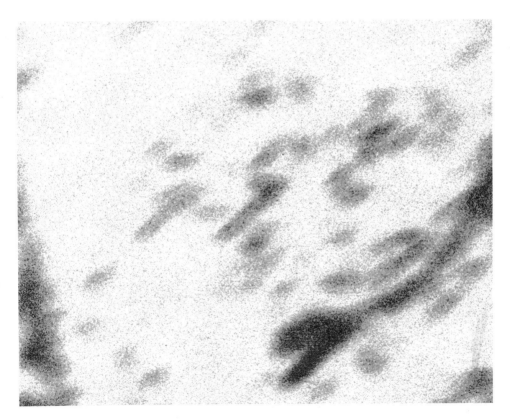

Figure 3 Time-of-Flight SIMS map of the distribution of the intensity of the ion -255 amu originating from stearic acid in lubricant residue at the surface of a cold-rolled steel sample; magnification approximately $500\times$; primary ions $^{69}Ga^+$ of 25 kV.

coefficient only. Peak intensities can therefore be more easily converted to concentrations using sensitivity factors only. SNMS can thus be expected to become the foremost technique for quantitative depth profiling in the near future, because it is fast, very sensitive, quantitative, and has superior depth resolution.

For static SIMS of organic materials, postionization (in this case using lasers in a nonresonant mode) can also be performed. Compared with the normal SIMS spectrum, the spectrum then contains several extra peaks if postionization is applied. These are due to the monomeric repeating unit of the polymer or to entire small molecules that are emitted. These extra ions are very useful for the identification and characterization of the materials.

Time-of-Flight SIMS.   The introduction of TOF analyzers, along with postionization, is one of the major developments in the SIMS techniques of the last few years. TOFSIMS is essentially static SIMS in which the quadrupole mass spectrometer has been replaced with a time-of-flight spectrometer, which gives the technique unique capabilities [28,29]. Ions are extracted at high voltage (e.g., 3 kV) and then enter into a field-free flight tube 1–2 meters in length where they are separated according to their flight time, which depends on their mass (Fig. 4). Advantages of this type of SIMS are that both the transmission and the mass resolution of the mass analyzer are considerably higher than those of the quadrupole analyzer. As a consequence, TOFSIMS has higher sensitivity (so that spectra can be recorded with lower total ion dose) and a mass resolution that enables peaks with nominally the same mass to be separated. The actual resolution obtained depends on the quality of the ion gun, which has to deliver pulses at the nanosecond level rather than being run continuously, as in quadrupole SIMS.

Because of the high mass accuracy, peak identities can now also be identified uniquely. An example is given in Fig. 5. This capability has removed most of the guesswork from SIMS analysis and has opened up many applications in materials science that are impossible in quadrupole SIMS because of peak overlap. For instance, studies of most practical metals, e.g., Al, are very difficult in quadrupole SIMS because the metal ions of interest almost always overlap with organic ions at nominally the same mass. Another important advantage of TOFSIMS is the much higher mass range than in quadrupole SIMS, which enables very large ions to be detected. An example is illustrated in Fig. 6, which shows the distribution of oligomers present in the surface of a polymer sample [23]. However, such results cannot be obtained with bulk materials. The polymer has to be present as an extremely thin film on an active metal (usually Ag). In the SIMS process the oligomers, which are emitted as neutral molecules, then become cationized by Ag^+ ions.

The spectra in TOFSIMS are similar to those in quadrupole SIMS, but not identical. The time-of-flight analyzer can detect ions with a much greater kinetic energy spread than the quadrupole. Therefore, inorganic ions or low-mass organic fragments, such as C^+ and CH^+, which are emitted with high kinetic energies, appear in higher intensities in TOFSIMS spectra than in quadrupole spectra.

TOFSIMS is rapidly gaining popularity as a tool for studying chemistry and orientation at organic surfaces, such as polymers and polymer blends. The problem of surface charging (the surface charges up positively as a result of electron emission) has been satisfactorily solved by the development of pulsed electron sources, which neutralize the charge. Imaging (as shown in Fig. 3) can be performed routinely with high sensitivity and submicron resolution. A limitation of the technique is, however, that standard spectra of many materials are not yet available. Although much can be derived from the chemistry of

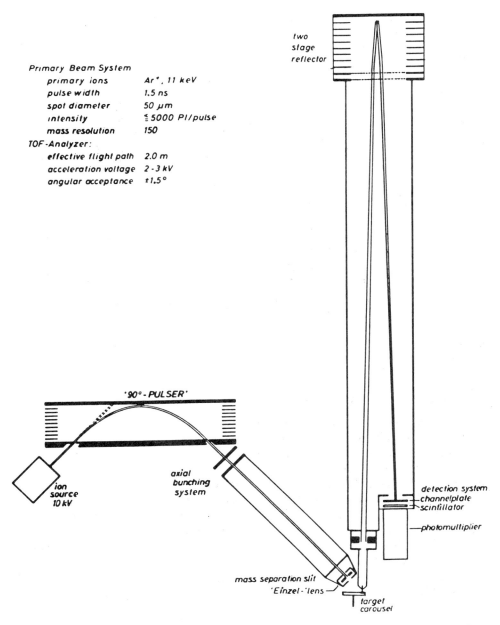

Figure 4 Principle of Time-of-Flight SIMS showing mass-separated primary beam of 10 kV Ar^+ ions and 2 meter flight tube with two-stage ion reflector system. (From Ref. 7.)

the material (if known!) and the exact mass of the ions, in many cases it is not possible to identify exactly the composition or structure of the material. Much more work needs to be done in this area before the technique can be applied routinely by unskilled analysts. Fortunately, several databases (for quadrupole SIMS) have been published [19,20] and others (for TOFSIMS) are being prepared.

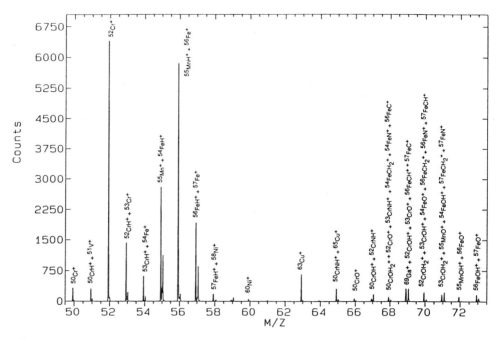

Figure 5 Part of positive, high mass-resolution Time-of-Flight SIMS spectrum of a polished 304L stainless steel surface; primary ions $^{69}Ga^+$ of 25 kV.

2. Some Selected Applications

In this section some selected examples of XPS, AES, and SIMS are discussed, which may illustrate the capabilities of these techniques for adhesives-related applications.

In the example shown in Fig. 7, a thin film of plasma-polymerized trimethylsilane had been deposited on cold-rolled steel as a pretreatment for improved adhesion and corrosion [30a]. The film thickness was determined by ellipsometry to be 500 Å. The composition was characterized by AES, XPS, and TOFSIMS. AES gave information on the bulk composition, surface enrichment, and interfacial oxide (Fig. 7a). Note that the C/Si ratio of the bulk of the film, after equilibrium sputtering conditions have been reached, is approximately 3, i.e., identical to that of the monomer from which the film was made. XPS was performed on the same samples at two different takeoff angles. Lowering the takeoff angle increases the surface sensitivity (sampling depth) of the technique. Shown in Fig. 7b are the Si2p lines from which the conclusion can be drawn that the surface of the film is enriched in Si-O bonds, whereas the bulk has a higher concentration of Si-C bonds. In Fig. 7c, parts of the TOFSIMS spectra are shown of the same film before and after solvent cleaning. This rinse was performed to check on the presence of low-molecular weight materials at the film surface, which are known to form in plasma polymerization. This rinsing treatment had practically no effect on the XPS and AES results, but the TOFSIMS spectra before and after rinsing are quite different. Before rinsing, the spectrum resembles that of polydimethylsiloxane [19,20]; after cleaning the surface is similar to that of SiO_2. The spectrum indicates a high concentration of silanol groups, as can be concluded from the high intensity of the peak at $+45$ amu ($SiOH^+$). The peak identification and the fit between observed and calculated masses are shown in

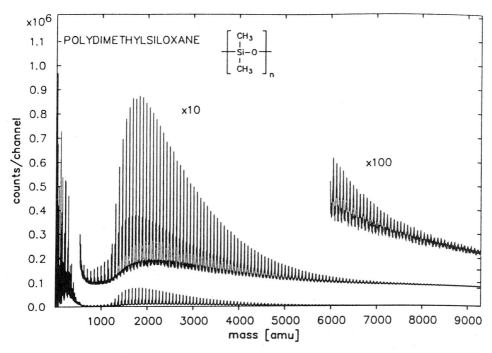

Figure 6 Positive Time-of-Flight SIMS spectrum obtained from a thin film of polydimethylsilox-ane on a silver substrate showing the polymer fragmentation in the 0–500 amu range and the oligomer distribution in the higher mass ranges. (From Ref. 7.)

Table 3 for a silane film on CRS [30b]. Only those compositions were accepted that had a deviation of less than 0.01 amu from the calculated mass. It is clear from this example that application of the three techniques discussed here yields complementary information, enabling a more detailed description of the film structure than any of the three techniques alone.

An example of the characterization of a thin film of a coupling agent adsorbed on a steel surface is given in Fig. 8 [24]. The coupling agent was vinylbenzylaminoethyl aminopropyltrimethoxy silane (SAAPS). Following hydrolysis in a mixture of water and alcohol, the silane was applied by dipping the metal in a very dilute solution. The figure shows a very high intensity of the peak at $+117$ amu after short immersion times, and a high Si^+ and lower $+117$ amu intensity after longer immersion times. The peak at $+117$ amu was uniquely identified as $CH_2 = CH - C_6H_4 - CH_2 +$, i.e., the end group of the coupling agent. This example thus demonstrates that the silane is highly oriented if applied as a monolayer, but this preferred orientation is absent after deposition of several monolayers. Knowledge of this orientation is important for the optimization of surface pretreatments by means of coupling agents.

An example of the use of deuterated materials in TOFSIMS studies is given in Fig. 9. Blends of polystyrene and fully deuterated polystyrene were prepared in experiments in which segregation effects were investigated [31]. Variables were the ratios between the two polymers and the molecular weights. The most characteristic peak in polystyrene is $+91$ amu ($C_7H_7^+$, tropyllium) and the corresponding ion for d^8-polystyrene is at $+98$

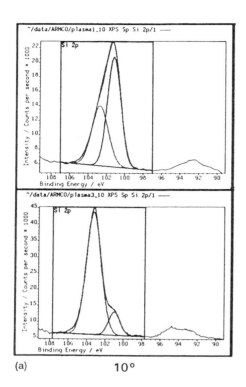

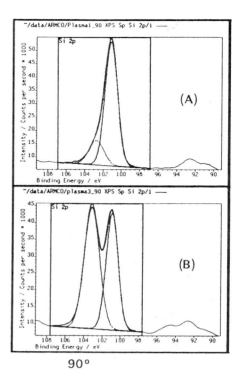

(a)   10°   90°

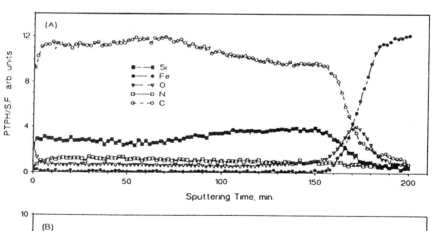

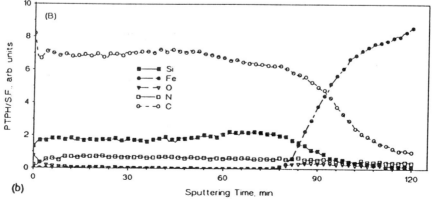

(b)

198

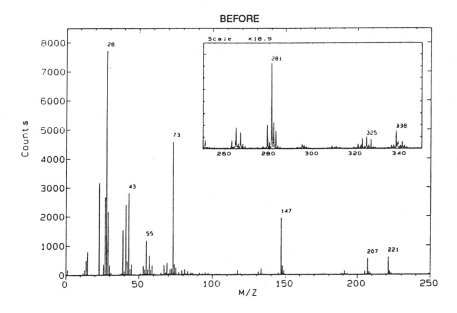

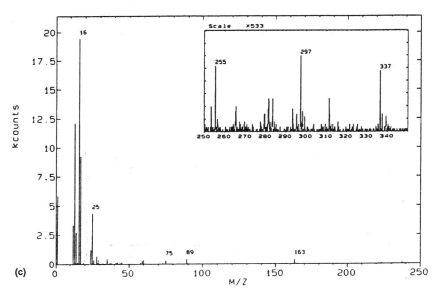

Figure 7 (a) XPS spectra at take-off angles 10° and 90° (between sample surface and normal) of the surface of plasma-polymerized trimethyl silane films on steel substrates; film (A) was deposited in non-reducing conditions, film (B) was prepared in reducing conditions. The spectra demonstrate the presence of highly oxidized Si in the extreme surface layers; (b) depth profiles by AES of the films of Fig. 7a showing regions with different elemental composition in the films and the presence of an iron oxide in film (A); the sputtering rate was 5 Å/min.; (c) positive (top) and negative (bottom) Time-of-Flight SIMS spectra of film (A) of Fig. 7a before and after (see pg. 200) rinsing in methanol. The changes in the spectra indicate that low-molecular weight soluble components were present in the as-deposited film. (From Ref. 30a.)

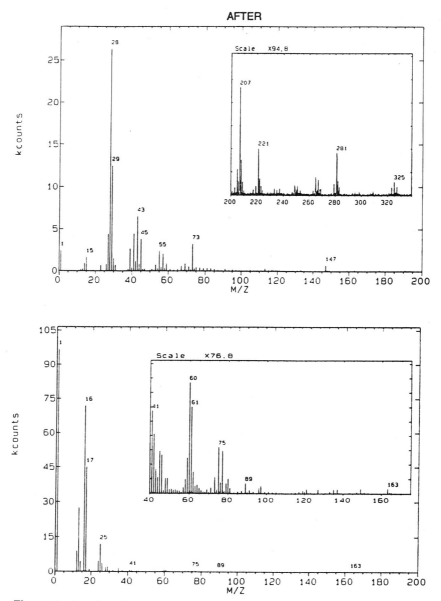

Figure 7 (continued)

amu. The spectrum in Fig. 9 illustrates that in this example of 1:1 PS/d^8-PS ratio of equal molecular weights, the deuterated material is enriched at the surface. This type of application, i.e., monolayer surface sensitivity with organic structural information capability and separation of all isotopes, is unique to SIMS. There appears to be no other technique, except perhaps SERS (to be discussed below), that could identify this phenomenon.

The final example is a combined application of TOFSIMS and XPS, which was used to characterize the interface between a metal and a polymer system [32]. The polymer system was a cathodic electroprimer that is widely applied in automotive applications over

Table 3 TOFIMS Peak Identification of Silane Film on Steel

Ion	Composition	pH 10.5		pH 8.0	
		Δm[b]	Counts/yield[c]	Δm[b]	Counts/yield[c]
−1	H^-	0	50.57	0	28.92
−16	O^-	2	16.70	3	12.60
−17	OH^-	2	10.55	1	5.56
+17	NH_3^+	1	0.41	1	1.42
+18	NH_4^+	2	0.80	2	2.65
+28	Si^+	0	23.30	0	19.29
+31	CH_5N^+	1	0.80	3	3.42
−35	Cl^-	—[a]	—	0	28.60
+39	[d]$C_2HN^+/C_3H_3^+$	1	1.85	2	4.18
−41	CHN_2^-	7	0.13	5	0.07
+41	[d]$C_3H_5/C_2H_3N^+$	3	3.30	3	5.90
+42	[d]$C_2H_4N^+/C_3H_6^+$	3	2.06	2	2.48
+43	[d]$C_3H_7^+/C_2H_5N^+$	4	2.73	1	2.60
+44	SiO^+	10	3.01	10	1.35
+45	$Si(OH)^+$	2	2.15	8	2.86
+55	$C_4H_7^+$	5	1.14	2	1.03
+56	$Fe+$	0	4.76	3	0.44
−60	SiO_2^-	2	0.26	1	0.27
−61	$HSiO_2^-$	11	0.15	10	0.13
+62	$Si(OH)_2^+$				
−77	$HSiO_3^-$	6	0.07	5	0.065
+79	SiO_3^+	6	0.15	3	0.36
−89	$C_2H_7NSiO^-$	1	0.01	1	0.01
+100	$C_2H_2NSiO_2^+$	3	0.60	3	0.10
+102	$C_2H_4NSiO_2^+$	1	0.13	2	0.15
+105	$C_3H_{11}NSiO^+$				
−118	$C_3H_{10}N_2SiO^-$	4	0.03	4	0.008
+121	$C_3H_{11}NSiO_2^+$				
−121	$HSi_2O_3NH_2^-$	0	0.013	2	0.009
−127	$C_2H_3SiOFe^-$	—	—	10	0.028
−137	$HSi_2O_4NH_2^-$	5	0.008	5	0.005
+147	$(CH_3)_5Si_2O^+$	—	—	5	0.05
+149	$C_3H_{11}NFeO_2^+$	—	—	4	0.17
+163	$C_4H_{13}NSi_2O_2^+$	—	—	5	0.06
+207	$C_5H_{15}Si_3O_3^+$				
+221	$C_7H_{21}Si_3O_2^+$				
+281	$C_7H_{21}Si_4O_4^+$				

[a]Ions listed without Δm and counts/yield values are for methanol-cleaned sample only.
[b]Difference between listed composition and measured mass in milli-amu.
[c]Ratio of counts in peak area and total ion yield.
[d]At pH 10.5/pH 8.0.
Source: Ref. 30b.

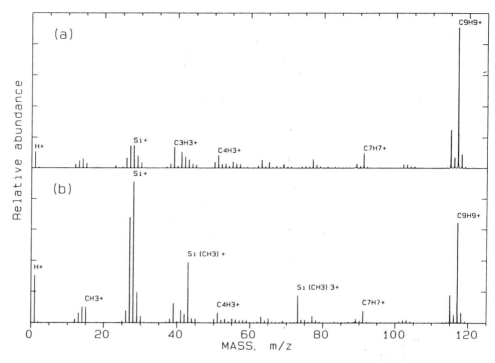

Figure 8 High mass-resolution positive Time-of-Flight SIMS spectra of films deposited on a zinc substrate of the hydrolyzed silane SAAPS (styryl aminoethyl aminopropyl trimethoxy silane); (a) immersion time 30 s; (b) immersion time 390 s. The spectra demonstrate a highly oriented film in (a) and a loss or orientation in film (b). The ion $C_9H_9^+$ is the styryl end group of the silane molecule. (From Ref. 24.)

the zinc phosphate conversion coating. Knowledge of the chemistry at the interface between the phosphate and the primer is important for the understanding and optimization of the adhesion and corrosion performance of the entire paint system. One aspect is, for instance, the degree of the cure of the primer, which may vary among different parts of the automobile. Therefore, an example is also given of TOFSIMS analysis of the primer surface/interface after undercure and overcure conditions.

In Fig. 10, XPS maps are shown of the C1s photoelectron line and ZnLLM Auger line recorded at the metal surface of a paint-galvanized steel system following exposure in a corrosion test. The panels showed several small circular spots where corrosion had occurred. The paint had been removed after the test [32]. The distributions of the two elements in the small corrosion spots are complementary, indicating that in these local areas organic debris was covering the metal surface. SIMS analysis detected in these areas high concentrations of Na^+ ions. Areas with high sodium concentrations are normally the cathodes of the corrosion cells, the counterions being the cathodically generated OH^- ions. This experiment thus demonstrated that the local areas shown in Fig. 10 were formed at cathodic sites, probably as a result of decomposition of the polymer by alkaline attack. This is a good example of a system where the XPS mapping capability is useful. AES, despite its higher spatial resolution, is not very useful for organic surfaces and

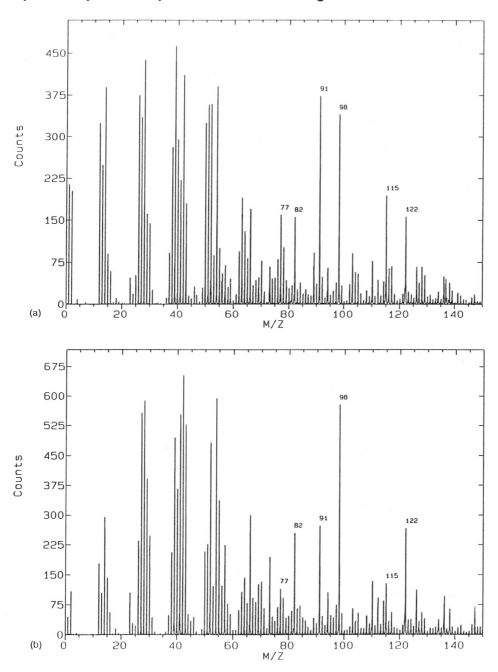

Figure 9 Positive Time-of-Flight SIMS spectra of 50/50 (by weight) blends of high molecular weight polystyrene and fully deuterated polystyrene (M_w = 1,950,000 for both polymers) before (a) and after (b) annealing for 24 hours at 200°C in an inert gas. The labeled ions 82, 98 and 122 amu are the deuterated analogs formed by the deuterated polymer of the ions 77, 91 and 115 amu formed by polystyrene. The spectra demonstrate the increase of the surface concentration of the deuterated polystyrene upon annealing. Courtesy F. J. Boerio and P. P. Hong, University of Cincinnati, to be published.

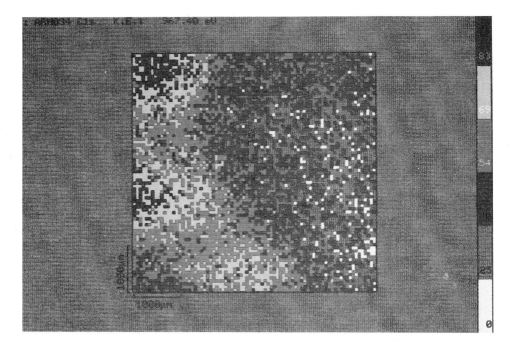

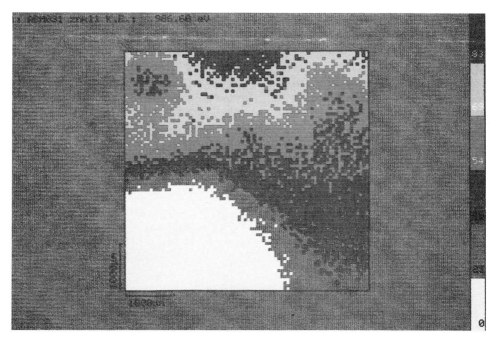

Figure 10 Element maps of carbon (C1s line, top) and zinc (ZnLMM line, bottom) recorded by imaging XPS at a corroded area on a galvanized steel surface. The area shown is 5×5 mm^2. (From Ref. 32.)

SIMS, with its much higher surface sensitivity, would also detect mainly organic material in the Zn-rich regions.

In Fig. 11, TOFSIMS spectra are presented of the two sides of the interface between a cathodic E-coat and a phosphated cold-rolled steel substrate. A variable in this experiment was the cure temperature of the paint [24]. The paint system is epoxy-based and can, after the cure, be described as an epoxy-urethane as the crosslinker is a blocked diisocyanate. The spectra indicate that in both cases the separation is very close to the metal (i.e., zinc phosphate) surface, but the difference in the spectra indicates differences in the cure conditions of the paint. Several high mass peaks, indicated on the spectra, decrease or increase with cure temperature. Other peaks that are marked demonstrate the presence of the crosslinker at the interface. Other conclusions drawn from these paint studies using TOFSIMS and imaging XPS were that the degree of cure is generally higher at the metal/coating and that the surface of the paint is always in a lower state of cure as a result of interfering oxidation reactions [32].

In similar studies in which the metal/coating interface was investigated as a function of immersion time in a dilute salt solution, it was found by TOFSIMS that paint degradation always occurred in regions with high Na^+ concentration [33]. The counterions Cl^- were not detected at the interface. These results are very important in that they prove that cation diffusion through an organic coating is a critical step in the complex series of events leading to corrosion beneath organic coatings and that alkaline paint attack plays a role in the propagation mechanism. The high sensitivity of SIMS, along with the mapping and peak identification capabilities, makes this technique very powerful for the study of metal-organic interfaces.

B. Vibrational Spectroscopies

The major spectroscopic techniques for use in adhesive bonding technology that are based on vibrational principles are several forms of infrared spectroscopy, Raman spectroscopy, and the more recent technique HREELS, the vibrational version of EELS used in electron microscopes. These techniques will be discussed in this chapter and some recent developments and applications of the techniques in adhesion studies will be given. Raman and IR spectroscopies are optical techniques and therefore they generally do not require a UHV requirement, which is a great advantage. Molecules adsorbed at surfaces or at interfaces can be observed *in situ*, without the need to destroy the sample. HREELS can also be used to study interfaces *in situ*, but does require an UHV environment, because excitation is performed by means of a low-energy electron source.

1. Principles and Applications

Several recent overviews of principles and applications of Raman, FTIR, and RHEEL spectroscopies are available in the literature [34–37]. The use of all major surface and interface vibrational spectroscopies in adhesion studies has recently been reviewed [38]. Infrared spectroscopy is undoubtedly the most widely applied spectroscopic technique of all methods described in this chapter because so many different forms of the technique have been developed, each with its own specific applicability. Common to all vibrational techniques is the capability to detect *functional groups*, in contrast to the techniques discussed in Sec. III.A., which detect primarily elements. The techniques discussed here all are based in principle on the same mechanism, viz., when infrared radiation (or low-energy electrons as in HREELS) interact with a sample, groups of atoms, not single elements, absorb energy at characteristic vibrations (frequencies). These absorptions are

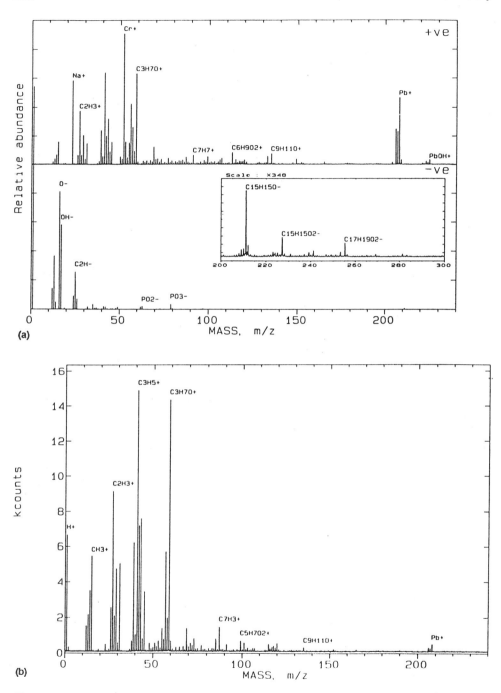

Figure 11 (a) positive and negative Time-of-Flight SIMS spectra of a $100 \times 100 \ \mu m^2$ area of the metal side of the interface between an epoxy cataphoretic paint system and phosphated cold-rolled steel. The paint was slightly overcured; (b) as in (a) but positive spectrum of paint side; (c) as in (a) but paint was undercured; metal side; (d) as in (c) but positive spectrum of paint side. (From Ref. 24.)

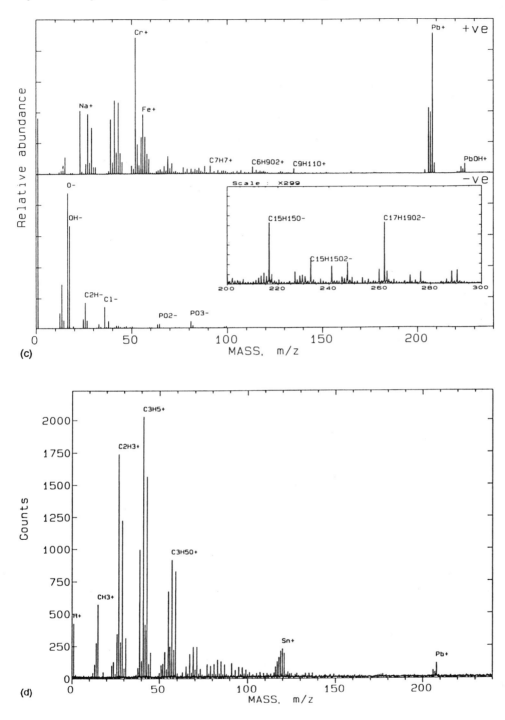

(c)

(d)

mainly used for qualitative identification of functional groups in the sample, but quantitative determinations are possible in many cases.

(a)   IR Techniques.   Infrared spectroscopy in its original dispersive form, in which the absorption of infrared light (the intensity of transmitted energy) when passing through a sample is measured by scanning through the spectrum, has been in use for a long time. The introduction of the Fourier transform principle in IR spectroscopy has prompted an interest in the use of the technique for surface analysis. The Fourier transform (FTIR) instruments use mirrors instead of slits. Each scan thus gives information over the entire spectrum resulting in higher throughput, sensitivity, and signal-to-noise ratio. The many different (and sometimes confusing!) acronyms used in IR mainly refer to different forms of sampling techniques that have been developed in recent years and all use the FT (or interferometric) principle. The most common of these acronyms are listed in Table 1. This list is not complete because some researchers use their own acronyms.

Basically, there are two categories of FTIR spectroscopies: reflection and nonreflection techniques [38]. The latter class comprises either acoustic detection or emission from the sample itself. The techniques recognized here are photoacoustic spectroscopy (PAS), emission spectroscopy (EMS), and photothermal beam deflection spectroscopy (PBDS). These techniques will not be considered further in this chapter. The reader is referred to the literature [39–42]. For adhesion studies the reflection techniques (SRIRS) are more important. The major classes of sampling techniques in SRIRS are:

Internal reflection IR (IRS); commonly known as MIR (multiple internal reflection) or ATR (attenuated total reflection).

External reflection IR (ERS); the techniques in this category can be a single reflection setup (reflection-absorption IR, RAIR, or grazing incidence reflection IR, GIR) or a multireflection setup (MRAIR). The single reflection technique is also frequently referred to as specular reflectance IR.

Diffuse reflectance IR (commonly called DRIFT).

The principles of these three major categories are shown in Fig. 12. ATR is useful primarily for identification of polymer films, liquids, or other materials that can be coated onto a crystal of high refractive index, such as Ge or TlBr/TlI. The sample must be in very good contact with the crystal for good spectra to be obtained. IR light is shone into the crystal at an angle of incidence that is higher than the critical angle of reflection for the crystal, so internal reflection of the radiation occurs and its intensity is attenuated as a result of absorption by the sample. Mostly rectangular crystals are used, but circular crystals have recently been developed for the study of aqueous solutions, films, and fibers [43–46]. Although the requirement of intimate contact between crystal and sample is severely limiting, the technique has the useful capability of providing a depth profile by varying the angle of incidence or by using crystals with different optical densities. As an example, the surface crystallinity or orientation of fibers with respect to their axis has been reported [46]. Further, the technique is uniquely suited for the study of solid/aqueous interfaces.

The technique of choice for studying thin films on metals (or certain other substrates) directly is single reflection RAIR [47–54]. The limitation here is that the substrate must be very smooth, but this can be easily achieved by polishing the metal before deposition of the film. Characterizations of thin organic layers on metal (oxide) surfaces, such as occurs in lubricants, corrosion inhibitors, adhesives, polymers, paints, and so forth, are specific applications of this rather recent form of FTIR. It should be noted that the relative band

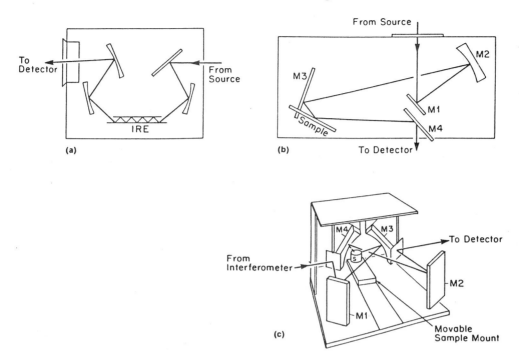

Figure 12 Attachments for IR analysis of surfaces; (a) attenuated total reflectance (ATR); (b) reflection-absorption (RAIR); (c) diffuse reflectance (DRIFT). (From Ref. 2.)

positions and shapes may be different in this technique than in conventional transmission IR. The spectrum may also change with the thickness of the organic film, which implies that polymer/metal interactions are in principle observed [47,51]. The technique is so surface sensitive that oxidation of metals can be determined in situ [51] and the packing structure of monolayers of organic molecules or Langmuir-Blodgett films can be studied [52,53]. In such studies the metallic substrate must have a high reflectivity. Ideal substrates are thus silver and copper. The technologically important substrates aluminum and steel have lower reflectivity or sensitivity.

In diffuse reflectance IR spectroscopy (DRIFT) light impinges on a solid powdered sample and is scattered in all directions. This light is collected and redirected to a detector. The powder must be very fine and is mixed with or dispersed in a suitable matrix, such as KCl or KBr with a particle size of less than 10 μm. It is not suitable for large powders or lumps. By using special sample preparation techniques, e.g., by placing the KBr powder over the sample, monolayers of adsorbed silanes have been studied [55], or water adsorbed on polymer surfaces has been detected [56].

In summary of this section, it can be stated that there are now numerous FTIR spectroscopy techniques, which as a result of their enhanced sensitivity and S/N have contributed immensely in recent years to the understanding of molecular phenomena at surfaces and interfaces as they are related to adhesion. The development of new sampling techniques still continues, and much activity can be expected to occur in the near future in the FTIR arena, along with another promising vibrational spectroscopy tool, Raman spectroscopy, to be discussed below.

(b) Raman Techniques. This vibrational spectroscopy is related and complementary to FTIR. Although its usage is currently not so widespread as that of the various IR techniques, primarily because of its much lower sensitivity, this may well change in the future as some exciting new developments have recently been published.

Raman spectroscopy is a long-established technique for the study of bulk materials. In principle, the technique is straightforward [57–63]. A small region of a transparent sample is illuminated by a monochromatic laser beam, and light that is scattered at a 90° angle with the incident beam is collected and directed into a spectrometer. Most of the scattered light is elastically scattered and has the same frequency v_o as the incident light. This is known as Rayleigh scattering. A small fraction of the scattered light, however, is inelastically scattered and thus contains new frequencies $v_o \pm v_k$. These frequency differences are associated with the transitions between the various vibrational levels in the sample molecules; hence the frequencies observed are in many cases similar to the wave numbers in FTIR. The lines with the lower frequencies are the Stokes lines; those with the higher frequencies are referred to as the anti-Stokes lines. The former series is usually measured in the Raman experiment. The intensities, frequencies, and polarization character of these lines can be determined. The lines are observed in a direction perpendicular to the incident beam, which is plane-polarized.

Raman spectroscopy is useful for studying aqueous solutions, e.g., of polymers, because the spectra are hardly affected by the presence of water. The two major problems of the technique are: (1) the low intensities of the Stokes lines (hence data acquisition is slow), and (2) laser-induced fluorescence effects, which may be so intense that they can completely wipe out all Raman scattering of interest.

The recent developments in the Raman technique, which are very promising and which may lead to its emergence as one of the most surface-sensitive techniques available in the near future, are the following.

Development of Fourier transform [64–66] and Hadamer transform [67] spectrometers as detectors in Raman spectroscopy, similar to FTIR; it has recently been shown that with the proper masking of the Rayleigh scattered radiation and special optical filters, fluorescence-free Raman spectra can be obtained [64].

Resonance Raman spectroscopy (RRS); if the wavelength of the incident radiation is chosen so that it coincides with an absorption band of the scattering molecules, the resonant Raman scattering cross-sections may be up to 10^6 times the cross-sections for normal Raman scattering. In such cases it is possible to detect monolayers (e.g., of dye molecules) at surfaces. This has indeed been demonstrated [68,69]. Recently RRS has found many new applications, mainly in biological studies.

Surface-enhanced Raman spectroscopy (SERS); in this resonance scattering the enhancement is caused by the substrate rather than by the adsorbed molecules.

SERS is thus a surface technique that has so far been restricted to only a few substrate metals, viz. copper, silver, and gold [70–75], but it is applicable to almost any adsorbate. It is also noteworthy that the two mechanisms in RRS and SERS are additive, so that certain dyes on SERS substrates can be detected at extremely low coverage [76]. It is generally accepted now that two mechanisms contribute to the strong enhancement (up to a factor of 10^6) of Raman scattering by molecules adsorbed at the roughened surface of Cu, Ag, or Au. One factor is the generation of large oscillating dipoles and therefore oscillating electrical fields at the metal surface. The other factor is the enhancement of the polarizability of the adsorbed molecules at optical wavelengths by the substrate, i.e., a charge-transfer mechanism [35]. In addition to the substrates already mentioned, enhancement has been observed for the alkali metals, aluminum, indium, palladium, platinum, and even some oxides (NiO and TiO_2), though with lower intensities than for the metals Ag, Cu, or Au.

The range at which molecules contribute significantly to the SERS signal varies between 5 and 50 Å, i.e., comparable to the sampling depth in XPS. The SERS signal does not depend linearly on adsorbate coverage, making quantitative analysis difficult. Further, the restriction to certain substrates is a serious limitation. Recently it has been reported that the strong enhancement effects by substrates such as silver spill over for a few nanometers into an adjacent phase. Therefore, if silver is coated with a very thin film of, for instance, SiO_2, SERS can be observed for molecules adsorbed onto the silica film [77]. Other approaches include the evaporation of silver overlayers to study surfaces of thin or thick films. An example is shown in Fig. 13. In this study the surface segregation in blends of polystyrene with deuterated polystyrene was investigated. SERS was used to investigate the interface with a roughened silver substrate, but also to study the surface by means of a thin overlayer. At both interfaces the technique detected an enrichment of the deuterated form after vacuum annealing [78]. The surface segregation was confirmed by TOFSIMS. In another application the specific adsorption of one of the components of an adhesive system was studied. The results are summarized in Fig. 14 [79]. These various intricate and ingenious sample

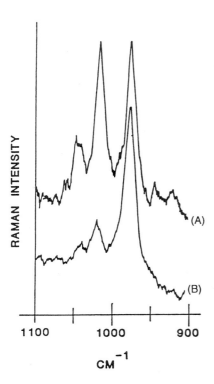

Figure 13 Surface-enhanced Raman spectra (SERS) obtained from thin films of blends of polystyrene and deuterated polystyrene before (A) and after annealing (B). Samples and conditions as in Fig. 9.

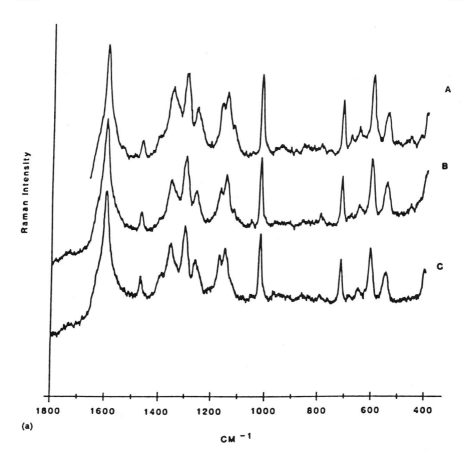

(a)

CM^{-1}

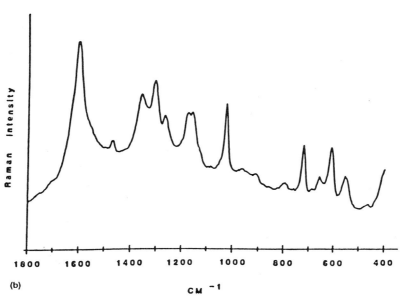

(b)

CM^{-1}

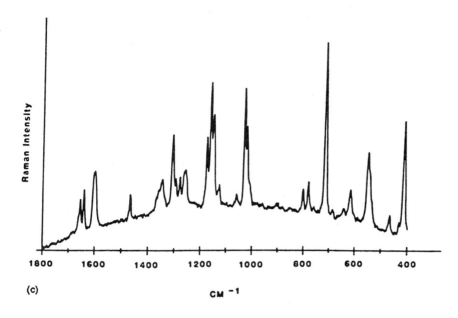

(c)

CM⁻¹

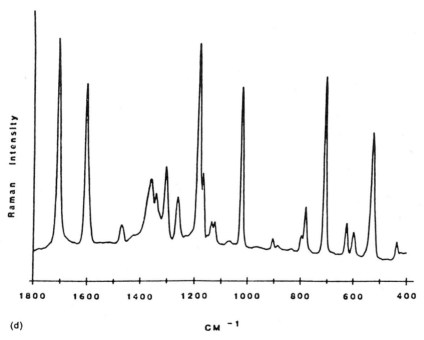

(d)

CM⁻¹

Figure 14 (a) Surface-enhanced Raman spectra (SERS) obtained from an acrylic adhesive spin-coated onto silver island films from (A) 1% and (B) 5% solutions in acetone and (C) from the bulk adhesive; (b) SERS spectrum from saccharin spin-coated onto silver island films from a 0.1% acetone solution; (c) normal Raman spectrum of the sodium salt of saccharin; (d) normal Raman spectrum of pure saccharin. The results demonstrate that saccharin, a component of the adhesive, is adsorbed on the substrate surface and this adsorption is detected non-destructively by SERS. (From Ref. 79.)

preparation methods that are currently being developed by several laboratories show great promise for an extension of the types of substrates and materials to which the SERS technique can be applied.

Extremely low-noise integrating multichannel detectors; the very recent introduction of these very special devices has made it possible for the first time to record Raman spectra from monolayers of organic materials (e.g., fatty acids) on substrates that do not give surface enhancement, for instance a water-air interface [80]. Conventional Raman spectroscopy may now have become a viable non-UHV surface analysis tool. Advantages of conventional RS over SERS are that the scattering cross-sections are independent of the type of substrate surfaces. Further, the intensity varies more linearly with coverage and the cross-sections do not vary a great deal for different molecules. Hence the conventional Raman technique is more quantitative than SERS.

We may thus begin to see more applications of conventional Raman spectroscopy, e.g., of monolayers on well-defined substrate surfaces. On the other hand, SERS will remain a powerful technique in its own right because of its greater sensitivity and surface specificity. It remains practically the only technique available to determine nondestructively and in a non-UHV environment the orientation of molecules in situ in a dense medium, since the contribution to the observed signal by molecules farther than about 50 Å away from the interface is negligible compared with the enhanced signal from the interfacial molecules.

(c) High-Resolution Electron Energy Loss Spectroscopy (HREELS). The specific attribute of this vibrational technique, compared with the ones discussed above, is that it can provide functional group information on the surface of polymers. In addition, one can study the interactions between these functional groups and thin films of metals evaporated onto these polymers. The technique, which has gone through some recent instrumental developments, is thus important for the furthering of our understanding of metal-polymer adhesion mechanisms.

The basic experiment in HREELS in the backscattering geometry is straightforward [37]. A monochromatized electron beam of 1–10 eV is directed toward the surface and the energy distribution of the reflected electrons is measured in an electron analyzer with a resolution of up to 7 meV. The spectrum consists of the elastic peak and peaks due to energy losses to the sample surface by the excitation of molecular vibrations. If plotted as wave numbers, these vibrations are very similar to those observed in IR techniques. The resolution achievable in this technique is, however, considerably less than in IR, which becomes clear if one considers that 1 meV $= 8.066$ cm^{-1}, so the spectral resolution in HREELS is of the order of 100 cm^{-1} (in IR the resolution is typically around 4 cm^{-1} or better). Detection of crystallinity or other high-resolution details as is possible in IR is therefore currently not achievable in HREELS.

The most meaningful information in this technique is obtained by varying the electron impact energy and the scattering geometry (angles Θ_i and Θ_r) by rotating the sample holder or the electron analyzer.

A major problem that has hindered applications of the technique to bulk polymer surfaces until recently is the surface charging of insulators. At the low incident electron energies used in this technique, the secondary electron emission is high, so that a positive charge develops. Therefore, this problem can be overcome by using low current defocused flood guns of 1–2 keV electrons [81]. Further, although spectra can now be taken of the surface of polymers, these usually show broad peaks or bands, nowhere near the resolution obtained in IR. It is remarkable that heating (or annealing) polymers in vacuum for a few minutes at temperatures in the range 200–250°C sometimes results in a pronounced sharpening of the peaks. No satisfactory mechanism for this effect has been put forward yet.

The major strength of the technique lies in the fact that vibrational information of organic surfaces can be obtained with absorption bands that are identical to those observed in IR. In this respect the technique bridges a gap between XPS and IR. Information can be obtained on the polymer surface itself or on molecules segregated to or adsorbed on the polymer surface. Aliphatic and aromatic groups can be distinguished and hydrogen is also detected via functional groups. The adsorption of water on polymer surfaces is also detected easily. On the other hand, because of the high surface sensitivity, it is difficult to prepare clean model surfaces for HREELS studies. The similarity between HREELS and IR bands facilitates spectral interpretation immensely. There is no systematic difference between peak positions in the two techniques. The intensities of the bands in the two techniques are vastly different, however, because electron and photon excitation of molecular vibrations seem to follow different selection rules. At present, the excitation mechanism in HREELS is not completely understood, making quantification impossible. No practical theory is available to quantify electron-induced vibrational spectra of polymers.

Some polymer surface studies that have been reported recently are the detection of the molecular orientation at polymer film surfaces. For instance, the spectrum of isotactic polystyrene is different from that of the atactic material [82]. The spectra of thin films of PMMA cast on Au, Al, or Cu were also different; especially the intensity of the $C=O$ stretching band at 1710 cm^{-1} varied considerably [83]. Thus HREELS seems to be capable of identifying molecular long-range order in polymeric surfaces.

Other recent and very interesting studies are those in which monolayers of metals have been evaporated onto polymeric substrates. This allowed a conclusion as to the nature of the interaction and the molecular sites that are preferably attacked by the metal atoms. Systems that have been studied recently are Cu-polyimide, Cr-polyimide, and Al-polyimide [84–86]. Very recently the interfaces formed between vacuum-evaporated Al and polyvinyl alcohol, polyacrylic acid, and polyethylene terephthalate (PET) were investigated [87]. Figure 15 shows the HREELS spectrum of PET. The vibrational assignments and their comparison with IR intensities are shown in Table 4. The HREELS spectra observed during PET metallization are shown in Fig. 16. The bands that disappear are those of $O-CH_2$ and $C=O$, and the new band that forms has been ascribed to the formation of a $C-O-Al$ group and eventually a carboxylate.

In summary, the HREELS vibrational spectroscopy technique appears to be a new and promising highly surface-sensitive tool for the study of interactions between metals and polymers related to adhesion phenomena. Current limitations are the lack of quantification, the lack of knowledge on the sampling depth, the difficulty to prepare clean polymer surfaces, the need for a clean vacuum (as opposed to the optical vibrational techniques), and the lack of suitable fingerprint spectra of clean surfaces of technological interest. However, as mentioned for Raman spectroscopies, we can expect much activity in the near future and the technique may be capable of carving its own niche in the gamut of available surface spectroscopies.

C. Miscellaneous Spectroscopies

The ion/electron and vibrational spectroscopies are the most widely used techniques in adhesive bonding technology. However, several lesser known spectroscopies are worth mentioning here briefly also, because they can provide unique information on metal-

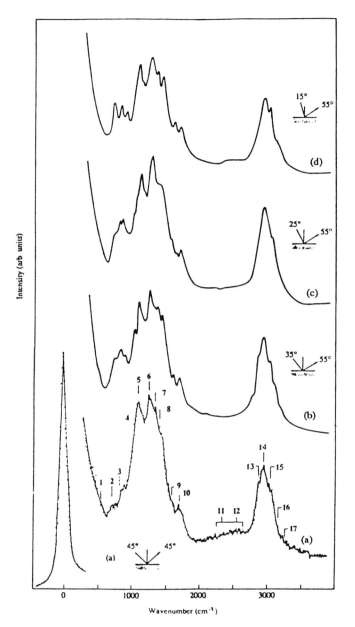

Figure 15 HREEL (three-point smoothed) spectra of a PET sample in different geometrical configurations: ($i =$ incidence angle; $r =$ reflection angle) (a) $i = 45°$, $r = 45°$; (b) $i = 35°$, $r = 55°$; (c) $i = 25°$, $r = 55°$; (d) $i = 15°$, $r = 55°$. (From Ref. 87.)

polymer interfaces in certain cases. A more complete list with suitable references is given in Table 1. The techniques that are discussed in this section are ellipsometry, electrochemical impedance spectroscopy (EIS), glow discharge optical spectroscopy (GDOS), and Rutherford backscattering spectroscopy (RBS).

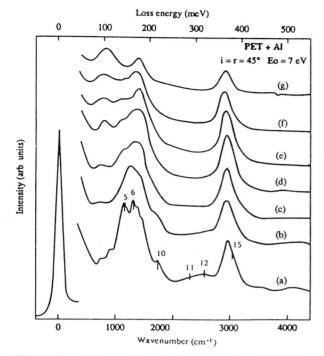

Figure 16 High-resolution electron energy loss (HREEL) spectrum of a metallized PET sample; (a) clean PET; (b)–(g) increasing coverage of the PET by Al atoms. (From Ref. 78.)

Table 4 Comparison of IR and HREELS Vibrational Bands of PET

Band	Infrared bands[a] cm^{-1}	Intensity	HREELS loss energy meV	cm^{-1}	Assignment
1	502	m	63	510	$\gamma_w(C = O)$
2	730	s	89	720	$\nu11(B_{1u})$ benzene
3	875	m	107	860	$\nu17B(B_{1u})$ benzene
4	1020	m	128	1030	$\nu18A(B_{2u})$ benzene
5	1120	s	138	1110	$\nu(O - CH_2)$
6	1263	s	157	1270	$\nu(O - C=)$
7	1343	m	169	1360	$\gamma_w(CH_2)$
	1370	w			
8	1455	m	181	1460	$\delta(CH_2)$
	1470	w			
9	1580	w	198	1600	$\nu8A(A_1), \nu8B(B_1)$
	1617	w			
10	1725	s	213	1720	$\nu(C = O)$
11			Unlocalized		$\nu(O - H)$, hydrogen-bonded water
12			Unlocalized		$\nu(O - H)$, hydrogen-bonded water
13	2890	w	360	2900	$\nu_s(CH_2)$ amorphous or $\nu_a(CH_2)$ crystalline
14	2970	m	367	2960	$\nu_a(CH_2)$ amorphous
15	3068	w	381	3070	$\nu_{20A}(B_{2u})$ and $\nu_{20B}(B_{3u})$ benzene
16			393	3170	$\nu(O - H)$?
17	3440	w	409	3500	$\nu(O - H)$

[a]Spectrum measured in this work; s = strong; m = medium; w = weak.
Source: Ref. 87.

1. Ellipsometry

This is a widely used optical technique for measuring film thicknesses. The technique uses polarized light at oblique incidence and yields information on the optical constants of materials and the thickness of overlayers. The signal is generated by interference of the components of the primary beam, which are reflected at phase boundaries between the substrate and the film. Best results are obtained when the films are no thicker than only a few multiples of the wavelength of the light. Discontinuous or island film structures can also be studied. An ellipsometric analysis gives two numbers delta and psi (Δ and Ψ), which together define the so-called complex reflection ratio ρ of the reflection coefficient in the directions parallel and perpendicular to the plane of incidence. More information on the principles can be found in the literature. There is a wealth of literature on the subject [88–91], both concise introductions and more comprehensive discussions or conference proceedings.

For applications in adhesive bonding research or technology, ellipsometry is useful for the quantitative determination of film thicknesses. Especially aluminum is a metal that has been studied extensively. It lends itself well to oxide thickness measurements because Al_2O_3 is transparent, which is a requirement. The thickness of the oxide formed in certain media can be determined [92]. Other studies reported on the use of ellipsometry to investigate the corrosion or rate of oxide film dissolution in certain environments in situ. As the film dissolves, the formation of pores and differences between the densities of different layers in the oxide film can be distinguished and related to the conditions of the anodizing process [93].

2. Electrochemical Impedance Spectroscopy (EIS)

This technique has rarely been used in adhesive bonding studies, but is well suited to determine the environmental degradation of adhesively bonded systems. Currently it is frequently used to study the degradation processes of painted metals immersed in an electrolyte, such as 5% NaCl [94]. Essentially the technique measures the total impedance of a sheet metal coated with a film of high resistivity. The metal is immersed in the electrolyte, held at its corrosion potential, and a small AC voltage perturbation is applied. The current response is measured and plotted as impedance |Z| versus the frequency. The usual spectrum is a plot of this impedance versus frequency over the range $0.1–10^5$ Hz (Bode plot), although there are other formats as well. The spectrum is recorded at several intervals, for instance once a week or month.

From the Bode plot the total capacitance of the coating system and the total resistivity of the system can be determined directly from the graph by extrapolation to zero frequency. With the use of a so-called equivalent circuit, a hypothetical electrical circuit that would give the same response over the frequency range, the capacitance and resistivity can be broken down into the individual components for the metal, the interface, and the coating [95,96]. Parameters such as coating capacitance, double-layer capacitance, pore resistance, and polarization resistance can all be derived from the appropriately chosen model. These can then be converted to physical properties such as percentage of water uptake, diffusion rates of ions and liquid water through the coatings, numbers of pores, corrosion reactions (anodic and/or cathodic) that take place at the interface, blister resistance (related to adhesion of the coating), and so forth. These parameters are normally determined as a function of immersion time so that the best system of metal pretreatment and type of coating can be determined and further optimized quickly and in the laboratory without having to rely on lengthy exposure in outdoor tests.

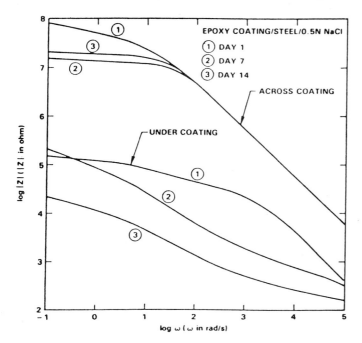

Figure 17 Electrochemical impedance data (Bode plots) for epoxy-coated steel as a function of immersion time in 0.5 N NaCl solution. (From Ref. 98.)

Excellent correlations between EIS analyses and field exposure of painted metals have been reported [97]. The coatings can be very thick (such as paint systems or adhesives), but very thin films, such as oxide, plasma coatings, conversion coatings, or other pretreatments, can also be investigated. As an example, Fig. 17 shows EIS curves as a function of immersion time in 3% NaCl of a cold-rolled steel sample coated with a film of plasma-polymerized trimethyl silane [98]. The variable in this study was the plasma cleaning procedure applied to the metal prior to film deposition. Table 5 shows a comparison of the parameters that can be derived from the impedance data as a function of pretreatment using an equivalent circuit consisting of two resistances and two capacitors plus the resistivity of the electrolyte. It is seen that all parameters show the same ranking order and that order is identical to that observed in actual corrosion exposure. This example thus exemplifies that EIS data are quick and reliable and can be obtained in a fraction of the time required for actual exposure in a corrosion test. Further, the results of this study indicate that the pretreatment of the metal is by far the most important step in the plasma polymerization process. The actual composition of the deposited coating or its thickness is of secondary importance [98].

EIS and other electrochemical methods appear to be useful for the study of the performance of metal pretreatments (cleaning processes, anodization, phosphating, chromating, etc.) prior to adhesive bonding. A quick comparison of methods can be achieved, and because the method is fast and straightforward, it can be used as a quality control method. On adhesively bonded system, EIS could be performed in more fundamental studies that would provide information on the nature and locus of degradation processes, when immersed in aggressive solutions.

Table 5 EIS Parameters Versus Corrosion Performance of Cold-Rolled Steel Coated with a Thin Film of Plasma-Polymerized Trimethylsilane

Sample	C_p (F)	R_{po} (Ω)	C_{dl} (F)	R_p (Ω)	R_1	% Rust
A	4.3×10^{-6}	196	3.0×10^{-5}	1.0×10^3	0.49	60
B	1.5×10^{-6}	618	1.2×10^{-6}	9.8×10^4	1.22	0
C	1.7×10^{-6}	182	1.6×10^{-6}	6.5×10^3	1.02	20
D	6.0×10^{-6}	69	8.6×10^{-5}	9.2×10^2	0.33	80
Control (15 min)	—	—	5.9×10^{-4}	3.6×10^2	—	100
Control (1 h)	—	—	1.6×10^{-3}	1.8×10^3	—	100

C_p = capacitance of coating; R_{po} = pore resistance of coating; C_{dl} = double-layer capacitance (interface); R_p = polarization resistance (interface); R_1 = slope of impedance curve in range 10^2–10^4 Hz.
Source: Ref. 98.

3. Glow Discharge Optical Spectroscopy (GDOS)

This technique and a variation, GDMS (glow discharge mass spectrometry), are essentially depth-profiling techniques [99]. However, there is a major difference between GDOS and depth profiling in typical surface analysis techniques such as AES and SIMS. The rate of profiling in GDOS is of the order of 1–5 μm/min. A high-intensity argon lamp (DC gas discharge) is used for sputtering the material. The sputtered elements are detected in the plasma by a spectrochemical analysis of the glow light via an optically transparent window. Because of the requirement of fast, simultaneous detection, a grating spectrometer is used, so each element requires its own photomultiplier.

Sputtering in GDOS occurs by the positive ions generated in the gas discharge glow from where they are accelerated toward the sample. The electrically conductive sample is pressed against the cathode of which it forms a part. The high sputtering rate allows depth profiling through metallic coatings of industrial interest, e.g., zinc coatings on galvanized steel [100]. Residual elements such as carbon can be detected below the coating. Also, entire conversion coatings can be profiled rapidly, such as phosphates and chromates or anodization layers, which are beyond the scope of AES or SIMS depth profiling. The technique could thus be used in quality control, for instance for determining the homogeneity of electrodeposits across the width of the strip, or for control of the oxide thickness in metal pretreatment operations.

The technique requires only a moderate vacuum and has a spatial resolution of a few millimeters. Currently, two commercial instruments are available. Most elements can be detected, including hydrogen. Limitations are the quantification, which requires suitable standards, and the sample has to be electrically conductive. Nonconductive samples can be analyzed only by grinding them and mixing them with a metallic or graphite powder.

As a typical example, Fig. 18 shows the depth profile of a several microns thick metallic coating on steel [100].

4. Rutherford Backscattering Spectroscopy (RBS)

This is another widely used technique that provides concentration depth profiles of materials. Detailed recent reviews of this technique are available [101–103]. Essentially, in RBS a sample is bombarded with a collimated beam of high-energy light ions, (typically $^4He^+$ of 1–2 MeV, although proton beams are also used). The scattered ions are energy-analyzed and counted using solid-state detectors. The scattering angle is very

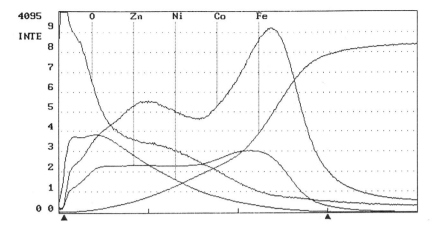

Figure 18 Glow-discharge optical spectroscopy (GDOS) depth profile of an electrodeposited metallic coating system consisting of 1 µm Zn-1%Co base layer and 3 µm Ni-20%Zn top layer on cold-rolled steel substrate. Sputtering time is in seconds.

high, usually >90° (typically 170°). The technique provides a depth profile and compositional analysis and is quantitative and nondestructive. A major drawback is, however, the need for a large particle accelerator.

Basically, what is measured in RBS is the distribution of the energy of the backscattered ions. The energy scale can, however, be converted to a mass scale because for a given projectile/target combination and scattering geometry it can exactly be calculated how much energy the primary particle loses in collisions with an atom of a particular mass. Energy loss to heavy atoms is less than to light atoms. The positions on the energy scale are thus related to collisions with surface atoms. However, surface atoms scatter only a small fraction of the primary particles. Below the surface, inelastic energy losses occur also. Therefore, the mass scale can also be converted to an energy scale. These principles are schematically shown in Fig. 19.

In addition to providing a depth profile, RBS can also give crystallographic information if the instrument is equipped with a precision goniometer. In certain directions of single crystals the phenomenon of channeling occurs. The backscatter yield is reduced when the beam is aligned with the major low index axes [104]. By aligning both incident and reflected beams along such directions, the relative positions of surface atoms can be determined with high precision. Also, the sites of impurity atoms can be determined, because for interstitial positions the backscatter yield is the same as for the random direction value.

RBS can be considered an almost ideal tool for the analysis of thin films of a few microns thickness if stoichiometric information is required. All elements > H are detected. Quantitative measurements of film thickness and composition and concentrations of impurities can be performed. A single RBS spectrum can show the amount of an impurity and its distribution throughout a thin film.

In general, however, only rather simple structures and compositions can be investigated successfully. The depth range that can be accessed varies but is of the order of 1–2 µm for proton and alpha particle beams. This depth depends on the primary particle energy, which has an upper limit of a few MeV above which nuclear reactions may occur.

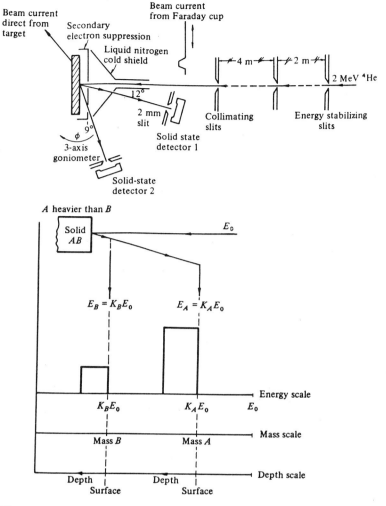

Figure 19 Experimental arrangement for Rutherford backscattering spectroscopy (RBS) (top) and schematic of Rutherford backscattering from a solid composed of elements of mass A and B (bottom). (From Ref. 101.)

The depth resolution is very high, e.g., 10–300 Å, and can be improved for very low depths by tilting the sample.

Table 6 summarizes some of the parameters of RBS and Fig. 20 shows an example of the quantitative measurement of the epitaxial growth of an oxide film [101].

IV. CONCLUDING REMARKS

This chapter summarizes the principles of some of the many spectroscopic techniques that are available for the analysis or study of aspects of adhesive bonding science and technology. As indicated in Table 1, there are dozens of techniques and new acronyms appear almost on a daily basis. The number of instrumental spectroscopies available today to the scientist is bewildering, especially the many techniques for surface characterization. Therefore, it is likely that some techniques have been missed, although it was attempted to

Table 6 Parameters in Rutherford Backscattering Spectrometry

Probe beam	H^+, $^4He^+$, other light ions
Probe energy	1–3 MeV
Beam diameter	~0.5–1.0 mm (~2 μm with microbeam)
Beam current	~2–20 nA
Analysis time	~5–30 min
Integrated charge	~1–40 μC ($6 \times 10^{12} - 2.5 \times 10^{14}$ ions)
Scattering angle	170°
Energy analyzer	Surface barrier detector 15–25 keV energy resolution
Probing depth	~1–2 μm
Depth resolution	20–30 nm (3–4 nm with tilted targets)
Mass resolution	Isotope resolution up to ~40 amu
Sensitivity	10^{-2}–10^{-4} monolayers for heavy surface impurities
	10^{-1}–10^{-2} monolayers for light surface impurities
Accuracy	3–5% (typical)

Source: Ref. 101.

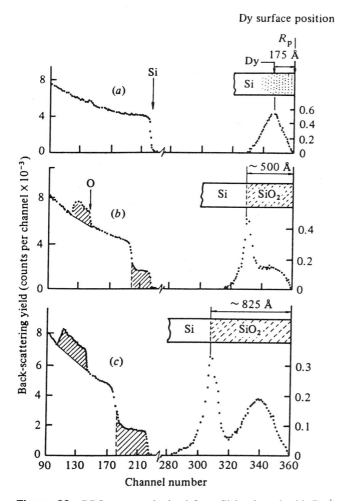

Figure 20 RBS spectra obtained from Si implanted with Dy^+ ions; (a) after implantation; (b) after growth of 500 Å of SiO_2; (c) after growth of 825 Å of SiO_2. (From Ref. 101.)

cover them all, at least in Table 1. The choice of techniques from that listing that were actually discussed in this chapter had to be limited and was in some cases somewhat arbitrary and subjective. However, some emphasis was put on techniques that can be used in the study of the science of adhesive bonding technology. Techniques for routine analysis, e.g., NMR or the various mass spectrometries, were not discussed in depth.

It is clear that tremendous developments have been made in recent years with certain techniques. A large number of spectroscopic techniques are now available that can be adapted, with little or no adaptation, to the study of various aspects of adhesive bonding. Important questions, such as adhesion mechanisms, failure mechanisms, locus of failure, and so forth, can now be addressed successfully by several experimental methods. The application of such spectroscopies has evidently led to a better understanding of the performance of materials used in adhesive bonding and such knowledge has triggered the development of new or improved materials.

In the choice of a particular spectroscopic technique, one should be well aware that no single technique, no matter how sophisticated, can solve all problems or answer all questions. Therefore, some insight into the principles and capabilities of the many techniques that one could choose is required for anyone who wishes to tackle problems in materials science. Providing this insight, and referring the interested reader to the appropriate literature for further studies, was the main objective of this chapter.

REFERENCES

1. E. A. Ledbury, A. G. Miller, P. D. Peters, E. E. Peterson and B. W. Smith, *Proc. 12th Natl. SAMPE Techn. Conf. Series*, SAMPE, Azusa, CA, 1980, p. 935.
2. J. A. Filbey and J. P. Wightman, in *Adhesive Bonding* (L-H. Lee, ed.), Plenum Press, New York, 1991, p. 175.
3. G. D. Davis, in *Adhesive Bonding* (L-H. Lee, ed.), Plenum Press, New York, 1991, p. 139.
4. C. Klauber and R. S. T. Smart, in *Surface Analysis Methods in Materials Science*, (D. J. O'Connor, B. A. Sexton and R. S. T. Smart, eds.), Springer-Verlag, Berlin, 1992, p. 13.
5. R. J. MacDonald and B. V. King, in *Surface Analysis Methods in Materials Science*, (D. J. O'Connor, B. A. Sexton and R. S. T. Smart, eds.), Springer-Verlag, Berlin, 1992, p. 117.
6. J. C. Vickerman, in *Methods of Surface Analysis* (J. M. Walls, ed.), Cambridge University Press, Cambridge, 1989, p. 169.
7. A. Benninghoven, F. G. Rüdenauer and H. W. Werner, *Secondary Ion Mass Spectrometry*, Wiley, New York, 1987.
8. A. W. Czanderna and D. M. Hercules, *Ion Spectroscopies for Surface Analysis*, Plenum Press, New York, 1991.
9. J. C. Vickerman, A. Brown and N. M. Reed, eds., *Secondary Ion Mass Spectrometry*, Oxford University Press, Oxford, 1989.
10. D. Briggs and M. P. Seah, eds., *Practical Surface Analysis by Secondary Ion and Neutral Mass Spectrometry*, Wiley, Chichester, 1990.
11. W. J. van Ooij and R. S. Michael, in *Metallization of Polymers* (E. Sacher, J-J. Pireaux and S. P. Kowalczyk, eds.), ACS Symposium Series No. 440, American Chemical Society, Washington, D.C., 1990, p. 60.
12. C. H. Becker, in *Ion Spectroscopies for Surface Analysis* (A. W. Czanderna and D. M. Hercules, eds.), Plenum Press, New York, 1991, p. 273.
13. C. J. Powell, D. M. Hercules and A. W. Czanderna, in *Ion Spectroscopies for Surface Analysis* (A. W. Czanderna and D. M. Hercules, eds.), Plenum Press, New York, 1991, p. 417.
14. D. Briggs and M. P. Seah, eds., *Practical Surface Analysis by Auger and X-ray Photoelectron Spectroscopy*, 2nd ed., Wiley, Chichester, 1990.

15. D. G. Armour, in *Methods of Surface Analysis* (J. M. Walls, ed.), Cambridge University Press, Cambridge, 1989, p. 263.

16. E. Taglauer, in A. W. Czanderna and D. M. Hercules, *Ion Spectroscopies for Surface Analysis*, Plenum Press, New York, 1991, p. 363.

17. G. C. Smith, *Quantitative Surface Analysis for Materials Science*, Institute of Metals, London, 1991.

18. H. E. Bishop, in *Methods of Surface Analysis* (J. M. Walls, ed.), Cambridge University Press, Cambridge, 1989, p. 86.

19. D. Briggs, A. Brown and J. C. Vickerman, *Handbook of Static Secondary Ion Mass Spectrometry (SIMS)*, Wiley, Chichester, 1989.

20. J. G. Newman, B. A. Carlson, R. S. Michael, J. F. Moulder, W. J. van Ooij and T. A. Hohlt, *Handbook of Static SIMS Spectra of Polymers*, Perkin Elmer Corporation, Eden Prairie, MN, 1991.

21. A. J. Swift and J. C. Vickerman, in *Surface Analysis Techniques and Applications* (D. R. Randall and W. Neagle, eds.), The Royal Society of Chemistry, Cambridge, 1990, p. 37.

22. I. V. Bletsos, D. M. Hercules, D. van Leyen, B. Hagenhoff, E. Niehuis and A. Benninghoven, *Anal. Chem. 63*: 1953 (1991).

23. I. V. Bletsos, D. M. Hercules, D. van Leyen, A. Benninghoven, C. G. Karakatsanis and J. N. Rieck, *Macromolecules 23*: 4157 (1990).

24. W. J. van Ooij and A. Sabata, *Surf. Interface Anal. 19*: 101 (1992).

25. D. Briggs and B. D. Ratner, *Polymer Commun. 29*: 6 (1989).

26. P. A. Cornelio and J. A. Gardella, Jr., in *Metallization of Polymers* (E. Sacher, J-J. Pireaux and S. P. Kowalczyk, eds.), ACS Symposium Series No. 440, American Chemical Society, Washington, D.C., 1990, p. 379.

27. D. E. Sykes, in *Methods of Surface Analysis* (J. M. Walls, ed.), Cambridge University Press, Cambridge, 1989, p. 216.

28. E. Niehuis, *Proc. Eighth Intern. Conf. on Secondary Ion Mass Spectrometry (SIMS VIII)*, Wiley, Chichester, 1992, p. 269.

29. P. M. Lindley, J. A. Chakel and R. W. Odom, *Proc. Eighth Intern. Conf. on Secondary Ion Mass Spectrometry (SIMS VIII)*, Wiley, Chichester, 1992, p. 219.

30. (a) A. Sabata, W. J. van Ooij and H. K. Yasuda, *Surface Interface Anal. 20*: 845 (1993). (b) W. J. van Ooij and A. Sabata, *J. Adhesion Sci. Technol. 5*: 843 (1991).

31. W. J. van Ooij and F. J. Boerio, unpublished work.

32. W. J. van Ooij, A. Sabata and A. D. Appelhans, *Surf. Interface Anal. 17*: 403 (1991).

33. W. J. van Ooij, A. Sabata and R. J. Koch, *J. Adhesion Sci. Technol. 7*: 1153 (1993).

34. J. T. Yates, Jr. and T. E. Madey, eds., *Vibrational Spectroscopy of Molecules on Surfaces*, Plenum Press, New York, 1987.

35. J. A. Creighton, in D. R. Randall and W. Neagle, eds., *Surface Analysis Techniques and Applications*, Royal Society of Chemistry, Cambridge, 1990, p. 13.

36. N. K. Roberts in D. J. O'Connor, B. A. Sexton and R. S. T. Smart, eds., *Surface Analysis Methods in Materials Science*, Springer-Verlag, Berlin, 1992, p. 187.

37. J-J. Pireaux, Ch. Grégoire, M. Vermeersch, P. A. Thiry, M. Rei Vilar and R. Caudano, in *Metallization of Polymers* (E. Sacher, J-J. Pireaux and S. P. Kowalczyk, eds.), ACS Symposium Series No. 440, American Chemical Society, Washington, D.C., 1990, p. 47.

38. M. W. Urban, *J. Adhesion Sci. Technol. 7*: 1 (1993).

39. P. R. Griffiths, ed., *Transform Techniques in Chemistry*, Plenum Press, New York, 1978.

40. M. W. Urban and J. L. Koenig, in *Vibrational Spectra and Structure* (J. Durig, ed.), Vol. 18, Ch. 3, Elsevier, Amsterdam, 1990.

41. M. W. Urban, S. R. Gaboury, W. T. McDonald and A. M. Tiefenthaler, in *Adv. Chem. Series*, No. 227, Ch. 17 (C. Craver and T. Provder, eds.), American Chemical Society, Washington, D.C., 1990.

42. M. W. Urban, in M. W. Urban and C. D. Craver, eds., *Adv. Chem. Series*, No. 236, Ch. 1, American Chemical Society, Washington, D.C., 1992.

43. J. E. Bertie and H. E. Eysel, *Appl. Spectrosc. 39*: 392 (1985).
44. M. W. Urban, J. L. Koenig, S. B. Shih and J. Allaway, *Appl. Spectrosc. 41*: 590 (1987).
45. M. W. Urban and J. L. Koenig, *Appl. Spectrosc. 41*: 1028 (1987).
46. A. M. Thiefenthaler and M. W. Urban, *Appl. Spectrosc. 42*: 163 (1988).
47. J. D. Swalen and J. F. Rabolt, in *Fourier Transform Infrared Spectroscopy: Applications to Chemical Systems* (J. R. Ferraro and L. J. Basile, eds.), Academic Press, New York, 1979, p. 283.
48. W. G. Golden, in *Fourier Transform Infrared Spectroscopy: Applications to Chemical Systems* (J. R. Ferraro and L. J. Basile, eds.), Academic Press, New York, 1979, p. 315.
49. D. L. Allara, A. Baca and C. A. Pryde, *Macromolecules 11*: 1215 (1978).
50. T. Nguyen, D. P. Bentz and W. E. Byrd, *Proc. ACS, PMSE Div. 59*: 459 (1988).
51. D. K. Otteson and A. S. Nagelberg, *Thin Solid Films 73*: 347 (1980).
52. D. L. Allara and J. D. Swalen, *J. Phys. Chem. 86*: 2700 (1982).
53. W. Knoll, M. R. Philpott and W. F. G. Golden, *J. Chem. Phys. 77*: 219 (1982).
54. J. F. Rabolt, F. C. Burns, N. E. Schlotter and J. D. Swalen, *J. Chem. Phys. 78*: 946 (1983).
55. S. R. Culler, M. T. McKenzie, L. J. Fina, H. Isihda and J. L. Koenig, *Appl. Spectrosc. 38*: 791 (1984).
56. E. G. Chatzi, H. Ishida and J. L. Koenig, *Appl. Spectrosc. 40*: 682 (1986).
57. J. A. Creighton, in *Advances in Spectroscopy*, (R. J. H. Clark and R. E. Hester, eds.), Wiley, Chichester, 1988, p. 37.
58. M. Moskovits, *Rev. Mod. Phys. 57*: 783 (1985).
59. A. Otto, in *Light Scattering in Solids* (M. Cardona and G. Gunterodt, eds.), (Topics in Applied Physics), Vol. 54, Springer, Berlin, 1984, p. 289.
60. A. Wokaun, in *Solid State Physics* (H. Ehrenreich, F. Seitz and D. Turnbull, eds.), Academic Press, New York, 1984, Vol. 38, p. 223.
61. R. K. Chang and B. L. Laube, *CRC Crit. Rev. Solid State Mater. Sci. 12*: 1 (1984).
62. R. P. Cooney, M. R. Mahoney and A. J. McQuillan, *Advances in Infrared and Raman Spectroscopy*, (R. J. H. Clark and R. E. Hester, eds.), Heyden, London, 1988, Vol. 9, p. 188.
63. I. Pockrand, in *Springer Tracts in Modern Physics*, Springer, Berlin, 1984, Vol. 104, p. 1.
64. T. Hirschfeld and D. B. Chase, *Appl. Spectrosc. 40*: 133 (1986).
65. D. B. Chase, *J. Am. Chem. Soc. 108*: 7485 (1986).
66. S. F. Parker, K. P. J. Williams, P. J. Hendra and A. J. Turner, *Appl. Spectrosc. 42*: 796 (1988).
67. D. C. Tillota, R. D. S. Freeman and W. G. Fately, *Appl. Spectrosc. 41*: 1280 (1987).
68. M. Fujhira and T. Osa, *J. Am. Chem. Soc. 98*: 7850 (1976).
69. H. Yamada, T. Amamiya and H. Tsubomura, *Chem. Phys. Lett. 56*: 591 (1978).
70. R. K. Chang and T. Furtak, *Surface-Enhanced Raman Scattering*, Plenum Press, New York, 1982.
71. H. Baltruschat and J. Heibaum, *Surf. Sci. 166*: 113 (1986).
72. Q. Feng and T. M. Cotton, *J. Chem. Phys. 90*: 983 (1986).
73. H. Ishida and A. Ishitani, *Appl. Spectrosc. 37*: 450 (1983).
74. T. Watanabe, O. Kawanami, H. Katoh and K. Honda, *Surf. Sci. 158*: 341 (1985).
75. T. Vo-Dinh, M. Y. K. Hiromote, G. M. Begun and R. L. Moody, *Anal. Chem. 56*: 1667 (1984).
76. B. Pettinger, K. Krisher and G. Ertl, *Chem. Phys. Lett. 151*: 151 (1988).
77. D. J. Walls and P. W. Bohn, *J. Phys. Chem. 93*: 2976 (1989).
78. F. J. Boerio, unpublished work.
79. F. J. Boerio, P. P. Hong, P. J. Clark and Y. Okamoto, *Langmuir 6*: 721 (1990).
80. T. Kawai, J. Umemura and T. Takenaka, *Chem. Phys. Lett. 162*: 243 (1989).
81. M. Liehr, P. A. Thiry, J-J. Pireaux and R. Caudano, *Phys. Rev. B. 33*: 5682 (1986).

82. M. Rei Vilar, M. Schott, J-J. Pireaux, Ch. Grégoire, R. Caudano, A. Lapp, J. Lopes da Silva and A. M. Botelho do Rego, *Surf. Sci. 211/212*: 782 (1989).
83. N. J. Dinardo, J. E. Demuth and T. C. Clarke, *J. Chem. Phys. 85*: 6739 (1986).
84. N. J. Dinardo, J. E. Demuth and T. C. Clarke, *Chem. Phys. Lett. 121*: 239 (1985).
85. J. J. Pireaux, M. Vermeersch, Ch. Grégoire, P. A. Thiry, R. Caudano and T. C. Clarke, *J. Chem. Phys. 88*: 3353 (1988).
86. N. J. Dinardo, J. E. Demuth and T. C. Clarke, *J. Vac. Sci. Technol. A4*: 1050 (1986).
87. Y. Novis, N. Degosserie, M. Chtaïb, J-J. Pireaux, R. Caudano, P. Lutgen and G. Feyder, *J. Adhesion Sci. Technol. 7*: 699 (1993).
88. R. Greef, in *Surface Analysis Techniques and Applications* (D. R. Randall and W. Neagle, eds.), Royal Society of Chemistry, Cambridge, 1990, p. 27.
89. P. S. Hauge, *Surf. Sci. 96*: 108 (1980).
90. R. Greef, in *Comprehensive Chemical Kinetics* (R. G. Compton, ed.), Elsevier, Amsterdam, 1989, Vol. 29, p. 453.
91. Proc. Fourth International Conference on Ellipsometry, *Surf. Sci. 96* (1980).
92. R. Greef and C. F. W. Norman, *J. Electrochem. Soc. 132*: 2362 (1985).
93. U. Künzelmann and G. Reinhard, *Wiss. Z. Tech. Univ. Dresden 38*: 111 (1989).
94. M. W. Kendig and F. Mansfeld, *Mat. Res. Soc. Symp. Proc. 125*: 293 (1988).
95. U. Rammelt and G. Reinhard, *Prog. Org. Coat. 21*: 205 (1992).
96. J. R. Sully, D. C. Silverman and M. W. Kendig, eds., *Electrochemical Impedance: Analysis and Interpretation*, ASTM STP 118, ASTM, Philadelphia, 1993.
97. E. Frechette, C. Compère and E. Ghali, *Corros. Sci. 33*: 1067 (1992).
98. W. J. van Ooij, A. Sabata and Ih-Houng Loh, *Proc. Eur. Symp. on Modification of the Passive Film*, Paris, France, February 15–17, 1993.
99. T. Miyama and I. Fikui, *1987 Pittsburgh Conf. on Anal. Chem. and Appl. Spectrosc.*, paper No. 314.
100. Y. Furunushi, T. Suzuki and M. Shimizu, *Proc. Intern. Conf. on Zinc and Zinc Alloy Coated Steel Sheet—GALVATECH '89*, Iron and Steel Institute of Japan, Tokyo, 1989.
101. W. A. Grant, in *Methods of Surface Analysis* (J. M. Walls, ed.), Cambridge University Press, Cambridge, 1989, p. 299.
102. S. H. Sie, in *Surface Analysis Methods in Materials Science* (D. J. O'Connor, B. A. Sexton and R. S. T. Smart, eds.), Springer Verlag, Berlin, 1992, p. 203.
103. L. C. Feldman, in *Ion Spectroscopies for Surface Analysis* (A. W. Czanderna and D. M. Hercules, eds.), Plenum Press, New York, 1991, p. 311.
104. B. R. Appleton and G. Foti, in *Ion Beam Handbook for Material Analysis* (J. W. Mayer and E. Rimini, eds.), Academic Press, New York, 1977, p. 67.
105. B. Schueler and R. W. Odom, *J. Appl. Phys. 61*: 4652 (1987), and *Thin Solid Films 153*: 1 (1987).
106. D. G. Armour, in *Methods of Surface Analysis* (J. M. Walls, ed.), Cambridge University Press, Cambridge, 1989, p. 263.
107. D. J. O'Connor, in *Surface Analysis Methods in Materials Science* (D. J. O'Connor, B. A. Sexton and R. S. T. Smart, eds.), Springer-Verlag, Berlin, 1992, p. 245.
108. M. Aono and R. Souda, *Nucl. Instrum. Methods Phys. Res. B27*: 55 (1987).
109. P. Avouris, F. Boszo and R. E. Walkup, *Nucl. Instrum. Methods Phys. Res. B27*: 136 (1987).
110. J. T. Yates, Jr., M. D. Alvey, K. W. Kolasinkski, and M. J. Dresser, *Nucl. Instrum. Methods Phys. Res. B27*: 147 (1987).
111. S. H. Sie, in *Surface Analysis Methods in Materials Science* (D. J. O'Connor, B. A. Sexton and R. S. T. Smart, eds.), Springer Verlag, Berlin, 1992, p. 203.
112. H. D. Hagstrum, in *Electron and Ion Spectroscopy of Solids* (L. Fiermans, J. Vennik, and W. DeKeyser, eds.), Plenum, New York, 1978, p. 273.

113. C. Webb and P. M. Williams, *Surface Sci. 53*: 110 (1975).
114. A. B. Christie, in *Methods of Surface Analysis* (J. M. Walls, ed.), Cambridge University Press, Cambridge, 1989, p. 127.
115. R. Leckey, in *Surface Analysis Methods in Materials Science* (D. J. O'Connor, B. A. Sexton and R. S. T. Smart, eds.), Springer-Verlag, Berlin, 1992, p. 291.
116. M. M. El-Gomati and J. A. D. Matthew, *J. Microsc. 147*: 137 (1987).
117. M. P. Seah and C. G. Smith, *J. Mater. Sci. 21*: 1305 (1986).
118. R. L. Gerlach, *J. Vac. Sci. Technol. 8*: 599 (1971).
119. P. S. Turner in *Surface Analysis Methods in Materials Science* (D. J. O'Connor, B. A. Sexton and R. S. T. Smart, eds.), Springer-Verlag, Berlin, 1992, p. 79.
120. L. C. Feldmand and J. W. Mayer, *Fundamentals of Surface and Thin Film Analysis*, North-Holland, New York, 1986, p. 247.
121. J. L. Cocking, J. A. Sprague and J. R. Reed, *Surf. Coat. Technol. 36*: 133 (1988).
122. J. L. Cocking and G. R. Johnson, in *Surface Analysis Methods in Materials Science* (D. J. O'Connor, B. A. Sexton and R. S. T. Smart, eds.), Springer-Verlag, Berlin, 1992, p. 371.
123. D. Norman, *J. Phys. C:19*: 3273 (1986).
124. N. K. Roberts, in *Surface Analysis Methods in Materials Science* (D. J. O'Connor, B. A. Sexton and R. S. T. Smart, eds.), Springer-Verlag, Berlin, 1992, p. 187.
125. J. Comyn, A. J. Kinloch, C. C. Horley, R. R. Mallik, D. P. Oxley, G. R. Pritchard, S. Reynolds, and C. R. Werrett, *Int. J. Adhesion Adhesives 5*: 59 (1985).
126. C. W. White, E. W. Thomas, W. F. van der Weg and N. H. Tolk, in *Inelastic Ion-Surface Collisions* (N. H. Tolk, J. Tully, W. Heiland, and C. W. White, eds.), Academic Press, New York, 1977, p. 201.
127. G. Betz, *Nucl. Instrum. Methods Phys. Res. B27*: 104 (1987).
128. P. J. Jennings, in *Surface Analysis Methods in Materials Science* (D. J. O'Connor, B. A. Sexton and R. S. T. Smart, eds.), Springer-Verlag, Berlin, 1992, p. 275.
129. G. L. Price, in *Surface Analysis Methods in Materials Science* (D. J. O'Connor, B. A. Sexton and R. S. T. Smart, eds.), Springer-Verlag, Berlin, 1992, p. 263.
130. D. P. Woodruf and T. A. Delchar, *Modern Techniques of Surface Science*, Cambridge University Press, Cambridge, 1986, p. 14.
131. in ref. 130, p. 150.
132. in ref. 130, p. 157.
133. L. C. Feldman and J. W. Mayer, *Fundamentals of Surface and Thin Film Analysis*, North-Holland, New York, 1986, p. 31.
134. A. Vértes and I. Czakó-Nagy, *Electrochimica Acta 34*: 721 (1989).

11

Ultraviolet Stabilization of Adhesives

Douglas Horsey *CIBA-GEIGY Corporation, Ardsley, New York*

I. INTRODUCTION

Antioxidants (including hindered phenols, phosphites, and thioethers) are commonly used in hot-melt adhesive (HMA) and sealant formulations [1–4]. Relatively high levels of antioxidants are used (typically, 0.5 to 1.0%) in some adhesives (e.g., hot melts) to protect the polymers from oxidation when they are subjected to high temperatures during compounding and processing. The presence of these additives further protects the adhesive during the final end-use application. Antioxidants are also used in adhesives subjected to less severe processing (e.g., solvent- and water-based adhesives to protect the adhesives during storage and end use. The thermal stabilization of adhesives is addressed in Chapter 12.

Some applications require exposure of sealants and adhesives to ultraviolet (UV) radiation. While the conventional antioxidants mentioned above will provide some protection against UV-initiated oxidation, many applications necessitate the incorporation of "light stabilizers" to reduce photoinduced polymer degradation. Two general classes of light stabilizers are used, UV absorbers (UVA) and hindered amine light stabilizers (HALS). In this chapter we discuss the stabilization mechanism of each of these classes and the relative importance of each of these stabilizers in preventing degradation of a thin-film adhesive versus a "thick" sealant. Although we focus on the fundamental aspects of the UV stability of adhesives, several examples are given to illustrate how these principles can be put into practice.

II. PHOTODEGRADATION OF ADHESIVES AND SEALANTS

The general autooxidation scheme for hydrocarbons shown in Fig. 1 represents the core reaction for hydrocarbon degradation [5]. As adhesives are mixtures of polymers and oligomers, this scheme can be used to understand the oxidative degradation of adhesives. Several aspects of the scheme are worth emphasizing before we examine the effect of UV light on polymer degradation. First, the formation of a peroxyradical (ROO•) from a polymer free radical (R•) is an extremely fast (diffusion-controlled) reaction in the

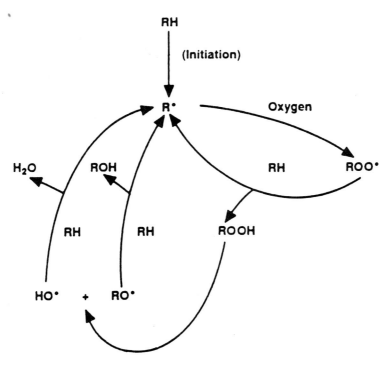

Figure 1 Autooxidation scheme for polymer degradation showing the cyclic nature of the process.

presence of oxygen. Second, the hydroperoxides (ROOH) formed from the peroxyradicals can further decompose to generate two new free radicals (RO•, HO•), which can initiate further reactions. This autocatalytic mechanism results not only in the cyclic nature of autoxidation but in an *exponential* increase in the number of free radicals available to initiate new degradation reactions. Although the same scheme is the basis for photooxidation of polymers, the presence of UV light exacerbates the situation, especially with regard to the initiation reactions.

Thermolysis of hydroperoxides (ROOH) becomes significant only above 100°C. Although this high-temperature reaction is a major cause of degradation in hot-melt adhesives, it is less important at ambient temperatures. However, *photolysis* of ROOH does occur to a significant extent at ambient temperatures and is a major cause of photoinduced polymer degradation [6]. Other photoinduced reactions can occur due to light absorption by trace levels of carbonyl impurities resulting from thermal oxidation of the polymer during manufacture, storage, and processing.

Although we focus here on these photooxidative pathways, it is important to mention the purely photolytic degradation of adhesives. Polymers that are chromophores in the UV range can directly absorb UV light without first undergoing oxidation. These polymers may undergo photolytic rearrangements, providing another nonoxidative pathway for loss of physical or chemical properties. Examples of these polymers are aromatic polyurethanes or polyesters.

III. ULTRAVIOLET ABSORBERS

The simplest and most direct method to reduce photooxidation is to block the UV radiation from reaching the polymer, much as a sunscreen is used to prevent sunburn. Fillers (carbon black, pigment, talc, TiO_2, etc.) can potentially provide an improvement in UV stability by this mechanism, but may cause other complications (e.g., additive adsorption or deactivation, metal impurities, etc.). Chemical ultraviolet absorbers are used routinely to prevent the photooxidation of polymers. There are several major requirements for chemicals to function as ultraviolet absorbers: (1) they must absorb strongly in the UV region (290 to 400 nm) but must have a sharp cutoff in the visible region (>400 nm) so as not to contribute color to the polymer, (2) they must be photostable, and (3) they must dissipate the photoexcitation energy in a harmless way. Representative structures of two common classes of UVA that fulfill these requirements, the 2-hydroxybenzophenones and the 2-hydroxyphenylbenzotriazoles, are shown in Fig. 2.

The absorption of light follows Beer's law, which says that the absorption of light equals a constant (extinction coefficient) multiplied by the UVA concentration multiplied by the path length:

absorption = const. × concentration × path length

in which

$$absorption = \log \frac{I_0}{I}$$

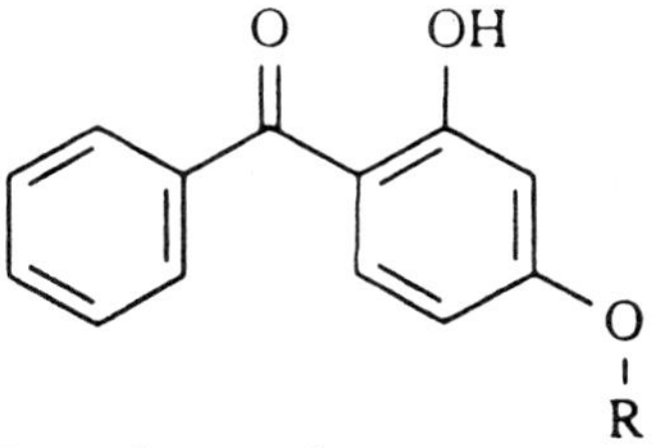

2-hydroxybenzophenones

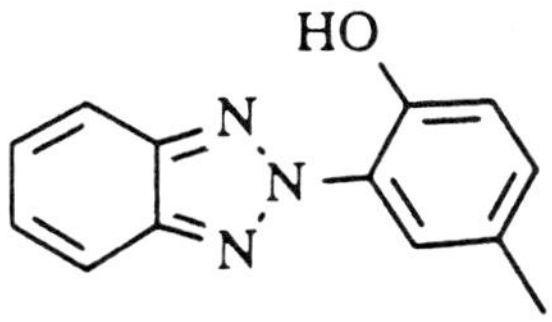

BTZ-1 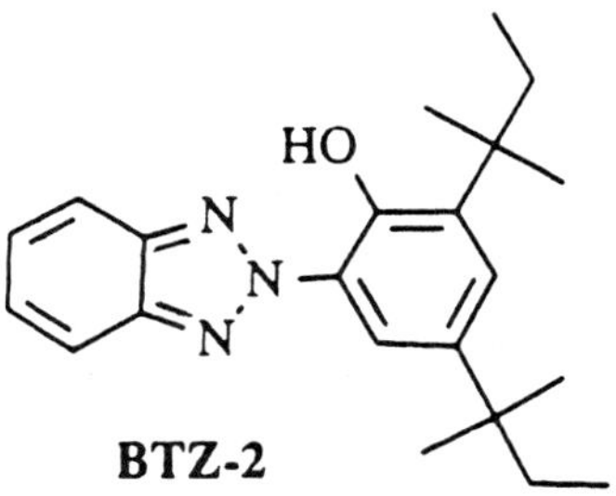 **BTZ-2**

2-hydroxyphenylbenzotriazoles

Figure 2 Representative structures of ultraviolet absorbers.

where I_0 is the intensity of incident light and I is the intensity of light having passed through the sample. This relationship implies that both the type and concentration of the absorbing species are important as well as the sample thickness.

To illustrate the function and limitations of these UVAs, we will look at the optical properties of a typical 2-hydroxyphenylbenzotriazole (BTZ). The UV absorption properties are shown in Fig. 3 for several concentrations of a solution of BTZ-2. Although the actual amount of light absorbed is a function of wavelength, for this discussion we can use an approximate value of 90% UV light absorption at a BTZ concentration of 0.0025% and a path length (sample thickness) of 1.00 cm. Beer's law tells us that if 90% of the light is absorbed by a 0.0025% BTZ solution in 10 mm (0.4 in.), 90% of the light is also absorbed by a 0.025% BTZ solution in 1.0 mm (0.04 in.), and 90% of the light will be absorbed by a 0.25% BTZ solution in 0.10 mm (0.004 in., 4 mils). A BTZ concentration of 0.25 to 1.0% is typical for protection of a polymer.

Carrying this further, if at this concentration of 0.25% BTZ, 90% of the light is absorbed in the first 0.10 mm, then 99% of the light is absorbed in the first 0.20 mm and 99.9% is absorbed in the first 0.30 mm of the sample. This explains why UV absorbers are very effective at protecting all polymers of sufficient thickness but are not effective at protecting sample surfaces or very thin films. This is demonstrated clearly in Fig. 4, where a sample of polypropylene exposed to UV from sunlight is microtomed and analyzed for degradation (loss of molecular weight and hence viscosity) [8]. With no light stabilizer, UV radiation is at a high intensity throughout the transparent sample. As oxygen diffuses in from both sample surfaces, the UV light accelerates oxidation. In the presence of a UVA, the bulk of the sample is protected except for the surface exposed to the UV light. (The effect of the HALS is explained below.) This dependence on sample thickness also explains why UV absorbers are widely used in sealant application but have

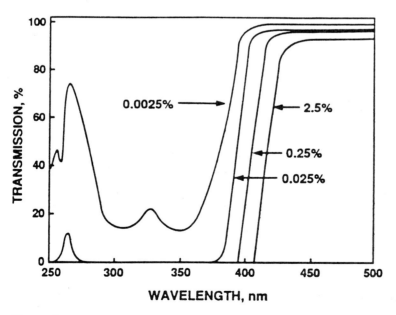

Figure 3 UV absorption spectra for various concentrations of a solution of BTZ-2 (1.0-cm cell in cyclohexane).

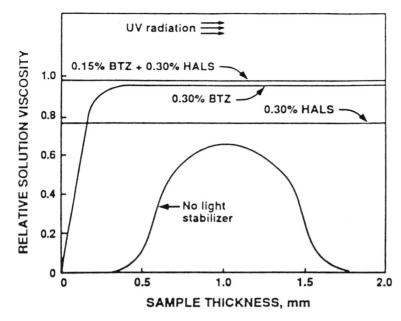

Figure 4 Effect of UV exposure on a 2-mm-thick polypropylene at various depths.

limited efficacy in adhesive films, where typical thickness may be about 0.025 mm (1 mil).

IV. HINDERED AMINE LIGHT STABILIZERS

As alluded to above, a second class of light stabilizers, the hindered amine light stabilizers (HALS), provide additional stability to polymers and can function even in thin film or in sample surfaces. HALS chemistry is based on the 2,2,6,6-tetramethylpiperidine derivatives (Fig. 5). Unlike UVA, HALS have no significant UV absorption and therefore have no effect on the incident radiation. The mechanism of protection by HALS is still being investigated [6,9], but the primary mechanism involves the HALS acting as an extremely efficient hydroperoxide decomposer and free-radical scavenger, involving a cyclic mechanism that regenerates the active HALS species (Fig. 6).

Referring back to Fig. 1, the HALS functions not by blocking the offending UV radiation but by destroying the unstable and radical-initiating hydroperoxides and by scavenging free radicals before they can be involved in the propagation reactions. Thus, by attacking a second pathway, the addition of a HALS to a polymer can provide additional stability beyond what can be achieved with a UVA alone. More important, since HALS efficacy is not dependent on competitive absorption of light, HALS protect the surface layer of a polymer and can protect thin films. Again this is demonstrated in Fig. 4, where even the exposed surface of the polymer is protected from degradation. The term *hindered amine light stabilizer* is a somewhat limiting misnomer, as this class of additive is also finding wide use as a thermal stabilizer. Unfortunately, HALS, functioning as antioxidants, cannot protect aromatic polymers from the nonoxidative photolytic processes described above.

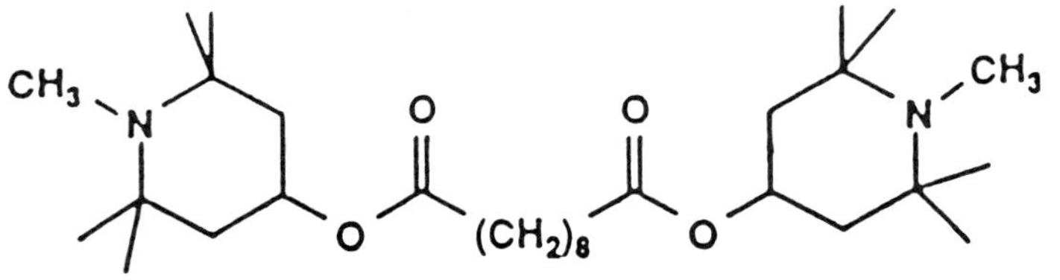

HALS - 2

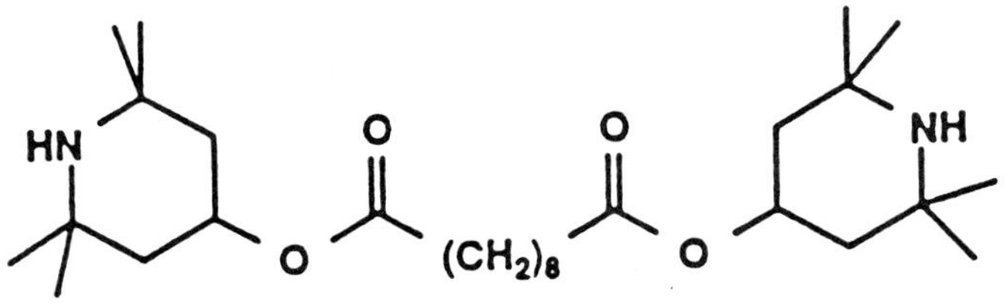

HALS - 1

Figure 5 Representative structures for hindered amine light stabilizers.

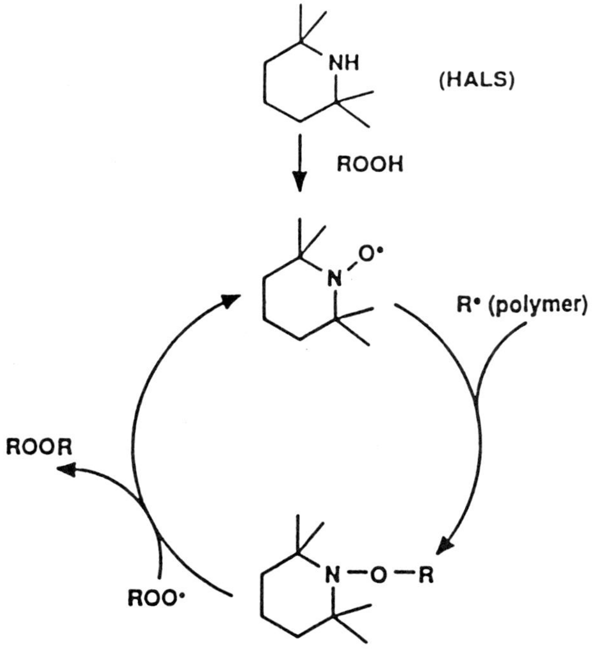

Figure 6 Regenerative mechanism for hindered amine light stabilizers.

V. ACCELERATED AGING OF POLYMERS

While the most meaningful exposure testing of polymers is done under actual exposure to sunlight, accelerated aging devices are used routinely to provide predictive and reproducible data in a reasonable amount of time [10,11]. The UV output of several commercial weathering devices are shown in Fig. 7. While many of these instruments are equipped to regulate temperature, humidity, and water spray, the most important environmental parameter to duplicate is the spectral distribution of sunlight. For instance, a carbon-arc lamp emits a high intensity of low-wavelength, high-energy radiation, which may initiate photoreactions not seen in actual use. The xenon-arc lamp more closely matches the spectral distribution of sunlight and has been found to correlate well with outdoor weathering. Other exposure devices, including the Q-U-V weathering tester [11], have shown good correlation with outdoor exposure when used with a UV-340 bulb.

VI. UV ABSORPTION BY HYDROCARBON POLYMERS

The photostability of most hydrocarbon polymers are generally not a function of their tendency to absorb light but rather of their propensity toward oxidation (free-radical formation). Although it is well known that the inherent UV stability of an unsaturated polymer is inferior to that of a saturated analog, this effect cannot be attributed to an increase in light absorption by the unsaturated system. For example, the UV spectra of a styrene–butadiene–styrene (SBS) block copolymer compared to the saturated styrene–

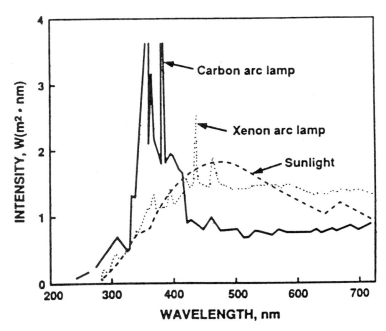

Figure 7 Comparison of sunlight versus artificial weathering device's spectral distribution.

ethylenebutylene–styrene (SEBS) analog shows no significant absorption in the important 300 to 400 nm UV range for sunlight. Thus for many polymers the relative UV stability correlates well with the thermal stability. Obviously, this may not hold where polymer photolysis becomes important.

VII. UV STABILITY OF ADHESIVE FILMS

With the considerations noted above, we can now look at the performance of UV stabilizers in several illustrative examples. Table 1 shows the effectiveness of BTZ-2 and HALS-1 at preventing the degradation of a natural rubber/hydrocarbon tackifier water-based adhesive when exposed to UV light. Natural rubber undergoes chain scission upon oxidation, resulting in a loss of molecular weight and hence a loss of cohesive strength. Samples were coated on polyester film, adhered to a glass plate, and exposed in a carbon-arc weatherometer (CAW). The extent of degradation was monitored by following discoloration, and by peeling the tape from the glass and observing cohesive failure. As shown by the data and as would be predicted from the arguments above, a HALS is more effective than the BTZ at protecting the adhesive. The BTZ/HALS combination was as effective as the HALs alone, which may indicate that a lower concentration of HALS would be sufficient under these exposure conditions.

In the second example (Table 2), two acrylic adhesive formulations are compared. The first contains a hydrogenated rosin ester tackifier, which shows relatively good UV stability. This tackifier can be replaced by one of lower stability (and lower cost), such as the nonhydrogenated rosin ester, by the incorporation of the appropriate light stabilizers. As these data show, some improvement in stability is seen when a BTZ is used at 0.5% (or a combination of BTZ/HALS at 0.25%/0.25%), but the best stability is achieved when the adhesive is stabilized with a HALS at the same level (0.5%). It should be pointed out that BTZ are commonly used in films and coatings (e.g., automotive clear coats) when the objective is to protect the material below the film. Similar applications can be imagined

Table 1   UV Stability of Natural Rubber Adhesive After Accelerated Aging in a Carbon-Arc Weatherometer (1.0 Mils Thick)

		Exposure in carbon-arc weatherometer	
	Unaged	8 h	16 h
Yellowness index[a]			
Unstabilized	6	9	10
BTZ-2, 2.0%	4	7	8
HALS-1, 2.0%	4	4	4
BTZ-2/HALS-1, 1/1%	4	4	4
Cohesive failure[b]			
Unstabilized	Okay	Fail	Fail
BTXZ-2, 2.0%	Okay	Okay	Fail
HALS-1, 2.0%	Okay	Okay	Okay
BTZ-2/HALS-1, 1/1%	Okay	Okay	Okay

[a]Higher numbers indicate more color development.
[b]Visual evaluation after peeling polyester film from glass slides.

Table 2 UV Stability of Tackified Acrylic Adhesive Film After
Accelerated Aging in a Xenon-Arc Weatherometer (1.0 Mils Thick)[a]

| | Yellowness index:[b] exposure in xenon-arc weatherometer | | | |
Stabilizer	0 h	50 h	100 h	250 h
Hydrogenated tackifier[c]				
Unstabilized	7	9	10	10
Nonhydrogenated tackifier[d]				
Unstabilized	9	17	18	20
BTZ-2, 0.5%	9	10	12	13
HALS-1, 0.5%	8	8	9	10
BTZ-2/HALS-1, 0.25/0.25%	9	10	11	12

[a]Formulation contains 70% acrylic polymer and 30% tackifier.
[b]Higher numbers indicate more discoloration.
[c]Hydrogenated rosin ester tackifier.
[d]Nonhydrogenated rosin ester tackifier.

where UVA-containing adhesive films are used to screen and protect other substrates. Conversely, in cases where the adhesive is very difficult to stabilize, the UVA can be incorporated into the substrate exposed to UV to protect the adhesive.

To illustrate the effect of sample thickness, we can compare adhesives to sealants, which in many cases can be viewed as thick adhesives. As the sample thickness is increased, the benefit of the BTZ is clearly demonstrated. In Table 3 a SEBS/hydrogenated hydrocarbon tackifier sealant formulation was prepared as a hot melt and poured into shallow petri dishes. Although both of these polymers have good inherent stability, sealant applications may require extended exposure to UV radiation. The discoloration data show that the BTZ prevents yellowing of the sealant. However, examination of the sealant surface shows surface crazing and cracking when not protected by incorporation of the HALS. The combination of the two classes of light stabilizers provides the best overall performance.

As shown above regarding acrylic adhesive formulation, the use of light stabilizers may allow the incorporation of a less stable (and less expensive) tackifier in sealant

Table 3 UV Stability of SEBS Sealant Film After Accelerated Aging
in a Xenon-Arc Weatherometer (2.0 Mils Thick)[a]

| | Apparent Gardner color: hours XAW exposure | | | |
Stabilizer	0	500	1000	2000
Unstabilized	1	5	6	6
BTZ-1, 0.5%	1	2	3	3
HALS-1, 0.5%	1	4	4	5
BTZ-2/HALS-1, 0.25/0.25%	1	3	3	3

[a]Formulation contains 70% SEBS and 30% hydrogenated resin tackifier. Compounded as hot melt.

Table 4 UV Stability of SEBS Sealant After Accelerated Aging
in a Xenon-Arc Weatherometer (2.0 Mils Thick)[a]

	Yellowness index:[b] hours XAW exposure				
	0	100	500	750	1000
Hydrogenated HC tackifier[c]					
BTZ-2/HALS-1, 0.2%/0.2%	9	9	11	12	17
Hydrogenated HC/rosin ester[d]					
BTZ-2/HALS-1, 0.0%/0.0%	6	55	110	120	125
BTZ-2/HALS-1, 0.2%/0.2%	6	25	80	90	100
BTZ-2/HALS-1, 0.4%/0.4%	7	9	60	80	80
BTZ-2/HALS-1, 0.8%/0.8%	7	8	16	28	30
BTZ-2/HALS-1, 1.5%/1.5%	7	7	9	10	11

[a]Formulation contains 70% SEBS and 30% tackifier. Solvent-based solutions coated onto glass slides.
[b]Higher numbers indicate more discoloration.
[c]Hydrogenated hydrocarbon tackifier.
[d]Containing a mixture of 50% hydrogenated hydrocarbon and 50% pentaerythritol rosin ester tackifier.

formulations. The data in Table 4 compare two SEBS sealants, one containing a hydrogenated hydrocarbon tackifier and the second containing a mixture of the same hydrocarbon tackifier with a rosin ester. Comparable stability to the "stable tackifier" can be achieved in a mixed formulation with a combination of BTZ and HALS.

VIII. SELECTION OF LIGHT STABILIZERS

In the preceding discussion we dealt with BTZ and HALS as classes of light stabilizer without much attention to structure variations within each class. As demonstrated by the numerous chapters of this book, many polymers are used for adhesive applications. Each of these materials may have unique photostability concerns. It is beyond the scope of this chapter to address the stabilization of specific adhesives. Similarly, it is beyond our scope to discuss the many structural variations of UVA and HALS chemistry. Many derivatives of these compounds are commercially available, and some adhesive and sealant formulations and/or applications may dictate the use of one stabilizer over another.

In the case of BTZ, the UV absorption properties of the compounds are relatively similar, with some variations in absorptivity and absorption maxima (λ_{max}). Since sufficient BTZ is typically used to remove essentially all the light in the surface layer of the polymer, the performance characteristics of BTZs are usually comparable. BTZ selection is therefore based on physical properties, not photochemical properties. Properties such as volatility (molecular weight) and physical form are usually considered, as well as economic concerns.

In the case of HALS, the activity is dependent on the amine functionality concentration. Therefore, the highest-activity HALS are generally those with a high portion of the molecular weight contributed by an amine such as HALS-1 and HALS-2. Additional structure built into the molecules may dilute and decrease the activity, but in some applications may increase polymer compatibility or decrease volatility. Oligomeric HALS have found extensive utility in high-surface-area applications such as films and fibers, due to their reduced volatility. Almost by definition, adhesives have high surface areas, but they also are between two substrates, which will limit volatility. As in the case of BTZ, physical properties are also important. Additionally, due to the basicity of the amine functionality, the use of conventional HALS is intolerable in some applications, and structural modifications of the HALS functionality are used to reduce interactions.

ACKNOWLEDGMENT

The author thanks CIBA-GEIGY Corporation, Ardsley, New York, for permission to publish these results.

REFERENCES

1. A. Patel and R. Thomas, *Tappi.* *70*(6): 166 (1988).
2. P. Rota Graziosi, G. Knobloch, H. Martin, and A. Patel, *Tappi.* *72*(12): 71 (1989).
3. D. Horsey, G. Knobloch, A. Patel, and H. Martin, *Tappi Hot Melt Symposium Proc.* 1990.
4. For a review of antioxidants, see P. Klemchuk, *Ullmann's Encyclopedia of Industrial Chemistry*, 5th ed., VCH Publishers, Deerfield Beach, Fla., 1985, Vol. A3, p. 91; or F. Gugumus, in *Plastics Additives Handbook*, 3rd ed. (R. Gachter and H. Muller, eds.), Oxford University Press, New York, 1990, p. 1.
5. J. Bolland and G. Gee, *Trans. Faraday Soc., 42*: 236, 244 (1946).
6. P. P. Klemchuk, *Encyclopedia of Material Science and Engineering*, Pergamon Press, Elmsford, N.Y., 1988, p. 5200.
7. F. Gugumus, in *Plastics Additives Handbook*, 3rd ed. (R. Gachter and H. Muller, eds.), Oxford University Press, New York, 1990, p. 129.
8. A. Patel and J. Usilton, in *Stabilization and Degradation of Polymers*, Advances in Chemistry Series 16 (D. L. Allara and W. J. Hawkins, eds.), American Chemical Society, Washington, D.C., 1978, p. 116.
9. P. P. Klemchuk, M. E. Gande, and E. Cordola, *Polymer Degradation Stabilization 27*: 65 (1990).
10. J. Wypych, *Weathering Handbook*, Chemtec Publishing, Toronto, 1990.
11. G. Fedor and P. Brennan, *Adhesives Age, 33*(5): 22 (1990).

12
Thermal Stabilization of Adhesives

Neal J. Earhart and Ambu Patel *CIBA-GEIGY Corporation, Ardsley, New York*

Gerrit Knobloch *CIBA-GEIGY Corporation, Basel, Switzerland*

I. INTRODUCTION

Adhesives are, in general, produced by the compounding of several different components: polymers, tackifier resins, and waxes or oils. The hydrocarbon-based components are susceptible to thermal oxidation and degradation [1]. Many adhesives are exposed to elevated temperatures as a result of compounding, storage, and end use. Hot-melt adhesives (HMAs), for example, are extremely prone to oxidation due to their high compounding and application temperatures. Each hydrocarbon-based component in an adhesive formulation follows a similar oxidation mechanism. The result of degradation, however, can be very different and can greatly influence the overall thermal stability of the adhesive formulation.

II. MECHANISM OF THERMAL OXIDATION
AND DEGRADATION

The basic scheme for the autooxidation of polymers is detailed in Table 1 [2]. The scheme can be broken down into several distinct reactions: initiation, propagation, and termination. The first step in the oxidation mechanism, initiation, occurs when a polymeric free radical (R•) is formed by exposure to heat, light, shear, or impurities. The propagation reactions involve the very rapid reaction of the polymer free radicals with oxygen-forming peroxy radicals (ROO•). The peroxy radicals can then react with the polymer to produce hydroperoxides (ROOH), which can decompose further to form two new free-radical species (RO• and HO•), which, in turn, can participate in other propagation reactions. The mechanism of autooxidation is cyclic in nature and leads to an exponential growth of free radicals. The termination mechanisms of these free-radical reactions are cross-linking and chain scission. The ramifications of degradation via cross-linking include hardening, skinning, gel formation, a decrease in tack, and an increase in viscosity. Degradation via chain scission results in softening, a viscosity decrease, an increase in tack, and a loss of cohesive strength. In addition to the change in physical properties as a result of degradation, discoloration is also possible. Although sample discoloration, in itself, may have a

Table 1 Major Reactions: Polymer Oxidation, Chain Scission, and Cross-Linking

Initiation

$$ROOH \xrightarrow{\text{heat}} RO\cdot + HO\cdot$$

$$2ROOH \longrightarrow RO\cdot + ROO + H_2O$$

$$RH \xrightarrow{\text{heat/light}} R'\cdot + H\cdot$$

$$R'\cdot + O \longrightarrow R'OO\cdot$$

$$RO\cdot + R'H \longrightarrow ROH + R'\cdot$$

Propagation

$$ROO\cdot + R'H \longrightarrow ROOH + R'\cdot$$

$$RO\cdot + R'H \longrightarrow ROH + R'\cdot$$

$$HO\cdot + R'H \longrightarrow H_2O + R'\cdot$$

$$R'\cdot + O_2 \longrightarrow R'OO\cdot$$

$$R'OO\cdot + R'H \longrightarrow R'\cdot + R'OH$$

Termination and cross-linking

$$R'\cdot + R'\cdot \longrightarrow R'\text{-}R'$$

$$R'O\cdot + R'\cdot \longrightarrow R'OR'$$

$$R'OO\cdot + R'\cdot \longrightarrow R'OOR'$$

$$\left. \begin{array}{l} R'OO\cdot + R'O\cdot \longrightarrow \\ R'OO\cdot + R'OO\cdot \longrightarrow \end{array} \right\} \begin{array}{l} \text{nonradical} \\ \text{products} \end{array}$$

Chain scission

$$\sim CH_2 - \underset{\underset{OOH}{|}}{\overset{\overset{CH}{|}}{C}} - CH_2 \sim \; \rightarrow HO\cdot + \sim CH_2 \quad \underset{\underset{O\cdot}{|}}{\overset{\overset{CH}{|}}{C}} - CH_2 \sim$$

$$\sim CH_2 - \overset{\overset{O}{\|}}{C} - CH_3 + \sim \cdot CH_2$$

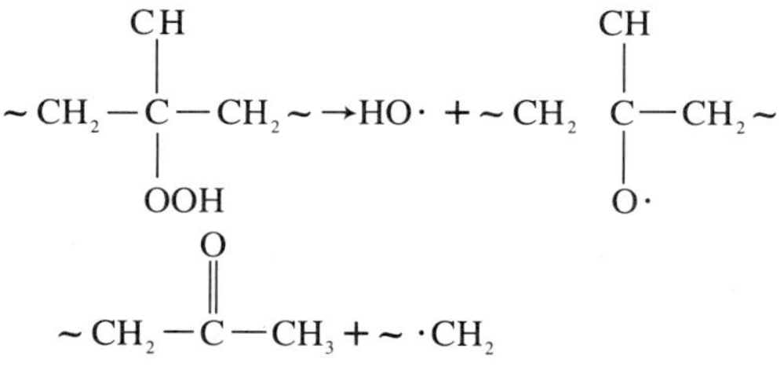

ROOH = peroxide (impurities)

R'H = polymer

minimal effect on the physical properties of an adhesive, the color formation results in an undesirable appearance which may be interpreted as being that of an inferior product.

The tactics used to inhibit the autooxidation process involve the obstruction of one or more of the degradation pathways. To inhibit or prevent the undesirable effects of degradation as a result of thermal oxidation, antioxidants are used in adhesive formulations [3,4]. Antioxidants are generally added to each component of an adhesive formulation and, in most cases, are also added to the final formulation.

III. ANTIOXIDANTS

There are two major classes of antioxidants and they are differentiated based on their mechanism of inhibition of polymer oxidation: chain-terminating or primary antioxidants and hydroperoxide-decomposing secondary antioxidants [5]. Primary or free-radical scavenging antioxidants inhibit oxidation via very rapid chain-terminating reactions. The majority of primary antioxidants are hindered phenols or secondary aryl amines. Generally, hindered phenols are nonstaining, nondiscoloring, and are available in a wide range of molecular weights and efficiencies. Amine-based primary antioxidants are very effective. However, they tend to interfere with peroxide cross-linking, and they discolor and stain. They are used primarily in situations where their color addition is unimportant or can be hidden.

A typical hindered phenol primary antioxidant is AO-1 (2,6-di-*tert*-butyl-*para*-cresol) (Table 2). Stabilization is achieved through proton donation from the — OH of AO-1 to a peroxy or alkoxy radical (Fig. 1). This reaction is in favorable competition with proton donation from a polymer carbon atom. Important to note, however, is that the resulting phenoxy radical is stable and does not abstract a proton from the polymer chain. This would be an undesirable effect, because the antioxidant would then be acting as a chain transfer agent. Due to its low cost, AO-1 is widely used as antioxidant. In adhesive formulations, where exposure to high temperatures is possible [e.g., hot-melt adhesives (HMAs)], the high volatility level of AO-1 renders it virtually useless. To solve this problem, state-of-the-art lower-volatility antioxidants are used to achieve the necessary level of stabilization (Table 2). In addition to their lower volatility, these antioxidants show higher activity, compatibility, and a resistance toward the formation of colored by-products during compounding and application temperatures of HMAs.

The performance of a primary antioxidant can be improved by the use of a secondary antioxidant. Secondary antioxidants or peroxide decomposers do not act as radical

Table 2 Key to Antioxidants

AO-1	2,6-di-*t*-butyl-*p*-cresol
AO-2	Tetrakis[methylene(3,5-di-*t*-butyl-4-hydroxylhydrocinnamate)]methane
AO-3	2,4-bis(*n*-octylthio)-6-(4'-hydroxyl-3',5'-di-*t*-butylanilino)-1,3,5-triazine
AO-4	2,4-bis[(octylthio)methyl]-*o*-cresol
AO-5	Triethyleneglycol-bis-3-(3'-*t*-butyl-4'-hydroxy-5'-methylphenyl)propionate
AO-6	Butylated reaction product of *p*-cresol and dicyclopentadiene
AO-7	Hindered phenol

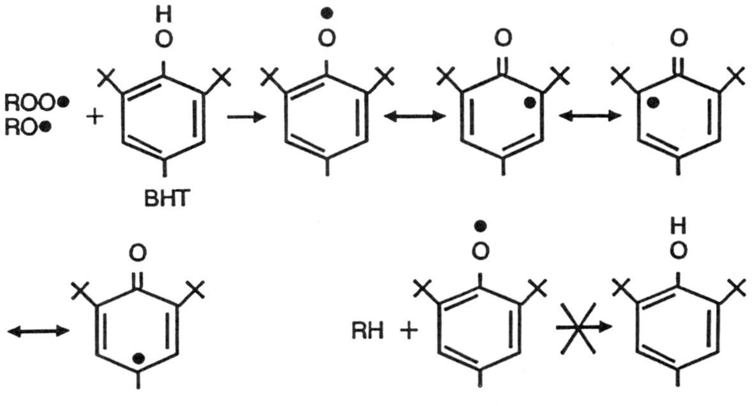

Figure 1 Radical stabilization using hindered phenols.

Phosphites:

$$(RO)_3P + R'OOH \longrightarrow (RO)_3P{=}O + R'OH$$

Thiosynergists:

$$R{-}S{-}R + R'OOH \longrightarrow R{-}S{=}O + R'OH \quad (+ \text{ further oxidation})$$
$$\underset{\displaystyle R}{\overset{\displaystyle |}{}}$$

Figure 2 Radical stabilization with secondary antioxidants.

scavengers but undergo redox reactions with hydroperoxides to form nonradical stable products (Fig. 2). This class of antioxidants (Table 3) includes phosphites such as tris(nonylphenyl)phosphite (PS-1) and thiosynergists or thioesters such as dilauryl thiodipropionate (TS-1). Phosphites reduce hydroperoxides to alcohols as they are oxidized to phosphate. Phosphites are generally highly effective process stabilizers and are nondiscoloring. Thiosynergists or thioesters are also nondiscoloring and are used with primary antioxidants to achieve long-term heat stability. Secondary antioxidants are generally used exclusively in combination with primary antioxidants. Thus secondary antioxidants are referred to as "synergists" because the overall level of stability achieved when a primary and a secondary antioxidant are used in combination is much greater than if either were used alone.

A variety of factors, including compounding or processing conditions, end use, and expected performance, should influence the selection of an appropriate antioxidant system. These factors include compatibility with the polymer and other additives or components, antioxidant mobility and volatility, discoloration, resistance to hydrolysis, extraction resistance, radical trapping efficiency, toxicity, and cost-effectiveness.

As mentioned previously, adhesives are primarily a blend of several hydrocarbon-based components, each of which is susceptible to thermoxidative degradation. The stabilization of adhesive compounds against thermooxidative degradation is complex. Typically used to inhibit or prevent degradation, antioxidants are added to each hydrocarbon component [6] as well as to the final adhesive formulation. To gain a better understanding of the stabilization of an adhesive formulation, it is beneficial to examine the degradation mechanisms of the individual components.

IV. STABILIZATION OF ADHESIVE COMPONENTS (RAW MATERIALS)

A. Stabilization of Polymers (Elastomers)

Typically, the polymers used in adhesive formulations have only a minimal level of stabilization to endure isolation/coagulation, drying/finishing, and warehouse storage. As an adhesive producer, it is most desirable to understand the level of performance of a stabilized polymer and determine whether additional stabilization is needed to provide the necessary level of performance.

Table 3   Key to Secondary Antioxidants

PS-1	Tris(nonylphenyl)phosphite
TS-1	Dilauryl thiodipropionate

1.   Stabilization of Ethylene–Vinyl Acetate Copolymers

Ethylene–vinyl acetate (EVA) copolymers are used in HMAs. The EVA acts as the binder, contributing cohesive strength to the adhesive formulation. Typically, an EVA used in a HMA is approximately 18 to 28 mol % vinyl acetate. In an EVA copolymer, the crystalline polyethylene (PE) region provides strength, compatibility with the wax, and the desired high-temperature properties. The amorphous region containing both VA and PE provides compatibility with the tackifier.

    Figure 3 shows the performance during static oven aging at 170°C (338°F) of a stabilized EVA polymer. The base stabilization of the EVA polymer by the producer using AO-1 provided an unsatisfactory level of stability. The presence of skinning and a more pronounced level of discoloration in the base AO-1–stabilized EVA requires additional antioxidant to meet the performance needs of a HMA. Upon the addition of AO-2 to the base polymer, it is clear that the stability of the EVA is improved significantly, skin formation is not observed, and color development is reduced substantially. The formation of insoluble gel as a result of cross-linking is also reduced dramatically with the addition of AO-2, as shown in Fig. 4.

2.   Stabilization of SIS Thermoplastic Elastomers (Styrene–Isoprene–Styrene Block Copolymer)

Thermoplastic elastomers (TPEs) with blocks of polydiene rubber are subject to degradation at the carbon–carbon double-bond sites and require proper stabilization. In SIS block copolymers, chain scission is the predominate degradation mechanism. In a SIS block copolymer, the addition of a more effective stabilizer, AO-3, alone or blended with a

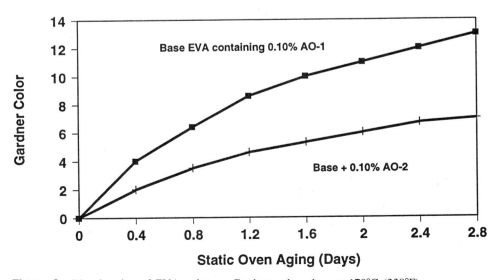

Figure 3   Discoloration of EVA polymer: Gardner color, days at 170°C (338°F).

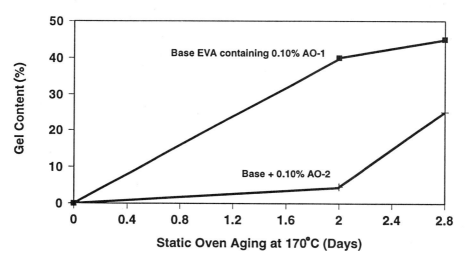

Figure 4 Gel formation in EVA polymer: gel content, % insoluble in toluene at 25°C (77°F).

secondary antioxidant, PS-1, can provide a significantly superior performance over AO-1 alone or with PS-1. Resistance to discoloration after static oven aging at 80°C (176°F) is improved dramatically (Fig. 5). Viscosity stabilization (melt flow index stability) (Fig. 6) is also improved drastically using AO-3/PS-1.

3. Stabilization of SBS Thermoplastic Elastomers (Styrene–Butadiene–Styrene Block Copolymer)

SBS block copolymers degrade primarily via cross-linking. The thermal stability of SBS copolymers can be improved with the addition of a more effective stabilizer system. Figures 7 and 8 show gel formation and discoloration of a SBS copolymer stabilized with different stabilizer systems after static oven aging at 70°C (158°C). As seen previously in

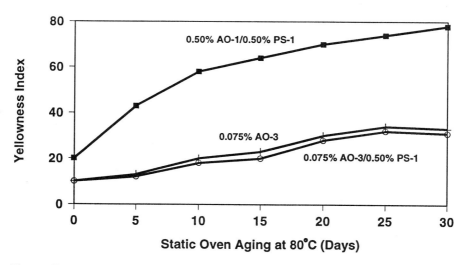

Figure 5 Discoloration of SIS polymer: yellowness index, days at 80°C (176°F).

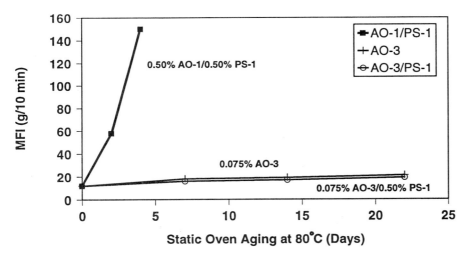

Figure 6 Viscosity stabilization of SIS polymer: melt flow index, days at 80°C (176°F). MFI (g/10 min)-200°C/5 kg.

SIS copolymers, a higher-molecular-weight, more effective stabilizer such as AO-3, alone or in combination with a secondary antioxidant, PS-1, provides superior color stability and resistance to gel formation than does the AO-1/PS-1 stabilizer system.

4. Stabilization of Carboxylated Styrene–Butadiene (X-SBR) Latices

Carboxylated SBR latices are used as adhesives in applications where durability and flexibility are desired. Some of the major uses for X-SBR latex are in tufted carpet backing, paper coatings, wall and vinyl floor tile adhesives, and pressure-sensitive adhesives. Typically, discoloration is the first measure of the degradation of an X-SBR latex. Discoloration of a dried latex film can often be related to a loss of the physical

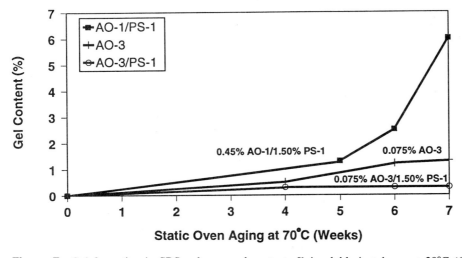

Figure 7 Gel formation in SBS polymer: gel content, % insoluble in toluene at 20°C (68°F).

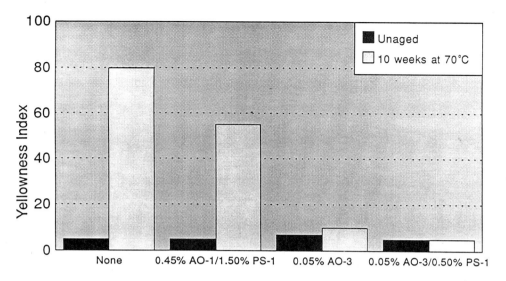

Figure 8 SBS polymer oven aging at 70°C (158°F): color formation, yellowness index.

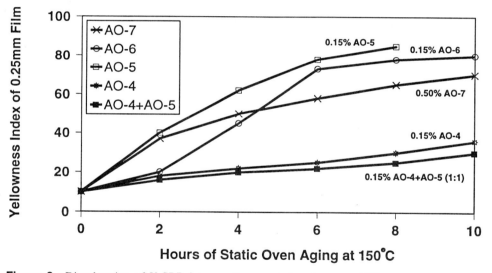

Figure 9 Discoloration of X-SBR latex: yellowness index, hours at 150°C (300°F).

properties and subsequently, to inferior performance in an adhesive formulation. Figure 9 illustrates the effects of adding an effective antioxidant system to an X-SBR latex on the level of discoloration as a result of static oven aging at 150°C (300°F). The addition of AO-4 and AO-5 alone an in a 1:1 ratio dramatically increases the resistance to discoloration compared to that of a traditional stabilization system such as AO-6 or AO-7.

B. Stabilization of Tackifier Resins

The adhesion or performance of nonpolar elastomers to various substrates can be increased by the addition of a tackifier resin. The tackifier modifies the elastomer by improving wettability, modifying the viscoelastic properties, and increasing polarity.

Tackifier resins are susceptible to thermooxidative degradation. It is not uncommon for tackifier resins to degrade rapidly at room temperature. The integrity of the tackifier resin is very important to the final properties of an adhesive formulation. Tackifier resins must be effectively stabilized to prevent degradation during storage in order to maintain their properties until production of the adhesive formulation. Degradation leads to changes in both the chemical and physical properties of the tackifier, such as compatibility, melt viscosity, and discoloration. The mechanism of tackifier oxidation follows the autooxidation as outlined in Table 1. The degradation of tackifiers can be followed by determining the hydroperoxide formation, color development, and viscosity changes during static oven aging at warehouse storage temperatures.

1. Stabilization of Rosin Ester Tackifier Resins

In static oven aging at 40°C (104°F) of a rosin ester tackifier (Fig. 10) the rate of hydroperoxide formation was reduced significantly using AO-2, with even better results using AO-3. The hydroperoxides are fairly stable at room temperature. At temperatures associated with hot-melt compounding or drying of solvent- and water-based formulations, hydroperoxides decompose spontaneously. The decomposition products initiate further reactions, which can result in the formation of color species. The addition of AO-2 and AO-3, which reduces the level of hydroperoxides formed, subsequently reduces the level of tackifier discoloration after oven aging (Fig. 11).

Melt viscosity relates to stability during processing and to end-use performance. A stable melt viscosity is a very important property of a rosin ester tackifier. The melt viscosity can also be related to the hydroperoxide content of the tackifier after aging. Figure 12 shows that the addition of AO-2 and AO-3 can significantly reduce the increase in the melt viscosity of the rosin ester during static oven aging at 40°C (104°F).

2. Stabilization of C_5-Hydrocarbon Tackifier

As shown in rosin ester–based tackifiers, the addition of AO-2 can greatly reduce the hydroperoxide formation of a C_5-hydrocarbon–based tackifier during static oven aging at 40°C (104°F) (Fig. 13).

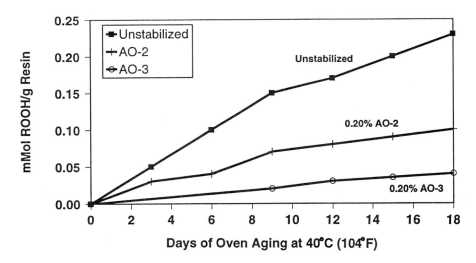

Figure 10 Stabilization of rosin ester tackifier: hydroperoxide formation during oven aging at 40°C (104°F).

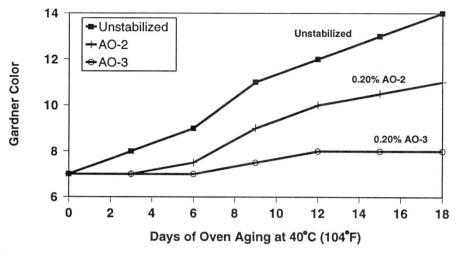

Figure 11 Stabilization of rosin ester tackifier: Gardner color after oven aging at 40°C (104°F).

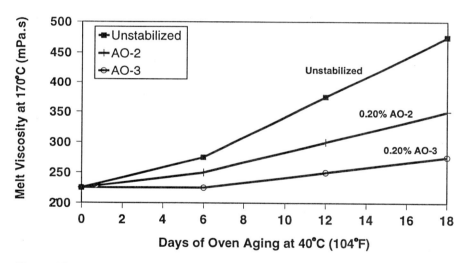

Figure 12 Stabilization of rosin ester tackifier: melt viscosity [at 170°C (338°F)] after oven aging at 40°C (104°F).

3. Effects of Tackifier Stability on the Physical Properties of a HMA

The prolonged effects due to thermal oxidation of a tackifier during storage correlates directly with the level of discoloration and viscosity changes in the HMA formulation. An effectively stabilized tackifier will produce a HMA with good color and controlled viscosity. When used in a HMA formulation, an unstabilized tackifier will result in a HMA with a high degree of discoloration and an unstable viscosity. Stabilizing a HMA formulation, however, will not correct for the addition of an unstabilized or preoxidized tackifier. The best performance can be achieved with the addition of an effective stabilizer to both the tackifier and the HMA.

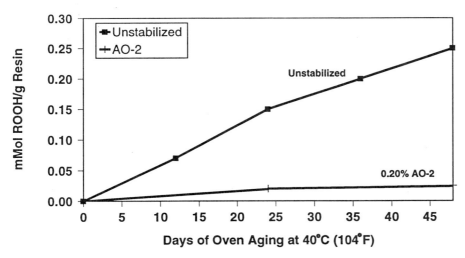

Figure 13 Stabilization of C₅-hydrocarbon tackifier: hydroperoxide formation during oven aging at 40°C (104°F).

C. Stabilization of Waxes

Waxes are used primarily in hot-melt adhesive formulations. Waxes are generally highly crystalline hydrocarbons. Waxes are added to HMA formulations to lower cost and decrease viscosity. Some of the HMA properties that are affected by wax content are the softening point and open time. Typically, waxes are thermally stable. During high-temperature storage and compounding, however, waxes can discolor very rapidly. Degradation of a wax can result in a reduction in the thermooxidative stability of the overall HMA formulation. As shown in Fig. 14, the addition of an antioxidant during static oven

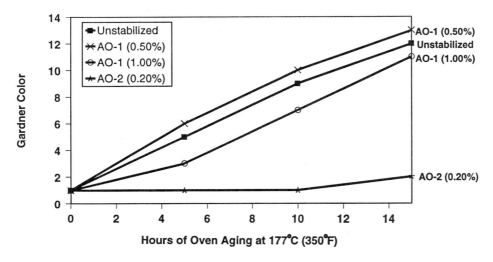

Figure 14 Stabilization of a microcrystalline wax: color formation during oven aging at 177°C (350°F).

aging at 177°C (350°F) of a microcrystalline wax can greatly reduce the degree of discoloration. The less volatile AO-2, at lower concentrations, significantly outperforms the more volatile AO-1. This reinforces the importance of volatility when an antioxidant is selected for a high-temperature application.

D. Influence of Stabilized Raw Materials on Adhesive Properties During Production and Storage

It has been demonstrated that the hydrocarbon-based raw materials of an adhesive formulation do undergo thermooxidative degradation and that the addition of antioxidants is necessary to maintain the integrity of these components during storage and compounding. In this section we illustrate the effects of using unstabilized raw materials in adhesive formulations in comparison with stabilized raw materials.

1. EVA Hot-Melt Adhesives

The effects of storage time of an unstabilized and a stabilized rosin ester tackifier on the properties of an EVA HMA are illustrated in Figs. 15 to 17. Significant effects on the initial color of the EVA HMA (Fig. 15) are observed when using an unstabilized tackifier. An increased level of hydroperoxides is also noted. In this situation, the addition of an antioxidant to the HMA will not correct the problem. However, the addition of an antioxidant to the HMA may reduce further discoloration during compounding or end-use applications.

Figure 16 illustrates the effects of an unstabilized tackifier on color formation as a result of high-temperature aging of the EVA HMA formulation. In this scenario the tackifier was aged for 18 days at 50°C and then combined with the other components at

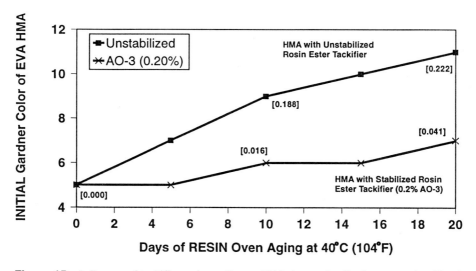

Figure 15 Influence of tackifier resin quality on EVA: hot-melt adhesive properties. Numbers in brackets represent ROOH content of the resin before compounding.

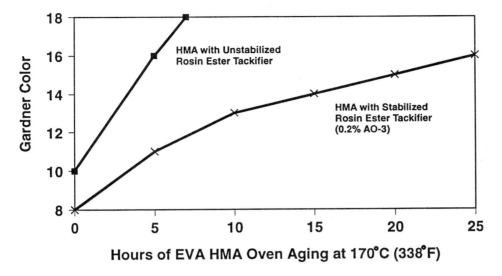

Figure 16 Influence of tackifier resin quality on EVA: hot-melt adhesive properties—discoloration. Resin aged for 18 days at 50°C before compounding.

177°C (350°F). The final HMA formulation was then aged at 170°C (338°F). Use of the unstabilized tackifier results in a darker initial color and a more rapid rate of discoloration than that of HMA using the stabilized tackifier.

A consistent viscosity of the EVA HMA during high-temperature aging can be achieved using a stabilized tackifier (Fig. 17). The EVA HMA using the unstabilized

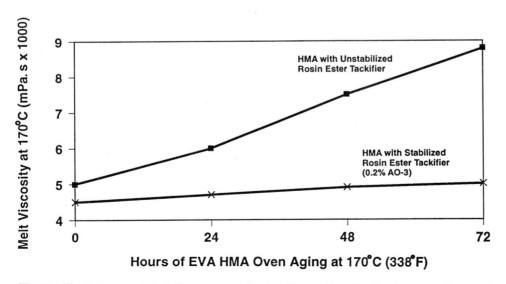

Figure 17 Influence of tackifier resin quality on EVA: hot-melt adhesive properties—melt viscosity. Resin aged for 18 days at 50°C before compounding.

tackifier has a higher initial viscosity and a more severe increase in viscosity during aging than those of the EVA HMA prepared with the stabilized tackifier.

E. Stabilization of Adhesive Formulations

In addition to stabilization of the raw materials, increased performance of an adhesive can also be achieved by the addition of antioxidants to the completed formulation. Typically, the antioxidant is added during the compounding stages of an adhesive.

1. Stabilization of SIS Hot-Melt Adhesives

SIS-based HMAs degrade under high-temperature compounding and application conditions. Degradation of the bulk SIS HMA results in skin formation, viscosity changes, and discoloration. In a pressure-sensitive adhesive (PSA) film application, the film can degrade under moderate storage conditions, resulting in loss of tack and peel strength. Improved performance of both the bulk adhesive and the PSA film can be achieved with the addition of AO-2 (Figs. 18 and 19).

2. Stabilization of EVA Hot-Melt Adhesives

EVA hot-melt adhesives are widely used in the packaging and bookbinding marketplace due to their superior adhesion to most substrates, their versatility, and their ease of formulation. EVA-based HMAs degrade under high-temperature processing and application conditions. Degradation usually results in discoloration, viscosity changes, and skin formation. EVA HMA can be effectively stabilized against discoloration and viscosity changes using AO-2 (Figs. 20 and 21).

V. CONCLUSIONS

The hydrocarbon-based raw materials used by the adhesives industry are prone to thermooxidative degradation. This degradation can occur during isolation, storage, compounding, and end use. Raw material degradation can effect the performance of the final

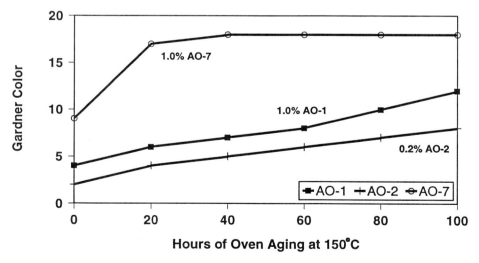

Figure 18 Stabilization of an SIS HMA: color development during oven aging at 150°C (300°F).

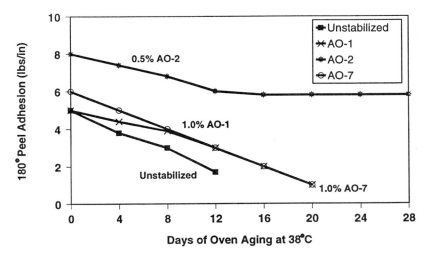

Figure 19 Stabilization of an SIS HMA: peel adhesion after oven aging at 38°C (100°F).

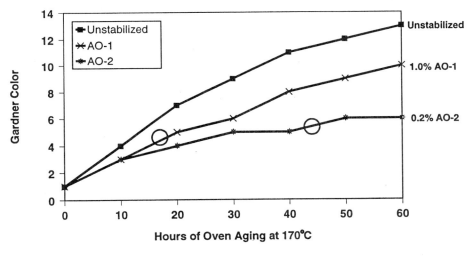

Figure 20 Stabilization of an EVA HMA: color development during oven aging at 170°C (338°F). (O), onset of skinning.

adhesive formulation detrimentally. Raw materials that have been stabilized with effective stabilizers in the early stages of their production will have increased storage life while maintaining consistent quality. Effective stabilization of the raw materials will result in an adhesive formulation with improved physical properties. The use of antioxidants only in the final formulations cannot "reverse" the effect of using predamaged unstabilized raw materials. The performance of the final adhesive formulation can be improved significantly by the further addition of effective antioxidants. Selection of the most effective antioxidant system to meet the performance demands will produce an adhesive that will maintain its physical properties during storage, compounding, and end use.

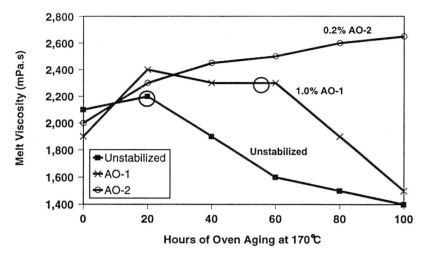

Figure 21   Stabilization of an EVA HMA: melt viscosity during oven aging at 170°C (338°F). (O), onset of skinning.

ACKNOWLEDGMENTS

The authors would like to thank CIBA-GEIGY Corporation, Ardsley, New York, and CIBA-GEIGY Ltd., Basle, Switzerland, for permission to publish this paper.

REFERENCES

1. J. Bolland and G. Gee, *Trans. Faraday Soc. 42*: 236 (1946).
2. A. Patel, in *Modern Plastics Encyclopedia*, McGraw-Hill, New York, 1984.
3. A. Patel and R. Thomas, *Tappi. 70*(6): 166 (1988).
4. D. Horsey, G. Knobloch, A. Patel, and H. Martin, *Tappi Hot Melt Symposium*, Tappi Press, Atlanta, June 1990, p. 21.
5. P. Klemchuk, *Ullmann's Encyclopedia of Industrial Chemistry*, 5th ed., VCH Publishers, Deerfield Beach, Fla.; 1985, Vol. A3, p. 91.
6. P. Rota Graziosi, G. Knobloch, H. Martin, and A. Patel, *Tappi. 72*(12): 71 (1989).

III
Adhesive Classes

13

Protein Adhesives for Wood

Alan L. Lambuth *Boise Cascade Corporation, Boise, Idaho*

I. INTRODUCTION

Human beings have apparently had a propensity for gluing things together since the dawn of recorded history and probably before. By experimentation, we learned over time that certain natural products or extractives could develop bonds in wood of sufficient strength to break in the material being glued rather than in the glue film itself. These observations ultimately led to papyrus laminating, decorative wood veneering, and furniture and musical instrument assembly, practiced by early Chinese and Egyptians and virtually every other civilization since that time using nature's own glues [1–3].

By today's standards, the ancient adhesive raw material choices were limited. Starch, blood, and collagen extracts from animal bones or hides were the principal early sources. Somewhat later, milk protein and fish skin extracts were discovered and included. Interestingly, vegetable proteins appear *not* to have been utilized as adhesives until recent times [4]. Tree pitch and petroleum bitumen were known and exploited as weatherproof coatings and caulks but not as adhesives, due to their plastic-flow behavior [5]; that is, the joints would creep and thus could not be used for structural support.

Adhesive durability has always been a problem. Although starch and protein glues were able to maintain long-term adhesive strength when kept completely dry, none were to any significant degree resistant to water or mold. Heat-cured blood glues and casein glues eventually provided some moisture resistance, but that was as durable as the technology of the times permitted. This fact limited the uses of adhesives strictly to interior or at least covered exterior applications from early historical times down through the Industrial Revolution and nearly to the present. From the middle of the nineteenth century onward, knowledge regarding more efficient means of dispersing and denaturing protein adhesives began to accumulate [6,7]. The results were significant improvements in protein glue working properties, bond strength, and water resistance, but still well short of true exterior durability. The urgencies of World War I brought protein-based adhesives to their nearest attainment of this goal in the form of chemically denatured, heat-set blood glues. These were used to assemble, among other things, laminated wooden airplane propellers and structural components in support of the war effort [8].

In the early 1920s the first of the phenol–formaldehyde and urea–formaldehyde resin adhesives were developed [9,10]. These early resins were slow to cure and somewhat difficult to use, but they ushered in the era of ''thermoset'' polymeric adhesives and true exterior durability. The urgencies of World War II forced the rapid development of these and other synthetic resin glues for their water and weather resistance but left protein adhesive technology to evolve and flourish in applications for which ''interior'' durability was still adequate [11]. In the years following the war, the greatly expanded and largely idled petrochemical industry went looking for appropriate new markets. Among the opportunities identified were synthetic resin adhesive applications. Within a decade, these were converted from costly wartime adhesive specialties to bulk commodity glues. Throughout this synthetic resin expansion, however, protein adhesives held onto their ''interior durable'' product applications, due in large part to their unique combination of low cost and cold-curing capability or, alternatively, very fast hot press curing (about twice as fast as phenolics) [12,13]. This situation continued into the 1960s, at which time the price of petrochemically based adhesives had become so low that they literally displaced protein adhesives from their traditional interior markets. Specifically, phenol and urea–formaldehyde resins replaced blood, soybean, and starch glues in all plywood and composite wood panels; resorcinol–formaldehyde resins replaced casein glues in lumber laminating and millwork applications; and poly(vinyl acetate) and acrylic emulsion glues replaced virtually all collagen adhesives (animal and fish bone/skin derived) from furniture, musical instruments, and general interior wood assembly [14]. Limited and specialized applications for protein glues, mainly in combination with synthetic resin polymers, continue to the present, however. These are discussed in the following sections. It is worth noting that protein glue technology in each of the application areas described above remains fully useful and industry approved in a backup sense and could be reintroduced rapidly if world petroleum resources were to be threatened again through events beyond national control, as in 1973.

In this chapter we address three of the most widely used families of protein-based adhesives for wood: soybean, blood, and casein. The technology presented is drawn primarily from the years 1930 through about 1960, when the consumption and technical refinements of these adhesives were at their peak. Soybean glues are discussed first because they were often utilized in combination with blood or casein to yield adhesives of intermediate performance properties as well as being used alone.

II. SOYBEAN ADHESIVES

A. Raw Material Source and Preparation

Soybeans are legumes, the seeds of a low-growing field vine. These vines are ancient in culture; the written record of their domestication in China dates back almost 5000 years [15]. From that time until now, soybeans have remained a very important agricultural crop for almost every temperate-climate civilization because of their unusually high content of both triglyceride oil and edible protein.

To process soybeans into these useful products, the beans are dehulled and the oil is removed by crushing at very high pressure or by solvent extraction. If the resulting dry soybean meal is intended for food, it is heated to 70°C or higher to coagulate the proteins and caramelize the carbohydrates, thus improving their nutritional qualities. If the soybean meal is intended for adhesive use, it is processed at temperatures below 70°C to preserve the alkaline solubility of the proteins [16].

The protein content of oil-free soybean meal ranges from about 35 to 55% on a worldwide basis. However, the industrial grades are generally blended to yield a uniform protein content of 44 to 52%, depending on the source. The other principal constituents of soybean meal are carbohydrates, totaling about 30%, and ash at 5 or 6% [17]. The moisture content after processing is quite low, usually less than 10%.

Production experience has shown that to perform well as a protein glue, adhesive-grade "untoasted" soybean meal must be ground to an extremely fine flour [18]. Typically, the dry extracted meal is ground or milled until at least 40%, and preferably 60 to 80%, will pass through a 46-μm (325-mesh) screen. For easier quality control with flours of this fineness, an alternative specific surface test method is available that determines average particle size in terms of surface area per gram [19]. For the range of mesh sizes recommended above, the corresponding specific surface values are about 3000 to 6000 cm^2/g.

B. Formulation

Soybean flour will wet and swell in plain water but will not disperse to yield useful adhesive properties. For this purpose, treatment with a soluble alkaline material is necessary. Almost any organic or inorganic alkali will disperse wetted soybean flour to some degree. However, soybean wood glues of maximum bonding efficiency require dispersion with several percent of a strong alkali such as sodium hydroxide or trisodium phosphate [20]. The effect of this strongly alkaline treatment is to break the internal hydrogen bonds of the coiled protein molecules, literally unfolding them and making all their complex polar structure available for adhesion to wood. Although essential for adhesion, this alkaline dispersion process exposes the protein structure to gradual destruction by alkaline hydrolysis. Thus a dispersed soybean glue has a definite useful life, slowly losing viscosity and adhesive functionality over a storage period of 6 to 12 h.

Although these strongly alkaline soybean glues are nearly colorless in an applied film, they cause a reddish-brown stain on wood surfaces as they cure, due to alkali burn of the cellulose itself [5]. If a colorless glue line on wood is desired, the wetted soybean flour must be dispersed with a less strongly alkaline material, such as hydrated lime or ammonia [21]. However, the adhesive bond strength of these low-color, mildly dispersed soybean glues is considerably less than that obtained with fully dispersed, highly alkaline formulations. Typical high- and low-alkali soybean glue formulations are listed in Tables 1 and 2.

The additions of hydrated lime and sodium silicate solution in the high-alkali mix (Table 1) accomplish two purposes: (1) they help maintain a level glue viscosity for a longer adhesive working life, and (2) they improve the water resistance of the cured glue film by forming some insoluble proteinates [22].

The starchy constituents of soybean flour also disperse in the presence of strong alkali to become useful adhesive molecules contributing to dry bond strength. However, this starchy fraction also retains its well-known sensitivity to water and is considered primarily responsible for limiting the performance of soybean glues with respect to water resistance [23].

The final addition of preservative shown in this formulation is essential in virtually all protein glues to provide mold resistance in high-humidity service. Without this protection, even heat-cured soybean adhesives *will* mold as the moisture content of the bonded wood approaches 20% [24]. Copper-8-quinolinolate, copper naphthenate, and orthophenyl phenol are among the few remaining preservatives permitted to be used in the United

Table 1 High-Alkali Soybean Glue: Ingredients and Mixing Procedure

	Amount (kg)
Water at 16–21°C	87.5
Adhesive-grade soybean flour[a]	48.5[b]
Pine oil or diesel oil defoamer: mix 3 min or until smooth in a counterrotating mixer	1.5[b]
Water at 16–20°: mix 2 min or until smooth	72.5
Fresh hydrated lime (as a slurry in)	6.0
Water at 16–21°C: mix 1 min	12.0
50% Sodium hydroxide solution: mix 1 min	7.0
Sodium silicate solution: mix 1 min	12.5[c]
Orthophenyl phenol: mix 10 min	2.5

[a]44% protein, specific surface 3000–6000 cm^2/g.
[b]Normally dry-blended for easier handling and dust control.
[c]8.90% Na_2O, 28.70% SiO_2, 41% Baumé.

States at this time for wood products. Where the use of chlorinated phenols is still permitted, they are also very effective preservatives for protein glues at the addition level shown. In this case, the sodium hydroxide content of the glue formulation converts the water-insoluble chlorinated fungicides to their soluble sodium salts.

Large quantities of this high-alkali soybean glue formulation were used to bond interior grades of softwood plywood between about 1940 and 1960 [13]. It was also used to some extent for assembling prefabricated wooden building components [11]. Its primary advantages were very low cost and the capability to bond almost any dry wood surface. It also offered real versatility in a bonding process because it could be hot pressed or cold pressed to promote cure. Appropriate pressing schedules for each curing mode are provided in tables that appear later in this chapter.

The low-alkali formulation (Table 2) has been used widely as a briquetting binder for wood, charcoal, and other absorbent particles. It is particularly suitable for paper and softboard laminating, where a colorless glue line and minimum swelling of the glue film on high-humidity exposure are desired [25]. It is not recommended for structural uses such as sheathing plywood because of its lower degree of protein dispersion and thus

Table 2 Low-Alkali Soybean Glue: Ingredients and Mixing Procedure

	Amount (kg)
Water at 16–21°C	112.5
Adhesive-grade soybean flour[a]	48.5[b]
Pine oil or diesel oil defoamer: mix 3 min or until smooth in a counterrotating mixer	1.5[b]
Water at 16–21°C: mix 2 min or until smooth	75.0
Fresh hydrated lime (as a slurry in)	15.0
Water at 16–21°C: mix 5 min	25.0

[a]44% protein, specific surface 3000–6000 cm^2/g.
[b]Normally dry-blended for easier handling and dust control.

lower bonding strength. Substituting borax or monosodium phosphate for the hydrated lime dispersing agent will yield similar nonstaining glues.

Over the years a number of denaturants or cross-linkers have been added to soybean glues to improve their water resistance, working life, and consistency. These may be roughly categorized as formaldehyde donors, sulfur compounds, and inorganic complexing salts. Each of these groups of compounds appears to react with the starchy constituents of whole soybean flour as well as the dispersed protein molecules. Formaldehyde itself acts too rapidly and thus is difficult to control. Instead, such compounds as dialdehyde starch, dimethylol urea, sodium formaldehyde bisulfite, and hexamethylenetetramine have been used successfully to toughen the cured glue film and improve its water resistance [26,27]. Similarly, carbon disulfide, thiourea, and ethylene trithiocarbonate, among the sulfur compounds [28,29], and the soluble salts of cobalt, chromium, and copper have been used to improve soybean glue working properties and adhesive performance [30,31]. These modifiers are generally added last when preparing the glue. The range of addition of all such denaturants is 0.1 to 1.0% based on the weight of soybean flour. Also, 5 to 20% of an aliphatic epoxy resin has been added to soybean glues, yielding significantly improved durability, but the cost is high [32]. Similarly, alkaline phenol–formaldehyde (PF) resins have been incorporated for both durability and mold resistance. Proportions have ranged from a straight soybean glue with a minor resin addition to a primarily PF resin glue containing a moderate quantity of soybean flour [33,34]. Soybean protein is presently being evaluated as a cost-reducing adhesive constituent in phenolic and urea–formaldehyde resin binders for wood particle and cellulosic waste reconstituted products.

C. Mixing, Application, and Pressing

Soybean glues are very easy to mix, provided that they are wetted with plain water as a first step. (If any form of alkali is present in the first mixing water, the dry soybean flour will form permanent lumps.) As with all protein glues, the first mix is kept thick to break down any lumps of dry powder that may be present. The division of water additions in both formulations listed earlier demonstrates this mixing procedure. Once the soybean flour particles have been uniformly wetted, further dilution and dispersion steps can follow without difficulty. Water additions are adjusted to yield a mixed glue viscosity in the broad range of 500 to 25,000 cP at 25°C, depending on purpose. Briquetting and paper laminating glues would typically be 500 to 1000 cP, while cold-press plywood glues should be 10,000 to 20,000 cP for best performance [35]. Hot-press formulations would be midrange. Particularly because of the heavy first mixing stage and the high final viscosities, an appropriate soybean glue mixer should have relatively strong, slow-turning blades plus a counterrotating scraper for continuous removal of glue buildup from the mixer walls.

Low-viscosity soybean briquetting adhesives are generally applied by spray. Paper and softboard laminating glues are usually applied by curtain coater, knife, or indirect roller. High-viscosity plywood and lumber assembly formulations are fairly well limited to application by spreader roll or extrusion. For very small assembly jobs, soybean glue can easily be applied by brush.

One of the real advantages of protein glues generally is their ability to be cured (under pressure) either hot or cold. Typical commercial schedules for each mode of cure are given in Tables 3 and 4. The cold-pressing schedule in Table 4 is the result of an interesting laboratory observation and subsequent industry-wide patent [35]. Alkaline protein glues, particularly soybean glues, lose water quite rapidly into adjacent dry wood

Table 3   Soybean Glue: Hot-Pressing Process for Interior Douglas Fir
Plywood

(a) Pressing schedule

Rough panel thickness (mm)	Number of plies	Panels per press opening	Platen temperature (°C)	Pressing time at full pressure (min)
4.8	3	2	110	3
6.4	3	2	116	3
7.9	3	2	121	3½
7.9	3	3	127	5
9.5	3	2	127	4
11.1	3	1	116	3
11.1	3	2	132	4¾
12.7	5	1	110	3½
12.7	5	2	121	6
14.3	5	1	110	3¾
14.3	5	2	121	6
15.9	5	1	116	4
17.5	5	1	121	4
19.0	5	1	127	4¼
20.6	5	1	132	4½
20.6	7	1	127	5
23.8	7 and 9	1	132	5½
27.0	7 and 9	1	132	6
30.2	7 and 9	1	132	7
33.4	9	1	132	7½
36.5	9	1	132	8
39.7	9 and 11	1	132	9

(b) Glue application rates

Core thickness (mm)	Mixed glue per single glue line (g/m^2)
1.59	195
1.54, 2.82	208
3.18	220
3.63, 4.23, 4.76. 6.35	232
All constructions 5 or more plies, 20.6 mm and thicker	245

On rough or warm veneer, add at least 12 g extra glue spread.

(c) Other conditions

1. Total assembly time per press load. 15 min.
2. Veneer temperature not to exceed 43°C.
3. Veneer moisture content not to exceed 8%.
4. Not less than 14 kg/cm^2 uniform hydraulic pressure.

Table 4 Soybean Glue: Cold-Pressing (No-Clamp) for Interior Douglas
Fir Plywood

(a) Glue application rates

Core thickness (mm)	Mixed glue per single glue line (g/m^2)
2.54	305–318
2.82	313–323
3.18	318–330
3.63	325–337
4.23	330–342
4.76	330–342
All constructions 5 or more plies, 20.6 mm and thicker	367

For rough or warm veneer, add an extra 20 g over these spreads.

(b) Other conditions

1. Hold press load 5 min after assembling last panel before applying pressure.
2. Total assembly time limit per press load, 25 min.
3. Veneer temperature not to exceed 43°C.
4. Veneer moisture content not to exceed 8%.
5. Use 12–14 kg/cm^2 uniform hydraulic pressure.
6. Pressing time to be measured after gauge reaches full pressure.
7. Pressure to be retained for 15 min.

surfaces. As a result, they gain sufficient gel strength in 15 to 20 min to permit removal of a glued wood assembly from its clamping device without loss of intimate contact between the glued surfaces. Cure is then completed over the next 6 to 12 h simply by placing the bonded products in storage at ambient temperatures with minimum handling. This method of cold pressing, called the no-clamp process, was used throughout the softwood plywood industry for many years. Prior to its introduction, all protein-bonded cold-press plywood was clamped for 6 to 8 h with bulky steel beams and turnbuckles.

Because protein glues develop bond strength primarily by water loss over time, roll pressing has proved unsuccessful as a clamping method for wood products. The short, intense period of pressure simply squeezes the still-fluid mix off the glue line without affording sufficient time for water loss and gelation. An exception is the soft rubber roll or brush roll lamination of paper to paper or paper to wood [36]. This can be accomplished at production speeds because of the extreme rapidity with which dry paper removes water from a protein glue film.

D. Blended Formulations

As mentioned earlier, the single largest commercial use of soybean flour in wood glues during the recent past has been as a blend with other adhesive proteins, mainly blood and casein, for bonding interior-grade plywood, doors, and millwork. These blended formulations exploit several unique properties of the soybean glues themselves and incorporate useful adhesive characteristics from the other protein materials.

1. Soybean–Blood Glues

For example, a blend of soybean flour with spray-dried soluble animal blood, a fairly expensive but very efficient adhesive protein, yields a glue with the best properties of each material [37]: namely, the cost becomes moderate and the consistency ideal for wood product assembly (slightly granular) because of the soybean flour. The hot-press curing time is very short and the cured glue bonds are considerably more water resistant because of the blood's thermosetting properties. Fortunately, both proteins require the same neutral wetting procedure and strongly alkaline dispersion steps. They are otherwise compatible in all proportions, yielding a series of cost/performance-related adhesives. Soybean–blood blend glues were by far the most widely used protein hot-press adhesives for interior structural plywood from the early 1940s until about 1960 [13]. Also, when the oil embargo of 1973 quickly placed phenolic resin adhesives on allocation through petrochemical restrictions, the plywood industry immediately returned to the use of soybean–blood hot-press glues for interior structural grades. Typical examples of low- and high-blood-content soybean-blend glues are described in Tables 5 and 6.

Both glues are ready to use when mixed and have a working life of 6 to 8 h at inside temperatures. Several points of difference between these glues should be noted:

1. The water content of the high-blood-glue formulation is much larger, which offsets most of the material cost increase. This is possible because the "water requirement" of alkaline-dispersed blood is much higher than that of soybean flour.
2. The order of addition of alkaline dispersing agents in the high-blood mix is partially reversed. Experience has shown that this helps impart a more granular consistency to the dispersed blood, which is otherwise very slick and smooth.
3. The final addition of hexamethylenetetramine illustrates the use of a formaldehyde donor to partially denature or cross-link the dispersed proteins. This adds further

Table 5 Low-Blood-Content Soybean Blend Glue: Ingredients and Mixing Procedure

	Amount (kg)
Water at 16–21°C	100.0
Adhesive-grade soybean flour[a]	36.0[b]
Dried soluble animal blood	7.5[b]
74-μm Wood flour[c]	5.0[b]
Pine oil or diesel oil defoamer: mix 3 min or until smooth in a counterrotating mixer	1.5[b]
Water at 16–21°C: mix 2 min or until smooth	110.0
Fresh hydrated lime (as a slurry in)	4.0
Water at 16–21°C: mix 1 min	8.0
Sodium silicate solution: mix 1 min	20.0[d]
50% Sodium hydroxide solution: mix 5 min	5.0
Orthophenyl phenol: mix 5 min	2.5

[a]44% protein, specific surface 3000–6000 cm^2/g.
[b]Normally dry-blended for easier handling and dust control.
[c]0.074 mm (200 mesh) and finer.
[d]8.90% Na_2O, 28.70% SiO_2, 41° Baumé.

Table 6 High-Blood-Content Soybean Blend Glue: Ingredients
and Mixing Procedure

	Amount (kg)
Water at 16–21°C	80.0
Dried soluble animal blood	35.0[a]
Adhesive-grade soybean flour[b]	8.5[a]
74-μm Wood flour[c]	5.0[a]
Pine oil or diesel oil defoamer: mix 3 min or until smooth in a counterrotating mixer	1.5[a]
Water at 16–21°C: mix 2 min or until smooth	19.5
Fresh hydrated lime (as a slurry in)	4.0
Water at 16–21°C: mix 1 min	8.0
Sodium silicate solution: mix 1 min	22.5[d]
50% Sodium hydroxide solution: mix 5 min	8.0
Powdered hexamethylenetetramine: mix 3 min	1.0

[a]Normally dry-blended for easier handling and dust control.
[b]44% protein, specific surface 3000–6000 cm^2/g.
[c]0.074 mm (200 mesh) and finer.
[d]8.90% Na_2O. 28.70% SiO_2, 41° Baumé.

granular character to the mixed glue, lengthens its working life, and improves the water and mold resistance of the cured adhesive film.

4. As with straight soybean glues, the low-blood-content formulation requires a mold-inhibiting ingredient to meet plywood performance standards, whereas the high-blood-content glue does not [24].

2. Soybean–Casein Glues

Blends of soybean flour with ground and screened casein also yield a very useful series of protein adhesives, in this case, mostly cured cold [38]. While alkaline-dispersed casein yields strong, water-resistant cold-cured bonds in wood, its sticky dispersed consistency does not permit the rapid water loss needed for quick-clamping procedures. By combining it with an appropriate amount of soybean flour, the cold-press no-clamp process can be used. Formulations of this type have proved so successful for bonding plywood faces onto wooden flush-door cores and frames in a short cold-pressing cycle that the entire industry has employed these protein adhesives from about 1950 to the present [39]. As a special performance property, the bonds of soybean–casein door glues maintain strong adhesion in a fire until the glue lines are literally charred away. Thus glues of this type are widely used to bond flush-design fire doors. They are also excellent adhesives for millwork in general [40]. Some current formulations also contain minor amounts of blood.

The formulation of the typical soybean–casein blend glue listed in Table 7 is quite different from any protein adhesive described thus far in that all ingredients, dispersing agents included, are dry-blended into a single packaged composition that requires only the addition of water to prepare. The uniform oiling of all ingredients during the blending operation is a key, since it slows down the solution of the alkaline ingredients long enough for the soybean flour and casein to wet out under reasonably neutral conditions. Then the alkaline agents dissolve. Highly alkaline dispersing conditions are provided by reaction of the sodium salts with lime to yield sodium hydroxide in situ plus insoluble calcium salts

Table 7 Soybean–Casein Dry Glue: Ingredients and Mixing Procedure

	Amount (kg)
Adhesive-grade soybean flour[a]	29.0
250-μm Lactic acid casein[b]	9.5
Fresh hydrated lime	3.5
74-μm Wood flour[c]	2.5
Granular sodium carbonate	2.5
Granular sodium fluoride	1.0
Granular trisodium phosphate	0.5
Pine oil or diesel oil defoamer	1.5
Water at 16–21°C	100
Dry glue: mix 2 min, until smooth and thick; let stand 15 min or until thinning has occurred; mix until smooth	50
Water at 16–21°C: mix 2 min or until smooth	25

[a]44% protein, specific surface 3000–6000 cm^2/g.
[b]0.250 mm (60 mesh).
[c]0.074 mm (200 mesh) and finer.

[6]. (The sodium hydroxide needed for this strongly alkaline dispersion step could not be included in a one-package composition because of its hygroscopic behavior.) These sequential solution and dispersion reactions require some time for completion, which is the reason for the 15-min pause in glue preparation.

The dry ingredients are blended intensively in an appropriate dry powder mixer while the defoamer is sprayed in to provide uniform distribution. Mixing directions are given in Table 7. The second water addition may be increased or decreased to obtain the final viscosity desired. (A normal range is 4000 to 8000 cP.) Working life is 4 to 6 h at inside temperatures. Application rates must be determined by experience but will generally range from 245 to 345 g of mixed glue per square meter of a single glue line.

While the soybean–casein blend glue can be used according to the short-cycle no-clamp process on dry softwood, it will require 4 to 6 h of clamp time to cure to machining strength when used on dense hardwoods. Water removal from the glue film is simply too slow on hardwoods to develop adequate early gelation. However, the ultimate bond strength is excellent. Note that this formulation represents about the maximum casein content at which short-cycle clamping is possible for softwood flush-door or millwork assembly. Above this level water loss is too slow, interfering with normal production rates through the press.

III. BLOOD GLUES

Since soybean–blood blend glues were covered in the preceding section, in this portion of the chapter we deal only with all-blood adhesive compositions.

A. Raw Material Source and Preparation

Historically, animal blood could be used for adhesives only in reasonably fresh liquid form. These glues performed well on wood. However, the very rapid spoilage rate of liquid blood imposed real limitations on the general availability and use of this adhesive raw material. It was not until about 1910 that techniques were developed for drying whole blood in commercial quantities without denaturing its protein content, thus maintaining its water solubility [41]. As a result, blood could be collected, processed, and stored indefinitely for later use. The effect of this development was to stimulate rapid growth in the technology of blood-based adhesives, especially for wood, about the time of World War I.

Virtually all the proteins in animal blood can be dispersed into useful adhesive form. These include the serum albumin and globulin and even the red cell hemoglobin [42]. The fibrin clotting substance is sometimes removed before drying (by agitation or acidification) because of its instability in solution. Thus, except for residual moisture content, dried blood is essentially 100% active adhesive protein.

The principal North American bloods sold in quantity for adhesive uses are beef and hog, with lesser amounts from sheep and horses. Because of its high lysine content, poultry blood is utilized almost exclusively as a feed additive or binder and is seldom available otherwise. For adhesive purposes, there are significant viscosity differences relating to species among these dried bloods, beef being highest and poultry lowest [43]. Viscosity and water-holding properties are also influenced by animal age, diet, activity, and other factors. As a result, industrial-grade dried soluble blood is generally blended in large quantities to provide average and reproducible properties for adhesive formulating.

The method employed for drying blood is now entirely spray drying. (Formerly, a certain amount of vacuum pan–dehydrated blood was also available.) Spray-drying conditions relating to temperature, dwell time, and humidity can be adjusted to produce a wide range of blood solubilities [44]. Also, chemical denaturants such as glyoxal can be added to the blood solution prior to drying to further modify its adhesive characteristics [45]. Solubilities from about 20 to 95% can be prepared with ± 5% control. (Dried bloods below 20% solubility can only be redissolved in strongly alkaline solutions, which destroy a significant portion of the adhesive proteins.) This controllable range of solubilities permits the formulation of blood glues with a variety of handling and performance properties.

Generally speaking, the lower the solubility of a dried blood product, the more granular and water holding is its alkaline-dispersed form [46]. For instance, blood glues of 20 to 40% solubility make excellent cold-press formulations (which *must* have a granular consistency) [42]. They also yield the most water- and mold-resistant (near-exterior) glue bonds when cured hot. By comparison, highly soluble bloods in the range 85 to 93% yield very slick and livery alkaline dispersions of somewhat lower water-holding capacity. Soybean flour is normally blended with these highly soluble bloods to produce appropriately granular glues. If soybean flour is not used, they must be combined with a particulate cellulosic filler such as wood flour or nutshell flour to develop this functional consistency. Examples of these glue types are provided in the following section.

B. Formulation

As with the soybean glues discussed previously, dried blood adhesives must initially be wetted or redissolved in plain water and then be subjected to one or more alkaline dispersing steps. Unlike vegetable proteins, however, high-solubility blood proteins can be dispersed and rendered strongly adhesive by more moderate alkaline agents such as hydrated lime or ammonia [47]. Especially with a denaturing compound added, these glues represented the most water-resistant adhesives available until the advent of phenol–formaldehyde resins [5]. An example is shown in Table 8. This mix is unique in the quantity of denaturant it employs. The aldehyde reaction actually causes the blood protein

Table 8 Blood Glue: Ingredients and Mixing Procedure

	Amount (kg)
90% Soluble spray-dried animal blood	50.0
Water at 16–21°C: mix 3 min or until smooth	40.0
Water at 16–21°C: mix until smooth	30–60
Ammonium hydroxide, sp. gr. 0.90: mix 3 min	3.0
Powdered paraformaldehyde (sift in slowly while mixing)	7.5
Allow mix to stand 30 min; mix briefly until glue is fluid and smooth.	

to gel for a short period before thinning out again to a working viscosity level. The useful life is 6 to 8 h. This formulation can be cured hot or cold, but hot pressing yields the most durable bonds.

The next resurgence of blood glue technology came during and after World War II. By that time, the highly alkaline multistep dispersing systems of soybean glues had become well established and were employed successfully with blood glues. Two examples utilizing low-solubility blood in typical plywood glue formulations are shown in Tables 9 and 10. The second mix (Table 10) demonstrates the use of hot water to coagulate the blood and lower its solubility *during* the mixing procedure [48]. Both these glues are excellent adhesives for interior-grade plywood when cured either hot or cold. Preservative or denaturant additions are not normally required to meet plywood performance standards.

As a point of interest, blood glues are not affected by many of the protein denaturants used to improve the performance of soybean glues: specifically, sulfur compounds and complexing salts [49]. However, they are very sensitive to aldehyde-acting compounds, and these have been employed at levels of 0.1 to 1.0% to yield improved consistency and water resistance. Typical of these are glyoxal, paraformaldehyde, methylol ureas, and formaldehyde addition compounds such as dialdehyde starch and sodium formaldehyde bisulfite.

Table 9 Dry Heat-Treated Blood Glue: Ingredients and Mixing Procedure

	Amount (kg)
Water at 16–21°C	150
20% Soluble spray-dried animal blood	37.5[a]
74-μm Wood flour[b]	10.5[a]
Pine oil or diesel oil defoamer: mix 3 min or until smooth	2.0[a]
Water at 16–21°C: mix 2 min or until smooth	165.0
Fresh hydrated lime (as a slurry in)	5.0
Water at 16–21°C: mix 1 min	10.0
50% Sodium hydroxide solution: mix 10 min	8.0
Sodium silicate solution: mix 5 min	17.5[c]

[a]Normally dry-blended for easier handling and dust control.
[b]0.074 mm (200 mesh) and finer.
[c]8.90% Na_2O, 28.70% SiO_2, 41° Baumé.

Table 10 Hot Water–Coagulated Blood Glue: Ingredients
and Mixing Procedure

	Amount (kg)
Water at 63°C	100.0
90% Soluble spray-dried animal blood	40.0[a]
74-μm Wood flour[b]	9.0[a]
Pine oil or diesel oil defoamer: mix 10 min	1.0[a]
Water at 10–16°C	175.0
Pine oil or diesel oil defoamer: mix 2 min or until smooth	1.0
Fresh hydrated lime (as a slurry in)	3.5
Water at 10–16°C: mix 2 min	7.0
50% Sodium hydroxide solution: mix 2 min	7.5
Sodium silicate solution: mix 5 min	17.5[c]

[a]Normally dry-blended for easier handling and dust control.
[b]0.074 mm (200 mesh) and finer.
[c]8.90% Na_2O, 28.70% SiO_2, 41° Baumé.

A special class of blood protein denaturants, used primarily with higher levels of blood solubility, are the alkaline phenol–formaldehyde (PF) resins. Low-molecular-weight, low-alkali PF resins cause granulation of dispersed blood protein without much effect on viscosity, usually a reduction [50]. Highly advanced, high-alkali PF resins such as those used as plywood adhesives generally cause rapid thickening and gelation of dispersed blood glues if not employed with care [33,51]. Resins of intermediate advancement and alkalinity are almost passive to dispersed blood. These interactions have been exploited to formulate blood–resin glues for different hot-press applications at almost every level of combination. Two examples at the extremes of the range will suffice (Tables 11 and 12). In the case of the low-resin-content formulation (Table 11), PF resin addition also functions as a preservative agent and was used widely for the purpose. This formulation is for hot pressing only. The bond-durability level can be characterized as "mid-exterior."

For the high-resin-content formulation (Table 12), a partial addition of the PF resin can be made just after the initial mix and before sodium hydroxide addition if more fluidity is needed for propeller-type stirring. This formulation is more properly termed a blood-fortified exterior PF resin adhesive for hot-pressing plywood or laminated veneer lumber [52]. Even in quantities this small, the effect of the animal blood is to reduce the hot-press curing time by 20 to 30% over that of phenolic resins used alone. For purposes of adhesive solids calculation, the blood content can legitimately be included with the phenolic resin solids.

A special application for which 80% soluble blood is particularly suited is its use in phenolic resin glues as a foaming agent to produce "air-extended" PF adhesives [53]. These are currently used to manufacturer plywood on automated production lines. For this purpose, the mixed adhesive containing blood is put through a special high-speed stirring and air-injection system that lowers the specific gravity of the adhesive from about 1.0 to 0.2 with very fine air bubbles. The low-density adhesive foam is then extruded onto passing veneer surfaces, which are assembled and hot pressed to produce exterior grades of plywood. (Recycled glue is defoamed and recirculated.) The primary advantage of this

Table 11 Low-Resin Blood Glue: Ingredients and Mixing Procedure

	Amount (kg)
Water at 16–21°C	87.5
20% Soluble spray-dried animal blood	25.0[a]
90% Soluble spray-dried animal blood	12.5[a]
74-μm Wood flour[b]	11.0[a]
Pine oil or diesel oil defoamer: mix 3 min or until smooth	1.5[a]
Water at 16–21°C: mix 2 min or until smooth	200.0
Fresh hydrated lime (as a slurry in)	3.0
Water at 16–21°C: mix 1 min	6.0
Sodium silicate solution: mix 1 min	22.5[c]
45–50% Solids low to intermediate advancement PF resin: mix 3 min	13.5[d]

[a]Normally dry-blended for easier handling and dust control.
[b]0.074 mm (200 mesh) and finer.
[c]8.90% Na_2O, 28.70% SiO_2, 41° Baumé.
[d]Georgia-Pacific 3195, Borden Cascophen 335-I, Neste CB 118.

Table 12 High-Resin Blood Glue: Ingredients and Mixing Procedure

	Amount (kg)
Water at 16–21°C	250.0
74-μm Nutshell flour[a]	75.0
Winter wheat flour	25.0
90% Soluble spray-dried animal blood	17.5
Diesel oil defoamer: mix 5 min or until smooth	2.5
50% Sodium hydroxide solution: mix 2 min	28.0
Granular sodium carbonate: mix 15 min	10.5
43% Solids highly advanced PF resin: mix 5 min while cooling the glue at 21–27°C	610.0[b]

[a]0.074 mm (200 mesh) and finer.
[b]Georgia-Pacific 5763, Borden Cascophen 318-G, Neste CB 303.

kind of adhesive is lowered cost: for example, savings up to 25% over that of conventionally applied phenolic adhesives. A typical foamable glue mix is described in Table 13.

C. Mixing, Application, and Pressing

As the formulations show, straight blood and soybean–blood blend glues are prepared in generally the same sequence and manner as outlined for soybean glues. Finished glue viscosity ranges are somewhat lower, typically 5000 to 8000 cP for hot-press formulations and 8000 to 20,000 cP for the thicker and grainier cold-press glues. Glue life at room temperature is 4 to 8 h: the cooler, the longer.

With respect to application methods, blood glues can be spread on wood surfaces by most conventional means. These include roller, knife, and extrusion but do not include

Table 13 Foamable Glue: Ingredients and Mixing Procedure

	Amount (kg)
Water at 16–21°C	170
Industrial wheat flour	50
80% Soluble spray-dried animal blood: mix 7 min	20
PF plywood resin	110[a]
50% Sodium hydroxide solution: mix 15 min	12
PF plywood resin	275[a]
50% Sodium hydroxide solution: mix 2 min	5
Surfactant: mix 2 min or until smooth	1[b]

[a]43% solids phenolic resin; Borden Cascophen 3136, Neste CB 305, Georgia-Pacific 4922.
[b]Emersol or equivalent.

curtain coating or spray, for which the glues must be thinned below practical film retention levels.

The major advantage of alkaline-dispersed blood glues over all other wood glues except resorcinol-based synthetic resin adhesives is their sensitivity to heat, resulting in extremely fast hot-press curing times [54]. This property is of sufficient importance to warrant reproducing an entire plywood hot-pressing schedule for purposes of comparison. The commercial blood glue pressing times shown in Table 14 are literally half those of current phenolic plywood resin adhesives. They are also significantly faster than those listed earlier for soybean glues. This hot-pressing schedule is suitable for all straight blood adhesive formulations and also for soybean–blood blend glues containing at least half blood as the active protein ingredient.

Blood and and soybean–blood blend glues of appropriate high viscosity and granular consistency can be pressed cold according to the schedule shown for soybean glues. For this purpose they must contain a terminal addition of about 5% preservative based on dry glue weight in order to meet product standards for mold resistance [24].

IV. CASEIN GLUES

A. Raw Material Source and Preparation

As with blood, the adhesive qualities of casein curd from milk were recognized in relatively ancient times. Mixed with a simple alkali such as lime, casein protein became an important adhesive for furniture and paint pigments and the preferred sizing agent for the canvas of Renaissance paintings [6,55].

Medieval furniture assembly was divided between casein glues and animal gelatin adhesives made from boiled extracts of bone and hide. The gelatin glues were applied to joints as a hot solution and gained bond strength almost immediately on cooling. The casein glues required much longer clamping times to develop adhesion by water loss and insoluble caseinate formation. However, the casein glues had true water resistance, whereas the animal gelatin rules remained forever sensitive to even slight moisture and heat exposure [5]. Thus casein tended to be used where durability was required. This

Table 14 Blood Glue Hot-Pressing Schedule for Interior Douglas Fir Plywood

Rough panel thickness (mm)	Number of plies	Panels per press opening	Glue spread per single glue line (g/m^2)	Minimum stand time[a] (min)	Pressing time (min at full pressure)			
					110°C	116°C	127°C	138°C
6.4	3	2	171	3	3	2¾	—	—
7.9	3	1	183	3	1¾	1½	1¼	1
7.9	3	2	183	3	3½	3	2¾	2¼
7.9	3	3	195	3	—	7	6	5¼
9.5	3	1	195	3	1¾	1¾	1½	1¼
9.5	3	2	195	3	—	4¼	3¾	3
11.1	3	1	208	3	2	1¾	1¾	1½
11.1	3	2	208	3	—	4½	4¼	4
11.1	5	1	171	3	3	—	—	—
11.1	5	2	171	3	—	6	—	—
12.7	5	1	183	3	2¾	2¼	2	1¾
12.7	5	2	195	3	—	6	5½	5
14.3	5	1	195	3	2¾	2½	2¼	2
14.3	5	2	208	3	—	7	6	5¼
15.9	5	1	195	3	3½	3	2½	1¼
15.9	5	2	208	3	—	7½	6½	5½

17.5	5	1	208	3	4	3½	3	2½
17.5	7	1	208	3	4¼	4	3¾	3¼
19.0	5	1	208	3	—	4¼	3¾	3
19.0	7	1	208	3	—	5	4¼	3¾
20.6	5	1	208	3	—	4½	3¾	3¼
20.6	7	1	208	3	—	6	4¼	4
22.2	7	1	208	3	—	—	4¼	4
23.8	7 and 9	1	220	4	—	—	6	5
25.4	7 and 9	1	220	4	—	—	7	5½
27.0	7 and 9	1	232	4	—	—	7	6
28.6	7 and 9	1	232	4	—	—	7½	6¼
30.2	7 and 9	1	232	4	—	—	8	6½
33.4	9	1	244	5	—	—	9	7
36.5	9	1	244	5	—	—	10	8
39.7	9 and 11	1	244	5	—	—	12	9

Other conditions:
1. Total assembly time limit per press load, 16 min.
2. Veneer temperature not to exceed 43°C.
3. Veneer moisture content not to exceed 8%.
4. On rough or warm veneer, add at least 12 g of extra glue spread.
5. Not less than 12 kg/cm² uniform hydraulic pressure.

[a]Stand time after assembling last panel before loading press.

association with water resistance has remained a favorable performance factor for casein glues down to the present.

Casein protein is recovered from skim milk by acid precipitation to pH 4.5. Mineral acids may be used or the milk can be cultured with bacteria that convert lactose (milk sugar) to lactic acid, which in turn precipitates the casein. The precipitated protein curd is washed free of acid with hot or cold water and is then dried and ground. The commercial designation for casein often includes its method of acid precipitation (e.g., ''lactic acid or sulfuric acid casein'').

Since industrial casein competes directly with the worldwide food uses of milk and its proteins, the price of casein tends to vary widely as the supply/demand economics of milk products rise and fall. In recent years, the cost has remained well over $2 (U.S.) a pound. Even at this price, however, certain casein blend and specialty glues continue to hold a significant place in current markets.

B. Formulation

For adhesive uses, the particle size of ground casein is normally controlled within the range of about 250 to 500 μm [56]. Particles coarser than 500 μm (30 mesh) may not dissolve and disperse completely during glue preparation. Those much finer than 250 μm (60 mesh) tend to form immediate lumps on wetting, even if oiled. For single-package casein glues (by far the most widely used type), preliminary oiling of the dry ingredients is a very important manufacturing step. It helps prevent the pickup of atmospheric moisture by alkaline salts in the dry composition followed by premature attack on the casein during storage. Oiling also slows down the solution of these salts in water at the time of glue mixing, thus allowing the casein particles to become wetted and lump-free in reasonably neutral water.

The lime content of casein glues is similarly important. A high percentage of lime (above 30% of dry casein weight) ensures maximum water resistance of the cured glue film but sharply reduces mixed adhesive working life. A lime content below 10% provides a long working life and adequately strong dry bonds on wood but significantly reduces moisture resistance. Most commercial adhesive formulations balance these properties by utilizing lime additions in the range 15 to 25% [57].

As with blood and soybean flour, the maximum adhesive capability of casein is attained only by complete aqueous dispersion of the folded protein molecules with a strongly alkaline inorganic salt such as sodium hydroxide [56]. Since sodium hydroxide cannot be incorporated successfully into a dry adhesive composition, it is quickly produced on mixing through a double decomposition reaction between calcium hydroxide and strongly ionized but less alkaline salts such as sodium fluoride, sodium carbonate, and trisodium phosphate. (The residues from this reaction are insoluble calcium compounds.)

The viscosity and consistency of casein glues can be altered substantially by reaction with most of the classic protein denaturants such as sulfur compounds, formaldehyde donors, and complexing metal salts [6,56]. One or more of these are frequently used as manufacturing control to offset the natural variability of casein and produce glues of uniform properties. The water resistance of cured casein glues is also improved by moderate denaturing.

Finally, to provide mold resistance adequate for interior and covered exterior structural requirements, a fungicide must be added to casein glues [40,58]. In this case there is no excess of sodium hydroxide in the glue composition to convert a water-insoluble

Table 15 Casein Dry Glue: Ingredients and Mixing Procedure

	Amount (kg)
500-250 μm Lactic acid casein	15.0
500-250 μm Sulfuric acid casein	15.0
74-μm Wood flour[a]	5.0
Fresh hydrated lime	6.5
Granular trisodium phosphate	4.0
Granular sodium fluoride	2.0
Powdered dimethylol urea	0.05
Diesel oil defoamer	1.45
Sodium orthophenylphenate[b]	1.0
Water at 16–21°C	100
Dry glue: mix 2 min, until smooth and thick	50
Let stand 15 min or until thinning has occurred; then mix 2 min or until smooth.	

[a]0.074 mm (200 mesh) or finer.
[b]Dowicide A, Dow Chemical Co.

fungicide to its soluble sodium salt. Therefore, it is added as a prepared soluble salt in order not to upset the fairly precise alkaline balance in the dry glue composition needed to fully disperse the casein. Sodium orthophenylphenate and sodium pentachlorophenate are examples.

The casein adhesive formulation described in Table 15 embodies all the foregoing technology. The dry ingredients are intensively blended in an appropriate dry powder mixer while the defoamer is sprayed in to provide uniform distribution. The dimethylol urea addition, a protein denaturant, is variable for glue viscosity control. The small adjustment is made in the defoamer. Mixing directions are provided in Table 15. The finished glue viscosity should be in the range 4000 to 8000 cP at room temperature, thickening gradually over several hours and attaining a firm gel overnight.

In a totally different area of application, casein adhesives for paper sizing, chipboard laminating, and label gluing are more nearly "casein solutions" [59,60]. They are simple dispersions with ammonia or borax at moderate pH and low viscosity. They are frequently combined with latexes or soluble rosin derivatives for special performance improvements [59].

C. Mixing, Application, and Pressing

Casein glues for wood pass through an early thick-consistency stage that requires fairly strong agitation to reduce them to a uniform and lump-free state. The mixer should be equipped for sidewall scraping to work thickened glue continuously back into the stirred composition. Counterrotating paddle mixers and bread dough mixers have proved ideal for this purpose.

Because of their thick, sticky consistency, casein glues are generally applied to only one of a mating pair of wood surfaces by roller, knife, or extrusion. Adequate adhesive wetting and transfer occur when the wood surfaces are brought together.

The stickiness of alkaline-dispersed casein glue provides two of its best performance attributes: long assembly-time tolerance and wipe resistance (difficulty of removal). A film of casein glue on dry lumber, for example, may allow an open/closed assembly time of 1 to 2 h before clamping is required. This property is especially useful in the timber laminating industry, where it permits many pieces of lumber to be stacked over each other, adjusted for position, and assembled into large, complex laminated beams [61]. This long assembly tolerance plus the gap-filling and wipe-resistant capabilities of casein glues made them the outstanding choice for laminated structural wood products from the mid-1930s onward. Today's phenol–resorcinol–formaldehyde laminating adhesives, which ultimately displaced casein glues on the basis of exterior durability, could still use a large measure of these working properties of casein glues.

While casein glues can be heat cured and were employed in the past to make hot-press plywood, most of the high-volume bonding applications have involved cold pressing. Casein glue films are adequately cured by water loss and insolubilizing of the proteins through various chemical reactions at room temperature [62]. Heating does not yield significantly improved water resistance. Except for soybean–casein blend glues, which take on the granular consistency of the soybean constituent, the inherent stickiness of straight casein glues dictates a fairly long clamping time to bring about water loss and adhesive hardening. Progressive shear tests have shown that these glues develop about half their dry strength in 3 h and substantially all of it in 6 to 8 h at room temperature. However, moisture resistance continues to improve for several days [63].

As mentioned previously, another performance attribute of casein glues that recommends their use in structural wood laminates is fire resistance [64]. While all three of the proteins discussed in this chapter burn to a char before losing bond strength, casein adhesives appear particularly durable in this respect. Thus casein glues remain the adhesives of choice for the economical assembly of wood-based fire doors of flush and panel designs. In yet another attribute, the combined adhesive strength and toughness (as opposed to brittleness) of casein glue films has made them an ideal bonding agent for wooden sporting equipment and other applications required to withstand flexing, vibration, and shock, such as racquets, hockey sticks, and fishing rods [6,65].

V. OTHER PROTEINS

Reference was made to the historical use of collagen glues derived from the gelatin extracts of animal bones and hides. This does not properly indicate the true importance to the wood industry of these materials. From ancient times to the present, animal glues have in fact remained one of the primary assembly adhesives for wooden furniture, cabinets, and musical instruments [66]. Applied as a hot, viscous solution to furniture joints, they rapidly develop gel strength on cooling that permits the prompt removal of clamping pressure. On subsequent drying, these glues cure to resilient, high-strength bonds between wood surfaces, especially those involving end grain. Animal glue bonds are strong and permanent as long as they are kept dry and reasonably cool but are subject to softening and fungicidal attack when moistened or heated. Water and temperature resistance can be improved through the incorporation of most of the protein denaturants listed earlier [67,68].

Animal glues are used widely in a variety of ways with paper. For example, they have been the dominant adhesive for rewettable gummed paper tapes, labels, and envelope seals [69]. They are an important coadditive with synthetic wet-strength resins and rosin sizes for coated paper products [70]. They have been a primary binder for the grit that forms sandpaper [71].

In contrast to their widespread use in furniture and paper products, animal glues have not proved useful as structural adhesives for wood. When used as the principal protein constituent, their water sensitivity is excessive compared with other available proteins. When combined with soybean, blood, or casein, animal gelatin glues are completely hydrolyzed and destroyed by the strong alkalies required to disperse these proteins. In addition, they soften when severely heated, which, by law, prohibits their use in structural wood products [5].

Although in recent years animal glues have been replaced substantially by the newer synthetic adhesives, particularly the vinyl and acrylic emulsions, large quantities are still sold in dry and stable liquid forms for furniture assembly or repair and paper bonding applications.

Generally similar comments can be made with respect to fish skin adhesive extracts regarding these and other wood gluing applications. Fish skin glues are normally prepared in stable liquid form through mild acid hydrolysis and are frequently combined with animal glues for improved rewettability, tack, and adhesion of paper to glass or metal surfaces [72]. They differ from animal glues in one important respect: namely, they will not soften at elevated temperatures, especially when treated with aldehydic or polyvalent metal ion cross-linkers. Thus, in addition to the more conventional paper bonding applications, this property has created a major field of use for fish gelatin extracts as durable but temporary protective coatings and light-convertible photoresist films [73].

Other vegetable protein sources are occasionally mentioned as substitutes for soybeans. These have included cottonseed meal, peanut flour, Alaska pea, and rapeseed meal, to name just a few. Although they do contain 25 to 35% useful protein substance, they have never made significant inroads on soybean flour for wood-gluing applications on the basis of comparable performance. However, they can be used and are prepared for adhesive purposes in the same manner as soybean flour itself.

REFERENCES

1. A. Lucas, *Analyst 51* (1926).
2. H. S. Shrewsbury, *Analyst 51* (1926).
3. Pliny, *Natural History*, Book XVI, Ch. 83, Bostock and Riley's translation, London, 1855.
4. O. Johnson, U.S. patent 1,460,757 (July 3, 1923).
5. T. R. Truax, The gluing of wood, *Bulletin 1500*, U.S. Department of Agriculture, Washington, D.C., 1929.
6. H. K. Salzberg, in *Handbook of Adhesives* (I. Skeist, ed.), Reinhold, New York, 1962, Chap. 9.
7. T. D. Perry, (1944), *Modern Wood Adhesives*, New York, Pitman.
8. Glues used in airplane parts, *Report 66*, U.S. Department of Agriculture Forest Products Laboratory, Madison, Wis., 1920; *Western Flying 3* (1927).
9. A. Knop and W. Scheib, *Chemistry and Application of Phenolic Resins*, Springer-Verlag, New York, 1979.
10. K. J. Saunders, *Organic Polymer Chemistry*, Chapman & Hall, London, 1973.
11. G. N. Arneson, Glues and gluing in prefabricated house construction, *Project 575B*, U.S. Department of Agriculture Forest Products Laboratory, Madison, Wis., 1946.
12. L. B. Lane and J. J. Frendreis, *Adhesives Age* (May 1961).
13. *Census of Manufactures, 1942 Through Preliminary 1958: U.S. Tariff Commission Reports of 1948*, U.S. Department of Commerce, Washington, D.C.

14. *Wood Handbook: Wood as an Engineering Material*, Agriculture Handbook 72, U.S. Department of Agriculture Forest Products Laboratory, Madison, Wis., 1987.

15. Pen Ts'ao Kong Mu, *The Records of Chinese Emperor Sheng-Nung*, 2838 B.C.

16. Prosoy, *Technical Bulletin*, Centra Soya Company, 1948.

17. R. S. Burnett, *Soybeans and Soybean Products*, Wiley-Interscience, New York, 1951.

18. G. Davidson, U.S. patent 1,724,695 (Aug. 13, 1929).

19. F. M. Lea and R. W. Nurse, (1939). *Trans. J. Soc. Chem. Ind. 58* (1939).

20. G. H. Brother, A. K. Smith, and S. J. Circle, *Soybean Protein*, U.S. Department of Agriculture, Bureau of Agricultural Chemistry, Washington, D.C., 1940.

21. G. Davidson and I. F. Laucks, U.S. patent 1,813,387 (July 7, 1931).

22. I. F. Laucks and G. Davidson, U.S. patents 1,689,732 (Oct. 30, (1928) and 1,691,661 (Nov. 13, 1928).

23. I. F. Laucks, *Chemurgic Digest 2* (1943).

24. *Mold Resistance of Plywood Made with Protein Adhesives*, Douglas Fir Plywood Association (now American Plywood Association), Tacoma, Wash., Feb. 15, 1952.

25. N. J. Sheeran, U.S. patent 2,788,305 (Apr. 9, 1957).

26. D. French and J. T. Edsall, *Adv. Protein Chem. 2* (1945).

27. T. Satow, U.S. patent 1,994,050 (Mar. 12, 1935).

28. J. Bjorksten, *Adv. Protein Chem. 6* (1951).

29. I. F. Laucks and G. Davidson, (1939), U.S. patent 2,150,175 (Mar. 14, 1939).

30. I. F. Laucks and G. Davidson, (1930), U.S. patents 1,786,209 (Dec. 23, 1930) and 1,805,773 (May 19, 1931).

31. D. M. Wood, U.S. patents 2,297,340 and 2,297,341 (Sept. 29, 1942).

32. A. L. Lambuth, U.S. patent 3,192,171 (June 29, 1965).

33. J. R. Ash and A. L. Lambuth, U.S. patent 2,817,639 (May 12, 1954).

34. Monsanto Company, British patent 688,222 (Mar. 4, 1953).

35. H. Galber and A. J. Golick, U.S. patent 2,402,492 (June 18, 1946).

36. J. R. Stillinger and W. Williams, Jr., *Timberman* (Jan. 1955).

37. C. N. Cone and H. Galber, U.S. patent 1,976,435 (Oct. 9, 1934).

38. L. Bradshaw and H. V. Dunham, U.S. patents 1,829,258 and 1,829,259 (Oct. 27, 1931).

39. Hollow core flush doors, *Report 1983*, U.S. Department of Agriculture Forest Products Laboratory, Madison, Wis., 1954.

40. M. L. Selbo, *Forest Products J. 3* (1949).

41. W. Eichholz, German patent 199,093 (Aug. 6, 1907).

42. K. H. Meyer, *Natural and Synthetic High Polymers*, Wiley-Interscience, New York, 1950.

43. C. E. Drugge and J. M. Hine, U.S. patent 2,963,454 (Dec. 6, 1960).

44. A. L. Lambuth, U.S. patent 3,324,103 (June 6, 1967).

45. S. A. Karjala and F. K. Dering, U.S. patent 3,301,692 (Jan. 31, 1967).

46. N. J. Sheeran, U.S. patent 2,870,034 (Jan. 20, 1959).

47. Blood albumin glues: their manufacture, preparation and application, *Report 281–282*, U.S. Department of Agriculture Forest Products Laboratory, Madison, Wis., 1936, revised 1938 and 1955.

48. Weldwood L1-R plywood, *Sales Bulletin*, American Institute of Architects File 19-F, U.S. Plywood Corporation, Feb. 1952.

49. R. A. Jarvi, U.S. patent 2,705,680 (Apr. 5, 1955).

50. O. C. Carmichael, U.S. patent 2,375,195 (May 8, 1945).

51. C. N. Cone, U.S. patent 2,895,928 (July 21, 1959).

52. PF 3097 phenolic resin adhesive for exterior type softwood plywood, *Product Bulletin*, Schedule PG 1976, Monsanto Company, Seattle, Wash., Dec. 6, 1968.

53. S. Nylund, *Proc. Forest Products Research Society Conference on Structural Wood Composites*, Madison, Wis., Nov. 1987.

54. J. M. Gossett and M. H. Estep, Jr., and M. J. Perrine, U.S. patent 2,874,134 (Feb. 17, 1959).

55. J. R. Spellacy, *Dried Casein and Condensed Whey*, Lithotype Process Co., San Francisco, 1953.

56. E. Sutermeister and F. L. Brown, *Casein and Its Industrial Applications*, 2nd ed., Reinhold, New York, 1939.

57. H. G. Higgins and K. F. Plomley, *Australian J. Appl. Sci. 1* (1950).

58. Adhesives casein-type, water- and mold-resistant, *Federal Spec. MMM A-125*, U.S. General Services Administration, Washington, D.C., 1955.

59. S. Jones, in *Adhesion and Adhesives* (S. Clark, J. E. Rutzeler, and R. C. Savage, eds.), Wiley, New York, 1954.

60. R. J. Lodge, Casein adhesives, *Modern Packaging—Encyclopedia Issue* (1959).

61. J. G. Mark, U.S. patent 2,279,256 (Apr. 7, 1942).

62. Casein glues: their manufacture, preparation and application, *Report D 280*, U.S. Department of Agriculture Forest Products Laboratory, Madison, Wis., 1950.

63. Control of conditions in gluing with protein and starch glues, *Report R 1340*, U.S. Department of Agriculture Forest Products Laboratory, Madison, Wis., 1950.

64. Adhesives: their use and performance in structural lumber products, *Report 2199*, U.S. Department of Agriculture Forest Products Laboratory, Madison, Wis., 1960.

65. B. L. Lambuth, *Angler's Workshop*, ISBS Champoeg Press, Eugene, Oreg., 1979.

66. R. C. Gill, *Furniture Manuf., 79*(2) (Feb 1957).

67. J. R. Hubbard, U.S. patent 2,043,324 (June 9, 1936).

68. G. Stainsby, in *Recent Advances in Gelatine and Glue Research*, Pergamon Press, New York, 1958.

69. J. R. Hubbard, in *Handbook of Adhesives* (I. Skeist, ed.), Reinhold, New York, 1962, Chap. 7.

70. Protein and synthetic adhesives for paper coating, *Monograph 22*, Technical Association of the Pulp and Paper Industry, Atlanta, Ga., 1961.

71. N. E. Oglesby, (1943) U.S. patent 2,322,156 (June 15, 1943).

72. H. C. Walsh, in *Handbook of Adhesives* (Irving Skeist, ed.), Reinhold, New York, 1962, Chap. 8.

73. R. E. Norland, in *Coatings Technology Handbook* (D. Satas, ed.), Marcel Dekker, New York, 1991, Chap. 52.

14

Animal Glues and Adhesives

Charles L. Pearson *Swift Adhesives Division, Reichhold Chemicals, Inc., Downers Grove, Illinois*

I. INTRODUCTION

Animal glues have been used for thousands of years in traditional adhesive and sizing applications. In later times, they also found uses as protective colloids, flocculents, coatings, in composition, and as a component of compounded adhesives. The earliest known use of animal glue, for veneering, dates to the period 1500–2000 B.C. in ancient Egypt and has been referred to in literature from biblical times. Commercial manufacture dates back to about A.D. 1690 in England and Holland. Numerous patents relating to the manufacturer of animal glues were issued during the period 1754–1844. The first manufacture of animal glue in the United States was early in the nineteenth century [1].

Until 1940, natural materials, including animal glues, were the only adhesive materials available. Animal glues were the adhesives of choice for such uses as woodworking, paper manufacture and converting, bookbinding, textile sizing, abrasives, gummed tape, matches, and a variety of other applications. The acceptance of animal glues as adhesives is based on their unique ability to deposit a viscous, tacky film from a hot aqueous solution, which forms a firm gel while cooling, and provides an immediate, strong, initial bond. Subsequent drying provides the final bond of high strength and resiliency. Ease of preparation, high tack, fast set, ready application, and good machining properties, even in high-speed operations, are characteristics of importance.

After introduction of emulsion-based adhesives, and later hot melts, animal glues were displaced in part or completely from their traditional uses, but they have retained a relatively small but significant share of the adhesive market due to their unique properties. It is estimated that U.S. consumption of animal glues is in the range 30 to 35 million pounds, and production worldwide is about 90 million pounds annually (W.E. Blair, Swift Adhesives, private survey, 1992).

II. CHEMICAL COMPOSITION

Animal glues are derived by the hydrolysis of the protein constituent collagen of animal hides and bones. Collagen in its natural state is water insoluble and must be conditioned to solubilize the protein. Collagen molecules are triple helices of amino acid sequences and contain both nonpolar and charged acidic and basic side chains. The conversion of collagen to the soluble protein of animal glue (gelatin) involves breaking the intra- and intermolecular polypeptide bonds through the use of acid or alkali and heat. The collagen–glue (gelatin) transition has been described as a stepwise process involving the melting of the trihelical network to an amorphous form, followed by the sequential hydrolysis of various types of covalent bonds [2,3].

Glues and gelatins are described as hydrolyzed collagen with the following formula [5].

$$C_{102}H_{149}O_{38}N_{31} + H_2O \longrightarrow C_{102}H_{151}O_{39}N_{31}$$

The approximate chemical composition of glue (gelatin) protein is

Carbon   50.3%
Hydrogen   6.2%
Oxygen   25.6%
Nitrogen   17.8%

Animal glues are composed of α-amino acids joined in polypeptide linkages to form long-chain polymers [5–8]. A typical chain fraction with three amino acids:

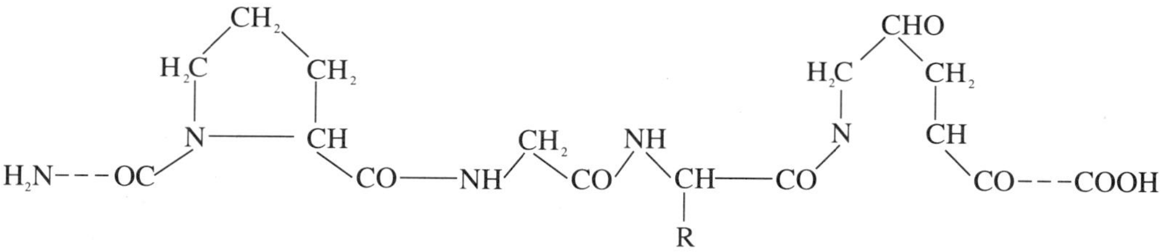

In aqueous solutions of animal glues, the polypeptide chains take up random configurations of essentially linear form. Studies have indicated that most glue molecules consist of single chains terminated at one end by an amino group and at the other end by a carboxyl group [7]. The molecules may also have side chains and contain cyclic structures. They contain cyclic structures and also have side chains. They may in part conform to the oriented chain in the original collagen. The polypeptide chains are of varying lengths and consequently, widely different molecular weights. A wide range of average molecular weights has been reported, ranging from approximately 10,000 to over 250,000. Molecular weight distribution is of equal importance in studying animal glue (gelatin) protein systems [3].

Amino acid studies corroborated by various analyses indicate that there are 18 different amino acids present in collagen and animal glue (gelatin) in varying amounts (see Table 1). The acidic and basic functional groups of the amino acid side and terminal groups confer polyelectrolyte characteristics to the protein chains. The chains contain both amine and carboxylic groups which are reactive and ionizable. These electrically charged sites affect the interactions among protein molecules and between protein molecules and water. These polar and ionizable groups are believed to be largely responsible for the gelation and characteristic rheological properties of animal glues. Cross-linkage between protein molecules is possible through hydrogen, ionic, and covalent bonds.

Table 1 Amino Acids Present in Collagen and Animal Glue

Amino acid	Average residues per 1000 total of all residues [9]	Character of R-radical [10]	
		Polarization	Ionic character
Alanine	103.2	Nonpolar	Neutral
Arginine	46.2	Polar	Basic
Aspartic acid	47.7	Polar	Acid
Glutamic acid	73.3	Polar	Acid
Glycine	339.8	Nonpolar	Neutral
Histidine	4.4	Polar	Basic
Hydroxyproline	97.8	Polar	Neutral
Hydroxylsine	6.2	Polar	Basic
Isoleucine	12.8	Nonpolar	Neutral
Leucine	24.4	Nonpolar	Neutral
Lysine	29.8	Polar	Basic
Methionine	5.4	Polar	Neutral
Phenylalanine	13.4	Nonpolar	Neutral
Proline	122.8	Nonpolar	Neutral
Serine	31.5	Polar	Neutral
Threonine	17.9	Polar	Neutral
Tyrosine	3.6	Weakly polar	Very weakly acid
Valine	20.7	Nonpolar	Neutral

Animal glues are amphoteric because the amine and carboxyl groups contained in the polypeptide protein chain are reactive and ionizable. In strongly acid solutions, the protein is positively charged and acts as a cation. In strongly alkaline solutions it is negatively charged and acts as an anion. The intermediate point, where the net charge on the protein is zero, is known as the *isoelectric point* (IEP) and is designated in pH units. The isoelectric point varies, depending on pretreatment of the collagen, whether acidic or alkaline. During processing, the acidic or alkaline treatments used hydrolyze the amide groups in the collagen to a greater or lesser extent, liberating the acid functions. Acid-processed glues (little amide group modification) have an isoelectric point near 9.0, and alkaline-processed glues (low residual amide groups) have an isoelectric point close to 4.8.

Gelation of aqueous solutions of animal glues upon cooling is an important characteristic. Gelation involves both intra- and intermolecular reorientation upon cooling of the solution. It is caused by the formation of random primary and secondary bonds. Intermolecular network formation is primarily the result of a cross-linking mechanism between molecular chains by hydrogen bonds [11].

III. TYPES OF ANIMAL GLUE

There are two major types of animal glue, hide glue and bone glue, differing in the type of raw materials used. Although process conditions may differ, both are obtained by hydrolysis of the collagen in the hide and of connective tissue or bone structure of the raw material. Both types are principally of cattle origin with tanning and meat-packing industries as the principal sources of raw materials.

IV. MANUFACTURE OF ANIMAL GLUE

Basic manufacturing procedures for animal glues generally involve alkaline pretreatment (for hide glues) or acidic pretreatment (for bone glues). The raw materials for hide glues include salted, limed, or pickled hide trimmings or splits, and chrome-tanned leather scrap. Tanned leather scrap requires special processing because of the chrome tannage.

Hide glues from hide trim and splits are prepared by initial washing with water. The stock is then soaked in lime (calcium hydroxide) and water for a period of weeks, which dissolves and removes extraneous protein-related materials, as well as conditioning the collagen for subsequent glue extraction by hydrolysis. The conditioned collagen is then washed with water, followed by acidulation with dilute acid such as sulfuric, hydrochloric, or sulfurous, for pH adjustment, followed by a final water rinse.

Chrome-tanned leather scrap for hide glues may be treated initially with lime or caustic, followed by a strong acid bath to remove the tannage. The stock is then soaked in magnesium hydroxide and rinsed prior to extraction. Alternatively, the chrome stock may be treated with a magnesium hydroxide soak only, prior to extraction, in which case the chromium tanning salts remain in the residue after extraction. The treated collagen is transferred to extraction kettles or tanks, where it is heated with water to convert the collagen and extract the glue. Several hot water extractions at progressively higher temperatures are made under carefully controlled conditions. Separate, successive dilute glue solutions are removed from the stock until the glue is completely extracted, usually in four extractions.

The dilute glue extractions, ranging from 2 to 9% glue solids, are filtered and concentrated by vacuum evaporation to 20 to 50% concentration prior to drying. In some plants the glue is chilled until it will gel, then dried in tunnel dryers which circulate heated air over gelled sheets stacked on wire nets with air space between, taking up to 48 h to dry. In newer installations, the concentrated glue solutions are cooled to the gelling point and are extruded in noodle form into a continuous dryer which completes the drying in 2 to 2½ h by circulating conditioned, filtered, heated air. The dried product, at 10 to 15% moisture content, is then ground to the desired particle size.

Bone glues fall into two categories, green bone or extracted bone. Green bone glues are prepared from fresh or ''green'' bones, which come primarily from the meat-packing industry. The bones are crushed, washed, and normally treated with dilute acid, either sulfuric or sulfurous, prior to extraction. Extracted bone glues are prepared from dry bones that have had a preliminary degreasing treatment with solvent prior to conditioning for extraction.

The glue is extracted from the conditioned bones by hydrolysis in pressure tanks by the successive application of steam pressure and hot water. Separate successive dilute glue solutions are removed from the tanks, followed by filtering or centrifuging to remove free grease and suspended particles. The dilute solutions are vacuum evaporated to high concentration, followed by drying and grinding, as described for hide glues. Modern animal glues, whether of hide or bone origin, contain adequate preservatives for protection against bacteria or mold growth under normal conditions of use, and may contain defoaming agents where foam control is desired in the end application.

V. PROPERTIES OF ANIMAL GLUES

Commercially available animal glues are sold in granular or pulverized form, and are dry, hard, odorless materials that vary in color from light amber to brown. Animal glues may be stored indefinitely in the dry form. The density of animal glues is approximately 1.27. A moisture content of 8 to 15% is considered commercially dry. An inorganic ash content of 2.0 to 5.0%, and a grease content of 0.2 to 3.0%, are in the normal range for commercial animal glues. The pH range of commercial glues is 5.5 to 8.0.

Animal glues are hydrophilic colloids and are soluble only in water. In cold water, the glue particles absorb water and swell, resulting in a jellylike sponge. Upon application of heat, the particles dissolve, forming a solution. Upon cooling, the solutions set to an elastic gel. The gelation is a thermally reversible reaction, and on application of heat the gel reverts to liquid form. The melting or gelling point will vary from below room temperature to over 120°F, depending on glue concentration, grade, and possible presence of modifiers.

An important characteristic of animal glues is their film forming and bonding properties. Dried films are continuous, noncrystallizing, permanent, and possess great strength and resilience. Tensile strengths in excess of 10,000 psi have been reported [12]. Animal glues are insoluble in oils, greases, waxes, alcohols, and other solvents. Being soluble only in water, the continuous films are ideally suited as barriers against these materials. With suitable practical methods, the films may be made moisture resistant.

Sensitivity to the effects of moisture may be reduced by the use of various insolubilizing agents (sometimes referred to as "tanning" or "hardening") which cross-link the protein molecules, rendering them less susceptible to hydration and solution. These agents include formaldehyde, paraformaldehyde, hexamethylene–tetramine, glyoxal, and dialdehyde starch. Metal salts will thicken, coagulate, and sometimes precipitate animal glue solutions. A degree of moisture resistance and raising of solution melting point may be obtained with these salts. Salts of aluminum, chromium, and iron have this effect on animal glue solutions.

Animal glues are readily compatible with and are frequently modified by water-soluble plasticizers such as glycerine, sorbitol, glycols, and sulfonated oils to increase film flexibility. They are also compounded with many other materials, such as dextrins, starches, sugars, various salts, pigments, poly(vinyl alcohols), and acetates, as well as some water-soluble solvents such as butyl cellosolve acetate for specific properties. Viscosities can be modified by compounding with thickeners, including compatible natural gums, alginates, and synthetic thickeners such as carboxymethyl cellulose.

Because of their amphoteric properties, animal glues possess electrical charges which unmodified or with suitable modification by simple chemical additives are highly effective as colloidal flocculents and as protective colloids in such applications as paper manufacturer, rubber compounding, ore and metal refining, and for water and industrial wastewater treatment.

A wide range of viscosities is possible, from almost water thin to in excess of 70,000 cP by variation of the dry glue concentration and test grade. Animal glues are available in a number of grades from low to high, varying in inherent viscosity and gelling properties. The gel property determines the grade and is the controlling factor in speed of set of a glue film for adhesive applications.

VI. GRADES AND TESTING

Animal glues are graded on the basis of gel strength (an arbitrary measure of the gelling property) and viscosity, which increases with an increase in gel strength. These properties have a marked bearing on glue application and end use. One of the earliest grading

Table 2 Glue Test Grades

Peter Cooper standard grade	National Association of Glue Manufacturers grade	Bloom (g)		Millipoise value (minimum)
		Range	Midpoint	
5A Extra	18	495–529	512	191
4A Extra	17	461–494	477	175
3A Extra	16	428–460	444	157
2A Extra	15	395–427	411	145
A Extra	14	363–394	379	131
#1 Extra	13	331–362	347	121
#1 Extra special	12	299–330	315	111
#1	11	267–298	283	101
1XM	10	237–266	251	92
1X	9	207–236	222	82
1¼	8	178–206	192	72
1⅜	7	150–177	164	62
1½	6	122–149	135	57
1⅝	5	95–121	108	52
	4	70–94	82	42
	3	47–69	58	
	2	27–46	36	
	1	10–26	18	

systems was introduced by Peter Cooper about 1844, establishing a basis for comparative values and market stability. The National Association of Glue Manufacturers adopted standard methods for testing animal glues in 1928. Table 2 shows animal glue test grades. Bone glues differ from hide glues, having a lower range of viscosities and gel strengths, due primarily to greater hydrolysis of the protein by the higher heat used during extraction. Table 3 shows the comparative properties of hide and bone glues.

The standard method for animal glues determines the viscosity by measuring the flow time in seconds of a 12.5% solution at 60.0°C through a standard pipette, and converting the results to millipoises. The method for gel strength (measured in grams) calls for cooling the 12.5% solution to 10.0°C and holding for 16 to 18 h, followed by determination of the weight in grams required to depress a 0.5 in.-diameter plunger a distance of 4 mm into the surface of the gelled sample using a bloom gelometer or comparable

Table 3 Comparative Properties of Hide and Bone Glues

Property	Hide glue	Bone glue
Gel strength (g)	50–512	50–220
Viscosity (mp)	30–200	25–90
pH	6.0–7.5	5.0–6.5
Moisture	10.0–14.0	8.0–11.0
Ash	2.0–5.0	2.0–4.0
Grease	0.3–1.0	0.4–4.0

instrument. The pH is determined electrometrically on a 12.5% solution at 40°C. The moisture content of dry glue is measured by drying a 10-g sample for 17 h at 105°C [13]. Compounded animal glue products made from known grades of animal glues are normally checked for viscosity using a Brookfield viscometer, and for solids by the oven-drying method or refractometer.

VII. PREPARATION OF ANIMAL GLUES

Since dry and compounded animal glues are used over a wide range of dilutions from 1% or lower to over 50%, no general ratio can be given. Dry glues are generally available in coarse (10 to 30 mesh) or fine (30 mesh and finer) granulation. Powdered glues (100 mesh and finer) are also available for special applications. Compounded glues are available as dry blends, in gelled cake form, or as liquid products.

Dry glues are readily prepared for use. The direct addition of dry glue to hot water in a jacketed, mechanically agitated mixer is recommended for fastest preparation. Thermostatic controls are recommended to hold the glue temperature at 140 to 145°F for use. Alternatively, the dry glue can be soaked in cold water until swollen, then transferred to a jacketed melting tank with agitator, melting, and stirring the glue into solution.

For best results, the dry glue should always be weighed. Water may be measured or weighed. Stainless steel mixing tanks are preferred. Water-jacketed mixers with low-pressure steam injection are recommended for large batches. Electrically heated mixers are usually preferred for small batches. Good agitation is important. Compounded dry blends are prepared in the same manner as dry glues. Cake glues (usually in 5 to 10-lb slabs) are placed in a suitable mixer, diluted as required with water, melted, and held at 140 to 145°F for use. Liquid glues are ready to use as is at room temperature. Cake glues contain modifiers such as corn syrup, sugars, magnesium sulfate (epsom salt), glycerine, sorbitol and other glycols, dextrins, clays and pigments, water-soluble organic solvents, and surfactants, depending on desired properties, as well as water, preservatives, and odorants.

VIII. FLEXIBLE AND NONWARP GLUES

Flexible and nonwarp glues are compounded animal glue–based products available in cake or dry blend form. Flexible cake glues are formulated to provide a permanent, flexible, resilient film. Glycerine is normally the primary plasticizer and may be modified with sorbitol and other glycols. Higher grades of animal glue are normally used to provide film strength and resiliency. The ratio of plasticizer to dry glue controls the film flexibility, and can be varied from a moderate degree of flexibility to a fully flexible film. These products are used in such applications as bookbinding, including hardcover books, directories, and catalogs, notebook binders, and soft-sided luggage.

Nonwarp cake glues are designed to provide a ''lay flat'' or nonwarp film with little or no curling upon drying. The nonwarp property is obtained through the use of various sugars and corn syrup. Tack and speed of set can be varied by choice of glue grade and ratio of dry glue to the sugars and other modifiers. Glues supplied in cake form have the advantage of ease of melting and minimal water dilution by the user (normally up to 20%). Dry blends are basically simple, nonwarp formulations supplied by the manufacturer in dry, fine mesh form, which the customer dissolves in hot water prior to use. These products have indefinite shelf life and lower shipping cost because of the absence of water.

Major areas of use for nonwarp glues are for set up boxes, casemaking for book covers, hard notebook binders, slipcases, looseleaf computer manuals, record covers, hard luggage, caskets, and laminating. Because of differences in equipment, application, and end-use requirements, compounded glues are formulated for each use and/or customer, and may be further modified to compensate for seasonal changes in humidity and temperature. Speed of set, for example, can vary from almost instantaneous to over 5 min. Properties of importance include viscosity at the recommended operating temperature, degree of tack and speed of set, and final bond characteristics.

IX. LIQUID ANIMAL GLUES

Liquid animal glues are modified dry glue solutions containing a gel depressant, usually urea, thiourea, ammonium thiocyanate, or dicyandiamide, so that they remain fluid at and somewhat below room temperature. Liquid animal glues can be modified with clays or calcium carbonate as fillers, and wetting and dispersing agents, plasticizers, and other modifiers as required. Solids are usually in the range 35 to 65%, with a viscosity range of 3000 to 5000 cP at room temperature.

X. GLUE APPLICATION

Recommended use temperature for dry animal glues and compounded products (except liquid glues) in most applications is 140 to 145°F (60 to 63°C). A range of 135 to 155°F (57 to 68°C) is acceptable but not preferred. Use at lower temperatures results in undesirable properties such as high viscosity, poor machinability, excessive glue use, and premature gelation, resulting in bond failure. Use at temperatures that are too high causes excessive water loss and degradation of the glue by heat hydrolysis. Excessive dilution of a glue solution to retard gelation is a bad practice that can result in a ''starved'' bond with insufficient glue to adhere. Use of a lower grade of glue or a gel depressant is indicated. Dry glues and compounded dry blends for adhesive use are usually prepared at 25 to 55% solids. Compounded cake glues are used as is, 'or diluted up to 20% with water. The amount of water used depends on glue grade, speed of operation, and type of material to be bonded.

Four basic steps for adhesive applications are as follows:

1. Deposit a thin, continuous glue film on one of the surfaces to be bonded.
2. Allow the glue film to become tacky (transition point from liquid to gel) before applying pressure.
3. Apply uniform pressure to ensure complete contact between surfaces to be bonded.
4. Continue pressure long enough to ensure a strong initial bond.

Grade selection for many uses involves consideration of the desired viscosity and gel properties for the specific end use. High test grades have the greatest water-taking properties, high viscosity, rapid gel formation and strength, fast speed of set, and greatest reactivity with insolubilizing materials. Low test grades have long tack life and open time, slow rate of gel formation and set, and best film-forming properties at high solids.

Table 4 Viscosity (cP) at 140°F for Dry Glues of Given Grade Test and Millipoise Value

Glue concentration (%)	High test (155 mp)	Medium-high test (102 mp)	Medium test (63 mp)	Low test (32 mp)
5	3.0	2.4	2.0	1.6
10	8.8	5.6	3.6	2.6
12.5	15.5	10.2	6.3	3.2
15	28.0	17.2	8.4	5.0
20	79.0	46.0	22.4	10.0
25	196	112	49.6	19.6
30	524	264	108	37.6
35	1,360	612	224	72.0
40	3,216	1,320	476	133
50	16,320	7,240	2,400	566

Source: Ref. 10, p. 146.

Medium test grades provide intermediate properties. Grade selection for adhesive use generally involves matching the gelation rate of the animal glue solution with the time from the application of the glue film to the bonding of the substrates (open time). The gelation rate and viscosity of animal glue solutions are usually closely related. Table 4 illustrates the viscosity in centipoise at 140°F for solutions of dry glues of given test and millipoise value [17].

XI. END USES

Current end uses for animal glues and compounded products include (see Fig. 1) bookbinding and directory/catalog binding, paper manufacture and converting, abrasives, ore and metal refining, paper box manufacture, matches, gummed tape, woodworking,

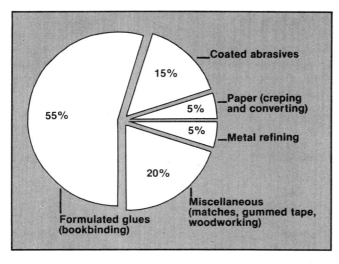

Figure 1 Use of animal glue product by end-use markets. (From Ref. 14.)

luggage, case covering and lining, rubber compounding, textiles, glass chipping, picture frame and decorative molding composition, and leather manufacturing.

A. Bookbinding

There are two areas where animal glue adhesives are used extensively: for hardcover books in casemaking and backlining operations, and for perfect binding of directories and catalogs. Animal glues, primarily compounded products, are used because of their good film flexibility, excellent tack and speed-of-set properties, and ease of cleanup. For hardcover books, the largest application is in casemaking for covers, where good tack when turn-ins are made, and a wide range of open times, which can be adjusted to meet the speed requirements of various types of casemaking equipment (from 20 to 120 cases per minute), are important properties. Reduction of scrap is also an important factor. The casemakers can be run at low speed, or can be stopped briefly for adjustment, without loss of covers. Film flexibility can be adjusted so that warpage of the cover is minimized. Machine cleanup is easy with hot water. Emulsion adhesives and hot melts have proved unsuitable, due to tack and open-time limitation, as well as difficult cleanup.

In addition to machine casemaking there are various related applications, including manual and semiautomated casemaking operations for such applications as looseleaf binders and sample books, where animal glue products are used because of good tack and open-time properties under a wide range of conditions. Animal glue products are also used for backlining and headbanding, a process that applies adhesive to the spine of the book to provide round and dimensional stability to the spine, and to adhere reinforcing cloth, decorative headbands, and paper liner. The modern backlining operation is usually in-line, with the backliner preceded by a rounder and backer that shapes the book, followed by a casing-in unit that applies the covers. These units operate in the range of 40 books per minute. Animal glue adhesives compete with hot melts for this application but are used widely because of their good film flexibility, tack and speed-of-set properties, and adhesion to gluing of adhesives, cloth, and paper. Wide latitude during short machine stops and ease of cleanup are also factors.

Books bound with animal glues range from inexpensive children's books to textbooks, which must withstand rough handling and continuous usage during their lifetime, to encyclopedias and reference books, which must retain their functionality for an extended period of time. Directories and catalogs are perfect bound with animal glue adhesives. In perfect binding, the book signatures (sections) are assembled and the spine is cut and leveled with saws and knives to separate the signatures into individual pages. Books range from less than ¼ in. to over 3 in. in thickness. The binder first applies an animal glue–based primer to the spine of the book. This adhesive penetrates to bind the cut edges of the pages tightly into the book, as well as to provide flexibility when the book is opened. The primer application is followed in line by the cover adhesive, which adheres the cover to the book and also provides a flexible film that will not crack when the book is opened. The perfect-bound book relies solely on the adhesive to bind the pages and cover strongly into one unit, in contrast to stapled books. Perfect-bound books are trimmed to final size after binding and cover application.

With binder speeds up to 200 to 250 books per minute, important factors are fast tack and speed of set. High bond strength of pages to cover is a requirement. When used properly, animal glue adhesives provide high ''page pull'' values, with paper failure in many cases. Flexibility of the adhesive film under a wide range of temperature and humidity conditions is also of great importance. A major factor in the use of animal glue adhesives in this application is their repulpability (the film is water soluble and biodegradable), which permits an economic return on the scrap trimmed from the book. Current competing hot melts are not repulpable, resulting in loss of trim for recycling.

B. Paper

Animal glues, in both dry and liquid forms, have been used extensively in the manufacture of paper and paper products. Attributes of importance include film forming, colloidal, amphoteric, and adhesive properties. Animal glues have application in the creping of various types of tissue, such as facial, toilet, and toweling. Creping typically involves machine speeds up to 5000 ft/min and dryer temperatures in the range of 600°C. Creping glues can be added at the paper machine head box, or by spray application to the surface of the dryer, typically a large single steam-heated cylinder known as a Yankee dryer. The creping glue adheres the paper tissue to the face of the cylinder, where a doctor blade peels it off, thereby forming the familiar crinkled pattern as the paper leaves the dryer. Solutions of 1 to 3% glue concentration, alone or in combination with poly(vinyl alcohols), release oils, and plasticizers, are typical. The amount of animal glue applied is generally not over 0.1% on the dry paper furnish.

Animal glues provide good adhesion to the dryer face, without which the paper sheet lifts or blows from the dryer surface with little or no crepe effect. Animal glue provides the final tissue with a fine, evenly textured crepe with a soft feel or hand. Longer doctor blade life has been reported when using animal glue. Repulpability of animal glues when recycling the tissue trim is an added benefit. Maintaining the required glue film on the face of the dryer is easy, as is removal during cleanup.

Animal glues have been used for both internal and surface paper sizing. Glue is a protective colloid for rosin used in beater sizing of fine, rag-content papers, particularly in hard-water conditions. In practice, the glue is added near the end of the beater operation after the pH of the furnish has been adjusted. The paper fiber is positively charged, so the rosin–glue particle is attracted to and binds with the paper fiber. The addition of small quantities of animal glue (typically 0.3 to 0.6% on the paper furnish) enhances the internal sizing strength, density, formation, and resistance to scuffing and erasure, particularly on bond papers. Animal glue can also be used as a wet-strength additive in conjunction with resins that are aldehyde donors, which renders the animal glue insoluble.

Animal glues are used for the surface sizing of fine papers such as bond, currency bank note, blueprint and reproduction, and chart and map papers. When animal glues are used alone or in combination with starch sizing, higher grades of glue are generally used. Application is either by immersion in tub sizing, or by surface application through a size press. The function of the sizing is to provide a protective film on the surface of the sheet, laying down the fibers and closing the pores to improve ink holdout, increase surface strength, and resist picking of paper fibers. By proper use of hardening agents such as aldehyde donors or metal salts such as aluminum, a good degree of moisture resistance can be imparted to the sheet. The good film-forming properties of animal glues, as well as resistance to oil, grease, and fats, are beneficial factors. Animal glues were used extensively for the retention and recovery of paper fiber, both on machine and in save-alls, because of their amphoteric, flocculating ability, but have largely been supplanted by synthetic resins.

C. Paper Converting

An example of the use of animal glue adhesives in the paper converting field is for the covering of rigid boxes and containers with paper, typically lithographed decorative labels in setup box manufacturing. Animal glue adhesives are preferred because of their controlled tack, nonwarp properties, and permanence of adhesion. In this operation, a thin paper cover is fed through a roller applicator and coated with adhesive. The paper is delivered to a belt and then to a spotting station, where the box is placed on the cover mechanically or by hand. The machine then wraps the cover material around the box and ejects the finished box. Animal glue adhesives are also used for record albums, paper laminating, convolute and spiral cans such as oil and juice cans, and other specialty box operations, such as cosmetics and cigarettes.

D. Abrasives

The coated-abrasive industry consumes a large volume of animal glue in such applications as sandpaper and cloth in rolls, sheets, and various specialties. Animal glue is also used for ''setup'' abrasive wheels, belts, disks, cones, and bobs, as well as in greaseless abrasive composition, which is basically a mixture of animal glue, water, and abrasive in stick form for application to polishing wheels. Abrasives are involved in the production and finishing of wood products and for fine metal finishing, which is of importance in the production of machinery, automobiles, household equipment, appliances, and similar products.

In the production of coated abrasives, the backing material, especially in the case of cloth, is sized as required with low- to medium-grade animal glue or glue and starch, and is applied in the concentration range 30 to 40% to impart body and strength. The base or ''make'' coat is applied to the paper or the presized cloth by a roller applicator which deposits the glue film in a thickness determined by the grit size: the larger the grit, the thicker the film. The base coat takes advantage of the gelling property of the animal glue. Glue concentration is in the range of 25 to 40% and may contain extenders such as calcium carbonate. Higher-grade glues are used for silicon carbide and aluminum oxide, medium grades for emery and garnet, and the lower grades for flint.

The adhesive grain is applied uniformly to the tacky glue film on the moving backing sheet from a gravity-fed hopper. The grain is held in place as the glue gels and can be electrostatically oriented into a vertical position as the sheet passes through the initial drying stage. The partially dried sheet is passed through a second coating machine, which applies a thin film of animal glue (or in some cases a resin), usually in the concentration range 10 to 15%, to lock the grit firmly in position. The abrasive cloth or paper is then dried in a controlled hot-air dryer and wound into jumbo rolls for storage and subsequent slitting and cutting to the desired shape and form.

Closely allied with coated abrasives is the use of animal glues by the end user for set up, polishing wheels, belts, disks, and bobs. The user prepares these by applying an animal glue base coat and an appropriate abrasive grain, and sometimes a top coat, followed by drying. The principles, grade selection, and handling are similar to those for coated abrasives. Applications include metal finishing, glass grinding, granite finishing, and leather finishing.

A major advantage of animal glues over phenolic resins used for the same purpose is film resilience. Phenolic films tend to be brittle, which can result in scratching and other undesirable finishing problems, especially with the finer grades of abrasives. A common practice is to use an animal glue make coat and a phenolic resin size coat, to take advantage of the adhesive and film properties of animal glue while imparting moisture resistance.

E. Woodworking

Animal glues were the traditional adhesives for woodworking, finding acceptance for edge gluing, assembly, veneering, inlays, and finishing. They have been largely supplanted by ready-to-use, synthetic, water-based emulsion adhesives, and to a smaller degree by hot melts for some applications. Animal glues are still used in high-quality furniture and for critical applications such as pianos.

Edge gluing includes the joining of wood pieces for table and desk tops, chair bottoms, core stock construction, and posts and blocks for turning. Assembly gluing is the putting together of panels and parts in case good, cabinet, chair, and drawer construction. For high-speed edge gluing, assembly gluing, post lamination, and operations with short assembly times, medium to higher testing grades are used. Low to medium grades are usually preferred for slower-speed assemblies, veneering, and edge banding operations. For medium to high test grades, water/glue ratios vary from $1\frac{1}{2}$ to $1\frac{3}{4}$:1 for rapid assembly times in the range 0 to 15 s closed time, to 2:1 to $2\frac{1}{2}$:1 for slow assembly times in the range of 1 min closed time until pressure is applied. For low to medium grades, water/glue ratios vary from $1\frac{1}{8}$ to $1\frac{3}{8}$:1 for rapid times to $1\frac{1}{4}$ to $1\frac{3}{4}$:1 for slow times.

In edge and assembly gluing, a thin uniform coat of glue is applied to the surface to be bonded. The surfaces are lined up, joined, and then clamped under adequate pressure (in the range 100 to 150 psi). The pieces are then removed for final air drying until the full strength of the bond is developed. Animal glues possess high tensile strength—greater than that of the wood—in addition to their tack and set properties. Also, the film does not creep. Liquid animal glues, usually 50 to 60% solids, find application in panel or frame assembly of hard board to wood for cabinets, drawer assembly, corner blocking, trailer assemblies, and other applications where a ready-to-use adhesive with a relatively long open time is desired.

As less commonly known use of animal glues is as a size for the finishing of high-quality furniture surfaces. In this procedure, a dilute, warm glue solution is applied to the surface of the wood and is allowed to dry. The compression grain of the wood is raised and the glue fills the exposed, porous wood surface. On sanding, a glasslike surface is obtained which takes a uniform, lasting stain at reduced levels of finish.

F. Matches

Animal glues continue to be used for the production of book and wooden stick matches. Efforts to replace it have been without success. The production of matches is dependent on the presence of the animal glue in the head. It not only serves as the binder for the ignition chemicals, combustible fillers and inert materials, but because of its air-entraining properties ensures a match head of proper density and burning characteristics.

The animal glue content of the match head is normally 10 to 12% of the dry ingredients. High-grade glues are used for book matches, and medium- to high-grade glues are used for safety and "strike anywhere" wood stick matches. Glues for matches require a low grease content, usually 0.3% or below, and controlled foam properties. The foam creates a uniform, porous head containing oxygen to promote ignition and combustion.

G. Ore and Metal Refining

In ore refining, the flocculating or suspending properties of animal glues, which depend on the relative electrical charges, are used to separate valuable metal-containing ores, including uranium, zinc, lead, and others, from base materials such as clay. The glue acts as a protective colloid (floc stabilizer) to inhibit precipitation or "sanding out" of fines during leaching, and as a flocculent to aid settling of fines and subsequent filtration.

Animal glues are used in the electrolytic refining and winning of such metals as copper, zinc, lead, tin, gold, and silver. Electrorefining is a process where metal is dissolved from an anode containing impurities, and replated on a cathode. In electrowinning, the metal is recovered from solution by deposition on a cathode. Glue is added at a level of 0.03 to 0.15% to the acid electrolyte solution and produces a uniform high current density, resulting in the production of a smooth, dense cathode deposit in which nodular growths, spines, and needle growths, which can ultimately short circuit the electrolytic cell, are avoided.

H. Gummed Tape

Low-test animal glues, both bone and hide, were used extensively in gummed tape manufacture. Animal glues have been displaced to a considerable extent by dextrins and starches. They are still used as an additive to dextrins and starches to improve adhesive properties, and also alone or with dextrins for specialty paper or cloth tapes. Animal glues possess a high, aggressive tack, good open time and set properties, permanent bond, and excellent machinability. Properties are frequently modified as to tack, open time, set, and wettability by the addition of modifiers such as wetting agents, plasticizers, and gel depressants, in addition to dextrins.

The gumming adhesive is generally prepared at 50% concentration and applied at a dry adhesive deposit of about 25% on the basis weight of the paper or cloth backing. The adhesive is applied to the backing by roller applicator at 140 to 145°F and either passed over drying rolls or through a tunnel dryer. It is stored in rolls for subsequent slitting, printing, and cutting into standard rolls for use.

I. Miscellaneous Applications

1. Textiles

Animal glues have been used for many years as a warp size to protect yarn from breaking and chafing during weaving. These glues have particular applications on rayon, acetate, and viscose, and cotton, nylon, Orlon, and Dynel have also been sized successfully. Sizing solutions commonly contain 2.5 to 8% glue solids using various grades of glue, depending on the operation. The glue is usually modified with plasticizers and softening agents such as sulfonated oils, glycerin, and wax emulsions as well as wetting agents. The modifiers are generally used at a level of 20 to 45% of the glue solids. These agents act to lubricate and provide flexibility to the size film on the yarn.

In the process of weaving crepe fabrics, the yarn is held in a tight twist by a glue film during weaving. The fabric is washed after weaving to remove the glue, which releases the twist and forms the characteristic crepe effect. Animal glues are also used in textile finishing as dye leveling agents, and in silk screen printing.

2. Rubber Compounding

Animal glues are used in rubber compounding to modify the physical properties of the rubber. In particular applications such as textile rollers, cots, and aprons used in weaving cloth, the glue prevents static buildup during weaving, which causes fiber lapping or winding of fibers around rollers. The static electricity is discharged through the rollers, due to the electrolytic properties of the glue and the humidity in the air. Because of their amphoteric electrolytic properties, animal glues are also used as suspending or coagulating agents in rubber–water emulsions.

3. Luggage and Case Covering

Large volumes of compounded animal glue adhesives are used for soft- and hard-sided luggage, case, casket, and table pad covering and lining. These are basically hand operations where the glue is applied by roller applicator to the covering or lining material, usually uncoated or coated cloth or leather, and the material is positioned on the piece and glued into place. This is generally followed by riveting or sewing to finish the unit. These products are characterized by good tack to hold the material in place, and long open time to allow movement of the material into final position. For soft-sided luggage, dry film flexibility is also important, to prevent glue cracking. The adhesive should also have a low reducible sulfur content, to minimize tarnishing of metal fittings.

4. Gaskets

Animal glue–impregnated gasket materials are used as seals between two joined surfaces. They are water, oil, and solvent resistant, flexible, compressible, economical, and do not bond to the sealed surface. In the production of fiber or paper gaskets, the saturating procedure is carried out by drawing the fiber or paper stock through a shallow tank containing a glue–plasticizer solution. The saturated sheet then passes through squeeze rolls to remove excess liquid. It next passes through a second bath of hardening or curing agent (a aldehyde donor) and plasticizer. The sheet is then slowly dried under controlled conditions and cut into widely varied shapes.

5. Glass Chipping

An unusual application for animal glues is in glass chipping, recently the subject of revived interest. Chipped glass is used for decorative panels in such items as doors, windows, and lampshades. The frost or fernlike design is produced by pouring an animal glue solution of about 30 to 35% solids on a clean, sandblasted glass surface, and allowing the glue to dry under controlled conditions. The glue adheres to the glass, and as the film dries, it contracts, pulling particles of glass from the surface and creating the design. The pattern is controllable, within limits, by choice of glue grade. Use of higher grades makes larger designs, while lower grades produce a finer, more even texture.

6. Other Applications

Animal glues are used in leather manufacturing as a component of fat liquoring compounds which are used to impart softness to the leather, and as a filling agent to give added body to the leather. They are also used in finishing.

Low-grade animal glues are employed as a component with whiting, linseed oil, and rosin in composition for picture frames and decorative moldings on wood. High-grade glues are used to make molds for plaster casting. Medium- to low-grade animal glues with good foaming properties and adhesive qualities are utilized to produce strong, lightweight, cellular concrete.

Selected animal glues, also known as technical gelatins, are used in the production of protein hydrolysates for use as a source of protein in such products as cosmetics, shampoos, and skin care lotions and creams. Generally, the glue protein is hydrolyzed to a molecular weight below 2000 by use of enzyme, acid, or alkaline treatment.

REFERENCES

1. R. H. Bogue, *The Chemistry and Technology of Gelatin and Glue*, McGraw-Hill, New York, 1922.
2. A. Veis and J. Cohen, *J. Am. Chem. Soc. 78*: 244 (1956).
3. A. Veis, *The Macromolecular Chemistry of Gelatin*, Academic Press, New York, 1964.
4. F. Hofmeister, *Z. Physiol. Chem. 2*: 299 (1878).
5. J. E. Eastoe, *Biochem, J. 61*: 589 (1955).
6. R. E. Newman, *Arch. Biochem. 24*: 289 (1949).
7. J. Pourdier, *Sci. Ind. Photogr. 19*: 81 (1948).
8. A. G. Ward, *J. Photogr. Sci. 3*: 60 (1955).
9. J. E. Eastoe and A. A. Leach, *Recent Advances in Gelatin and Glue Research* (G. Stainsby, ed.), Pergamon Press, New York, 1958, pp. 173–178.
10. J. R. Hubbard, in *Animal Glues*, 2nd ed. (I. Skeist, ed.), Van Nostrand Reinhold, New York, 1977, p. 140.
11. G. Stainsby, ed., *Recent Advances in Glue and Gelatin Research*, Pergamon Press, New York, 1958.
12. J. L. Schwendeman, et al., Gelatin as a possible structural material for space use, *Report ASD-TDT-63-444*, OTS, U.S. Department of Commerce Washington, D.C., July 1963.
13. *Standard Methods for the Sampling and Testing of Animal Glues*, National Association of Glue Manufacturers, New York, 1962.
14. W. E. Blair and C. L. Pearson, *Adhesives Age 30*(6): 32–35 (1987).

15

Carbohydrate Polymers as Adhesives

Melissa G. D. Baumann and Anthony H. Conner *Forest Products Laboratory, USDA–Forest Service, Madison, Wisconsin*

I. INTRODUCTION

Carbohydrates in the form of polysaccharides are readily available from all plants, the exoskeletons of various marine animals, and some microorganisms. Because up to three-fourths of the dry weight of plants consists of polysaccharides, it is not surprising that many polysaccharides are readily available at low cost. Polysaccharides, especially from plant sources, have served a variety of uses in human history, ranging from basic necessities, such as food, clothing, and fuel, to paper and adhesives.

Three major carbohydrate polymers are readily obtained from biomass and are commercially available. These polysaccharides are cellulose, starch, and gums. The use of each of these types of carbohydrate polymers in and for adhesives is discussed in this chapter.

II. ADHESIVES FROM CELLULOSE

Cellulose is the principal structural material in the cell wall of all plants and is also found in algae, bacteria, and animals (tunicates). Approximately 10^{11} tons of cellulose is formed each year; this puts cellulose among the most important renewable resources in the world [1].

A. Cellulose Structure

Cellulose is a homopolymer of β-D-anhydroglucopyranose monomeric units that are linked via ether linkages between C-1 of one monomeric unit and C-4 of the adjacent monomeric unit (Fig. 1). As illustrated, every other monomeric unit is rotated approximately 180° about the long axis of the cellulose chain when compared to its two neighboring monomeric units. Because of this rotation, cellobiose is usually considered to be the repeat unit of the cellulose polymer. Chain lengths in cellulose can range from 700

The Forest Products Laboratory is maintained in cooperation with the University of Wisconsin. This article was written and prepared by U.S. government employees on official time, and it is therefore in the public domain and not subject to copyright.

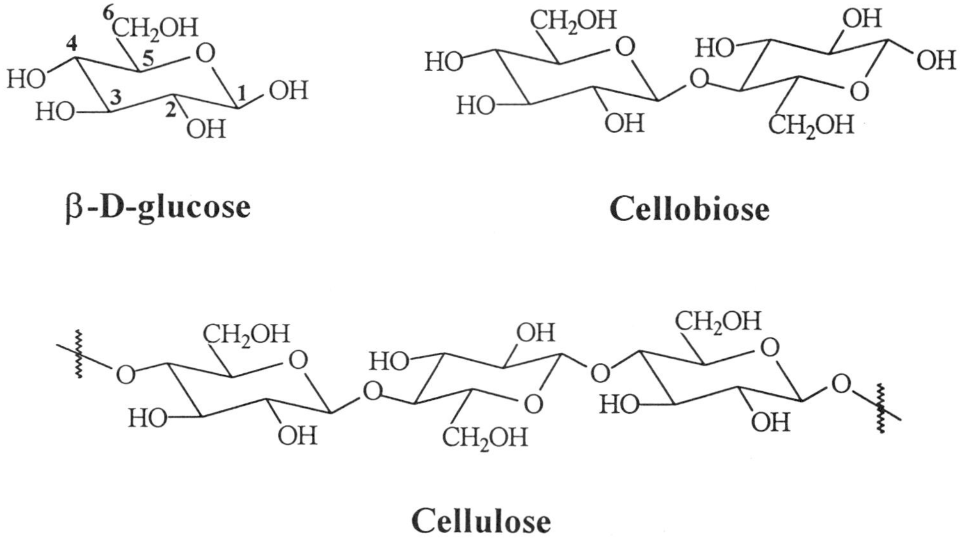

β-D-glucose

Cellobiose

Cellulose

Figure 1 Molecular structures of β-ᴅ-glucose, cellobiose, and cellulose. β-ᴅ-Glucose is the main building block of cellulose, while cellobiose is the actual repeating unit.

to 25,000 glucose units, depending on the source [2]. The two most common sources for cellulose are cotton linters and wood pulp, with approximate chain lengths of 1000 to 5000 and 500 to 2100 glucose units, respectively.

As a result of the large number of hydroxyl groups, cellulose molecules readily form hydrogen bonds with other cellulose molecules to give highly crystalline structures. Because the hydrogen bonding between cellulose molecules is not easily disrupted, cellulose does not dissolve in most common solvents. As a consequence, cellulose itself is not useful as an adhesive. Instead, cellulose is converted to various derivatives that can be used in the formulation of adhesives. Both esterification and etherification reactions can be carried out at the hydroxyl groups of cellulose. These cellulose esters and ethers can readily be dissolved into aqueous or organic solvents. Certain derivatives are thermoplastic and thus have been used in plastics and as hot-melt adhesives.

B. Modification of Cellulose

Each glucose unit in a molecule of cellulose has three hydroxyl groups that can be used to derivatize the cellulose by reactions common to all alcohols. It is uncommon, and for some derivatives impossible, to achieve a degree of substitution (DS) of 3. Most important derivatives of cellulose have a DS that is somewhat below that value. For a given derivative, the DS must be specified since the properties of the derivative depend almost as much on DS as they do on the substituting agent.

1. Esters

Esterification of cellulose to give cellulose trinitrate was discovered by Schonbein in 1846 using a mixture of sulfuric and nitric acids. The resultant compound was so flammable that its first use was as smokeless gunpowder. By the end of the nineteenth century, cellulose nitrates had been prepared with a lower DS, and they could safely be used for

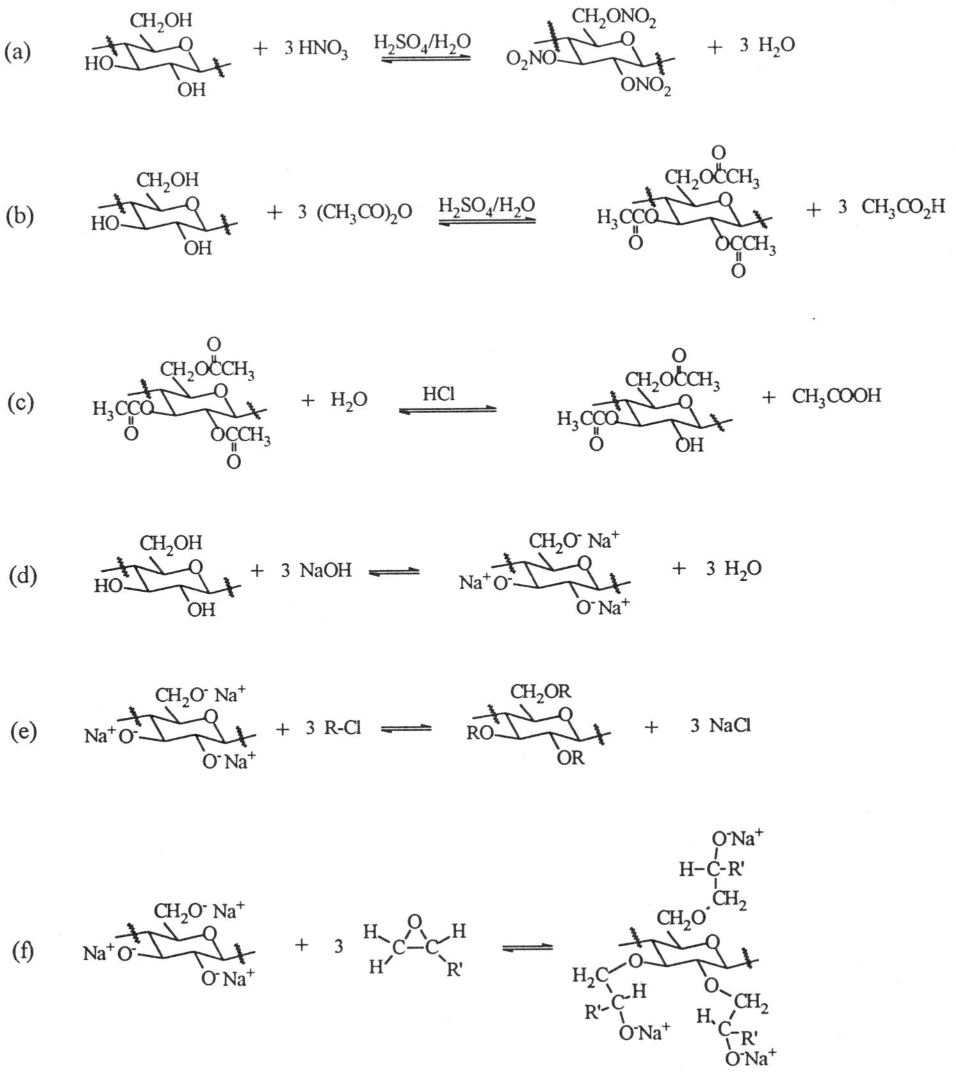

Figure 2 Major reactions of cellulose: (a) nitration by nitric acid; (b) esterification by acetic anhydride; (c) hydrolysis of tertiary cellulose acetate by hydrochloric acid; (d) deprotonation of cellulose, the first step in etherification; (e) etherification of sodium cellulose by an alkyl halide; (f) etherification of sodium cellulose by an epoxide.

other purposes. All cellulose nitrates are prepared by Schonbein's method, in which an aqueous slurry of cellulose is reacted with nitric acid in the presence of sulfuric acid (Fig. 2a). The reaction is in equilibrium and thus the removal of water during the reaction forces the reaction to completion [3]. The relative concentrations of the reacting species determine the ultimate DS that can be obtained. Boiling the nitrated product in water removes sulfate groups that can make the cellulose nitrate unstable, and digestion may be the final preparation step if a lower-viscosity material is desired.

The discovery that cellulose esters could be prepared with organic substituents led to the development of cellulose derivatives that had decreased flammability compared to that of cellulose nitrate. The most important organic ester is cellulose acetate. It is prepared by the reaction of acetic anhydride on cellulose in the presence of sulfuric acid. Acetic acid is used as the solvent and the reaction is carried out for about 8 h to yield the triester (defined as having a DS greater than 2.75) (Fig. 2b). The derivatives with lower DS values are obtained by the hydrolysis of the triester by hydrochloric acid to yield the desired substitution (Fig. 2c).

Esters of other aliphatic acids are prepared in a similar manner using the appropriate anhydride. Industrially important esters include cellulose propionate, cellulose butyrate, cellulose acetate propionate (CAP), and cellulose acetate butyrate (CAB). The mixed esters, CAP and CAB, are prepared by using a mixture of anhydrides in the desired ratios, or by reacting cellulose with propionic or butyric acid and acetic anhydride. Both reactions require sulfuric acid as a catalyst.

2. Ethers

Cellulose ethers are prepared by replacing the hydrogen on the cellulose hydroxyl groups with an alkyl group. The substitution reaction first involves the removal of the hydrogen by sodium hydroxide to make sodium cellulose (Fig. 2d). The sodium cellulose is then reacted with the appropriate alkyl halide or epoxide. Reaction with an alkyl halide yields the cellulose ether plus sodium halide (Fig. 2e). The epoxide reaction involves opening the epoxide ring (Fig. 2f), yielding a hydroxyl group on the substituent, which is deprotonated in the strongly basic reaction medium. Cellulose ethers that have been used as adhesives include methyl, ethyl, carboxymethyl, hydroxyethyl, and benzyl cellulose.

Methyl cellulose and ethyl cellulose are prepared using methyl chloride and ethyl chloride, respectively, as the alkyl source. Carboxymethyl cellulose (CMC) is the most important ether prepared from cellulose. It is commonly available as its sodium salt, which is prepared in an alcoholic solvent using either chloroacetic acid or sodium chloroacetic acid as the substituting agent. Hydroxyethyl cellulose (HEC) and hydroxypropyl cellulose are both prepared using epoxides: ethylene oxide for HEC and propylene oxide for HPC. The reaction is carried out in a water-miscible solvent, such as the lower alcohols. Because the hydroxyethyl and hydroxypropyl substituents have alcohol groups, further reaction can occur at these groups. Therefore, it is possible to get more than three substituents per anhydroglucose unit. For this reason, a quantity called molar substitution (MS) is defined to be the average number of hydroxyethyl or hydroxypropyl groups per ring, including both those attached directly to the ring and those attached to the ether substituent.

C. Properties and Uses

Adhesives derived from cellulose are used in a wide variety of applications that require an economical means for bonding porous substrates. Specific uses and formulations of various cellulose ester and ether adhesives are discussed in the following sections.

1. Cellulose Nitrate

Cellulose nitrates with DS values of 1.8 to 2.3 are used in plastics, lacquers, coatings, and adhesives. The most common use as an adhesive is as general-purpose household cement, in which the nitrate and a plasticizer are dissolved in a mixed ketone–ester organic solvent. Upon application to the substrate, the solution rapidly loses solvent to form tough, moisture-resistant, clear films. These thermoplastic films are prone to discoloration upon exposure to sunlight, and, as might be expected, are very flammable.

2. Cellulose Acetate

Cellulose acetate is the most important ester produced from cellulose; however, its use in adhesives is limited. Both the triacetate (DS greater than 2.75) and secondary acetate (DS of 2.4 to 2.6) are used industrially in plastics and textiles. The triacetate is soluble in mixtures of organic solvents, and the secondary acetate is soluble in acetone. Cellulose acetate is more heat resistant than cellulose nitrate but is less water resistant and tends to become brittle with age.

3. Cellulose Acetate Butyrate

Use of the mixed ether cellulose acetate butyrate (CAB) helps to overcome some difficulties associated with using cellulose acetate as an adhesive. CAB is soluble in a greater range of organic solvents than is the pure acetate, and it is more compatible with common plasticizers. It can be applied either as a hot-melt adhesive or in a solvent solution. Because CAB is grease resistant, it has been used in paper sizing and coating to make the paper more resistant to staining.

4. Methyl Cellulose

Methyl cellulose with a low DS (0.4 to 0.6) is soluble in dilute aqueous sodium hydroxide. As substitution is increased, the methyl cellulose becomes soluble in water (DS 1.3 to 2.6), then in organic solvents (DS 2.4 to 2.6). The most commonly used derivatives have a DS of 1.2 to 2.0, which results in cold-water solubility and solution stability for pH 2 to 12. Upon drying, solutions of methyl cellulose give clear, odorless, tasteless films that are resistant to oils and organic solvents. Methyl cellulose is used for paper coating and sizing to impart grease resistance, in ceramics as a binder, as a non-staining paste for wallpaper, and in adhesives for leather drying. The last application takes advantage of the fact that upon heating, methyl cellulose thickens reversibly. Thus hides that are attached to the platen during the drying process are easily removed when dry. Table 1 shows a formulation for a leather adhesive [4].

5. Ethyl Cellulose

In contrast to methyl cellulose, ethyl cellulose is commonly prepared in its organic soluble state with a DS of 2.3 to 2.6. Films of ethyl cellulose are thermoplastic, and they resist alkali and salts. Because of their organic solubility, the films tend to swell very little in the presence of water. The resistance of ethyl cellulose to chemical degradation has led to its

Table 1 Formulation of Leather Adhesive with Methyl Cellulose

Component	Amount (part)
Methyl cellulose (4 Pa · s)	2.0
Plasticizer	0.2
Casein	0.3
Water	97.5

use in films, lacquers, and adhesives. In adhesives, ethyl cellulose may either be applied in a solvent or as a hot melt.

6. Carboxymethyl Cellulose

Carboxymethyl cellulose, generally as its sodium salt, is the most widely used cellulose ether. CMC is water soluble with a DS of 0.4 to 1.2 in common applications. It was originally used as a replacement for natural gums in adhesives, but it has since developed many uses of its own. Sodium carboxymethyl cellulose is readily water soluble because of its ionic nature; its largest use is in laundry detergents, where it helps to suspend soil particles during washing. This resistance to greases and soil has made CMC useful in fabric sizing as a soil repellant. Because CMC is completely nontoxic, it is used in many food applications, where its affinity for water prevents drying of the product. Although it is generally no longer used as an adhesive, per se, it is still used in adhesives as a thickener.

7. Hydroxyethyl Cellulose

Hydroxyethyl cellulose with an MS value greater than 1.6 is soluble in hot or cold water; with an MS value of 2.3 to 2.6, it is soluble in organic solvents. HEC forms oil- and grease-resistant films that retain clarity over time. It is used as an adhesive in billboards, corrugated board, plywood, and wallpaper, and as a sizing and binding agent in paper products. Although not approved for direct use in food, HEC is used as an adhesive in packaging materials for foodstuffs.

III. ADHESIVES FROM STARCH

Starch is produced by plants as a way to store the chemical energy that they produce during photosynthesis. Starch is found primarily in the seeds, fruits, tubers, and stem pith of plants, most notably corn, wheat, rice, sago, and potatoes. In 1985 alone, more than 1.6 billion kilograms of starch was used in applications involving bonding of materials [5]. Clearly, this makes starch a very important adhesive material.

A. Starch Structure

Like cellulose, starch is a naturally occurring polymer of glucose. It differs from cellulose in two significant aspects: the glucose rings are in the α-D configuration rather than the β-D configuration, and starch can be differentiated into two types of polymers. One polymer, amylose, consists of α-D-anhydroglucopyranose monomeric units combined linearly through 1–4 linkages with little or no branching. The other polymer, amylopectin, is linked through 1–4 linkages but also has branches that form at the primary alcohol group on C-6 (Fig. 3). Careful analyses of various starches have shown that there is also an intermediate fraction that is thought to be an infrequently branched amylopectin [6].

The amount of amylose and amylopectin in a starch depends on the source of the starch. Most starches contain 20 to 30% by weight of amylose, although certain hybrids can contain more than 80% amylose. The most commonly available industrial starches are waxy cornstarch, regular cornstarch, high-amylose cornstarch type V, and high-amylose cornstarch type VII, with amylose concentrations of 0, 28, 55, and 70%, respectively [7].

Starch alone suspended in cold water is essentially unable to act as an adhesive because the starch is so tightly bound in granules. The granules consist of crystalline regions where straight-chain molecules and straight sections of branched molecules are

α-D-glucose   **Amylose**

Amylopectin

Figure 3   Molecular structures of α-D-glucose and the two major molecules that make up starch, amylose, and amylopectin.

Figure 4   Outer surface of starch granule showing radial arrangement of crystallites. Crystallites are shown as darker lines. [Adapted from K. H. Meyer, *Adv. Colloid Sci. 1*: 168–169 (1942).]

aligned. The crystallite regions are linked together by more amorphous areas where the molecules are not aligned. Within the starch granule, the molecules and crystallites are arranged radially in concentric layers (Fig. 4). It is these granules that must be opened to obtain adhesive bonding.

B.   Modification of Starch

Starch must be modified before it can be used as an adhesive. Methods for opening the starch granules include heating, alkali treatment, acid treatment, and oxidation.

1. Heat Treatment

The simplest method of breaking up starch granules is well known to the cook. To thicken a sauce, cornstarch must be heated. During the heating process, the starch granules first swell and then burst with a coincident thickening of the suspension. The temperature at which this thickening occurs is called the gelation temperature. For starches in pure water, gelation occurs between 57 and 72°C [8]. Observation of gelled starch under a polarizing microscope indicates that the crystallinity of the starch granule is lost during the gelling process. However, the starch is not truly in solution but rather in a colloidal suspension. Suspensions of amylose and high-amylose starches have a tendency to harden and become solid upon cooling. This process is called retrogradation or setback and is a result of the tendency of linear molecules to align with one another. This aligning effect also means that at the same solids content, suspensions with a higher ratio of amylose to amylopectin have a higher viscosity.

2. Alkali Treatment

The gelation temperature can be lowered by the addition of sodium hydroxide to a starch suspension. If sufficient alkali is added, the starch can be induced to gel at room temperature.

3. Acid Treatment (Thin-Boiling Starches)

Acid modification of starch is achieved by heating the starch to 49 to 54°C with small amounts of aqueous mineral acid, followed by neutralization with base. The acid acts mainly on the amorphous regions of the starch granules, leaving the x-ray and bi-refringence patterns of the crystalline regions essentially unchanged. Dried acid-modified starch appears very similar to its unmodified counterpart; however, upon heating a suspension to the gelation temperature, the differences become obvious. The acid-modified starch tends to give a much thinner solution at the same solids content when compared to unmodified starch. This makes the modified starch useful in applications that require a higher solids content.

4. Oxidation

Oxidized starch is commonly obtained by aqueous alkaline hypochlorite treatment. A starch suspension at pH 8 to 10 is treated with hypochlorite (5 to 10% Cl based on starch) for long enough to produce the desired viscosity. Acid is liberated during the reaction, so base must be added to maintain the pH for optimum reactivity. The resultant starch contains a mixture of carboxyl and carbonyl oxygens. Some shortening of chain length is observed during the reaction, but as in acid modification, there appears to be little change in the crystalline region of the starch. Dried oxidized starch is generally whiter than unmodified starch since the oxidation and subsequent rinsing tend to remove impurities that may be present in native starch. Oxidized starches behave similarly to the acid-modified starches upon gelling. However, the oxidized starches have greater tack and adhesive character, and thus they are used more frequently in adhesive preparations.

C. Dextrins

Dextrins are the product of dry-roasting starch in the presence of an acid catalyst. Although potato, tapioca, and sago starches are the easiest to convert to dextrins, the low cost and ready availability of cornstarch make it the most commonly used starch. Dextrins are generally divided into three categories: white dextrins, canary or yellow dextrins, and

British gums. Their differences are determined by the roasting time and temperature and the amount of catalyst used.

1. White Dextrins

White dextrins are produced at low temperatures (120 to 130°C) and roasting times (3 to 7 h) in the presence of a high concentration of catalyst. The primary reaction that occurs during the formation of white dextrin is hydrolysis of the starch molecules [9]. This reaction initially acts at the 1–6 linkages, and continues with the 1–4 linkages as the concentration of 1–6 linkages decreases. Eventually, repolymerization occurs, which yields small, highly branched dextrins. Very little repolymerization occurs in white dextrins, resulting in a white or buff powder with a degree of polymerization of approximately 20. The solubility of the white dextrins can range from 1 to 95% in water, with the lower-solubility grades resembling starch in their characteristics. The higher-solubility grades are more similar to the lower-conversion yellow dextrins. The small amount of repolymerization in the white dextrins makes the suspensions susceptible to retrogradation, and thus the suspensions must be used soon after preparation.

2. Yellow Dextrins

Yellow dextrins are prepared at higher temperatures (135 to 160°C) and longer roasting times (8 to 14 h) in the presence of less acid catalyst than are the white dextrins. These conditions promote further repolymerization, yielding a yellow or tan powder with a degree of polymerization between 20 and 50. The yellow dextrins are, for the most part, water soluble; less than 1 part water to 1 part dextrin is required for a working suspension. Yellow dextrin suspensions exhibit good viscosity stability, so retrogradation is less of a problem.

3. British Gums

In British gums, the repolymerization reaction is allowed to proceed to the greatest extent. Dry roasting is carried out for 10 to 24 h at temperatures between 150 and 180°C, and a very small amount of acid catalyst is used. These dextrins tend to be the darkest in color, which ranges from yellow to dark brown. As with the white dextrins, the British gums exhibit a wide range of solubilities in water depending on the exact reaction conditions used. However, the British gums are not prone to retrogradation, and they tend to give a more viscous suspension at the same concentration.

D. Additives and Formulation Variables

The formulation of starch and dextrin adhesives can be viewed more as an art than the result of rigorous scientific study. Not surprisingly, the purpose for which the adhesive is to be used and the method by which it will be applied greatly determine the properties needed in the resin. Factors that must be controlled include viscosity, solids content, stability, tack, slip, substrate penetration, drying rate, flexibility, water and microbial resistance, and cost. Some of these are determined by the type of starch or modification, while others require the addition of an additive to give the adhesive the desired properties. In sections below we discuss some of the most common additives that are used with starch-derived adhesives.

1. Borax

Borax (sodium tetraborate) in the presence of small amounts of sodium hydroxide is the most widely used additive to starch-based adhesives. It is commonly used in dextrin adhesives, where it increases the viscosity and acts as a tackifier and viscosity stabilizer. These effects are particularly important in machine application of adhesive to substrate. When used in adhesives, borax is often added in amounts up to 10% based on dry starch before the starch is cooked. Enough sodium hydroxide is added to convert the borax to sodium metaborate, which is the active boron species in thickening. The metaborate is able to hook two starch molecules together, forming a complex (Fig. 5) [10]. If additional sodium hydroxide is added, the complex will dissociate; the viscosity of the suspension will begin to decrease with increasing sodium hyroxide [11].

2. Plasticizers

Plasticizers are used to control brittleness of the glue line and to regulate the speed of drying. Commonly used plasticizers act in one of three ways: by forming a solid solution with the dried adhesive, by controlling the moisture in the film, and by lubricating the layers within the dried adhesive. Plasticizers that form a solid solution, such as urea, sodium nitrate, salicylic acid, and formaldehyde, tend to decrease the viscosity of the adhesive preparation. Urea is the most commonly used of these additives, and it may be added at a level of 1 to 10% based on dry starch. Hygroscopic plasticizers such as glycerol and ethylene glycol are commonly used to decrease the drying rate of the film and ensure that the film does not become brittle. Soaps, polyglycols, and sulfonated oil derivatives are used in small amounts as lubricating adhesives to impart permanent flexibility to the glue line, since they are not affected by changes in humidity.

3. Additives to Increase Water Resistance

Starch-based adhesives used in any application that requires water resistance must contain additives that resist water. Commonly used additives of this type are urea–formaldehyde, melamine–formaldehyde, and resorcinol–formaldehyde precondensates, poly(vinyl alcohol), and poly(vinyl acetate). The greatest water resistance is imparted by the formaldehyde-based precondensates; poly(vinyl alcohol) and poly(vinyl acetate) are used for adhesives that are resistant to cold water but can dissolve in hot water.

4. Viscosity Stabilizers

As mentioned previously, one problem encountered in starch-based adhesives is retrogradation. Colloid stabilizers such as soaps and sodium chloride are used to retard this tendency. Borax, sodium hydroxide, and several common plasticizers also perform this function to some extent, so viscosity stabilizers may not be necessary if these additives are used for other purposes.

5. Fillers

Fillers, in amounts of 5 to 50%, are used to control the penetration of the adhesive into the substrate and to control the setting of the glue. Mineral fillers such as clay and bentonite are commonly used.

6. Other Additives

Other additives that may be found in starch-based adhesives include preservatives to retard microbial growth, bleaches to remove colored impurities and prevent discoloration of the glue over time, defoamers to prevent foaming during processing, and organic solvents to enhance bonding to waxed surfaces. The most commonly used preservative is formaldehyde, and common bleaches include sodium bisulfite, hydrogen and sodium peroxide, and sodium perborate. Defoaming agents and solvents to be used in starch ad-

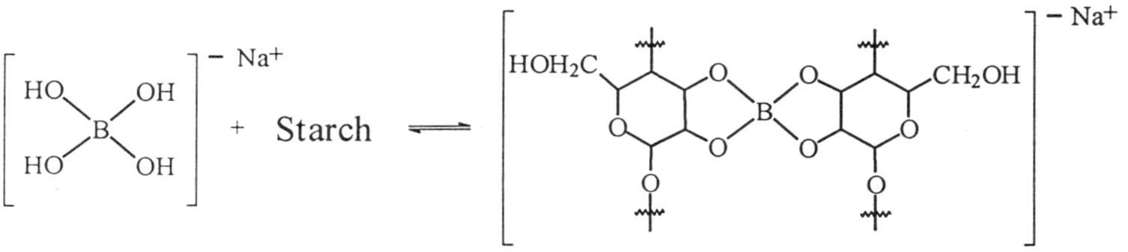

Figure 5 Complexation of starch molecules by the borax in basic solution.

hesives must be matched to the type and use of the adhesive, with special attention to compatibility and toxicity of the components.

E. Applications

The majority of starch-derived adhesives are used in the paper and textile industries as binders and sizing materials. However, the discussion in this chapter is limited to glues and pastes, since paper and textile uses have been covered thoroughly elsewhere [12]. This section provides an overview of several gluing applications, with special attention to the properties required of the glue in each case.

1. Corrugating Adhesives

Corrugated board is produced by the adhesion of a fluted layer of paper to a flat layer. A two-phase starch adhesive is commonly used to join the two layers. The liquid phase, a gelled mixture of starch and sodium hydroxide in water, is called the carrier. The solid phase of ungelled starch and borax is suspended in the carrier phase. The mixture is applied to a warm fluted sheet, which is then placed into contact with a hot flat sheet. The ungelled starch gels from the heat of the sheets. Table 2 shows a recipe for a typical corrugating adhesive [13].

Table 2 Recipe for Corrugating Adhesive

Starch adhesive phase	Amount (part)
Liquid (carrier)	
Water	11.9
Starch	3.4
40% Aqueous NaOH	2.0
Cook at 71°C for 15 min	—
Cold water	8.5
Solid (suspension)	
Water	56.6
Borax decahydrate	0.54
Stir to dissolve	—
Starch	16.9
Stir to disperse	—
Add carrier solution	—

2. Bag Manufacture

Three types of adhesives are used in the manufacture of paper bags and sacks: side-seam adhesive, bottom paste, and cross paste. During production of a single-layer bag, the paper is first formed into a long tube held together by side-seam adhesive. To be compatible with machinery used in this process, the adhesive needs a viscosity in the range 2 to 4 Pa·s with a solids content of 20 to 22%. Generally, the side-seam adhesive is made using high-soluble white dextrin or acid-modified starch, so the viscosity of the suspension remains low. Bottom paste adhesive is used to close the bottoms of the paper bags; it is much more viscous and has greater tack than side-seam adhesive. The high tack is necessary to keep the bags from opening after they have been formed. Bottom paste adhesives are usually composed of white dextrins or starch; viscosity may be as high as 140 Pa·s at 25°C. Cross paste adhesive is used only for the manufacture of multiwall bags. It is similar to side-seam adhesive, with the added requirement that it must not bleed through the paper to other layers. This is achieved by the addition of mineral fillers or poly(vinyl acetate). Water resistance of bag adhesives can be improved by the addition of urea–formaldehyde or poly(vinyl alcohol). Typical formulations for these adhesives are listed in Table 3.

3. Laminating Adhesives

Laminates prepared from paper or paperboard may be bonded with carbohydrate adhesives. The properties of the adhesive used with each will differ greatly depending on the surface of the material and the equipment to be used for laminating. However, all laminating adhesives should exhibit high tack, low penetration into the substrate, and

Table 3 Typical Formulations for Paper Bag Adhesives

Adhesive and component or process	Amount (part)
Side-seam adhesive	
Water	49.7
White dextrin (94% soluble)	39.7
White dextrin (13% to 15% soluble)	5.0
Borax decahydrate	4.0
Preservative	0.01
Antifoam agent	0.03
Cook to 85°C for 20 to 30 min	—
Cool to room temperature	—
50% Aqueous NaOH	1.2
Bottom paste	
Water	76.7
Starch	19.2
Borax decahydrate	3.1
Preservative	0.2
Soap	0.8
Cook to 93°C for 20 to 30 min	—
Adjust solids to 30%	—

Source: Ref. 14, Chap. 26, p. 14.

noncurling behavior. White dextrin is often used at about 20% of the total adhesive; a typical formulation is given in Table 4.

4. Carton Sealing.

Cartons from corrugated board are most often sealed with liquid glues or hot melt adhesives. However, dextrin adhesives are still used as a result of their low cost and ready availability. Adhesives for sealing cartons must have low and stable viscosity, be able to form strong bonds, and set up quickly. If staining of the carton is not a concern, sodium hydroxide may be added to increase adhesion. A formulation for a white dextrin adhesive is given in Table 5.

5. Remoistening Adhesives

Adhesives for remoistening applications such as tapes, stamps, labels, and envelope flaps must be capable of rewetting. Such adhesives should also be glossy, noncurling, and nonblocking under humid conditions. This requires that the formulation contain acid-modified starch or high-soluble white and yellow dextrins to allow a high solids content. A plasticizer is added to prevent brittleness in the final film. In tape and stamps, the paper with adhesive has traditionally been run over a bar to produce microcracks in the adhesive layer, thus rendering it noncurling. A similar effect has been achieved by solvent

Table 4 Typical Formulation for a Laminating Adhesive

Component or process	Amount (part)
Water	54.6
White dextrin (high soluble)	20.2
Clay	13.5
Urea	6.7
Borax decahydrate	5.0
Cook to gel	

Source: Ref. 15, Chap. 26, p. 4.

Table 5 Formulation of Adhesive for Sealing Cartons

Component or process	Amount (part)
Water	110
White dextrin	80
Preservative	2
Borax	12
Antifoam agent	0.125
Cook at 85°C for 20 min	—
Cool to 50°C	—
Water	10
50% NaOH	0.12

Source: Ref. 14, Chap. 26, p. 4.

Table 6   Formulations for Remoistening Adhesives

Adhesive and component or process	Amount (part)
Envelope adhesive	
Yellow dextrin (95% soluble)	65.7
Water	32.9
Tributyl phosphate	0.2
Heat to 88°C for 30 min	—
Cool to 60°C	—
Corn syrup	1.1
Gummed tape adhesive	
Water	50
Waxy starch (acid-modified)	44
Urea	6

Source: Refs. 15 and 16.

deposition of the adhesive. In this process, a cold-water-soluble dextrin is suspended in an organic solvent, where it is insoluble. When this suspension is applied to paper and dries, it leaves behind a layer of discrete dextrin particles, which will not curl. Formulations for two remoistening adhesives are given in Table 6.

IV. ADHESIVES FROM NATURAL GUMS

Gums are hydrophobic or hydrophilic polysaccharides derived from plants or microorganisms that upon dispersing in either hot or cold water produce viscous mixtures or solutions. Natural gums include plant exudates (gum arabic, gum ghatti, gum karaya, gum tragacanth), seed gums (guar gum, locust bean gum, tamarind), plant extracts (arabinogalactan from larch; agar, algin, funoran from seaweed), and the extracellular microbial polysaccharides (xanthan gum, dextran). Gums are used for many industrial purposes, as shown in Table 7. In recent years, synthetic polymers and microbially produced gums increasingly have replaced plant-derived gums.

Historically, several adhesives have been derived from natural carbohydrate polymers. In a few cases, such polymers have been utilized because of their own particular adhesive properties. However, natural carbohydrate polymers are usually used as modifiers for more costly synthetic resins, especially as thickeners, colloidal stabilizers, and

Table 7   Applications for Natural Gums

Product	Amount (%)
Detergents, laundry products	16
Textiles	14
Adhesives	12
Paper	10
Paint	9
Food	8
Pharmaceuticals, cosmetics	7
Other	24

flow controllers. Adhesive uses for natural gums include pressure-sensitive tape, denture adhesives, pharmaceutical tablet binders, household products, and label pastes [17].

V. CONCLUSIONS

Carbohydrates, in the form of polymers such as cellulose, starch, and natural gums, are available in large quantities, especially from plant sources. Each of these has potential for utilization as adhesives and in adhesive formulations. This has been true historically and will be increasingly true in the future as petroleum-derived polymeric materials become scarce and their prices rise. However, because the bonds formed by carbohydrate polymer adhesives are generally sensitive to water, future applications of these adhesives will increasingly depend on modifying the natural polymer to give components that can undergo further cross-linking to form water-insensitive bonds.

REFERENCES

1. V. P. Karlivan, in *CHEMRAWN I: Future Sources of Organic Raw Materials* (L. E. St.-Pierre and G. R. Brown, eds.), Pergamon Press, Elmsford, N.Y., 1980.
2. D. N.-S. Hon, in *Adhesives from Renewable Resources*, ACS Symposium Series 385 (R. W. Hemingway and A. H. Conner, eds.), American Chemical Society, Washington, D.C., 1989, p. 291.
3. L. C. Wadworth and D. Daponte, in *Cellulose Chemistry and Its Applications* (T. P. Nevell and S. H. Zeronian, eds.), Ellis Horwood, Chichester, West Sussex, England, 1985, p. 349.
4. W. D. Paist, in *Handbook of Adhesives* (I. Skeist, ed.), Van Nostrand Reinhold, New York, 1977, pp. 212–221.
5. H. M. Kennedy, in *Adhesives from Renewable Resources*, ACS Symposium Series 385 (R. W. Hemingway and A. H. Conner, eds.), American Chemical Society, Washington, DC, 1989, p. 272–273.
6. R. L. Whistler and W. M. Doane, *Cereal Chemistry 38*: 251 (1961).
7. L. Kruger and N. Lacourse, in *Handbook of Adhesives* (I. Skeist, ed.), Von Nostrand Reinhold, New York, 1990, pp. 153–166.
8. J. Delmonte, *Technology of Adhesives*, Reinhold, New York, 1947, pp. 277–302.
9. F. Wood, in *Recent Advances in the Chemistry of Starch and Cellulose* (J. Honeyman, ed.), Interscience, New York, 1959, pp. 285–306.
10. F. A. Cotton and G. Wilkinson, *Advanced Inorganic Chemistry*, Wiley, New York, 1980, p. 298.
11. J. A. Radley, in *Industrial Uses of Starch and Its Derivatives* (J. A. Radley, ed.), Applied Science, London, 1976, p. 37.
12. A. H. Zijderveld and P. G. Stoutjesdijk, in Ref. 9, pp. 199–228; M. J. Mentzer, in *Starch: Chemistry and Technology* (R. L. Whistler, J. N. BeMiller, and E. F. Paschall, eds.), Academic Press, New York, 1984, pp. 543–574; Starch and starch products in paper coating, *Tappi Monograph 17*, Technical Association of the Pulp and Paper Industry, New York, 1957.
13. W. O. Kroeschell, ed., *Preparation of Corrugating Adhesives*, Tappi Press, Atlanta, GA., 1977, pp. 19–20.
14. A. Blair, in *Handbook of Adhesive Bonding* (C. V. Cagle, ed.), McGraw-Hill, New York, 1973.
15. H. M. Kennedy and A. C. Fischer, Jr., in *Starch: Chemistry and Technology* (R. L. Whistler, J. N. BeMiller, and E. F. Paschall, eds.), Academic Press, New York, 1984.
16. J. Lichman, in *Cellulose Chemistry and Its Applications* (T. P. Nevell and S. H. Zeronian, eds.), Ellis Horwood, Chichester, West Sussex, England, 1985, pp. 679–691.
17. A. H. Conner, in *Adhesives from Renewable Resources*, ACS Symposium Series 385 (R. W. Hemingway and A. H. Conner, eds.), American Chemical Society, Washington, DC, 1989, p. 272–273.

16

Natural Rubber-Based Adhesives

Sadhan K. De *Indian Institute of Technology, Kharagpur, India*

I. INTRODUCTION

Natural rubber adhesives can be classified into two principal types: latex adhesives and solution adhesives. Natural rubber latex is obtained by tapping the tree *Havea brasiliensis*. The latex consists of about 35% solids. Ammonia is added to the latex immediately after the tapping to prevent bacterial attack and coagulation. Before marketing, the latex is concentrated to a total solids content of about 60 to 70%.

Solution adhesive is obtained from solid rubber obtained by coagulation of latex as obtained from the tree. The coagulation is effected by dilute aqueous solution of organic acids. The solid rubber can be graded in terms of dirt content, ash content, nitrogen content, and volatile matter. It can also be categorized by plasticity retention index and Mooney viscosity. To maintain uniformity it is necessary to choose solid natural rubber of known characteristics.

II. LATEX ADHESIVES

Latex adhesives are made from natural rubber latex by adding stabilizers, wetting agents, and other components. They are applied to the substrate by brush, spray, doctor knife, or reverse roll coater. The adhesive is dried to film near room temperature. The adhesive strength can be improved by vulcanizing the system. The applications of latex adhesives are to porous substances such as paper, leather, and textiles.

Latex adhesives can be handled easily because they are more fluid due to lower solid content. They have little incendiary risk because of the absence of solvent. They are cheaper because they do not contain solvent, which is lost on drying. However, natural rubber solution adhesives possess intrinsically greater adhesion to the substrate than do latex adhesives. Following are some examples of nonvulcanizing natural rubber latex.

1. *Self-seal envelope adhesives*

	Parts by weight (wet)
60% Natural rubber latex	167
10% Potassium hydroxide solution	2
50% Aqueous dispersion of zinc diethyldithiocarbamate (accelerator)	1

2. *Leather adhesive*

	Parts by weight (wet)
60% Natural rubber latex	167
20% Ethylenediaminetetracetic acid solution	2.5
50% Aqueous dispersion of poly(2,2,4-trimethyl 1,2-dihydroquinoline) (antioxidant)	1

3. *General-purpose pure gum adhesive*

	Parts by weight (wet)
60% Natural rubber latex	167
50% Dispersion of zinc dibutylthiocarbamate	2
10% solution of ammonium caseinate	10

4. *Adhesives for the carpet industry.* Formulations for backing carpets of hessian and polypropylene staple construction are given below.

	Parts by dry weight	
	A: Primary backing	B: Secondary backing
60% Natural rubber latex	100	100
Stabilizer/wetting agent [sodium lauryl ether sulfate (27%)]	1.5	1.0
10% Aqueous dispersion of thiourea	1.0	1.0
50% Aqueous dispersion of N-phenyl, N'-cyclohexyl p-phenylenediamene (antioxidant)	1.0	1.0
Water	To give 75 % solid	
Whiting	400	250
50% Dispersion of polyacrylate thickener	0.2	0.3

Ingredients are to be mixed in the order given above. Stabilizer, thiourea, water, and antioxidant must be completely dispersed in the mix before adding whiting. Whiting must be completely dispersed in the mix before adding thickeners.

5. *Tire cord adhesive*

	Parts by weight (wet)
60% Natural rubber latex	125
40% vinyl pyridine latex	62.5
Water	507
RF resin solution	266.3
Water	240
Sodium hydroxide	0.3
Resorcinol	11
40% Formaldehyde solution	15

Mature for 6 h at 25°C

III. SOLUTION ADHESIVES

Solution adhesives consist of solid rubber dissolved in a solvent such as toluene, naphtha, or trichloroethane. The solvent used will depend on the drying and flammability considerations in the application. Milled raw rubber can be shredded and agitated in the solvent until a clear solution is obtained. Other components are added at this stage and mixed uniformly. Alternatively, the solid rubber can be compounded with other components and the mix dissolved in the solvent.

Solution adhesives are applied at 10 to 25% total solids content by spray, doctor knife, reverse roll coater, roller, or by spatula or trowel. The adhesive is dried at room temperature or in air ovens, care being taken to eliminate the fire and health hazards associated with some solvents.

The toughness and durability of the bond may be improved by using vulcanizable rubber solution adhesives. These are normally supplied as two components which are mixed prior to use. Component A is prepared as follows:

	Parts by weight
Natural rubber	10
Zinc oxide	1
Antioxidant	0.1
Sulfur	0.1
Solvent	80

Component B is prepared as follows: To 100 parts of component A, add 4 parts of a 10% solution of a dithiocarbamate accelerator. The mixed adhesive will vulcanize at room temperature or at a higher temperature, if required. Typical vulcanizing time at room temperature is 3 to 4 days.

IV. PRESSURE-SENSITIVE ADHESIVE TAPES

Solid natural rubber is widely used in adhesive formulations for making electrical insulation tapes, packaging tape, and surgical tapes and plasters. Backing may be made from paper crepe, cloth, or synthetic materials, depending on the application. The rubber can be applied from solvents or calendered directly onto the backing. The following formulations are for packaging tape and surgical tape adhesives:

	Parts by weight	
	Packing tape	Surgical tape
Natural rubber	100	100
Ester gum (tackifier)	175	100
Lanolin (wool fat)	25	25
Antioxidant	1	1
Zinc oxide	50	100
Solvent (toluene, naphtha, or trichloroethylene)	400	200

Initial shredding or granulation of the rubber aids dissolution. Alternatively, lightly mill the rubber and zinc oxide, then add solvent. Blend the swollen mixture in a Z-blade mixer until homogeneous, then add ester gum, antioxidant, and finally, lanolin.

ACKNOWLEDGMENT

The author acknowledges permission of the Rubber Research Institute of Malaysia, Kuala Lumpur, to cite the information on adhesive formulations.

BIBLIOGRAPHY

NR Technology Rubber Developments Supplement, No. 2, Natural Rubber Producers' Research Association, Brickendonbury, U.K., 1970.

NR Technology Rubber Developments Supplement, No. 2, Natural Rubber Producers' Research Association, Brickendonbury, U.K., 1972.

17

Polysulfide Sealants and Adhesives

Naim Akmal* *University of Cincinnati, Cincinnati, Ohio*

A. M. Usmani *Firestone, Carmel, Indiana*

I. INTRODUCTION

High-sulfur-containing polymers known as polysulfides were introduced by Thiokol in 1928. The solid polysulfide polymers contain 37 to 82% bound sulfur, but the more important liquid polymers containing about 37% sulfur find application in high-performance sealants and adhesives. These conventional polysulfide polymers are now available from Morton International (United States), Toray Thiokol (Japan), and Chemiewerke (Germany). During the 1960s and 1970s, new mercaptan-terminated polymers were introduced that have varying polymer backbones. The properties, especially the chemical resistance of the polymers, depend on their backbone structures.

Polysulfide sealants account for about 43 million pounds of the total 500 million pounds of the U.S. sealant market. Major fields of polysulfide sealants are aircraft, automotive, construction, and marine. End uses of polysulfides are very diversified: for example, sealing integral fuel tanks, sealing pressurized cabins, potting electrical connectors, sealing bolted steel tanks, glazing of windshields, glazing of rear automotive lights, recreational vehicles, vibration damping in trailers, gas tank liners, curtain walls, building exterior joints, highway joints, airfields, insulated glass, swimming pools, flight decks, decks of pleasure craft, solid-rocket fuel binder, relief maps, printing rolls, dental impressions, hoses and gaskets.

In this chapter we describe the chemistry and technology of polysulfide polymers; processing and manufacture of polysulfide sealants and adhesives, including their formulations; curing reactions; and characterization and testing. Adhesion considerations are also discussed briefly.

II. CHEMISTRY OF POLYSULFIDE POLYMERS

A. Preparation of Conventional Polysulfide Polymers

Preparation of liquid polysulfide polymers has been discussed extensively [1–4]. The polymer is synthesized by reacting aqueous sodium polysulfide with bis(2-chloroethyl)

Current affiliation: Teledyne Analytical Instruments, City of Industry, California formal. Cross-linking is introduced by including a small amount of 1,2,3-trichloropropane. The reactions of polymerization done at 100°C produce a mixture of chain lengths in which sulfur is present as $-C-S_2-C-$ or $-C-S_3-C-$. Dihalide monomer is added to aqueous polysulfide solution containing bischlorobutyl formal and bis-4-chlorobutyl ether are used in small amounts where improvement in low-temperature performance is required. Specific emulsifying and nucleating agents (alkyl naphthalene sulfonate and magnesium hydroxide sol) are used in the polymerization. The polymerization reaction is as follows:

$$n\text{ClCH}_2\text{CH}_2\text{OCH}_2\text{OCH}_2\text{CH}_2 \;\; \text{Cl} \;+\; n\text{Na}_2\text{S}_{2.25} \longrightarrow$$

$$\leftparen\!\!-\text{CH}_2\text{CH}_2\text{OCH}_2\text{OCH}_2\text{CH}_2\text{S}_{2.25}\!\!-\rightparen_n \;+\; 2n\text{NaCl}$$

The high-molecular-weight solid polysulfide polymer is reduced by sodium sulfite, resulting in splitting of the polymer into segments that are simultaneously terminated by mercaptan groups as follows:

$$\text{RSSH} + \text{NaSH} + \text{NaHSO}_3 \longrightarrow 2\text{RSH} + \text{Na}_2\text{S}_2\text{O}_3$$

The concentration of the splitting salts regulates the molecular weight from 1000 to 8000. The extent of cross-linking depends on the mole percent of trichloropropane, a cross-linking agent used in the initial reactions, usually in the range 0.05 to 2.0 mol %. The amount of cross-linking agent regulates modulus and elongation: a decrease gives the lower modulus and higher elongation desirable in applications involving greater movement.

The typical structure of a liquid polysulfide (e.g., Morton's LP-32) is

$$\text{HS}\!-\!\!\leftparen\!\!\text{CH}_2\text{CH}_2\text{OCH}_2\text{OCH}_2\text{CH}_2\text{SS}\!\!-\rightparen_{23}\!\!\text{CH}_2\text{CH}_2\text{OCH}_2\text{OCH}_2\text{SH}$$

Use of a mixed dihalide monomer feed will produce random copolymers. During copolymerization, interchange takes place, resulting in randomization. Therefore, block copolymers cannot be prepared by stepwise addition of dihalide monomers. To prepare block copolymers, homopolymers should be made, blended in the desired proportion, and then the blend co-cured by a conventional technique. When cured by mixing an oxidizing agent, water is split off, with the hydrogen coming from the mercaptan groups of two polymer molecules and the oxygen being supplied by the oxidizing curative. The molecules are joined at the sulfurs.

B. Modified Polysulfide

Recent advances have resulted in chemical modification of conventional liquid polysulfide polymers in a one-step reaction with dithiol, as follows:

$$\text{HS}\!-\!\!\leftparen\!\!\text{CH}_2\text{CH}_2\text{OCH}_2\text{OCH}_2\text{CH}_2\!-\text{S}\!-\!\text{S}\!\!-\rightparen_n\!\!\text{CH}_2\text{CH}_2\text{OCH}_2\text{OCH}_2\text{CH}_2\text{SH} \;+\; \text{HSRSH}$$

Liquid polysulfide polymer   Dithiol

$$\downarrow \text{catalyst}$$

$$\text{HS} \!\!-\!\!\!\left[\!\!-\text{CH}_2\text{CH}_2\text{OCH}_2\text{OCH}_2\text{CH}_2\!\!-\!\!\text{S}\!\!-\!\!\text{S}\!\!-\!\!\right]_{n-m}\!\!\!\text{CH}_2\text{CH}_2\text{OCH}_2\text{OCH}_2\text{CH}_2\text{SH}$$

Segmented lower-molecular-weight liquid polysulfide

+

$$\text{HS} \!\!-\!\!\!\left[\!\!-\text{CH}_2\text{CH}_2\text{OCH}_2\text{OCH}_2\text{CH}_2\!\!-\!\!\text{S}\!\!-\!\!\text{S}\!\!-\!\!\right]_{m}\!\!\!\text{CH}_2\text{CH}_2\text{OCH}_2\text{OCH}_2\text{CH}_2\!\!-\!\! \quad \text{SRSH}$$

Dithiol-modified-liquid polysulfide polymer

This new dithiol-modified liquid polysulfide polymers have lower viscosity and improved compatibility with formulating ingredients (e.g., plasticizers, pigments, and fillers). Thus dithiol-modified liquid polysulfide polymers give more latitude in the formulation of high-performance sealants, adhesives, and coatings. The volatile organic content (VOC) of such products is lower than that of products formulated by conventional liquid polysulfide polymers.

C. Other Mercaptan-Terminated Polymers

In the 1960s Diamond Alkali marketed several polymers having polyether backbone terminated with mercaptan groups [5,6]. Polymers of poly(oxyalkalene)polyol backbone can be esterified with thio-substituted organic acids to produce terminal mercaptan end groups. These polymers were cured similar to conventional polysulfides, but gave generally poorer properties and were therefore withdrawn from the market.

In the mid-1970s, polysulfide polymer chemistry was advanced when Products Research and Chemical Corporation introduced polyoxypropylene urethane backbone with mercaptan terminal groups [7]. Molecular weight regulation and minimization of side reactions are important features. The backbone is significantly different from that of conventional polysulfide polymer, yet the curing chemistry is the same and the cured product is a polysulfide rubber. A typical structure of this type of resin is shown below.

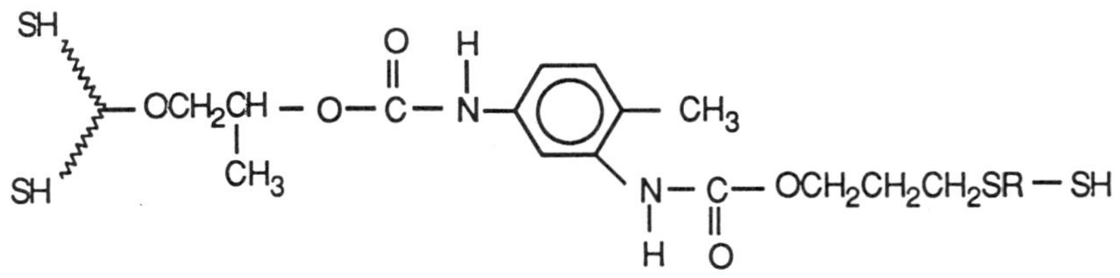

Mercaptan-terminated polyoxypropylene urethane

D. Polythioether Polymers

This polymer does not contain S–S or formal linkages, and these are the weak links in conventional polysulfides. The polythioether polymer has excellent resistance to fuel and organic solvents and has better thermal stability than that of conventional polysulfides. Polythioethers can be terminated with mercaptan, hydroxyl, silyl, and nonreactive end groups. The curing chemistry can thus be varied based on the terminal groups. A typical structure of this class of polymer is

$$+(OCH_2CH_2SCH_2CH_2)_2+(OCHCH_2SCH_2CH_2)]_n$$
$$|$$
$$CH_3$$

Polythioether

III. PROPERTIES OF POLYSULFIDE POLYMERS

Liquid polysulfide polymers are available in a series of viscosities and cross-link densities. In general, polymers in the range 400 to 500 P are used in sealants and adhesives, while lower-viscosity polymers are used for coatings and casting compounds. The tensile properties of unfilled polysulfide polymers are poor but are improved by suitable reinforcement with pigments and fillers. Both the molecular weight of the liquid polysulfide polymer and oxidative curing influence the physical properties. Higher tensile strength is obtained with higher-molecular-weight materials.

Cured liquid polysulfide compositions have excellent resistance to many oils and solvents (e.g., hydrocarbons, esters, ketones, dilute acids and alkalis). Systems must be properly formulated and cured to obtain maximum solvent resistance. Swelling tests on cured, filled polysulfides have been reported by Usmani et al. by measuring weight gain versus immersion time [8]. With jet reference fuel (JFR), an equilibrium was quickly reached. During early immersion in water, the weight gain was linear with time. Later, a square-root weight gain versus immersion time was found to exist.

The glass transition temperature (T_g) of polysulfides depends on the hydrocarbon moiety and the length of the polysulfide chain. The amount of cross-linking monomer is small, and therefore it did not influence T_g. Generally, the greater the hydrocarbon content, the lower the T_g. Higher-ranking polysulfides have higher T_g. The thermal stability of polysulfide polymers depends on the polymer backbone and the curative used to vulcanize the polymer. Commercially available polysulfides have an ethyl formal disulfide backbone, and this regulates the upper temperature limits. In an acid-catalyzed hydrolytic attack, formaldehyde is released, which in turn reduces the disulfide bond to mercaptan. The formic acid so generated catalyzes hydrolysis of the formal group. The terminal mercaptan group can react with a hydroxyl group to give a monosulfide bond. The degradation results in weight loss and loss of flexibility due to the monosulfide structure formation. Disulfide and formal groups provide a flexibilizing effect due to free rotation. Calcium oxide can neutralize formic acid and absorb water and is therefore an effective stabilizer. Practical cure rates cannot be achieved in anhydrous formulations by metal dioxide curing agents. Thermal instability can also arise when the mercaptan group reacts with the metal oxide. Sulfur mitigates formation of the mercaptide groups. Polysulfide sealants cured using manganese dioxide and chromate salts provide continuous service at 250°F.

Tobolsky has studied extensively the viscoelastic properties of polysulfide polymers [9]. Polysulfide polymers have the unique ability to relieve internal stress or stress between mercaptan and disulfide linkages. The stress decay of cross-linked polysulfide elastomer follows the equation

$$F(t) = F(0) + e^{-t/T}$$

where $F(t)$ is the final stress, $F(0)$ the initial stress, t the time, and T the relaxation time. The relaxation time (n hours) for polysulfide polymers at 80°C for several curing agents are 0.68 for lead oxide at 7.3 parts by weight of resin (phr), 32 for manganese dioxide at

18.9 phr, and 200 for 2,4-toluene diisocyanate plus N-methyl-2-pyrolidone at 7.0 phr. The ability of polysulfide to relieve stress is extremely valuable in maintaining adhesion in joints subjected to joint movement.

IV. COMPOUNDING, PROCESSING, AND MANUFACTURE OF POLYSULFIDE SEALANTS

The polysulfide sealant must be specifically formulated to meet the desired requirements and to obtain optimum properties. Pot life, working properties, and sealant properties should be properly adjusted. Suitable fillers should be dispersed and suitable additives should be incorporated into the formulation. Curing agent, curing modifier, filler, plasticizer, and adhesion additive are discussed briefly below.

Mercaptan-terminated liquid polysulfide polymers are polymerized to rubbery solids by oxidizing agents (e.g., lead dioxide, activated manganese dioxide, calcium peroxide, cumene hydroperoxide, alkaline dichromates, and p-quinonedioxime). The curing process involves the oxidation of the terminal mercaptan groups in the polysulfide polymers to form the corresponding disulfide.

$$M^{n+} + H \left[SCH_2CH_2OCH_2OCH_2CH_2S \right]_n S \longrightarrow$$

$$M^{(n-2)+} + \left[SCH_2CH_2OCH_2OCH_2CH_2S \right]_m$$

The process can simply be written as

$$M^{n+} + 2RSH \longrightarrow M^{(n-2)+} + RSSR + 2H^+$$

and it involves the net transfer of two electrons per molecule of disulfide. Dichromate oxidation of mercaptans in an aqueous medium may proceed via the following pathway:

$$Cr^{VI} + RSH \rightleftharpoons RSCr^{VI}$$

$$RSCr^{VI} + RSH \longrightarrow Cr^{VI} + RSSR$$

$$RSCr^{VI} \longrightarrow Cr^{VI} + RS\cdot$$

Mercaptan oxidation can proceed through coordination of the mercaptan to manganese dioxide, with manganite ion providing a base for deprotonation of the mercaptan.

$$-MnO_2 + 2RSH \longrightarrow$$

O\\Mn//O with SR and SR

$$\downarrow$$

$$MnO \qquad 2RS\cdot \longrightarrow RSSR$$

In general, the pH of the system governs the curing rate. Acidic materials retard, whereas alkaline materials accelerate the cure. Thus examples of retarders are stearic acid and metallic stearates. Typical accelerators are amines, inorganic bases, water, dinitrobenzene, and sulfur.

Table 1 Polysulfide Sealant Formulations (wt %)

Ingredients	Sealant				
	One-part	Building	Insulating glass	Aircraft	Casting compound
Polysulfide polymer	20	35	30	65	35
Fillers	50	40	50	25	35
Plasticizers	25	20	15	5	27
Adhesion additives	2	2	2	2	—
Curing agents	3	3	3	3	3

Source: Ref. 10.

Fillers increase the strength, impart needed rheological properties, and reduce the cost of sealants. Tensile properties are increased significantly, depending on type of filler, its particle size, and the type of cure. Improper filler selection can ruin the performance of a polysulfide sealant. Calcium carbonates (wet or dry ground limestone, precipitated), carbon blacks (furnace, thermal), calcined clays, silica and silicate fillers, and rutile titanium dioxide are typical fillers used in polysulfide sealants. Generally, combination of fillers are used in formulation. Plasticizers improve the working properties while lowering the modulus of the sealant. The plasticizer must be compatible with cured sealant, should have low volatility, and must be safe. Polymeric and esteric types are commonly used plasticizers.

Sealant adhesion is improved by the incorporation of additives. Typical phenolic resin additives are Methylon AP-108, Durez 16674, Bakelite BRL 2741, and Resinox 468. Epoxies are also good adhesion promoters. Silanes (e.g., A-187 and A-189) are known to increase adhesion. Table 1 lists five types of sealant formulations suggested by Panek that are useful in several end applications [10]. Generally, the integral fuel tank polysulfide sealant consists of two parts: 9 parts by weight of sealant base components mixed with 1 part by weight of accelerator. A typical composition is shown in Table 2.

Most sealants, especially those used in building applications, contain adhesion additives. Occasionally, to obtain good bonding, primers are used. For metals, a dilute

Table 2 Chemical Composition of a Typical Aircraft Sealant

Ingredient	Function(s)	Percent
Base component		
Calcium carbonate	Filler, reinforcer	26.15
Titanium dioxide	Filler, opacifier	3.10
Liquid polysulfide polymer	Vehicle	58.50
Volatile diluent	Viscosity adjuster	2.25
Accelerator component		
Manganese dioxide	Curing agent	5.53
Processing oil	Modulus adjuster	3.95
Diluent	Viscosity adjuster	0.51

Source: Ref. 11.

solution of silanes in organic solvents has been used. A film-forming primer is required for porous surfaces. Masonary primers generally contain a chlorinated rubber or a modified phenolic resin either alone or in combination with additional plasticizer. Thin layers of silanes give good polysulfide adhesion to metals, glass, and ceramic substrates.

Production of polysulfide sealant basically involves mixing and dispersion. Therefore, equipment used in coating and ink manufacture is applicable to sealant manufacture. Viscosities of up to 60,000 P are common in polysulfide sealant, and therefore heavy equipment is generally used in its manufacture. Typical useful equipment includes sigma blade mixers, kneader-extruders, and high-speed dispersators. A three-roll paint mill gives excellent processing. Transferring and packaging of sealants requires heavy-duty displacement pumps.

V. POLYSULFIDE SEALANT CHARACTERIZATION AND TESTING

Typical examples of physical, rheological, and mechanical characterization of sealants include specific gravity, percent solid, viscosity, flow, application time, working life, tack-free time, standard curing rate, liquid-immersed curing rate, resistance to rupture, low-temperature flexibility, peel strength, resistance to solvents, tensile strength, elongation modulus, chalking, accelerated storage stability, and hydrolytic stability. Usmani has proposed the use of parallel-plate rheometry (PPR) and dynamic mechanical analysis in the characterization and curing of polysulfide sealants [11–13].

Mechanical and physical quality control tests on polysulfide sealants can produce erroneous and misleading results, especially in predicting long-term performance. Problems such as poor adhesion, inadequate cure, and short walking life can frequently occur, resulting in tedious and costly repair.

The composition of polysulfide sealants can be determined by centrifuging thinned polysulfide sealants and resolving them into components. Both quantitative and qualitative analysis can be performed [3,8]. The filler fraction can be analyzed by x-ray analysis and scanning electron microscopy (SEM). Nuclear magnetic resonance (NMR) spectroscopy and gel permeation chromatography can be used for vehicle analysis. Mazurek and Silva have described a SEM method of analysis for cured polysulfide sealants [14]. Paul has studied the effects on environment on the performance of polysulfide sealants [15]. Numerous data can be obtained by monitoring the effects of various environmental conditions. Understanding chemical processes will assist in improving properties, however.

VI. POLYSULFIDE/EPOXY ADHESIVES

Fettes and Gannon were first to report reactions of liquid polysulfide polymers with epoxy resins [16]. In these adhesives, the epoxy resin is the major component. LP-3 in which n = 8 has been used most extensively. This polymer has a molecular weight of about 1000 with viscosity in the range 7 to 12 P. Epoxy resins most widely used in adhesive formulations have viscosity in the range 80 to 200 P with an epoxy equivalent of 175 to 210. Typical resins are Epon 820 and 828 (Shell), ERL-3794 (Union Carbide), and Araldite 6020 (CIBA). The epoxy–polysulfide reaction is prompted by organic amines (e.g., diethylenetriamine and benzyldimethylamine). The general reaction of a poly-sulfide polymer, epoxy resin, and amine hardness is

$$— RSH \; + \; \underset{\underset{O}{\diagdown\diagup}}{C—C} — R' — \underset{\underset{O}{\diagdown\diagup}}{C—C} \; + \; R''NH_2 \longrightarrow$$

$$—RS — \underset{\underset{OH}{|}}{C} — C — R' — \underset{\underset{OH}{|}}{C} — C \; \underset{\underset{H}{|}}{N} \; R''—$$

These adhesives are used in the construction, electrical, and transportation industries because of flexibility, adhesion to many substrates, and chemical resistance. They find application as adhesives for steel, aluminum, ceramics, wood, and glass; in crack repair for concrete and patch repair for concrete; as a grouting compound; and as an automotive body solder. Fillers (e.g., calcium carbonate, graphite, milled glass fibers, silica, and talc) can be added to polysulfide–epoxy adhesives to extend pot life, reduce exotherm, and increase rigidity and impact strength. Formulations of liquid polysulfide–epoxy concrete adhesives are shown in Table 3.

VII. ADHESION CONSIDERATIONS

Interfacial aspects and adhesion of polysulfides have been studied extensively and reported by us [3,17]. The epoxy-modified polysulfide has improved adhesion due to chemical reactions that increase electronic attraction forces. Water has been found to be

Table 3 Liquid Polysulfide–Epoxy Concrete Adhesive Formulations

Parts by weight	Formulation 1	Formulation 2
Part I		
Thiokol LP-3	100	100
Silica (HDS-100)	80	—
Hydrite 121	—	140
EH 330	20	20
Toluene	—	65
Part II		
Epon 820	200	200
Hydrite 121	—	105
Toluene	—	5
Working and curing properties		
Brush life (h)	0.2	1.4
Trowel life (h)	0.25	—
Setting time for 0.25 mm (h)	0.8	3.5
Tack-free time for 0.25 mm (h)	5.0	—
Cure time for 0.25 mm (h)	—	24
Ratio of part I to part II	1:1 (wt)	1:1 (vol)

the most potent debonding agent in cured polysulfides. Formation of thiourethane is responsible for excellent adhesion of polysulfide onto polyurethane coatings.

$$R-N{=}C{=}O \quad + \quad R-SH \quad \longrightarrow \quad R-NH-\overset{\overset{\displaystyle O}{\displaystyle \|}}{C}-S-R'$$

Contained in cured   Polysulfide   Thiourethane
polyurethane   sealant

Titanates used as primers interact with polysulfide to produce a tough layer of Ti–S, resulting in enhanced adhesion [17]. We have employed SEM extensively to study polysulfide debonding and events occurring at the interface and have found it to be a very useful tool.

REFERENCES

1. J. C. Patrick and H. R. Furguson, U.S. patent 2,466,963, to Thiokol Chemical Corp. (1946).
2. M. B. Berenbaum and J. R. Panek, in *Polyalkylene Sulfides and Other Polythioethers* (N. G. Gaylord (ed.), Interscience, New York, 1962, pp. 43–224.
3. A. M. Usmani, *Polymer Plastics Technol. Eng. 19*: 165 (1982).
4. N. D. Ghatge, S. P. Vernekar, and S. V. Lonikar, *J. Rubber Chem. Technol. 54*: 198 (1981).
5. G. M. La Fave and F. Y. Hayashi, U.S. patent 3,278,496, to Diamond Alkali (1966).
6. G. M. La Fave, F. Y. Hayashi, and A. W. Fradkin, U.S. patent 3,258,496, to Diamond Alkali (1966).
7. L. Morris, R. E. Thompson, and I. P. Seegman, U.S. patent 3,431,239, to Products Research and Chemical Corp. (1969).
8. A. M. Usmani, R. P. Chartoff, J. M. Butler, I. O. Salyer, and D. E. Miller, *AFWAL-TR-80-4095*, 1981.
9. A. V. Tobolsky and W. J. Macknight, *Polymeric Sulfur and Related Polymers*, Wiley-Interscience, New York, 1965.
10. J. R. Panek, in *Handbook of Adhesives* (I. Skeist, ed.), Van Nostrand Reinhold, New York, 1990.
11. A. M. Usmani, *Polymer Plastics Technol. Eng. 19*: 185 (1982).
12. A. M. Usmani, R. Chartoff, W. Werner, I. Salyer, J. Butler, and D. Miller, *J. Rubber Chem. Technol. 54*: 1081 (1981).
13. A. M. Usmani, *Polymer News 10*: 23 (1985).
14. W. Mazurek and V. M. Silva, *MRL-R-1096*, Australia, 1987.
15. D. B. Paul, *Polymer Mater. Sci. Eng. 56*: 468 (1987).
16. E. M. Fettes and J. A. Gannon, U.S. patent 2,789,958, to Thiokol (1957).
17. A. M. Usmani, in *Adhesive Joints: Formation, Characterization, and Testing* (K. L. Mittal, ed.), Plenum Press, New York, 1984.

18

Phenolic Resin Adhesives

A. Pizzi* *University of the Witwatersrand, Johannesburg, South Africa*

I. INTRODUCTION

Phenolic resins are the polycondensation products of the reaction of phenol with formaldehyde. Phenolic resins were the first true synthetic polymers to be developed commercially. Notwithstanding this, even now their structure is far from completely clear, because the polymers derived from the reaction of phenol with formaldehyde differ in one important aspect from other polycondensation products. Polyfunctional phenols may react with formaldehyde in both the ortho and para positions to the hydroxyl group. This means that the condensation products exist as numerous positional isomerides for any chain length. This makes the organic chemistry of the reaction particularly complex and tedious to unravel. The result has been that although phenolic resins were developed commercially as early as 1908, were the first completely synthetic resins ever to be developed, have vast and differentiated industrial uses today, and great strides have been made in both the understanding of their structure and technology and application, several aspects of their chemistry are still only partially understood.

It may be argued with some justification that such a state of affairs is immaterial, because satisfactory resins for many uses have been developed on purely empirical grounds during the last 80 years. However, it cannot be denied that gradual understanding of the chemical structure and mechanism of reaction of these resins has helped considerably in introducing commercial phenolic resins designed for certain applications and capable of performances undreamed of in formulations developed earlier by the empirical rather than the scientific approach. Knowledge of phenolic resin chemistry, structure, characteristic reactions, and kinetic behavior remains an invaluable asset to the adhesive formulator in designing resins with specific physical properties. The characteristic that renders these resins invaluable as adhesives is their capability to deliver water, weather, and high-temperature resistance to the cured glue line of the joint bonded with phenolic adhesives, at relatively low cost.

*Current affiliation: Université de Nancy, Epinal, France

II. CHEMISTRY

Phenols condense initially with formaldehyde in the presence of either acid or alkali to form a methylolphenol or phenolic alcohol, and then dimethylolphenol. The initial attack may be at the 2-, 4-, or 6-position. The second stage of the reaction involves methylol groups with other available phenol or methylolphenol, leading first to the formation of linear polymers [1] and then to the formation of hard-cured, highly branched structures.

Novolak resins are obtained with acid catalysis, with a deficiency of formaldehyde. A novolak resin has no reactive methylol groups in its molecules and therefore without hardening agents is incapable of condensing with other novolak molecules on heating. To complete resinification, further formaldehyde is added to cross-link the novolak resin. Phenolic rings are considerably less active as nucleophilic centers at an acid pH, due to hydroxyl and ring protonation.

However, the aldehyde is activated by protonation, which compensates for this reduction in potential reactivity. The protonated aldehyde is a more effective electrophile.

The substitution reaction proceeds slowly and condensation follows as a result of further protonation and the creation of a benzylcarbonium ion that acts as a nucleophile.

Resols are obtained as a result of akaline catalysis and an excess of formaldehyde. A resol molecule contains reactive methylol groups. Heating causes the reactive resol molecules to condense to form large molecules, without the addition of a hardener. The function of phenols as nucleophiles is strengthened by ionization of the phenol, without affecting the activity of the aldehyde.

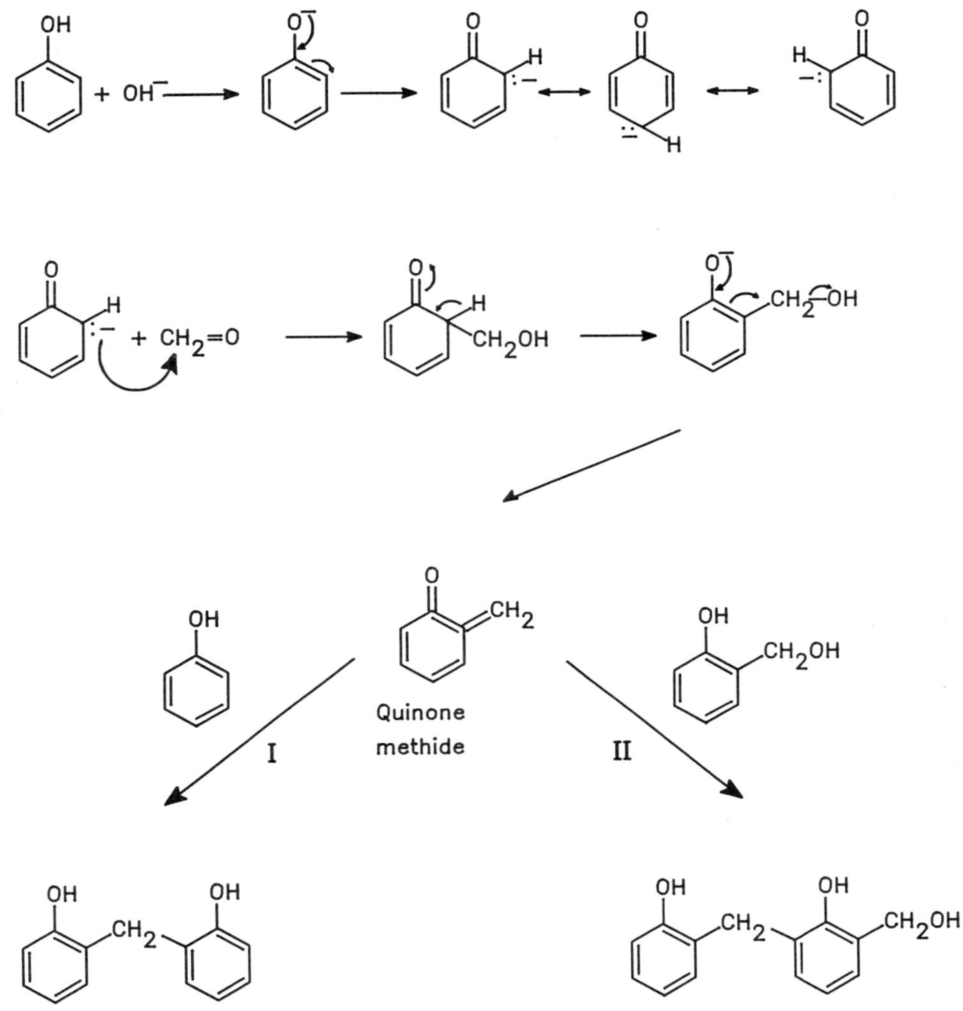

Quinone methide

I

II

Megson [2] states that reaction II (in which resols are formed by the reaction of quinone methides with methylolphenols or other quinone methides) is favored during alkaline catalysis. A carbonium ion mechanism is, however, more likely to occur. Megson [2] also states that phenolic nuclei can be linked not only by simple methylene bridges but also by methylene ether bridges. The latter generally revert to methylene bridges if heated during curing, with the elimination of formaldehyde.

The difference between acid-catalyzed and base-catalyzed processes is (1) in the rate of aldehyde attack on the phenol, (2) in the subsequent condensation of the phenolic alcohols, and to some extent (3) in the nature of the condensation reaction. With acid catalysis, phenolic alcohol formation is relatively slow. Therefore, this is the step that determines the rate of the total reaction. The condensation of phenolic alcohols and phenols forming compounds of the dihydroxydiphenylmethane type is, instead, rapid. The latter are therefore predominant intermediates in novolak resins.

Novolaks are mixtures of isomeric polynuclear phenols of various chain lengths with an average of five to six phenolic nuclei per molecule. They contain no reactive methylol groups and consequently cross-link and harden to form infusible and insoluble resins only when mixed with compounds that can release formaldehyde and form methylene bridges (such as paraformaldehyde or hexamethylenetetramine).

In the condensation of phenols and formaldehyde using basic catalysts, the initial substitution reaction (i.e., the formaldehyde attack on the phenol) is faster than the subsequent condensation reaction. Consequently, phenolic alcohols are initially the predominant intermediate compounds. These phenolic alcohols, which contain reactive methylol groups, condense either with other methylol groups to form ether links, or more commonly, with reactive positions in the phenolic ring (ortho or para to the hydroxyl group) to form methylene bridges. In both cases water is eliminated.

Mildly condensed liquid resols, which are the more important of the two types of phenolic resins in the formulation of wood adhesives, have an average of fewer than two phenolic nuclei in the molecule. The solid resols average three to four phenolic nuclei but with a wider distribution of molecular size. Small amounts of simple phenol, phenolic alcohols, formaldehyde, and water are also present in resols. Heating or acidification of these resins causes cross-linking through uncondensed phenolic alcohol groups, and possibly also through reaction of formaldehyde liberated by the breakdown of the ether links.

As with novolaks, the methylolphenols formed condense with more phenols to form methylene-bridged polyphenols. The latter, however, quickly react in an alkaline system with more formaldehyde to produce methylol derivatives of the polyphenols. In addition to this method of growth in molecular size, methylol groups may interact with one another, liberating water and forming dimethylene–ether links ($—CH_2—O—CH_2—$). This is particularly evident if the ratio of formaldehyde to phenol is high. The average molecular weight of the resins obtained by acid condensation of phenol and formaldehyde decreases hyperbolically from over 1000 to 200, with increases in the molar ratio of phenol to formaldehyde from 1.25:1 to 10:1.

A. Acid Catalysis

Consideration must be given to the possibility of direct intervention by the catalyst in the reaction. Hydrochloric acid is the most interesting case of an acid catalyst, as is ammonia of an alkaline catalyst. When the phenol–formaldehyde reaction is catalyzed by hydrochloric acid, two mechanisms may come into operation. Vorozhtov has proposed a reaction route that passes through the formation of bischloromethyl ether (Cl—CH₂—O—CH₂—Cl) [3]. Ziegler [4,5] has suggested a route through the formation of a chloromethyl alcohol (Cl—CH₂—OH) as intermediate. The second route appears to be the more probable. Both hypotheses agree that chloromethylphenols are the principal intermediates. The chloromethylphenols have been prepared and isolated by various means. They are highly reactive compounds which, with phenols, form dihydroxydiphenylmethanes and complex methylene-linked multiring polyphenols. Reaction is highly selective and takes place in the para position.

B. Alkaline Catalysis

Different mechanisms of alkaline catalysis have been suggested according to the alkali used. When caustic soda is used as the catalyst, the type of mechanism which seems the most likely is that which involves the formation of a chelate ring similar to that suggested by Caesar and Sachanan [6]. The chelating mechanism may initially cause the formation of a sodium–formaldehyde complex or of a formaldehyde–sodium phenate complex and is similar in concept to the mechanisms advanced for metal ion catalysis of phenolic resins in the pH range 3 to 7.

When ammonia is used as a catalyst, the resins formed are very different in some of their characteristics from other alkali-catalyzed phenol–formaldehyde resins. The reaction mechanism appears to be quite different from that of sodium hydroxide–catalyzed resins. An obvious deduction is that intermediates containing nitrogen are formed. Several such intermediates have been isolated from ammonia-catalyzed phenol–formaldehyde reactions by various researchers [7–9]. Similar types of intermediates are formed when amines or hexamethylenetetramine are used instead of ammonia. In the case of ammonia, the main intermediates are dihydroxybenzylamines and trihydroxybenzylamines.

These intermediates contain nitrogen and have polybenzylamine chains. They react further with more phenol, causing splitting and elimination of the nitrogen as ammonia or amines and producing nitrogen-free resins. However, this requires a considerable excess of phenol and a high temperature. With phenol–hexamethylenetetramine resins of molar ratio 3:1, the nitrogen content of the resin cannot be reduced to less than 7% when heated at 210°C. When the rate is increased to 7:1, the nitrogen content on heating at 210°C can be reduced to less than 1%.

Ammonia-, amine-, and amide-catalyzed phenolic resins are characterized by greater insolubility in water than that of sodium hydroxide–catalyzed phenolic resins. The more ammonia used, the higher the molecular weight and melting point that are obtained without cross-linking. This is probably due to the inhibiting effect of the nitrogen-carrying groups (i.e., —CH$_2$—NH—CH$_3$ or —CH$_2$—NH$_2$), which is caused by their slow rate of subsequent condensation and loss of ammonia. Ammonia, amines, and amides are sometimes used as accelerators during the curing of phenolic adhesives for wood products.

C. Metallic Ion Catalysis and Reaction Orientation

In the pH range 3 to 7 the higher rate of curing of phenolic resins prepared by metal ion catalysis is due to preferential ortho methylolation [10] and therefore also to the high proportion of ortho–ortho links of the uncured phenolic resins prepared by metal ion catalysis. The faster curing rates of phenolic resins prepared by metallic ion catalysis is then due to the higher proportion of the free higher-reactive para positions available for further reaction during curing of the resin. The mechanism of the reaction [11] involves the formation of chelate rings between metal, formaldehyde, and phenols or phenol nuclei in a resin.

The rate of metal exchange in solution [11,12] and the instability of the complex formed determine the accelerating or inhibiting effect of the metal in the reaction of phenol with formaldehyde. The more stable that complex II is, the slower the reaction proceeds to the formation of resin III. A completely stable complex II should stop the reaction from proceeding to resin III. If complex II is not stable, the reaction will proceed to form phenol–formaldehyde resins of type III. The rate of reaction is directly proportional to the instability or the rate of metal exchange in solution of complex II. The acid catalysis due to the metal ion differs only in degree from that of the hydrogen ion [13].

The effect of the metal is stronger than that of hydrogen ions, because of higher charge and greater covalence, since its interaction with donor groups is often much greater

[13]. This allows phenolic resin adhesives to set in milder acid conditions. Most covalent metal ions accelerate the phenol–formaldehyde reaction. The extent of acceleration depends on the type of metal ion and the amount of it that is present. The capability of acceleration in order of decreasing acceleration effectiveness has been reported to be [11] Pb^{II}, Zn^{II}, Cd^{II}, $Ni^{II} > Mn^{II}$, Mg^{II}, Cu^{II}, Co^{II}, $Co^{III} > Mn^{III}$, $Fe^{III} >> Be^{II}$, $Al^{III} > Cr^{III}$, Co^{II}. The most important conclusion to be drawn is that the accelerating effect is indeed present in both the manufacture of phenol–formaldehyde resin and its curing. Therefore, the fast rate of curing of high-ortho phenolic resins can be ascribed only partially to the high proportion of para positions available. The other reason for the fast rate of curing is that the metallic ion catalyst is still present, and free to act, in the resin at the time of curing. In such a resin, a considerable number of ortho positions (especially of methylol groups in ortho positions to the phenolic hydroxyls) are still available for reaction and capable of complexing.

III. CHEMISTRY AND TECHNOLOGY OF APPLICATION OF PHENOLIC RESIN ADHESIVES FOR WOOD

A. General Principles of Manufacture

A typical phenolic resin is made in batches in a jacketed, stainless steel reactor equipped with an anchor-type or turbine-blade agitator, a reflux condenser, vacuum equipment, and heating and cooling facilities. Molten phenol, formalin (containing 37 to 42% formaldehyde, or paraformaldehyde), water, and methanol are charged into the reactor in molar proportions between 1:1.1 and 1:2, and mechanical stirring is begun. To make a resol-type resin (such as that used in wood adhesive manufacture), an alkaline catalyst such as sodium hydroxide is added to the batch and it is then heated to 80 to 100°C. Reaction temperatures are kept under 95 to 100°C by applying vacuum or by cooling water in the reactor jacket. Reaction times vary between 1 and 8 h according to the pH, the phenol/formaldehyde ratio, the presence or absence of reaction retarders (such as alcohols), and the temperature of the reaction.

Since a resol can gel in the reactor, dehydration temperatures are kept well below 100°C, by applying vacuum. Tests have to be done to determine first, the degree of advancement of the resin, and second, when the batch should be discharged. Examples of methods of such tests are the measurement of the gel time of a resin in a 150°C hot plate or at 100°C in a water bath. Another method is measuring the turbidity point, that is, precipitating the resin in water or solutions of a certain concentration.

Resins that are water soluble and have a low molecular weight are finished at as low a temperature as possible, usually around 40 to 60°C. It is important that the liquid, water-soluble resols retain their ability to mix with water easily when they are used as wood adhesives. Resols based on phenol are considered to be stable for 3 to 9 months. Properties of a typical resin are a viscosity of 100 to 200 cP at 20°C, a solids content of 55 to 60%, a water mixibility of a minimum of 2500%, and a pH of 7 to 13, according to the application for which the resin is destined.

Phenol–formaldehyde (PF) resins present lower reactivity at a pH of about 4. The accepted effect of the pH and of the phenol/formaldehyde molar ratio on the rate of polymerization and rate of hardening of phenolic resols is shown in Fig. 1. Recently, however [14], the concepts expressed in the graph have been found to be only partially

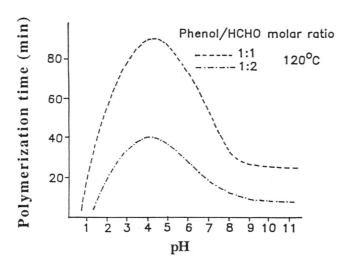

Figure 1 Rate of polymerization as a function of pH for phenolic resols of different molar ratios at 120°C (accepted wisdom).

correct, at least with regard to the dependence of the PF adhesive rate of curing as a function of pH. The expected asynthotic acceleration expected over pH 7 to 8 and due to the formation of phenate ions has been proven not to be the only effect present. At first, acceleration occurs, but after a pH of approximately 8 to 9 the rate of hardening of the resin slows down considerably [14], contrary to accepted wisdom, as shown in Fig. 2. There are several reasons for this behavior [14,15]. The probable reason why it was not noticed earlier appears to be due to the slow gel times of PF resin, which make it very tedious to check reactivity effectively.

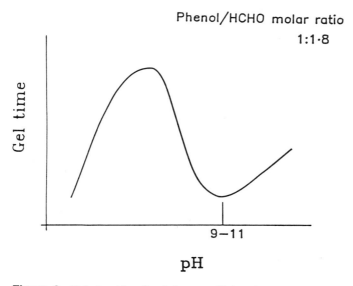

Figure 2 Relationship of gel time to pH for phenolic resols.

B. Curing Acceleration Under Alkaline Conditions

The so-called α- and β-set acceleration of curing for very alkaline PF resins for foundry core binders was pioneered in the early 1970s [16], although it had been discovered in the early 1950s [16]. In this application the addition of considerable amounts of such esters in liquid form (α-set) or as a gas (β-set), such as propylene carbonate, methyl formate, glyceroltriacetate, and others, were found to accelerate resin curing to extremely short times. This technique is now used extensively around the world for foundry core PF binders [16] and is being considered for wood adhesives [14,15] and rigid alkaline PF foams. The technique is applicable in the approximate pH range 7 to 14. The mechanism that makes PF curing acceleration possible has only been explained recently [14] and is based on the carbanion behavior of the aromatic nuclei of phenate ions, leading to a more complex variant of the Kolbe–Schmitt reaction. The ester, or residue of its decomposition, attacks the negatively charged phenolic nuclei, and its reaction is not limited to the ortho and para sites, transforming the phenolic nuclei in a polycondensation reagent of functionality higher than 3, leading to much earlier gelling. Furthermore, polycondensation occurs not only according to the phenol–formaldehyde mechanism but also according to a second reaction superimposed on it. The reaction of propylene carbonate, according to this mechanism, superimposed onto the normal phenol–formaldehyde reaction has been shown to be as follows [14,15]:

C. Physical Properties of Phenol–Formaldehyde Resins

Hardened phenol–formaldehyde resins have a specific gravity of approximately 1.2 to 1.3, a refractive index of 1.6, and a specific heat of 0.5. They are typically brown in color, and novolaks are lighter than resols. Resols are dark yellow, orange reddish, or brownish, even when made with pure raw materials. However, if the alkali is neutralized, resols become almost colorless. The best results were obtained with citric, lactic, and phosphoric acids. Pale-colored, hardened resins can be prepared with them [17]. Phenolic resins are relatively stable up to about 200 to 250°C, although oxidative degradation takes the form of attack at the methylene bridges to produce substituted dihydroxy-benzophenones [18]. Above this temperature, they begin to char slowly, and at higher temperatures charring is more rapid. At about 400°C decomposition is rapid, yielding phenols and aldehydes and leaving a cokelike residue.

In the A stage, simple phenol–formaldehyde resins are readily soluble in alcohol, esters, ketones, phenols, and some ethers, and insoluble in hydrocarbons and oils. As a class, resols tend to be more soluble in alcohols and water, and novolaks tend to be more soluble in hydrocarbons. In the early stages of condensation, resols are often soluble in water, owing to the presence of methylolphenols, especially polyalcohols. This is more pronounced with resols that are derived from phenol. Cresilic resols are less soluble, and xylenolic resols are almost insoluble in water. The solubility of A-stage resins in dilute aqueous sodium hydroxide or in mixtures of water and alcohols follows the same trend.

Solubility in alcohols and insolubility in hydrocarbons appear to go together. The alcohol and water solubility can be reduced only by using acetaldehyde, or other aldehydes in the place of formaldehyde, and by introducing hydrocarbon chains, particularly in the ortho or para position, in the aromatic ring. B-stage resins are soluble in only a few solvents, such as boiling phenols, and acetone, aqueous sodium hydroxide, and deca- and tetrahydronaphtalenes. Resins in the hardened or C stage are very resistant to most chemical reagents. They are unaffected by all ordinary organic solvents and water, although a few percent of water may be absorbed in filled material, mainly by the filler, thus causing slight swelling. The C-state resins dissolve slowly in boiling phenols such as naphthols. Resins from the simplest phenols can also be broken down and dissolved by hot, strong, alkali solutions.

Simple phenol–formaldehyde resins are readily attacked by sodium hydroxide. However, cresol–formaldehyde, and especially xylenol–formaldehyde, are much less susceptible to attack. Resins are often more resistant to strong alkaline solutions (i.e., 15 to 20%) than to dilute solutions (i.e., 5%). The filler has a considerable influence on the chemical resistance of the resins. Inert mineral fillers have a better resistance than cellulosic fillers. C-stage resins are resistant to most acids, except sulfuric acid stronger than 50%, formic acid, and oxidizing acids such as nitric and chromic acids. The insolubility of hardened resins in acetone is used to test the degree of cure of the resin. The curing temperature influences the amount of matter that is insoluble in acetone after prolonged heating [19]. The higher the hardening temperature, the lower the amount of acetone extractives. The mechanical properties of hardened phenol–formaldehyde resins are greatly influenced by the moisture content. This applies even more to resins containing fillers, plasticizers, and other ingredients. The rate of water absorption decreases with time, but thick samples may not reach an equilibrium even after several months in water. Therefore, in measuring the mechanical properties of resins, it is necessary to condition the test pieces under carefully controlled temperature and humidity prior to making the tests. In many cases, the mechanical properties of hardened resins are largely dependent on the type and orientation of the filler. This applies particularly to water absorption, tensile strength, and impact strength. It also applies to shear strength, with the condition that in the plane of the laminations the shear strength depends on the adhesion between the laminae of sheet material. The properties of the resin are more important than those of the filler in determining the compression strength.

IV. APPLICATIONS

A. PF Wood Binders

Phenolic resins are used as binders for exterior-grade plywood and particleboard, which need the superior water resistance provided by these resins. In the manufacture of plywood, the phenolic resins adhesive is usually applied to the wood veneers by roller or extrusion coating. The coated veneer is then cross-grained, stacked, and cured in a multidaylight press for 5 to 10 min at 120 to 130°C and at 11 to 16 kg/cm^2. In the manufacture of particleboard, the phenolic resin adhesives are sprayed onto the wood chips, or sprayed plus spread by continuous blenders. The glued wood chips are formed in a mat and then pressed for 2 to 12 s/mm, according to thickness and press temperature moisture content, at 190 to 230°C and 25 to 35 kg/cm^2.

The only type of phenolic resins used commercially for this application are resol-type resins, which have the following structure:

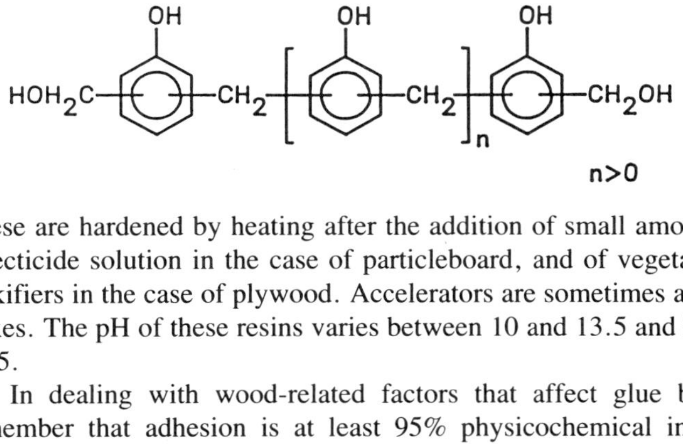

$$n > 0$$

These are hardened by heating after the addition of small amounts of wax emulsion and insecticide solution in the case of particleboard, and of vegetable or mineral fillers and tackifiers in the case of plywood. Accelerators are sometimes added in both types of glue mixes. The pH of these resins varies between 10 and 13.5 and is generally between 12 to 12.5.

In dealing with wood-related factors that affect glue bonds, it is important to remember that adhesion is at least 95% physicochemical in nature. The mechanical aspects of bond formation (such as keying cured adhesive solids into the wood surface) contribute negligibly to the bond strength or wood failure. The main chemical forces in thermosetting resin adhesion are primary valence bonds, covalent bonds, and hydrogen bonds, plus secondary forces such as van der Waals and London forces and any other types of electrostatic, dipolar, and associative forces. It is therefore essential that the resin contain significant numbers of functional groups and that the wood surface present a significant number of reactive sites to enable the resin to bond. Any factors that limit resin functionality or block reactive sites on the wood structure necessarily impede adhesion.

1. Properties of Phenolic Adhesives for Plywood [20]

Certain attributes of phenolic resins have been designed to give the strongest and most durable plywood bonds. Laboratory and field experience have demonstrated that certain types of phenol–formaldehyde plywood resins perform significantly better on veneers than do others. These superior resins have several properties in common:

1. They are relatively low in alkali content, generally about one-third to not more than one-half molar.
2. They have a lower molecular weight for hardwood veneers than do phenolic resins designed for softwood gluing.
3. They are high in methylol group content. Alternatively, they may contain free formaldehyde or require a matching catalyst that contains paraformaldehyde.
4. Even in dried adhesive carrier films or powdered resins, phenolic resins for hardwoods share the B-stage characteristic of reliquefying briefly under heat and pressure to allow transfer and flow on the glue lines of a plywood panel. This liquefaction can occur without water. This is unlike all but the most recently developed softwood phenolic resins.
5. Phenolic resins for hardwoods have higher thermal softening points than those of many other conventionally prepared resins. This indicates a network that has more cross-links after final cure and also greater durability.
6. They have a 40 to 45% solids content and 150 to 600 cP of viscosity at 25°C.

In general, lower resin alkali content and lower molecular weight are associated with slower cure, which explains why softwood plywood adhesive resins are both more alkaline and more condensed. Adequate exterior-grade adhesion can be attained on softwood veneer if these considerably condensed phenolic resins, which also cure more quickly, are used. Conversely, the increased functionality of hardwood adhesive resins partially compensates for their inherently slower cure rate. Hardwood phenolic resins required about 30% longer press times for adequate cure.

Notwithstanding their good adhesive capability, phenolic resins for hardwood gluing carry one distinct disadvantage: they do not prepress as well as softwood phenolic resins. Their lower condensation and longer flow are not assets. Prepressing is important because it is done to minimize face veneer losses and to reduce precure times on hot platens. Therefore, other means can be employed to obtain the required tackiness. Additives and adhesive formulations for this purpose are available. To increase the prepress capacity of low-alkali, long-flow phenolic resins suitable for hardwood gluing, small amounts of starch or poly(vinyl alcohol) can be added to the resin glue mix just before use.

Finished resin viscosity is increased to allow for the thickening effect of these additives without reducing the average molecular weight of the phenolic resin. Water-soluble thickeners (such as hydroxyethyl cellulose, polyethylene glycols, and maleic anhydride copolymers) contribute to prepress tack. However, they cause a large increase in resin viscosity, and the amount added must be small. Consequently, the benefit to prepressing is limited. Most animal- and vegetable-based thickeners, such as gum arabic, are subject to hydrolysis in alkaline phenolic resins, and lose effectiveness in a matter of hours.

2. Additives

A number of additives and modifiers contribute useful properties to phenolic resins used for wood gluing. Multipurpose additives are the aminoresins and urea–formaldehyde and melamine–formaldehyde polymers. These include not only urea and melamine plywood adhesive resins, but also dimethylol urea, trimethylol melamine, and hexamethylol melamine. Added in amounts from 5 to 15% of phenolic resin solids, they improve resin tack and prepressing, increase long-assembly-time tolerance, shorten pressing times, and enhance resin functionality. This results in stronger bonds on wood veneers. As long as they are used in limited quantities, they have a negligible effect on long-term phenolic bond durability. They appear to be well protected from hydrolytic degradation by the cured phenolic polymer network. The dispersion of the aminoresin molecules in the alkaline medium of the phenolic resin inhibits their curing reaction, which is acid catalyzed. It causes them to function as methylolated cross-linking units for the phenol–formaldehyde polymer.

Formaldehyde in liquid solution or solid form, and formaldehyde-generating compounds, are also phenolic resin additives that improve functionality and decrease curing times. Paraformaldehyde is used most frequently, but hexamethylenetetramine, formaldehyde/sodium bisulfite complexes, tris(hydroxymethyl)nitromethane, and glyoxal are also used. Significant effects are obtained when 3 to 5% is added, based on phenolic resin solids. Further reduction in curing time are possible if 1 to 2% resorcinol is added or resorcinol-acting natural extractives such as wattle-tannin extract. This should be mixed with paraformaldehyde and added to the liquid phenolic resins glue mix.

Formaldehyde additions overcome the effect of phenolic extractives in certain hardwood species, which prevent proper cure or adhesion of phenol–formaldehyde resins. Free formaldehyde appears to react rapidly with these phenolic extractives before they can interfere with the phenolic resin curing mechanism. With certain wood species that are rich in extractives, this technique has been used to increase bond durability from interior-grade to true exterior-grade performance.

Natural phenolic compounds are used as both replacements for substantial portions of synthetic phenol in plywood adhesive resins and as glue mix additives to improve performance; 4 to 6% is added, based on phenolic resin solids. They bring about improvements in assembly time tolerance and flow with no significant change in adhesion. Glue mix additions of wattle extract with additional formaldehyde produce faster hot-pressing cycles. However, some assembly-time tolerance and pot life has to be sacrificed in the process, but full exterior-grade durability is retained.

The lignin residues from wood pulp production are another class of extractives currently receiving attention as phenolic resin additives. Substituted phenols, such as cresols and xylenols, have been used as glue mix additives for phenolic adhesive resins to improve assembly-time tolerance. They are also used as solvents to remove oleoresinous deposits on the surface of pitchy softwood veneers. They can be used as flow promoters in phenolic hardwood adhesives. To avoid interference with the rate of resin curing, the amount added should not exceed 3 to 4% of the phenolic resin solids content.

Complexing additives commonly include the soluble salts of boron, chromium, zinc, cobalt, lead, manganese, magnesium, and others. When added to phenolic resin adhesives, some of these compounds have been successful in reducing pressing times and in improving prepress performance [21]. Borax is widely used in North America to shorten the prepress cycles of phenolic plywood glues for softwoods. However, these compounds tend to increase the molecular weight of a phenolic resin by complexing several molecules together through their phenolic hydroxyl groups. The gain in the resin molecular weight and prepress tack is sometimes accompanied by a reduction in assembly-time tolerance and the loss of the B-stage melt-flow behavior. In hardwood gluing, this is sometimes not advantageous, and the addition of complexing salts should be approached with caution.

Mixed borate salts are very effective as a treatment for the preservation of wood products against fungi and most insects. However, the boron salts, which become localized in high concentration on the veneer surface, tend to gel the phenolic resin before it can reach the wood surface and bond to it. However, very dilute aqueous solutions of borates (i.e., 0.25%) applied to softwood veneers in their green state decrease their thermal degradation during high-temperature drying and preserve their reactive sites for bonding with phenolic adhesives.

3. Formulation of Plywood Glue Mixes

The guiding principles for the preparation of plywood adhesive glue mixes are:

1. To maintain the highest possible phenolic solids content in the mixed glue (preferably in the range 30 to 40%).
2. To incorporate a cellulosic filler, such as a nutshell flour of 200 mesh or finer, in about 20 to 40% of phenolic solids. Nonabrasive inorganic fillers may also be satisfactory.
3. Alternatively, to add about half this amount of unrefined starchy material, such as wheat flour.
4. To add no alkali, or at the most only 1 to 2%, to disperse and stabilize the starchy material.
5. To add only enough water to produce a glue viscosity that can be handled by the gluing equipment. The preferred viscosity range is 1500 to 2500 cP, measured at 250°C.
6. To ensure proper wetting of the veneer by the glue film, a surface-active agent should be added (about 0.1 to 0.25% on resin solids).

Examples of glue mixes incorporating these principles are listed in Table 1.

4. General Observations on Particleboard Manufacture [21]

In the case of the application of phenolic adhesives to the manufacture of exterior-grade particleboard, the closest attention must be focused on the application of the resin rather than on its formulation. A good phenolic resin for plywood can be used successfully for the manufacture of particleboard once the various conditions of application have been understood. The pressing time of the board varies according to the type of adhesive, its reactivity, and the moisture content of the glued particles. In many cases a light water spray is applied to the top surfaces of the board before prepressing to shorten the pressing time. The light film of water covering the surface is vaporized when it comes in contact

Table 1 Examples of Glue Mixes

Material	Compositions		
40–45% Solids PF resin	100	100	100
300—Mesh coconut (or walnut) shell flour	12	14	10
Industrial wheat flour	6	—	—
50% Sodium hydroxide	2	—	—
Surfactant	0.1	0.1	0.1
Water	10	5	—
Total parts	130.1	120.1	110.1
Phenolic solids	31–35%	33–37%	36–41%

with the hot caul sheet of the press and migrates from the surfaces toward the core of the panel, causing a faster increase in temperature and a faster cure.

The water spray prevents precuring of the adhesive on the surface of the board during closure of the press before contact with the hot top caul sheet. The wood undergoes a partially irreversible plastic deformation during pressing, caused by the combined action of pressure and heat. Different products can be obtained by varying the type of pressing cycle of the board. Different pressing procedures and diagrams are available. A diagram for industrial three-layer boards may read as follows:

1. *Maximum pressure.* 23 to 27 kg/m^2 is reached as fast as possible after pressure closure (i.e., 35 to 50 s; other processes use pressure as high as 35 kg/cm^2).
2. *Contact with the gauge bars.* As a rule, contact is made after 60 to 120 s from the start of the press closure. The higher the density, the longer the time it takes. This, however, can be reduced by increasing the moisture content of the glued mat.
3. *Steam escape.* It is expected to begin 1 to 3 min after making contact with the gauge bars.
4. *Pressure decrease.* After approximately 1½ to 2 min of maximum pressure, the pressure is slowly decreased until the final pressure on the panel is as low as 2 to 3 kg/cm^2. This takes place toward the end of the cycle, just before press opening.

This pressing diagram produces a board with high-density face layers, and the shortest possible pressing time, at a given temperature and a low power consumption. The main properties of panels with high-density face layers are the stiffness of the panel; better warp resistance; high dimensional stability; hard, glossy, and shockproof surfaces that need less adhesive for subsequent veneering; and narrow thickness tolerances.

Considerable variation in the properties of the final board can be obtained by varying the moisture contents of surface and core layers, and by using faster resins in the core layer and slower reacting resins in the surface layer. This can also be achieved by varying geometry and sizes of the wood chips, the density of the board, and so on. Small variations in the manufacture and characteristics of the phenolic resin used do not affect the property of the finished particleboard as extensively as do the factors listed above. Experiments [21] on the correlation of curing and bonding properties of particleboard glued with resol-type phenolic resins by differential scanning calorimetry show that resols tend to reach two endotherm peaks, the first at 65 to 80°C and the second at 150 to 170°C. Resols used for particleboard have been shown to begin curing at lower temperatures than those for novolak resins. Resol-glued particleboard shows no bond formation at 120°C. At 130°C the resol-glued panels show internal bond strengths of 0.55 to 0.7 MPa. The internal bond strength for the wet tests increases as the board core temperatures goes over 150°C during pressing. The normal temperatures for 12- to 13-mm-thick board glued with phenolic adhesive are 170 to 230°C. The pressing time is 18 to 12 s/mm. Typical results obtained using phenol–formaldehyde adhesives for particleboard are shown in Table 2.

5. Dry-Out Resistance

One of the more common difficulties in bonding pine veneers and chips is adhesive dry-out. Dry-out is associated with the high liquid absorbancy of pine sapwood and it appears especially during long assembly times. This problem can be overcome by using phenolic resins modified through reaction with alkylated phenols, especially 3,4-xylenol [22]. Another technique used to achieve similar results is the manipulation of synthesis procedures used in preparing a standard phenol-formaldehyde resin [22]. The dry-out

Table 2 Results Obtained Using PF Adhesives for Particleboard

| Density (g/cm^3) | Swelling after a 2-h boil | | Internal bond strength | | Cold-water swelling (%) |
	Measured wet (%)	Measured dry (%)	Dry (kg/cm^2)	After a 2-h boil (kg/cm^2)	
±0.700	±15	2–3	10–11	5–8	9–11

resistance imparted by alkylated phenols is due to an initial semithermoplastic character in the resin. This is derived from their monomer bifunctionality and linear polymer that is consequently formed.

If a linear and essentially non-cross-linking prepolymer is prepared from phenol and formaldehyde, it can be coreacted with a nonlinear and cross-linking prepolymer to form a resin. The latter resin will have some initially semithermoplastic or dry flow character but will be primarily a thermosetting one. The product is an alkaline novolak–resol copolymer. Evaluation of this copolymer concept has shown that many resins possess a controlled initial semithermoplastic character which improves resistance to dry-out. Good dry-out resistance is achieved without loss of press-time efficiency or broad-range bonding ability. Such resins perform noticeably better than other types of resins which are resistant to dry-out.

Such a resin of the alkaline novolak–resol type can be prepared by coreacting a prepolymer, prepared by reacting formaldehyde and phenol in the molar ratio 2.6 : 1.0, and a prepolymer obtained by reacting formaldehyde and phenol in a molar ratio of 1 : 1. The two prepolymers are then mixed in 50 : 50 proportions by mass and coreacted.

B. Foundry Sand Binders and Mineral Fiber Binders

Phenolic resins are also extensively used in the binding of foundry molds. Both resols and novolak resins are used for this application. The sand is coated with the phenolic resin at a rate of 3 to 4%. The PF resin can be used both as an organic solvent solution and in powder form. Coating of the substrate can be done both at ambient or at higher temperature. In higher-temperature coatings novolaks are the preferred resins and in this application, waterborne resins (75% resin) can also be used. Hexamethylenetetramine as well as wax are added. Hexamine is often added separately from the resin to avoid precuring.

Another equally important field of application of phenolic resins is in the binding of mineral fibers such as glass fiber and rock wool. These are used for thermal and acoustic insulation at densities in the range 2.5 to 70 kg/m^3. Both powder and liquid resins, generally in water solution, are used for this purpose. Liquid resins are generally applied at about 10% concentration in water; the water evaporates, cooling the fiber and avoiding decomposition of the resins; and the resinated mat is then cured in a hot-air circulation oven at 175 to 200°C for 2 to 5 min.

C. Binders from PF Copolymers with Other Resins

The characteristics of PF resins and the reactive chemical groups they present render them particularly suitable for the preparation of binders by coreaction with other resins. This is still a relatively young field, and the most interesting and relevant co-resins that are used or explored in this respect are the aminoplastic resins, in particular urea–formaldehyde and melamine–formaldehyde (the copolymerization with the latter being a somewhat older use) and the diisocyanates.

While melamine–formaldehyde resins have been known for a long time to be able to form true copolymers with PF resins, this has not been the case for urea–formaldehyde resins. Until quite recently, copolymerization between PF and UF resins or urea was not thought to be likely [23], the system curing as a polymer blend only. Such deduction was based on the lack of detection of any methylene bridge between a phenolic nuclei and the amido group of urea. Recently, phenol–urea–formaldehyde resins of two different types and for two different purposes were instead shown to be able to copolymerize. First, Tomita [24] has shown that copolymerization between PF and urea resins occurs under acid conditions, the driving force of this work being the aim to produce a phenol–urea–formaldehyde copolymer in which the methylol groups are on the amido group of urea, and thus able to cure rapidly at very mild acidic pH values as a UF resin while retaining a level of water resistance of the resin. Second, in the alkaline pH range PF resins were shown to be able to copolymerize rapidly with urea, doubling the PF linear degree of polymerization while presenting full water resistance of the cured resin [25].

Both behaviors are easily explained by the relative rates of PF condensation and of urea hydroxymethylation and subsequent combination. Thus in Fig. 3 the relative gel times of PF and UF resins, as well as the relative rate constants of PF autocondensation and urea hydroxymethylation and self-condensation, are shown, indicating quite clearly in which pH ranges copolymerization is possible and with which species [26]. From Fig. 3, urea and PF resin, with little free formaldehyde or methylol ureas, will easily copolymerize in the pH ranges higher than 7; instead, PF and UF resins copolymerize in the range 6–9.

PF resols in water solution have been shown [27] to react rapidly and readily with polymeric MDI (4,4′-diphenyl methane diisocyanate) with minimal deactivation of the

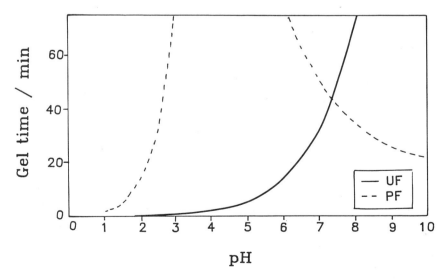

Figure 3 Relationship of gel time to pH for PF and UF resins.

isocyanate groups by the water. This peculiar behavior is based on the much faster rate of reaction of the isocyanate group with the PF methylol groups (hydroxybenzyl alcohols groups) than with water [27]. Such adhesives are now used industrially to a limited extent for bonding difficult to glue hardwood veneer species into exterior-grade plywood [28], and present exceptional adhesion, ease of bonding, adequately long pot life, and very high adhesive strength.

REFERENCES

1. A. Pizzi, R. M. Horak, D. Ferreira, and D. G. Roux, *Cell. Chem. Technol. 13*: 753 (1979).
2. N. J. L. Megson, *Phenolic Resin Chemistry*, Butterworth, London, 1958.
3. V. N. Vorozhtov and E. N. Yurvigina, *J. Gen. Chem. USSR 1*: 49 (1931).
4. E. Ziegler, *Oesterr. Chem.-Ztg. 49*: 92 (1948).
5. E. Ziegler, *Mh. Chem. 78*: 334 (1948).
6. C. Caesar and A. N. Sachanan, *Ind. Eng. Chem. 40*: 922 (1948).
7. T. Shono, *Proc. World Eng. Congr. Tokyo 31*: 533 (1931).
8. K. Hultzsch, *Chem. Ber. 82*: 16 (1949).
9. A. Zinke, *J. Appl. Chem. 1*: 257 (1951).
10. D. A. Fraser, R. W. Hall, and A. J. L. Raum, *J. Appl. Chem. 7*: 676 (1957).
11. A. Pizzi, *J. Appl. Polymer Sci. 24*: 1247, 1257 (1979).
12. R. W. Kluiber, *J. Am. Chem. Soc. 82*: 4839 (1960).
13. A. E. Martell and M. Calvin, *Chemistry of Metal Chelate Compounds*, Prentice Hall, Englewood Cliffs, N.J., 1952.
14. A. Pizzi and A. Stephanou, *J. Appl. Polymer Sci. 49*: 2157 (1993).
15. A. Pizzi and A. Stephanou, *Holzforschung* (in press, 1993).
16. P. H. R. B. Lemon, *Int. J. Mater. Prod. Technol. 5*(1): 25 (1990).
17. A. A. Vanscheidt and O. N. Simonova, *Plastmassy Sb. Stat. 129* (1939).
18. R. T. Conley and J. F. Bieron, *J. Appl. Polymer Sci. 8*: 103 (1963).
19. H. G. K. Pritchett, *Trans. Plastics Inst. London 18*: 30 (1950).
20. A. L. Lambuth, *IUFRO Conference on Wood Gluing Working Party S5.04.07*, Merida, Venezuela, Oct. 1977.
21. A. Pizzi, *Wood Adhesives Chemistry and Technology*, Vol. 1, Marcel Dekker, New York, 1983.
22. C. M. Chen and J. T. Rice, *Forest Prod. J. 26*: 17 (1976).
23. *Encyclopedia of Polymer Science*.
24. B. Tomita, *Adhesives for Tropical Woods Symposium*, Taipei, Taiwan, May 1992.
25. A. Pizzi and A. Stephanou, *J. Appl. Polymer Sci.* (in press, 1993).
26. N. Meikleham, M.Sc. thesis, University of the Witwatersrand, Johannesburg, South Africa, 1993.
27. A. Pizzi and T. Walton, *Holzforschung 46*: 541 (1992).
28. A. Pizzi, J. Valenzuela, and C. Westermeyer, *Holzforschung 47*: 69 (1993).

19

Natural Phenolic Adhesives I: Tannin

A. Pizzi* *University of the Witwatersrand, Johannesburg, South Africa*

I. INTRODUCTION

The loose term *renewable resources adhesives* has been used to identify polymeric compounds of natural, vegetable, origin that have been modified and/or adapted to the same use as some classes of purely synthetic adhesives [1]. At present two classes of these adhesives exist: one already extensively commercialized in the southern hemisphere and the other on the slow way to commercialization. These two types of resins are tannin-based adhesives [2] and lignin adhesives [3–6]. Both types are aimed primarily at substituting synthetic phenolic resins. In some aspects, such as performance, they closely mimic, or are even superior to, synthetic phenolic adhesives, while in others they behave vastly different from their synthetic counterparts. In this chapter we focus primarily on tannin-based adhesives because they have already been in extensive industrial use in the southern hemisphere, in certain fields of application, for the last 20 years. These adhesives are of some interest not only for their excellent performance in some applications but also for their mostly environment-friendly composition. Lignin adhesives are treated briefly here and in detail in chapter 20.

II. LIGNIN ADHESIVES

Lignin is a phenolic polymer that is one of the main polymeric constituents of wood. It is generally produced in great quantities as waste from paper pulp mills. It is composed of repeating phenylpropane units.

Current affiliation: Université de Nancy, Epinal, France

Considerable research has been carried out on lignin adhesives and binders. While for certain applications, such as binders for nontarred rough rural roads, lignin derivatives have been used for many years, in the main area of potential application, wood adhesives, industrial use has been lagging. A variety of effective lignin adhesives formulations exist and have already been reviewed extensively [3], some of them having been used for periods of up to 3 years in some particleboard or plywood mills. All these are not used today because all of them always come against two broadline problems: or the formulation tends to be corrosive or hard on equipment in the plant, or the lignin in the formulation tends to noticeably slow down panel pressing time, with consequent loss of mill productivity. In North America there are now encouraging indications that premethylolated lignin (prereacted with formaldehyde) can be added up to 20 to 30% of synthetic phenolic resins for plywood without lengthening panel pressing times, and one or two mills already appear to be using such a system [4]. Considering their very limited application, the reader is referred to more extensive reviews on this subject [3,6].

III. TANNIN-BASED ADHESIVES

The word *tannin* has been used loosely to define two different classes of chemical compounds of mainly phenolic nature: hydrolizable tannins and condensed tannins. The former, including chestnut, myrabolans (*Terminalia* and *Phyllantus* tree species), and dividi (*Caesalpina coraria*) extracts, are mixtures of simple phenols such as pyrogalol and ellagic acid and of esters of a sugar, mainly glucose, with gallic and digallic acids [2]. They can and have been used, successfully, as partial substitutes (up to 50%) of phenol in the manufacture of phenol–formaldehyde resins [7,8]. Their chemical behavior toward formaldehyde is analogous to that of simple phenols of low reactivity, and their moderate use as phenol substitutes in the above-mentioned resins does not present difficulties. Their lack of macromolecular structure in their natural state, the low level of phenol substitution they allow, their low nucleophilicity, limited worldwide production, and higher price somewhat decrease their chemical and economical interest.

Condensed tannins, on the other hand, constituting more than 90% of the total world production of commercial tannins ($\pm$ 200,000 tons per year), are both chemically and economically more interesting for the preparation of adhesives and resins. Condensed tannins and their flavonoid precursors are known for their wide distribution in nature and particularly for their substantial concentration in the wood and bark of various trees. These include various *Acacia* (wattle or mimosa bark extract), *Schinopsis* (quebracho wood extract), *Tsuga* (hemlock bark extract), and *Rhus* (sumach extract) species, from which commercial tannin extracts are manufactured, and various *Pinus* bark extract species. Where bark and wood of trees were found to be particularly rich sources of condensed tannins, commercial development ensued through large-scale afforestation and/or industrial extraction, mainly for use in leather tanning. The production of tannins for leather manufacture reached its peak immediately after World War II and has since progressively declined. This decline of their traditional market, coupled with the increased price and decreased availability of synthetic phenolic materials due to the advent of the energy crisis of the early 1970s, stimulated fundamental and applied research on the use of such tannins as a source of condensed phenolics.

A. Condensed Tannins

The structure of the flavonoid constituting the main monomer of condensed tannins may be represented as follows:

This flavonoid unit is repeated 2 to 11 times in mimosa tannin, with an average degree of polymerization of 4 to 5, and up to 30 times for pine tannins, with an average degree of polymerization of 10 to 11.

The nucleophilic centers on the A ring of a flavonoid unit tend to be more reactive than those found on the B ring. This is due to the vicinal hydroxyl substituents, which cause general activation in the B ring without any localized effects such as those found in the A ring.

Formaldehyde reacts with tannins to produce polymerization through methylene bridge linkages at reactive positions on the flavonoid molecules, mainly the A rings. The reactive positions of the A rings are one of positions 6 or 8 (according to the type of tannin) of all the flavonoid units and both positions 6 and 8 of the upper terminal flavonoid units. The A rings of mimosa and quebracho tannins show reactivity toward formaldehyde comparable to that of resorcinol [12]. Assuming the reactivity of phenol to be 1 and that of resorcinol to be 10, the A rings have a reactivity of 8 to 9. However, because of their size and shape, the tannin molecules become immobile at a low level of condensation with formaldehyde, so that the available reactive sites are too far apart for further methylene bridge formation. The result may be incomplete polymerization and therefore weakness. Bridging agents with longer molecules should be capable of bridging the distances that are too long for methylene bridges. Alternatively, other techniques can be used to solve this problem.

In condensed tannins from mimosa bark the main polyphenolic pattern is represented by flavonoid analogs based on resorcinol A rings and pyrogallol B rings. These constitute about 70% of the tannins. The secondary but parallel pattern is based on resorcinol A rings and catechol B rings [2,9]. These tannins represent about 25% of the total of mimosa bark tannin fraction. The remaining parts of the condensed tannin extract are the "nontannins" [9]. They may be subdivided into carbohydrates, hydrocolloid gums, and small amino and imino acid fractions [2,9]. The hydrocolloid gums vary in concentration from 3 to 6% and contribute significantly to the viscosity of the extract despite their low concentration [2,9]. Similar flavonoid A- and B-ring patterns also exist in quebracho wood extract (*Schinopsis*, *Balansae*, and *Lorentzii*) [10–12], but no phloroglucinol A-ring pattern, or probably a much lower quantity of it, exists in quebracho extract [12–14]. Similar patterns to wattle (mimosa) and quebracho are followed by hemlock and Douglas fir bark extracts. Completely different patterns and relationships do instead exist in the case of pine tannins [15–17] which present instead only two main patterns: one represented by flavonoid analogs based on phloroglucinol A rings and catechol B rings [15,17]. The other pattern, present in much lower proportion, is represented by phloroglucinol A rings and phenol B rings [15,17]. The A rings of pine tannins then possess only the phloroglucinol type of structure, much more reactive toward formaldehyde than a resorcinol-type structure, with important consequences in the use of these tannins for adhesives.

In condensed polyflavonoid tannin molecules the A rings of the constituent flavonoid units retain only one highly reactive nucleophilic center, the remainder accommodating the interflavonoid bonds. Resorcinolic A rings (wattle) show reactivity toward formaldehyde comparable to, though slightly lower than, that of resorcinol [18]. Phloroglucinolic A rings (pine) behave instead as phloroglucinol [19]. Pyrogallol or catechol B rings are by comparison unreactive and may be activated by anion formation only at relatively high pH [13]. Hence the B rings do not participate in the reaction except at high pH values (pH 10), where the reactivity toward formaldehyde of the A rings is so high that the tannin–formaldehyde adhesives prepared have unacceptably short pot lives [18]. In general tannin adhesives practice, only the A rings are used to cross-link the network. With regard to the pH dependence of the reaction with formaldehyde, it is generally accepted that the reaction rate of wattle tannins with formaldehyde is slowest in the pH range 4.0 to 4.5 [23]; for pine tannins, the range is between 3.3 and 3.9.

Formaldehyde is generally the aldehyde used in the preparation, setting, and curing of tannin adhesives. It is normally added to the tannin extract solution at the required pH, preferably in its polymeric form of paraformaldehyde, which is capable of fairly rapid depolymerization under alkaline conditions and as urea–formalin concentrates. Hexamethylenetetramine (hexamine) may also be added to resins due to its formaldehyde-releasing action under heat. Hexamine is, however, unstable in acid medium [20] but becomes more stable with increased pH values. Hence under alkaline conditions the liberation of formaldehyde might not be as rapid and efficient as wanted. Also, it has been fairly widely reported, with a few notable exceptions [21], that bonds formed with hexamine as hardener are not as boil resistant [22] as those formed by paraformaldehyde. The reaction of formaldehyde with tannins may be controlled by the addition of alcohols to the system. Under these circumstances some of the formaldehyde is stabilized by the formation of hemiacetals [e.g., $CH_2(OH)(OCH_3)$] if methanol is used [19]. When the adhesive is cured at an elevated temperature, the alcohol is driven off at a fairly constant rate and formaldehyde is progressively released from the hemiacetal. This ensures that less formaldehyde is volatilized when the reactants reach curing temperature and that the pot life of the adhesive is extended. Other aldehydes have also been substituted for formaldehyde [18,23,25].

In the reaction of polyflavonoid tannins with formaldehyde, two competitive reactions are present:

1. The reaction of the aldehyde with tannin and with low-molecular-weight tannin–aldehyde condensates, which are responsible for the aldehyde consumption.

2. The liberation of aldehyde, available again for reaction. The latter reaction is probably due to the passage of unstable $-CH_2-O-CH_2-$ ether bridges initially formed to $-CH_2-$ linked compounds.

It is interesting to note that while $-CH_2-O-CH_2-$ ether bridged compounds have been isolated for the phenol–formaldehyde [24] reaction, their existence for fast-reacting phenols such as resorcinol and phloroglucinol has been postulated, but they have not been isolated, as these two phenols have always been considered too reactive with formaldehyde. They are detected by a surge in the concentration of formaldehyde observed in kinetic curves due to methylene ether bridges decomposition [16].

When heated in the presence of strong mineral acids, condensed tannins are subject to two competitive reactions. One is degradative leading to lower-molecular-weight products, and the second is condensative as a result of hydrolysis of heterocyclic rings (*p*-hydroxybenzyl ether links) [13]. The *p*-hydroxybenzylcarbonium ions created condense randomly with nucleophilic centers on other tannin units to form "phlobaphenes" or "tanner's red" [13,26–28]. Other modes of condensation (e.g., free-radical coupling of B-ring catechol units) cannot be excluded in the presence of atmospheric oxygen. In predominantly aqueous conditions, phlobaphene formation or formation of insoluble condensates predominates. These reactions, characteristic of tannins and not of synthetic phenolic resins, must be taken into account when formulating tannin adhesives.

Sulfitation of tannin is one of the oldest and most useful reactions in flavonoid chemistry. Slightly sulfited water is sometimes used to increase tannin extraction from the bark containing it. In certain types of adhesives, the total effect of sulfitation, while affording the important advantages of higher concentration of tannin phenolics in adhesive applications due to enhanced solubility and decreased viscosity, and of higher moisture retention by the tannin resins, allowing slower adhesive film dry-out, hence longer assembly times [12], also represents a distinct disadvantage in that sulfonate groups promote sensitivity to moisture with adhesive deterioration and bad water resistance of the cured glue line even with adequate cross-linking [29–32].

IV. TECHNOLOGY OF INDUSTRIAL TANNIN ADHESIVES

The purity of vegetable tannin extracts varies considerably. Commercial wattle bark extracts normally contain 70 to 80% active phenolic ingredients. The nontannins fraction, consisting mainly of simple sugars and high-molecular-weight hydrocolloid gums, does not participate in the resin formation with formaldehyde. Sugars reduce the strength and water resistance in direct proportion to the amount added. Their effect is a mere dilution effect of the adhesive resin solids, with consequent proportional worsening of adhesive properties. The hydrocolloid gums, instead, have a much more marked effect on both original strength and water resistance of the adhesive [2,19,33]. If it is assumed that the nontannins in tannin extracts have a similar influence on adhesive properties, it can be expected that unfortified tannin–formaldehyde networks can achieve only 70 to 80% of the performance shown by synthetic adhesives.

In many glued wood products, the demands on the glue line are so high that unmodified tannin adhesives are unsuitable. The possibility of refining extracts has proved fruitless largely because the intimate association between the various constituents makes industrial fractionation difficult. Fortification is in many cases the most practical approach to reducing the effect of impurities. Fortification generally consists of copolymerization of the tannin with phenolic or aminoplastic resins [18,19,34]. It can be carried out during manufacture of the adhesive resin, during glue mix assembly, just before use, or during adhesive use. If added in sufficient quantity, various synthetic resins have been found effective in reducing the nontannin fraction to below 20% and in overcoming other structural problems [18,19]. The main resins used are phenol–formaldehyde and urea–formaldehyde resols with a medium to high methylol groups content. These resins can fulfil the functions of hardeners, fortifiers, or both. Generally, they are used as fortifiers between 10 and 20% of total adhesive solids, and paraformaldehyde is used as a hardener. Such an approach is the favorite one for marine-grade plywood adhesives. These fortifiers are particularly suitable for resorcinolic types of condensed tannins, such as mimosa. They can be copolymerized with the tannins during resin manufacture, during use, or both [18,34]. Copolymerization and curing are based on the condensation of the tannin, with the methylol groups carried by the synthetic resin. Since tannin molecules are generally large, the rate of molecular growth in relation to the rate of linkage is high, so that tannin adhesives generally tend to have fast gelling and curing times and shorter pot lives than those of synthetic phenolic adhesives. From the point of view of reactivity, phloroglucinol tannins such as pine tannins are much faster than mainly resorcinol tannins such as mimosa. The usual ways of slowing them down and, for instance, to lengthen adhesive pot life are:

1.   To add alcohols to the adhesive mix to form hemiacetals with formaldehyde and therefore act as retardants of the tannin–formaldehyde reaction
2.   To adjust the adhesive's pH to have the required pot life and rate of curing
3.   To use hexamine as hardener, which under the current conditions gives very long pot life at ambient temperature but still fast curing time at higher temperatures

The viscosity of bark extracts is strongly dependent on concentration. The viscosity increases very rapidly above a concentration of 50%. Compared to synthetic resins, the tannin extracts are more viscous at the concentrations normally required in adhesives. High viscosity of aqueous solutions of condensed tannins is due to the following causes, in order of importance:

1.   *Presence of high-molecular-weight hydrocolloid gums in the tannin extract* [33]. The viscosity is directly proportional to the amount of gums present in the extract [33].
2.   *Tannin–tannin, tannin–gum, and gum–gum hydrogen bonds.* Aqueous tannin extract solutions are not true solutions, but rather, colloidal suspensions in which water access to all parts of the molecules present is very slow. As a consequence, it is difficult to eliminate intermolecular hydrogen bonds by dilution only [33].
3.   *Presence of high-molecular-weight tannins in the extract* [33].

The high viscosity of tannin extracts solutions has also been correlated with the proportion of very high molecular weight tannins present in the extract. This effect is not well defined. In most adhesive applications such as in plywood adhesives, the viscosity is not critical and can be manipulated by dilution.

In the case of particleboard adhesives decrease of viscosity is, instead, an important prerequisite. When reacted with formaldehyde, unmodified condensed tannins give adhesives having characteristics that do not suit particleboard manufacture: namely, high viscosity, low strength, and poor water resistance. The most commonly used process to eliminate these disadvantages in the preparation of tannin-based particleboard adhesives consists of a series of subsequent acid and alkaline treatment of the tannin extract, causing hydrolysis of the gums to simple sugars and some tannin structural changes, thus improving viscosity, strength, and water resistance of the unfortified tannin–formaldehyde adhesive [33]. Furthermore, such treatments may cause partial rearrangement of the flavonoid molecules that causes liberation of some resorcinol in situ in the tannin, rendering it more reactive, allowing better cross-linking with formaldehyde, and ultimately yielding an adhesive which without addition of any fortifier resins gives truly excellent performance for exterior-grade particleboard [1,2,33].

This modification cannot be carried out too extensively, but only to a limited extent, to avoid precipitation of the tannin from solution by the formation of ''phlobaphenes.'' Typical results obtained are shown in Table 1.

Particular gluing and pressing techniques have been developed for tannin particleboard adhesives [35,36] to achieve pressing times much faster than those obtained with synthetic phenol–formaldehyde adhesives. Pressing times of 7 s/mm of panel thickness have been achieved and press times of 9 s/mm at 190 to 200°C press temperature are in daily operation: these are pressing times comparable to what is obtainable with urea–formaldehyde or melamine–formaldehyde resins. The success of these simple types of particleboard adhesives relies heavily on industrial application technology rather than just on the preparation technology of the adhesive itself [33,37]. A much higher moisture content of the resinated chips is tolerable with these adhesives than with any of the synthetic PF and aminoresin adhesives.

Table 1 Unfortified Tannin–Formaldehyde Adhesives Obtained by Acid–Alkali Treatment, for Exterior-Grade Particleboard: Example of Industrial Board Results

	Swelling after a 2-h boil				
Panel	Measured wet (%)	Measured dry (irreversible swelling) (%)	Original internal bond (IB) tensile perpendicular (kP/cm^2)	IB after a 2-h boil[a] (kP/cm^2)	Cyclic test after five cycles measured (%)
0.700	11.0	0.0	13.0	9.0	3.0

[a]*Source:* Ref. 2. From Ref. 34.

The best adhesive formulation for phloroglucinolic tannins such as pine tannin extracts is, instead, a comparatively new adhesive formulation that is also capable of giving excellent results when using resorcinolic tannins such as a wattle tannin extract [38]. The adhesive glue mix consists only of a mix of an unmodified tannin extract 50% solution to which has been added paraformaldehyde and polymeric nonemulsifiable 4,4′-diphenyl methane diisocyanate (commercial MDI) [38]. The proportion of tannin extract solids to MDI is 70:30 based on mass, but can be lower in MDI. This adhesive is based on the following peculiar mechanism, by which the MDI, in water, is hardly deactivated to polyureas [39]:

Reaction Scheme 1

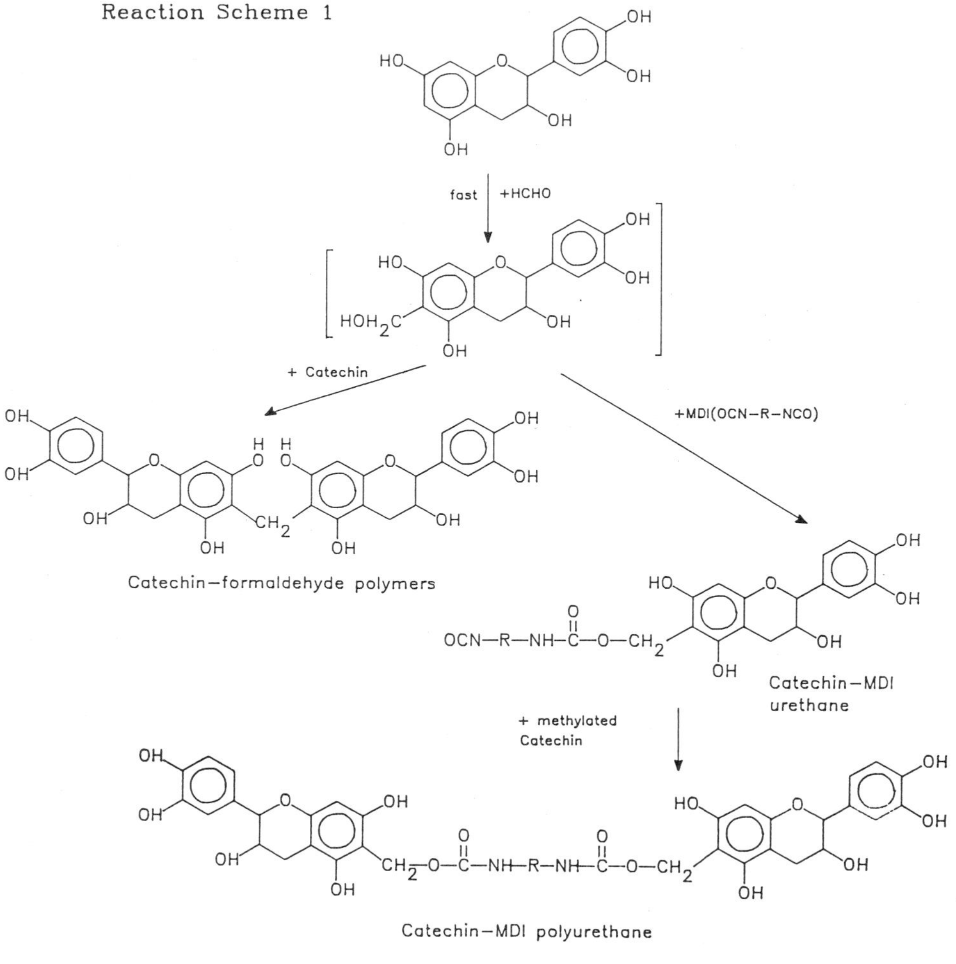

Catechin–MDI polyurethane

The properties of the particleboard manufactured with this system using pine tannin adhesives are listed in Table 2. The results obtainable with this system are hence quite good and not too different from the results obtainable with some of the other tannin adhesives already described. In the case of a phloroglucinolic tannin extract being used, no pH adjustment of the solution is needed.

One point that was given close consideration is the deactivating effect of water on the isocyanate group of MDI. It has been found that the amount of deactivation by water of

Table 2 Properties of Particleboard Manufactured Using Pine Tannin Adhesives

| Density | Swelling after a 2-h boil | | Original internal bond (IB) tensile perpendicular (kP/cm^2) | IB after[a] 2-h boil[a] (kP/cm^2) | IB retention after a 2-h boil (%) |
	Measured wet (%)	Measured dry (irreversible swelling) (%)			
0.690	15.0	4.3	8.4	4.3	51

[a]*Source:* From Ref. 34.

this group when in a concentrated solution (50% or over) of a phenol is much lower than previously thought [38–41]. This is the reason that aqueous tannin extract solutions and MDI can be reacted without substantial MDI deactivation by the water present.

A. Corrugated Cardboard Adhesives

The adhesives developed for the manufacture of damp-ply-resistant corrugated cardboard are based on the addition of spray-dried wattle extract, urea–formaldehyde resin, and formaldehyde to a typical Stein–Hall starch formula of 18 to 22% starch content [42,43]. The wattle tannin–urea–formaldehyde copolymer is formed in situ, and any free formaldehyde left in the glue line is absorbed by the wattle extract. The wattle extract powder should be added at a level of 4 to 5% of the total starch content of the mix (i.e., carrier plus slurry). Successful results can be achieved in the range of 2 to 12% of the total starch content, but 4% is the recommended starting level. The final level is determined by the degree of water hardness and desired bond quality. This wattle extract–UF fortifier system is highly flexible and can be adopted to dampproof a multitude of basic starch formulations.

B. Cold-Setting Laminating and Fingerjointing Adhesives for Wood

A series of different novolaklike materials are prepared by copolymerization of resorcinol with resorcinolic A rings of polyflavonoids, such as condensed tannins [44–46]. The copolymers formed have been used as cold-setting exterior-grade wood adhesives, complying with the relevant international specifications. Several formulations are used. The system most commonly used commercially relies on the simultaneous copolymerization of resorcinol and of the resorcinolic A rings of the tannin, due to their comparable reactivities toward formaldehyde.

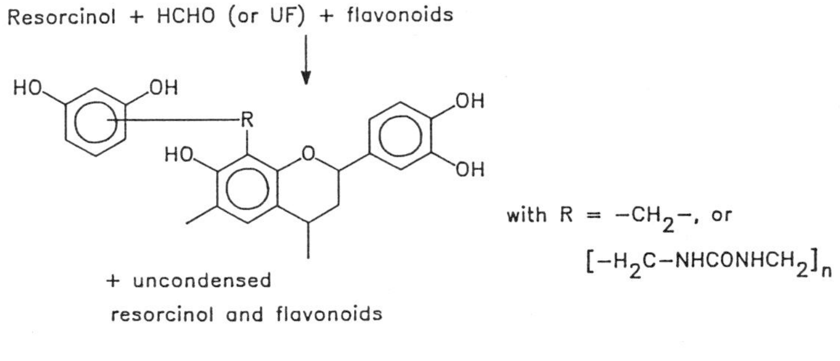

with R = –CH$_2$–, or

[–H$_2$C–NHCONHCH$_2$]$_n$

Table 3 Results of Tannin–Resorcinol–Formaldehyde Cold-Setting Adhesives Used on Beech Strips

	Dry	After a 24-h cold-water soak	After a 6-h boil
Tensile strength (N)	3200–3800	2300–2900	2200–2800
Wood failure (%)	90–100	75–100	80–100

Source: Ref. 47.

The final mixture of the products of this system is an adhesive that can be set and cured at ambient temperature by the addition of paraformaldehyde. Other cold-set systems exist and are described in the more specialized literature [2,44–46]. The typical results obtainable with these adhesives are indicated in Table 3.

A particularly interesting system now used extensively in several southern hemisphere countries is the so-called "honeymoon" fast-setting, separate-application system [48,49]. In this system one of the surfaces to be mated in the joint is spread with a standard synthetic phenol–resorcinol–formaldehyde adhesive plus paraformaldehyde hardner. The second surface is spread with a 50% tannin solution at pH 12. When the two surfaces are jointed, fingerjoints develop enough strength to be installed within 30 min and laminated beams (glulam) need to be clamped for only 2½ to 3 h instead of the traditional 16 to 24 h, with a consequent considerable increase of factory productivity. This adhesive system also provides full weather- and boil-proof capabilities.

C. Tire Cord Adhesives

Another application of condensed tannin extracts that has proved technically successful is as tire cord adhesives. Both thermosetting tannin formulation [50] and tannin–resorcinol–formaldehyde formulations have been experimented with successfully.

V. ANALYSIS

Various methods of analysis are available for the determination of tannin content. These methods can generally be grouped into two broad classes:

1. *Methods aimed at the determination of tannin material content in the extract.* The classical method of this type still used is the hide-powder method. These methods were devised to determine which percentage of the extract would participate in leather tanning. Their main drawback for their use for adhesives is their incapability of detecting and determining the approximate 3 to 6% of monoflavonoids and biflavonoids, or phenolic "nontannins," present in the extract, which do not contribute to tanning capacity but which do definitely react with formaldehyde and contribute to adhesive preparation.
2. *Methods aimed at the determination of phenolic material present in the extract that can be reacted with formaldehyde.* These methods were devised particularly for tanning extracts used in adhesives and are all based on the determination of some of the products of reaction of the flavonoids with formaldehyde.

Accepted methods of the first type comprise the hide-power method [51], the refractometric method, and various visible, ultraviolet, and infrared spectrometric methods. Accepted methods of the second type include the Stiasny–Orth method [52,53] and its modifications and the Lemme [54] sodium bisulfite backtitration method.

REFERENCES

1. A. Pizzi, *Holzforsch. Holzverwert. 43*(4): 83 (1991).
2. A. Pizzi, in *Wood Adhesives Chemistry and Technology*, Vol. 1 (A. Pizzi, ed.) Marcel Dekker, New York, 1983, Chap. 4; *J. Macromol. Sci. Chem. Ed. C18*(2): 247 (1980).
3. H. H. Nimz, in *Wood Adhesives Chemistry and Technology*, Vol. 2 (A. Pizzi, ed.), Marcel Dekker, New York, 1983, Chap. 5.
4. P. M. Tahir and T. Sellers, Jr., *Proc. Division 5, 19th IUFRO World Congress*, Montreal, Quebec, Canada, 1990, pp. 207–214.
5. N. G. Lewis and T. R. Lantzy, in *Adhesives from Renewable Resources*, ACS Symposium Series 385 (R. W. Hemingway, A. H. Conner, and S. J. Branhome, eds.), American Chemical Society, Washington, D.C., 1989, Chap. 2.
6. W. G. Glasser and S. Sarkanen, *Lignin, Properties and Materials*, ACS Symposium Series 397, American Chemical Society, Washington, D.C., 1989.
7. E. Kulvik, *Adhesives Age 18*: 3 (1975).
8. E. Kulvik, *Adhesives Age 19*: 3 (1976).
9. D. G. Roux, *Modern Applications of Mimosa Extract*, Leather Industries Research Institute, Grahamstown, South Africa, 1965, pp. 34–41.
10. H. G. C. King and T. White, *J. Soc. Leather Traders' Chem. 41*: 368 (1957).
11. D. G. Roux and E. Paulus, *Biochem. J. 78*: 785 (1961).
12. H. G. C. King, T. White, and R. B. Huges, *J. Chem. Soc.*, 3234 (1961).
13. D. G. Roux, D. Ferreira, H. K. L. Hundt, and E. Malan, *Appl. Polymer Symp. 28*: 335 (1975).
14. J. W. Clark-Lewis and D. G. Roux, *J. Chem. Soc.*, 1402 (1959).
15. R. W. Hemingway and G. W. McGraw, *Appl. Polymer Symp. 28* (1976).
16. D. du T. Rossouw, A. Pizzi, and G. McGillivray, *J. Polymer Sci. Chem. Ed. 18*: 3323 (1990).
17. L. J. Porter, *N. Z. J. Sci. 17*: 213 (1974).
18. A. Pizzi and H. O. Scharfetter, *J. Appl. Polymer Sci. 22*: 1945 (1978).
19. H. O. Scharfetter, A. Pizzi, and D. du T. Rossouw, *IUFRO Conference on Wood Gluing*, Merida, Venezuela, Oct. 1977.
20. H. M. Saayman, *LIRI Research Bulletin 466*, Leather Industries Research Institute, Grahamstown, South Africa, 1967.
21. F. W. Herrick and L. H. Bock, *Forest Prod. J. 8*(10): 269 (1958).
22. H. McLean and J. A. F. Gardner, *Pulp Paper Mag. Can.* (Aug. 1952).
23. K. F. Plomley, *Paper 39*, Division of Australian Forest Products Technology, 1966.
24. N. J. L. Megson, *Phenolic Resins Chemistry*, Butterworth, Sevenoaks, Kent, England, 1958.
25. A. Pizzi and H. O. Scharfetter, *CSIR Special Report HOUT 138*, Pretoria, South Africa, 1977.
26. R. Brown and W. Cummings, *J. Chem. Soc.*, 4302 (1959).
27. K. Freudenberg and J. M. Alonso de Lama, *Annalen 612*: 78 (1958).
28. R. Brown, W. Cummings, and J. Newbould, *J. Chem. Soc.*, 3677 (1961).
29. A. Pizzi, *Colloid Polymer Sci. 257*: 37 (1979).
30. L. K. Dalton, *Aust. J. Appl. Sci. 1*: 54 (1950).
31. L. K. Dalton, *Aust. J. Appl. Sci. 4*: 54 (1953).
32. J. R. Parrish, *J. S. African Forest Assoc. 32*: 26 (1958).
33. A. Pizzi, *Forest Prod. J. 28*: 12 (1978); *J. Appl. Polymer Sci.*, (in press, 1993).
34. Specification for particleboard, Deutschen Normenausschuss, V100 and V313, *DIN 68761*, Part 3, 1967.

35. A. Pizzi, *Adhesives Age 20*(12): 27 (1977).
36. A. Pizzi, *Adhesives Age 21*(9): 32 (1978).
37. A. Pizzi, *Holzforsch. Holzverwert. 31* (4): 85 (1979).
38. A. Pizzi, *J. Macromol. Sci. Chem. Ed. A16*(7): 1243 (1981); *Holz Roh-Werkstoff 40*: 293 (1982).
39. A. Pizzi, E. P. Von Leyser and J. Valenzuela, *Holzforschung, 47*: 69 (1993).
40. A. Pizzi, *J. Appl. Polymer Sci. 25*: 2123 (1980).
41. A. Pizzi and T. Walton, *Holzforschung, 46*: 541 (1992).
42. A. E. McKenzie and Y. P. Yuritta, *Appita 26* (1974).
43. P. A. J. L. Custers, R. Rushbrook, A. Pizzi, and C. J. Knauff, *Holzforsch. Holzverwert. 31* (6): 131 (1979).
44. A. Pizzi and D. G. Roux, *J. Appl. Polymer Sci. 22*: 1945 (1978).
45. A. Pizzi and D. G. Roux, *J. Appl. Polymer Sci. 22*: 2717 (1978).
46. A. Pizzi, *J. Appl. Polymer Sci. 23*: 2999 (1979).
47. Specification for synthetic adhesive resins for wood, *British Standard BS1204*, Parts 1 and 2, 1965.
48. A. Pizzi, D. du T. Rossouw, W. Knuffel, and M. Singmin, *Holzforsch. Holzverwert. 32*(6): 140 (1980).
49. A. Pizzi and F. A. Cameron, *Forest Prod. J. 34*(9): 61 (1984).
50. K. H. Chung and G. R. Hamed, in *Chemistry and Significance of Condensed Tannins* (R. W. Hemingway and J. J. Karchesy, eds.), Plenum Press, New York, 1989.
51. D. G. Roux, *J. Soc. Leather Trades' Chem. 35*: 322 (1951).
52. E. Stiasny and F. Orth, *Collegium 24*: 50, 88 (1924).
53. A. Wissing, *Svensk Papperstid 20*: 745 (1955).
54. A. Lemme, U.S. patent 3,232,897 (1966).

20
Natural Phenolic Adhesives II: Lignin

A. Pizzi* *University of the Witwatersrand, Johannesburg, South Africa*

I. INTRODUCTION

The occurrence of lignin as a waste product in pulp mills has made it an attractive raw material for adhesives ever since the beginning of the sulfite pulping of wood. The first patents dealing with the application of spent sulfite liquor (SSL) as an adhesive for paper, wood, and other lignocellulosic materials date back to the end of the nineteenth century [1], and since then have shown an ever-growing number. On the other hand, technical utilization of lignin on a large scale is still at a very low level for the amount produced worldwide. Presently, most of the spent liquors in pulp mills are burned. Only about 20% is used for various purposes, such as dispersants, oil-well-drilling muds, pelletizing materials, molding stabilizers, and concrete grinding additives.

As a major wood component, native lignin is neither hygroscopic nor soluble in water. However, during technical sulfite pulping, lignin becomes soluble in water, due to partial degradation and introduction of sulfonic acid groups ($-SO_3H$). In applying SSL as an adhesive, it has to be converted to an insoluble state during the curing period. Cross-linking in lignin can be achieved either by condensation or by radical coupling reactions. A great number of patents have become known during the past three decades [2] dealing with the application of SSL as a wood adhesive, in which the lignin is cross-linked by condensation reactions. However, either high temperatures and long heating times or mineral acids are required for these condensation reactions, which cause structural changes or charring in the wood particles. Recently, cross-linking of the lignosulfonate molecules by radical combinations, which avoids mineral acids and high temperatures, has been developed, but this presents disadvantages as well, as the use of peroxides is not favored in wood processing plants, for a variety of reasons. The use of lignin by polycondensation reaction with formaldehyde also presents the disadvantage of slower pressing time in their application to panel products.

II. CHEMICAL BACKGROUND OF THE CURING REACTION OF LIGNIN

Lignin is composed of phenylpropane (C_9) units that are linked together by carbon-to-carbon as well as carbon-to-oxygen (ether) bonds. Our present knowledge of lignin

---

Current affiliation: Université de Nancy, Epinal, France

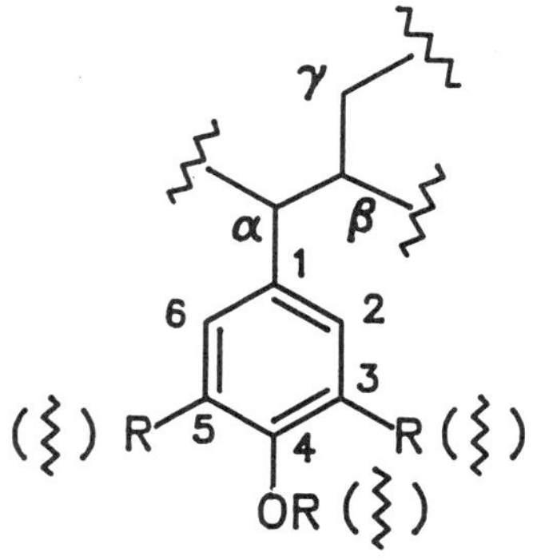

Figure 1   Phenyl propanoid units of lignin: R, R_2 = H, OCH_3; R_3 = H, CH_3, CH_2; = possible linkage to other phenyl propanoid units.

structure is based on the assumption that it is formed from p-hydroxycinnamyl alcohols by oxidative coupling [2,3] oxidized by hydrogen peroxide and peroxidase to a phenoxy radical (R). The unpaired electron in R is delocalized and reacts at three different sites of the radical, indicated by the four resonance structures R_a to R_d, leading to the dilignols A to F (Fig. 1).

A. Cross-Linking by Condensation Reactions

When lignosulfonate is treated with strong mineral acids at elevated temperatures or heated at temperatures above 180°C, condensation reactions leading to diphenyl methanes and sulfones take place. The reactivity of lignosulfonates depends to some extent on the cation. Of the four lignosulfonates obtained technically, the calcium-based exhibit the lowest and the ammonium-based the highest reactivity; the sodium and magnesium lignosulfonates show a medium reactivity.

Hydrobenzyl alcohol groups as well as sulfonic acid groups on the carbon α to the aromatic rings of some of the phenyl propane units of the random polymer react in the presence of strong mineral acids with the aromatic nuclei of other phenylpropane units. This reaction, leading to diphenylmethanes, is of the same type as the formation of phenolic resins from phenol and formaldehyde. Lignin also reacts with formaldehyde and can be cross-linked by it in the same manner as that of synthetic polyphenolic resins.

B. Cross-Linking by Oxidative Coupling [2]

Lignosulfonic acid in technical SSL contains about 0.4 of a free phenolic hydroxy group per C_9 unit. Therefore, like the formation of lignin in plants, cross-linking of lignosulfonate is possible by oxidative coupling. Oxidants such as hydrogen peroxide, and catalysts such as sulfur dioxide or potassium ferricyanide, are most effective. Treatment of a 50% technical SSL with this redox system leads to a very vigorous exothermic reaction under evaporation of water. The yield of the resin under certain conditions exceeds 70%, indicating that some carbohydrates must also have been enclosed in the resin. The advantage of this type of cross-linking compared with condensation reactions is that it needs neither mineral acids nor high temperatures, due to the recombination of radicals, for which the activation energy is very low. The strongly exothermic reaction causes a uniform temperature profile during pressing of particleboard without external heat.

III. APPLICATION OF LIGNIN AS AN ADHESIVE FOR PARTICLEBOARD, PLYWOOD, AND FIBERBOARD

According to its structure as a polyphenol, lignin as an adhesive should be similar to phenol–formaldehyde resins. This is true for native lignin in wood; to transfer them into insoluble resins, technical lignins (lignosulfonate and black liquor) have to be additionally cross-linked. However, condensation reactions in lignin by heat or mineral acids cannot be as effective as in synthetic phenol-formaldehyde (PF) resins, due to the lower number of free positions in the aromatic nuclei of lignin and their considerably lower reactivity than in PF resins. First, there is only 0.5 of a free 5-position (ortho to the phenolic groups) per C_9 unit; positions 6 and 2 are less reactive. Second, there is less than one benzyl alcohol or ether group per C_9 unit in lignin, whereas in synthetic PF resins up to three methylol groups can be introduced into one phenolic ring. Finally, the aromatics in lignin are considerably less reactive toward hydroxybenzyl alcohol groups than is phenol, due to the presence of methoxy or methoxy-equivalent groups rather than hydroxy groups on the lignin aromatic rings. For these reasons, lignin in technical spent liquors cannot be as effectively cross-linked as synthetic PF resins. At least higher press temperatures at longer heating times or higher acid concentrations are necessary.

Quite a number of patents have been pending during the past three decades dealing with lignin as adhesive for particleboard, plywood, and fiberboard in the absence of conventional PF or urea–formaldehyde (UF) adhesives [2]. Besides lignin, in most cases additional cross-linking agents for lignin are necessary, such as epoxides, polyisocyanates, polyols, polyacrylamides, polyethyleneimine, aldehydes, maleic anhydride, amines, proteins, melamine, hydrazine, and so on. So far, these procedures, for different reasons, have not led to any major practical application. Very few procedures, such as those of Pelikan et al. (1954), Pedersen and Jul-Rasmussen (1963), Shen (1973), Nimz et al. (1972), and others, use lignosulfonates or SSL without integrated cross-linking chemicals [2]. The patent of Pelikan et al. describes ways of using lignin as an adhesive for floor layers by cross-linking it oxidatively with chromium trioxide. The mechanism of cross-linking is the same as with hydrogen peroxide (Nimz, 1972) [2], but it is much less effective. Its applicability to particleboards has been tried, but the boards exhibited low tensile strengths and disintegrated in water at 20°C in less than 2 h.

A. Curing Lignin Boards by Long Pressing Time and Postheating Treatment

According to Pedersen and Jul-Rasmussen [2], wood chips are mixed with 20 to 25% of their weight with a 50% technical SSL and pressed at 185°C for 30 min, giving a 12-mm-thick board that must be postheated at 195°C for 80 min in an autoclave. The pH value of the SSL had been adjusted to 3 by citric acid. The particleboard obtained has a bending strength of 230 kP/cm^2, a tensile strength perpendicular to the grain of 5.3 kP/cm^2, and a density of 0.7 g/cm^3. Press temperatures may vary between 170 and 235°C and temperatures of the autoclave between 170 and 210°C.

High press and autoclave temperatures as well as long heating times are necessary for effective cross-linking by condensation reactions in lignin, as pointed out above. The color of the boards is dark, due to decomposition reactions and charring caused by the high temperatures, and the density of the boards usually reaches values at around 0.8 g/cm^3 if the required tensile strength is to be obtained. The temperature in the core layer during pressing reaches 140°C. This may also cause condensation reactions between wood and SSL, as well as chemical and physical changes in the wood particles [5].

The relatively high dimensional stability of the particleboard toward water may be caused by these changes. Pedersen and Jul-Rasmussen (1963) found a thickness expansion after a 2-h soaking in water at 20°C of only 1.5% and 13.8% water absorbance. Open-air tests, extending over 5 years, carried out by the wood panel products laboratory of the Technical Research Centre of Finland [6] revealed that SSL boards, obtained according to the Pedersen procedure, were superior in strength and in surface properties to UF as well as to PF particleboard. Roffael [7] has shown that water absorption of Pedersen SSL particleboard at different air humidities is only about half as high as with conventional PF particleboard. Weathering for 1 year gave a nearly constant humidity at around 6% for SSL particleboard, while PF boards gave humidities between 12 and 15%. Also, after soaking the boards in water at 20°C for 24 h, the LS boards lost only 25% of their initial tensile strength, while that of conventional PF particleboard decreased by 70%. In contrast, the mechanical strength properties of SSL particleboard were inferior to those of PF board [8].

The Pedersen procedure has been applied to mill-scale tests in Denmark, Switzerland, and Finland, but has been discontinued in all cases. One reason for this failure is the high cost caused by the two-stage heating treatment. The autoclave must consist of refined steel, due to the evolution of corrosive gases such as sulfur dioxide, causing additional high costs. Another reason was the long pressing and curing time needed for manufacture. However, one of the main reasons for the discontinuation of the procedure was the frequent fires induced by the high pressing and posttreatment temperatures [2].

As mentioned above, the condensation rate of lignosulfonates depends on the cation, with ammonium ions exhibiting the highest reactivity. Shen and Calvé [4] used fractionated ammonium-based SSL as binder for particleboard and found the highest reactivity, leading to the best mechanical board properties, with a low-molecular-weight fraction. Unexpectedly, the tensile strength of dry particleboard obtained with a low-molecular-weight ammonium-based SSL fraction increased with the sugar content of the SSL. Best board properties were obtained with 6% of a low-molecular-weight (0 to 5000) ammonium-based SSL fraction having 50 to 60% sugar. In this case, a pressing time of 8 min at 210°C was sufficient for manufacture of 11-mm-thick waferboard to meet the Canadian standard requirements for exterior-grade particleboard.

Obviously, the sugars take part in the condensation reactions of lignosulfonate by production of furfural. While the bending strength of dry boards increased steadily, with the sugar content of the SSL going up to 80%, the bending strength after a 2-h boiling of the boards reached a maximum at about 50 to 60% carbohydrates, indicating that the condensation between lignin and carbohydrates leads to better water resistance than that between carbohydrates only.

B. Curing Lignosulfonate Particleboard with Sulfuric Acid

In 1973, K. C. Shen of the Eastern Forest Products Laboratory in Ottawa, Canada, proposed sulfuric acid as a curing agent for SSL waferboard. The pressing conditions were the same as those of conventional PF particleboard, when poplar wafers were first sprayed with 1% of 15 to 20% sulfuric acid and then with 4 to 5% SSL powder, which adheres at the surface of the wet wood wafers. Later [9], concentrated sulfuric acid (9%)

was added to the SSL before spray drying, and the powder adhered to the wax-coated wafers. High pressing temperatures of about 205°C were also necessary, the catalytical effect of the sulfuric acid merely reducing the pressing time to that of industrial conditions for PF particleboard.

The strength properties of the boards, having an average specific gravity of 0.67 g/cm^3, were measured by the torsion shear at the center plane of 1×1 in. specimens, from which the internal bond strength (tensile strength perpendicular to the surface) can be obtained by multiplication with the factor 0.7 [10]. Values obtained for internal bond strength and modulus of rupture (MOR) for dry samples as well as after 2 h of boiling met the Canadian standards for particleboard [11]. The torsion shear strength and MOR of dry boards were independent of the pressing time, while the wet strength increased proportionally with the pressing time. This means that for exterior-grade requirements, distinct pressing times are necessary.

Best board properties were obtained with 1% concentrated sulfuric acid, based on dry wood particles. At higher acid concentrations the strength of dry boards decreased, while that of wet boards showed a further increase. However, charring of the wood particles takes place at acid concentrations higher than 0.9%. The thickness expansion of SSL boards obtained with 1% sulfuric acid lay between 26 and 46%, after soaking in water at 20°C for 1 week, and between 51 and 66% after 2 h of boiling. These values are considerably worse than those of exterior-grade PF particleboard.

The acidity of the particleboard was found to be pH 3, after disintegration of the boards in 10 times their weight of water [12]. It has been reported that the acidity had no longer-term influence on the mechanical board properties, checked by conventional accelerated aging treatments [12]. In this case, 11-mm boards had been pressed for 6 min and a part of them had been postheated at 149°C for 2 h. Of the accelerated aging treatments only 20 days of heating at 149°C showed faster aging of SSL than of PF boards, which is due to the higher acidity of the SSL boards.

On the whole, the SSL particleboard obtained by Shen cannot be compared technically with exterior-grade PF particleboard. The Shen procedure, however, has found no practical application yet, as the results obtained are still far from those obtainable with PF particleboard. In 1977 Shen mentioned that a short production trial run had been carried out at a waferboard plant; although the preliminary results were promising, addition work was still required to modify the binder formulation and production parameters to meet the requirements of plant operation, and to obtain results comparable to those of PF particleboard.

C. Curing SSL Boards with Hydrogen Peroxide

The drawback inherent in the Pedersen and Shen procedures—high pressing temperatures and long pressing times or strong mineral acids—can be avoided if cross-linking of the lignin molecules is achieved by radical coupling instead of condensation reactions (Nimz et al. 1972 [2]). In this case, the formation of new carbon–carbon as well as carbon–oxygen bonds between two radicals is a very fast reaction with a low activation energy, which needs no external heating or mineral acids as catalyst. This means that the reaction is very specific, and side reactions such as decompositions and charring can be avoided, while linkages between wood and SSL may also occur.

The essential radicals are formed from phenolic groups in the lignosulfonate molecules by oxidation with hydrogen peroxide in the presence of a catalyst. Out of a number of catalysts, sulfur dioxide (SO_2) has been proven to be the most effective [13]. A 50% calcium-based SSL containing about 1% SO_2 at pH 2 reacts vehemently with a 35% hydrogen peroxide solution in a strongly exothermic reaction, forming an insoluble gel. The reaction time is less than 1 min but depends on the source and composition of SSL. At higher pH values, the reaction takes some minutes or needs heating up to about 70°C, but after reaching 70°C, the reaction is also very fast.

It has been reported [13] that the SSL containing the catalyst and the hydrogen peroxide solution have to be sprayed separately on the wood chips. Under certain conditions, the hydrogen peroxide can be mixed with the SSL and sprayed together on the wood chips, in a single operation [13]. Another possibility consists of adding half of the SSL as spray-dried powder, lowering the humidity of the blended chips to about 13%, which is the upper limit according to German standards. The powder may either be mixed together with the liquid SSL and the hydrogen peroxide or added separately after the wood chips have been sprayed with the mixture of SSL solution and hydrogen peroxide. The humidity of the wood chips can thus be adjusted to predetermined values. The pot life of the blended wood chips, which is the assembly time between spraying and pressing, would then be extended. Medium-density interior-grade particleboard can be obtained from wood chips with 20% SSL, based as dry material on dry wood chips, at pressing temperatures between 100 and 120°C under otherwise conventional manufacturing conditions for UF particleboard.

There are several reasons why this system has not found industrial favor: (1) the unfavorable situation due to the presence of a peroxide in wood panel plants, such as possible machinery corrosion, and other problems, and (2) the fact that the produced board is often relatively very soft immediately out of the particleboard press, rendering particularly problematic its early handling.

IV. LIGNIN IN COMBINATION WITH PHENOL– FORMALDEHYDE ADHESIVES

The number of patents pending on lignin as a substitute or extender for phenolic wood adhesives during the past three decades is high [2]. Under certain conditions, up to 40% of phenol–formaldehyde (PF) adhesive can be replaced by lignosulfonate or black liquor without significantly extending the curing time or worsening board properties. Lignin–PF formulations have been used in manufacturing particleboard, fiberboard, and plywood. The reason for their application has to be seen in the lowering of costs, resulting from the difference in cost between phenol–formaldehyde and lignin. However, in most cases the lignin has to be pretreated by deionization, ultrafiltration, or cation change. Two recent procedures that have become better known are discussed next in more detail.

A. Lignin–PF Formulations

In 1971, Roffael and Rauch [14,15] claimed that the curing time of SSL particleboard could be reduced and, according to Pedersen, the postheating treatment in an autoclave avoided when phenolic resins of the novolak type were added to the SSL [15]. Due to coagulations between calcium-based SSL and phenol–formaldehyde, the calcium-based SSL has to be transferred into sodium-based SSL. The board properties are strongly dependent on the pH value of the glue: for example, 10% SSL, 4% novolak, and 2.1%

hexamethylenetetramine were applied to dry pine wood chips to prepare 9-mm-thick boards at a pressing time of 12 min.

While the highest bending and tensile strengths were obtained between pH 5 and 7, the percentage swelling in water at 20°C, after 24 h, had a minimum at pH 3.5. For this reason, a pH value of 4.7 has been suggested by the authors as a compromise [15]. Both the mechanical strength and the dimensional stability of the particleboard can be improved by higher ratios of novolak in the glue formulation or increasing pressing temperatures up to 250°C. Besides conventional contact heating in a flat press, high-frequency heating was applied, raising the temperatures in the core layer during pressing to 220°C for 1 min, which diminished the pressing time. In contrast to their publication in 1971 [15], Roffael and Rauch found in 1973 [16] that phenolic resins of the resol type also improve the binding properties of SSL in particleboard, and the percentage swelling can be improved to meet the German standard specification [18] (6% after 2 h in water at 20°C) by applying a postheating treatment at 200°C for 1 h.

The postheating treatment could be avoided when higher amounts of resol-type resin were used. In conventional PF particleboard the PF resin amounts to about 8%, based on dry wood particles. It has been found [16] that up to 33% of the resol-type adhesive in conventional PF particleboard can be substituted in the surface layers of a three-layer 22-mm board by sodium-based lignosulfonate without major deterioration in the mechanical board properties. In 20-mm one-layer particleboards at pH 9, up to 25% of the phenol–formaldehyde resin could be replaced by sodium–lignosulfonate under conventional pressing conditions, leading to particleboard meeting the German standards specification [7]. Furthermore, 10% of the PF resin in beech/plywood could be substituted for by sodium-based lignosulfonate at a pressing temperature of 165°C, and up to 30% at 190°C [8]. The highest shear strength of the plywood was obtained with an adhesive formulation of pH 12 to 13.

B. Karatex Adhesive

According to Forss and Fuhrmann (1972) [6,17], the amount of lignin in lignin–PF adhesives for particleboard, plywood, and fiberboard can be increased to 40 to 70% if a high-molecular-weight fraction (MW > 5000) of either lignosulfonate or black liquor, obtained from alkaline pulping of wood, is applied. Fractionation of SSL or black liquor can be achieved by ultrafiltration [6,17]. According to the authors, the higher effectivity of high-molecular-weight lignin molecules is due to their higher level of cross-linking, which requires less phenol–formaldehyde for the formation of an insoluble copolymer than do low-molecular-weight lignin molecules. However, bearing in mind the findings of Shen [4] that low-molecular-weight ammonium-based SSL is more effective, Forss and Fuhrmann appear not to have checked the influence of inorganic salts in SSL or black liquor that are separated off during ultrafiltration [2]. Forss and Fuhrmann assume that condensates between smaller lignin molecules and phenol–formaldehyde "are unable to contribute to the three-dimensional network" [6], which is unlikely because low-molecular-weight lignin molecules are more reactive than are high-molecular-weight molecules.

In the manufacture of particleboard either high-frequency (HF) heating or combined contact HF heating has been applied. In the latter case, the press platen temperature has been 180°C. German standard requirements for weather-resistant particleboards were met at pressing times between 10 and 12 s/mm and 8 to 12% adhesive, based on dry wood particles. One advantage inherent in the fractionation by ultrafiltration is that the lignin becomes more uniform and less dependent on variations in pulping conditions and wood source, which sometimes cause serious problems in the application of technical lignins. Full-scale plywood mill tests, some of them running continuously for several weeks, appear to have been performed in two Finnish plywood mills. Again, this procedure does not appear to be in operation anymore.

C. Methylolated Lignins

The fundamental problem of lignin, slowing of the pressing time obtainable with phenol–formaldehyde resins, was partly eliminated by Sellers (1990) [19] and by Calvé (1990) [20], who first reacted lignin with formaldehyde in a reactor for a few hours. A methylolated lignin (ML) equivalent to a phenol–formaldehyde resol was obtained. As in this case the reactivity of the methylol groups of lignin introduced depend on the reactivity of phenolic nuclei available for reaction, mixing with a synthetic phenol–formaldehyde resin ensures that the reactivity of the PF resin is not impaired. In this manner up to 30% methylolated lignin could be used to substitute for the phenol–formaldehyde adhesive, with no drop in performance and pressing times. In plywood industrial plant trials with such PF–ML systems, Sellers and Calvé both obtained excellent results. It is believed that at least one Canadian plywood mill is using such a system industrially today. As plywood pressing time is not the really critical variable in a plywood mill, this system did not itself prove suitable for application to particleboard mills, where the shortness of the pressing time that can be obtained is the determining variable.

Attempts were made to use more reactive lignins, such as bagasse (sugarcane waste) lignins, which present 0.7 to 0.9 of a reactive position for each phenylpropane unit, using the same approach. Although good particleboard could be obtained with a mixture of 67% methylolated bagasse lignin and 33% PF resin, these could be obtained only at pressing times of 37 to 50 s/mm, still far too long to be of any interest to a particleboard mill [21]. Thus, for particleboard, the low reactivity of lignins toward formaldehyde and the limited number of sites available for reaction with formaldehyde on most aromatic nuclei of the phenylpropane units of lignin are clearly the limiting factors to utilization of this material.

It then became clear that a different but equally or more efficient cross-linking route to be employed in parallel to formaldehyde cross-linking had to be used if feasible pressing times for particleboard mills were to be achieved. Two parallel approaches toward this end have proved successful. First, methylolated bagasse lignin (MBL) and methylolated kraft lignin (MKL) were reacted in water with diisocyanate according to a new reaction and the mechanism observed for PF resins [22]. Combinations of polymeric MDI (4,4′-diphenylmethanediisocyanate), synthetic PF, and methylolated lignins yielded particleboard with full exterior-grade properties at pressing times as fast as 20 s/mm when using up to 55% methylolated bagasse lignin [23]. Pressing times using methylolated kraft lignin were also faster but still too slow [23]. Second, as a consequence of the elucidation of PF α-set acceleration mechanisms [24], pressing times as short as 7.5 s/mm for MBL and 10 s/mm for MKL were obtained, at a lignin content of the resin as high as 65% of total adhesive [25]. These pressing times are faster than for synthetic phenolic resins and almost of the same order of magnitude as for UF resins. Industrial plant trials are being held for this system, which appears for the first time to have eliminated the main problem of lignin in wood adhesives for particleboard, that is, the problem of too long pressing times.

V. LIGNIN IN COMBINATION WITH UREA–FORMALDEHYDE RESINS

In 1965, W. Arnold [2] found that the pressing time of SSL particleboard obtained with sulfuric acid as catalyst (see the Shen procedure) can be reduced by 50% and the specific gravity of the boards by 7 to 10% if 10 to 30% of the SSL-blended wood chips are replaced by UF-blended wood chips. However, at pressing times of 0.6 to 1 min/mm board thickness, the pressed particleboard still has to be posttreated at high temperatures to meet German standards for mechanical strength properties. Again, the necessary posttreatment of the boards has hindered the practical application of this finding. On the other hand, small amounts (up to 10% of the UF resin) of SSL improve the cold adherence of blended UF particleboard (Schmidt-Hellerau, 1973, 1977) [2]. This has found practical applications in current industrial use in some western European particleboard mills.

Roffael [7] has shown that 20% of the UF in the surface layers of UF particleboard can be replaced by ammonium-based lignosulfonate without significantly worsening the mechanical board properties. The release of formaldehyde decreased only slightly, which was attributed to the reduction of UF resin rather than to a reaction of formaldehyde with the lignin. The binding of formaldehyde by lignin in UF particleboard is claimed in three Japanese patents [2], together with other patents dealing with lignin UF formulations as wood adhesives. Other improvements achieved by lignosulfonate in UF resins are decrease of adhesive viscosity, increased wettability of wood particles, and improved water resistance of finished boards.

According to a recent report [20], substitution of up to 15% of the UF adhesive in particleboards by SSL does not cause major impairment in particleboard properties. This can be seen from the properties of 17.7-mm-thick one-layer particleboard obtained with 8% UF binder (F/U = 1.27), replaced partially by 10 to 30% magnesium-based SSL. The adhesive contained 0.5% paraffin emulsion and 3% ammonium chloride, Pressing time was 10 s/mm at 200°C [2]. It is obvious, however, that substitution of 20% or more of the UF binder by magnesium-based SSL worsens both the strength and water resistance of the boards, while the gelling time (pot life) of the adhesive is increased. When three-layer 20-mm particleboard was manufactured with 15% of calcium-based (A), sodium-based (B), or ammonium-based (C and D) lignosulfonate and 85% UF binder, with pressing times of 9 s/mm at 200°C and 10.5% adhesive in the surface layer and 8.5% in the core layer, the board properties were different.

At board densities of about 0.7 g/cm^3, the bending and tensile strengths of the UF–LS boards are not decreased compared to boards prepared with 100% UF binder, while the percentage of swelling is increased. The formaldehyde release is considerably decreased by ammonium (C and D)- rather than by calcium- and sodium-based SSLs, indicating that the ammonium ions react with formaldehyde under the conditions existing in the boards, but not the lignin. In the case of calcium- and sodium-based SSLs, the reduction in formaldehyde release lies between 10 and 18%, which corresponds to the amount of SSL in the UF–SSL formulation.

In a recent patent by Edler (1978) [2] it has been claimed that about 33% of UF binder in particleboards can be replaced by ammonium-based SSL if certain conditions are maintained. First, the UF resin should have a relatively high number of methylol (CH$_2$OH) groups, characterized by a Witte number of 1 to 1.8, preferably 1.6, which leads to better compatibility between UF and SSL. Second, the concentration of ammonium ions has to be adjusted to 0.2 to 4%. The ammonium ions react with free formaldehyde, forming less reactive hexamethylenetetramine, which leads to excessive sulfonic acid groups in lignin. If the ammonium-ion concentration is higher than 4%, based on dry lignosulfonate, the acidity becomes too high, resulting in very fast curing. The latter causes soft board surfaces and diminished strength properties. On the other hand, if the ammonium ion concentration is below 0.2%, the curing time becomes too long. The properties of four types of particleboard, obtained with three different types of adhesive—A having a Witte number of 1.58, B of 1.50, and C of 1.02, the latter prepared using only UF resin, while types A and B contained 33% lignosulfonate and 67% UF—Pressing times were about 15 s/mm at 160°C. Wood chips consisting of 67% pine and 33% Douglas fir gave different results. The mechanical strength properties of boards obtained with resins of types A and B (33% lignosulfonate) show no major impairment compared with those of conventional UF boards (resin type C). Significant improvements in the water resistance are gained, due to the polyphenolic structure of lignin [2].

REFERENCES

1. F. Melms and K. Schwenzon, *Verwertungsgebiete der Sulfitablauge*, VEB Deutscher Verlag für Grunstoffindustrie, Leipzig, 1967.
2. H. H. Nimz, in *Wood Adhesives Chemistry and Technology*, Vol. 1 (A. Pizzi, ed.), Marcel Dekker, New York, 1983, Chap. 5.
3. K. Freudenberg, *Science 148*: 595 (1965); K. Freudenberg, in *Constitution and Biosynthesis of Lignin* (K. Freudenberg and A. C. Neish, eds.), Springer-Verlag, Berlin, 1968.
4. K. C. Shen and L. Calvé, *Adhesives Age*, 25 (Aug. 1980).
5. R. J. Mahoney, 1980, *Proc. 14th International Particleboard Symposium*, Washington State University, Pullman, Wash., 1980.
6. K. G. Forss and A. Fuhrmann, *Paperi Puu 11*: 817 (1976).
7. E. Roffael, *Adhaesion 11*: 334; 12: 368 (1979).
8. E. Roffael and W. Rauch, *Holz-Zentralbl. 43/44* (Apr. 10, 1974).
9. D. P. C. Fung, K. C. Shen, and L. Calvé, Spent sulphite liquor–sulphuric acid binder: its preparation and some chemical properties, *Report OPX 180 E*, Eastern Forest Products Laboratory, Ottawa, Ontario, Canada, 1977.
10. K. C. Shen and M. N. Carroll, *Forest Prod. J. 19*(8): 17 (1969); K. C. Shen, *Forest Prod. J. 24*(2): 38 (1974).
11. Canadian standard for particleboard, *CSA 0188/68*.
12. K. C. Shen, *Forest Prod. J. 27*(5): 32 (1977).
13. H. H. Nimz and G. Hitze, *Cell. Chem. Technol. 14*:371 (1980).
14. E. Roffael and W. Rauch, *Holzforschung 26*: 197 (1972).
15. E. Roffael and W. Rauch, *Holzforschung 25*: 149 (1971).
16. E. Roffael and W. Rauch, *Holzforschung 27*: 214 (1973).
17. K. G. Forss and A. Fuhrmann, *Forest Prod. J. 29*(7): 39 (1979).
18. *LCHW-Ligninsulfonate in Holzspanplatten*, Studie 2 and 3, Lignin-Chemie Waldhof-Holmen, Dusseldorf, Germany, 1979.
19. P. Md. Tahir and T. Sellers, Jr., *19th IUFRO World Congress*, Montreal, Quebec, Canada, Aug. 1990.
20. L. Calvé, *19th IUFRO World Congress*, Montreal, Quebec, Canada, Aug. 1990.
21. A. Pizzi, F. A. Cameron, and G. H. van der Klashorst, in *Adhesives from Renewable Resources*, ACS Symposium Series 385 (R. W. Hemingway, A. Conner, and S. J. Branham (eds.), American Chemical Society, Washington, D.C., 1988, Chap. 7.
22. A. Pizzi and T. Walton, *Holzforschung 46*(6): 541–547 (1993).
23. A. Stephanou and A. Pizzi, *Holzforschung 47*: 439 (1993).
24. A. Pizzi and A. Stephanou, *J. Appl. Polymer Sci. 49*: 2157 (1993).
25. A. Stephanou and A. Pizzi, *Holzforschung 47*: 501 (1993).

21
Resorcinol Adhesives

A. Pizzi* *University of the Witwatersrand, Johannesburg, South Africa*

I. INTRODUCTION

Resorcinol–formaldehyde (RF) and phenol–resorcinol–formaldehyde (PRF) cold-setting adhesives are used primarily in the manufacture of structural glulam, finger joints, and other exterior timber structures. They produce bonds not only of high strength but also of outstanding water and weather resistance when exposed to many climatic conditions [1,2]. PRF resins are prepared mainly by grafting resorcinol onto the active methylol groups of the low-condensation resols obtained by the reaction of phenol with formaldehyde. Resorcinol is the chemical species that gives to these adhesives their characteristic cold-setting behavior. At ambient temperature and on addition of a hardener, it provides accelerated and improved cross-linking not only to the resorcinol–formaldehyde resin but also to the phenol–formaldehyde resins onto which resorcinol has been grafted by chemical reaction during resin manufacture. Resorcinol is an expensive chemical, produced in very few locations around the world (to date only three commercial plants are known to be operative: in the United States, Germany, and Japan), and its high price is the determining factor in the cost of RF and PRF adhesives. It is for this reason that the history of RF and PRF resins is closely interwoven, by necessity, with the search for a decrease in their resorcinol content, without loss of adhesive performance.

In the past decades, significant reductions in resorcinol content have been achieved: from pure resorcinol–formaldehyde resins, to PRF resins in which phenol and resorcinol were used in equal or comparable amounts, to the modern-day commercial resins for glulam and finger jointing in which the percentage, by mass, of resorcinol on liquid resin is on the order of 16 to 18%. A step forward has also been the development and commercialization of the "honeymoon" fast-set system [3], coupled with the use of tannin extracts, which in certain countries are used to obtain PRFs of 8 to 9% resorcinol content without loss of performance and with some other advantages. This was a "system" improvement, not an advance on the basic formulation of PRF resins.

II. CHEMISTRY OF RF RESINS

The same chemical mechanisms and driving forces presented for phenol–formaldehyde resins apply to resorcinol resins. Resorcinol reacts readily with formaldehyde to produce

---

Current affiliation: Université de Nancy, Epinal, France

resins which harden at ambient temperatures if formaldehyde is added. The initial condensation reaction, in which A-stage liquid resins are formed, leads to the formation of linear condensates only when the resorcinol/formaldehyde molar ratio is approximately 1:1 [4]. This reflects the reactivity of the two main reactive sites (positions 4 and 6) of resorcinol [5]. However, reaction with the remaining reactive but sterially hindered site (2-position) between the hydroxyl functions also takes place [4]. In relation to the weights of resorcinol–formaldehyde condensates which are isolated and on a molar basis, the proportion of 4- plus 6-linkages relative to 2-linkages is 10.5:1. However, cognizance must be taken of the fact that the first-mentioned pair represents two condensation sites relative to one. The difference in reactivity of the two types of sites (i.e., 4- or 6-position relative to the 2-position) is then 5:1 [4]. Linear components always appear to form in preference to branched components in A-stage resins [4]; that is, terminal attack leads to the preferential formation of linear rather than branched condensates. This fact can be attributed to:

1. The presence of two reactive nucleophilic centers on the terminal units, as opposed to single centers of doubly bound units already in the chain.
2. The greater steric hindrance of the available nucleophilic center (nearly always at the 2-position) of the doubly bound units as opposed to the lower steric hindrance of at least one of the nucleophilic centers of the terminal units (a 4- or 6-position always available). The former is less reactive as a result of the increased steric hindrance. The latter are more reactive.
3. The lower mobility of doubly bound units, which further limits their availability for reaction.

most common type of resorcinol/formaldehyde "tetramer"

The absence of methylol (—CH$_2$OH) groups in all six lower-molecular-weight compounds isolated [4] reflects the high reactivity of resorcinol under acid or alkaline conditions. It also shows the instability of its *para*-hydroxybenzyl alcohol groups and their rapid conversion to *para*-hydroxybenzyl carbonium ions or quinone methides. This explains how identical condensation products are obtained under acid or alkaline reaction conditions [4]. In acid reaction conditions methylene ether–linked condensates are also formed, but they are highly unstable and decompose to form stable methylene links in 0.25 to 1 h at ambient temperature [6,7].

From a kinetic point of view, the initial reaction of condensation to form "dimers" is much faster than the later condensation of these "dimers" and higher polymers. The reaction of resorcinol with formaldehyde, on an equal molar basis and under identical conditions, also proceeds at a rate which is approximately 10 to 15 times faster than that of the equivalent phenol–formaldehyde system [13]. The high reactivity of the resorcinol–formaldehyde system renders it impossible to have these adhesives in resol form. Therefore, only resorcinol–formaldehyde "novolaks," thus resins not containing methylol groups, can be produced. Thus all the resorcinol nuclei are linked together through methylene bridges with no methylol groups and methylene–ether bridges.

The reaction rate of resorcinol with formaldehyde is dependent on the molar ratio of the two constituents, the concentration of the solution, pH, temperature, presence of various catalysts, and amount of certain types of alcohols present [8–11]. The effect of pH and temperature on the reactivity of the resorcinol–formaldehyde system is shown in Fig. 1 [11,12]. Methanol and ethanol slow down the rate of reaction. Other alcohols behave similarly, the extent of their effect being dependent on their structure. Methanol lengthens gel time more than that of other alcohols; higher alcohols are less effective. The retarding effect on the reaction is due to temporary formation of hemiformals between the alcohols and the formaldehyde. This reduces the reaction rate because of the lower concentration of available formaldehyde [9,10]. Other solvents also affect the rate of reaction by forming complexes or by hydrogen bonding with the resorcinol [9,12].

In the manufacture of pure resorcinolic resins, the reaction would be violently exothermic unless controlled by the addition of alcohols. Because the alcohols perform other useful functions in the glue mix, they are left in the liquid glue. PRF adhesives are generally prepared by reaction of phenol with formaldehyde to form a polymer that has been proved to be in the greatest percentage, and often completely, linear [4]. This can be represented as follows:

$$m \geq 0 \text{ in integer numbers}$$
$$0 \leq n \leq 2 \text{ in integer numbers}$$

I

In the reaction, the resorcinol chemical is added in excess, in a suitable manner, to polymer I to react with the $-CH_2OH$ groups to form polymers of the following type, in which the terminal resorcinol groups can be resorcinol chemical or any type of resorcinol–formaldehyde polymer.

$$p \geq 1 \text{ in integer numbers}$$

Where straight resorcinol adhesives are not suitable, resins can be prepared from modified resorcinol [12]. Characteristic of these types of resins are those used for tire cord adhesives, in which a pure resorcinol–formaldehyde resin is used, or alternatively, alkyl resorcinol or oil-soluble resins suitable for rubber compounding are obtained by prereaction of resorcinol with fatty acids in the presence of sulfuric acid at high temperature followed by reaction with formaldehyde. Worldwide, more than 90% of resorcinol adhesives are used as cold-setting wood adhesives; the most notable other application is use as tire cord adhesives, which constitutes less than 5% of the total use.

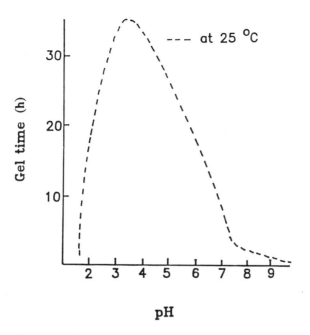

Figure 1   Effect of pH and temperature on the reactivity of the resorcinol–formaldehyde system.

III. WOOD LAMINATING AND FINGER JOINTING ADHESIVES

Various adhesive formulations can be used for the manufacture of laminated wood beams and finger joints for structural purposes. Only those adhesive formulations that at some time or other have been used in industrial applications will be described. All these formulations are based totally or partially on resorcinol, and the hardening process is carried out at ambient temperatures up to 50°C [14].

A. Adhesive 1: Resorcinol–Formaldehyde Adhesive

A resorcinol–formaldehyde novolak is produced accordingly to the following schematic reaction:

$$n \geqslant 0$$

If paraformaldehyde and fillers, generally wood and nutshell flours, are added, the resin becomes capable of setting in 2 to 3 h and curing in 16 to 24 h at ambient temperature.

B. Adhesive 2: Phenol–Resorcinol–Formaldehyde Adhesive and Powder Hardener

A phenol–formaldehyde resol is prepared and resorcinol is grafted onto it according to the following schematic reaction:

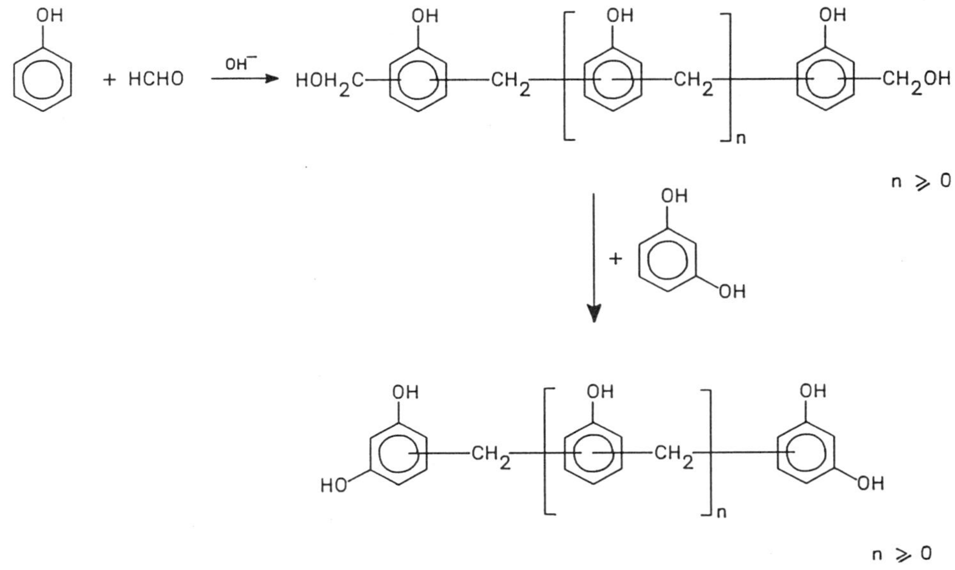

$$n \geqslant 0$$

$$n \geqslant 0$$

The resin produced is capable of setting in 2 to 3 h and cures in 16 to 24 h at ambient temperature once paraformaldehyde and flour fillers have been added.

C. Adhesive 3: Urea–Resorcinol–Formaldehyde Adhesive

A urea–formaldehyde resin is prepared and resorcinol is grafted onto it as terminal units. The urea–formaldehyde resin can also be purchased ready-made. The schematic reaction is as follows:

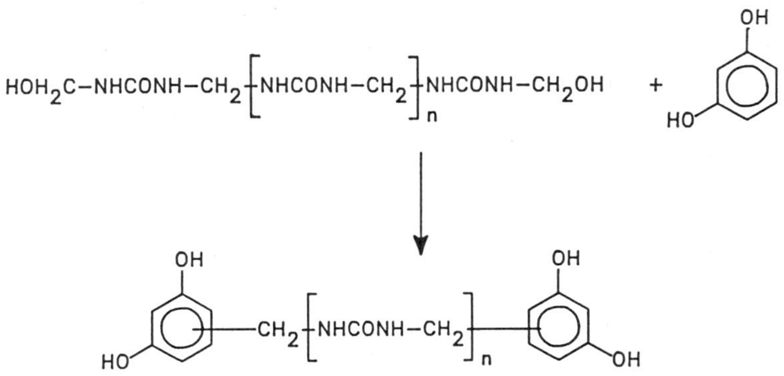

The behavior of this adhesive is identical to that of adhesive 2, although larger amounts of resorcinol are used. These adhesives have a high urea/formaldehyde ratio, are water resistant, and are capable of radio-frequency curing. They can also be used for plywood manufacture.

D. Adhesive 4: Phenol–Resorcinol–Formaldehyde and Liquid Hardener

A phenol–formaldehyde resol is prepared which constitutes the resin. Resorcinol chemical or resorcinol formaldehyde novolaks, in aqueous or water–alcohol solutions, are used as hardeners according to the following scheme:

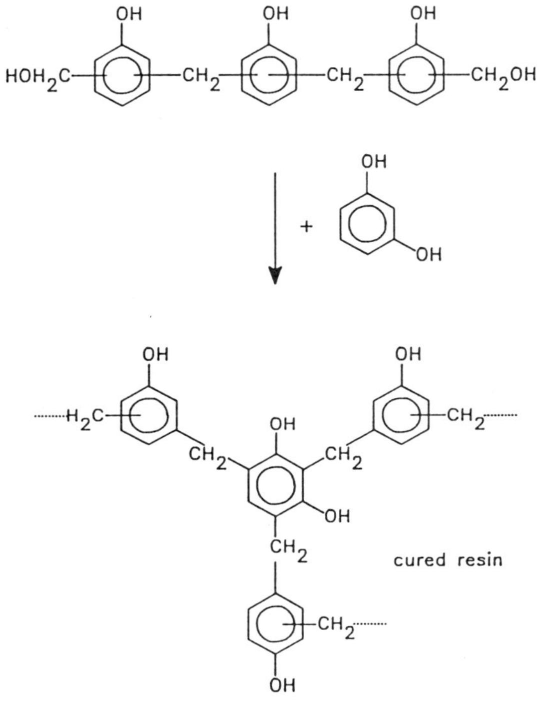

E. Adhesive 5: Phenol–Resorcinol–Formaldehyde and Liquid Hardener

A phenol–resorcinol–formaldehyde similar to adhesive 2 is prepared and a phenol–formaldehyde resol of the same type of adhesive 4 is used as hardener. The schematic curing reaction is the following:

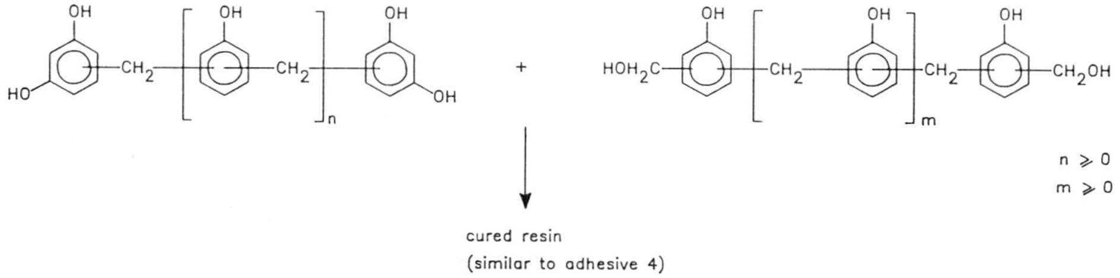

Adhesives of type 2, in which a liquid phenol–resorcinol–formaldehyde adhesive and a powder hardener are used are currently the most commonly used industrially. Adhesives of type 1 were used extensively up to a few years ago. They fell into disfavor because of the high price of the resorcinol chemical needed to make them. They are still used industrially when particularly difficult wood-gluing problems arise. Adhesives of types 4 and 5 have been used quite extensively in the past, but have been superseded in the last few years by adhesives of type 2, which have several handling advantages.

Adhesives of type 3, although good, have not really caught on commercially and were developed as an alternative to phenol–resorcinol–formaldehyde adhesives of type 2 due to the ever-increasing price of oil-derived phenol. Adhesives of types 1, 2, and 3 are mixed before use in a mass/mass ratio of liquid adhesive resin (50 to 60% solid content) to powder hardener of 5:1. The hardener is generally a mixture of 10 parts paraformaldehyde and 10 fillers. It is comprised of 200-mesh wood flour or a mixture of wood flour and nutshell flour, also 200 mesh. Adhesives of types 4 and 5 have a liquid resin/liquid hardener ratio of 1:1 by mass. (This is because the hardener is also a resin.)

All properly formulated resorcinol-based adhesives must have a viscosity low enough in aqueous-alcoholic solutions to flow with ease into all the interstices of the wood surface. Wetting ability is promoted by the alcohol. The paraformaldehyde used as hardener is an addition polymer composed of a few to over 100 formaldehyde monomers. It dissolves slowly in water, by depolymerization to formaldehyde monomers. The rate of depolymerization depends on the degree of polymerization of the paraformaldehyde, the size of the particles, and the pH. Therefore, the working life or pot life of a glue mix can be adjusted by selecting the type of paraformaldehyde and the pH correctly, as its success also depends on the rate of supply of formaldehyde monomer. Fillers are added to give consistency to the glue mix, to control viscosity and thixotropic characteristics, to form a fibrous reinforcement of the adhesive film, and to lessen the cost. Wood flour is used as a filler to obtain better gap filling where rough or uneven surfaces must be bonded, or where low bonding pressures must be used. Nutshell flours, such as coconut shell flour, walnut shell flour, peach pip shell flour, macadamia nutshell flour, or even olive stone flour are used as fillers to lower absorptive qualities and to provide smooth-flowing mixtures. Clays and silica smokes can be used in very small amounts to control the thixotropic consistency of the glue mix.

As the formaldehyde reacts with the resorcinol-based resin, condensation occurs, with the formation of high-molecular-weight polymers. There is considerable secondary-forces interaction between the growing resorcinol polymers and the noncrystalline hemi-cellulosic and lignocellulosic molecules. The highly polar methylol groups and the phenolic hydroxy groups link to cellulose and lignin groups by van der Waals, hydrogen, and electrostatic bonds. The growing adhesive polymers continue to interact to form colloidal particles and then a gelatinous film. This mechanism depends strongly on the moisture content of the wood, which determines the rate of water and solvent absorption.

The advantage of ambient-temperature curing is that the moisture escapes gradually from the hard film formed on curing, inducing a minimum of residual stresses on the joint and allowing the glue line to assume the aspect of a molecularly porous solid. As a consequence, the hard film is able to transpire in the same way as wood, which minimizes checking or crazing and allows the glued joint to survive exposure to the extremes of humidity cycles. Typical level of strength and wood failure results obtained in specified standard tests are shown in Table 1.

Table 1 Typical Tensile Strength and Percentage Wood Failure Results Obtainable with Synthetic PRF Resins[a]

Dry test		24-h cold-water soak		6-h boil test	
Strength (N)	Wood failure (%)	Strength (N)	Wood failure (%)	Strength (N)	Wood failure (%)
3000–3500	90–100	2600–3200	75–100	2500–3000	75–100

[a]Test according to Refs. 24 and 25.

IV. SPECIAL ADHESIVES OF REDUCED RESORCINOL CONTENT

A. Fast-Setting Adhesives for Finger Jointing and Glulam

Together with the more traditional finger jointing adhesives that have just been discussed, a series of ambient-temperature fast-setting separate application systems have also been developed. These eliminate the long delays caused by the use of more conventional phenol–resorcinol–formaldehyde adhesives, which require lengthy periods to set. These types of phenolic adhesives are applied separately. They were first developed in the United States [15–18] to glue large components where presses were impractical. Kreibich [18] describes these separate applications or "honeymoon" systems as follows: "Component A is a slow-reacting phenol–resorcinol–formaldehyde resin with a reactive hardener. Component B is a fast-reacting resin with a slow-reacting hardener. When A and B are mated, the reactive parts of the component react within minutes to form a joint which can be handled and processed further." Full curing of the slow-reacting part of the system takes place with time. m-Aminophenol is a frightfully expensive chemical, and for this reason these systems were discarded and not used industrially [14]. In their original concept, component A is a traditional phenol–resorcinol–formaldehyde cold-setting adhesive at its standard pH of 8 to 8.5 to which formaldehyde hardener has been added. Flour fillers may be added or omitted from the glue mix. Component B is a phenol/meta-aminophenol/formaldehyde resin with a very high pH (and therefore a high reactivity), which contains no hardener or only a very slow hardener.

More recently, a modification of the system described by Kreibich has been used extensively in industry with good success. Part A of the adhesive is again a phenol–resorcinol–formaldehyde (PRF) cold-setting adhesive, with powder hardener added at its standard pH. Part B can be either the same PRF adhesive with no hardener and the pH adjusted to 11, or a 50 to 55% tannin extract solution at a pH of 11 to 12, provided that the tannin is of the condensed or flavonoid type, such as wattle, quebracho, hemlock, or pine extract, with no hardener [3,19]. The results obtained with these two systems are good and the resin not only has all the advantages desired but also as a result of the use of vegetable tannins and of the halving of the resorcinol content of the entire adhesive system, is considerably cheaper [3,19,20].

The adhesive works in the following manner: Once the component A glue mix is spread on one finger-joint profile and the two profiles are joined under pressure, the reaction of component B with the hardener of part A is very fast. In 30 min at 25°C, finger joints prepared with these adhesives generally reach the levels of strength that finger joints glued with more conventional phenolic adhesives are able to reach only after 6 h at 40 to 50°C or in 16 to 24 h at 25°C [3,20]. Clamping of laminated beams (glulam) bonded with these fast-set honeymoon adhesives is in average of only 3 h at ambient temperature compared with the 16 to 24 h necessary with traditional PRF resins [19,20].

B. Branched PRF Adhesives

Recently, another step forward has been taken in the formulation of PRF adhesives of lower resorcinol content. Liquid resorcinol or PRF resins appear to be mostly linear [4]. The original concept in "branching" erroneously maintained that if a chemical molecule capable of extensively branching (three or more effective reaction sites with an aldehyde) the PF and PRF resins is used after, before, or during, but particularly during or after, the preparation of the PF resin, the polymer in the branched PRF adhesive has (1) higher molecular weight than in normal PRF adhesives where branching is not present, and (2) higher viscosity in water or water–solvent solutions of the same composition and of the same resin solids content (concentration). It also needs a much lower resorcinol amount on total phenol to present the same performance of normal, linear PRF adhesives. This can be explained schematically as follows:

resorcinol $-CH_2-[phenol-CH_2-]_n$ resorcinol

resorcinol $-CH_2-[phenol-CH_2-]_n$ resorcinol

resorcinol $-CH_2-[phenol-CH_2-]_n$ resorcinol

n in integer numbers

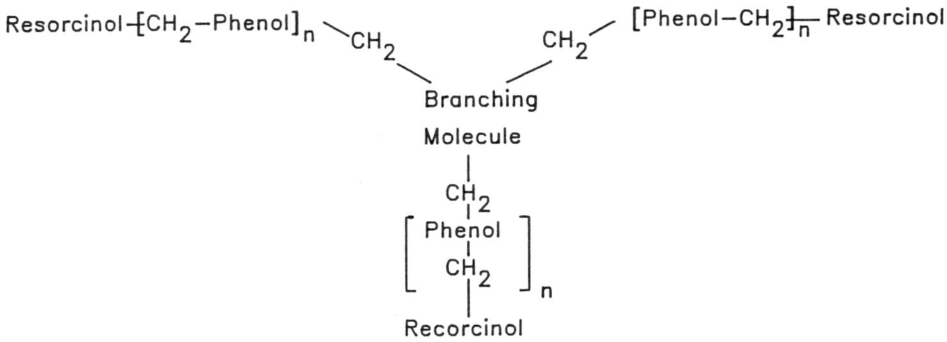

Resorcinol$-[CH_2-Phenol]_n$ CH_2 CH_2 $[Phenol-CH_2-]_n$ Resorcinol

Branching

Molecule

$$\left[\begin{array}{c} CH_2 \\ Phenol \\ CH_2 \end{array} \right]_n$$

Recorcinol

with $n \geq 1$ and an integer number and comparable to, similar to, or equal to n in the preceding scheme for the production of PRF resins.

When comparing linear and branched resins, for every n molecules of phenol used in the particular schematic examples shown, 2 molecules of resorcinol are used in the case of a normal, traditional, linear PRF adhesive, whereas only 1 molecule of resorcinol for n molecules of phenol is used in the case of a "branched" PRF adhesive. The amount of resorcinol has then been halved or approximately halved in the case of the branched PRF resin. A second effect caused by the branching is a noticeable increase in the degree of polymerization of the resin. This causes a considerable increase in the viscosity of the liquid adhesive solution. Because PRF adhesives must be used within fairly narrow viscosity limits, to return the viscosity of the liquid PRF adhesive within these limits, the resin solids content in the adhesive must be lowered considerably, with a consequent further decrease in total liquid resin of the amount of resorcinol and of the other materials, except solvents and water. This decreases the cost of the resin further, without decreasing its performance.

Thus, to conclude, the decrease of resorcinol by branching of the resin is based on two effects:

1. A decrease of resorcinol percentage in the polymer itself, hence in the resin solids, due to the decrease in the number of the phenol–formaldehyde terminals onto which resorcinol is grafted during PRF manufacture
2. An increase in molecular weight of the resin, which by the need to decrease the percentage resin solids content to a workable viscosity, decreases the percentage of resorcinol on liquid resin (not on resin solids)

It is clear that in a certain sense a branched PRF will behave as a more advanced, almost precured phenolic resin. While the first effect described is a definite advance on the road to better engineered PRF resins, the second effect can also be obtained with more advanced (reactionwise) linear resins. The contribution of the second effect to the decrease in resorcinol is not less marked than that of the first effect. It is, however, the second effect that accounts for the differences in behavior between branched and linear PRF adhesives.

Branching molecules could be resorcinol, melamine, urea, and others [21]. Urea is the favorite, because it is much cheaper than the others and needs to be added in only 1.5 to 2% of total resin. When urea is used as a brancher, the adhesive assumes an intense and unusual (for resorcinol resins) blue color after a few days, hence its nickname, "blue glue." However, later work [22,23] has shown that tridimensional branching has very little to do with the improved performance of these low-resorcinol contact adhesives, with tridimensionally branched molecules contributing, at best, no more than 8 to 9% of the total strength [22,23]. In reality, addition of urea causes the reaction as foreseen, not in three points branching but only in two sites of the branching molecule. This is equivalent to saying that most of the resin doubles linearly in molecular weight and degree of polymerization, while the final effect, good performance at half the resin resorcinol content, is maintained [22,23]. This effect is based on the relative reactivity for phenolic methylols of urea and of unreacted phenol sites, and thus while the macro effect is as wanted, at the molecular level it is only a kinetic effect due to the different relative reactivities of urea and phenol under the reaction conditions used. Thus

$$\text{resorcinol} - CH_2 - \left[\!- \text{phenol} - CH_2 -\right]_{\!n} \text{resorcinol}$$

$$\text{resorcinol} - CH_2 - \left[\!- \text{phenol} - CH_2 -\right]_{\!n} \text{resorcinol}$$

$$\downarrow$$

$$\text{(with urea)}\quad \text{resorcinol} - CH_2 - \left[\!- \text{phenol} - CH_2 -\right]_{\!n} \text{urea} - \left[\!- CH_2 - \text{phenol} -\right] CH_2 - \text{resorcinol}$$

Halving of the resorcinol content is still obtained, but from 90 to 100% of the polymers in the resin are still linear.

It is noticeable that the same degree of polymerization and "doubling" effect cannot be obtained by lengthening the reaction time of a PF resin without urea addition [22,23]. These liquid resins then work at a resorcinol content of only 9 to 11%, considerably lower than that of traditional PRF resins. These resins can also be used with good results for honeymoon fast-setting adhesives in PRF–tannin systems, thus further decreasing the total content of resorcinol in the total resin system to a level as low as 5 to 6%.

V. FORMULATION FOR ADHESIVES OF TYPE 2 [14]

A basic formulation capable of giving more than adequate results is presented here so that a starter in the field can get acquainted with these types of adhesives. The procedure for the preparation of this resin can be modified in many ways by varying catalysts, concentration, molar ratios, and condensation conditions.

Phenol, 110 parts by mass + 22 parts water
First formalin 37% solution, 49 parts by mass
H_2SO_4 10% solution, 22 parts by mass
First NaOH 40% solution, 4.5 parts by mass
Second NaOH 40% solution, 9.25 parts by mass
Second formalin 37% solution, 90 to 93 parts by mass
Methanol or methylated spirits, 30 parts by mass (at start of reaction)
Resorcinol, 71 parts by mass
Tannin extract, 19 parts by mass

Phenol, water, methanol, and the first amount of formalin solution are charged on the reaction vessel and heated mildly until the phenol is dissolved. H_2SO_4 is added and the temperature increased to reflux. The mixture is refluxed for 3.5 to 4 h (generally ±4 h). It is cooled to 50 to 60°C and the following is added: the two amounts of NaOH 40% solution (slowly) and the second amount of formalin solution. The mixture is refluxed for 4.5 to 4.75 h and then resorcinol is added. The mixture is refluxed for a further 30 to 50 min. Spray-dried wattle tannin extract is added immediately before or during cooling to adjust viscosity. The pH must be adjusted to 8.5 to 9.5 according to the pot life required. The hardener is 50:50 mixture of paraformaldehyde 96% (usually a fast grade) and 180 to 200-mesh wood flour (60:40 mass proportion).

REFERENCES

1. J. M. Dinwoodie, in *Wood Adhesives Chemistry and Technology*, Vol. 1 (A. Pizzi, ed.), Marcel Dekker, New York, 1983, pp. 1–58.
2. R. E. Kreibich, in *Wood Adhesives: Present and Future* (A. Pizzi, ed.), Applied Polymer Symposium 40, 1984, pp. 1–18.
3. A. Pizzi, D. du T. Rossouw, W. Knuffel, and M. Singmin, *Holzforsch. Holzverwert. 32*(6): 140 (1980).
4. A. Pizzi, R. M. Horak, D. Ferreira, and D. G. Roux, *Cell. Chem. Technol. 13*: 753 (1979); *J. Appl. Polymer Sci.*
5. R. A. V. Raff and B. M. Silverman, *Ind. Eng. Chem. 43*: 1423 (1951).
6. D. du T. Rossouw, A. Pizzi, and G. McGillivray, *J. Polymer Sci. Chem. Ed. 18*: 3323 (1980).
7. A. Pizzi and P. van der Spuy, *J. Polymer Sci. Chem. Ed. 18*: 3477 (1980).
8. R. A. V. Raff and B. H. Silverman, *Can. Chem. 29*: 857 (1951).

9. A. R. Ingram, *Can. Chem. 29*: 863 (1951).
10. C. T. Liu and T. Naratsuka, *Mozukai Gakkaishi 15*: 79 (1969).
11. P. H. Rhodes, *Mod. Plastics 24*(12): 145 (1947).
12. R. H. Moult, in *Handbook of Adhesives*, 2nd ed. (I. Skeist, ed.), Reinhold, New York, 1977, pp. 417–423.
13. G. G. Marra, *Forest Prod. J. 6*: 97 (1956).
14. A. Pizzi, in *Wood Adhesives Chemistry and Technology*, Vol. 1 (A. Pizzi, ed.), Marcel Dekker, New York, 1983, pp. 105–178.
15. G. F. Baxter and R. E. Kreibich, *Forest Prod. J. 23*(1): 17 (1973).
16. R. W. Caster, *Forest Prod. J. 23*(1): 26 (1973).
17. H. Ericson, *Papper Tra 1*: 19 (1975).
18. R. E. Kreibich, *Adhesives Age 17*: 26 (1974).
19. A. Pizzi and F. A. Cameron, *Forest Prod. J. 34*(9): 61 (1984).
20. A. Pizzi and F. A. Cameron, in *Wood Adhesives Chemistry and Technology*, Vol. 2 (A. Pizzi, ed.), Marcel Dekker, New York, 1989, pp. 229–306.
21. A. Pizzi, in *Wood Adhesives Chemistry and Technology*, Vol. 2 (A. Pizzi, ed.), Marcel Dekker, New York, 1989, pp. 190–210.
22. E. Scopelitis, M.Sc. thesis, University of the Witwatersrand, Johannesburg, South Africa, 1992.
23. E. Scopelitis and A. Pizzi, *J. Appl. Polymer Sci. 47*: 351 (1993); *48*: 2135 (1993).
24. Specification for synthetic resin adhesives for wood, Part 2; Close contact joints, *British Standard BS 1204-1965*.
25. Standard specification for phenolic and aminoplastic adhesives for laminating and fingerjointing of timber, and for furniture and joinery, *SABS 1349-1981*, South African Bureau of Standards, 1981.

22

Urea–Formaldehyde Adhesives

A. Pizzi* *University of the Witwatersrand, Johannesburg, South Africa*

I INTRODUCTION

The urea–formaldehydes (UF) are the most important and most used class of aminoresin adhesives. Aminoresins are polymeric condensation products of the reaction of aldehydes with compounds carrying aminic or amidic groups. Formaldehyde is by far the primary aldehyde used. The advantage of UF adhesives are their (1) initial water solubility (this renders them eminently suitable for bulk and relatively inexpensive production), (2) hardness, (3) nonflammability, (4) good thermal properties, (5) absence of color in cured polymers, and (6) easy adaptability to a variety of curing conditions [1,2].

Thermosetting amino resins produced from urea are built up by condensation polymerization. Urea is reacted with formaldehyde, which results in the formation of addition products such as methylol compounds. Further reaction and the concurrent elimination of water leads to the formation of low-molecular-weight condensates which are still soluble. Higher-molecular-weight products, which are insoluble and infusible, are obtained by further condensing the low-molecular-weight condensates. The greatest disadvantage of the amino resins is their bond deterioration, caused by water and moisture. This is due to the hydrolysis of their aminomethylenic bond. Therefore, UF adhesives are used only for interior applications.

II. CHEMISTRY OF UF RESINS: UREA–FORMALDEHYDE CONDENSATION

The reaction between urea and formaldehyde is complex. The combination of these two chemical compounds results in both linear and branched polymers, as well as tridimensional networks, in the cured resin. This is due to a functionality of 4 in urea (due to the presence of four replaceable hydrogen atoms) (in reality, only trifunctional) and a fuctionality of 2 in formaldehyde. The most important factors determining the properties of the reaction products are (1) the relative molar proportion of urea and formaldehyde, (2) the reaction temperature, and (3) the various pH values at which condensation takes place. These factors influence the rate of increase of the molecular weight of the resin. Therefore the characteristics of the reaction products differ considerably when lower and

---

Current affiliation: Université de Nancy, Epinal, France

higher condensation stages are compared, especially solubility, viscosity, water retention, and rate of curing of the adhesive. These all depend to a large extent on molecular weights.

The reaction between urea and formaldehyde is divided into two stages. The first is alkaline condensation to form mono-, di-, and trimethylolureas. (Tetramethylolurea has never been isolated.) The second stage is the acid condensation of the methylolureas, first to soluble and then to insoluble cross-linked resins. On the alkaline side, the reaction of urea and formaldehyde at room temperature leads to the formation of methylolureas. When condensed, they form methylene–ether links between the urea molecules. The alkaline products from urea and formaldehyde, and from mono- and dimethylolureas, are as follows:

The reaction also produces cyclic derivatives: uron, monomethyloluron, and dimethyloluron.

On the acid side, the products precipitated from aqueous solutions of urea and formaldehyde, or from methylolureas, are low-molecular-weight methyleneureas [3]:

$$H_2NCONH(CH_2NHCONH)_nH$$

These contain methylol end groups in some cases, through which it is possible to continue the reaction to harden the resin.

The monomethylolureas formed copolymerize by acid catalysis and produce polymers and then highly branched and cured networks.

$$HOH_2C-NH-CO-NH_2 \xrightarrow{H^+} HO\overset{H}{\underset{+}{-}}CH_2-NH-CO-NH_2 \longrightarrow$$

X

$$CH_2-NH-CO-NH_2 \\ \overset{+}{} \\ \ddots \\ \overset{+}{NH_2}-CO-NH-CH_2$$

$$\overset{+}{H_2C}-NH-CO-NH-CH_2\left[NH-CO-NH-CH_2\right]_n NH-CO-NH_2$$

The kinetics of the formation and condensation of mono- and dimethylolureas and of simple urea–formaldehyde condensation products has been studied extensively. The formation of monomethylolurea in weak acid of alkaline aqueous solutions is characterized by an initial fast phase followed by a slow bimolecular reaction [4,5]. The reaction is reversible. The rate of reaction varies according to the pH with a minimum rate of reaction in the pH range 5 to 8 for a urea/formaldehyde molar ratio of 1:1 and a pH of ± 6.5 for a 1:2 molar ratio [6] (Fig. 1). The 1:2 urea/formaldehyde reaction has been proved to be three times slower than the 1:1 molar ratio reaction [7].

The rapid initial addition reaction of urea and formaldehyde is followed by a slower condensation, which results in the formation of polymers [7]. The rate of condensation of urea with monomethylolurea to form methylenebisurea (or UF ''dimers'') is also pH

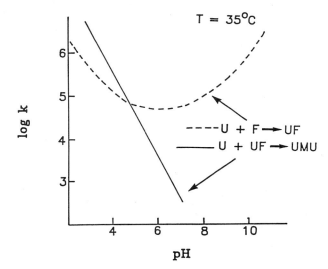

Figure 1 Influence of pH on the addition and condensation reactions of urea and HCHO. U, urea; F, HCHO; M, —CH$_2$—.

dependent. It decreases exponentially from a pH of 2 to 3 to neutral pH value. No condensation occurs at alkaline pH values.

The initial addition of formaldehyde to urea is reversible and is subject to general acid and base catalysis. The forward bimolecular reaction has an activation energy of 13 kcal/mol. The reverse unimolecular reaction has an activation energy of 19 kcal/mol [5]. The rates of introduction into the urea molecule of one, two, and three methylol groups has been estimated to have the ratio 9:3:1, respectively. The formation of *N,N'*-dimethylolurea to monomethylolurea is three times that of monomethylolurea to urea.

Methylenebisurea and higher oligomers undergo further condensation with formaldehyde [8] and monomethylolurea [9], behaving like urea. The capability of methylenebisurea to hydrolyze to urea and methylolurea in weak acid solutions (pH 3 to 5) indicates the reversibility of the aminomethylene link and its lability in weak acid moisture. It explains the slow release of formaldehyde over a long time in particleboard and other wood products manufactured with UF resins.

III. GENERAL PRINCIPLES OF MANUFACTURE AND APPLICATION

It is very important in the commercial production of urea–formaldehyde (UF) resins to be able to control the size of the molecules by the condensation reaction, since their properties change continuously as they grow larger. The most perceptible change is the increase in viscosity. Low-viscosity syrups are formed first. These change into high-viscosity syrups, which are clear to turbid. Molecular weight may vary from a few hundred to a few thousand, with a wide range of molecular size. These molecules are built up by water splitting off at random between reactive groups of neighboring molecules, thereby increasing their size. Once their solubility, viscosity, pH, concentration, and so on, have been determined, they constitute the resins available commercially. The most important factors influencing the final properties of aminoplastic resins in industrial manufacture are the purity of the reagents, the molar proportions of the materials used, the preparation process used, and the pH variation and control.

The most common method of preparation for commercial UF resin adhesives is the addition of a second amount of urea during the preparation reaction. This consists of reacting urea and formaldehyde in more than equivalent proportions. Generally, an initial urea/formaldehyde molar ratio of 1:2.0 to 2.2 is used. Methylolation can in this case be carried out in a much shorter time, by using temperatures of up to 90 to 95°C. The mixture is then maintained under reflux. When the exotherm subsides (usually after 10 to 30 min), the methylol compounds have formed, and the reaction is completed under reflux by adding a trace of an acid to decrease the pH to the UF polymer-building stage (pH 5.0 to 5.3). As soon as the right viscosity is reached, the pH is increased to stop polymers building and the resin solution is cooled to about 25 to 30°C. More urea (called second urea) is added to consume the excess of formaldehyde, until the molar ratio of urea to formaldehyde is in the range 1:1.1 to 1:1.7. After this addition of urea, the resin is left to react at 25 to 30°C for as long as 24 h. The excess water is eliminated by vacuum distillation until a resin solids concentration of 64 to 65% is reached, and the pH adjusted to achieve suitable shelf life or storage life.

The final addition of urea can be done in one operation, or the urea may be added at suitable intervals in smaller lots. Second or further ureas can be added at a temperature slightly higher than ambient or can be added at higher temperatures, 60 to 90°C, according to the type of final resin wanted [10–13]. Increasing second or further urea additions tend to improve bond quality, especially at low formaldehyde/urea molar ratios [10–13]. Higher-molar-ratio resins tend to exhibit an overall better initial bond quality [11], but present an exponentially increased formaldehyde emission problem [13], most often disqualifying them from many, or most, modern uses. Some UF resins used for joinery are also produced without a final, or second, urea addition. The pH used during the condensation reaction (not the methylolation) is generally in the range 4.8 to 5.3.

Control of the average molecular size of the finished resin is essential for correct flow in plywood and particleboard applications while in the hot press prior to curing. Too low a level of condensation (i.e., low-molecular-weight resins) may give too much flow; the resin ''runs away'' from the wood or sinks into it rapidly under pressure, leaving ''starved'' glue lines. This can be corrected by lowering the pH by adding an acid or acid-producing substance, usually a curing agent, hardening catalyst, or simply, hardener. If a resin of too high a condensation stage (i.e., high-molecular-weight resins) is on hand, its flow under normal pressure and temperature may be too low to produce good results. This can usually be corrected by adding flow agents to it, provided that at least some flow is left in the resin. It is generally an advantage to produce resins with ample flow in the factory. Their storage life is longer and finishing can be done at any time, at short notice, to specification, particularly by adjusting the flow and speed of cure.

Resins that have lost part of their flow during manufacture or storage must be corrected by the addition of a flow agent. The simplest means is often the addition of water sprayed on the compound and mixed in well. If a resin is still capable of flowing, this procedure produces a resin with properties that are still acceptable. In cases where moisture content control is critical, it may be necessary to allow a little more time for ''heating'' to let the added moisture escape. However, if the flow is very low, and large quantities of water must be used to bring the flow back to normal, this method is not recommended. The large amount of water would cause longer ''breathing'' times to be necessary due to excessive volatile components, and excessive shrinkage may take place, causing excessive stress on the glue lines. It must be kept in mind that excessive water addition causes UF resin precipitation. The best way to correct flow in these cases is to mix the resin with large amounts of an equal resin of the same quality that has a higher flow. Any proportion may be used to bring the flow back to normal. If increased flow is desired, 0.5 to 2.0% of spray-dried UF or MF resin can also be added to function as a flow agent. Methylol compounds, such as dimethylolurea, also increase flow, but they increase the water released during reaction more than do spray-dried resins. Lubricating agents such as calcium stearate are also able to give a fair degree of flow increase.

Many substances have been suggested as curing agents. These include the following acid products: (1) boric acid, (2) phosphoric acid, (3) acid sulfates, (4) hydrochlorides, (5) ammonium salts of phosphoric or polyphosphoric acid, (6) sodium or barium ethyl sulfate, (7) acid salts of hexamethylenetetramine, (8) phtalic anhydride, (9) phtalic acid, (10) acid resins such as poly(basic acid)–poly(hydric alcohol), (11) oxalic acid or its ammonium salts, and many others. However, the most widely used curing agents in the wood products industry are still ammonium chloride or ammonium sulfate. Their effect can be altered by retarding the reaction of the resin. This is done by the simultaneous addition of small amounts of ammonia solution (which is eliminated during hot curing) to lengthen the pot life of the glue mix. Latent catalysts that produce acid only on heating may also be used, such as dimethyloxalate and other easily hydrolizable esters, or halogenated substances such as 0.1 to 0.2% of bromohydrocinnamic acid and others.

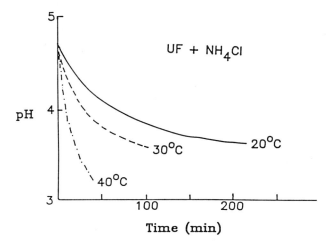

Figure 2 pH change of UF resin with ammonium chloride hardener as a function of temperature and time. (From Ref. 2.)

The driving force in the use of these salts as hardeners is their capacity to release acid, which decreases the pH of the resin and thereby accelerates curing. The speed of the reaction between the ammonium salt and formaldehyde (or ammonia and formaldehyde when this is present) also determines, together with the amount of heat supplied, the rate of acid release and therefore the rate of curing:

$$4NH_4Cl + 6HCHO \rightarrow 4HCl + (CH_2)_6N_4 + 6H_2O$$
$$\text{Hexamethylene-}$$
$$\text{tetramine}$$

Ammonium chloride is a better hardener than hydrochloric acid, as the latter produces weaker joints. The effect of a fixed amount of ammonium chloride on the pH change and on the rate of resin curing as a function of time and temperature is shown in Fig. 2.

Often, particularly in cold-setting UF resins for joinery, hardeners consisting of mixtures of a salt such as ammonium chloride or ammonium sulfate with an acid such as phosphoric acid, citric acid, or others are used to regulate pot life and rate of curing. Both pot life and rate of curing of the resin can then be regulated by (1) varying the concentration of the hardener in the resin, (2) by changing the relative proportions of acid and salt, and (3) by changing the type of acid and/or salt composing the hardener. Acting on these three principles, setting times of between a few minutes and several hours can easily be obtained.

A. Plywood Adhesives

The UF adhesives for plywood generally contain less than 2 mol of formaldehyde per mole of urea, and most of them are condensed to a slightly viscous, hydrophilic stage and are quite soluble in water. The degree of polymerization, and hence the viscosity under comparable conditions, of UF resins for plywood is generally higher than those of UF resins for particleboard.

The application of urea–formaldehyde resins for gluing purposes is based on the excellent control that can be exercised on the condensation reaction by varying the pH, a procedure easily applicable to a production scale. A small amount of an acid as hardener is added at ambient temperature. This produces no visible change at first, or possibly for a few hours; finally, thickening sets in, and the resins gels and hardens into an insoluble material. While the adhesive is still in liquid form, it can be spread on the wood surfaces, which have to be glued and joined under pressure. These have to be cured either at room temperature for a few days or at elevated temperature for a few minutes. Solutions of ammonium salts, generally ammonium chloride or ammonium sulfate, or mixtures of ammonium chloride with urea, are generally used as resin hardeners. Often, ammonia solution is added to lengthen the usable life of the glue mix. Hydraulic presses with multiple openings are generally used for the production of plywood or flat veneer work. They can operate at pressures of 10 to 16 kg/cm^2, but generally operate in the range 12 to 14 kg/cm^2, according to the wood species, to avoid destruction of the porous structure of the wood.

The temperature is usually in the range of 120 to 160°C according to the type and moisture content of the veneers. It is chosen according to its capacity for the fastest pressing time and its ability to produce a good joint without blisters. Different pressing conditions are used in different countries, and the resin must be manufactured keeping the differences in the conditions of application in mind. There is quite a difference, for example, between a UF adhesive and glue mix, which is capable of giving good joints at 5 to 8% moisture content of the veneer and a pressing temperature of 120° C, and a UF resin and glue mix usable at a veneer moisture content of 0 to 1% and pressing temperatures of 140 to 160°C. The former needs better flow characteristics and faster curing under standard measured conditions than does the latter if optimum pressing times and production schedules are to be maintained. Lower temperatures lengthen the curing time of the resins considerably but have the advantage that when the cured plywood sheets are taken out of the hot press, they tend to warp less on cooling or drying.

The use of fillers with plywood UF adhesives has important economical consequences and is necessary for technical reasons, because the fillers produce "body" in the glue solution and therefore prevent joint "starvation" in porous wood. Without filler it would be difficult to prevent part of the adhesive from flowing away or flowing into the open pores of the wood; or in the gluing of medium to thin veneers, from flowing through them to the other side, thereby causing undesirable resin patches on the outer veneer surfaces. As a rule, 20 to 50% filler is used for joinery and up to 100% for plywood. The most common fillers are wheat flour, corn flour, rye flour, very fine hardwood flour, and gypsum. If gypsum is used, it must be free of calcium hydroxide, because this interferes with the acid curing agent.

B. Particleboard Adhesives

A very important application for UF adhesives is in the manufacture of particleboard. The glue mix is generally composed of a liquid resin to which water has been added to decrease viscosity and to facilitate spraying, plus small amounts of ammonium chloride or sulfate and small amounts of ammonia solution. Small quantities of insecticides, wax emulsion, and fire-retarding agents (such as ammonium phosphates) are added before spraying the adhesive onto the wood chips. Pressing temperatures and maximum pressures used in the cycle are in the range of 150 to 200°C and 2 to 35 kg/cm^2, respectively.

The moisture content of glued furnish chips is 7 to 8% for the board core and 10 to 12% for the surface. The resin contents used (i.e., solids) are 6 to 8% for board core and 10 to 11% for board surfaces.

It must be realized that on curing, the viscosity of UF resins changes, not only at a different rate but also in a different manner according to the temperature. The viscosity gradually increases, with temperatures of up to $\pm$ 50°C. Above 60°C the viscosity quite rapidly reaches a maximum, and then decreases. This indicates that the resin tends to degrade under prolonged heating at high temperatures. To avoid this problem, UF-bonded particleboard must never be pressed for too long and must never have a "hot-stack" or "postcure" period after pressings. They must preferably be cooled after manufacture to avoid deterioration in strength and quality. The cured UF resins degrade rapidly at any temperature at a pH below 2. The viscosity for a good particleboard resin is on the order of 100 to 450 cP (at 20°C) [14].

As important as viscosity is resin flow, which reflects viscosity under hot-pressing conditions. Resin flow is a determining factor in manufacturing good particleboard. Excessive flow causes the resin to soak into the wood particles and causes glue-line starvation; insufficient flow causes insufficient contact surface. The gel time generally used at 100°C for glue mixes of UF-bonded particleboard is 3 to 12 min, with 30 s to 5 min for board faces and cores, respectively. The actual gel time in the press depends on the press temperature and is considerably shorter.

C. UF Adhesives for Low-Formaldehyde-Emission Particleboard

UF resins in their cured state are nontoxic. Urea itself is also harmless. However, free formaldehyde and formaldehyde generated by slow hydrolysis of the aminoplastic bond are highly reactive and combine easily with proteins in the human body. This may cause a painful inflammation of the mucous membranes of the eyes, nose, and mouth [15]. Even a low concentration of formaldehyde vapor in the air can cause disagreeable irritations of the nose and eyes. However, such irritations usually disappear in a short time without permanent damage. Occasionally, allergic or anaphylactic reactions develop and complete removal from exposure is necessary.

High temperatures and high relative humidity can result in odor problems in a room containing particleboard manufactured with UF resins [15]. The release of formaldehyde from UF particleboard is caused by two factors. It can be due to free formaldehyde present in the board that has not reacted, and it can be due to formaldehyde formed by hydrolysis of the aminoplastic bond as a result of temperature and relative humidity [2,15]. While the first type of release lasts only a short time after manufacture of the particleboard, the second type of release can continue throughout the entire working life of the board. A considerable number of variables influence the emission of formaldehyde from a UF-glued particleboard. The main ones are the molar ratio of urea to formaldehyde (which influences both types of release), the press temperature, and in service, the ambient temperature and relative humidity.

UF resins for particleboard with urea/formaldehyde molar ratios of 1:1.45, 1:1.32, and 1:1.25 have free formaldehyde contents of 0.8%, 0.3%, and less than 0.2%, respectively [12]. While the current tendency internationally is to use UF resins that have a urea/formaldehyde molar ratio of 1:1.3 or much lower, which release much less formaldehyde, these resins perform less well in the production of UF-glued particleboard

Table 1 Comparison of Particleboard Prepared with UF Resins of Various Molar Ratios

U/F molar ratio	Approximate density (g/cm)	Internal bond (MPa)	Percent water swelling (2h)	Percent HCHO released, perforator method (mg HCHO/100 g board)
1:1.4–1.5	0.680	0.7 – 0.8	4	50–70
1:1.3–1.35	0.680	0.6 – 0.7	4–5	25–30
1:1.1–1.25	0.680	0.45 – 0.55	5	5–20

Source: Ref. 2.

[12,14]. In particular, they do not allow as much flexibility in particleboard production as do resins with higher formaldehyde/urea molar ratios. This fact stresses the need of greater control and supervision of the production at particleboard plants where UF resins of low molar ratio are used. An example of the variation in properties between particleboard manufactured with different molar ratio resins is given in Table 1.

It is also necessary to use more hardener when working with UF resin of a lower formaldehyde/urea molar ratio, as the gel time of the resin is slower. Up to 5% urea can sometimes be added to the glue mix to decrease the amount of formaldehyde released during pressing and to decrease the initial amount of free formaldehyde present in the finished board immediately after manufacture. Strict norms have been established in many countries with regard to limits of formaldehyde emission from particleboard bonded with UF resins [15,19,20]. Recent work indicated that good E1-type UF resins of urea/formaldehyde molar ratio lower than 1:1.1 can be prepared in a variety of ways [13,16–18]. Although the theoretical basis of this finding has been discussed in part elsewhere [13], to be able to advance a tentative theory for low-formaldehyde-emission UF resins, it is of interest to apply these findings to the formulation and preparation of UF resins of low formaldehyde emission, first in the laboratory and then at the industrial level. First, these resins can be divided into two broad classes: (1) those resins based on addition of melamine or melamine–formaldehyde (MF) resins to the UF resin, and (2) those UF resins in which very low formaldehyde emission capability is obtained exclusively by the manipulation of their manufacturing parameters. The former class is nothing but a subset of the second [13].

The underlying principle of a low-formaldehyde-emission UF resin is that a certain amount of free urea needs to be present to (1) mop up a large amount of the free formaldehyde that may be present at the end of the preparation, and (2) to mop up the greater part of the free formaldehyde that may be generated during hot curing of the resin. A third possible requirement would be that some free monomeric urea species should still be left to mop up, over a long period of time, some of the formaldehyde that may be liberated during the service life of the board.

Such requirements of a UF resin are fundamentally quite divergent and extreme. They mean that addition of great amounts of urea are needed, possibly at the end of the reaction; such urea will react with the free HCHO present or generated during hot curing, but will also react with the active methylol groups present on the urea resin itself, severely limiting the possibility of cross-linking of the resin and ultimately affecting adversely, and diminishing, its cured strength. These two sets of divergent requirements indicate that in general, a low-formaldehyde-emission UF formulation must be a compromise between strength and emission requirements. Once this basic conflict of requirements is understood, it can be overcome to attain formulations that give both good strength and low formaldehyde emission. A UF resin is a mixture of molecular species: namely, methylolurea, UF polymers, and methylolated UF polymers. It has already been proven, both theoretically [16,17] and by applied means [11,17,18] that while monomeric and polymeric methylolated species contribute more to the adhesion of the resin to the wood substrate, it is the polymeric fraction (methylolated and nonmethylolated) that contributes most to the cohesion of the resin. Thus a resin to which great amounts of final urea are added will have a proportionally high amount of urea and monomeric methylolated species, giving both good adhesion and low formaldehyde emission, and proportionally a lower amount of prebuilt polymeric species, giving poor cohesion, hence lower strength. Conversely, a resin of final higher urea/formaldehyde molar ratio such as the classic UF resins used for the last few decades, will have a large number of polymeric species, will still be heavily methylolated, most of the methylolated species will be polymeric however, and will still have a considerable amount of free and potentially free formaldehyde. These resins will have good cohesion and good adhesion, hence good strength, but very high HCHO emission.

The logical manner to avoid the conflicting requirements of the two properties wanted is then to prebuild in some easy and convenient manner the particular mixture of species that will give the correct balance of strength and emission for the applications required. Thus, although UF resins of very low urea/formaldehyde ratios [13] can be prepared by adding great amounts of second and third ureas, the high predominance of urea and other monomeric species in relation to polymer proportions will give boards of poor strength, albeit of very low HCHO emission. The required balance of chemical species and of properties can then be achieved more easily by preparing two or more UF resins, and/or preresins, which are mixed in various amounts to yield the desired balance of acceptable strength and low emission [16].

D. Other UF Adhesives Applications

Although particleboard and plywood are the major users of UF adhesives, two other applications, although consuming much lower proportions of these resins, are also worthy of note. The first is in the furniture and joinery industry, including the manufacture of hollow-core doors. While, in the latter application thermosetting resins with characteristics and glue mixes similar to those for plywood are used, often (but not always) cured by radio frequency, the former can be simpler resins of higher urea/formaldehyde molar ratio to which cold-setting capability and different pot lives are given by variety of hardener types; in these, hardeners formed by an acid plus a salt are the norm. The second application of note is in foundry applications as sand core binders. In this application UF resins compete with phenolic and furanic resins. In general, however, the resins used for the hot-box process are UF resins modified with 20 to 50% furfuryl alcohol to obtain a UF–furanic resin copolymer, and phenol–formaldehyde resins modified with urea. Small amounts of paraffin wax and corn flour are often added to facilitate mixing of the resin with the sand (generally between 1 and 2.5% resin on sand).

IV. ANALYSIS

Methods of formaldehyde analysis include the iodometric, sulfite [21], and mercurimetric [22,23] methods. The sulfite method measures only the formaldehyde present, whereas the iodometric method can also estimate the methylol groups. Another method is based on the partition of formaldehyde between water and isoamyl alcohol [24]. Estimation of the formaldehyde in the alcohol phase of a mixture of an aqueous solution of the resin and isoamyl alcohol allows deduction of the amount of free formaldehyde. This procedure has the advantage that no risk of reaction arises between free formaldehyde and the resin components.

Kappelmeier [25] has suggested the use of aniline, benzylamine, and phenylethylamine as reagents for the identification and analysis of urea in UF resins. He has provided evidence that the methylene–ether groups form a bridge between urea residues in urea–formaldehyde resins. The use of benzylamine, in particular (which yields dibenzylurea from urea derivatives), has been developed as a method of analysis. In determining the ratio of urea to formaldehyde in UF resins, the benzylamine method has been coupled with a process of formaldehyde estimation which involves depolymerization with phosphoric acid, followed by distillation into alkaline potassium cyanide solution [26].

Steiner [11] advocates the use of bromination in CCl_4 and subsequent x-ray analysis to determine available reactive methylol groups. High-resolution nuclear magnetic resonance (NMR) has also been used to analyze UF resins and trace their kinetic behavior [27]. Particularly useful is $^{13}CNMR$ analysis of liquid UF resins, where clear identification of monomeric species, methylolated or not, methylol ureas, methylol groups on the polymer, methylene–ether linkages, methylene bridges, sites of branching, uron, free formaldehyde and other features can be achieved easily and rapidly [28,29]. For example, this technique makes it possible easily to estimate the probable bonding ability and approximate emission class to which boards bonded with it are likely to belong [16].

V. FORMULATION

An introduction to the typical resin synthesis of a UF resin used as an adhesive for wood products and in industrial applications is given below. It constitutes a handy formulation for whose who want to work in this field. It is not a low-formaldehyde-emission formulation. To 1000 parts by mass of 42% formaldehyde solution (methanol < 1%) are added 22% NaOH solution to pH 8.3 to 8.5, 497 parts by mass of 99% urea, and the temperature raised in ± 50 min from ambient to 90°C while maintaining pH in the range 7.3 to 7.6 by small additions of 22% NaOH. The temperature is maintained at 90 to 91°C until the turbidity point is reached (generally another 15 to 20 min). The pH is then corrected to 4.8 to 5.1 by addition of 30% formic acid, and the temperature is raised to 98°C. The water tolerance point is reached in ± 18 min and the pH is then adjusted to 8.7. Vacuum distillation of the reaction water with concomitant cooling is then initiated. After distillation of the wanted amount of water to reach a resin content of 60 to 65%, the resin is cooled to 40°C, 169 parts by mass of second urea is added, the pH is adjusted to 8.5 to 8.7, and the resin is allowed to mature at 30°C for 24 to 48 h; resin characteristics: solids content, 60%; density, 1.268 g/cm^3; free HCHO, 0.4%; viscosity 200 cP; pH, 8.

REFERENCES

1. J. M. Dinwoodie, in *Wood Adhesives Chemistry and Technology*, Vol. 1 (A. Pizzi, ed.), Marcel Dekker, New York, 1983, pp. 1–58.
2. A. Pizzi, in *Wood Adhesives Chemistry and Technology*, Vol. 1 (A. Pizzi, ed.), Marcel Dekker, New York, 1983, pp. 59–104.

3. G. Zigeuner, *Fette Seifen Austrichmittel, 56*: 973 (1954); *57*: 14, 100 (1955).
4. L. E. Smythe, *J. Phys. Colloid Chem. 51*: 369 (1947).
5. G. A. Growe and C. C. Lynch, *J. Am. Chem. Soc. 70*: 3795 (1948); *71*: 3731 (1949); *75*: 574 (1953).
6. L. Bettelheim and J. Cedwall, *Svenska Kem. Tidskr. 60*: 208 (1948).
7. G. Smets and A. Borzee, *J. Polymer Sci. 8*: 371 (1952).
8. J. I. de]ong and J. de Jonge, *Recl. Trav. Chim. Pays-Bas 72*: 207 (1953).
9. J. I. de Jong and J. de Jonge, *Recl. Trav. Chim. Pays-Bas 72*: 213 (1953).
10. K. Horioka, M. Noguchi, K. Moriya, and A. Oguro, *Bull. Gov. Forestry Exp. Sta. Tokyo 113*: 20 (1959).
11. S. Chow and P. R. Steiner, *Forest Prod. J. 23*(12); 32 (1973).
12. B. Sundin, *Proc. FESYP International Particleboard Symp.*, Hamburg, Germany, 1978, p. 112.
13. A. Pizzi, L. Lipschitz, and J. Valenzuela, *Holzforschung* (in press, 1994).
14. R. Marutzky and L. Ranta, *Holz Roh-Werkstoff 37*: 389 (1979).
15. R. Marutzky, in *Wood Adhesives Chemistry and Technology*, Vol. 2 (A. Pizzi, ed.), Marcel Dekker, New York, 1989, pp. 307–388.
16. D. Levendis, A. Pizzi, and E. Ferg, *Holzforschung 46*: 263 (1992); *J. Appl. Polymer Sci. 50*: 907 (1993).
17. A. Pizzi, *J. Adhesion Sci. Technol. 4*(7): 573, 589 (1990).
18. A. Pizzi, *Holzforsch. Holzverwert 43*(3):63 (1991).
19. H.-J. Deppe, in *Luftqualitat in Innenraumen* (K. Aurand, B. Seifert, and J. Wegner, eds.), Fischer Verlag, Stuttgart, Germany, 1982, pp. 91–128.
20. H.-J. Deppe, *Holz-Kunststoffverarb. 20*(7/8): 12 (1985); *21*(7/8):12 (1986).
21. J. I. de Jong and J. de Jonge, *Recl. Trav. Chim. Pays-Bas 71*: 890 (1952).
22. J. I. de Jong, *Recl. Trav. Chim. Pays-Bas 72*: 653 (1953).
23. A. Petz and M. Cherubim, *Holz Roh-Werkstoff 13*: 70 (1955).
24. G. Widmer, *Kunststoffe 46*: 359 (1956).
25. C. P. A. Kappelmeier, *American Chemical Society Meeting*, Boston, 1951.
26. P. P. Grad and R. J. Dunn, *Anal. Chem. 25*: 1211 (1953).
27. B. Tomita and Y. Hirose, *J. Polymer Sci. 14*: 387 (1976).
28. J. R. Ebdon and P. E. Heaton, *Polymer 18*: 971 (1977).
29. R. M. Rammon, W. E. Johns, J. Magnuson, and A. K. Dunker, *J. Adhesion 19*: 115 (1986).

23
Melamine–Formaldehyde Adhesives

A. Pizzi* *University of the Witwatersrand, Johannesburg, South Africa*

I. INTRODUCTION

Melamine–formaldehyde (MF) and melamine–urea–formaldehyde (MUF) resins are among the most used adhesives for exterior and semiexterior wood panels and for the preparation and bonding of both low- and high-pressure paper laminates and overlays. Their much higher resistance to water attack is their main distinguishing characteristic from urea–formaldehyde (UF) resins. MF adhesives are expensive. For this reason, MUF resins which have been cheapened by addition of a greater or lesser amount of urea are also often used. Notwithstanding their widespread use and economical importance, the literature on melamine resins is only a small fraction of that dedicated to urea–formaldehyde resins. Often MFs and MUFs are described in the literature as a subset of UF amino resins. This is not really the case, as they have peculiar characteristics and properties all of their own which in certain respects are very different from those of UF adhesives.

II. USES FOR MF RESINS

MF resins are used as adhesives for exterior and semi-exterior-grade plywood and particleboard. In this application their handling is very similar to that of urea–formaldehyde resins for the same use, with the added advantage of their excellent water and weather resistance. Melamine–formaldehyde resins are also used for the impregnation of paper sheets in the production of self-adhesive overlays for the surface of wood-based panel products and of self-adhesive laminates. In this application the impregnation substrate, α-cellulose paper, is thoroughly impregnated by immersing it in the resin solution, squeezing it between rollers, and drying without curing it to proper flow by passing it through an air-draft tunnel oven at 70 to 120° C at $\pm$ 10 m/s. The dry MF-impregnated sheets can then be bonded by one of two main processes:

1. The sheets of MF-impregnated paper, consisting of one surface layer or a few surface layers, are bonded together and with a substrate of paper sheets impregnated with phenolic resins to form laminates of variable thickness. In the impregnated papers is

---

Current affiliation: Université de Nancy, Epinal, France

the dry but still active MF resin, which functions as the adhesive of the MF-impregnated sheet to both MF-impregnated sheets and at the interface between MF-impregnated and PF-impregnated layers. These laminates are high-pressure laminates.

2. The MF in an impregnated paper sheet is not completely cured but still has a certain amount of residual activity and is applied directly in a hot press, in a single sheet, on a wood-based panel, to which it bonds by completing the MF adhesive curing process.

Press platens are made from stainless steel or chromium-plated brass and copper. The chromium layer preserves surface quality longer than does ordinary steel. The MF laminates exhibit a remarkable set of characteristics. Because of their unusual chemical inertness, nonporosity, and nonabsorbance, they resist most substances, such as mild alkalies and acids, alcohols, solvents such as benzene, mineral spirits, natural oils, and greases. No stains are produced on MF surfaces by these substances. In addition to almost unlimited coloring and decorating possibilities, this remarkable resistance has resulted in the extensive use of MF laminated wood-based panel products for tabletops, sales counters, laboratory benches, heavy-duty work areas in factories and homes, wall paneling, and so on.

III. CHEMISTRY

A. Condensation Reactions

The condensation reaction of melamine (I) with formaldehyde (Fig. 1) is similar to but different from the reaction of formaldehyde with urea. As for urea, formaldehyde first attacks the amino groups of melamine, forming methylol compounds. However, formal-

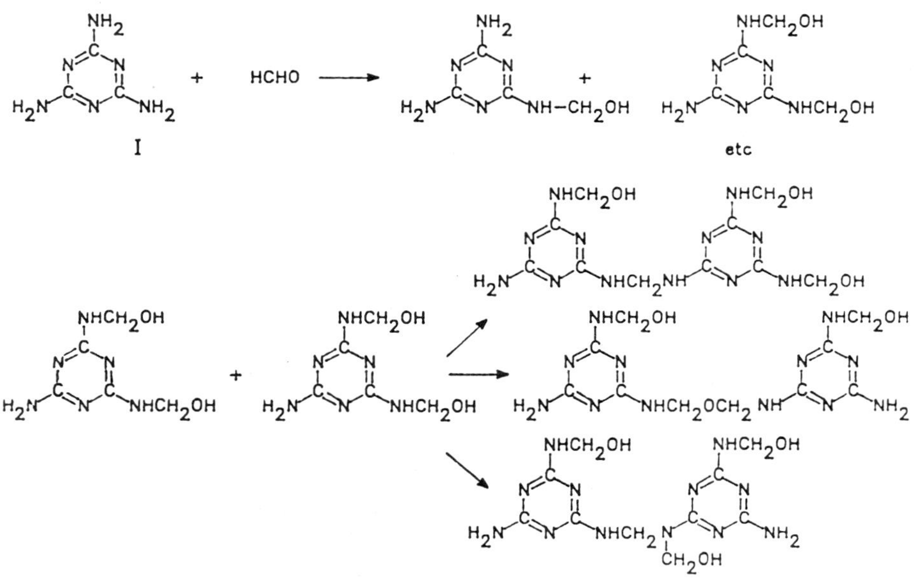

Figure 1 Methylolation (hydroxymethylation) and subsequent condensation reactions to form melamine–formaldehyde adhesive resins.

dehyde addition to melamine occurs more easily and completely than does addition to urea. The amino group in melamine accepts easily up to two molecules of formaldehyde. Thus complete methylolation of melamine is possible, which is not the case with urea [1]. Up to six molecules of formaldehyde are attached to a molecule of melamine. The methylolation step leads to a series of methylol compounds with two to six methylol groups. Because melamine is less soluble than urea in water, the hydrophilic stage proceeds more rapidly in MF resins formation. Therefore, hydrophobic intermediates of the MF condensation appear early in the reaction. Another important difference is that MF condensation to give resins, and their curing, can occur not only under acid conditions, but also under neutral or even slightly alkaline conditions. The mechanism of the further reaction of methylol melamines to form hydrophobic intermediates is the same as for UF resins, with splitting off of water and formaldehyde. Methylene and ether bridges are formed and the molecular size of the resin increases rapidly. These intermediate condensation products constitute the large bulk of the commercial MF resins. The final curing process transforms the intermediates to the desired MF insoluble and infusible resins through the reaction of amino and methylol groups which are still available for reaction.

A simplified schematic formula of cured MF resins has been given by Koehler [2] and Frey [3]. They emphasize the presence of many ether bridges besides unreacted methylol groups and methylene bridges. This is because in curing MF resins at temperatures up to 100°C, no substantial amounts of formaldehyde are liberated. Only small quantities are liberated during curing up to 150°C. However, UF resins curing under the same conditions liberate a great deal of formaldehyde.

At the condensation stage attention must be paid to the formation of hydrolysis products of the melamine before preparation starts. The hydrolysis products of melamine are obtained when the amino groups of melamine are gradually replaced by hydroxyl groups. Complete hydrolysis produces cyanuric acid.

melamine   ammeline

ammelide   cyanuric acid

Ammeline and ammelide can be regarded as partial amides of cyanuric acid. They are acid and have no use in resin production. They are very undesirable by-products of the manufacture of melamine because of their catalytic effect in the subsequent MF resin production, due to their acidic nature. If present, both must be removed from crude melamine by an alkali wash and/or crystallization of the crude melamine.

B. Mechanisms and Kinetics

The mechanism of the initial stages of the reaction of melamine with formaldehyde, leading to the formation of methylol melamines, is very similar to that of urea. The reaction mechanism of the acid-catalyzed condensation reactions of methylol melamines to form polymers and resins has been elucidated by Sato and Naito [4]. Melamine and formaldehyde react similarly to urea and formaldehyde, although basic differences are evident in the reaction rates and mechanism. The primary products of reaction are methylolmelamines, and evidence indicate that such compounds are formed only at ambient or higher temperature, except in acid pH ranges. The reaction is reversible throughout the pH range. Its forward rate is proportional to either [melamine][HCHO] or [melamine][H^+CHOH] or [melamine$^+$][HCHO], according to the pH used.

Methylolmelamine forms "dimers" by condensation with melamine under neutral and acid conditions (70°C). This process is irreversible. The initial hydroxymethylation is very rapid. Its rate is determined by the condensation of conjugated acids of methylolmelamines with melamine. The reaction rate is proportional to [melamine]^2[HCHO] [5]. When the [mineral acid]/[melamine] ratio is 0.0 to 1.0, the early-stage hydroxymethylation of melamine is dependent on the concentration of the melamine molecule (base species) MH and its conjugated acid MH_2^+ in the following manner [6]:

$$\text{rate} = k_{H_2O}[MH][HCHO] + k_H[MH_2^+][HCHO] + k_{MH_2^+}[MH_2^+][MH][HCHO] + k_{MH}[MH]^2[HCHO]$$

In the absence of added acid, when the ratio [mineral acid]/[melamine] is $= 0$, the rate of the reaction can thus be represented as

$$\text{rate} = k_{H_2O}[MH][HCHO] + k_{MH}[MH]^2[HCHO]$$

The condensation reaction has been studied by investigating the kinetics of the initial state of the condensation of di- and trimethylolmelamine (MF_2 and MF_3) in the pH range 1 to 9. Regardless of pH, the initial rate is equal to [4].

$$\text{rate} = k[MF_n]^2 \text{ (with } n = 2 \text{ or } 3)$$

In the presence of mineral acid, the main reaction at the early stage of the condensation is the reaction between the methylolmelamine molecule and its conjugated acid (MF_nH^+) [7]. This was found at an [acid]/[MF_n] ($n = 2$ or 3) ratio lower than 1.0 (pH 2.7). With an [acid]/[MF_n] ratio higher than 1.0 to 1.2 (pH < 2), the main condensation takes place between the conjugated acids themselves.

At equal pH values the condensation rate of trimethylolmelamine is considerably faster than that of dimethylolmelamine. This is the opposite of the rates of mono- and dimethylolurea. This means that while the nitrogen of the amino group in the case of urea is more reactive and therefore more nucleophilic than the nitrogen of the aminomethylol group, the opposite is true in the case of melamine. The reaction for MF_2 is primarily between the carbon of the methylol group next to the nitrogen in HM^+CH_2OH, and the nitrogen of the amino group in MCH_2OH. For MF_3, the condensation is mainly between the carbon of the methylol group next to the charged nitrogen in H^+MCH_2OH, and the nitrogen of the aminomethylol group in MCH_2OH [4]. The condensation rate therefore increases with the increasing electrophilicity of the carbon of the methylol group and the increasing nucleophilicity of the nitrogen of the amino group or aminomethylol group. Therefore, in MF_3 the carbon in HM^+CH_2OH is more electrophilic than the same carbon in MF_2. On the other hand, the nitrogen of the aminomethylol group in HM^+CH_2OH of MF_3 is less nucleophilic, and therefore less reactive, than the nitrogen of the amino group of MF_2. The effects of the carbon and nitrogen atoms are consequently opposite to each other in the MF_n condensation. Since the effect of the carbon is greater than the effect of the nitrogen on the reaction rate, MF_3 condenses faster than MF_2. At lower pH values the effect of the nitrogen becomes negligible and MF_3 is even faster than MF_2 in condensing to polymers.

The difference between the kinetic behavior of urea and melamine can be ascribed to the different effect of the nitrogen atom in the two compounds. With regard to the formation of methylol compounds as a result of hydroxymethylation, the functionality of melamine has been observed to be 6 against formaldehyde [29]. Similarly, melamine reacts easily with formaldehyde to form MF_3. It also forms MF_6 in concentrated formaldehyde [26,30]. For example, urea readily forms dimethylolurea, but forms trimethylolurea with marked difficulty [29,31] and never forms tetramethylolurea. These results suggest that the nitrogen of the aminomethylol group in methylolurea is considerably less nucleophilic than the nitrogen of the amino group in urea. However, the nitrogen of the aminomethylol group in methylolmelamine is not markedly less nucleophilic than the nitrogen of the amino group in melamine. Presumably, this is due to the difference in basicity between urea and melamine. The same is also true of their condensation reactions.

C. Mixed Melamine Resins

With regard to melamine–urea–formaldehyde, copolymers can be prepared which are generally used to cheapen the cost of MF resins, but which also show some worsening of properties. Copolymerization was proven by means of model compounds and polycondensates [8]. MUF resins obtained by copolymerization during the resin preparation stage are superior in performance to MUF resins prepared by mixing preformed UF and MF resins, especially because processing of such mixtures is quite difficult [9]. The relative mass proportions of melamine to urea used in these MUF resins is generally in the melamine/urea range 50:50 to 40:60. [24]. Melamine–phenol–formaldehyde resins, which in some respects show better properties than those of their corresponding MF and PF resins, have also been prepared [10–12]. Analysis of the molecular structure of those resins in both their uncured and cured states showed that no co-condensates of phenol and melamine form and that two separate resins coexist. This is due to the difference in reactivity of the phenolic and melamine methylol groups as a function of pH. Also, in their cured state an interpenetrating network of the separate PF and MF resins, as a polymer blend, is formed, not a copolymer of the two [13,16].

IV. RESIN PREPARATION, GLUE MIXING, AND HARDENING

Preparation of MF resins is not unduly difficult. The industrial equipment is standard: a stainless steel- or glass-lined reactor equipped with suitable cooling and heating coils or jacket, good mechanical stirring, reflux, and a vacuum distillation condenser for dewatering the reaction mixture toward the end of resin manufacture. Formulations that do not need dewatering are also used commercially. The molar ratio of melamine to formaldehyde is generally 1:2 to 3. Lower formaldehyde ratios can be used, depending on the application for which the resin is needed and on the system of manufacture.

Due to their characteristic rigidity and brittleness in their cured state, when MF resins are used for impregnated paper overlays, small amounts of modifying compounds, typically 3 to 5%, are often copolymerized with the MF resin during its preparation to give better flexibility to the finished product. Most commonly used are acetoguanamine, ϵ-caprolactame, and p-toluenesulfonamide.

acetoguonamine   E-coprolactame   p-toluene sulphonamide

The effect of these is to decrease cross-linking density in the cured resin due to the lower number of amidic or aminic groups in their molecules. Thus in resin segments where they are included, only linear segments are possible, decreasing the rigidity and brittleness of the resin. Acetoguanamine is most used for modification of resins for high-pressure paper laminates, while caprolactame, which in water is subject to the following equilibrium,

is used primarily for low-pressure overlays for particleboard. Small amounts of noncopolymerized plasticizers such as diethylene glycol can also be used for the same purpose. Due to the peculiar structure of the wood product itself, MF adhesives for particleboard generally do not need the addition of these modifiers. Often, a small amount of dimethylformamide, a good solvent for melamine, is added at the beginning of the reaction to ensure that all the melamine is dissolved and is available for reaction. Sugar is often added to lessen cost of the resin. The aldehyde group of sugars have been proven to be able to condense with the amide groups of melamine and hence to copolymerize in the resin. Their quantity in MF resins must be limited to very low percentages, and if possible, sugars should not be used at all, as with aging they tend to cause yellowing, crazing, and cracking of cured MF paper laminates and to have a bad effect on adhesive long-term water resistance in both plywood and particleboard.

MF adhesive resins for plywood and particleboard must be prepared to quite different characteristics than those for paper impregnation. The latter must have lower viscosity but still high resin solids content because they need to penetrate the paper substrate to a high resin load, to be dried without losing adhesive capability, and only later to be able to bond strongly to a substrate. Instead, MF adhesive resins for plywood and particleboard are generally more condensed, to obtain lower penetrability of the wood substrate (otherwise, some of the adhesive is lost by overpenetration into the substrate). The reverse applies for paper substrates, where the contrasting characteristics desired—good paper penetration and fast curing—can be obtained in several ways during resin preparation. These characteristics can be achieved by producing, for example, a resin with a lower degree of condensation and high-methylol-group content. Typically, a MF resin of a lower level of condensation with a melamine/formaldehyde molar ratio of 1:1.8 to 2 will give the desired characteristics. Its high methylol content and somewhat lower degree of polymerization will give low viscosity at a high resin solids content, favoring rapid wetting and impregnation of the paper substrate, while the high proportion of methylol groups will give it fast cross-linking and curing capabilities.

A second, equally successful approach is to produce a MF resin of lower methylol group content and higher degree of condensation to which a small second addition of melamine (typically, 3 to 5% total melamine) is effected toward the end of resin preparation. The shift to lower viscosity and higher solids content given by a second addition of melamine, shifting to lower values the average of the resin molecular mass distribution, yields a resin of rapid impregnation characteristics. Conversely, the higher degree of polymerization of the major part of the resin gives fast cross-linking and curing, due to the lower number of reaction steps needed to reach gel point. Typical total M/F molar ratios used in this system are 1:1.5 to 1.7.

Figure 2 shows typical temperature and pH diagrams for the industrial manufacture of MF resins. The important control parameters to take care of during manufacture are the turbidity point (the point during resin preparation at which addition of a drop of MF reaction mixture to a test tube of cold water gives slight turbidity) and the water tolerance or hydrophobicity point, which marks the end of the reaction. The latter is a direct measure of the extent of condensation of the resin and indicates the percentage of water or mass of liquid on the reaction mixture that the MF resin can tolerate before precipitating out. It is typically set for resins of higher formaldehyde/melamine ratios and lower

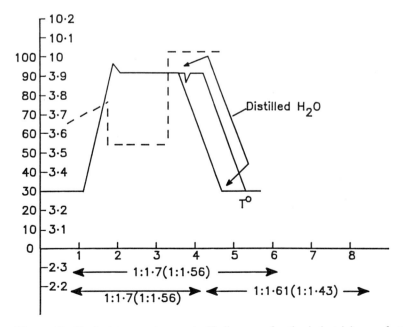

Figure 2 Typical temperature and pH diagrams for the industrial manufacture of MF resins.

condensation levels at around 170 to 190%, but for resins of lower formaldehyde/melamine molar ratios and higher condensation level it is set at around 120%. As can be seen from the diagrams in Fig. 2, once maximum reaction temperature is reached, pH is lowered to 9 to 9.5 to accelerate formation of the polymer. Once the turbidity point is reached, pH is again increased to 9.7 to 10.0, to slow down and more finely control the endpoint, determined by reaching of the wanted value of the water tolerance point. Industrial MF resins are generally manufactured to a 53 to 55% resin solids content with a final pH of 9.9 to 10.4 (but lower pH values are also used for low-condensation resins). To have acceptable rates of curing if higher pH values are used, higher quantities of hardener need to be used, which is clearly uneconomical. For typical MF resins for low pressure (particleboard), self-adhesive overlay pressing times of between 30 and 60 s at 170 to 190°C press temperature are required according to the type of resin used. Pressing conditions for particleboard and plywood adhesives are identical to those used for UF resins.

Glue mixing presents different requirements according to the final use of the MF resin. Hardeners are either acids or materials that will liberate acids on addition to the resin or on heating. In MF and MUF adhesives for bonding particleboard and plywood, the use of small percentages of ammonium salts, such as ammonium chloride or ammonium sulfate, is well established and is indeed identical to standard practice in urea–formaldehyde resins. In MF adhesives for low- and high-pressure self-adhesive overlays and laminates the situation is quite different. Ammonium salts cannot be used for the latter application for two main reasons. First, evolution of ammonia gas during drying and subsequent hot curing of the MF impregnated paper would cause high porosity of the cured MF overlay. Second, the stability of ammonium salts, in particular of ammonium chloride, might cause the MF-impregnated paper to cure and deactivate at ambient temperature after a short time in storage, causing the resin to have lost its adhesive capability by the time it is needed in hot curing. Third, the elimination of ammonia during drying and curing would leave the cured, finished, paper laminate essentially very acid due to the residual acid of the hardener left in the system. This badly affects the resistance to water attack of the cured MF surface defeating the primarily advantage for which such surfaces have justly become so popular. Thus a stable, self-neutralizing, non-gas-releasing hardener is needed for such an application. Several have been prepared. One of the most commonly used is the readily formed complex between morpholine and *p*-toluenesulphonic acid. Morpholine and *p*-toluenesulfonic acid readily react exothermically to form a complex of essentially neutral pH that is stable up to well above 65°C.

During heat curing of the MF paper overlay in the press, the complex decomposes, the MF resin is hardened by the acid that is liberated, morpholine is not vaporized and lost to the system, and on cooling the complex is re-formed, leaving the cured glue line essentially neutral.

In MF glue mixing for overlays and laminates, small amounts of release agents to facilitate release from the hot press of the cured bonded overlay are added. Small amounts of defoamers and wetting agents to further facilitate wetting and penetration of the resin in the paper are always added. A typical glue mix is shown in Table 1.

Table 1 Typical Paper Impregnation Glue Mix for Self-Adhesive Low-Pressure MF Overlays

Ingredient	Parts by mass
MF resin, 53% solids content	99.1
Release agent	0.08
Wetting agent	0.16
Hardener (morpholine/p-toluenesulfonic acid complex)	0.64
Defoamer	0.02

V. CHEMICAL AND PHYSICAL ANALYSIS

The analysis of these resins is difficult when unknown products, particularly fully cured articles, have to be tested for UF and MF resins. Widmer [17] offers a method for the identification of UF and MF resins in technical products. This involves preparing crystalline products of urea and melamine and identifying them under the microscope. Melamine (in the form of melamine crystals) and urea (in the form of long, crystalline needles of urea dixanthate) can be seen. This method allows one to distinguish between urea and melamine even in a cured adhesive joint.

Quantitative determination of MF resins is also rather difficult. A method was developed by Widmer [17] for the quantitative determination of melamine in MF condensation products. In this method the resins are destroyed under pressure by aminolysis, leaving the melamine intact. This is then converted to melamine picrate, which is easily crystallized and weighed. The Widmer method makes it possible to determine quantitatively the presence of urea and melamine in intermediate condensation products and in cured UF and MF resins (even when they have been mixed). Estimations are seldom in error by more than a few percent.

Hirt et al [18] have published an effective and rapid method for the detection of melamine by ultraviolet spectrophotometry. This method can be used for products containing MF. It makes use of the strong absorption of the melamine ion at 235 nm. The resin is extracted from comminuted melamine–formaldehyde samples by hydrolyzing it to melamine by boiling it under reflux in 0.1 N hydrochloric acid. Stafford [19] also gives a method for the identification of melamine in wet-strength paper.

Uncured MF resins analysis are carried out by GPC and ^{13}C NMR. GPC is an inconvenient method for MF resins. DMF, DMSO, or salts solutions are generally used as solvents with a differential refractometer as detector. Derivatives of the resin, as those obtained by sylilation, are generally used to decrease molecular association by hydrogen bonding. ^{13}C NMR is a more convenient technique, and the chemical shifts of the different structural groups in the resin can easily and readily be identified [20,21]. This method is also quite convenient in comparing MF resin structures obtained by different manufacturing methodology.

MF resins produce high-quality plywood and particleboard because their adhesive joints are boilproof [22.23]. Considerable discussion has occurred and many investigations have been carried out on the weather resistance of MF adhesives. Many authors uphold the good weather resistance of the more recently developed MF adhesives, especially those in which small amounts of phenol have been incorporated. The more

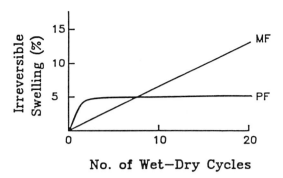

Figure 3 Irreversible thickness swelling characteristics of particleboard bonded with MF and PF adhesives during a wet–dry cycle.

general trend, however, is to consider the wood products manufactured with these resins as capable of resistance to limited weather and water exposure only, such as in flooring applications, rather than being capable of true exterior-grade weather resistance, for which phenolic adhesives are preferred. In Fig. 3 is shown the different behavior of MF-bonded and PF-bonded particleboard to a series of wet–dry cycles. Whereas PF-bonded boards initially deteriorate rapidly, then stabilize to a constant swelling value, MF-bonded particleboards have slower initial deterioration but never stabilize and continue to deteriorate with time and additional wet–dry cycles. This indicates that MF-bonded wood products are not completely impervious to further water attack, indicating the fundamental susceptibility of the aminoplastic bond to water. The rate of deterioration, and therefore bond hydrolysis, is faster as temperature rises. Considering the insolubility of melamine in cold water, this is quite understandable.

VI. FORMULATIONS

For starting to experiment in MF resins, the following two formulations are suggested.

A. Formulation for Exterior Particleboard

In a reaction vessel charge at 25°C, 44.4 parts by mass of water, add 30% NaOH solution to pH 11.2 to 12.0, followed by 15.5 parts of 91% paraformaldehyde prills, 34.8 parts of melamine powder, 2.8 parts of caprolactame, and 2.5 parts of N,N'-dimethylformamide while maintaining the temperature at 25°C; heat in $\pm$ 40 min. to 92 to 95°C. When the temperature reaches 80°C, adjust the pH with 30% NaOH solution, if necessary, to pH 9.9. At 93°C, cool to 90°C and maintain temperature there. Adjust pH to 9.55 to 9.65 with formic acid. Hold the pH at this value while checking, adjusting, and recording the pH value every $\pm$ 10 min. Check the turbidity point at 10-min intervals until the turbidity point is reached. At this time bring pH up to 9.95 to 10.05. Check, adjust, and record the pH every 10 min. Start distilling water under vacuum to a solid of $\pm$ 53 to 55%. Check the water tolerance at 10-min intervals until it is 170 to 180%. Then apply full vacuum and cool the resin to 30 to 35°C.

B. Formulation for Low-Pressure MF Paper-Impregnated Overlays

Follow the same procedure as for formulation A, but at the end of the water vacuum distillation add 1.7 to 1.9 parts by mass of the second melamine and heat the reaction mixture to 95°C again and maintain this temperature for 5 to 6 min. Then cool rapidly.

REFERENCES

1. A. Pizzi, in *Wood Adhesives Chemistry and Technology,* Vol. 1 (A. Pizzi, ed.), Marcel Dekker, New York, 1983, Chap. 2.
2. R. Koehler, *Kunststoffe Tech. 11*: 1 (1941); *Kolloid Z. 103*: 138 (1943).
3. R. Frey, *Helv. Chim. Acta 18*: 491 (1935).
4. K. Sato and T. Naito, *Oikym. J. (Japan) 5*(2): 144 (1973).
5. M. Akano and Y. Ogata, *J. Am. Chem. Soc. 74*: 5728 (1952).
6. K. Sato and S. Ouchi, *Polymer J. (Japan) 10*(1): 1 (1978).
7. A. Takahaski, *Chem. High Polymer (Japan) 7*: 115 (1950); *Chem. Abstr. 46*: 438 (1952); *Chem. High Polymer (Japan) 9:* 15 (1952); *Chem. Abstr. 48*: 1730 (1954).
8. D. Braun and H.-J. Ritzert, *Angew. Makromol. Chem. 156*: 1 (1988); *135*: 193 (1985).
9. D. Braun and H.-J. Ritzert, *Kunststoffe 77*: 1264 (1987).
10. A. Bachmann and T. Bertz, *Aminoplaste*, VEB Verlag fur Grundstoffindustrie, Liepzig, 1967, p. 81.
11. A. Knop and W. Scheib, 1979, *Chemistry and Application of Phenolic Resins*, Springer-Verlag, Berlin, 1979, p. 134.
12. K. Bruncken, in *Kunststoffhandbuch*, Vol. 10 (R. Vieweg and E. Becker, eds.), Hanser, Munich, 1968 p. 352.
13. D. Braun and W. Krausse, *Angew. Makromol. Chem. 108*: 141 (1982).
14. D. Braun and W. Krausse, *Angew. Makromol. Chem. 118*: 165 (1983).
15. D. Braun and H.-J. Ritzert, *Angew. Makromol. Chem. 125*: 9 (1984).
16. D. Braun and H.-J. Ritzert, *Angew. Makromol. Chem. 125*: 27 (1984).
17. G. Widmer, *Paint Oil Chem. Rev. 112*: 18, 26, 28, 30, 32–34 (1949); *Kunststoffe 46*(8): 359 (1956).
18. R. C. Hirt, F. T. King, and R. G. Schmitt, *Anal. Chem. 26*(8): 1273 (1954).
19. R. W. Stafford, *Paper Trade J. 120*: 51 (1945).
20. H. Schindlebauer and J. Anderer, *Angew. Makromol. Chem. 79*: 157 (1979).
21. B. Tomita and H. Ono, *J. Polymer Sci. Chem. Ed. 17*: 3205 (1979).
22. Synthetic resins adhesives for wood, *British Standard B5 1204*, 1965.
23. J. M. Dinwoodie, in *Wood Adhesives Chemistry and Technology*, Vol. 1 (A. Pizzi, ed.), Marcel Dekker, New York, 1983, Chap. 1.
24. W. Clad and C. Schmidt-Hellerau, *Proc. 11th International Particleboard Symposium*, Washington State University, Pullman, Wash., 1977, pp. 33–61.

24
Polyurethane Adhesives

Dennis G. Lay and Paul Cranley *The Dow Chemical Company,*
Freeport, Texas

I. INTRODUCTION

The development of polyurethane adhesives can be traced back more than 50 years to the
pioneering efforts of Otto Bayer and co-workers. Bayer extended the chemistry of
polyurethanes initiated in 1937[1] into the realm of adhesives about 1940[2] by combining
polyester polyols with di- and polyisocyanates. He found that these products made
excellent adhesives for bonding elastomers to fibers and metals. Early commercial
applications included life rafts, vests, airplanes, tires, and tanks [3]. These early developments were soon eclipsed by a multitude of new applications, new technologies, and
patents at an exponential rate.

The uses of polyurethane adhesives have expanded to include bonding of numerous
substrates, such as glass, wood, plastics, and ceramics. Urethane prepolymers were first
used in the early 1950s [4] to bond leather, wood, fabric, and rubber composites. A few
years [5] later one of the first two-component urethane adhesives was disclosed for use as
a metal-to-metal adhesive. In 1957 [6] the first thermoplastic polyurethane used as a hot-
melt adhesive (adhesive strips) was patented for the use of bonding sheet metal containers.
This technology was based on linear, hydroxy-terminated polyesters and diisocyanates.
Additional thermoplastic polyurethane adhesives began appearing in the 1958–1959
period [7,8]. During this period the first metal-to-plastic urethane adhesives were developed [9]. Waterborne polyurethanes were also being developed, with a polyurethane latex
claimed to be useful as an adhesive disclosed in 1961 by du Pont [10]. A commercial
urethane latex was available by 1963 (Wyandotte Chemicals Corporation) [11]. The
adhesive properties of urethane latexes were explored further by W.R. Grace in 1965
[12]. In the early 1960s, B.F. Goodrich developed thermoplastic polyester polyurethanes
that could be used to bond leather and vinyl [13]. In 1968 Goodyear introduced the first
structural adhesive for fiberglass reinforced plastic (FRP), used for truck hoods [14].

Polyurethane pressure-sensitive adhesives began appearing in the early 1970s [15].
By 1978 advanced two-component automotive structural adhesives (Goodyear) were
commercially available. Waterborne polyurethane adhesives received additional attention
during this period [16]. In 1984, Bostik developed reactive hot-melt adhesives [17].
Polyurethane adhesives are sold into an ever-widening array of markets and products,

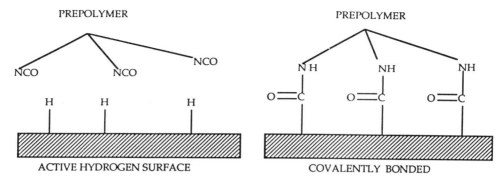

Figure 1 Typical mechanism for a urethane adhesive bonding covalently to a polar surface.

where they are known for their excellent adhesion, flexibility, low-temperature performance, high cohesive strength, and cure speeds that can readily be tailored to the manufacturer's demands [89].

Urethanes make good adhesives for a number of reasons: (1) they effectively wet the surface of most substrates (the energy level of very low energy surfaces such as polyethylene or polypropylene must be raised before good wetting occurs) [18], (2) they readily form hydrogen bonds to the substrate, (3) their small molecular size allows them to permeate porous substrates, and (4) they form covalent bonds with substrates that have active hydrogens. Figure 1 shows the typical mechanism for a urethane adhesive bonding covalently to a polar surface.

Polyurethane adhesive consumption has been estimated at 217 million pounds (1991) having a value of approximately $301 million (see Fig. 2). Applications contributing to this volume are shown in Table 1. It is interesting to note that while the packaging market is the fourth-largest market in terms of pounds of urethane adhesives sold, it is substan-

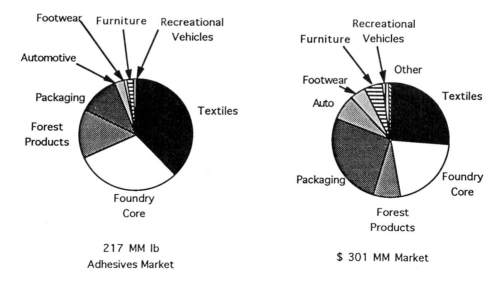

Figure 2 Polyurethane adhesive consumption, 1991.

Table 1 Sales Distribution for Polyurethane
Adhesives, 1991

Market segment	Volume (lb $\times$ 10^6)	Sales ($\times$ 10^6)
Textiles	82.7	$79.6
Foundry core binders	66.0	62.5
Forest products	30.7	22.6
Packaging	25.0	79.7
Automotive[a]	6.2	21.4
Footwear	1.6	13.2
Furniture	3.8	15.4
Recreational vehicles	1.6	5.0
Other	0.35	1.8

[a]Does not include windshield sealant volumes.

tially larger than the forest products market and the foundry core binder market in terms of dollars. Overall, the polyurethane adhesives market market grew at an annual rate of approximately 3% from 1986 to 1991. Specific market segments such as automotive and recreational vehicles easily surpassed the gross national product (GNP) growth rate. In the next few years a number of specific market segments are expected to grow at about 5% per year. These would include vehicle assembly (automotive and recreational vehicles), electronics, furniture, and curtain wall manufacture.

II. APPLICATIONS OVERVIEW

The textile market has traditionally been the largest consumer of polyurethane adhesives. There are a number of high-volume applications, including textile lamination, integral carpet manufacture, and rebonded foam. Textile lamination occurs through either a solution coating process or flame bonding. Flame bonding textile lamination is accomplished by melting a polyurethane foam by flame and then nipping the foam between two textile rolls while it is still tacky. Integral carpet manufacture describes carpeting that is manufactured by attaching either nylon, wool, or polypropylene tufts that are woven through a polypropylene scrim with a urethane adhesive to a polyurethane foam cushion in a continuous process. Rebonded foam is made using scrap polyurethane foam bonded together with a urethane prepolymer and is used primarily as carpet underlay. Durability, flexibility, and fast curing speeds are all critical parameters for these applications.

Foundry core binders are isocyanate-cured alkyd or phenolic adhesives used as binders for sand used to produce foundry sand molds. These sand molds are used to cast iron and steel parts. A fast, economical cure of the sand mold is required under ambient conditions.

Packaging adhesives are adhesives used to laminate film to film, film to foil, and film to paper in a variety of packaging constructions. A broad variety of products are sold to this market, with solvent-based, high solids, 100% solids, and waterborne adhesives all being used. Polyurethane adhesives are considered one of the high-performance products offered to this industry because of their excellence in adhesive properties, heat resistance, chemical resistance, and fast curing properties. Polyurethane adhesives can also be designed to meet U.S. Food and Drug Administration approval, a requirement for food packaging applications.

Solvent-borne adhesives represent the majority of the volume in the packaging market, with both one- and two-component systems being used. Waterborne polyurethane adhesives are a much smaller segment that has been driven by environmental considerations. Growth has slowed in recent years because of generally inferior performance compared to solvent-based adhesives and because most of the major converters have already made capital investments in solvent recovery systems.

Isocyanates are used in the forest products industry to adhesively bond wood chips, which are then pressed to form particleboard and oriented strandboard. Urethanes are also used to fill knotholes and surface defects in finished plywood boards (''plywood patch''). These filled systems must cure rapidly and be sanded easily.

The transportation market has used polyurethane adhesives for such diverse applications as bonding FRP and SMCs (sheet molding composites) panels in truck and car applications, polycarbonate headlamp assemblies, door panels, and weatherstrip flocking.

The construction market for polyurethane adhesives consists of a variety of applications, such as laminating thermal sandwich panels, bonding gypsum board to wood ceiling joists in modular and mobile homes, and gluing plywood floors. Early green strength, low shrinkage, and high bond strength are critical properties.

The furniture industry uses polyurethane adhesives to bond veneers of various composition to boardstock and metal substrates. Both waterborne and solvent-based adhesives are used.

Footwear is a sizable niche for polyurethane adhesives that are used to attach the soles. Polyurethane adhesives compete primarily with neoprene-based adhesives and have replaced much of the neoprene volume due to improved performance. However, the overall market has declined as U.S. manufacturers have moved production overseas.

III. BASIC URETHANE CHEMISTRY

Isocyanates react with active hydrogens as depicted in Fig. 3. This addition reaction occurs with the active hydrogen adding to the nitrogen atom and the electron-rich nucleophile (Lewis base) reacting with the carbonyl group. Generally, the stronger the base, the more readily it reacts with the isocyanate. Table 2 shows typical reaction rates of some active hydrogen-containing compounds.

As expected, the aliphatic amines and aromatic amines (the strongest bases in the table) react the fastest. The urethanes industry has taken advantage of this reactivity in two-component commercial processes, demanding fast cure by using specially designed metering equipment and spray heads.

Alcohols and water react readily at room temperature. Most urethane adhesives depend on the —NCO group reacting with either water or alcohols. Primary OH groups

$$R\text{-}N{=}C{=}O \quad + \quad H\text{-}A \quad \longrightarrow \quad R\text{-}\underset{H}{\overset{}{N}}\text{-}\underset{A}{\overset{}{C}}{=}O$$

Isocyanate   Active hydrogen   Adduct

Figure 3   Reaction of isocyanate with active hydrogen.

Table 2 Typical Reaction Rates for Selected Hydrogen-Containing Compounds

Active hydrogen compound	Typical structure	Relative[a] reaction rate
Aliphatic amine	$R — NH_2$	100,000
Secondary aliphatic amine	$R_2 — NH$	20,000–50,000
Primary aromatic amine	$Ar — NH_2$	200–300
Primary hydroxyl	$R — CH_2OH$	100
Water	$H — O — H$	100
Carboxylic acid	$R — CO_2H$	40
Secondary hydroxyl	$R_2CH — OH$	30
Urea proton	$R — NH — CO — NH — R$	15
Tertiary hydroxyl	$R_3C — OH$	0.5
Urethane proton	$R — NH — CO — OR$	0.3
Amide	$R — CO — NH_2$	0.1

[a]Uncatalyzed reaction rate, 80°C [1].

are two to three times as fast as sterically hindered secondary OH groups under equivalent conditions. The reaction rates shown in Table 2 reflect uncatalyzed reaction rates and should be used as an indication of relative reaction rates. Actual rates are dependent on solvent, temperature, and the presence of catalysts. Catalysts can significantly accelerate these reactions and can in some cases alter the order of reactivity [19].

A. Branching Reactions

There are a number of complex reactions that can occur besides the desired reaction of the polyol hydroxyl group with the isocyanate group to form a urethane, as shown in Fig. 4. Isocyanates can continue to react with undesirable consequences under conditions of high heat or strong bases. Basic impurities and excess heat catalyze branching reactions, leading to variations in prepolymer viscosity, gelation, and exotherms. Most basic impurities arise from the polyol, since polyols are typically produced under basic condition. As such, the net acidity of the overall system (contribution of acidic or basic components from the reactants) plays a critical role in determining the final viscosity achieved [20,21].

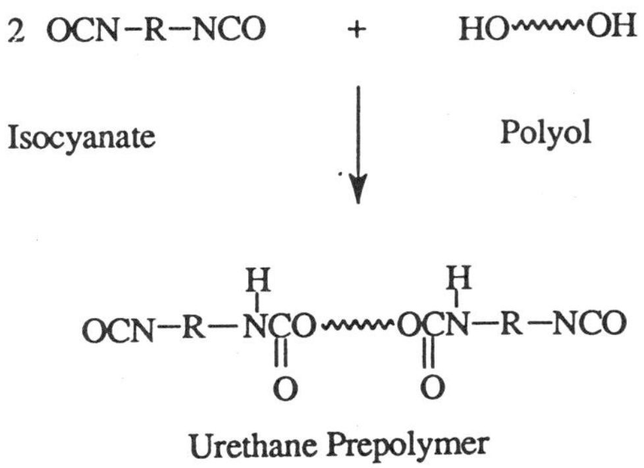

2 OCN–R–NCO   +   HO⌇⌇⌇OH

Isocyanate   Polyol

OCN–R–NCO⌇⌇⌇OCN–R–NCO

Urethane Prepolymer

Figure 4 Reaction of polyol hydroxyl group with isocyanate group to form a urethane.

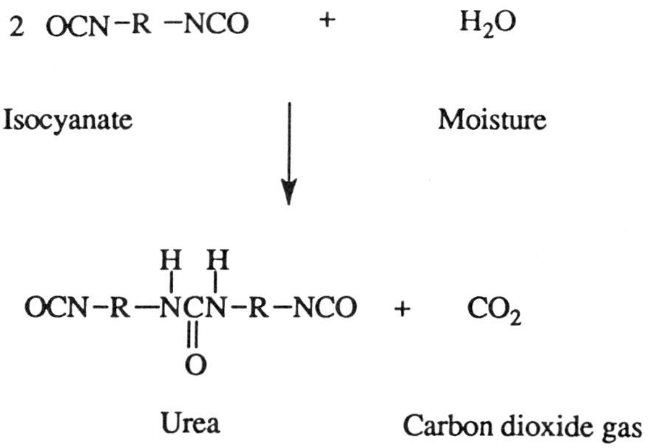

$$2 \text{ OCN}-\text{R}-\text{NCO} \quad + \quad H_2O$$

Isocyanate   Moisture

$$\text{OCN}-\text{R}-\overset{\overset{\displaystyle H}{|}}{N}\underset{\underset{\displaystyle O}{\|}}{C}\overset{\overset{\displaystyle H}{|}}{N}-\text{R}-\text{NCO} \quad + \quad CO_2$$

Urea   Carbon dioxide gas

Figure 5 Reaction of isocyanate with water.

The presence of water will lead to the formation of ureas and evolve CO_2 as shown in Fig. 5. This mechanism is thought to proceed through the formation of an unstable intermediate, carbamic acid, which then decomposes to give CO_2 and an aromatic amine. The amine will then react further with another isocyanate to give a urea linkage. All common moisture-cured urethanes give off CO_2 upon curing, which can pose problems if not properly controlled. Urea groups are known to cause high prepolymer viscosity because of increased hydrogen bonding and because of their ability to react further with excess isocyanate groups to form a biuret, as shown in Fig. 6.

At room temperature the biuret reaction proceeds very slowly; however, elevated temperatures and the presence of trace amounts of basicity will catalyze the biuret reaction as well as other branching reactions. These would include the formation of allophanate groups, as shown in Fig. 7 (due to the reaction of urethane groups with excess isocyanate groups), or trimerization of the terminal NCO group (to form an isocyanurate), as shown

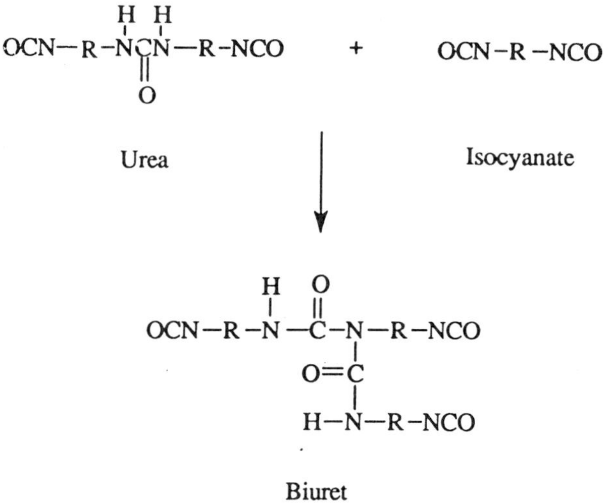

$$\text{OCN}-\text{R}-\overset{\overset{\displaystyle H}{|}}{N}\underset{\underset{\displaystyle O}{\|}}{C}\overset{\overset{\displaystyle H}{|}}{N}-\text{R}-\text{NCO} \quad + \quad \text{OCN}-\text{R}-\text{NCO}$$

Urea   Isocyanate

$$\text{OCN}-\text{R}-\overset{\overset{\displaystyle H}{|}}{N}-\overset{\overset{\displaystyle O}{\|}}{C}-\text{N}-\text{R}-\text{NCO}$$

Biuret

Figure 6 Reaction of urea with isocyanate.

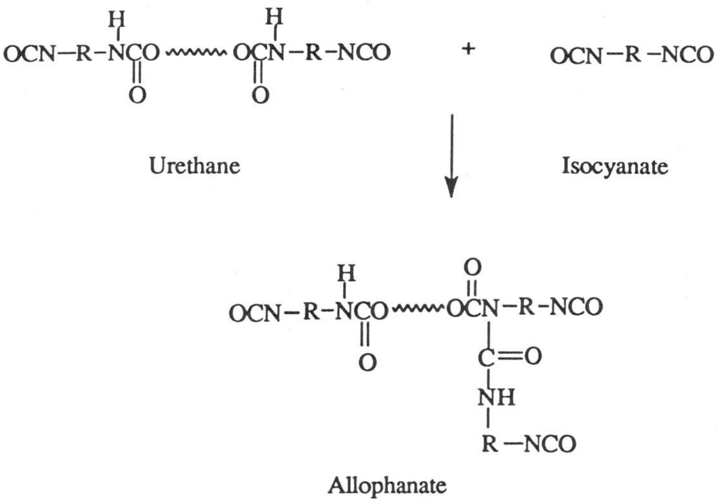

Figure 7 Reaction of urethane with isocyanate.

in Fig. 8. Biurets and allophanates are not as stable thermally or hydrolytically as branch points achieved through multifunctional polyols and isocyanates. The allophanates shown in Fig. 7 can continue to react with excess isocyanates to form isocyanurates (as shown in Fig. 8), a trimerization reaction that will liberate considerable heat. In most cases the desired reaction product is the simple unbranched urethane or a urea formed by direct reaction of an isocyanate with an amine. Ureas are an important class because they typically have better heat resistance, higher strength, and better adhesion. By controlling the reaction temperature (typically less than 80°C) stoichiometry, and using a weakly basic catalyst (or none at all), the reaction will stop at the urethane or urea product. Increasing the functionality of the polyol or the isocyanate will achieve branching or cross-linking in a more controlled fashion.

B. Catalysts

As noted previously, strong or weak bases that are sometimes present in the polyols will catalyze the urethane reaction. The effect of catalysts on the isocyanate reaction is well documented. Indeed, the first reported examples occur in the literature well before urethanes became a commercially significant class of compounds. The first use of a catalyst with an isocyanate was reported by Leukart in 1885 [22]. Other early reports were from French and Wirtel (1925), who used triethylamine to catalyze the reaction of phenols with 1-napthylisocyanate [23]. Baker and Holworth (1947) detailed the mechanism of the urethane reaction [24].

Commercial catalysts consist of two main classes: organometallics and tertiary amines. Both classes have features in common in that the catalytic activity can be described as a combination of electronic and steric effects. Electronic effects arise as the result of the molecule's ability to donate or accept electrons. For example, in the tertiary amines, the stronger the Lewis base, generally the stronger the polyurethane catalyst. Empty electronic orbitals in transition metals allow reactants to coordinate to the metal center, activating bonds and placing the reactants in close proximity to one another.

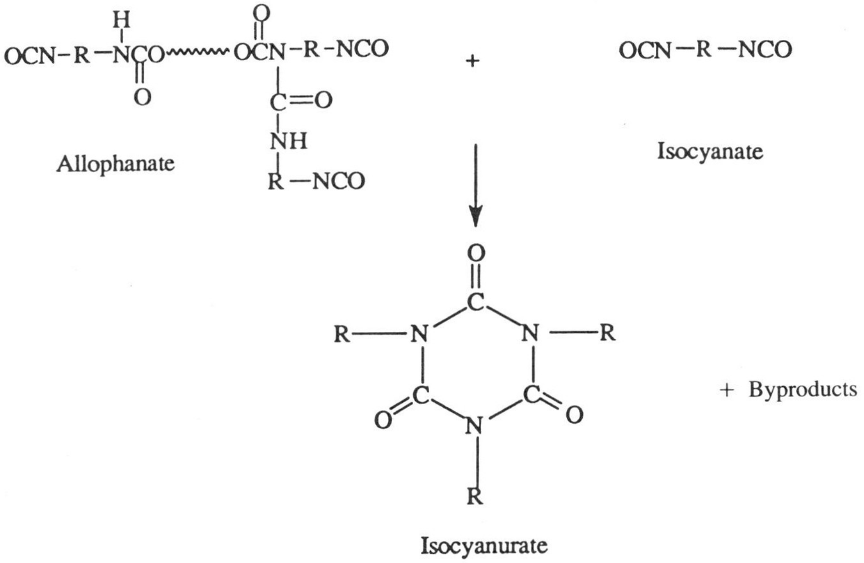

Figure 8 Reaction of allophanate with isocyanate.

Steric effects arise from structural interactions between substituents on the catalyst and the reactants that will influence their interaction. The importance of steric effects can be seen by comparing the activity for triethylenediamine to that of triethylamine. The structure of triethylenediamine (see Fig. 9) forces the nitrogens to direct their lone electron pairs outward in a less shielded position than is true of triethylamine. This results in a rate constant for triethylenediamine that is four times that of triethylamine at 23°C [1].

Organometallic complexes of Sn, Bi, Hg, Zn, Fe, and Co are all potent urethane catalysts, with Sn carboxylates being the most common. Hg catalysts have long induction periods that allow long open times. Hg catalysts also promote the isocyanate–hydroxyl reaction much more strongly than the isocyanate–water reaction. This allows their use in casting applications where pot life and bubble-free parts are critical. Bismuth catalysts are replacing mercury salts in numerous applications as the mercury complexes have come under environmental pressure.

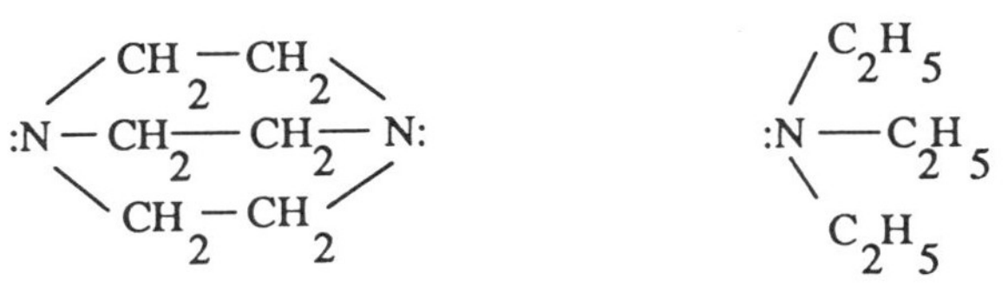

(a)   (b)

Figure 9 Structure of (a) triethylenediamine and (b) triethylamine.

Table 3 Gelation Times (min) at 70°C

Catalyst	TDI	Isocyanate m-xylene diisocyanate	Hexamethylene diisocyanate
None	>240	>240	>240
Triethylamine	120	>240	>240
Triethylenediamine	4	80	>240
Stannous octoate	4	3	4
Dibutyltin di(ethylhexoate)	6	3	3
Bismuth nitrate	1	0.5	0.5
Zinc napthenate	60	6	10
Ferric chloride	6	0.5	0.5
Ferric 2-ethylhexoate	16	5	4
Cobalt 2-ethylhexoate	12	4	4

Source: Ref. 19.

Catalysts will not only accelerate reaction rates but may also change the order of reactivity. Table 3 illustrates this behavior. These data indicate that amines do not affect the relative reactivities of different isocyanates and show that Zn, Fe, and Co complexes actually raise the reactivity of aliphatic isocyanates above aromatic isocyanates.

IV. URETHANE POLYMER MORPHOLOGY

One of the advantages that a formulator has using a polyurethane adhesive is the ability to tailor the adhesive properties to match the substrate. Flexible substrates such as rubber or plastic are obvious matches for polyurethane adhesives because a tough elastomeric product can easily be produced. Polyurethanes derive much of their toughness from their morphology.

Polyurethanes are made up of long polyol chains that are tied together by shorter hard segments formed by the diisocyanate and chain extenders if present. This is depicted schematically in Fig. 10. The polyol chains (typically referred to as soft segments) impart low-temperature flexibility and room-temperature elastomeric properties. Typically, the lower-molecular-weight polyols give the best adhesive properties, with most adhesives being based on products of molecular weight less than 2000. Generally, the higher the soft segment concentration, the lower will be modulus, tensile strength, hardness, and tear strength, while elongation will increase. Varying degrees of chemical resistance and heat resistance can be designed by proper choice of the polyol.

Short-chain diols or diamines are typically used as chain extenders. These molecules allow several diisocyanate molecules to link forming longer-segment hard chains with higher glass transition temperatures. The longer-segment hard chains will aggregate together because of similarities in polarity and hydrogen bonding to form a pseudo-cross-linked network structure. These hard domains affect modulus, hardness, and tear strength and also serve to increase resistance to compression and extension. The hard segments will yield under high shear forces or temperature and in fact determine the upper use

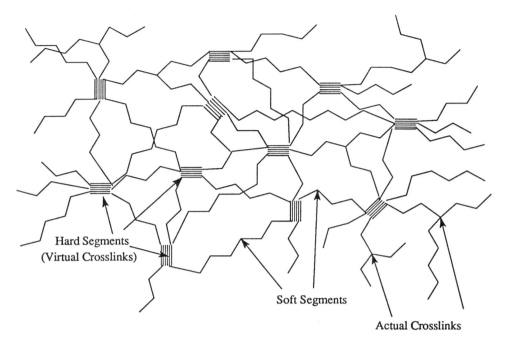

Figure 10   Polyol-chain structure of polyurethane.

temperature of the product. Once the temperature or shear stress is reduced, the domains will re-form.

The presence of both hard segment and soft segment domains for polyurethanes gives rise to several glass transition temperatures, one below $-30°C$ which is usually associated with the soft segment, transitions in the range 80 to 150°C, and transitions above 150°C. Transitions in the range 80 to 150°C are associated with the breakup of urethane hydrogen bonds in either the soft segment or the hard segment. Transitions higher than 150°C are associated with the breakdown of hard segment crystallites or aggregates. Linear polyurethane segmented prepolymers can act as thermoplastic adhesives which are heat activated. A typical use for this type of product is in the footwear industry.

By proper choice of either the ioscyanate or the polyol, actual chemical cross-links can be introduced in either the hard or soft segments that may be beneficial to some properties. The effectiveness of these cross-links is offset by a disruption of the hydrogen bonding between polymer chains. Highly cross-linked polyurethanes are essentially amorphous in character exhibiting high modulus, hardness, and few elastomeric properties. Many adhesives fall into this category.

V.   PREPOLYMER FORMATION

Most urethane adhesives are based on urethane prepolymers. A prepolymer is made by reacting an excess of diisocyanate with a polyol to yield as isocyanate-terminated urethane as shown in Fig. 11. Prepolymers may have excess isocyanate present (''quasi-prepolymers'') or they may be made in a 2:1 stoichiometric ratio to minimize the amount of free isocyanate monomer present. Most moisture-cured prepolymers are based on 2:1

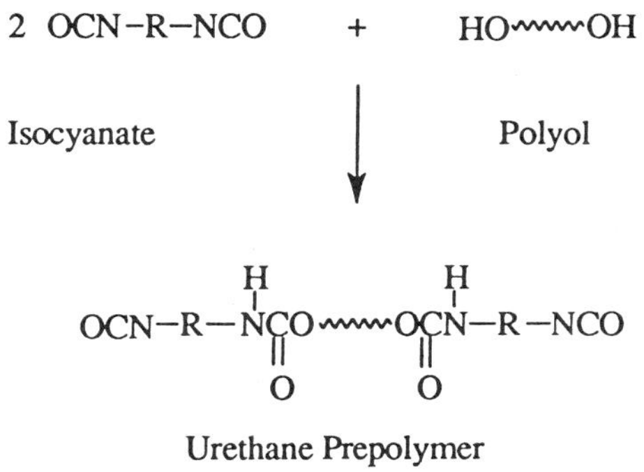

2 OCN−R−NCO   +   HO〰〰OH

Isocyanate   Polyol

H   H
| |
OCN−R−NCO〰〰OCN−R−NCO
‖   ‖
O   O

Urethane Prepolymer

Figure 11 Reaction of isocyanate with polyol.

stoichiometric ratios. Two-component adhesives generally are based on quasi-prepolymers, which use the excess isocyanate to react with either chain extenders present in the other component or with the substrate surface.

Prepolymers are ioscyanates and react like isocyanates, with several important differences. Prepolymers typically are much higher in molecular weight, higher in viscosity, are lower in isocyanate content by weight percent, and have lower vapor pressures. Prepolymers are important to adhesives for a number of reasons. The desired polymeric structure of the adhesive can be built into the prepolymer, giving a more consistent structure with more reproducible physicals. In addition, since part of the reaction has been completed, reduced exotherms and reduced shrinkage are normally present. For two-component systems, better mixing of components usually occurs, since the viscosity of the two components more closely match. In addition, the ratios of the two components match more closely. Side reactions such as allophanate, biuret, and trimer are lessened. Finally, prepolymers typically react more slowly than does the original diisocyanate, allowing longer pot lifes.

VI. ADHESIVE RAW MATERIALS

Polyols for adhesive applications can be generally broken into three main categories: (1) polyether polyols, (2) polyester polyols, and (3) and polyols based on polybutadiene. Polyether polyols are the most widely used polyols in urethane adhesives because of their combination of performance and economics. They are typically made from the ring-opening polymerization of ethylene, propylene, and butylene oxides, with active proton initiators in the presence of a strong base as shown in Fig. 12.

Polyether polyols are available in a variety of functionalities, molecular weights, and hydrophobicity, depending on the initiator, the amount of oxide fed, and the type of oxide. Capped products are commercially available as well as mixed-oxide feed polyols, as shown in Fig. 13. Polyether polyols typically have glass transitions in the $-60°C$ range, reflecting the ease of rotation about the backbone and little chain interaction. As one would expect from such low glass transition temperatures, they impart very good low-temperature performance. The polyether backbone is resistant to alkaline hydrolysis, which make them useful for adhesives used on alkaline substrates such as concrete. They are typically very low in viscosity and exhibit excellent substrate wetting. In addition, their low cost and ready availability from a number of suppliers add to their attractiveness.

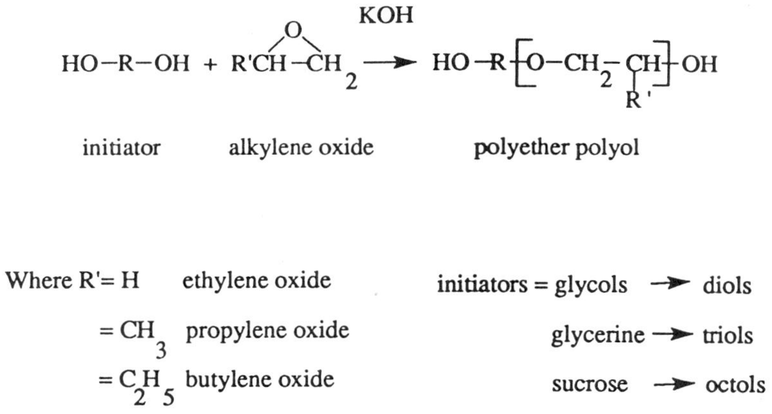

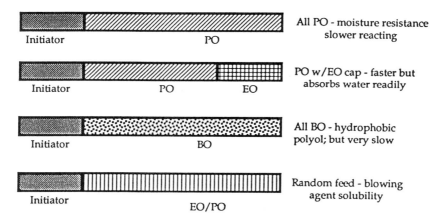

KOH
HO–R–OH + R'CH–CH2 → HO–R[O–CH2–CH]–OH
R'
initiator
alkylene oxide
polyether polyol
Where R'= H ethylene oxide
= CH3 propylene oxide
= C2H5 butylene oxide
initiators = glycols → diols
glycerine → triols
sucrose → octols

Figure 12 Ring-opening polymerization to form polyether polyols.

Initiator
PO
All PO - moisture resistance slower reacting
Initiator
PO
EO
PO w/EO cap - faster but absorbs water readily
Initiator
BO
All BO - hydrophobic polyol; but very slow
Initiator
EO/PO
Random feed - blowing agent solubility

Figure 13 Various commercially available capped products and mixed-oxide feed polyols.

The more commonly used polyether polyols range in molecular weight from 500 to 2000 for diols and 250 to 3000 for triols. Lower-molecular-weight, higher-functionality polyols are traditionally used in rigid-foam applications but have also been used as cross-linkers for two-component, fast-curing urethane adhesives. Polytetramethylene glycols (PTMOs; see Fig. 14) can be considered a subset of polyether polyols. They offer improved physical properties compared to polyethers based on ethylene oxide, propylene oxide, or butylene oxide, combining high tensile strength (due to stress crystallization) with excellent tear resistance. They are also noted for their excellent resistance to hydrolysis. They are typically priced at a premium to other polyols.

Polyester polyols are used widely in urethane adhesives because of their excellent adhesive and cohesive properties. Compared to polyether-based polyols, polyester-based adhesives have higher tensile strengths and improved heat resistance. These benefits come

$$HO-(CH_2-CH_2-CH_2-CH_2-O)_n-H$$

Figure 14 Structure of polytetramethylene oxide.

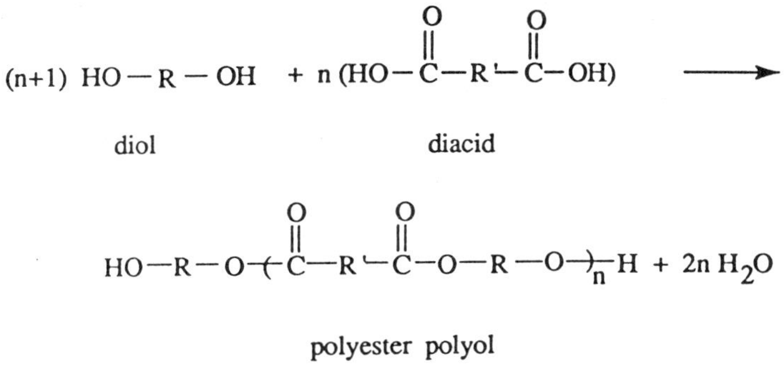

$$(n+1) \; HO-R-OH \; + \; n \; (HO-\overset{O}{\overset{\|}{C}}-R'-\overset{O}{\overset{\|}{C}}-OH) \longrightarrow$$

diol   diacid

$$HO-R-O-(\overset{O}{\overset{\|}{C}}-R'-\overset{O}{\overset{\|}{C}}-O-R-O)_n-H \; + \; 2n \; H_2O$$

polyester polyol

Figure 15 Reaction of diol with diacid to form polyester polyol.

at the sacrifice of hydrolytic resistance, low-temperature performance, and chemical resistance. One of the more important application areas for these products is in the solvent-borne thermoplastic adhesives used in shoe sole binding. These products are typically made from adipic acid and various glycols (see Fig. 15).

Some glycerine or trimethylolpropane may be used to introduce branching structures within the polyester backbone. Phthalic anhydride may also be used to increase hardness and water resistance. Inexpensive terephthalic acid–based polyesters from recycled polyethlyeneterephthalate (PET) resins have more recently become popular.

Polycaprolactones (see Fig. 16), another type of polyester polyol, offer improvements in hydrolysis resistance and in tensile strength (can stress crystallize) over adipic acid–based polyester polyols. They are typically higher in viscosity and higher in cost than polyether polyols of comparable molecular weight. When moisture resistance is critical, urethane adhesives incorporating polybutadiene polyols are used. These products are hydroxy-terminated, liquid polybutadiene resins. The hydrocarbon backbone greatly decreases water absorption, imparting excellent hydrolytic stability. Polybutadiene compounds also have exceptional low-temperature properties, with glass transition temperatures being reported below $-70°C$ [25]. These products are priced at a 40 to 50% premium over comparable polyether polyols. The structure of polybutadiene polyols is shown in Fig. 17.

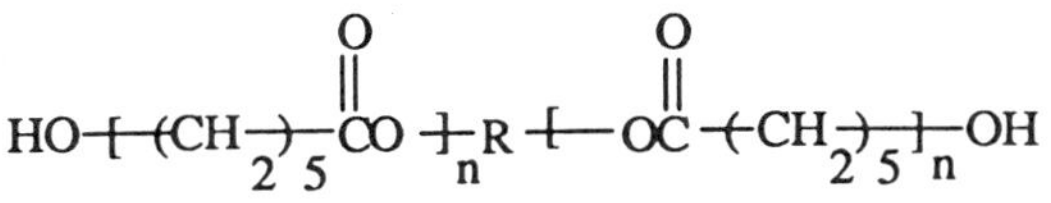

$$HO-[(CH_2)_5-\overset{O}{\overset{\|}{C}}O]_n-R-[-O\overset{O}{\overset{\|}{C}}-(CH_2)_5]_n-OH$$

Figure 16 Structure of polycaprolactone diol.

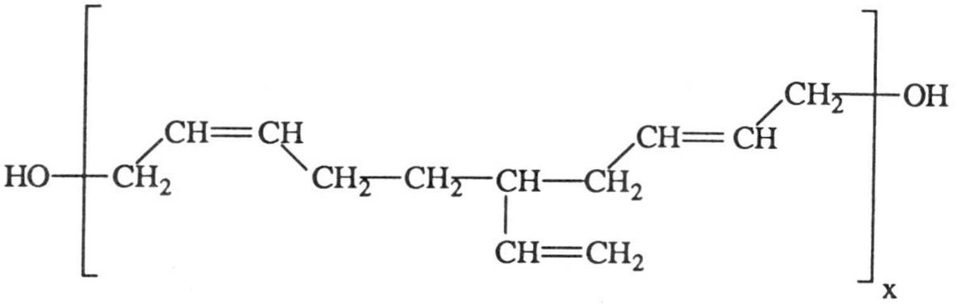

Figure 17 Structure of polybutadiene polyol.

A. Isocyanates for Adhesive Applications

Toluene diisocyanate (TDI) is a colorless, volatile, low-viscosity liquid commonly used in the adhesives area to manufacture low-viscosity prepolymers for flexible substrates. The structure of TDI is shown in Fig. 18. TDI is typically supplied as an 80:20 mixture of the 2,4 and 2,6 isomers, respectively, with two grades of acidity available. Type I TDI is low in acidity (10 to 40 ppm); type II TDI is higher (80 to 120 ppm). Type II TDI is generally used for prepolymer applications because the additional acidity is available to neutralize trace bases found in polyether polyols. These trace bases can cause branching reactions during prepolymer cooks, causing high viscosities and even gelations if not properly controlled (see Section V). The extra acidity present also serves to stabilize the prepolymer, extending the shelf stability. In addition, since TDI is predominately the 2,4 isomer, a reactivity difference is noted for the isocyanate groups. Since the less hindered site reacts first, the sterically hindered site is left when prepolymers are formed, leading to prepolymers that are more shelf stable. TDI prepolymers are used in adhesives for the textile and food packaging laminates industry, where a fit is found for their low viscosity and low cost. The volatility of TDI and additional handling precautions that must be taken when using TDI has limited its growth in adhesive applications.

Methylene diphenyl diisocyanate (MDI) is used where high tensile strength, toughness, and heat resistance are required. MDI is less volatile than TDI, making it less of an inhalation hazard. The acidity levels in MDI are very low, typically on the order of 0 to 10 ppm, so the trace base levels in the polyols is much more critical in prepolymer production than with TDI. The structure of MDI is shown in Fig. 19. There are several commercial

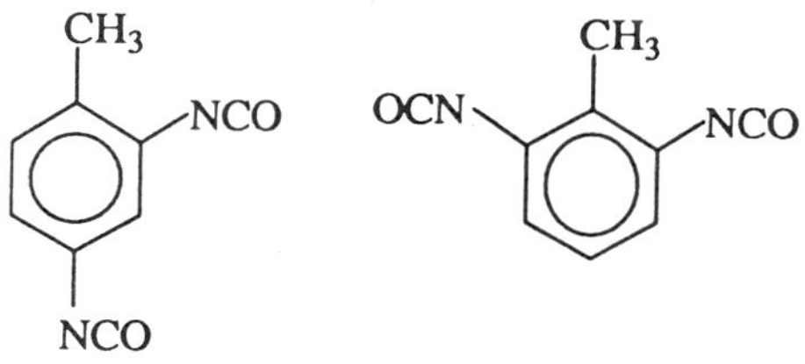

2, 4 TDI   2, 6 TDI

Figure 18 Structure of the 2,4 and 2,6 isomers of toluene diisocyanate.

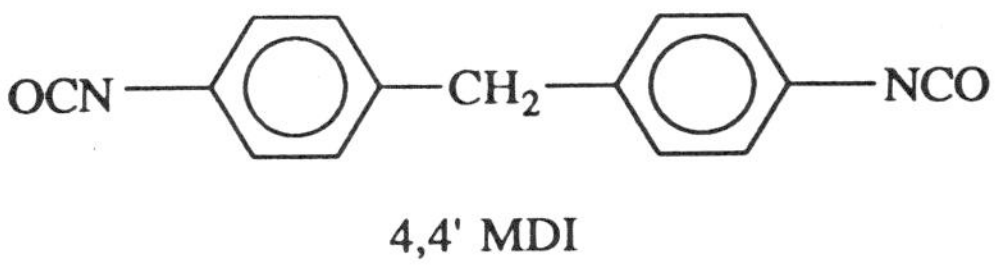

4,4' MDI

Figure 19 Structure of methylene diphenyl diisocyanate.

suppliers of MDI that typically supply grades with 98% or better 4,4' isomer. MDI is a solid at room temperature (melting point 38°C, 100°F), requiring handling procedures different from those for TDI. MDI should be stored as a liquid at 115°F or frozen as a solid at (-20°F) to minimize dimer growth rate. MDI reacts faster than TDI, and because the NCO groups in MDI are equivalent, they have the same reactivity, a contrast to TDI. MDI is used in packaging adhesives, structural adhesives, shoe sole adhesives, and construction adhesives.

Several MDI products have been introduced that address the inconvenience of handling a solid. They are seeing increased usage in the adhesives industry and are expected to experience a higher growth rate. Most MDI producers offer a uretonimine-modified form of MDI that is a liquid at room temperature. The uretonimine structure is shown in Fig. 20. In addition, several producers have introduced MDIs containing elevated levels of the 2,4' isomer, as shown in Fig. 21. At approximately the 35%, 2,4' isomer level, the product becomes a liquid at room temperature, greatly increasing the handling ease. A number of advantages are seen: slower reactivity, longer pot life, lower-viscosity prepolymers, prepolymers with lower residual monomeric MDI, and improved shelf stability.

Polymeric MDIs are made during the manufacturing of monomeric MDI. These products result as higher-molecular-weight oligomers of aniline and formaldehyde get phosgenated. A typical structure for these products is shown in Fig. 22. These oligomers average 2.3 to 3.1 in functionality and contain 30 to 32% NCO. Much of the hydrolyzable chlorides and color bodies produced in the manufacturing process of MDI is left behind in

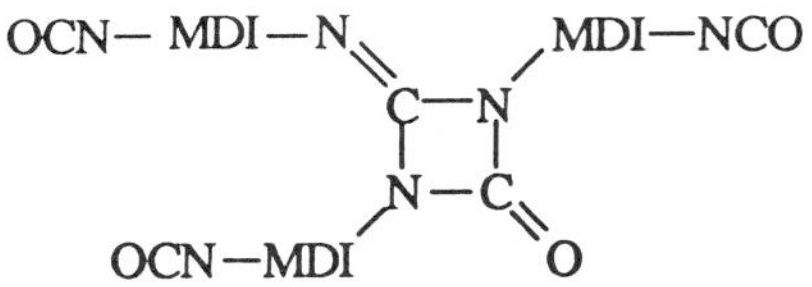

Figure 20 Structure of uretonimine.

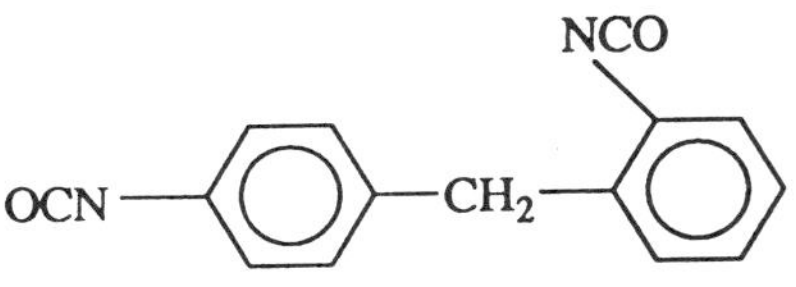

Figure 21 Structure of the 2,4' isomer of MDI.

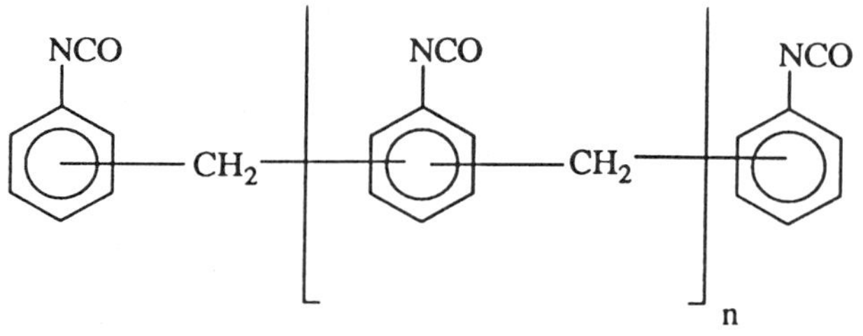

Figure 22   Structure of polymeric MDI.

these products. The acidity levels can be 10 to 50 times the level found in pure MDI, and the products are dark brown in color. The higher acidity level decreases reactivity; however, this decrease is offset somewhat by the higher functionality.

Polymeric MDIs are typically lower in cost than pure MDI and because of the increased asymmetry have a lower freeze point (liquids at room temperature). They are less prone to dimerization, and as a consequence are more storage stable than are pure MDI and derivatives. Polymeric MDIs are used whenever the color of the finished adhesive is not a concern. They are generally not used for prepolymers because high-viscosity branched structures typically result. They are widely used as adhesives in the foundry core binder area, in oriented strandboard or particleboard, and between rubber products and fabric or cord. It is interesting to note that the polymeric isocyanates used commercially today are structurally very similar to the Desmodur R (trademark, Bayer) products used over 40 years ago [2].

Aliphatic isocyanates are used whenever resistance to ultraviolet light is a critical concern. Examples of aliphatic isocyanates are hexamethylene diisocyanate, hydroge-nated MDI, isophorone diisocyanate, and tetramethylxylene diisocyanate. Structures for these molecules are shown in Fig. 23. The aliphatic isocyanates are usually more expensive than aromatic isocyanates and find limited use in adhesive applications. UV resistance is usually not a critical concern in adhesives because the substrate shields the adhesive from sunlight.

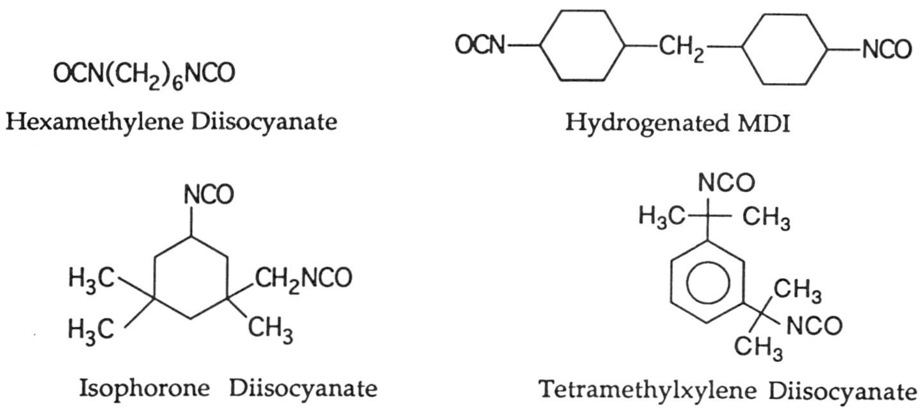

Figure 23   Structure of various aliphatic isocyanates.

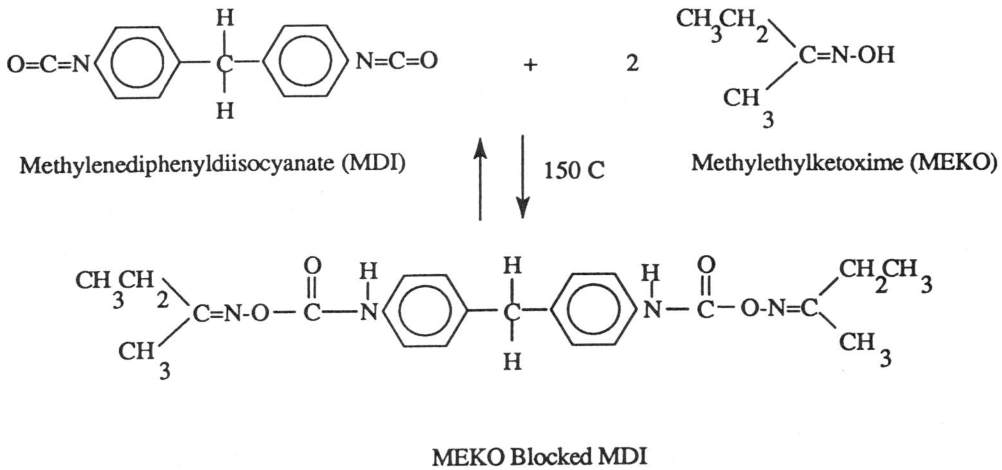

Figure 24 Preparation of a methylethylketoxime blocked isocyanate.

Blocked isocyanates are also used in urethane adhesives. Blocking or "masking" of the isocyanates refers to reacting the isocyanate groups with a material that will prevent the isocyanate from reacting with active hydrogen-containing species at room temperature but will allow that reaction to occur at elevated temperatures. Blocked isocyanates are easily prepared and their chemistry has been developed extensively since their inception by Bayer and co-workers during the early 1940s [26–28]. As an example, the preparation of a methylethylketoxime blocked isocyanate is shown in Fig. 24.

Blocked isocyanates offer a number of advantages to unblocked isocyanates. The traditional concern for moisture sensitivity can be addressed by blocking the isocyanate. Heat activation is then required, but most commercial adhesive applications can meet this requirement. Water-based dispersions and dispersions of the isocyanate in the polyol or other reactive media become possible using blocked isocyanates. There are a number of blocked isocyanates commercially available that could be used in adhesive applications. Miles (Bayer) produces a series of aromatic and aliphatic blocked isocyanates marketed for primers, epoxy flexibilizers, wire coatings, and automotive topcoat applications. Blocked isocyanates are widely patented for fabric laminating adhesives [29], fabric coating adhesives [30–33], and tire cord adhesives [34–39].

B. Toxicology

Polyether polyols are generally considered to be low in toxicity with respect to eye and skin irritation; however, amine-initiated polyether polyols have been found to be more irritating to the skin and eyes. The manufacturer's material safety data sheet should always be consulted before use. Oral toxicity is generally a secondary concern in an industrial environment. The vapor pressure of polyols is generally negligible; thus vapor inhalation is not a usually not a concern [40]. Low-molecular-weight glycols (chain extenders) are considered more problematical than polyether polyols. While generally the vapor pressure of these products is low, there are processes that could potentially result in vapor concentrations close to the exposure limits [40]. The exposure guidelines for chain extenders may be written to differentiate between aerosols and vapors. For more specific handling information the manufacturer should be consulted.

The toxicology of isocyanates is a primary concern when developing or using polyurethane adhesives. Respiratory effects are the primary toxicological manifestation of repeated overexposure to diisocyanates [41–45]. In addition, most of the monomeric isocyanates are eye and skin irritants. Precautions should be taken in the workplace to prevent exposure. The risk of overexposure is primarily (but not limited to) allergic sensitization with asthma-type symptoms. Manufacturers' guidelines (MSDS) should be consulted for the most current information and legal requirements.

C. Fillers and Additives

Fillers are used in adhesives to improve physical properties, to control rheology, and to lower cost. The most common polyurethane filler are calcium carbonate, talc, silica, clay, and carbon black. A more rigorous treatment of this subject can be found in Katz and Milewski [46]. Fumed silicas and carbon blacks are used primarily as thixotropes in application areas that require a nonsagging bead. Calcium carbonates, clays, and talcs are used to improve the economics of an adhesive formulation. A major concern using fillers with urethane prepolymers is the moisture content associated with the fillers. Fillers typically must be dried prior to use with urethane prepolymers or isocyanates. Hygroscopic fillers should be avoided, as moisture introduced by the filler can lead to poor shelf stability of the finished product.

Pigments are sometimes used in polyurethane adhesive systems, but since most adhesives are generally hidden from view, pigments do not play major roles. Pigments may be used to color the adhesive to match the substrate. Pigments are more typically used to color one side of a two-component system to help the user distinguish between the isocyanate and the polyol. They are also sometimes used as an aid to judge mix ratios. Carbon black and titanium dioxide are two commonly used pigments.

Plasticizers can also be used in polyurethane adhesives to lower viscosity, improve filler loadings, improve low-temperature performance, and to plasticize the polyurethane adhesive. Phthalate esters, benzoate esters, phosphates, and aromatic oils are common examples [47]. Plasticizers should be used sparingly, as adhesion will generally decrease as levels increase.

VII. SURFACE PREPARATION AND PRIMERS

Proper surface preparation is the key to obtaining good adhesive bonds having a predictable service life. Substrate surfaces may have dirt, grease, mold-release agents, processing additives, plasticizers, protective oils, oxide scales, and other contaminants that will form a weak boundary layer. When the adhesive fails it is usually through this region, giving a low-strength bond. Some form of surface treatment is necessary to obtain optimum bond strength. The primary goal of surface treatment is to remove any weak surface boundary layer on the substrate [48]. A large number of surface treatments have been developed, with many targeted toward specific substrates. These would include mechanical abrasion, etching, solvent cleaning, detergent washing, flame treatments, chemical treatments, and corona discharges [18,49–54].

Primers are also used in conjunction with a surface treatment to either improve adhesive performance or to increase production flexibility in a bonding operation. Isocyanates have been used for over 50 years as primers on substrates such as rubber, plastic, fibers, and wood [55]. Isocyanates will react with polar groups on the surface and promote bonding.

Silane coupling agents are commonly used as primers for glass, fiber composites, mineral-filled plastics, and cementacious surfaces. The silane coupling agents have been found to be especially effective with glass substrates. One end of the coupling agent is an alkoxysilane that condenses with the silanol groups on the glass surface. The other end of the coupling agent is an amino, mercapto, or epoxy functionality that will react with the isocyanate group in the adhesive. Epoxy silanes have also been used as additives to adhesives to improve water resistance [56]. Other organometallic primers are based on organotitanates, organozirconates, and some chromium complexes [48].

VIII. COMMON ADHESIVE TYPES

A. One-Component Adhesives

The oldest types of one-component polyurethane adhesives were based on di- or triisocyanates that cured by reacting with active hydrogens on the surface of the substrate or moisture present in the air or substrate. The moisture reacts with the isocyanate groups to form urea and biuret linkages, building molecular weight, strength, and adhesive properties. Prepolymers are also used either as 100% solids or solvent-borne one-component adhesives. Moisture-cured adhesives are used today in rebonded foam, tire cord, furniture, and recreational vehicle applications.

A second type of one-component urethane adhesive comprises hydroxypolyurethane polymers based on the reaction products of MDI with linear polyester polyols and chain extenders. There are several commercial suppliers of these types of thermoplastic polyurethanes. The polymers are produced by maintaining the NCO/OH ratio at slightly less than 1:1 to limit molecular weight build to the range 50,000 to 200,000 with a slight hydroxy content (approximately 0.05 to 0.1%). These are typically formulated in solvents for applications to shoe soles or other substrates. After solvent evaporation heat is used to melt the polymer (typically 50 to 70°C; at these temperatures the polymers reach the soft, rubbery, amorphous state), so the shoe upper can be press fit to the sole. Upon cooling, the adhesive recrystallizes to give a strong, flexible bond [57]. More recently, polyisocyanates have been added to these to increase adhesion and other physical properties upon moisture curing. In Section IX.B we discuss this in more detail.

The use of waterborne polyurethane adhesives has grown in recent years as they have replaced solvent-based adhesives in a number of application areas. There are a number of papers and patents covering the use of waterborne polyurethanes in shoe soles, packaging laminates, textile laminates, and as an adhesive binder for the particleboard industry [58–61]. Because waterborne polyurethane adhesives have no VOC (volatile organic content) emissions and are nonflammable, they are environmentally friendly. Typically they can be blended with other dispersions without problems and exhibit good mechanical strength. Water-based systems are fully reacted, linear polymers that are emulsified or dispersed in water. This is accomplished by building hydrophilicity into the polymer backbone with either cationic or anionic groups or long hydrophilic polyol segments or, less frequently, through the use of external emulsifiers. Figure 25 illustrates the more common functional groups that can be built into the urethane molecule that will confer hydrophilicity.

A typical example of how these groups are built into the polymer backbone is shown in Fig. 26. A urethane prepolymer is reacted with chain extenders containing either carboxylates or sulfonates in a water-miscible solvent (e.g., acetone). The reaction

ANIONIC GROUP

$$COO^- \ R_3NH^+$$

NONIONIC GROUP

$$(CH_2CH_2O)\text{-}R_n$$

CATIONIC GROUP

$$-N^{\pm}-$$
H A⁻ ... R

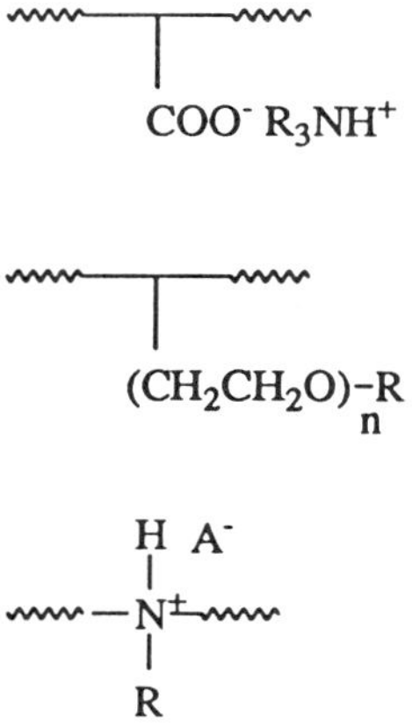

Figure 25 Common functional groups that confer hydrophilicity in the urethane molecule.

product is an isocyanate-terminated polyurethane or polyurea with pendant carboxylate or sulfonate groups. These groups can easily be converted to salts, which as water is added to the prepolymer–solvent solution, allows the prepolymer to be dispersed in water. The solvent is then stripped, leaving the dispersed product. There are variations on this theme that allow lower solvent volumes to be used [62]. Long hydrophilic polyol segments can also be introduced. Chain extenders with hydrophilic ethylene oxide groups pendant to the backbone are reacted with the prepolymer to form a nonionic self-emulsifying polyurethane. This reaction is also carried out in a water-miscible solvent that can later be stripped from the solvent–water solution.

Blocked isocyanates can also be considered a one-component adhesive. The use of a blocking agent allows the isocyanate to be used in a reactive medium that can be heat activated. One-component adhesives based on blocked isocyanates are thus not amenable to room-temperature curing applications. The chemistry of these products is covered in more detail in Section VI.A.

B. Two-Component Adhesives

The second major classification of common polyurethane adhesives is the two-component system. Two-component polyurethane adhesives are widely used where fast cure speeds are critical, as on OEM (original equipment manufacturers) assembly lines that require quick fixture of parts, especially at ambient or low bake temperatures. Two-component urethanes are required in laminating applications where no substrate moisture is available or where moisture cannot penetrate through to the adhesive bond. Two-component urethanes are also useful where CO_2 (generated by a one-component moisture cure) or a volatile blocking agent would interfere with the adhesive properties.

Two-component adhesives typically consist of a low-equivalent-weight isocyanate or prepolymer that is cured with a low-equivalent-weight polyol or polyamine. They may be 100% solids or solvent borne. Since the two components will cure rapidly when mixed, they must be kept separate until just before application. Application is followed quickly by mating of the two substrates to be bonded.

Efficient mixing of the two components is essential for complete reaction and full development of designed adhesive properties. In-line mixing tubes are adequate for low-

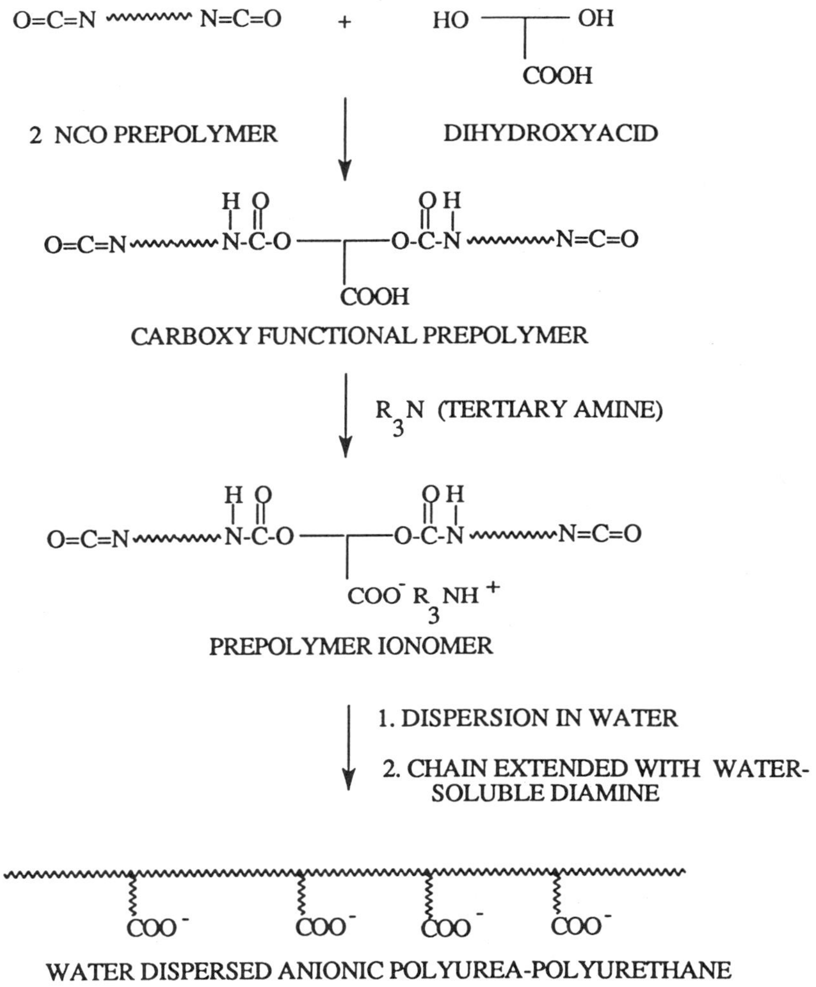

$$O=C=N \,\text{⌇⌇⌇⌇}\, N=C=O \quad + \quad HO-\overset{\displaystyle |}{\underset{\displaystyle COOH}{}}-OH$$

2 NCO PREPOLYMER   DIHYDROXYACID

$$O=C=N\text{⌇⌇⌇⌇}\overset{H}{\underset{}{N}}\overset{O}{\underset{}{-C}}-O-\overset{}{\underset{COOH}{}}-O-\overset{O}{\underset{}{C}}\overset{H}{\underset{}{-N}}\text{⌇⌇⌇⌇}N=C=O$$

CARBOXY FUNCTIONAL PREPOLYMER

R_3N (TERTIARY AMINE)

$$O=C=N\text{⌇⌇⌇⌇}\overset{H}{\underset{}{N}}\overset{O}{\underset{}{-C}}-O-\overset{}{\underset{COO^- R_3NH^+}{}}-O-\overset{O}{\underset{}{C}}\overset{H}{\underset{}{-N}}\text{⌇⌇⌇⌇}N=C=O$$

PREPOLYMER IONOMER

1. DISPERSION IN WATER

2. CHAIN EXTENDED WITH WATER-
   SOLUBLE DIAMINE

$$\text{⌇⌇⌇⌇⌇⌇⌇⌇⌇⌇⌇⌇⌇⌇⌇⌇⌇⌇⌇⌇}$$
$$COO^- \qquad COO^- \qquad COO^- \qquad COO^-$$

WATER DISPERSED ANIONIC POLYUREA-POLYURETHANE

Figure 26 How functional groups are built into the polymer backbone of urethane.

volume adhesive systems. For larger-volume demands, sophisticated meter mix machines are required that will mix both components just prior to application. Commercial systems for delivering two-component adhesives are segmented based on the viscosity ranges of the components. The ranges can be broken down into low, middle, and high viscosity, with, for example, Liquid Control Corp., Sealant Equipment and Engineering Inc., and Graco Inc., respectively, supplying equipment for the three ranges [88].

In present-day high-speed assembly line operations, adhesives are applied robotically. The adhesive bead is applied quickly and evenly to parts on a conveyor line just prior to being fitted. These operations, especially the need to handle the adhered substrates soon after assembly, demand fast-curing adhesive systems [63]. Two-component adhesives are used to bond metals to plastics in automobiles, to laminate panels in the construction industry, to laminate foams to textiles, to laminate plastic films together, and to bond poly(vinylidene chloride) films to wood for furniture. A commercial waterborne two-component adhesive is sold by Ashland under the trademark ISOSET. This system is used for exterior sandwich panels by recreational vehicle manufacturers and is composed of a water-emulsifiable isocyanate and a hydroxy-functionalized emulsion latex.

IX. RECENT DEVELOPMENTS

A. Hybrid Adhesives

Over the last three decades there have been a number of attempts to wed the unique benefits of polyurethane adhesives with the benefits of other adhesive systems. These attempts have led to the reporting of a variety of urethane hybrids. Early work focused on simple blends; for example, in 1964 Union Carbide blended organic isocyanates with ethylene–vinyl acetate copolymers [64]. These blends were used as an adhesive interlayer in glass laminations, particularly safety glass laminates. Similarly, polyurethane–epoxy blends for safety glass laminates were reported in 1970 [65].

More recent efforts have focused on developments that create true hybrids. For example, blocked isocyanate prepolymers have been mixed with epoxy resins and cured with amines [66–68]. These blocked prepolymers will react initially with the amines to form amine-terminated prepolymers that cross-link the epoxy resin. Several blocked isocyanates are commercially available. The DESMOCAP (trademark, Miles, Inc.) 11A and 12A products are isocyanates (believed to be blocked with nonylphenol) used as flexibilizing agents for epoxy resins. ANCAREZ (trademark, Pacific Anchor, Inc.) 2150 is a blocked isocyanate epoxy blend used as an adhesion promoter for vinyl plastisols. A one-package, heat-cured hybrid adhesive was reported consisting of isophorone diisocyanate, epoxy resin, and a dispersed solid curative based on the salt of ethylenediamine and bisphenol A [69]. Urethane amines are offered commercially that can be used with epoxy resins to develop hybrid adhesive systems [70].

Urethane acrylic hybrids have been reported based on several approaches. Pacific Anchor has developed a urethane acrylate that is commercially available (ANCAREZ 300A). Acrylic polyols have been synthesized in the presence of polyether polyols by Saunders for use in two-component structural adhesives with improved tensile and impact strength [71,72]. Pressure-sensitive acrylic prepolymers with hydroxyl groups have been formulated with isocyanate prepolymers to give adhesives with improved peel strength [73,74]. Aqueous-based vinyl-to-fiberboard adhesives were reported by Chao using water-dispersible MDI with a functionalized acrylic latex and an aqueous dispersion polyurethane to give improved shear and hot peel strength [75]. Acrylonitrile dispersion graft polyether polyols have also been used in two-component SMC adhesives [76].

Urethanes have also been used to toughen vinyl-terminated acrylic adhesives for improved impact resistance. Thus rubber-toughened urea–urethane acrylates [77,78] water-dispersible urethane acrylates [79], and high-temperature-performance urethane–acrylate structural adhesives have been reported [80]. Polyurethanes terminated with acrylic functionality are also used for anaerobic or radiation-cured adhesives with improved toughness [81].

B. Reactive Hot Melts

Polyurethane reactive hot melts are 100% solid, hot-melt thermoplastic prepolymers that moisture cure slowly after application. Conventional hot melts are known for their quick setting, excellent green strength, ease of application, and low toxicity. Their primary limitation is low heat resistance (at elevated temperatures, adhesive will soften and flow) and poor adhesion to some substrates, due to insufficient wetting. The use of a polyurethane prepolymer with low levels of free isocyanates as a hot melt offers distinct advantages: initial green strength is still achieved, and in addition, the isocyanate will moisture cure slowly, converting the thermoplastic adhesive to a thermoset. There are a number of recent patents on reactive hot melts [82–85]. The tensile strength of the adhesive increases, heat resistance is improved, and the final cured adhesive will not flow at elevated temperatures [86]. A limitation of this technology is the need for porous substrates or bond designs that will allow the diffusion of moisture into the adhesive so that moisture curing will occur. The adhesive itself must be protected from moisture prior to use. This technology should be applicable to assembly line operations which require an adhesive that gets high initial green strength.

C. Pressure-Sensitive Adhesives

The use of polyurethanes in the pressure-sensitive adhesives market has been relatively small. Polyurethanes have been somewhat limited to being used as additives to pressure-sensitive adhesives to improve their cohesive strength. Recent developments in the institutional carpet backing or automotive carpet floor mat markets suggest that pressure-sensitive urethanes can succeed commercially [87].

X. SUMMARY

Polyurethane adhesives as a class can no longer be perceived as new raw materials. From a base of 217 million pounds, double-digit growth can no longer be expected. Even so, significant growth will continue. Formulators are taking advantage of the tremendous flexibility of urethane chemistry in designing new adhesive products. Specialty niches such as waterbornes and reactive hot melts, for example, will continue to emerge and fuel growth. Exciting times lie ahead for innovative formulators of polyurethane adhesives!

REFERENCES

1. J. Saunders and K. Frisch, *Polyurethanes: Chemistry and Technology*, Pt. 1, Interscience, New York, 1963.
2. J. M. DeBell, W. C. Goggin, and W. E. Gloor, *German Plastics Practice*, DeBell and Richardson, Cambridge, Mass., 1946.
3. B. A. Dombrow, *Polyurethanes*, Reinhold, New York, 1957.
4. Farbenfabriken-Bayer, U.S. patent 2,650,212 (1953).
5. Stoner-Mudge Co., U.S. patent 2,769,826 (1956).
6. J. F. Anderson and L. F. Fiedler, U.S. patent 2,801,648, to B.F. Goodrich (1957).
7. C. S. Schollenberger, H. Scott, and G. R. Moore, *Rubber World 137*:549 (1958).
8. B.F. Goodrich, U.S. patent 2,871,218 (1959).
9. French patent 1,192,267 (1959).
10. J. E. Mallone, U.S. patent 2,968,575, to E.I. du Pont de Nemours and Co. (1961).
11. *Experimental Urethane Latex E-204*, Bulletin, Wyandotte Chemicals Corporation, July 15, 1963.
12. S. P. Suskind, *Appl. Polymer Sci 9*: 2451 (1965).
13. B. F. Goodrich, U.S. patent 3,015,650 (1962).
14. M. E. Kimball, *Adhesives Age*, p. 21 (June 1981).
15. Continental Tapes, U.S. patent 3,802,988 (1974).

16. D. Dieterich and J. N. Rieck, *Adhesives Age 21*(2):24 (1978).
17. H. Von Volthenberg, *European Adhesives Sealants 1*(4): 28 (1984).
18. R. A. Bragole, *Urethanes in Elastomers and Coatings*, Technomic, Westport, Conn., 1973, p. 136.
19. J. W. Britain and P. G. Gemeinhardt, *J. Appl. Polymer Sci. 4*: 207 (1960).
20. H. G. Scholten, J. G. Schumann, and R. E. TenHoor, *Rubber Technol. 5*(3): 395 (1960).
21. *Preparation of Prepolymers*, Form 109-008-95-290-SAI, Dow Chemical Co., Freeport, Texas (1990).
22. R. Leuckart, *Ber. Chem. Ges.* p. 873 (1885).
23. H. E. French and A. F. Wirtel, *J. Am. Chem. Soc. 48*: 1736 (1926).
24. J. W. Baker and J. B. Holdsworth, *J. Chem. Soc.*, p. 713 (1947).
25. *Hydroxyl Terminated Poly BD Resins, Functional Liquid Polymers*, Electrical applications brochure, Atochem North America, Inc., Oct. 1990.
26. Z. W. Wicks, *Progr. Org. Coatings 3*: 73 (1975).
27. Z. W. Wicks, *Progr. Org. Coatings 9*: 3 (1981).
28. S. Petersen, *Liebigs Ann. Chem. 562*: 205 (1949).
29. Y. Miura and K. Amamitsu, Japanese patent 7,010,636; *C.A. 73*: 78229 (1970).
30. J. F. Levy and J. Kusean, U.S. patent 3,705,119; *C.A. 78*: 59887 (1973).
31. K. H. Weber, H. Streigler, G. Berndt, and R. Voigt, British patent 1,186,649; *C.A. 73*: 44790 (1970).
32. H. L. Elkin, U.S. patent 3,384,506; *C.A. 69*: 20349 (1968).
33. C. J. Amirsakis, German Offen. 1,803,038; *C.A. 71*: 71755 (1969).
34. G. W. Rye, R. S. Bhakuni, J. L. Cormany, and T. E. Evans, German Offen. 1,921,672; *C.A. 73*: 46449 (1970).
35. E.I. du Pont, British patent 987,600; *C.A. 63*: 3148 (1965).
36. Farbenfabriken Bayer, French patent 1,525,628; *C.A. 38*: 4365 (1969).
37. K. Graehling, G. Angerer, and W. Gimplinger, German patent 1,470,776; *C.A. 73*: 36322 (1970).
38. T. Kigane, S. Yamada, Y. Isozaki, and Y. Yada, Japanese patent 70 18, 210; *C.A. 74*: 4486 (1971).
39. R. Miller, J. L. Witt, and M. L. Tidmore, U.S. patent 3,707,178; *C.A. 78*:73395 (1973).
40. V. K. Rowe and M. A. Wolf, in *Patty's Industrial Hygiene and Technology*, Vol. 2C (F. D. Clayton and F. E. Clayton, eds.), Wiley, New York, 1982.
41. T. D. Landry and C. A. Steffens in *Reaction Polymers* (W. F. Gum, W. Riese and H. Ulrich, eds.), Oxford University Press, Hanser Publishers, New York, 1992, p. 747.
42. R. J. Davies, *Clin. Immunol. Allerg. 4*(1): 103 (1984).
43. M. H. Karol, *CRC Crit. Rev. Toxicol. 16*(4): 349 (1986).
44. D. E. Banks, B. T. Butcher, and J. E. Salvaggio, *Ann. Allergy 57*: 389 (1986).
45. A. W. Musk, J. M. Peters, and D. H. Wegman, *Am. J. Ind. Med. 13*: 331 (1988).
46. H. S. Katz and J. V. Milewski, *Handbook of Fillers and Reinforcements for Plastics*, Van Nostrand Reinhold, New York, 1978.
47. D. Zalucha, *Polyurethane Adhesives*, Technomic Urethane Course, Atlanta, Ga. 1988.
48. A. J. Kinloch, *Adhesion and Adhesives*, Chapman & Hall, London, 1987, pp. 101–170.
49. J. J. Martin, in *Adhesion and Adhesives*, Vol. 2 (R. Houwink and G. Salomon, eds.), Elsevier, New York, 1967.
50. W. A. Dukes and A. J. Kinloch, *Sira Technical Review 3*, Sira, London, 1976.
51. R. C. Snogren, *Handbook of Surface Preparation*, Palmerton, New York, 1975.
52. Recommended Practice for the Preparation of Surfaces of Plastics Prior to Adhesive Bonding, *ASTM D 2093-69*, ASTM, Philadelphia, 1980.
53. D. M. Brewis, ed., *Surface Analysis and Pretreatment of Plastics and Metals*, Applied Science Publishers, London, 1982.
54. J. Shields, *Adhesives Handbook*, Butterworth, London, 1985.

55. Du Pont, U.S. patent 2,277,083 (1942).
56. A. F. Lewis, L. M. Zaccardo, and A. M. Schiller, U.S. patent 3,391,054, to American Cyanamid Company (1968).
57. G. Oertel, *Polyurethane Handbook*, Carl Hanser, Munich, 1985, p. 554.
58. L. Maempel, *Adhesion 5*: 14 (1988).
59. Japanese patent 80 08 344 (1980).
60. British patent 1,250,266 (1971).
61. J. M. Gaul, T. Nguyen, and J. S. Babiec, Jr., *J. Elastomers Plastics 16*: 206 (1984).
62. D. Dieterich, *Progr. Org. Coatings 9*: 281 (1981).
63. G. M. MacIver, *Structural Adhesives for Composite Bonding*, Adhesives and Sealants Council Mini Seminar, Apr. 1987.
64. B. O. Baum, U.S. patent 3,157,563; to Union Carbide Corporation (1964).
65. C. M. Roseland and V. E. Hamilton, U.S. patent 3,546,064; to McDonnell Douglas Corporation (1970).
66. R. Grieves, U.S. patent 4,623,702, to Pratley (1986).
67. D. G. Lay, paper presented to the *SPI Epoxy Resin Formulators Meeting*, San Francisco, Feb. 20–22, 1991.
68. D. G. Lay and T. G. Millard, paper presented at *ASC's Spring Meeting*, St. Louis, Mo., Apr. 14–17, 1991.
69. A. Goel, U.S. patent 4,737,565, to Ashland (1988).
70. J. Durig, S. Beinborn, and S. Sawant, paper presented at the *SPI Epoxy Resin Formulators Meeting*, San Francisco, February 20–22, 1991.
71. F. Saunders, U.S. patent 4,731,416, to Dow Chemical Co. (1988).
72. C. Bluestein, *ACS Symp. Ser. 172 (Urethane Chem. Appl.)*: 505 (1981).
73. W. DeVry et al., U.S. patent 4,145,514, to B.F. Goodrich (1979).
74. Y. Lee, U.S. patent 4,214,061, to B.F. Goodrich (1980).
75. Y. Chao, U.S. patent 4,636,546, to Rohm and Haas (1987).
76. T. Glamondi, U.S. patent 4,742,113, to Lord (1988).
77. R. Schappert, U.S. patent 4,721,751, to PPG (1988).
78. J. Saracsan, U.S. patent 4,452,964, to Ashland (1984).
79. A. Travati, U.S. patent 4,497,932, to Resem (1985).
80. T. Dawdy, U.S. patent 4,452,944, to Lord (1984).
81. L. Baccei, U.S. patent 4,309,526, to Loctite (1982).
82. F. Reischle et al., U.S. patent 4,585,819, to H.B. Fuller Co. (1986).
83. V. C. Markevka et al., U.S. patent 4,775,719, to H.B. Fuller Co. (1988).
84. V. C. Markevka et al., U.S. patent 4,820,368, H.B. Fuller Co. (1989).
85. H. Gilch, U.S. patent 4,618,651, to USM (1986).
86. M. Bowtell, *Adhesives Age*, p. 42 (Sept. 1987).
87. L. W. Mobley, paper presented to the *Polyurethanes World Congress*, Nice, Frances, Sept. 24–26, 1991.
88. Liquid Control Corp., North Canton, Ohio; Sealant Equipment and Engineering Inc., Oak Park, Michigan; Graco, Plymouth, Michigan.
89. P.E. Cranley in *Reaction Polymers* (W. F. Gum, W. Riese and H. Ulrich, eds.), Oxford University Press, Hanser Publishers, New York, 1992, p. 692.

25
Polyvinyl and Ethylene–Vinyl Acetates

Ken Geddes *Crown Berger Limited, Darwen, Lancashire, England*

I. INTRODUCTION

Poly(vinyl acetate) (PVA) and ethylene–vinyl acetate (EVA) copolymer adhesives have much in common, yet represent extremes in the degree of sophistication of their production processes. Both products are stable suspensions in water of a film-forming polymer, the particles of which are generally spherical. They are made by emulsion polymerization, which uses a free-radical addition mechanism to polymerize the monomer in the presence of water and stabilizers. Vinyl acetate is the sole or major monomeric raw material.

The major difference between the processes is the incorporation in EVAs of ethylene, an internally plasticizing monomer. The use of ethylene requires production equipment suitable for safe handling of a highly flammable, high-pressure gas. Despite the volatility and very low flash point of vinyl acetate, simple PVAs are manufactured successfully by many small-scale producers in developing countries; EVAs are made only at sophisticated and costly plants where ethylene gas, engineering skills, and a significant demand for the product come together.

II. CHEMISTRY

Vinyl acetate is characterized by having an activated double bond. While being an acceptably stable material under normal ambient storage, it is readily attacked by a free radical. This simple addition gives another free radical, and the addition of a series of monomer units results in a polymer chain. Thus

$$CH_2{=}CH{-}\underset{\underset{O}{\|}}{C}{-}CH_3 \quad + \cdot SO_4^- \quad \rightarrow \quad CH_2{-}\underset{\underset{SO_4^-}{|}}{CH}\cdot{-}\underset{\underset{O}{\|}}{C}{-}CH_3$$

Vinyl acetate monomer

$$+ \; CH_2{=}CH{-}\underset{\underset{O}{\|}}{C}{-}CH_3$$

$$\underset{\underset{SO_4^-}{|}}{CH_2}{-}CH\cdot{-}\underset{\underset{O}{\|}}{C}{-}CH_3 \longrightarrow \underset{\underset{SO_4^-}{|}}{CH_2}{-}\underset{\underset{\underset{CH_3}{|}}{\underset{C=O}{|}}}{CH}{-}CH_2{-}\underset{\underset{\underset{CH_3}{|}}{\underset{C=O}{|}}}{CH}\cdot \; etc.$$

III. RESIN PREPARATION

The essential property of any polymer used as an adhesive is that it should have good cohesion and stick to the surfaces it joins. This can occur only if the polymer forms a film on application or at some stage during the joining process. In the case of a water suspension such as a PVA or an EVA adhesive, this involves a smooth coalescence of the polymer particles as the water evaporates. For this to occur, the polymer must be above its minimum filming temperature (MFT). Some applications lend themselves to the application of heat. However, most rely only on the evaporation of water at ambient temperature, coupled with absorbence (or "wicking") into the substrate, if porous.

Poly(vinyl acetate) has a glass transition temperature (GTT) of about 30°C. Glass transition temperature marks the change in properties from a material with a glasslike nature to one with rubbery properties. In particular, particles above their GTT may deform, flow, and adhere. This results in GTT being the greatest factor in determining the minimum temperature for the formation of a coherent film. The GTT of any polymer is dependent on its structure. As the temperature rises, polymer chains vibrate under the influence of stretching and bending motion of individual bonds. At the GTT the steric and covalent locking of the chain is overcome, allowing the molecule as a whole to bend and rotate, subject to the special constraints of neighboring polymer chains. It is clear that any internally plasticizing copolymerized monomer or external solvent or plasticizer that assists this process by making the main chain of polymer more flexible, or eases its rotation by spacing adjacent chains of polymer, reduces the temperature at which film formation can take place. Ethylene acts directly as a polymer backbone plasticizer. A polymer such as poly(vinyl acetate) homopolymer:

$$-CH_2-CH-CH_2-CH-CH_2-CH-CH_2-\ etc.$$
$$\qquad\quad |\qquad\qquad\quad |\qquad\qquad\quad |$$
$$\qquad\quad OAc\qquad\quad OAc\qquad\quad OAc$$

becomes the much more flexible copolymer with ethylene:

$$-CH_2-CH-CH_2-CH_2-CH_2-CH-CH_2-\ etc.$$
$$\qquad\quad |\qquad\qquad\qquad\qquad\qquad\quad |$$
$$\qquad\quad OAc\qquad\qquad\qquad\qquad\quad OAc$$

In the latter structure the acetate groups are too widely spaced to influence rotation.

PVA adhesives are plasticized externally by smaller molecules that space the chains and minimize chain-to-chain interactions. The classic material for this purpose is dibutyl phthalate (DBP), but any compatible solvent may be chosen. Even water itself acts as a temporary coalescing solvent to some degree. Volatile materials gradually evaporate, allowing the PVA adhesive to harden and gain tensile strength. This is particularly important if strength at high ambient temperatures is required. Conversely, materials less volatile than DBP may be chosen to ensure that a degree of flexibility is retained in the joint, even after many years at high temperatures or repeated rinsing with water. For ultimate stability over time, copolymerization with ethylene must be chosen.

Lowering of the minimum filming temperature and increasing flexibility is not the only reason for introducing other polymerizable monomers into the preparation of the adhesive. A large variety of materials are available, having in common a double bond that is either activated or may become so on the approach of a free radical. Such monomers may modify the behavior of the final polymer, conferring resistance against alkalis, or to improve adhesion in wet conditions. Chain branching to increase molecular weight may be introduced. Stability of the polymer against freezing or mechanical shear are other desirable properties that may be gained by copolymerization.

Because of the simple addition of monomer to growing chains, and because the reaction takes place in the bulk of the polymer or on the surface of polymer particles rather than in solution, molecular weight is very high. Chain growth can cease only when two free-radical-terminated chains collide, an initiator fragment adds to and eliminates a growing chain, or a chain transfer reaction takes place with a small, volatile molecule. In practice, the chances of two growing chains mutually eliminating each other by addition are very low because of their high molecular weight and consequential lack of mobility. One recent theory relies on chain transfer to monomer and subsequent desorption of the small free radical as the only significant mechanism of chain termination. Molecular weight can be increased by the addition of small amounts of monomer with more than one double bond. These are known as chain branching agents. Conversely, molecular weight is decreased by the addition of a material containing a reactive hydrogen such as *n*-dodecyl mercaptan (1-dodecane thiol) as a chain transfer agent. The hydrogen is readily removed from the thiol (—SH) group, giving a terminated chain on the polymer but also a —S • free radical capable of starting a new chain. Formulation factors such as the level of the free-radical initiator used and process factors such as temperature of reaction and the amount of agitation also play a part. Low levels of initiator, low temperatures, and a carefully judged degree of agitation all favor high molecular weight.

Turning to practical considerations, choice of process and formulation is all important in successful production of PVA and EVA adhesives. Some formulation examples are given in a later section, but various aspects are discussed here.

A. Process

Addition polymerization is exothermic, and one of the major constraints to high production rates are the problems associated with heat removal. In processes using ethylene, the pressure of the gas determines solubility in the liquid phases (i.e., water and vinyl acetate monomer droplets) and in the polymer particles. This concentration of ethylene at the point of polymerization determines the ethylene content of the final polymer. Use of high pressures in such systems eliminates refluxing of the vinyl acetate, losing a very effective heat removal mechanism available to simple batch-process PVA production. Refluxing, however, gives condenser vapor losses. Great care has to be taken to ensure that the condenser is adequate to deal with the volume of vapor to be condensed. Returning condensate from an inverted condenser is also a very effective cooling agent, as it is often very much below the reaction temperature. The sensible heat removal adds to the evaporative cooling.

Care also has to be taken not to overcool the reaction. If so, a slowdown in the polymerization rate may occur, with excess free monomer, leading to an exotherm followed by foaming or an overloading of the condenser. A reduction in the monomer feed rate at this time is essential, but again care has to be exercised, as a sudden loss of cooling from the incoming monomer stream coupled with a drop-off in the reflux rate can give an uncontrollable exotherm.

The batch process uses a kettle fitted with an agitator. Other features are temperature probes and a cooling jacket. Simple processes operate at atmospheric pressure and use a condenser. If no condenser is fitted or ethylene gas is to be used, the reactor must be pressurized. On larger kettles, an external heat exchanger may be employed. This system is attractive, as it avoids the need to manage reflux, and the total cooling capability has flexibility through variable rates of pumping through the heat exchanger. Its disadvantage is the need to clean and maintain the heat exchanger, and manufacture of grades of adhesive with poor mechanical stability, high viscosity, or a tendency to foul surfaces can be difficult or uneconomic.

The batch process starts with filling the reactor with most of the water, much or all of the stabilizer [frequently poly(vinyl alcohol) in adhesives], and a small proportion of the monomer. On agitation and raising the temperature to above 65°C, addition of a water-soluble free-radical generator such as ammonium persulfate initiates polymerization. This establishes the number of particles and the average particle size of the emulsion polymer. A continuous stream of vinyl acetate is run or pumped in with additional initiator until the required concentration of polymer is obtained, this coinciding with the maximum working volume of the kettle. It follows that in this process there is a wide spread of residence times within the reactor. The initial polymer is present from the outset, but shells of fresh polymer built around the early particles have a relatively short period within the reactor. The water and the stabilizers are also present at the beginning, which gives maximum time for degradation and grafting reactions. It is, however, energetically inefficient to agitate such viscous solutions over the full period of the process.

One alternative is the *Loop process* [1–3]. This employs a rather simple principle. A small volume of reaction mixture is recirculated, while streams of monomer and water phase [a stabilizer solution such as aqueous poly(vinyl alcohol)] are pumped into the reactor in the correct proportions. The reactor is fully filled and a balancing volume of product is released through a pressure-sustaining valve. Any unreacted monomer remaining in the outlet stream polymerizes on the way to the cooling tank or over a few hours, prior to packing. The volume of this type of reactor is only 40 to 80 L in volume compared to 3000 to 100,000 L for a batch reactor.

The two types of reactor may be compared. The Loop reactor is more efficient energetically as the volume of reaction mixture to be agitated is so much less. It should be said, however, that the savings are not proportional to the volumes involved, as the diameter of the Loop pipes give greater frictional losses. As with many calculations involving viscosity in emulsion polymer production, complications arise due not only to pseudoplasticity of reaction mixtures, but because of different behavior at different shear rates and temperatures.

One of the greatest contrasts between the processes is the residence time within the reactor. It has already been noted that the poly(vinyl alcohol) in the batch process is usually present from the start. The residence time is therefore several hours. In contrast, the mean residence time of materials in the loop process is around 2 to 10 min. This has obvious advantages in terms of minimizing degradation of colloids but will also restrict grafting between colloid and monomer.

Aside from process comparisons, the main contrast between the systems is that of size, weight, and cost, especially for pressurized systems. Construction of batch reactors for use with ethylene at pressures of 1000 psi (70 atm) and upward has to be massive. The simple construction of the Loop process—just pumps and pipework—lends itself to use at high pressures. Apart from cost and weight, the small volume of the Loop reactor has obvious safety advantages. Despite these attractions, the Loop reactor system has so far been used successfully only for low-pressure systems such as poly(vinyl acetate) homopolymer for adhesives and copolymers for paint. Large-scale production of ethylene–vinyl acetate copolymers has yet to be demonstrated.

B. Formulation Factors

The majority of standard product formulations for PVA and EVA adhesives use poly(vinyl alcohol) as the main protective colloid and thickener. Poly(vinyl alcohol) can be obtained in a number of grades produced from poly(vinyl acetate) by hydrolysis. As a consequence, almost all products are effectively vinyl acetate/vinyl alcohol copolymers. Lower levels of hydrolysis, 88% most commonly, are readily water soluble in hot or cold water, although in practice care has to be taken in making solutions to avoid clumping of the grains, which can then be difficult to disperse. Eighty-eight percent hydrolyzed material is normally coupled with a "fully hydrolyzed" grade, in which the hydrolysis has been taken into the range 97 to 99.5% of the theoretical maximum.

These grades are not easily soluble in cold water, but once made into a solution by heating, are usually stable. Because of insolubility in cold water, the higher hydrolysis-grade materials have better water resistance as dried films. Their disadvantage is to give less viscous products for poly(vinyl alcohol) of the same molecular weight. This may be associated with the particle size of the polymer formed—the presence of vinyl acetate groups gives some surfactancy, especially if the acetate groups are in blocks. Finer average particle size—and hence more particles—fill the free space more effectively, increasing viscosity. Particle–particle interactions are also important, and particle size distribution has a profound effect on rheology.

Introduction of surfactants, especially anionic surfactants or in mixtures with non-ionics, give wider distributions that often lose viscosity on shearing. This makes pumping and application easier but may bring in the problem of overspreading. For consistency of application between different machines, a more Newtonian rheology is an advantage. Many commercial adhesives are further compounded by the addition of fillers, thickening agents of various kinds, plasticizers, and solvents, although the range of the latter that are acceptable is diminishing rapidly.

IV. ANALYSIS AND TESTING

PVA and EVA products are often sold with rather limited information. Often solids content, viscosity range, and pH are the only real specifications given. Minimum filming temperature (or in some cases, glass transition temperature) may be quoted, together with comonomer type, if any, and some brief application recommendations. Manufacturers may in production test for more properties than they publish, especially grit content, particle size, and unreacted monomer. Grit is the material retained by a standard sieve and comprised of oversized particles up to beads, skins, and pieces of reactor wall fouling which have found their way through the system. Most adhesives are filtered prior to packing, but this is less easy and less important than in lower-viscosity paint grades. The high viscosity of many adhesive products makes the use of fine screens an economic impossibility because of the slow speed of filtration.

Viscosity is a property sometimes difficult to assess, as figures can be measured on any one of several types of viscometer. One common type is the rotating disk viscometer, which must be used in a container big enough to eliminate wall effects. The main alternative is the cup-and-bob viscometer, where the viscous drag of the liquid between stationary and rotating concentric cylinders is indicated by a spring-loaded pointer moving over a dial.

For use with high-speed applicators, high shear cone and plate viscometer results may be quoted as secondary information. Many poly(vinyl alcohol)-stabilized products are comparatively insensitive to shear and give broadly similar results with different types of viscometer. This behavior under shear is known as Newtonian and is a feature, inter alia, of large particles with a narrow particle size distribution. High shear viscosity testing also indicates if there is sufficient mechanical stability to allow application by knife or roller, although this is not usually a problem with colloid-stabilized emulsion polymers and adhesives.

Particle size range is often from 200 to 4500 nm or more, with one or more peaks. Multipeak distribution may indicate agglomeration at some stage of the preparation, and microscopy can be used to show if the peaks are of single particles or of an agglomerated mass of smaller ones. Freeze–thaw resistance is called for in many countries. The key here is to avoid the higher degrees of hydrolysis in the poly(vinyl alcohol) (not greater than 98% hydrolyzed), and ensure that sufficient stabilizer is present to cover the surfaces. Nonionic surfactants will act as antifreezes and suppress the freezing point and it is often of value to quote the behavior of the adhesive at $-5°C$ and $-20°C$. Residual monomer should also be checked for quality control purposes and kept below the specified level. Gas–liquid chromatography is the favored method of analysis, but bromination is also widely used.

Application tests by their nature are often specific to the materials to be bonded and the application machinery in use. For packaging applications it is important that sufficient strength is generated within seconds of the bond being formed to hold the surfaces in position until the adhesive dries. Testers are available that apply a measured, standardized film to a series of kraft paper pieces. A second sheet of kraft paper (or whatever may be appropriate) is applied and the papers are peeled apart at specified time intervals. The time at which the surface of the paper is first torn off (as contrasted to the earlier tests, in which the surfaces are partially covered with adhesive) is noted. Short times are necessary for high-speed machinery. High viscosity, or suitable rheology to create resistance to parting of the adhesive layer, is obviously of value and is called wet tack or "grab." As most of the initial drying is through wicking of the water in the adhesive into the paper or board to be bonded, standardization of the paper used in this test is vital. Wood bond strength is more concerned with ultimate strength and the test pieces are usually allowed to dry thoroughly for 24 to 48 h or more. Small beechwood slips giving a controlled area of overlap are generally used, although end bonding of beech dowels is an alternative. In each case, bond strength is measured using a tensiometer and it should be noted that the overlap method gives the strength to resist shearing in a direction parallel to the wood surface, while the dowel method tests the resistance at right angles to the wood. In both cases the wood should fail before the adhesive, although if the adhesive fails, the strength measured may be a function of the amount of plasticizer or softening comonomer used in the preparation.

Ethylene–vinyl acetate adhesives are used for many of the same applications as externally plasticized PVAs but have especially good performance in the field of PVC lamination to hardboard and chipboard. Again, an adhesive is sought where the mode of failure is the cohesion of the wood. Drying is slow in this case because the PVC foils are largely impervious to water. Wicking into the substrate and vapor loss is the only mode of water removal. Hence it is essential that bonds be matured for a suitable period prior to test. Various test methods exist; the simplest uses 5-cm-wide strips of PVC foil laminated to plywood. This is dried and matured and suspended inverted with the plywood at 45° to the horizontal. Weights may be suspended from the width of the PVC, using a clamp or a firm clip to which weights are attached. This is simply extended to high-temperature peel strengths by placing the test piece within an oven, say at 70°C. The distance that the specimen has peeled after 30 min can be measured. Weights employed are usually 350 g or 500 g, but greater weights are possible. Ninety-degree peel tests can be conducted using tensile-strength testing equipment, but they are more difficult to conduct at temperatures above ambient.

V. FORMULATIONS

A. Poly(vinyl acetate) Homopolymers

1. Batch Process

A simple formulation by the batch process is as follows:

		Parts by weight (kg)	
1.	Initial reactor Charge		
	Process water	360.5	⎫
	Poly(vinyl alcohol) (88% hydrolyzed)	12.6	⎬ Presolution
	Nonyl phenol (15 M) ethyl oxide condensate	6.6	⎭
	Linear C_{12} sulfate, sodium salt	0.3	
	Sodium bicarbonate	1.6	
	Antifoam (nonsilicone)	0.4	
	Water	64.4	Rinse
2.	Initial monomer charge		
	Vinyl acetate	25.0	
3.	First initiator		
	Sodium persulfate	1.1	
	Water	6.0	
4.	Continuous monomer feed		
	Vinyl acetate	460.0	
5.	Continuous initiator feed		
	Sodium persulfate	0.2	
	Water	6.0	
6.	Final initiator feed		
	Sodium persulfate	0.1	
	Water	3.0	
7.	Plasticizer		
	Dibutyl phthalate	50.0	
8.	Preservative		
	Preservative	2.2	
		1000.0	

Process: Make a presolution of the poly(vinyl alcohol). Add to the polymerization kettle, agitate, and heat to 65°C, meanwhile adding the other ingredients of the initial reactor charge. At 65°C, add the initial vinyl acetate monomer and the first initiator. Heat cautiously to 80°C, during which time the initial vinyl acetate will polymerize (shown by the development of a blue color, a reduction or cessation of reflux, and a slight exotherm).

Start to add the continuous monomer and initiator feeds to go in over 4 h at a steady rate. Monitor temperature and reflex continuously, especially in the early stages of the reaction. Slow the feed of monomer if reflux is excessive or temperature cannot be maintained at 80 to 85°C. Ensure that the agitation is sufficient at all times to give a small vortex that blends in added monomer and condenser return smoothly but does not create foam or splashing. When feeds are complete, add final initiator and allow temperature to rise to 90 ± 2°C (heat if necessary). Hold for 20 min, then cool, adding the dibutyl phthalate at about 65°C and the preservative at 35°C or less.

2. Loop Process

A Loop continuous reactor uses a broadly similar formulation to the batch case, but the poly(vinyl alcohol)-containing solution is pumped in continuously rather than added to the reactor initially. Also, a redox initiator is used:

		Parts by weight (kg)
1.	Initial reactor filling	
	Process water	46.3
	Poly(vinyl alcohol) (88% hydrolyzed)	2.2
	Poly(vinyl alcohol) (98% hydrolyzed)	1.3
	Sodium acetate	0.1
	Sodium metabisulfite	0.1
		50.0
2.	Water/stabilizer feed	
	Process water	489.6
	Poly(vinyl alcohol) (88% hydrolyzed)	23.6
	Poly(vinyl alcohol) (98% hydrolyzed)	14.1
	Sodium acetate	1.1
	Sodium metabisulfite	1.0
3.	Monomer feed	
	Vinyl acetate	419.2
4.	Initiator feed	
	t-Butyl hydroperoxide	1.1
5.	Plasticizer	
	Dibutyl phthalate	50.0
6.	Preservative/antifoam	
	Preservative	0.2
	Antifoam	0.1
		1000.0

Process: Pump into the Loop reactor the initial fill. Start the circulation pump. Start to pump the water–stabilizer solution, followed by the monomer feed and the initiator, to give a total feed of raw materials of 500 kg/h. Allow the temperature to rise to 55°C and apply cooling to stabilize the temperature to 55 ± 1°C. Collect the product and add dibutyl phthalate after holding the completed required volume for 20 min at about 50°C. Cool and add the preservative and antifoam at 35°C or less.

B. Vinyl Acetate–Ethylene Copolymers

1. Batch Reactor Process

The following formulation illustrates the use of a redox initiator in a batch process. Also, the introduction of N-methylol acrylamide increases the molecular weight and chain cross-linking, minimizing the thermoplastic properties of the adhesive and the tendency to cold flow.

		Parts by weight (kg)
1.	Initial reactor charge	
	Deionized water	380.0
	Poly(vinyl alcohol) (88% hydrolyzed)	25.0
	Sodium bicarbonate	2.0
	Ammonium persulfate	5.0
2.	Initial monomer charge	
	Vinyl acetate	50.0
	Ethylene	(to 250 psi)
3.	Reducing initiator	
	Sodium formaldehyde sulfoxylate	1.5
	Deionized water	25.0
4.	Continuous monomer feed (1)	
	Vinyl acetate	446.0
5.	Continuous monomer feed (2)	
	N-Methylol acrylamide	4.5
	Deionized water	60.0
6.	Preservative, etc.	
	Preservative	0.5
	Antifoam	0.5
		1000.0*

*Excluding the weight of the combined ethylene.

Process: Load the reactor with the initial charge, using a presolution of the poly(vinyl alcohol) in 300 g of the water. Rinse in the sodium bicarbonate and ammonium persulfate with the remaining water. Switch on the agitator and purge with nitrogen. Then pump in the initial vinyl acetate and pressurize with ethylene gas. Raise the temperature to 35 to 40°C and maintain at this temperature. Start to add the reducing initiator feed to go in over about 8 h. After 1 h, start to add the continuous monomer feeds (1) and (2) to go in over about 7 h. At all times the unreacted monomer should be kept at 1 to 2% of the reaction mixture to ensure even copolymerization of the N-methylol acrylamide.

2. Loop Process

A Loop continuous reactor formulation is given below, where the redox initiator is of an unusual manganese type.

	Parts by weight (kg)
1. Initial reactor filling	
Water	50
2. Water/stabilizer solution feed	
(feed rate: 3.95 kg/min)	
Water	236.66
Poly(vinyl alcohol) (80% hydrolyzed)	11.36
Sodium bisulfite	3.79
3. Monomer feed	
(feed rate: 8.52 kg/min)	
Vinyl acetate	440.18
Ethylene	71.00
4. Initiator feed	
(feed rate: 3.95 kg/min)	
Manganese^{3+} sulfate/sodium	
pyrophosphate complex	0.35
Water	236.66
	1000.00
	(Items 2, 3, and 4)

Process: Fill the reactor with water and set the circulation pump to 800 rev/min. Start feeding the water–stabilizer solution at 3.95 kg/min, followed by the Mn^{3+} complex, also at 3.95 kg/min. Then start pumping in the vinyl acetate, finally beginning to feed the ethylene into the vinyl acetate feed at a pressure of 260 psi, the pressure-sustaining valve on the outlet of the Loop reactor being adjusted accordingly. This formulation was found to give a 95% conversion of ethylene with a 13.4% by weight incorporation in the final polymer.

C. Formulated Adhesive

	Parts (%)
PVA homopolymer (58% solids, 10% DBP)	82.6
Dibutyl phthalate (DBP)	5.7
Ethyl acetate	5.5
Calcium carbonate filler (micronized)	4.8
Preservative	0.2
Water	1.2
	100.0

Process: Blend all ingredients except the filler with the PVA. When complete add the filler slowly, blending well between additions.

VI. APPLICATIONS

Applications for poly(vinyl acetate) homopolymers include:

Wood adhesives
Packaging adhesive
General building adhesives; polystyrene tiles, hardboard; plasterboard
Ceramic tile adhesives
Remoistenable adhesives
Concrete patching adhesives
Bookbinding adhesive

Applications for vinyl acetate–ethylene copolymers include:

Bonding plastic foils and films in packaging
Lamination of PVC films to chipboard and plywood
Paper and board
Remoistenables for envelope flaps, paper labels, etc.
Bookbinding
Do-it-yourself (DIY) and household adhesives
Shoe and leather industry
General building adhesives
Nonwovens and flocking adhesives
Cigarette side-seam adhesives
Heat-sealable adhesives
Textiles

VII. SUMMARY

Vinyl acetate homopolymers are simply-made adhesive bases manufactured by addition polymerization in the presence of water and stabilizers. They are made commercially by the batch reactor process or by the Loop reactor continuous process. External plasticizers such as dibutyl phthalate are often added to confer flexibility and to lower the temperature at which they form a film on drying. Higher-quality products may be made by the copolymerization of ethylene with vinyl acetate to form an EVA. This involves the safe handling of ethylene gas under high pressure, and the plant required is more complex and considerably more costly. The Loop process has considerable attraction in the field of pressure polymerization.

REFERENCES

1. K. R. Geddes, in *Surface Coatings*, Vol. 3, A. D. Wilson, J. W. Nicholson, and H. J. Prosser, eds., Elsevier, New York, 1990, pp. 199–228.
2. M. B. Khan, *European Coatings J. 12*: 886–888, 891–892 (1991).
3. K. R. Geddes and M. B. Khan, European patent 0 417 893 A1 (1991).

26
Unsaturated Polyester Adhesives

A. Pizzi* *University of the Witwatersrand, Johannesburg, South Africa*

I. INTRODUCTION

Although unsaturated polyester resins are often regarded as casting plastics in at least one important use, glass-fiber lamination, they are used as adhesives. The method of binding glass-fiber mats with unsaturated polyesters started in 1942 by U.S. Rubber [1].

II. SYNTHESIS

A. Reaction Between Dicarboxylic Acids or Anhydrides and Diols

The synthesis of unsaturated polyesters usually involves a bulk reaction at elevated temperatures between dibasic acids or anhydrides and diols. A general reaction scheme for maleic anhydride and 1,2-ethanediol can be illustrated as follows:

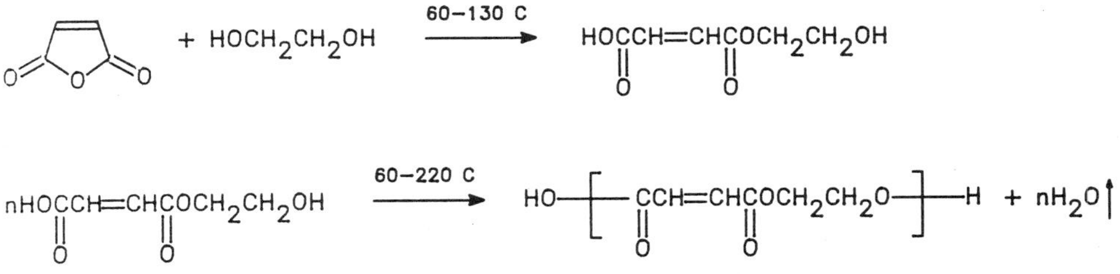

$$\text{maleic anhydride} + HOCH_2CH_2OH \xrightarrow{60-130\ C} HOCCH{=}CHCOCH_2CH_2OH$$

$$n\,HOCCH{=}CHCOCH_2CH_2OH \xrightarrow{60-220\ C} HO{-}\left[{-}CCH{=}CHCOCH_2CH_2O{-}\right]{-}H + nH_2O\uparrow$$

$$M_n = 1000 \text{ to } 3000$$

During this reaction most of the maleate groups are isomerized into fumarate groups. Since esterification is a reversible process, reaction water must be removed efficiently, especially in the last stages of the reaction, where the decrease in carboxyl group concentration is slow and the increase in viscosity is fast. These last stages are usually carried out under vacuum. However, to avoid losses of volatile reactants, an azeotropic distillation of reaction water in the presence of added organic solvent such as toluene or

---

*Current affiliation: Université de Nancy, Epinal, France.

443

xylene may be used [3]. The main drawbacks of this process are the longer reaction time and the difficulty in removing the last traces of solvent. Phthalic anhydride can be used to substitute maleic anhydride partially but extensively. The reaction scheme can be represented as follows:

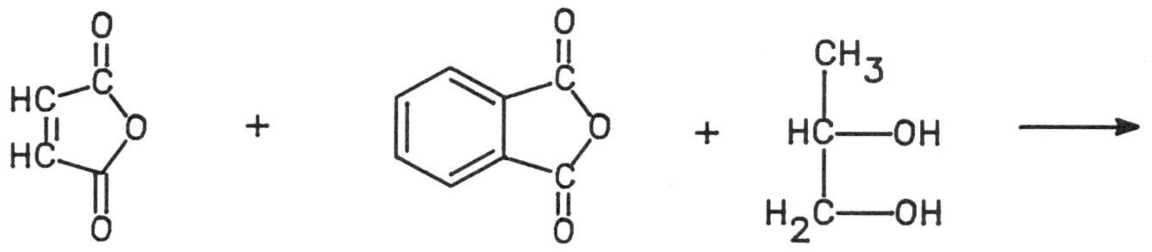

Maleic
Anhydride

Phthalic
Anhydride

Propylene
Glycol

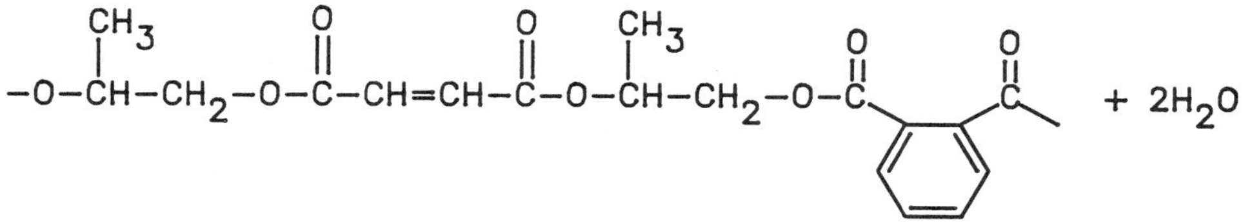

$+ 2H_2O$

B. Kinetics and Mechanisms

The theoretical analysis of the kinetic data for bulk polyesterification reactions is difficult because of the high concentrations of reactive end groups at the beginning of the reaction and because of the changes in dielectric constant of the medium during the reaction [2]. According to Flory [4], only the experimental results obtained for extents of reaction above 0.8 should be considered that is when the polarity no longer changes and when the reactive groups form a dilute solution in the polyester. Within these limits, experimental data show that both mono- and polyesterifications are third-order reactions [4], second order in acid and first order in alcohol. A reasonable mechanism involves nondissociated ion pairs and can be described, and with a protonic catalyst [3], as in the two following schemes:

$$2RCO_2H \underset{}{\overset{K}{\rightleftarrows}} RC(OH)_2^+ \ RCO_2^-$$

$$R, R' = \text{polyester chains} \qquad -d[RCO_2H]/dt = kK[RCO_2H]^2[R'OH]$$

$$AH \xrightleftharpoons{K_1} A^-H^+$$

$$RCO_2H + A^-H^+ \xrightleftharpoons{K_2} RC(OH)_2^+ \ A^-$$

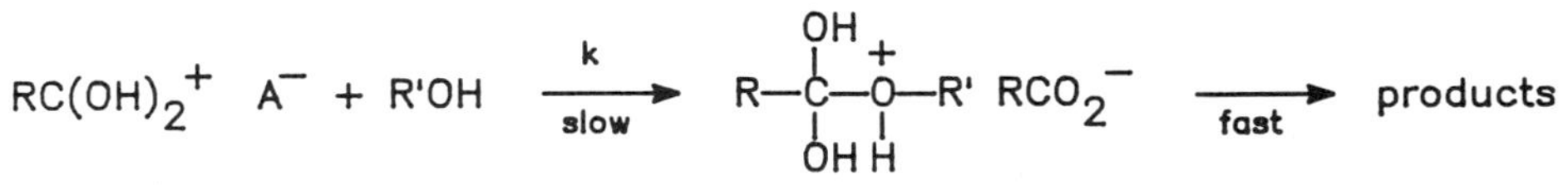

$$RC(OH)_2^+ \ A^- + R'OH \xrightarrow[\text{slow}]{k} \underset{\underset{OH\ H}{|\ \ |}}{R-\overset{\overset{OH}{|}}{\underset{|}{C}}-\overset{+}{O}-R'} \ RCO_2^- \xrightarrow{\text{fast}} \text{products}$$

R, R' = polyester chains $-d[RCO_2H]/dt = kK_1K_2[AH][RCO_2H][R'OH]$

According to classic organic chemistry, direct esterification can be catalyzed by either acidic or basic compounds, and Ingold [4] has proposed eight different mechanisms, four for acid-catalyzed and four for base-catalyzed processes. Basic compounds are seldom used as catalysts for polyesterification, and among the acid-catalyzed mechanisms, $A_{AC}2$ is by far the most frequently observed. Hundreds of compounds have been claimed as effective catalysts in the patent literature; strong protonic acids (H_2SO_4, benzene-, naphthalene-, and p-toluenesulfonic acids are the most popular), oxides or salts of heavy metal ions (acetates are often preferred for their higher solubility), and organometallic compounds of titanium, tin, zirconium, and lead are the catalysts most frequently reported.

The mechanisms proposed for direct esterification of low-molecular-weight esters have been investigated in detail by many workers and have been discussed in detail in a review by Bender [4]. According to these investigations, the following scheme is generally accepted for proton-catalyst reactions.

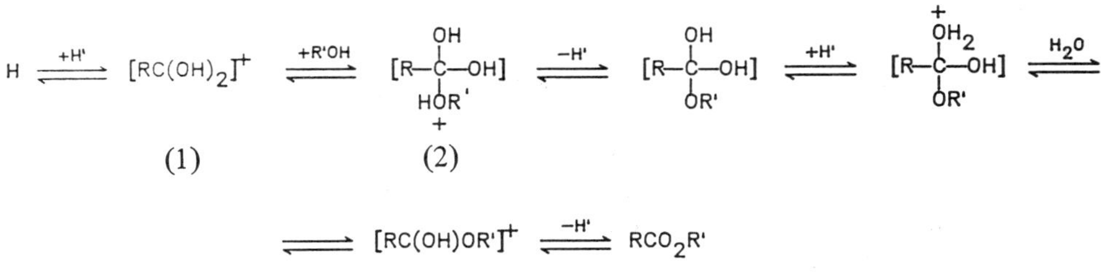

$$H \xrightleftharpoons{+H'} [RC(OH)_2]^+ \xrightleftharpoons{+R'OH} \underset{\underset{+}{\underset{HOR'}{|}}}{[R-\overset{\overset{OH}{|}}{\underset{|}{C}}-OH]} \xrightleftharpoons{-H'} \underset{\underset{OR'}{|}}{[R-\overset{\overset{OH}{|}}{\underset{|}{C}}-OH]} \xrightleftharpoons{+H'} \underset{\underset{OR'}{|}}{[R-\overset{\overset{\overset{+}{OH_2}}{|}}{\underset{|}{C}}-OH]} \xrightleftharpoons{H_2O}$$

(1)   (2)

$$\xrightleftharpoons{} [RC(OH)OR']^+ \xrightleftharpoons{-H'} RCO_2R'$$

In this scheme, the reaction of the protonated form (1) of the carboxylic acid with the hydroxy compound to give the addition intermediate (2) is usually taken as the rate-controlling step. This mechanism is usually extrapolated to proton-catalyzed direct polyesterification.

Owing to the low basicity of substrates such as carboxylic acids, the concentration of protonated species (1) can be extremely low, and alternative mechanisms, involving a nucleophilic attack assisted by compounds able to form hydrogen bonds in cyclic transition states such as (3) or (4), have also been considered [4].

(3)   (4)

C. Side Reactions

The chemical structure of unsaturated polyester is more complex than expected in view of the chemistry described above. The ^{13}C NMR spectra of unsaturated polyesters present many small peaks that cannot be assigned to carboxylic or hydroxylic end groups alone. These are due to a number of side reactions. Of these, the addition of hydroxyl groups to double bonds is one of the most important side reactions in the synthesis of unsaturated polyesters by polycondensation. It leads to the formation of side chains and a modification of the stoichiometry due to diol consumption [4].

D. Catalysts

Selection of the proper catalyst and the amount to be used for any application depends on the resin, the temperature at which the resin is to be cured, the required working or pot life, and the time of gelation. No catalyst is available that can meet all the requirements. Therefore, combinations of catalysts, or of catalysts and accelerators, must be used to obtain the best results.

When it is necessary to start and even cause a complete cure at lower temperatures so that the polymerization heat can readily be dissipated, methyl ethyl ketone peroxide (MEKP) is the catalyst generally used. It does not lead to a full cure by itself at ambient temperatures. However, with the addition of an accelerator, the catalyst will cause gelation and almost complete cure within short periods of time, depending on the percentage of each used with the resin. From 0.5 to 2.0% of MEKP and 0.1 to 1.0% of cobalt naphthenate accelerator can be used, depending on the desired working time of the resin.

It is important that special care be taken to avoid the contamination of organic peroxides with accelerators or promotors used in polymerization reactions. These materials should never be added directly to one another or consecutively to a resin unless one ingredient is thoroughly mixed in before adding the other. In some cases, vigorous or explosive decompositions may result if direct contamination occurs.

E. Resin Reactivity

The maximum exothermic temperature reached, the time required for the reaction to attain peak exothermic temperature, and the time of gelation are important factors to be considered when selecting a resin. Reactivity tests provide a method for determining the behavior, uniformity, and curing characteristics of a resin. The use of a resin for a specific application often depends on the reactivity of the resin. Measurements of reactivity are helpful in the evaluation of accelerator, catalysts, and other materials that must be considered for the correct use of the resin.

The inhibitor in the resin counteracts the catalyst which dissociates into free radicals to initiate polymerization during the induction period. As the inhibitor becomes completely consumed, near the end of the interval, the free radicals from the catalyst initiate polymerization. The beginning of the polymerization is evidenced by the exothermic reaction, which caused the temperature of the resin to rise above the ambient bath temperature and the gelation of the resin. Knowing the time it takes for gelation is very helpful in selecting the correct resin for a particular application.

The time period it takes for the temperature of the resin to rise from 5°F above the bath temperature to its maximum is the propagation interval. The rate of polymerization increases until the rate of heat evolution of the resin equals the rate of heat loss to the bath. Polymerization is complete after the peak exothermic temperature is reached. The maximum exothermic temperature together with the propagation interval indicates the rate at which cure is attained.

F. Cross-Linking Mechanism

1. Free-Radical Formation

The decomposition of initiators is induced by heat in the case of molding compounds or by accelerators at temperatures below the decomposition temperature of the initiator in the case of cast polyester resins. Two types of accelerators are used, metal salts—mainly cobalt salts—and amines. The oxidoreduction of metal salts by peroxides produces free radicals. The process is very efficient since both lower and higher valencies of cations participate in the reaction.

$$RO_2H + Co^{2+} \longrightarrow Co^{3+} + RO + OH^-$$

$$RO_2H + Co^{3+} \longrightarrow Co^{2+} + RO_2 + H^+$$

$$Co^{3+} + OH^- \longrightarrow Co^{2+} + OH$$

2. Free-Radical Cross-Linking

The free radicals first react with the chemical inhibitor which has previously been added to the resin, since the inhibitor material must be chemically dissipated before any reaction between free radicals and the $C=C$ double bonds can proceed [5]. Apparently, the free radicals serve to open the double bonds in the polyester linear chain to set in motion that portion of the polymerization process designated as initiation. Either the opened double bonds react with the vinyl groups of the monomer, or the free radicals serve to also open (add to) these latter unsaturated $C=C$ bonds, permitting them to perform their cross-linking function, uniting the polyester chains into a three-dimensional network. There is further evidence that free radicals may also, to some degree, react with the unsaturated monomer to form various products of decomposition [5].

Theoretically, the reaction of polyesters should go to completion with all the double bonds reacted upon by free radicals and complete cross-linking established under the most favorable conditions. However, in actual practice, as determined by iodometric analysis, the true amount of residual unsaturation (indicating how far the polymerization has not gone) has been traced in the actual curing of polyesters, and can be summarized as follows:

1. What may be considered as an optimum cure with full-properties potential realized occurs when 92 to 95% of the unsaturation has been converted. Neither extra catalyst nor postcuring will convert this slight amount of remaining unreacted material.
2. The failure of all unsaturated sites to become reacted during final cure accounts for the discoloration of polyesters upon weathering and long-term aging. The unreacted double bonds eventually take up oxygen due to the action of sunlight and other factors, and peroxides are formed, creating a yellowish or amber color.

III. STRUCTURE–PROPERTIES RELATIONSHIPS

For a given polyester formulation, the properties of the final compound are a function of its condensation (e.g., carboxyl and hydroxyl group concentration), viscosity and molecular weight distribution, and the structural features of the three-dimensional network obtained after free-radical copolymerization. An increase in the molecular weight of an unsaturated polyester improves its hardness, tensile and flexural strength, and its heat distortion temperature (HDT) until a plateau value is reached. Carboxyl end groups import higher viscosities and better physical properties to polyesters than do hydroxyl end groups.

Generally, both the physical and chemical properties of a polyester are affected by the ratio and type of the acid and diol components and of the copolymerizable monomer. To this effect higher proportions of maleic anhydride lead to a higher density of cross-linking and thus greater hardness and heat resistance of the cured resin. Conversely, phthalic anhydride is the most common reagent used to decrease the density of cross-linking, increasing the flexural strength. Equally, a variety of glycol can be used to obtain different resin properties. Propylene and ethylene glycols, diethylene glycol, and neopentyl glycol are commonly used.

Vinyl monomers are usually added to the polyester resin as solvents of the unsaturated polyester, this to decrease viscosity within manageable limits, as well as cross-linking reagents. Styrene is the vinyl monomer most commonly used. Thus the degree of cross-linking can be controlled not only by modifying the concentration of unsaturated acid residues in the resin backbone, but also by changing the proportion of vinyl monomer added to the resin. The length of the cross-links can be controlled to a certain extent by modifying the concentration and type of vinyl monomer used.

IV. GLASS-FIBER LAMINATION

One of the main uses of polyester resin is to function as the adhesive for glass-fiber lamination. The cross-linking reaction of unsaturated polyesters is exothermic; that is, it is accompanied by a rise in temperature. Indeed, one of the useful features of an inorganic adherend functioning also as a reinforcing agent in these resins is that the heat of reaction is dissipated efficiently, achieving better temperature control across the width of the laminate. Poor temperature control during curing often gives rise to one or several of the following defects: warpage, shrinkage, motley surface resulting from overcure, and blisters resulting from undercure [6].

Glass fibers are the preferred form of adherend for polyester resins since they provide the strongest laminates. The glass may be of various types: for example, electrical glass, a low-alkali borosilicate glass, or alkali glass with an alkali content of 10 to 15%. The former gives laminates with the best weathering and electrical properties, but the latter is cheaper. For good adhesion to be achieved between resin and glass it is necessary to remove any sizing (in the case of woven cloths) and then to apply a finish to the fibers. The function of a finish is to provide a bond between the inorganic glass and the organic resin. Today, the most important of these finishes are based on silane compounds. In a typical system vinyl trichlorosilane is hydrolyzed in the presence of glass fiber, and this condenses with hydroxyl groups on the surface of the glass [7].

The glass-fiber strands are converted into three basic forms: roving, filament yarn, and mat. These fibers have a high tensile strength and differ from natural fibers in that they have no inner cellular structure and therefore do not absorb moisture internally. They do absorb it on the surface, however, and can be wetted with organic liquids. The roving is in a twinelike form, prepared by twisting and collecting 60 simple yarns (12,240 filaments) on a spool. In this form glass is available for chopping into shorter fibers of varying lengths for use in preforming machines or for incorporation into molding compounds. The most common mat is obtained by cutting fibers into 5-cm lengths, collecting them by suction on a moving metal screen, applying a binder, baking, and collecting the bound mat in rolls. The 5-cm fiber length confers optimum properties in respect to manufacture, molding characteristics, and all-round strength properties [7].

REFERENCES

1. P. F. Bruins, *Unsaturated Polyester Technology*, Gordon and Breach, New York, 1976.
2. A. Fradet and E. Marechal, *Adv. Polymer Sci. 43*: 51 (1982).
3. A. Fradet and P. Arland, 1989, *Comprehensive Polymer Science*, Vol. 5, *The Synthesis, Characterization, Reactions and Applications of Polymers*, pp. 331–334.
4. G. C. Eastmond and A. Ledwith, 1989, in *Comprehensive Polymer Science*, Vol. 5, *The Synthesis, Characterization, Reactions and Applications of Polymers*, pp. 275–297.
5. S. S. Oleesky and J. G. Moir, *Handbook of Reinforced Plastics*, Van Nostrand Reinhold, New York, 1964, pp. 13–55.
6. K. W. Lem and C. D. Han, *J. Appl. Polymer Sci. 28*: 3185 (1983).
7. G. Lubin, 1986, *Handbook of Fibre Glass and Advanced Plastics Composites*, pp. 23–45.

27

Hot-Melt Adhesives

A. Pizzi* *University of the Witwatersrand, Johannesburg, South Africa*

I. INTRODUCTION

Hot melts are a widely used class of adhesives that are used for many applications but are rarely used for structural bonding, as seldom are they able to match the tensile strengths of other adhesive classes. Their primary uses are in packaging and in wood for edge veneering and veneer splicing. There are important reasons for employing hot-melt adhesive systems, such as:

1. Ease of application via high-speed equipment
2. Formation of strong, permanent, and durable bonds within a few seconds of application
3. No environmental hazard and minimal wastage because of 100% solid systems
4. Ease of handling
5. Absence of highly volatile or flammable ingredients
6. Excellent adhesion
7. Wide formulation possibilities to suit individual requirements (e.g., color, viscosity, application temperature, and performance characteristics)
8. Cost-effectiveness

Hot melts are 100% solid thermoplastic materials that are supplied in pellet, slug, block, or irregular-shaped chip form. They require heating via appropriate application equipment, which usually is fairly sophisticated in order to control the required temperature and coverage rate. Upon application, the heat source is removed and the thermoplastics set immediately (within a few seconds). Hot melts are thus well suited to high-speed continuous-bonding operations.

II. ETHYLENE–VINYL ACETATE HOT MELTS FOR EDGING

A. Physical Characteristics

Edge veneering requires use of a hot-melt adhesive that is relatively high in viscosity at application temperatures (usually around 200°C). The reasons for this are as follows:

Current affiliation: Université de Nancy, Epinal, France

451

1. The adhesive must have sufficient body to prevent flowing from vertical surfaces after application.
2. It must not penetrate the substrate surface too deeply, causing glue starvation.
3. It must have easy spreading and excellent wetting characteristics.

Viscosities of these hot melts are on the order of 50,000 to 60,000 mPa·s (cP) at 200°C. Viscosity is achieved through the correct selection of ethylene–vinyl acetate (EVA) copolymer grades, coupled with the quantity and type of reinforcing filler that is added to the system. The ball and ring softening point is an early indication of the degree of heat resistance of a particular hot melt. The softening point is influenced by the combination of ingredients, but to a large extent by the grade and quantity of EVA copolymer and tackifying resin contained in the system. Using a 5.1-g lead ball, the average softening points are between 90 and 105°C.

For optimum adhesion, the wetting characteristics (of the hot melt to substrates during application) are vital. Proper wetting is related to viscosity but is again largely influenced by resin selection and quantity. Stability of the adhesive is another important consideration. During prolonged periods at elevated temperature while contained in the hot-melt applicator, the hot melt must resist oxidation and thermal breakdown of components. This often leads to discoloration, charring, and inferior bonds. As a result of charred material, nozzle blockages can also be encountered.

B. Formulation Considerations

EVA hot melts consist basically of the following:

1. EVA copolymer
2. Tackifying and adhesion-promoting resins (e.g., hydrocarbon, rosin esters, coumarone–indene, terpene resins)
3. Fillers, usually barium sulfate (barytes) or calcium carbonate (whiting)
4. Antioxidants

1. EVA Copolymer

EVA copolymer is the main binder in the system and largely influences the following: (a) viscosity and rheology characteristics, (b) cohesive strength, (c) flexibility, and (d) adhesive strength. A variety of EVA grades are available, allowing the formulator a choice of varying vinyl acetate contents coupled with varying viscosities (melt index). Higher VA contents generate greater adhesion to plastics, coupled with increased flexibility. The higher the VA content, however, the higher the cost. Broadly speaking, EVA-based edge-veneering hot melts utilize grades averaging 28% vinyl acetate, and formulations usually contain 40% binder.

2. Resins

A certain percentage of resin is almost always incorporated into formulations, with resin content varying from 8 to 25%. Hydrocarbon resins are used most often, but rosin esters, terpenes, and indene resins, which are more heat stable, are also common. Resins provide better flow, hot-tack, adhesion, and wetting characteristics.

3. Fillers

The heavy fillers, such as barytes, are used at levels of up to 50% by weight, but more commonly at around 35 to 40%. The filler imparts cohesive strength and body to the adhesive, and also reduces the cost considerably. Barium sulfate is the filler chosen in most cases because of its high density and hence low pigment volume concentration. Barium sulfate grades vary from beige to dark brown, and this assists in formulating specific opaque colors to match color requirements. Finely ground calcium carbonate is sometimes used as a filler where very light colors are required. Titanium dioxide pigment is commonly used as a toner, at levels of 2 to 5%.

4. Antioxidants

Antioxidants are added to protect the organic components, especially resins, from oxidation/discoloration at high temperatures. A large choice exists. These materials are usually added at levels of 0.2 to 0.5%.

C. Production Technique and Equipment

Because of their relatively high melt viscosities, the EVA hot melts need special manufacturing equipment. For example, a Z-blade mixer such as a Baker Perkins or Winkworth with oil-heated jacketing is required. Mix temperatures are kept as low as possible ($\pm 110°C$) to keep bulk thick. The high-viscosity kneading action ensures rapid dissolution of EVA copolymer and resin. Fillers are easily dispersed and a homogeneous mix is achieved rapidly with this type of agitation. Upon completion, the molten product is extruded into ropes approximately 6 mm in diameter, which are cooled through a chilled water trough and then granulated into pellet form. Alternatively, hot-melt slugs are supplied where application equipment utilizes this form. It is essential to ensure that any residual moisture picked up during the cooling process is eliminated via an air-drying cyclone before packing.

III. POLYAMIDE HOT MELTS

The polyamide hot melts are high-performance systems and are used selectively where good heat resistance is required. Their high cost relative to EVA types makes them rather unattractive for general use. Polyamide resins offer high tensile strengths and high initial tack, often without the need for additional formulating. Their higher melt points ensure good heat-resistance qualities and are responsible for rapid setting on cooling. Their two main drawbacks are cost and the tendency to char easily if kept at high temperatures.

Hot-melt polyamide resins are obtained by the reaction of diamines with diacids. While in their simplest form polyamides are the reaction of a particular diamide with a particular diamine, most of the polyamides used in adhesive formulations are complex reaction products obtained by combining several diacids and diamine to obtain the particular properties required. The most common diacid used is a dibasic acid obtained by polymerizing oleic or linoleic acid or other unsaturated fatty acids. This acid can be represented as HOOC—R—COOH, where R is a hydrocarbon residue of 34 carbon atoms and of indeterminate configuration. Commercial forms of this dimeric diacid also contain preparations of products obtained by polymerization of three or more molecules of unsaturated fatty acids and thus contain varying quantities of trimeric acids and of higher homologs. Monomeric forms are also present. The most used diamine for this type of adhesive is ethylenediamine, $H_2N—(CH_2)_2—NH_2$, but other diamines are also used, responding to the general formula

$$\begin{array}{ccc} X & & X \\ | & & | \\ HN & \!\!\!\!-\!\!\!\!-\!\!\!\!-\ R\ -\!\!\!\!-\!\!\!\!-\!\!\!\! & NH \end{array}$$

where X and Y can be H or other chemical groups. Polyamides are then formed according to the schematic reaction

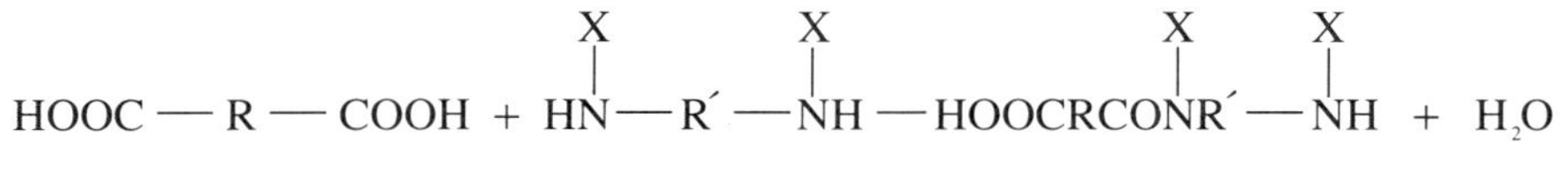

$$\underset{}{HOOC-R-COOH} + \overset{X}{\underset{|}{HN}}-R'-\overset{X}{\underset{|}{NH}}-HOOCRCON\overset{X}{\underset{|}{R'}}-\overset{X}{\underset{|}{NH}} + H_2O$$

or simply

$$n\,HOOC-R-COOH + n\,\overset{X}{\underset{|}{HN}}-R'-\overset{Y}{\underset{|}{NH}}-OCRCON\overset{X}{\underset{|}{R'}}-\overset{Y}{\underset{|}{N}}_n +$$

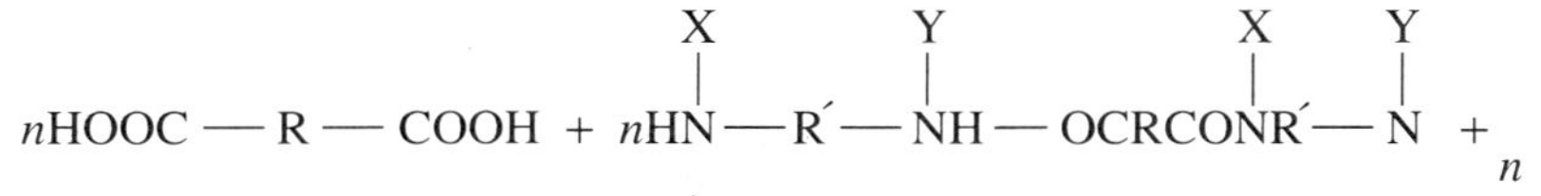

The reaction occurs with the elimination of water to form amide groups. The high polarity of the amide groups contributes to give, by formation or interchain hydrogen bonds, the characteristic polymer strength and adhesive properties to the polyamides.

The basic resins need some form of modification to achieve (1) suitable application viscosities, (2) flexibility, and (3) reduction in costs if possible. Suitable polyamide resins (those of the more flexible variety) are thus frequently modified by the addition of EVA copolymer (high-viscosity, high-melt-point grade). The amount of EVA that can be added is restricted to a maximum of 25% in most cases because of compatibility problems. The blend is then further modified with selected tackifying resin addition and small quantities of filler, to reach an optimum balance of performance properties. To achieve maximum adhesion, it is common for polyamide hot melts of this type to be used in conjunction with a polyamide resin solution primer system for edging material. The primer is invariably a dilute solution of the base polyamide resin.

IV. ADHESIVE APPLICATION GUIDELINES

In general, one should ensure that operation of the machine is in accordance with the manufacturer's instructions, being sure to set the machine according to the adhesive supplier's specifications for line speed, operating temperature, and adhesive coating weight. During application, the following guidelines should be observed:

1. Adhesive reservoir temperature: 204°C
2. Application roller temperature: 191°C (application roller to be 12 to 13°C lower than reservoir temperature)
3. Adhesive application weight: 200 to 250 g/m^2
4. Melting time of adhesive: 1½ to 3 h

Correct application weight and spread of the hot melt can be checked by bonding a transparent PVC strip and applying at a pressure of 2 to 4 kg/cm^2. If the correct pressure has been applied, the pattern caused by the applicator wheel on the adhesive should disappear, with little or no squeeze-out at the edges. The adhesive reservoir must be filled completely and the lid kept in place, to avoid heat loss. It is necessary to clean gluepots thoroughly at regular intervals, as well as filters, nozzles, and glue lines (ever 2 to 4 weeks). Suitable solid hot-melt cleaners, available from adhesive suppliers, have largely replaced the traditional use of cleaning solvents. The applicator wheel or roller is used at a pressure of 2 to 4 kg/cm^2.

V. APPLICATION AREAS

A. Veneer Splicing

Particleboard with a decorative wooden veneer surface usually employs a hot-melt adhesive to effectively mate veneer edges down the length of the joint. Polyamide hot-melt adhesives are widely used for this veneer splicing process. The adhesive is more often than not supplied as a thread and positioned as such in a zigzag configuration. A heated press is employed to activate the adhesive followed by rapid cooling and setting. The rapid set required is best achieved with polyamide hot-melt adhesives, since their setting temperatures are much higher than, for example, those for ethylene–vinyl acetate (EVA) types, and since the range between application and setting temperatures achievable with polyamides is narrow. Another important feature of polyamide resins in the context of this application is the low melt viscosities achievable, thus ensuring rapid spreading and wetting of the molten film.

B. Edge Veneering and Edge Banding

Edge veneering and edge banding constitute by far the main area employing hot-melt adhesives, which are based predominantly on EVA copolymer resins. For some applications, however, formulated polyamide hot melts are also used, particularly where exceptional heat resistance of the bond is required. Modern materials such as decorative surface board products, used in the manufacture of furniture components, require exposed edges to be covered with suitable edging materials. Most laminated surface board products consist of a decorative melamine or PVC layer bonded to a chipboard substrate. These board products have their own performance characteristics, which may influence the edge bond. The choice, application, and fabrication method of edging plays a very important part in the manufacturing and final application of the furniture produced. The choice of correct edging selection, therefore, depends very much on performance requirements and aesthetic value.

VI. GRAVURE APPLICATOR WHEEL TECHNOLOGY

A. Application Process

The gravure applicator wheel is the most important individual component on an edge-bander. Adhesive transfer to the substrate takes place when the substrate comes in contact with the gravure wheel applicator, which should rotate at the same speed as the moving track. Fresh adhesive is resupplied to the applicator wheel via the doctor blade in less than one revolution of the wheel.

By adding another doctor blade and reversing the direction of the rotation of the gravure applicator wheel, much more adhesive can be driven onto the substrate. This may be necessary when edging substrates with a low density, wide edges, or edges that require a lot of gap filling (plywood). Worn gravure applicator wheels should be replaced immediately. On replacement, a change in machine performance will take place and equipment adjustments should be carried out.

B. Heating

Where a cartridge heater is mounted in the center of the applicator wheel shaft, a high-temperature grease must be used as a heat-conducting medium between cartridge heater and applicator wheel; otherwise, there will be rapid cartridge heater burnout due to overheating. Adhesive temperature at the applicator wheel should be 12 to 13°C cooler than the adhesive temperature in the reservoir, to increase cartridge heater life.

C. Behavior of Hot-Melt Adhesives on Gravure Wheel Applicators

The single most important component on an edgebander is the gravure wheel applicator. All other components of the machine support the performance of the gravure wheel applicator. It is the gravure wheel application that controls the amount of hot-melt adhesive that is applied to the substrates, which, in turn, determines the number of calories of heat present to keep the hot-melt liquid until the time of bond formation.

Of the hot-melt adhesive that is located between the peaks and the doctor blade, only 1% is transferred to the substrate; the balance becomes nothing more than squeeze-out. The volume of adhesive that is found in the groove area is the actual material that is transferred to the substrate edge. The purpose of the doctor blade is not to act as an adjustment to increase or decrease adhesive transfer to the substrate; rather, it serves to replace in the grooves the adhesive that has transferred to the substrate. An incorrect doctor-blade setting will either cause excess squeeze-out or insufficiently fill the grooves, which will result in less than maximum adhesive transfer. The volume of adhesive that is transferred to the substrate is the single major controlling factor in determining the open time of the hot-melt adhesive. The volume of adhesive that does transfer is preset at the factory and is determined by the actual dimensions and geometry of the groove area.

If the adhesive is too cohesively strong (too cold) or too cohesively weak (too hot), the adhesive will break out of the top of the groove in the gravure applicator wheel, reducing the adhesive transfer. This is exactly why the open time of the hot-melt adhesive is determined by the amount of adhesive transferred, not by raising the application temperature. If the shear force is too low to move the adhesive from the gravure applicator wheel (slow line speed), adhesive transfer is reduced from maximum; too high a shear force (fast line speed) also reduces transfer. Shear force affects the cohesive strength of the hot melt exactly like temperature.

VII. FORMULATIONS FOR TYPICAL EVA EDGE-VENEERING HOT MELTS

Following are formulas for hot melts in various applications. Ingredients are listed in parts by weight.

A. General-Purpose Hot Melts for Both Wood and Plastic Veneers

	White	Natural	Brown
Hydrocarbon resin, 90°C m.p.	5.50	5.50	5.50
Rosin ester, 85°C m.p.	8.00	8.00	8.00
Coumarone indene resin, 105°C m.p.	5.00	5.00	5.00
Butylated hydroxytoluene (BHT) antioxidant	0.20	0.20	0.20

	White	Natural	Brown
Elvax 250	14.00	14.00	14.00
Elvax 210	10.00	10.00	10.00
Elvax 150	15.00	15.00	15.00
TiO$_2$ pigment	4.50	0.80	—
Superfine light barytes	37.80	41.50	—
Pink barytes	—	—	42.30
	100.00	100.00	100.00

B. Low-Cost Hot Melt for Wooden Veneer Only, Natural Color

Hydrocarbon resin, 90°C m.p.	12.00
Hydrocarbon resin, 100°C m.p.	10.00
Elvax 250	30.00
BHT antioxidant	0.20
Superfine light barytes	47.20
TiO$_2$	0.60
	100.00

C. Hot Melt for Difficult Plastic Surfaces (e.g., Deccon, Natural Color)

Polyterpene resin, 115°C m.p.	30.00
BHT antioxidant	0.20
Elvax 260	10.00
Elvax 250	35.00
CaCO$_3$ (15 μm)	24.80
	100.00

28

Reactive Acrylic Adhesives

Dennis J. Damico *Lord Industrial Adhesives, Erie, Pennsylvania*

I. INTRODUCTION

Acrylic adhesives were developed in Germany in the late 1960s as an outgrowth of poly(methyl methacrylate) chemistry. These early acrylic adhesives were developed primarily for bonding thermoplastics. Formulations that have developed since then have progressed beyond the ability to bond thermoplastics and now bond well to metals as well as a number of different substrates. Acrylic adhesives differ from urethane and epoxy adhesives in that they cure by addition polymerization involving free-radical reactions. The extent to which this type of chemistry affects the handling and performance of acrylic adhesives has an impact on the performance and handling of this type of adhesive and is discussed throughout this chapter.

II. CHEMICAL REACTIONS IN ACRYLIC ADHESIVES

Figure 1 depicts the most common means of initiating acrylic adhesive cures. The reaction involves the interaction of an aromatic amine with a peroxide at room temperature to form a free radical. This free radical reacts further with methacrylate monomers present in the adhesive formulation to form the final cured high-molecular-weight adhesive material.

Figure 2 shows the chemical structure of some of the more common acrylic monomers used to formulate acrylic adhesives. These monomers interact with the free radicals to fully cure acrylic adhesives. Adhesive formulators will select one of these monomers or a blend of these monomers to obtain specific adhesive properties.

Other mechanisms capable of forming free radicals that can initiate the cure of acrylic adhesives are:

Perester dibasic acid + metal ion (U.S. patent 4,348,503)
Saccharin salt + α-hydroxysulfone (U.S. patent 4,081,308)
Hydroperoxide + thiourea and metal salt (e.g., cobalt) (U.S. patent 4,331,795)
N,N-Dimethyl + saccharin p-toluidine (U.S. patent 3,658,624)

and

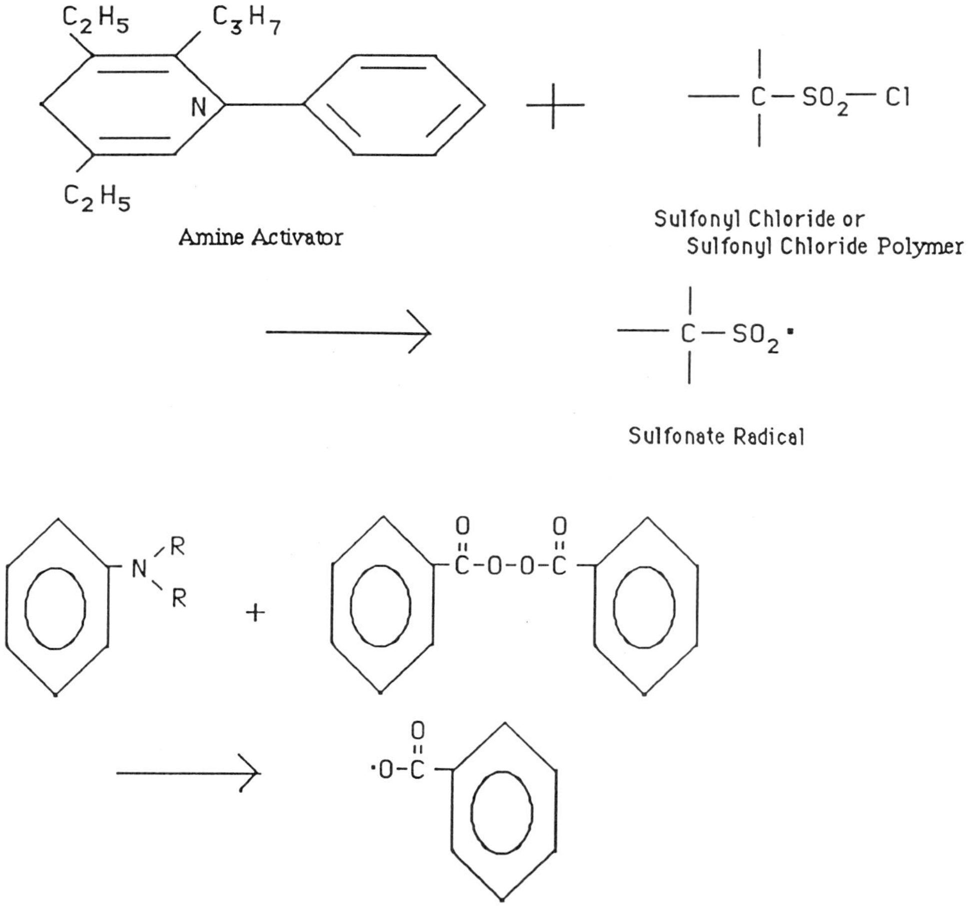

Figure 1 Free-radical formation acrylic chemistry.

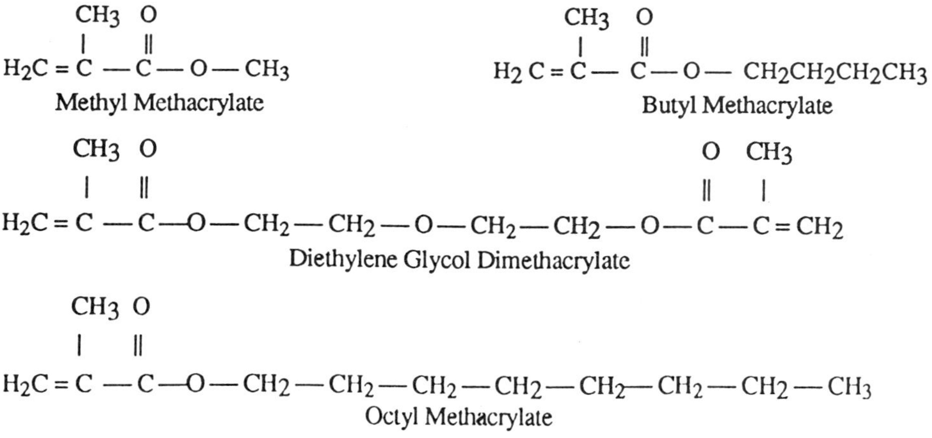

$$\begin{array}{cc} CH_3 & O \\ | & \| \\ H_2C = C - C - O - CH_3 \end{array}$$
Methyl Methacrylate

$$\begin{array}{cc} CH_3 & O \\ | & \| \\ H_2C = C - C - O - CH_2CH_2CH_2CH_3 \end{array}$$
Butyl Methacrylate

$$\begin{array}{cc} CH_3 & O \\ | & \| \\ H_2C = C - C - O - CH_2 - CH_2 - O - CH_2 - CH_2 - O - C - C = CH_2 \end{array}$$
Diethylene Glycol Dimethacrylate

$$\begin{array}{cc} CH_3 & O \\ | & \| \\ H_2C = C - C - O - CH_2 - CH_2 - CH_2 - CH_2 - CH_2 - CH_2 - CH_2 - CH_3 \end{array}$$
Octyl Methacrylate

Figure 2 Reactive species (monomers) for acrylics.

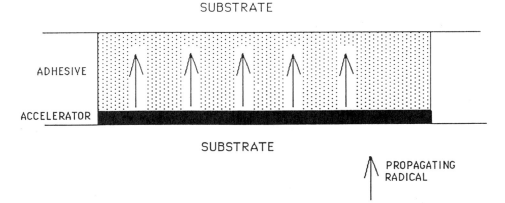

Figure 3 Accelerator cure.

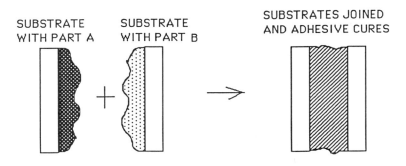

Figure 4 Honeymoon cure.

As mentioned in Section I, the chemistry depicted here enables the acrylic adhesive formulation to cure by addition polymerization. Figures 3 and 4 illustrate curing techniques made possible by the cure chemistry of acrylic adhesives. "Honeymoon" and accelerator lacquer cures are not possible with other condensation-cured adhesives in Fig. 3. Figure 5 shows that once cure is initiated, it proceeds very rapidly in contrast to other types of cures, such as in urethane and epoxy.

III. HANDLING OF ACRYLIC ADHESIVES: DO'S AND DON'TS WITH ACRYLICS

A number of handling characteristics follow from the cure chemistry of acrylics. One notable difference with acrylic adhesives is that it is not always necessary to mix the resin and hardener portion intimately at the time of use. In some cases, as is shown in Section II with the A/B method, the simple joining together of surfaces coated with the A and B portions of the adhesive will initiate cure.

Another important consideration with many types of acrylic adhesives is the mix ratio. Many acrylic adhesives must be mixed at mix ratios of 10:1 or 20:1. This requires

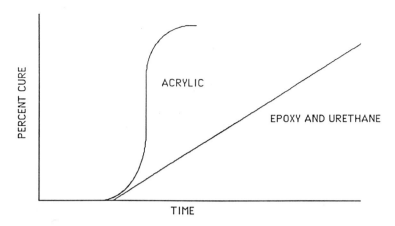

Figure 5 Percent cure–time curves for acrylic versus epoxy and urethane.

proper selection of meter/mix/dispense equipment, if used, to meter and control the mix ratio accurately.

A third consideration to be noted in working with acrylic adhesives is the gelation phase of the cure. Acrylic adhesives will often form a soft gel during the course of the cure. After this gel state is achieved it is important that bonded parts not be moved until a full cure is obtained. Finally, it is important to the adhesive user to be aware of the fact that acrylic cure is, to some degree, affected by atmospheric oxygen. This will often be visible for cure inhibition at the surface of adhesive resin in contact with atmospheric oxygen. This phenomenon is readily observable as surface softness or tackiness with some types of acrylic adhesives.

Safety considerations are also important in using acrylic adhesives. A potential user of acrylic adhesives should consult safety data sheets supplied by manufacturers as to the identity of methacrylate monomers used in recipes. Some acrylic adhesives are formulated with flammable monomers. Proper precautions should be taken when using this type of adhesive.

IV. BOND STRENGTHS AVAILABLE WITH ACRYLIC ADHESIVES

Generations of acrylics have evolved and been made available to the adhesive user. In the case of acrylic adhesives, the terms *second-generation acrylic adhesive*, *DH acrylic adhesive*, *acrylic adhesive hybrid*, *HP acrylic adhesive*, and *toughened acrylic adhesive* have all been used. Consequently, it is important for the adhesive user to consult performance and specification sheets supplied by adhesive companies to determine which type of acrylic adhesive is used and what types of performance are available. Tables 1 and 2 list some typical performance properties of the various types of acrylic adhesives that have been made available in recent years. As can be seen from examining these tables, considerably different performance characteristics are possible with different types of acrylic adhesives. Again, acrylic adhesive manufacturing specifications are the best source of specific values obtainable.

Table 1 Bond Performance of Acrylic Adhesives (U.S. Patent 3,994,764)

Substrate	Shear strength (psi)		MPa
Steel/steel (clean)	3800		26.2
Aluminum/aluminum	3100		21.4
Wood/wood	>1800	(wood failed)	12.4
Steel/glass	>800	(glass failed)	5.5
Steel/PVC	>2600	(PVC failed)	17.9
Steel/glass-reinforced plastic	2750		19.0
Steel/nylon	2980		20.5
Steel/polystyrene	2420		16.7
Steel/ABS	2480		17.1
Steel/natural rubber	600	(rubber failed)	4.1
Wood/ABE	600		4.1
Hot-dip galvanized steel/steel	420		2.9

Table 2 High-Performance Acrylic Adhesives for Bonding of Oily Metals[a]

Metal	Surface preparation	Shear strength (psi)	MPa
SAE 1010 cold-rolled steel	Oiled	6390	44.1
	Solvent wiped	6100	42.1
	Grit blasted	5940	41.0
6061-T6 aluminum	Oiled	5090	35.1
	Solvent wiped	4870	37.6
	Grit blasted	5030	34.7
2024-T3 Alclad aluminum	Oiled	4880	33.6
	Solvent wiped	4720	32.5
5052-0 aluminum	Oiled	2140	14.8
	Solvent wiped	2190	15.1

[a]Assemblies were made and tested according to ASTM D-1002-72 using 1.6-mm (0.0630-in.)-thick metal stock and a 0.25-mm (0.020-in.)-thick glue line.

V. TYPICAL FORMULATIONS

Listed here are a number of typical adhesive formulations. Listed in these tables are both the general composition of the resin portion as well as the composition of the hardener portion. Some useful patent references follow the formulations.

A. HP Acrylic Adhesive

	Parts
Polymer syrup 25% polymer in MMA (carboxylated butadiene acrylonitrile)	30–60
Methyl methacrylated (MMA)	10–30
Mathacrylic acid (MAA)	1–5
Methacryloxyethyl phosphate	1–10
Mixed aromatic amides	1–3
Calcium molybdate/zinc phosphate blend	2–12

B. Typical Formulation

	Parts
Styrene/methyl methacrylate	40
Methacrylic acid	9
Polymethyl methacrylate syrup	49
Diisopropanol p-toluidine	1.5
Toluhydroquinone	0.03

C. Heat-Resistant Epoxy Acrylic Hybrid

	Parts
Polymer monomer syrup	10–30
Elastomer solution in MMA	20–40
MMA	5–15
Inorganics	25–40
Methacryl phosphate	5
Stabilizers/initiators	3
Epoxy	Varies
BPO	4

D. U.S. Patents

3,658,624	4,081,308	4,574,142
3,725,504	4,223,115	4,467,071
3,832,640	4,263,419	4,703,089
3,873,640	4,331,795	4,769,419
3,890,407	4,348,503	4,855,001
3,970,709	4,386,036	5,096,962
3,994,764	4,452,944	

VI. SUBSTRATES

Table 3 is a partial list of substrates that can be bonded with acrylic adhesives. In addition to substrates, acrylic adhesives can be looked at in terms of potential end-use applications. They have been used in a number of applications, ranging from aerospace and aluminum bonding to a variety of military applications. They have also found use in bonding a number of thermoplastics in the transportation industry as well as metals of various types used in transportation and other areas. The electronics industry, in particular magnet bonding, has also been an area where acrylic adhesives have found use.

Due to some of the developments in the high-performance acrylic adhesives, very good adhesion has been obtained to unprepared metals. This is particularly important in the automobile area, where large areas are often bonded with little or no surface preparation. Acrylic adhesives for this type of application have also been developed that have much greater resistance to high-temperature cycles and long-term environmentals than that of other types of adhesives.

Table 3 Typical Substrates Bonded
with Acrylic Adhesives

Steel
Aluminum
Galvanized steel
Painted steel
Polycarbonate
Polymethyl methacrylate
ABS
Polyvinyl chloride
General thermoplastic bonding
Composites

BIBLIOGRAPHY

Bachmann, A. G., "*Aerobic*" *Acrylic Adhesives (New Technology in Acrylic Adhesives)*, Adhesives '84 Conference, Society of Manufacturing Engineers, Dearborn, Mich., 1984.

Bowen, R. L., and H. Argentar, *J. Appl. Polymer Sci.* 17 (1973).

Brendley, W. H., Jr., *Paint Varnish Prod.* (July 1973).

Damico, D. J., *Adhesives Age* (Oct. 1987).

Gatechair, L. R., and D. Wostratzky, *J. Radiat. Curing* (July 1973).

Lees, W. A., *Intern. J. Adhesion Adhesives* 1(5) (1981).

Lees, W. A., *Brit. Polymer J. 11*(June 1979).

Pasternack, G., *J. Radiat. Curing* (July 1982).

Rohm & Haas Company, *Catalysts for the Polymerization of Acrylic Monomers*, June 1959.

Seymour, R. B., *Introduction to Polymer Chemistry*, McGraw-Hill, New York, 1971.

Wood, D. W., and R. N. Lewis, *Mod. Plastics* (July 1974).

Yeames, J. E., *A New Adhesive for Structural Bonding of Engineering Materials*, Adhesives '84 Conference, Society of Manufacturing Engineers, Dearborn, Mich., 1984.

29

Anaerobic Adhesives

Richard D. Rich *Loctite Corporation, Newington, Connecticut*

I. INTRODUCTION

Anaerobic adhesives are mixtures of acrylic esters that remain liquid when exposed to air but harden when confined between metal surfaces. These mixtures can be used for a large number of industrial purposes, such as locking threaded fasteners, sealing threaded pipe connections, retaining cylindrical machine components, sealing flange joints, bonding structural components, sealing porous metal castings, welds and powdered metal parts, and many other applications that are still being found more than 40 years after the initial invention [1]. Several reviews have been published that describe anaerobic adhesives and their applications [2–6].

The first anaerobic adhesives were made at the General Electric Company by aeration of a polyethylene glycol dimethacrylate. This "Anaerobic Permafil" required continuous aeration to prevent hardening [7]. Although a number of internal applications had been identified, the problems associated with shipping and storage made the product so impractical that the company decided to discontinue its manufacture.

Vernon Krieble, chairman of the Chemistry Department at Trinity College in Hartford, Connecticut, learned about the product through his son, Robert Krieble, who was employed at General Electric. Vernon Krieble found a chemical solution to the problem by using cumene hydroperoxide (I) (see Section XII for all structures) as the initiator and packaging in half-filled oxygen-permeable polyethylene bottles [8]. He licensed the GE patent and in 1954 founded the American Sealants Corporation, which later became Loctite Corporation [9]. At the present time anaerobic adhesives and sealants are manufactured or sold on every continent by more than a dozen companies. Applications in virtually every industry, and technological innovation, as measured by patent activity, continue unabated.

II. FORMULATION

A large number of variations are possible for anaerobic curing products, but all will consist of the following components:

1. *Monomer*. Methacrylate esters are used in almost all anaerobic products. Acrylates, acrylic and methacrylic acids, and a few other vinyl polymerizable monomers may be used for special purposes.

2. *Initiator*. A hydroperoxide (typically, cumene hydroperoxide) is almost always used as an initiator, although there are some variations and it is also possible to take advantage of small amounts of ''native'' hydroperoxide present in the methacrylate monomers.

3. *Accelerator*. A large number of chemical accelerators have been developed which can catalyze the anaerobic cure and reduce the large differences in cure speed on different surfaces. The most commonly used accelerators are saccharin [benzoic sulfimide (II)] and aromatic amines such as *N*,*N*-dimethyl-*p*-toluidine (III) and 1,2,3,4-tetrahydroquinoline (IV).

4. *Stabilizers*. All methacrylate monomers must contain some free-radical inhibitor if they are to be shipped and stored safely. Hydroquinone and *p*-methoxyphenol are most commonly used for this purpose. Most formulations will also contain benzoquinone, naphthoquinone, and similar stabilizers. Since the anaerobic compositions are strongly catalyzed by traces of metals, many formulators have found it advantageous to add chelators such as tetrasodium ethylenediaminetetraacetic acid (EDTA) (V).

5. *Modifiers*. A very large number of modifications in the cured and uncured properties of anaerobic formulations can be brought about by the addition of components that have little or no effect on the fundamental anaerobic cure chemistry. These modifiers can increase the viscosity, control thixotropy, add color or fluorescence, increase sealing effectiveness, reduce strength, increase toughness, increase heat resistance, provide lubrication, and reduce settling of fillers.

6. *Surface activators*. In some applications anaerobic sealants will cure more rapidly if the surface is treated with a solution containing a metal salt or other chemical that will catalyze the polymerization. These substances will often be components that could not be added to the sealant without causing premature gellation.

III. REACTION MECHANISM

Anaerobic adhesives and sealants have been developed primarily in industrial laboratories, and most of the published literature are patents. A number of papers have been published within the last decade which discuss the reaction mechanisms of anaerobic adhesive cure [10–20].

A. Oxygen Inhibition

The polymerization mechanism of anaerobic adhesives is similar to that of other free-radical initiation systems except for the special ways in which the inhibiting effect of oxygen is used to delay the polymerization, and in the chemical activation that occurs at the metal surface.

Initiation:		I	$\rightarrow$	$I\cdot$
	$I\cdot + M$		$\rightarrow$	$M\cdot$
Inhibition:	$M\cdot + O_2$		$\rightarrow$	$MOO\cdot$ Weak free radical
Propagation:	$M\cdot + M$		$\rightarrow$	$M\cdot$
Termination:	$M\cdot + M\cdot$		$\rightarrow$	M_2 Free radical lost
	$M\cdot + I\cdot$		$\rightarrow$	MI Free radical lost

The reaction rate of oxygen with free radicals is very high and the peroxy radical formed is a relatively poor initiator. When the supply of oxygen is used up within a thin bond line, the propagation step can provide rapid development of adhesive strength. Although there is little disagreement about the importance of oxygen in the propagation step, the greatest interest and study has been directed to the initiation step in the process described above. The role of the bonding surfaces and the effect of different types of accelerators account for most of the literature on the cure mechanism of anaerobic adhesives.

B. Transition Metals

An important factor in the initiation of anaerobic adhesive cure is the redox reaction between a hydroperoxide and transition metals with adjacent oxidation states [10].

$$Fe^{2+} + ROOH \rightarrow Fe^{3+} + RO\cdot + OH^-$$
$$Fe^{3+} + ROOH \rightarrow Fe^{2+} + ROO\cdot + H^+$$

In the reactions above, other transition metals will react similarly, and copper is particularly active. Where cumene hydroperoxide (CHP) is used, $R = C_6H_5C(CH_3)_2$.

C. Accelerators

The use of saccharin and N,N-dimethyl-p-toluidine (DMPT) results in a substantial acceleration of the initiating reaction. Although each of these components itself is an accelerator, their combination has a strong synergistic effect. It has been suggested that a charge-transfer complex is formed by these materials [11]. It is not clear whether this complex is itself an initiator or whether it acts on other components to generate initiating species.

This same study indicated that the rate of anaerobic polymerization was nearly independent of the concentration of CHP and proceeded at a significant but slower rate with no hydroperoxide. This does not indicate that the hydroperoxide is not essential to the anaerobic cure but that it does not participate in the rate-determining step.

The use of 1-acetyl-2-phenylhydrazine [APH (VI)] and saccharin resulted in a somewhat slower reaction rate than with DMPT unless a catalytic amount of copper was added. In this study the concentration of CHP was found to be very important where the molar ratio of CHP/APH was less than 1. Where the ratio was greater than 1, the rate was independent of CHP [12]. These papers [11,12] have been reviewed [13]. The accelerating effect of the salts of saccharin and 6-methyl-1,2,3,4-tetrahydroquinoline (VII), 1,2,3,4-tetrahydroquinoline (IV), or 1,2,3,4-tetrahydroquinaldine (VIII) on the anaerobic polymerization of methyl methacrylate were studied. No organic peroxides were required for these polymerizations [14].

The effect on initiation reactions of charge-transfer complexes of *o*-benzosulfanilide (IX) and tertiary aromatic amines with various substituents was studied. The rates increased with increasing electron donor and decreasing acceptor properties of the substituents on the amine [15]. Aromatic tertiary amines mixed with benzosulfimides formed charge-transfer complexes whose decay products were effective catalysts for polymerization of polyethylene glycol dimethacrylate. The most effective catalyst system was DMPT with dibenzenesulfonimide (X). The effectiveness of the system was increased by the addition of CHP [16].

A study using a model reaction system consisting of cumene hydroperoxide (CHP), *N*,*N*-dimethyl-*p*-toluidine (DMPT), and *o*-benzoic sulfimide (saccharin) in toluene (without reactive acrylic monomer) showed conclusively that the DMPT was depleted significantly during the initiation step of an anaerobic reaction. The saccharin concentration was unchanged during this process. The initiating species may be radical anions derived from DMPT rather than reactive free radicals derived from the CHP [17].

A dilatometric study was carried out on the polymerization of an anaerobic system containing diethylene glycol dimethacrylate, DMPT, CHP, saccharin, and iron filings. DMPT was more effective than *N*,*N*-dimethylaniline (XI), triethylamine, or tributylamine. Saccharin was more effective than benzoyl cyanamide (XII), phthalimide (XIII), or succinimide (XIV) [18].

D.   Role of Saccharin

It has been suggested that one of the functions of the saccharin in anaerobic cure is to dissolve metal ions from the surfaces in order to catalyze the decomposition of CHP. X-ray photoelectron spectroscopy of a model anaerobic adhesive cured in contact with a metal surface indicates that trace amounts of metal or metal ions are found at the interface [19].

E.   Oxygen Absorption

Studies of the oxygen absorption of triethylene glycol dimethacrylate indicate that the reaction is catalyzed by DMPT and by DMPT with saccharin. This oxygen absorption appears to be continuous, although the active oxygen content of the system appears to level off due to decomposition of the peroxide formed. Oxalic acid substantially inhibits the absorption of oxygen [20].

IV.   MONOMERS

The first anaerobic patents mentioned only the polyglycol dimethacrylates with tetraethylene glycol dimethacrylate (XV) as the dominant example [7,8]. Neopentylglycol dimethacrylate (XVI) was first mentioned in a patent assigned to Borden [21]. The use of acrylic or methacrylic acids to improve adhesion to smooth surfaces was mentioned in a patent assigned to Loctite [22]. The use of diallylphthalate (XVII) and diallylphthalate prepolymer as comonomers with tetraethylene glycol dimethacrylate was claimed in a patent filed by Kalinowski [23].

A series of polyurethane polyacrylates were prepared by reaction of toluene diisocyanate (XVIII), other isocyanates, and isocyanate-terminated oligomers with hydroxyalkyl methacrylates [24]. These monomers could be tailored to provide the strength and toughness required for some structural adhesive applications. The incorporation of hard and soft segments into the polyurethane backbones provided significant improvements in toughness, cure-through-gap, and cryogenic strength properties [25].

The use of monomethacrylates in anaerobic formulations was disclosed in a patent assigned to Loctite. Specifically mentioned were hydroxyethyl (XIX), hydroxypropyl (XX), cyclohexyl (XXI), tetrahydrofurfuryl (XXII), dimethylaminoethyl (XXIII), and glycidyl methacrylates (XXIV) and cyanoethyl acrylate [26]. Methacrylate esters containing residual carboxylic acid groups were prepared by the reaction of hydroxyethyl methacrylate with phthalic anhydride (XXV), pyromellitic dianhydride (XXVI), and benzophenonetetracarboxylic acid dianhydride (XXVII). The residual acid provided improved adhesion [27,28]. The reaction product of hydroxyalkylmethacrylates with maleic anhydride (XXVIII) also produced monomers with residual acid as well as additional curable unsaturation [29]. The dimethacrylates of the bisglycol esters of dicarboxylic acids were used to formulate anaerobic adhesives. Among the dicarboxylic acids mentioned were phthalic (XXIX), maleic (XXX), fumaric (XXXI), and malonic (XXXII) [30].

Three Bond Company used trimethylolpropane trimethacrylate (XXXIII) [31] and ethoxylated bisphenol A dimethacrylate (XXXIV) [32] in anaerobic formulations. These monomers have some advantages in providing improved heat resistance. At Henkel & Cie. dicyclopentadienyl methacrylate (XXXV) was used in anaerobic formulations with high strength [33]. Rohm and Haas has disclosed the use of dicyclopentenyloxyethyl acrylate (XXXVI) and methacrylate in anaerobic formulations [34]. These monomers provide good cure strength on metal parts that have not been degreased and also have lower odor and volatility than do the corresponding dicyclopentadienyl esters. Silicone methacrylates have been formulated by Dow Corning [35] and Toshiba Silicones [36].

Methacrylate esters have been prepared by the reaction of methacrylic acid with epoxies such as the diglycidyl ethers of bisphenol A (XXXVII) [37]. Methacrylate esters suitable for anaerobic adhesives have also been prepared by the reaction of glycidyl methacrylate (XXXVIII) with a hydroxyl-terminated polyester [38]. The reaction of isocyanatoethyl methacrylate (XXXIX) with polyols resulted in monomers that could be formulated into anaerobic adhesives and sealants [39].

V. INITIATORS

The most commonly used initiator for anaerobic adhesives is cumene hydroperoxide. Many other hydroperoxides have been disclosed, such as *t*-butylhydroperoxide (XL), *p*-menthane hydroperoxide (XLI), diisopropylbenzene hydroperoxide (XLII), pinene hydroperoxide (XLIII), and methyl ethyl ketone hydroperoxide (XLIV) [40]. Some diperoxides, such as di-*t*-butylperoxide (XLV) and dicumylperoxide (XLVI), have been claimed, but these may function only because of hydroperoxide contamination [41].

Storage-stable anaerobic formulations can be prepared with no hydroperoxide if the methacrylate resin is aerated in the presence of an amide and a tertiary amine [42]. Anaerobic adhesives have been formulated with alkyl hydroxyethyl peroxides such as *t*-butyl-2-hydroxyethyl peroxide (XLVII) [43]. An adhesive formulated with *t*-butylperoxymaleic acid has improved surface adhesion (XLVIII) [44].

VI. ACCELERATORS

The first accelerator used in an anaerobic adhesive was tributylamine [45]. Saccharin was also found to be an effective accelerator [46] and the combination of saccharin and *N,N*-dimethyl-*p*-toluidine was particularly effective if properly stabilized [47]. A large number of patents have been issued on various accelerators and combinations thereof. *N*-Aminorhodanine (XLIX) [48], 1-acetyl-2-phenylhydrazine (VI) [49,50], benzenesulfonylhydrazide (L) [51], dibenzenesulfonamide (X) [52], and similar compounds have been disclosed.

The use of saccharin has been of particular interest and a number of compounds have been prepared that have a similar chemical structure and reactivity. The reaction of sulfuryldiisocyanate (LI) with acetic acid gives a disulfonamide that is an effective accelerator [53]. Similar reactions of *p*-toluenesulfonyl isocyanate (LII) and chlorosulfonyl isocyanate (LIII) can be used to prepare many different compounds which are active accelerators [54,55]. These methods allow the preparation of accelerators with improved solubility.

VII. STABILIZERS

The monomers used in anaerobic adhesives and sealants generally contain at least one free-radical stabilizer, such as hydroquinone or *p*-methoxyphenol. It was found that benzoquinone, naphthoquinone, and similar compounds provided improved shelf stability without retarding the anaerobic cure [56]. It was also found that anaerobic formulations could be stabilized with a stable nitroxide free radical such as di-*t*-butyl nitroxide (LIV) [57]. The use of a soluble metal chelating agent such as tetraasodium EDTA (V) was found to be an effective method of stabilizing an anaerobic formulation against small amounts of metal contamination [58].

VIII. MODIFIERS

The wide variety of applications of anaerobic adhesives and sealants is made possible by the modifications that make the viscosity appropriate to the application. An application that requires penetration into close-fitting parts should have very low viscosity, while a product used with large, loose-fitting parts should have a high viscosity. A styrene–acrylate copolymer could be used to increase the viscosity [59]. Polymethacrylates, cellulose esters, butadiene–styrene copolymers, acrylonitrile–butadiene–styrene copolymers, poly(vinyl chloride), copolymers of vinyl chloride and vinyl acetate, poly(vinyl acetate), cellulose ethers, polyesters, polyurethanes, and other thermoplastic resins have also been used to control the flow characteristics of anaerobic sealants [60]. The flow characteristics of anaerobic formulations can also be controlled by the addition of fumed silica and other solid additives which can impart "thixotropic" properties [61].

Many anaerobic adhesives and sealants may require a relatively low strength so that the components can be disassembled for repair or replacement. Many liquid plasticizers have been used for this purpose [62], but the use of a low-molecular-weight polyester has been found to be advantageous [63]. The "toughness" properties of anaerobic adhesives can be enhanced by the addition of a reactive elastomer [64]. The heat resistance of anaerobic adhesives and sealants can be enhanced by the addition of a bismaleimide (LV) [65]. These maleimide additives appear to be relatively unreactive during the initial anaerobic cure. As the adhesive is exposed to high temperatures the methacrylate backbone degrades and the methacrylates can then copolymerize with the maleimides, forming a more heat-resistant matrix [66].

The addition of dyes to anaerobic adhesives and sealants assists in identification and inspection of the products. Automated inspection procedures are made possible with dyes that fluoresce under ultraviolet light. Titanium dioxide pigments can make the sealants more visible.

Solid fillers are added to some anaerobic adhesives and sealants for various purposes. Mica, talc, and other mineral fillers can help to provide an instant seal capability to anaerobic pipe sealants. The sensitivity of the anaerobic cure system to metal contamination requires that these fillers be chosen very carefully.

Powdered graphite, polytetrafluoroethylene, and polyethylene can function as lubricants in pipe sealants and thread-locking compounds. This lubrication can prevent galling in close-fitting pipe threads [67]. Lubricating additives in thread-locking sealants can provide control of the clamping force exerted by a fastener at a given tightening torque [68].

IX. PRIMER/ACTIVATORS

The cure speed and adhesion of an anaerobic sealant can be increased by treatment of the surface with a solution containing activating chemicals. Early anaerobic thread lockers were strongly affected by part cleanliness, and degreasing the parts with a chlorinated solvent improved performance dramatically. The condensation product of an aldehyde and a primary or secondary amine, a sulfur-containing free-radical accelerator (LVI), or a compound of an oxidizable transition metal were some of the materials that could be added to activate the anaerobic cure [69–71]. Ferrocene (LVII), a derivative containing the ferrocene moiety, or a polymer incorporating ferrocene was an effective activator for anaerobic adhesives [72]. The copper, cobalt, manganese, or chromium salts of an acid phosphate acrylic monomer (LVIII) were found to be effective activators as well as adhesion promoters [73].

X. APPLICATIONS

1. *Thread locking.* The first applications for anaerobic adhesives were for locking threaded fasteners. Filling the "inner space" between a nut and bolt with a hard, dense material prevents self-loosening.

2. *Thread sealing.* The effect of filling the space between threaded parts or the space between inner and outer pipe threads provides a seal that can prevent the leakage of oil and other fluids in machinery as well as prevent corrosion of the threaded parts.

3. *Retaining.* Cylindrical press-fits and bearing assemblies can be retained with anaerobic adhesives, allowing accurate alignment and relaxed tolerances. Retaining and sealing of cup plugs and oil seals in castings is a major application.

4. *Impregnation.* Powdered metal parts, porous castings, and welds can be sealed against leakage of liquids or gases. This impregnation can also allow such parts to be plated and improves their machinability.

5. *Preapplied.* Thread lockers and sealants are made that can be coated on threaded parts in the form of a dry-to-the-touch film. These preapplied materials remain inert until assembly releases a quick-curing resin.

6. *Gasketing*. Anaerobic flange sealants can be applied manually or by automated methods such as tracing, stenciling, and screen printing. These products can eliminate a variety of preformed, precut gaskets and can also be used as a gasket dressing.

7. *Structural bonding*. Tough structural bonds can be achieved with some anaerobic adhesives for bonding components such as ferrite magnets, honing stones, identification tags, and decorative inserts.

Other applications include the fabrication of foundry molds [74,75] and the surface mounting of electronic components [76].

XI. STANDARDS AND SPECIFICATIONS

Testing standards and performance specifications for anaerobic adhesives and sealants have been established by government agencies and industrial organizations in several countries. In the United States there are military specifications for thread lockers, sealants, and retaining compounds. Mil-S-22473E, 12 April 1983, "Sealing, Locking and Retaining Compounds: (Single Component)" covers 15 of the earliest "letter grade" products. Specifications are set for color, viscosity, locking torque on ⅜-24 steel nuts and bolts, and fluid tightness. The effects of immersion in a number of fluids, heat aging at 149°C, and hot strength at 149°C (or 93°C for some grades) are also measured. This specification calls for measurement of an "average locking torque" after 90, 180, 270, and 360 degrees of turn.

Mil-S-46163A, 12 July 1983, "Sealing, Lubricating and Wicking Compounds: Thread-Locking, Anaerobic, Single-Component" covers nine grades of product for sealing (type I), lubricating (type II), and wicking (type III). Specifications are set for color, viscosity, locking torque (break and prevailing torque) on ⅜-16 steel, zinc- and cadmium-plated nuts and bolts, fluid tightness, lubricity, and "wicking" into pre-assembled fasteners. The immersion, heat aging, and hot strength tests are similar to those done in Mil-S-22473E.

Mil-R-46082B, 10 June 1983, "Retaining Compounds Single Component, Anaerobic" (Amendment 6, 9 January 1990) covers three types of retaining compounds, which are tested with a pin-and-collar compressive shear specimen. The three types vary primarily in viscosity, although there are also differences in heat resistance and strength. These products are subjected to immersion, heat aging, and hot strength tests similar to those described above.

In the United Kingdom the Ministry of Defence (MOD) has issued specifications DTD 5628–5633, which cover test procedures and performance requirements for a range of products. Five strength bands and four viscosities, from penetrating to thixotropic, are defined. The torque strengths are tested on M8 nuts and bolts and the shear strength in 12-mm pins and collars. The development of these specifications and the test procedures have been described by C. L. Brett at the MOD. The "breakloose" torque on nuts and bolts requires particular attention to a transient measurement where the first torsional motion is detected. Other products show somewhat different behavior, with no distinct "breakloose," and the torque at which the sealant begins to yield is not easily detected [77]. British Standard BS 5292 relates to the use of anaerobic sealants on gas appliances.

In Germany, standards have been published describing the "Compression Shear Test" (DIN 54452), "Dynamic Viscosity Determination of Anaerobic Adhesives by Rotational Viscometer" (DIN 54453), "Initial Breakaway Test at Bonded Threads"

(DIN 54455), and ''Torsion Shear Test'' (DIN 54455). DIN 54455 is particularly interesting since it is one of a very few tests in which a nut and bolt (M10) are seated to a measured torque before the anaerobic sealant is allowed to cure.

In the United States the Industrial Fastener Institute has published standards entitled ''Test Procedure for Locking Ability Performance of Non-metallic Locking Element Type Prevailing Torque Lock Screws'' and ''Test Procedure for the Locking Ability Performance of Chemical Coated Lock Screws.'' The American Society for Testing and Materials (ASTM) has published a ''Standard Test Method for Shear Strength of Adhesives Using Pin-and-Collar Specimen'' (ASTM D4562-90, October 1990). A subcommittee of ASTM Committee D-14 on Adhesives is actively studying torque strength tests and performance standards for anaerobic adhesives but has not yet reached any publishable conclusions (G. Litteral, personal communication).

The International Organization for Standardization (ISO) has circulated a Draft International Standard (ISO/DIS 10964, February 1992) ''Adhesives—Anaerobic Adhesives—Determination of Torque Strength of Anaerobic Adhesives on Threaded Fasteners.'' This standard is still subject to change and will not be adopted before September 1992. The draft describes testing procedures for liquid and preapplied sealants using manual and graphical procedures.

XII. LIST OF STRUCTURES

(XVI)

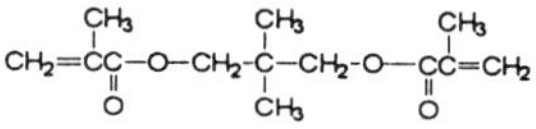

(XVII)

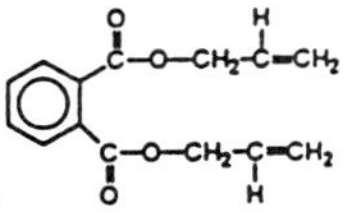

(XVIII)

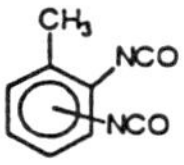

(XIX)

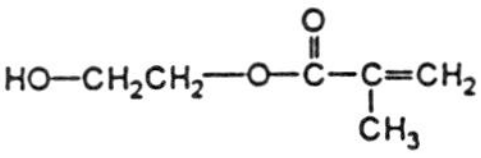

(XX)

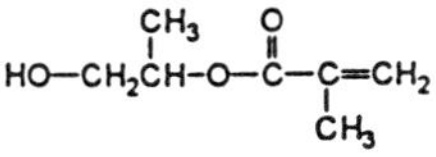

(XXI)

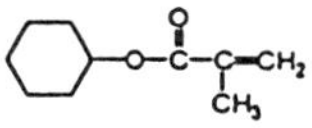

(XXII)

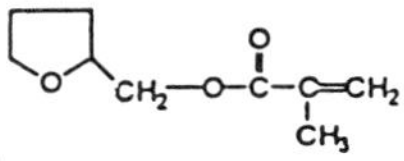

(XXIII)

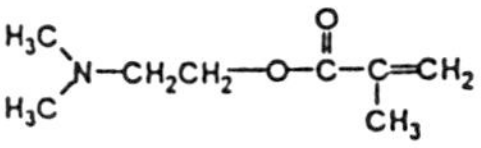

(XXIV)

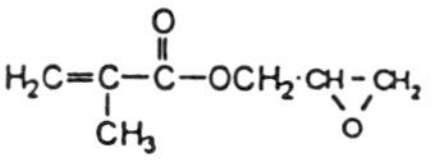

(XXV)

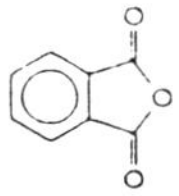

(XXVI)

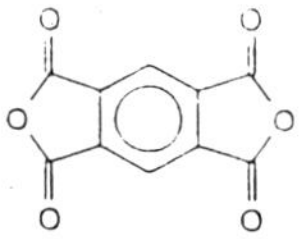

(XXVII)

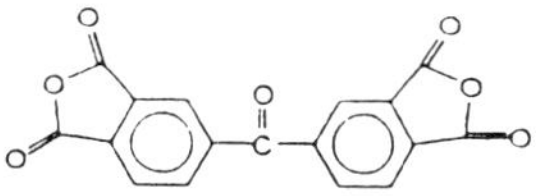

(XXVIII)

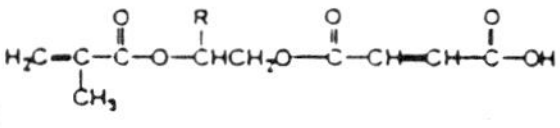

(XXIX)

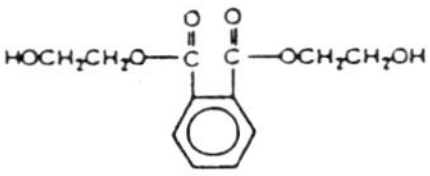

(XXX)

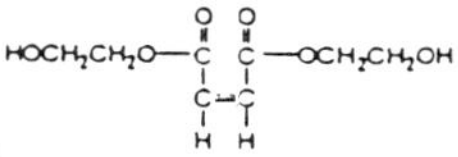

(XXXI)

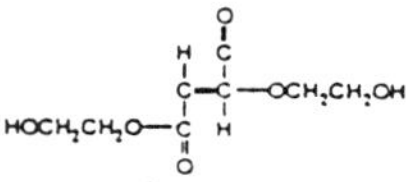

(XXXII)

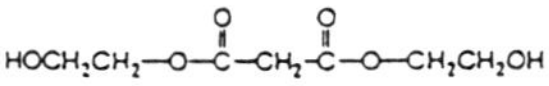

(XXXIII)

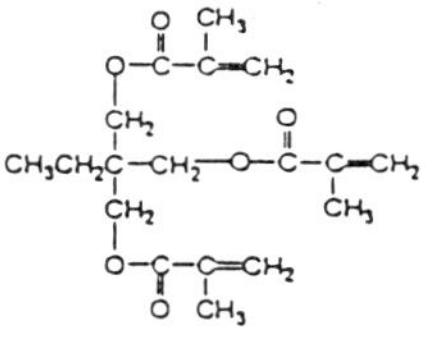

(XXXIV)

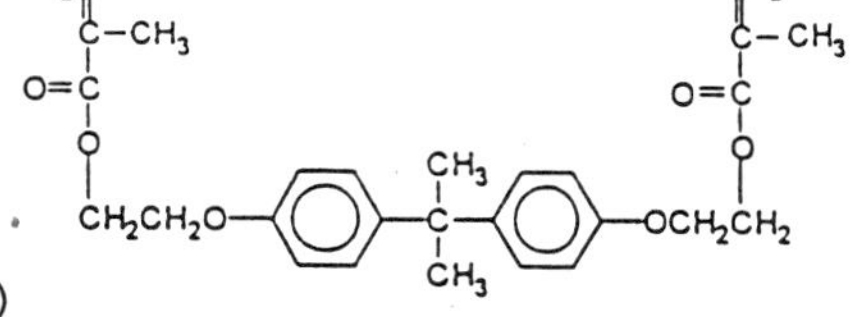

(XXXV)

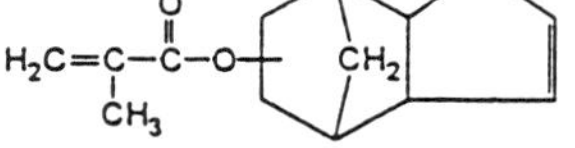

(XXXVI)

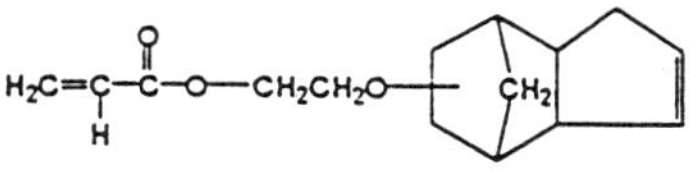

(XXXVII)

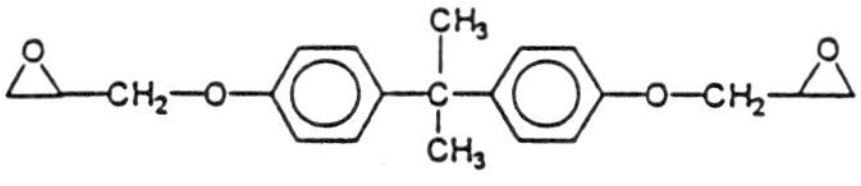

(XXXVIII)

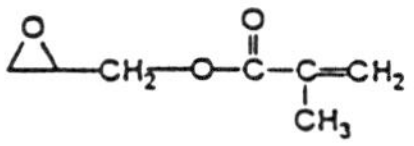

(XXXIX)

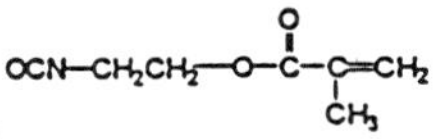

(XL)

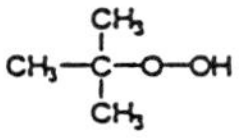

(XLI)

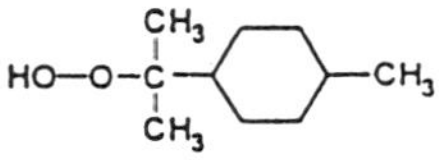

(XLII)

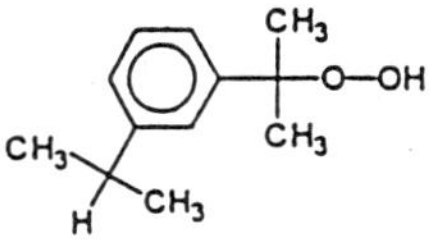

(XLIII)

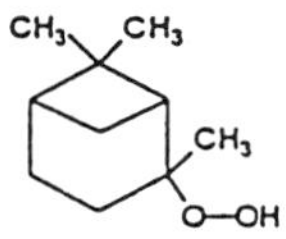

(XLIV)

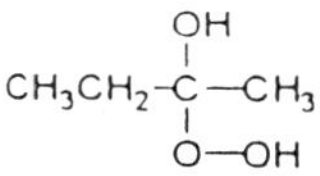

(XLV)

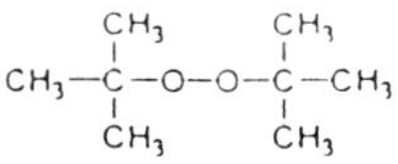

(XLVI)

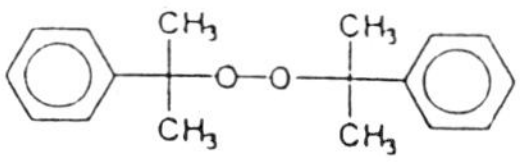

(XLVII)

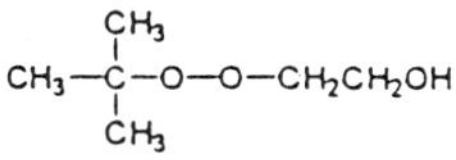

(XLVIII)

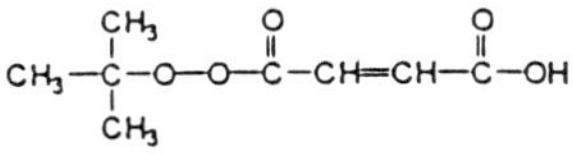

(XLIX)

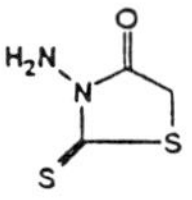

(L)

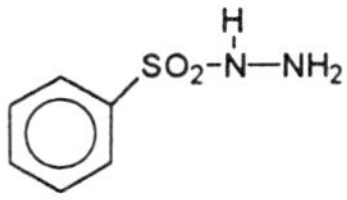

(LI) OCN—SO₂—NCO

(LII)

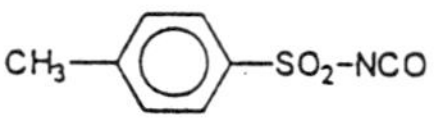

(LIII) Cl—SO₂—NCO

(LIV)

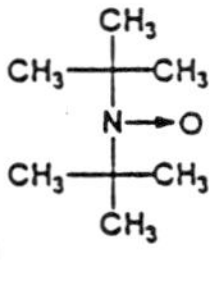

(LV)

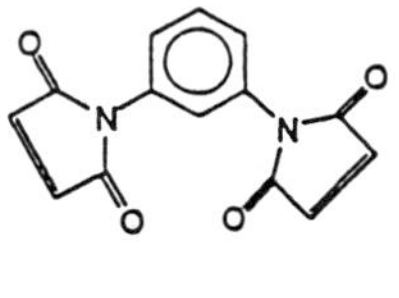

(LVI)

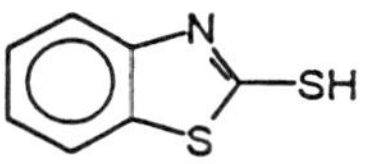

(LVII)

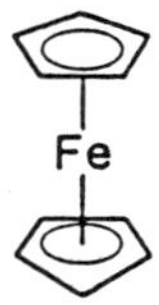

(LVIII)

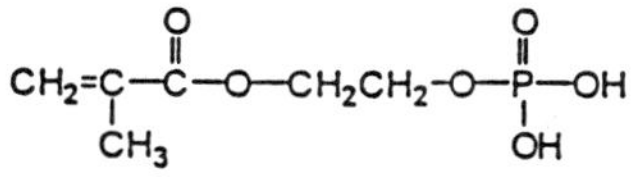

REFERENCES

1. G. S. Haviland, *Machinery Adhesives for Locking, Retaining and Sealing*, Marcel Dekker, New York, 1986.
2. L. J. Baccei and M. Hauser, in *Encyclopedia of Materials Science and Engineering*, Vol. 1, (M. B. Bever, ed.), Pergamon/MIT, Cambridge, Mass., 1986, pp. 47–51.
3. C. W. Boeder, in *Structural Adhesives Chemistry and Technology* (S. R. Hartshorn, ed.), Plenum Press, New York, 1986, pp. 217–247.
4. J. M. Rooney and B. M. Malofsky, in *Handbook of Adhesives*, 3rd ed. (I. Skeist, ed.), Van Nostrand Reinhold, New York, 1990, pp. 451–462.
5. W. A. Lees, *Brit. Polymer J.* 11: 64 (1979).
6. P. Penczek, *Adhaesion*, 32(4): 25 (1988).
7. R. E. Burnett and B. W. Nordlander, U.S. patent 2,628,178 (1953).
8. V. K. Krieble, U.S. patent 2,895,950 (1959).
9. E. S. Grant, *Drop by Drop: The Loctite Story*, Loctite Corporation, Newington, Conn., 1983.
10. D. J. Stamper, *Brit. Polymer J.* 15: 34 (1983).
11. Y. Okamoto, *J. Adhesion* 32: 227 (1990).
12. Y. Okamoto, *J. Adhesion* 32: 237 (1990).
13. R. A. Pike, *Chemtracts Macromol. Chem.* 2(6), 406 (1991).
14. T. Okamoto and H. Matsuda, *Nippon Setchaku Kyokaishi* 20(10): 468 (1984).
15. D. A. Aronovich, A. F. Murokh, E. Yu. Nikolaev, and A. P. Sineokov, *Plastmassy 1*: 51 (1990).
16. S. B. Meiman, D. A. Aronovich, E. G. Pomerantseva, G. I. Poduvalova, N. V. Fokeeva, and A. P. Sineokov, *Deposited Doc. SPSTL 918 Khp-D82*, 1982.
17. J. B. D. Smith, *J. Appl. Polymer Sci.* 45: 1 (1992).
18. M. Teodorescu, G. Hubca, C. Oprescu, and M. Dimonie, *Mater. Plastice*, 25: 170 (1988).
19. S. J. Hudak, F. J. Boerio, P. J. Clark, and Y. Okamoto, *Surface Interface Anal.* 15: 167 (1990).
20. V.-H. Chao, J.-F. Chung, and S.-T. Voong, *Kao Fen Tzu T'ung Hsun 2*: 80 (1978).
21. W. Karo and B. D. Halpern, U.S. patent 3,125,480 (1964).
22. J. W. Gorman and B. W. Nordlander, U.S. patent 3,300,547 (1967).
23. L. W. Kalinowski, U.S. patent 3,249,656 (1966).
24. J. W. Gorman and A. S. Toback, U.S. patent 3,425,988 (1969).
25. L. J. Baccei, U.S. patents 4,018,851 (1977), 4,295,909 (1981), and 4,309,526 (1982).
26. B. W. Nordlander, U.S. patent 3,435,012 (1969).
27. A. M. Brownstein, U.S. patents 3,428,614 and 3,451,980 (1969).
28. T. H. Shepherd and F. E. Gould, U.S. patent 3,595,969 (1971).
29. G. P. Werber, U.S. patents 4,209,604 (1980) and 4,431,787 (1984).
30. Y. Fukuoka, S. Kusayama, and M. Suzuki, U.S. patent 3,457,212 (1969).
31. S. Kiyono and R. Ogawa, Japanese patent 44,007,541 (1969) [*CA 71*: 125643j].
32. T. Saito, U.S. patent 3,890,273 (1975).
33. B. Wegemund and J. Galinke, U.S. patent 3,642,750 (1972).
34. W. D. Emmons and V. J. Moser, U.S. patent 4,234,711 (1980).
35. R. H. Baney and O. W. Marko, U.S. patent 4,035,355 (1977).
36. F. Tetsuo, T. Masahiro, and E. Isao, European patent 467,160 (1992).
37. B. Wegemund and G. Tauber, U.S. patent 3,660,526 (1972).
38. R. A. Pike and F. P. Lamm, U.S. patent 4,524,176 (1985).
39. D. K. Hoffman, U.S. patent 4,320,221 (1982).
40. K. Azuma, I. Tsuji, H. Kato, H. Tatemichi, A. Motegi, O. Suzuki, and K. Kondo, U.S. patent 3,925,322 (1975).
41. V. K. Krieble, U.S. patent 3,218,305 (1965).
42. W. A. Lees, D. J. Bennett, J. R. Swire, and P. Harding, U.S. patent 3,795,641 (1974).

43. B. M. Malofsky, U.S. patent 4,007,323 (1977).
44. P. J. Clark, U.S. patent 4,916,184 (1990).
45. V. K. Krieble, U.S. patent 3,041,322 (1962).
46. V. K. Krieble, U.S. patent 3,046,262 (1962).
47. V. K. Krieble, U.S. patent 3,218,305 (1965).
48. G. B. Bachman, U.S. patent 3,491,076 (1970).
49. D. P. Melody, D. A. Doherty, J. F. O'Grady, and R. D. Rich, U.S. patent 4,180,640 (1979).
50. R. D. Rich, U.S. patents 4,287,330 (1981) and 4,321,349 (1982).
51. W. Gruber, J. Galinke, and J. Keil, U.S. patent 3,984,385 (1976).
52. W. Gruber, J. Galinke, and J. Keil, U.S. patent 3,985,943 (1976).
53. K. Reich, U.S. patent 4,429,063 (1984).
54. A. F. Jacobine, U.S. patent 4,513,127 (1985).
55. A. F. Jacobine and D. M. Glaser, U.S. patent 4,622,348 (1986).
56. R. H. Krieble, U.S. patent 3,043,820 (1962).
57. D. J. O'Sullivan and D. J. Stamper, U.S. patent 3,682,875 (1972).
58. E. Frauenglass and G. P. Werber, U.S. patent 4,038,475 (1977).
59. J. W. Gorman and B. W. Nordlander, U.S. patent 3,300,547 (1967).
60. E. Frauenglass and W. E. Cass, U.S. patent 3,625,875 (1971).
61. E. Frauenglass, U.S. patent 3,547,851 (1970).
62. W. A. Lees and J. R. Swire, U.S. patent 3,419,512 (1968).
63. A. G. Bachman, U.S. patent 3,794,610 (1974).
64. T. R. Baldwin, D. J. Bennett, and W. A. Lees, U.S. patent 4,138,449 (1979).
65. B. M. Malofsky, U.S. patent 3,988,299 (1976).
66. L. J. Baccei, and B. M. Malofsky, *Polymer Sci. Technol.* 29: 589 (1984).
67. C. B. Fairey, E. Frauenglass, and L. W. Vincent, U.S. patent 4,813,714 (1989).
68. L. O'Connor, *Mech. Eng. (9): 52 (1991).*
69. *A. S. Toback and J. T. O'Connor, U.S. patent 3,591,438 (1971).*
70. *A. S. Toback, U.S. patent 3,616,040 (1971).*
71. *A. S. Toback and W. E. Cass, U.S. patent 3,625,930 (1971).*
72. *B. M. Malofsky, U.S. patent 3,855,040 (1974).*
73. *P. J. Clark, U.S. patent 4,990,281 (1991).*
74. *K.-H. Bruning, W. Kuhlgatz, E. Mekus, H.-U. Schubert, H. Schwarzer, and W. Schuh, British patent 2,004,788 (1979).*
75. *G. E. Green and J. L. Greig, U.S. patent 3,986,546 (1976).*
76. *S. Grant and J. Wigham, Hybrid Circuits, 8: 15 (1985).*
77. *C. L. Brett, Int. J. Adhesion and Adhesives, 2(1): 19 (1982).*

30

Aerobic Acrylics: Increasing Quality and Productivity with Customization and Adhesive/Process Integration

Andrew G. Bachmann *Dymax Corporation, Torrington, Connecticut*

I. INTRODUCTION

It is axiomatic that manufacturing productivity is increased and per unit cost is decreased by making assembly and automation processes more efficient. High-quality parts further lower costs by reducing rework and replacement liability costs. Environmental consciousness and regulatory compliance have become a permanent and increasing component of assembly costs.

Aerobic adhesives cure rapidly to form tough, durable bond lines with structural strength. They also produce highly adhesive protective coatings, sealants, and pottings *at the place and speed required by the assembly process*. Overall economic efficiency is improved because:

The complete lack of solvents allows easy compliance with environmental and worker safety regulations.
Aerobic adhesives improve productivity of automated assembly processes by curing ''instantly,'' but only on demand.
Profitability is increased by substantially eliminating the time delay between assembly and quality control procedures.
Tough, durable bond lines reduce replacement and liability costs.

Defined in very broad terms, most manufactured items are assembled by using:

Press or snap fits
Mechanical fasteners (screws, rivets, welding)
Chemical fasteners (adhesives)

''Aerobic'' adhesives were developed as an advance over anaerobic and cyanoacrylate adhesives. Anaerobic (threadblocking) adhesives are used to augment mechanical fasteners such as screws, bolts, or pres-fits. Cyanoacrylates tend to be used for nondurable bonding on rigid surfaces and for durable rubber bonding.

Aerobic adhesives, on the other hand, were designed to be used as the sole attachment device. They are as fast as many ''instant glues'' yet durable enough to be considered a "chemical fastener," replacing the need for mechanical fasteners, in many instances.

Chemical fasteners must be designed to meet the requirements of the assembly, the process, and environmental protection and worker safety.

Adhesives and sealants designed to complement and enhance all three stand the best chance of offering equipment designers and manufacturers the highest quality at the lowest per unit cost.

In addition, the twin challenges of increasing global competition and heightened environmental quality awareness mandate that manufacturers replace the commonly used, but simplistic, price per pound decision calculation with an in-depth understanding of the factors associated with obtaining the highest-quality product, the most efficient process, and having the least environmental impact.

Aerobic acrylic UV and activator curing adhesives were developed with the idea of meeting the requirements of maintaining or increasing product and environmental quality while delivering the ultimate in process efficiency.

II. HISTORICAL PERSPECTIVE/TECHNOLOGY REVIEW

A. Aerobic Acrylic Curing Technology—Activator and Heat Cures

Chemical or heat curing structural strength aerobic adhesives technology was first introduced to the assembly industry in the early 1980s (SME Technical Papers AD88-649 and AD84-580). High-performance characteristics derived from combining urethane oligomers with (meth)acrylic monomers and "elastomeric domain"' compounds. The meaning of the term "aerobic" and how it is different from anaerobic adhesives and 2nd-generation acrylic adhesives are explained in these and other articles [1–3,5].

Aerobic adhesives are composed of proprietary catalysts, elastomeric domain fillers, and low-vapor-pressure monomers. Formulations derived from this technology do not exhibit the severe sensitivity toward air inhibition shown by other acrylic adhesives. The result is that aerobic adhesives are usable on more porous surfaces and in wider gaps than was previously considered practical.

Of course, all vinyl polymerizations can be inhibited by air, hence the commonly seen "tacky" feel of many ultraviolet (UV) cured products. Therefore, those physical properties that are recognized as being affected by air inhibition mechanisms are compared in Appendix A.

One of the tests is maintaining strength through a gap. It can be affected by air inhibition, as well as by the diffusion of accelerators. In another test, air was beaten into several adhesive formulations. The aerobic adhesives were only marginally affected by air inclusion when cured between surfaces. Like anaerobics, fillets or adhesive squeezed outside of a bonded joint will remain uncured unless exposed to UV light.

In the last test, a porous material, pine wood, was chosen as a substrate because it assures that air is to remain in intimate contact with the curing adhesive. The ability to lock the threads of an iron nut and bolt without use of an activator is a recognized test for determining whether a formulation is capable of anaerobic cure.

B. Reactive Aerobic Acrylic Adhesives— Use and Cure Mechanism

1. Chemical Cure

Chemical bondings techniques for all acrylic adhesives are similar. First, the activator is applied to one of the surfaces to be joined as a thin film. While the permissible "ratio" of activator to adhesive is quite wide (10 to 50:1), too much activator will lower bond strengths. Activators are available in solvents to help assure thin film applications. Second, the adhesive is applied to the mating surface, and the two are brought together and held until handling strength is reached.

Application of activator to both substrates is recommended where one substrate is porous or where gaps of more than 20 mils are to be filled. Activator and adhesive may be applied to the same surface, but this initiates the bonding action so that the parts must be joined immediately.

(a) Advantages of Activator Curing Aerobic Acrylics

Simple no-mix application
Multiple curing methods available
Broad applicability to a wide variety of substrates (metal, plastic glass, ceramics) with minimal surface preparation
Tolerance for oil-contaminated surfaces
Rapid bonding at room temperature
Excellent bond flexibility
High peel strength, lap shear strength, and impact resistance
Excellent solvent resistance
Wide latitude in cure rate
Ability to bond heat-sensitive substrates
Low shrinkage during polymerization
Excellent reproducibility and reliability
Good low- and high-temperature properties
Wide formulating flexibility for specific end-use requirements

2. Heat Cure

Many aerobic formulations can be heat-cured, *but only between two surfaces*. Table 1 shows the heat curing properties in an adhesive used to mount surface mount devices on

Table 1 Time for Complete Cure

Temperature (°C)	Aerobic adhesive	Epoxy	Anaerobic adhesive
110	5 min	Unknown	Unknown
120	45 s	39 min	30 min
150	20 s	2–5 min	1–3 min
75	10 s	1 min	20 s

printed circuit boards compared to a typical epoxy and a typical anaerobic formulation used for the same purpose.

3. Curing by Ultraviolet Light

Aerobic adhesives are easily converted to UV curing formulations. This is not surprising as (meth)acrylics are commonly used as the base resin in UV inks and coatings. However, UV products usually have rather poor adhesion to hard, tough surfaces such as metal.

Many UV curing aerobic adhesives show the unique property of having structural properties regardless of the cure mechanism used, as shown below. A more detailed discussion of UV curing aerobic acrylics can be found in Sec. IV.

Figure 1 shows that complete cure using a preapplied activator requires 24–28 h. UV cure, however, is complete—reaching the same strength level in only seconds.

C. Increasing Product Quality Through Use of Aerobic Acrylic Adhesives

Aerobic acrylic UV and activator curing technology was developed with a view to maintaining the formulation flexibility necessary to meet a broad range of application requirements. Adhesives, sealants, and coatings providing improved product quality have been the result. In addition to increased formulation flexibility, improvements in resin

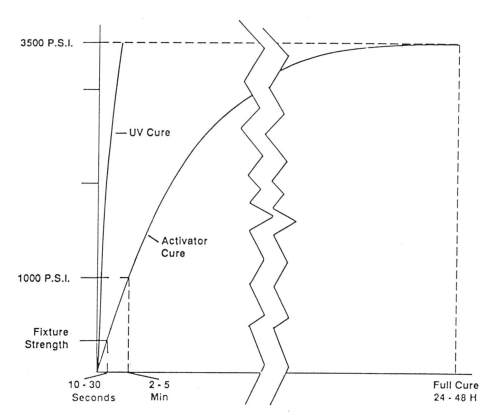

Figure 1 Typical adhesive tensile strength on cold-rolled steel.

characteristics possessing toughness and durability have resulted in assemblies of superior quality and performance [4,6,7].

D. Tough, Durable Bonds Yield Higher-Quality Assemblies

The essence of acceptable assembly quality is to eliminate production line rejects and failures over the life of the manufactured product. Aerobic acrylic activator curing adhesives and sealants have been in use in demanding automotive, medical, electrical, and electronic applications for the past 10 years. Designed to resist vibration, shock, and impact, they have withstood the severe environmental stresses of thermal, as well as physical, shock. Excellent for bonding closely fitting metal, glass, ceramic, filled plastic, and thermoset plastic parts, aerobic adhesives improve durability by also acting as shock-absorbing cushions. High tensile/shear strength to 3000 psi, combined with flexibility and the ability to absorb shock, results in a toughness developed to increase the useful life of the bonded or sealed part. Table 2 compares the toughness as measured in impact strength of aerobic acrylics with other adhesive products offered to OEM manufacturers. Appendix B tabulates a range of properties available from different aerobic adhesive formulations [1].

E. Adhesive Customization Improves End-Product Quality [1,5]

Aerobic adhesives have a wider latitude in formulating for specific end-use requirements than other acrylic adhesives. Flexibility, opacity, cure speed, and surface compatibility may be controlled over a wide range to produce adhesives for metal, glass, many plastics, ceramics, and hard woods. The cured properties of specific aerobic formulations are scarcely affected by efficient thickening agents such as fumed silica. As a consequence, it is possible for an end user to pick a set of cured properties and then have the adhesive's viscosity modified to specific requirements. The properties of three representative cured formulations are outlined in Appendix B.

Two of many excellent examples of how aerobic acrylic technology improves product quality are the following. A photocopier manufacturer uses an aerobic adhesive to attach large glass lenses to metal frames. The application requires an adhesive that can absorb the forces generated from the large differences in expansion characteristics between glass and metal. Type B formulated adhesives were shown to provide adequate bonds with

Table 2 Typical Torsional Impact[a]

Adhesive	Values (in.-lb)
Polyamide epoxy	7 + 5
Cyanoacrylate	2 ± 2
Anaerobic	12 ± 5
2nd-generation acrylic	20 ± 7
General-purpose aerobic acrylic (metal deformation)	40 ± 10
High-impact aerobic acrylic	60 +

[a]On degreased, sandblasted steel, 1/2-in. overlap.

tensile strength exceeding the deformation strength of the metal frames. However, when subjected to a 300°F temperature range, the glass lenses frequently broke. Increasing the bondline thickness to 10 mils and using a flexible type C formulation produced shock-absorbing bonds, and the lenses now pass the thermal cycle requirements.

An aerobic acrylic adhesive has also been used to increase both product performance and reliability while reducing assembly costs in the manufacture of motorcycle alternators. A major manufacturer of high-performance motorcycles had been holding ceramic magnets in place with a plastic retaining ring. The manufacturer had to increase the alternator's output from 17 to 22.5 amps at 3600 rpm without increasing the size of the alternator. It was known that replacing the retaining ring with a bonding process would permit more magnet mass in the same space.

F. Adhesive Customization Together with Rapid Fixture and/or Cure "On Demand" Improve Process Efficiency

One hundred percent solids, solvent-free aerobic acrylic adhesives are often customized to the needs of the assembly and the process. They produce permanent, structural strength bonded assemblies, pottings, and (electronic) coatings within seconds. Productivity is increased; assembly automation processes are made more efficient because the curing technology can be customized to fit the requirements of the process. The several curing methodologies are completed in seconds and can be installed at the time and place most convenient to the manufacturing process. Table 3 summarizes the advantages of using aerobic acrylic adhesive technologies.

Examples of how aerobic acrylics' formulation flexibility allows tailoring the adhesive to the process to maximize efficiency include the following:

1. A manufacturer of high-speed printers also requires 5 min of assembly time before fixture occurs in a 10 mil shimmed bondline gap. In this application, a number of ferrite magnets are bonded to metal frames and assembled in the proper geometrical

Table 3 Enhancing Quality and Processes Through Aerobic Adhesives and Coatings

Enhancing productivity	Productivity is increased by making assembly automation processes more efficient.
Enhancing quality	High-quality parts reduce replacement liability costs. Quality improvements due to toughness, durability, and structural strength of bondpottings and sealants and protective coatings lower liability exposure.
Bondline durability	Aerobic adhesives cure to form bonds with structural strength; produce highly adhesive practical coatings, sealants, and pottings at the speed required by an efficient assembly process.
Customized curing	Aerobic acrylics improve productivity. They facilitate automation by curing "instantly," but only on demand.
Profitability	Profitability can be increased by lowering per unit labor content.
Worker safety and regulatory compliance	Low-volatile, 100% solids means minimal EPA and OSHA impact.

arrangement in one step. This adhesive process replaces a combination of mechanical clamps, spacers, and holding screws.

2. A manufacturer of restaurant appliances fastens a stainless steel top to a combined counter/refrigerator/stove unit. The standard attachment technique included both welding and the use of self-taping screws. An aerobic acrylic adhesive has replaced these techniques resulting in a lower cost assembly requiring less skilled labor.

3. Aerobic acrylics also show enhanced cure speed, tensile strength, moisture resistance, and the ability to resist automotive bake cycles, which can be as high as 450°F. The operation also requires fixturing through a variable 10 to 20 mil gap in less than 30 s. Thixotropic adhesive is applied to the area to be bonded and activator sprayed directly on top of the adhesive. The metal is immediately folded to form a "hem" joint.

4. A fuse manufacturer replaced a 30-min epoxy used in a heat-cure process to assemble stainless steel inserts into phenolic holders. By utilizing an aerobic adhesive bonding process, the assembly was complete in 30 s at room temperature. The resulting increase in productivity more than offset the higher adhesive cost. A cyanoacrylate bonding process was even faster, but did not resist aging in a moist atmospheric environment or pass a drop test for impact resistance.

5. Many kinds of adhesives are utilized in the bonding of loudspeaker magnets, DC motor assemblies, and other types of magnet bonding. Some manufacturers still prefer to use mechanical clamps. The factors that contribute to the choice of a joining process include the cost of the adhesive, labor rates, speed of adhesive fixture, acceptability of odor level, and energy costs. Aerobic acrylic adhesives have been used primarily for their rapid fixture (as little as 12 s) resulting in faster assembly rates. Table 4 compares cure speed and other simple considerations used when choosing an adhesive for attachment of ferrite. The above applications are illustrated in Fig. 4.

The most familiar automotive use of activator curing types of aerobic acrylics is in DC motor assembly. This end use is particularly illustrative of the process streamlining and

Table 4 Some Adhesives for Magnet Bonding

Adhesive	Cure conditions	Gap	Disadvantages
Heat curing epoxy	30 min at 300°F	Any	Clean surfaces critical; high energy cost; cool-down time
Nylon hot melt	Induction heating of preapplied patch	Any	Clean surfaces critical; high energy cost; cool-down time; parts breakage
Anaerobic acrylic	30–60 s	10–20 mils	Clean surfaces critical; two-step process; cost/lb
2nd-generation acrylic	30–120 s	10–30 mils	Flammable vapors; noxious odors; stringy two-step process; OSHA/EPA problems
Aerobic acrylic	10–20 s	10–40 mils	Two-step process; cost/lb

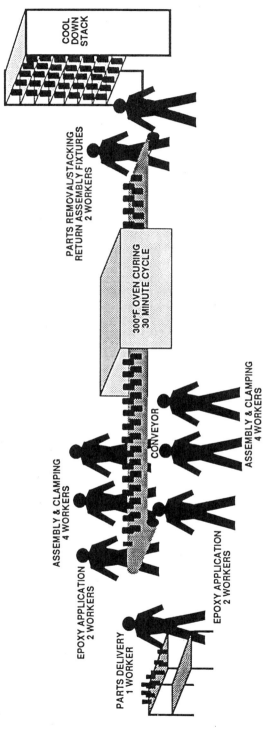

Figure 2 Heat-cured epoxy assembly method. Steps (nine workers): (1) one brings parts to conveyor belt. (2) Two apply epoxy adhesive. (3) Four assemble magnets into housing, attach stainless steel fixturing clamp, place on conveyor belt. (4) Parts are heated to 300°F for 30 min to cure the epoxy. (5) Two workers remove hot parts to cool down stack. Return fixtures to beginning of line. (6) After 24-hr cool down, parts go into the main assembly process.

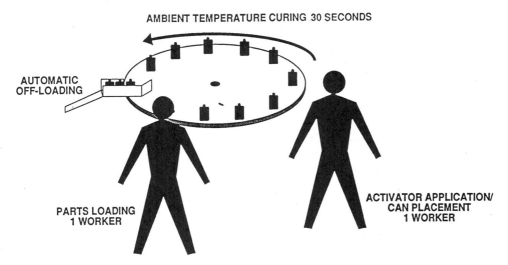

AMBIENT TEMPERATURE CURING 30 SECONDS

AUTOMATIC
OFF-LOADING

PARTS LOADING
1 WORKER

ACTIVATOR APPLICATION/
CAN PLACEMENT
1 WORKER

Figure 3 Dymax method. Steps: (1) Worker loads magnets to automatic feeder/adhesive dispenser; also brings motor cans to second worker station. (2) Second worker applies activator to can housing with a semiautomatic dispenser. (3) Second worker then places can over magnets that have been previously placed into fixtures robotically. (4) Turntable cycles, engaging fixture pressure mechanism. (5) Bonded parts are automatically off-loaded to continue in production.

cost savings opportunities offered by this unique technology. An epoxy method is shown in Fig. 2.

First used commercially in 1982, to bond ferrite magnets, cold bonding processes have increased in utility and sophistication, as indicated in Fig. 3. Productivity enhancing and quality improving uses are limited only by the imagination of the design engineer.

Table 5 shows the difference in labor, time, and, by extension, cost between using an epoxy and an aerobic acrylic adhesive.

III. AN EXAMPLE OF HIGH RELIABILITY BONDING WITH 100% QUALITY ASSURANCE TESTING

Figure 5 shows two ferrite magnets being held on a typical fixture prior to DC motor assembly. A special viscosity grade of rapid curing, moderate tensile, high-durability adhesive acrylic is depicted on the vertically held magnets. An easily dispensable viscosity was combined with a nondripping rheology as part of the customizing process for the manufacturer. After dispensing, the DC motor can be placed (usually robotically) over the ferrite fixture. A rotation of the timing table opens springs on the fixture and the magnets are pressed against the can with 20–50 lb pressure. Fixture strength occurs within 10–30 s, depending on the adhesive grade chosen, whereupon the bonded can can proceed to the next assembly step.

A 100% quality assurance check can be made within 2–5 min by pushing on the bonded ferrites at a predetermined force with a "go-no-go" gauge.

One vendor for DC motors experienced a rejected assembly about once in every 10,000 units. This simple go-no-go test finds this small number of rejects, which are then recycled.

(a)

(b)

Figure 4   Ferrite assemblies bonded in 10–30 s with aerobic acrylics. (a) Automotive DC motors. (b) Alternators. (c) Loudspeaker. (d) Stepping motors.

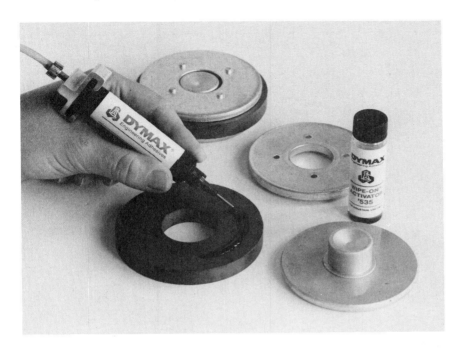

(c)

(d)

Table 5 Process Comparison

	Epoxy method[a]	Dymax method
Workers	9	2
Adhesive cost	5 cents/part	20 cents/part
Scrap rate	5%	0.1%
Process	Off line (24-hr delay)	In line—large Q.C. and W.I.P. savings
Other	Oven and energy costs	Capital depreciation

[a]Another method to cure epoxies or preapplied (nylon) adhesive patches is induction heating. Although it is a successful and attractive laboratory curing technique, at this writing, production line trials have yielded reject rates as high as 15%.

Figure 5 Two ferrite magnets being held on a typical fixture prior to DC motor assembly.

Parts can be easily cleaned and rebonded, eliminating concerns of waste disposal.

Figure 6 illustrates high adhesive strength combined with UV cure, e.g., a sealant being applied to a zinc-plated DC motor flange. UV cure is obtained in about 30 s.

The plastic headlamp is structurally bonded together in the same time frame. Two adhesives are used; one adheres to the nylon reflector, the other to the lens.

A capacitor potting compound needs aggressive adhesion to the metal can edges, providing a positive seal despite stresses induced during thermal changes.

The cost savings realized from process improvement, quality improvement, and worker environment have resulted in aerobic adhesives being widely used in DC motor assembly. One motor, entirely constructed using a combination of aerobic adhesives, press bits, and swagging, is illustrated in Fig. 7. Both activator and UV curing methods are used where these cure technologies maximize productivity and durability.

IV. UV CURING/A NEW TECHNOLOGY IN ACRYLIC ADHESIVES

A. Historical Perspective/Technology Review

The concept of cross-linking polymers with UV light attained full commercial acceptance in the coatings and ink industries in the late 1970s extending UV technology to joining smooth, nonporous substrates. These initial efforts proved disappointing as resins tended to give low-strength bonds to surfaces that were difficult to adhere to, including metals and other nonporous surfaces. The presence of oils, dirt, and grease was particularly harmful, as well. Plastics, which have low-energy surfaces, and glass were also difficult to bond structurally. Early attempts at formulating structural adhesives reflected these limitations. Consequently, in the 1970s and early 1980s, the use of UV adhesives gained little acceptance.

In the early 1980s, UV curing grades of aerobic acrylic adhesives, however, were developed to provide structural bonds within seconds at room temperature on a wide variety of substrates. Parts may be precisely positioned and then structurally joined ''on demand'' at both a place and time convenient to the assembly process.

B. Advantages of UV Curing Aerobic Acrylic

Ultraviolet curing grades of aerobic acrylics yield tough, impact-resistant, structural adhesives with tensile strengths up to 3000 psi when joining metal, plastic, glass, ceramics, and other substrates. Adhesive potting to ¼ in. can be produced in a single exposure to high-intensity UV light. Transparent optical adhesives are also derived from this technology.

Figure 1 showed steel adhesion tensile strengths of a typical aerobic adhesive cured with either activator or UV light. The ability to use multitype curing technologies to achieve structural adhesion strength greatly expands design capabilities and helps design in a more optimum, a more productive, assembly process.

Surface dryness and depth of cure are dependent on both the lamp used and the individual formulation. High intensities are required for the most rapid and deepest cures. A potting grade UV-aerobic will cure to ¼ in. in 30 s under a 100,000 microwave/cm^2 mercury vapor lamp. Table 6 shows curing times with various lights. The optical region of the electromagnetic spectrum is defined into UV, a narrow band of visible light, and a belt of infrared radiation, as shown in Fig. 8. An exploded view of the UV spectrum with absorbance curves for Dymax UV adhesives as compared to typical UV inks and coatings is also shown in Fig. 8.

C. Curing Lights and Mechanisms

Only initiators sensitive to long-wave UV light over 325 nm are used in UV aerobic acrylic adhesives. Long-wavelength UV light is sometimes referred to as ''black light'' and is generally considered harmless. Short-wave length radiation, on the other hand, can produce burns. (Sunburn from overexposure is a familiar example.) The high-intensity

Figure 6 High adhesive strength combined with UV cure.

lamps listed in Table 6 all require shielding to protect workers from the short wavelengths that they emit in addition to long-wave UV light.

Special long-wave light generators are available for curing aerobic adhesives utilizing fiber optics that will transmit UV light, and small "penlight" lamps have become available, too. These light sources are ideal for hand assembly of small parts such as electronic components. These light sources are illustrated in Fig. 9.

D. Bonding Categories

Categories of uses of UV aerobic acrylic adhesives are listed in Table 7. One category is the bonding and lamination of clear materials such as glass and plastics. Because UV aerobics exhibit high strength, toughness, and good structural properties, the second category has been designated "light welding": a process of joining parts by bridging the joint with cured adhesive, replacing long cure times or mechanical fasteners. A third category is adhesive potting and speciality coatings on smooth surfaces such as metal.

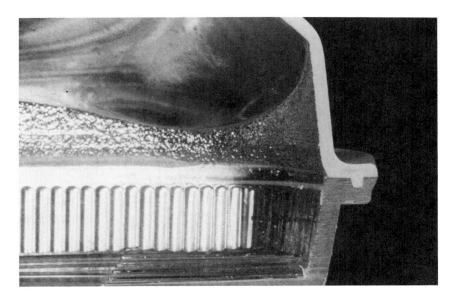

Figure 6 (continued)

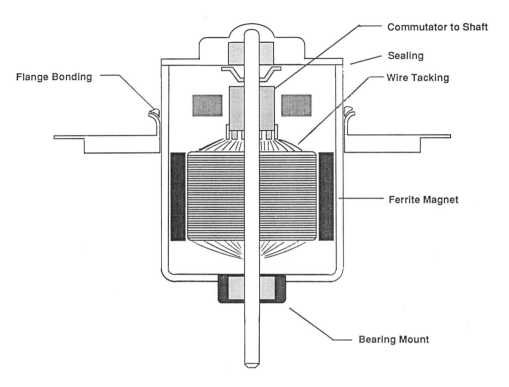

Commutator to Shaft

Sealing

Flange Bonding

Wire Tacking

Ferrite Magnet

Bearing Mount

Figure 7   Motor constructed using a combination of aerobic adhesives, press bits, and swagging.

Light Welding

The ability to bridge-bond many surfaces while retaining structural properties has led to a new concept for the joining of abutting surfaces. "Welding with light" is the process of laying a bead of liquid adhesive across the joint formed by abutting metal, glass, ceramic, thermoplastic, paper, and other surfaces and joining them by polymerizing (curing) the adhesive in a few seconds with exposure to long-wave UV light. Parts to be joined can be positioned to exacting tolerances. The adhesive may be applied either before or after positioning and then fixtured or "welded" on demand by exposure to UV light. The low shrinkage of UV aerobics helps assure nonmovement of parts due to the curing cycle. Small parts can be joined in 1–10 s at room temperature. Thick bondlines require longer times or more powerful light sources.

Some ferrite parts require bridge bonding because adhesive between the surfaces interferes with electrical properties. The common technique of using a paste epoxy requiring long time-cycles or heat-cures can be replaced by a fast-curing "lightwelding" technique. The parts are held in position, and a bead of thixotropic UV aerobic adhesive is applied across the joint and cured with UV light. A 200-W in. industrial light requires 2 s, while a low-power "black light" effects cures in 30 s. Figure 10 illustrates various UV applications.

Table 6 Some Sources of Light

Source	Wavelength	Intensity at 350–400 nm (watts/cm^2)	Approximate price range	Seconds to cure 1 (2-mil gap between glass slides)[c]	Seconds to cure 1 (1/8″ deep wire tack)[c]
Sunlight	Full spectrum	Very low	$75–$500	10–30	Not recommended
Mineral light	350–400 nm	100–7000		5–20	15–100 (higher intensities only)
Black light					
275-watt consumer style sunlamp[a]	Full spectrum + heat	5000–7000	$40–$75	1–10	5–30
Industrial-grade mercury arc lamp type I	300–400 nm + visible light	5000–10,000	$200–$500	1–15	5–30
Industrial-grade mercury arc lamp type II[b]	Full spectrum + heat	1,500,000	$800–$5000	0.5–4	2–10
Pulsed xeron arc lamps; solar simulators[b]	Full spectrum +	Very high	$50,000 +	Fast	1–5
Dymax "worker friendly" UV lamps	300–400 nm + visible light	50,000–150,000	$900–$3000	Fast	1–5

[a]Shielding Recommended
[b]Shielding Required
[c]Speed of cure varies according to formulation

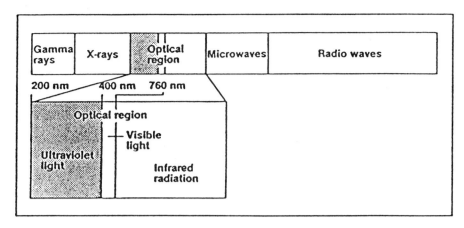

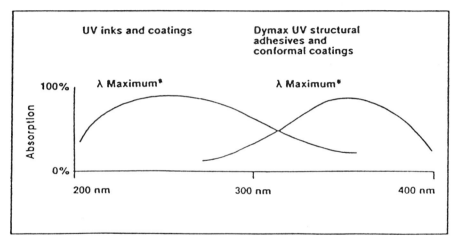

Figure 8 Optical region of the electromagnetic spectrum is defined into ultraviolet, a narrow band of visible light, and a belt of infrared radiation. An exploded view of the UV spectrum with absorbance curves for Dymax UV adhesives compared to typical UV inks and coatings. λ maximum = that point at which maximum absorption of UV energy occurs.

E. UV Aerobics—Improving Efficiency in the Assembly Process

Because of the relatively higher resin costs, process improvement evaluations are replacing unit or batch cost considerations. Many of the new coatings and adhesives provide downstream benefits, which, when factored into the overall cost of manufacturing, equal or reduce cost as compared to traditional coating technologies.

For example, a highly filled epoxy resin, for potting wires into a board connector, represents about one or two cents of the total component cost. The cost associated with mixing, heat curing, handling, and cooling hot pats adds about five cents to the assembly process cost. Finally, cost associated with downstream quality control and rejected parts adds, conservatively, another two or three cents.

Alternately, the newest generation of structural adhesives adds up to four cents material cost per component. The material's high efficiency and quality, however, lower assembly and quality costs. A company saves as much as 30–50% in overall component manufacturing cost and achieves a dramatic reduction in rejects.

Similarly, a 70% solvent-based or water-based conformal coating solution might cost as little as $75/gallon, versus solvent-free curing alternatives at $150/gallon. The energy expense to remove water, however, is high. The economic and environmental penalties to remove solvents are even greater. New OSHA and EPA regulations are expected to further restrict solvent usage. The seemingly prime disadvantages of infrared (IR) and UV alternatives are further offset when one considers that up to 70% of the lower-priced coating can be lost to evaporation.

The automotive industry leads the way in harsh environmental use of consumer electronics. Electronics represents over 10% of current costs to car makers. Much of the cost is for sensing and controlling components.

Because conventional UV resins cure only when exposed to UV light, resin underneath components will not cure and is unacceptable for automotive use. The aerobic UV coatings have a unique secondary heat curing ability (as low as 85°C), which makes them attractive. Simply raising the temperature of the boards for 5 min allows the uncured resin to "shadow cure."

One major automobile manufacturer was concerned about the use of urethane coatings with regard to solvent safety and emissions regulations. The cost and size of the complex coating supply and solvent evaporation equipment were other issues. Also, a 2-hr oven time limited the type of board it could coat and presented a severe handicap to short-run production and just-in-time delivery systems.

The company now sprays an aerobic coating onto the board. Because there are no solvents, boards move immediately beneath a high-intensity UV light for just 5 s. Short-run production is no longer a problem. Estimated total cost savings in this application is about $1.00/board.

As we have suggested, aerobic acrylic adhesive technology has proven remarkably well suited to cures effected by UV light. Aerobic acrylics cured by UV light have the potential for vastly broadening the market for acrylic adhesives.

As previously discussed, UV potting compounds used to be limited by their inability to cure through more than several mils. For larger gaps (most potting projects), slower curing epoxies have been the sole option for many manufacturers. Now, however, fast curing UV aerobic acrylic adhesives can pot depths of several inches in areas accessible to UV light.

F. UV Aerobics—Improving End-Product Quality

Adhesive and coating advances allow electronics design and assembly engineers to meet new quality demands. They bring new cost considerations to electronics production, too. For example, conformal coatings, tacking adhesives, and potting compounds used in military electronics have stringent specifications because of the demanding environments in which they are used.

Protection from high vibration, G-force stresses, thermal shocks, and moist or abrasive atmospheres is mandated, along with adhesion, strain relief, and overall stronger component attachment. Consumer products, from computers to appliances, are moving from "safe" office and home areas to more demanding environments as well.

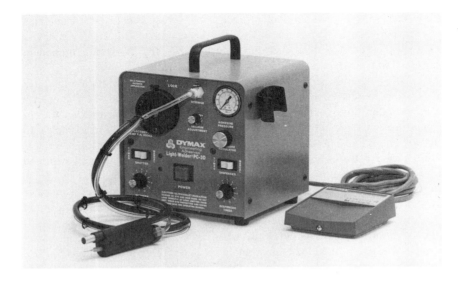

Figure 9 Long-wave light generators.

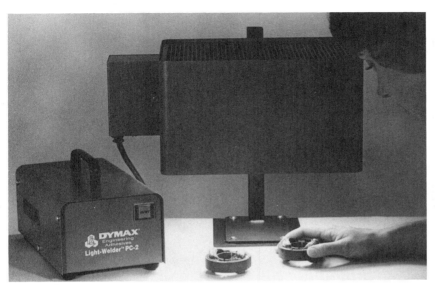

Figure 9 (continued)

In one example, a UV aerobic acrylic replaced an epoxy adhesive in an aircraft application. Epoxy adhesives frequently have been used to bond through-hole hybrid components onto PC boards. This prevents breaking of the fragile leads under common aircraft stresses such as temperature changes and vibration.

The manufacturer encountered drawbacks with this method, however. Cured epoxy is labor intensive and difficult to remove when rework is required. It is rigid and does not allow adequate expansion/contraction during temperature cycling. Cracks can occur in the bond, weakening the strength up to 50% of pretemperature cycling values. One must

Table 7 Bonding Categories Using UV Aerobic Acrylic Adhesives

Bonding	Welding	Potting/coating (other)
Glass window assembly	Chip bonding	Potting to 1/4-in. deep
Lens bonding	Speaker software assembly	Abrasion-resistant coatings
Polycarbonate shield assembly	Wire tacking	Solder mask
Plastic film laminations	Temporary bonding	Encapsulating
Costume jewelry	Ferrite magnet assembly	
Glass/plastic stemware	Electrical leads to lamp housing	
Lamp assembly	Sealing motor parts	
	Tamper proofing of mechanical fasteners	
	Costume jewelry	

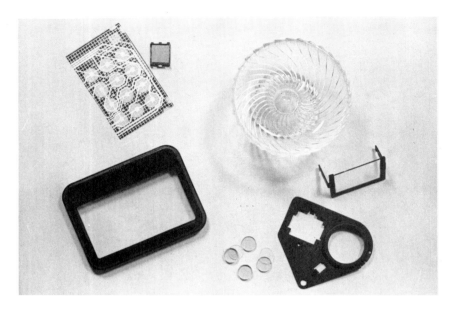

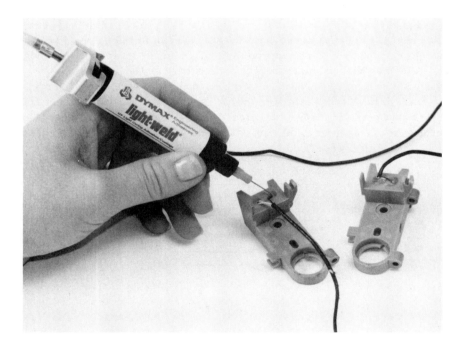

Figure 10   Various UV applications.

Figure 10 (continued)

measure and mix two components. Epoxy requires a minimum 15-min oven cure at 250°F, presenting the danger of circuit damage.

When a UV aerobic acrylic combining urethane and acrylic resins was substituted, the following characteristics were obtained: high strength before and after thermal shock testing, ease of rework, good bonding and processing, no stringing on application, and no runout during heating. The pattern spreads evenly and is cured with a single pass under a light. This application has been well documented in a paper given at the U.S. Navy Best Manufacturing Production Workshop, September 1989.

The same paper, authored by Dr. Olexander Hnojewyj and Mark Murdoch, Litton Industries, Applied Technology Division (Sunnyvale, CA), documents how a flexible grade of UV aerobic acrylic provided a quick cost-effective solution for supporting 46-flex and 26-flex pins, which previously had exhibited a higher-than-expected breakage rate.

The aerobic acrylic had excellent adhesion to the polyimide-flex-and-epoxy-glass laminate and supplied visual clarity for inspection of the reinforced pins. Cure was less than 1 min, and the material flowed well into through-hold cavities and around pins.

The material proved easy to apply both manually and with a pneumatic dispenser. UV acrylic reworkability with isopropyl alcohol or heat was quick and easy as well. In this case, the acrylic was more expensive than epoxy or silicone, but proved more cost-effective because of its virtual elimination of flex failures.

The new generation of acrylic adhesives show their versatility in other electronic assembly applications.

A surface mount aerobic acrylic was chosen by an Asian PC board manufacturer because the adhesive could be applied and cured faster, yielding an overall reduction of per-unit costs and improved productivity. Additionally, the aerobic acrylic yielded better green strength and UV fixturing before heat cure held chips in place and prevented skewing. The fast cure at higher assembly line throughout also helped prevent damage to temperature-sensitive components. Overall quality was improved.

Wire and component tacking with UV curing aerobic acrylic adhesives has been adopted by a manufacturer of flight systems boards where the need was to tack numerous

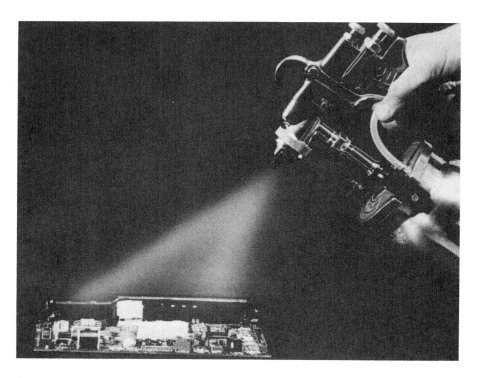

(a)

Figure 11 Application and curing of UV conformal coatings. (a) Spray coating. (b) Dip coating. (c) Table-top high conveyorized intensity UV lamp cures at 6–12 ft/min.

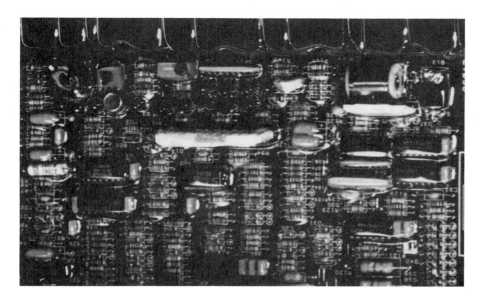

(b)

(c)

wires and components in place and reduce manufacturing costs at the same time. Two populated boards with fluorocarbon insulated wires, soldered in various configurations, were bonded in this test. Adhesives being compared were exposed to various mil-standard specifications for humidity, immersion, temperature and Humiseal coating.

The cyanoacrylates failed because they separated from the board. Although epoxies exhibited the necessary resistance to harsh environments, their lengthy cure time and a need to hold the wires in place during cure make their use inconvenient and impractical. The UV acrylic proved economical and effective, maintained integrity, and met all military standards.

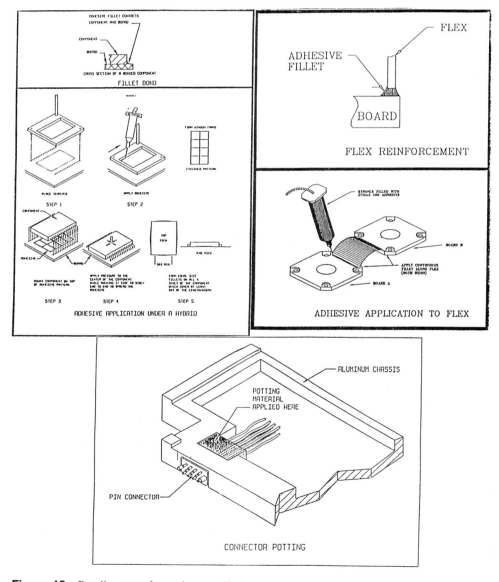

Figure 12 Bonding area for various applications.

Aerobic acrylic conformal coatings have provided major advantages to PCB manufacturers. Currently, solvent-based products dominate conformal coating processes, but safety, EPA, and processing considerations indicate their eventual decline. Although most solvents evaporate in ovens within a few hours, smaller volume manufacturers use systems requiring several days to complete solvent evaporation and/or cure of the resin systems.

UV aerobic coatings offer a solvent-free component that is 100% solids and requires no mixing. It is applied by spray or dip and then cured in as few as 5 s using high-powered UV lights. For low-volume applications, low-cost desktop lamp systems are available. This rapid cure translates into line speeds up to 15 ft/min. Figure 11 illustrates application and curing for conformal coating. Figure 12 portrays the actual bonding area for various applications.

V. SUMMARY

Acrylic adhesives provide substantial advantages for assembly operations such as fast, room-temperature fixtures, the ability to bond a wide variety of surfaces (even oily metal), bondline flexibility, and good environmental and solvent resistance. Aerobic acrylic adhesives, owing to their lack of noxious odors, low flammability, and lower toxicity parameters, overcome many of the disadvantages previously associated with acrylic adhesives. In addition, one-component UV light curing grades with superior structural properties and full cure in as little as 1 s, offer new opportunities in potting, bridge bonding, wire tacking, and glass plastic assemblies.

Because structural-strength aerobic adhesives and sealants cure at the speed and convenience of the assembly process, it is no longer necessary to pace assembly speed to an adhesives open time. Desire cures are effected in seconds. This ability to provide ''instant cure'' at the time and place demanded by the assembly process offers an attractive assembly and sealing alternative. The use of UV curable adhesives for sealing applications represents a new technology. With ever-increasing competition, cost savings due to automation have become the key to productivity. The potential of this new, easily adaptable technology are now being recognized by users in a variety of industries to lower costs and increase productivity while providing high-quality goods and incurring little environmental impact.

APPENDIX A: EFFECTS OF AIR INHIBITION

Tensile shear (steel) thickness (ASTM D-1002) 24-hr cure	Anaerobic structural adhesive	2nd-generation acrylic	Aerobic acrylic
2 mils	4500 psi	5000 psi	4000 psi
20 mils	2500 psi	3200 psi	3500 psi
60 mils	0	1000 psi	2500 psi
20 mils (with air beaten into the adhesive)	1000 psi	2000 psi	3000 psi
Cure on pine wood 40-mil gap	500 psi	Partial substrate failure	Substrate failure
Time to fixture steel nut and bolt at 75°F; no accelerator used	2–6 hr	4–24 hr	No fixture after 30 days

APPENDIX B: EFFECT OF AEROBIC FORMULATION VARIATIONS

	A	B	C
Fixture rate[a]	5 min	25 s	12 s
Complete room temperature cure	24–72 hr	4–24 hr	2–8 hr
Bondline appearance	White, hazy	White, hazy	Transparent
Gap filling	30 mils	20 mils	20 mils
Shore D hardness	50	75	10
Texture	Tough, resilient	Rigid	Very flexible
Typical end uses	Ferrite, bonding, shim "Plexiglas" fiberglass, assembly	Sheet metal, galvanized steel, bonding, graphite, glass	Lens bonding, Kevlar, window mounts
Tensile, cold-rolled steel	2000 psi	3500 psi	1800 psi

[a]Fixture rates are defined as the time required for two microscope slides, bonded in a ½-in. overlap, to resist movement from light finger pressure. Gaps estimated at 1–3 mils.
[b]Varies with ambient temperature and condition of the substrate.

REFERENCES

1. A. G. Bachmann, *Aerobic Acrylic Adhesives*, SME Technical Paper AD84-580, 1984.
2. A. G. Bachmann, *New Adhesives in Motor Construction*, Small Motor Manufacturer's Association, fall meeting, November 1987.
3. E. S. Stefanides, Adhesive allows 1200/hr. rate assembly of motor housings, *Design News*, December 1986.
4. G. Pfaff, Tagushi Symposium, Arlington, VA, May 1988.
5. A. G. Bachmann, *Aerobic Adhesives II: The Competitive Edge*, SME Technical Paper AD88-649.
6. A. G. Bachmann, Adhesives spur innovation, *Assembly Engineering*, February 1990.
7. O. Hnojewyj, *Ultraviolet Curable Materials for Military Electronic Applications*, Third Annual Navy Best Practices Workshop, San Diego, CA, September 1989.

31
Technology of Cyanoacrylate Adhesives for Industrial Assembly

William G. Repensek *National Starch and Chemical Company, Oak Creek, Wisconsin*

I. INTRODUCTION

Adhesives have been used since ancient times, but only since the advent of synthetic polymers (plastics) has their use become of major importance because manufacturers can now synthesize polymers and compound adhesive formulations to fit the requirements of the application. In other words, adhesives can be engineered for improved characteristics such as speed of cure, heat resistance, and impact resistance. Cyanoacrylates are just such an adhesive.

Cyanoacrylate adhesives can be defined as single-component chemically active liquids that react very rapidly with moisture or other weakly alkaline materials to form clear, hard solids. Their important characteristics are:

1. Very fast curing
2. Applied as liquids
3. Cured by a chemical reaction
4. Activated by alkaline materials
5. Form hard plastic materials after curing.

II. CHEMISTRY OF THE SYSTEM

The term *cyanoacrylate* comes from the chemical structure of these materials. Figure 1 depicts the general formula for cyanoacrylates. As the figure shows, cyanoacrylate adhesives are made up of carbon, hydrogen, oxygen, and nitrogen. The way these atoms are arranged in the molecule is important because the arrangement or configuration of the atoms affects the chemical properties of the molecule.

The CN group is called a nitrile or cyano group. In addition to giving these adhesives part of their name, the cyano group is highly polar and gives the molecule some of its strong adhesive character. The remainder of the molecule, the acrylate portion, is also polar and further enhances its adhesive character.

The letter ''R'' is used in organic chemistry to represent a part of a molecule that differs within a ''family'' of compounds. In the cyanoacrylate family, the R group is the

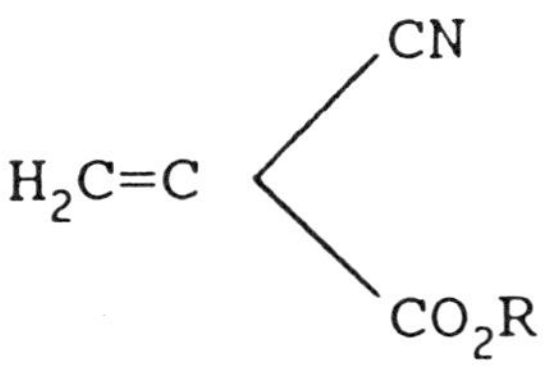

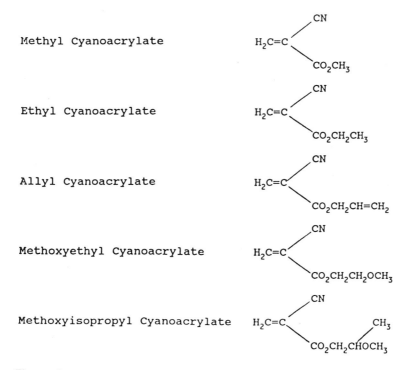

Figure 1 General formula for cyanoacrylate adhesives.

Methyl Cyanoacrylate $H_2C=C \begin{smallmatrix} CN \\ CO_2CH_3 \end{smallmatrix}$

Ethyl Cyanoacrylate $H_2C=C \begin{smallmatrix} CN \\ CO_2CH_2CH_3 \end{smallmatrix}$

Allyl Cyanoacrylate $H_2C=C \begin{smallmatrix} CN \\ CO_2CH_2CH=CH_2 \end{smallmatrix}$

Methoxyethyl Cyanoacrylate $H_2C=C \begin{smallmatrix} CN \\ CO_2CH_2CH_2OCH_3 \end{smallmatrix}$

Methoxyisopropyl Cyanoacrylate $H_2C=C \begin{smallmatrix} CN \\ CO_2CH_2CHOCH_3 \end{smallmatrix}$

Figure 2 Configuration of common cyanoacrylate adhesives.

alcohol that forms the ester with the acylic acid portion of the molecule. The type of alcohol used gives the cyanoacrylate adhesive its name. Thus methyl cyanoacrylate is the ester formed between methyl alcohol and cyanoacrylic acid. Figure 2 shows the configuration of some of the commercially available cyanoacrylate adhesives.

All of the molecules shown in Fig. 2 are thin, watery liquids at room temperature. In this low-viscosity state, these molecules are capable of flowing freely onto the surface of the parts to be bonded. However, such thin liquids would have little or no ability to bridge the gap between the mated parts. The carbon-to-carbon double bond shown in Fig. 2 to be common to all cyanoacrylate is capable of reacting such that adjacent molecules are linked together to form a large chain. The chains become so large that the adhesive changes from a liquid to a hard, tough solid. The chemical reaction involved, called polymerization, is depicted in Fig. 3.

Figure 3 Polymerization of cyanoacrylate adhesives.

The initiator that drives the polymerization or curing reaction of cyanoacrylate adhesives can be any of the chemicals that generate free radicals. Electromagnetic radiation in the form of heat or UV light can also trigger the reaction, but usually only enough to cause problems with product shelf life. The more likely route for cyanoacrylate polymerization is by an ionic initiation. Any molecule more alkaline than water can initiate the curing reaction. This type of polymerization is characteristically much faster than a free-radical type and is the reason that cyanoacrylates cure so rapidly. It is this cure speed more than any other property that makes these adhesives so popular on the production line. Many adhesives are stronger or more durable than cyanoacrylates, but none can cure as quickly and to such a wide variety of substrates as the cyanoacrylates.

Although it is not essential to understand the chemistry of cyanoacrylate polymerization to be able to use these adhesives, knowing that a chemical reaction is taking place helps the user to understand how application conditions affect their performance. Consider the fact that the common polymers, such as polyethylene, polystyrene, and poly(vinyl chloride), are made in sophisticated reactors. Parameters such as temperature, monomer concentration, and amount of activator are carefully controlled.

With cyanoacrylate adhesives, the reactor used to convert the liquid monomer to the hard solid is the space between the parts being bonded. When conditions vary in this space, the performance of the adhesive will vary. Such parameters as temperature, humidity, space between the parts, and the type of surface being bonded can vary considerably in a given application.

Figure 2 shows a number of different types of cyanoacrylic esters. There are subtle differences between them that can be utilized in specific applications. Methyl cyanoacrylate is a more polar compound than any of the others. This gives the cured adhesive a higher cohesive, or internal, strength. As a result, it has a higher shear strength which can be utilized on metal parts and other parts that are rigid enough themselves to benefit from the strength of this hard, brittle polymer.

Ethyl cyanoacrylate is a little less polar than methyl cyanoacrylate, and has the ability to wet plastic surfaces more readily, and is a better solvent for plastics. With this added ability to make intimate contact with the surface, the bonds on plastic are stronger with ethyl cyanoacrylate than with the methyl ester. This difference in performance gives rise to the adage that *m*ethyl is for *m*etal and *e*thyl is for *e*verything else. Sometimes this difference can be utilized in reverse to good advantage to avoid stress cracking on such sensitive plastics as polycarbonate and polyacrylate.

Also shown in Fig. 2 is an allyl cyanoacrylate. This molecule contains a second double bond that can be made to react after the initial polymer chain is formed. This secondary bonding can occur between adjacent polymer chains, causing cross-linking of the chains. Such cross-linked polymer chains are more heat resistant than is the uncrosslinked polymer.

Table 1 High-Temperature Resistance of Allyl Cyanoacrylate

	Shear strength[a]	
Temperature (°C)	Allyl	Methyl
25	3000	3600
100	3900	250
120	900	0
150	1500	0
250	1700	0

[a]Steel/steel per ASTM D1002. Aged 1 week at temperature indicated, cooled and measured at 25°C.

The data presented in Table 1 compare the heat resistance of allyl cyanoacrylate and methyl cyanoacrylate determined by heating a steel lap shear specimen for 1 week at various temperatures. At room temperature the two types are essentially equal in strength. The slightly higher strength of the methyl cyanoacrylate caused by its higher polarity can be seen clearly in the higher value obtained in the test cured and aged at room temperature.

The effect of exposure to 100°C shows up in the loss of strength for the uncrosslinked methyl cyanoacrylate and a higher strength for the allyl cyanoacrylate because of the extra strength contributed by the reaction of the double bond in the allyl cyanoacrylate. After exposure to 120°C, all the strength of the straight-polymer-chain methylcyanoacrylate is lost. The strength of the cross-linkable allyl cyanoacrylate is also reduced, suggesting the loss of the contribution to its strength by intermolecular association. Only the contribution from the allyl group survives.

Higher temperature causes the allyl double bonds to react faster. As a result, more cross-linking can take place and more of the strength is retained. Resistance to temperatures above 120°C is possible provided that the parts are clamped during the curing process. Because of the extensive cross-linking, the resultant polymer is very brittle and it is recommended only for metal and other rigid, high-temperature-resistant substrates.

The fourth type of cyanoacrylates presented in Fig. 2 are the alkoxyalkyl esters. Methoxyethyl cyanoacrylate and methoxyisopropyl cyanoacrylate esters have all the desirable properties of the methyl, ethyl, and allyl cyanoacrylates, with the added advantage of low vapor pressure. As a result, these monomers have little or no odor, which makes them popular for use in environments where ventilation is a problem. The low vapor pressure also reduces the fogging of adjacent parts so often seen with ''regular'' cyanoacrylates on damp days, a problem discussed in more detail below.

In addition to the benefits of low odor and reduced fogging, these adhesives form stronger bonds to low-energy substrates such as EPDM rubber, natural rubber, and other difficult-to-bond plastics. This property seems to be a function of the solvent action of the uncured adhesive, so care must be taken to avoid stress cracking when the adhesive is used on sensitive substrates such as polycarbonates and polyacrylates.

While the alkoxy cyanoacrylates cure by the same mechanism as regular cyanoacrylates, the cure speed is a bit slower and the overall strength is about 20% lower than that of ethyl cyanoacrylates. The strength is well in excess of the strength of most plastic substrates however, the 20% reduction in strength is not significant.

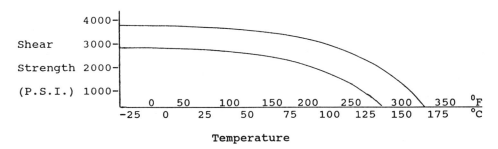

Figure 4 Hot strength of allyl cyanoacrylate, 100 cP Brookfield, bond cured at room temperature for 24 h, heated for 2 h at temperature and tested hot.

Table 2 Temperature Performance of Allyl Cyanoacrylates

	Lap shear strength (psi) (steel)
Room temperature	3112
100°C (212°F)	3353
250°C (480°F)	1493

III. HOT STRENGTH

In general, standard industrial cyanoacrylates do not operate effectively above 180°F (see Fig. 4). However, the new allyl types of cyanoacrylates can operate as high as 480°F before the bond looses sufficient strength to be operationally effective (see Table 2). Allyl cyanoacrylates for metal-bonding applications have proven effective in wave solder and under-hood (automotive) applications. In Fig. 4, bonded assemblies are cured at room temperature for 24 hrs. The assemblies are heated for 2 hrs and tested hot.

IV. SPEED OF CURE

A typical cure curve for 100-cP methyl cyanoacrylate is shown in Fig. 5. The speed of cure is influenced by the thickness of the bond, the activity of the surface, and the designed speed of the adhesive.

1. *Versatile.* There are few materials to which cyanoacrylate will not bond/adhere, even dissimilar materials (polyethylene, polypropylene, and Teflon are some materials that require pretreatment).
2. One component without solvent, therefore easily dispensed.
3. *Economical.* A small drop goes a long way. It is important to apply the least amount for maximum strength (see Table 3).

The following equivalents are for application of the adhesive to actual parts.

1 lb = 30,000 bonds 1 in. square and 1 mil thick
1 oz =   1,875 bonds 1 in. square and 1 mil thick
2 g =   132 bonds 1 in. square and 1 mil thick

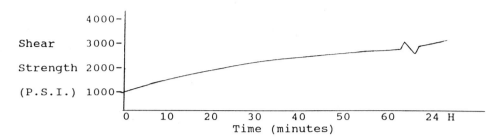

Figure 5 Typical cure curve for 100-cP methyl cyanoacrylate.

Table 3 Approximate Conversion Chart for
Estimating Use[a]

1 drop =	0.006 g
2 drops =	0.012 g
5 drops =	0.030 g
10 drops =	0.060 g
50 drops =	0.600 g (⅓ g)
167 drops =	1.00 g (⅔ g)
330 drops =	2.00 g
75,500 drops =	1 lb
4,666 drops =	1 oz

[a]Based on the assumption that 1 drop = 0.006 g.

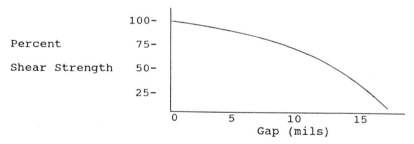

Figure 6 Gap-filling capability of cyanoacrylate.

Cyanoacrylate adhesives have limited gap-filling capability (see Fig. 6). Low-viscosity grades can fill gaps to approximately 2 mils, medium viscosity up to 10 mils and high viscosity up to 15 to 20 mils. However, note that the speed of cure will be slower and the ultimate strength lower as the gap increases. For example, when the gap is 15 mils it requires 1 h for sufficient handling strength, but with a gap of 1 mil it takes only 10 s. Cure speed can be increased by the use of a chemical accelerator. Moisture and water resistance is limited. Resistance against water, although poor when two rigid parts are bonded, is markedly improved when at least one of the two parts is flexible.

Table 4 Typical Fixture Time with Activator[a]

Gap (in.)	With activator	Without activator
0.000	1 s	10 s
0.005	50 s	600 s

[a]Steel/steel, 25°C, ethyl cyanoacrylate.

V. ACTIVATORS

Since bases are catalysts for the curing reaction and acids are stabilizers for the cyanoacrylates, the pH value of the surface will control the cure speed. Surfaces that tend to be acidic will cure slowly compared to a neutral surface, which in turn will cure more slowly than an alkaline surface. In most applications the objective is to speed the cure; therefore, all the commercially available activators are weak bases dissolved in a volatile carrier. Applying an activator to a surface places a layer of the weak base in position to initiate the cure. Since they are stronger bases than moisture, they are able to neutralize the stabilizer systems in the adhesive more effectively, and thus they can tolerate larger gaps than would be possible with moisture alone. In general, the fixturing time is 10 times faster with activator than without it. Even with activator, the effect of the gap is clear (see Table 4).

Another advantage of the activator is the ability to cure a film or drop of the adhesive on a surface. The activator can be applied to either the surface being coated or to the top surface of the adhesive. The adhesive cures to a clear, hard, dry plastic that can be used to locate parts in position or to form a protective coating. Because the curing reaction is ionic, it is not sensitive to oxygen as are free-radical reactions, and it will cure through to the surface without the tackiness associated with free-radical curing systems.

VI. IMPROVED COMMERCIAL CYANOACRYLATE COMPOUNDS

A. New Flexible Cyanoacrylates

New cyanoacrylate compounds exhibit good adhesion to various plastics and elastomeric surfaces, such as Mylar, copper foil, and vinyl films. These products show better impact resistance and good flexibility compared to standard cyanoacrylates, good resistance to cracking under flexing or bending, and a longer open time than that of standard products.

B. New Cure-Through-Gap Cyanoacrylates

Through changes in the manufacturing process, cyanoacrylates can now be highly purified. This added purification step has led to the development of a group of materials that can cure through a gap without the typical reduction in shear strength and overall performance. More traditional formulations show a dramatic reduction in shear strength as the gap increases (see Fig. 6). Table 5 shows cure through a gap and the corresponding shear strength for these new, highly purified cyanoacrylates.

Table 5 Percent Cure Through 0.008-in. Gap

Time (h)	Gap-filling cyanoacrylate	Regular cyanoacrylate
2	25	13
24	100	40
72	100	100

Table 6 Set Time Across a 0.004-in. Gap

Product viscosity (cP)	Time (s)	
	Surface-insensitive cyanoacrylate	Ethyl cyanoacrylate
100	6–8	55
500	7–12	>60
1500	8–12	>60
8000	8–12	n.a.[a]

[a]Not applicable.

C. New Ultrafast-Cure Surface-Insensitive Cyanoacrylates

The new range of surface-in sensitive cyanoacrylates provides ultrafast cures independent of gap. In addition, these cyanoacrylates will rapidly bond acidic and low-energy surfaces. The fast cure also minimizes the occurrence of frosting and fogging. Table 6 shows a comparison of these new surface-insensitive materials compared to a standard ethyl-grade cyanoacrylate. These products are also suited to bonding various wood substrates and porous surfaces without the use of activators.

D. New Low-Odor Cyanoacrylates

Due to unique manufacturing processes and chemical reformulations, cyanoacrylates can now be completely odor free. These products, which have almost no detectable odor, improve worker comfort and acceptance. Second, there is no unsightly fogging (chlorosis) of expensive parts adjacent to the bond line as can be seen with other types of cyanoacrylate adhesives on hot, humid days. This improves production rates of acceptable parts.

VII. SIGNIFICANT CYANOACRYLATE CHARACTERISTICS

A. Polypropylene and Polyethylene Bonding

Polyolefin bonding has been advanced using cyanoacrylates through the use of surface primers. These primers promote adhesion to untreated polyethylene (PE), polypropylene (PP), and EPDM rubber. Table 7 shows comparison bonds using standard industrial-grade cyanoacrylates.

Table 7 Comparison Bonds Using Standard Industrial-Grade Cyanoacrylates

	PE/PE		PP/PP	
Lap shear (psi)	Primed	Unprimed	Primed	Unprimed
After 15 min, room temperature	90	50	185	52
After 24 h, room temperature	108	50	200	52

B. Medical-Grade Materials

When fully cured, cyanoacrylates will meet class VI standards for plastics, the highest class of safety and biocompatibility as defined by the *United States Pharmacopoeia* (USP).

C. Thermal Conductivity

Cyanoacrylates are essentially thermally nonconductive materials. The value for a typical methyl-grade cyanoacrylate is 2.1 (Btu-in.)/(hr-ft^2-°F).

D. Durability

Assemblies joined with cyanoacrylate adhesives exhibit good long-term durability, particularly when the materials are somewhat flexible, such as rubbers and most plastics (see Fig. 7). Bonded lap shear specimens have been aged outdoors for 7 years with good retention of strength (see Table 8).

When impact resistance and/or strength is required, it can be improved dramatically by the "rubber sandwich" technique. In this case, a rubber sheet bonded between the rigid plastic or metal substrates will absorb all peel and impact forces. It is also useful in absorbing stresses when thermal expansion and contraction occur.

VIII. CHLOROSIS

There are times when various environmental factors can affect bonding results. One such phenomenon is *chlorosis*, in which white particles appear on the bonded parts. Chlorosis

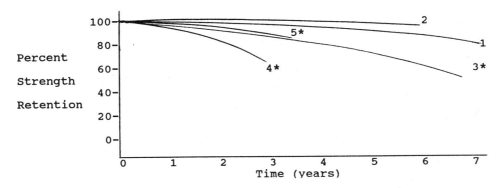

Figure 7 Durability of cyanoacrylate adhesives. Asterisk denotes substrate failure.

Table 8 Strength Retention of Cyanoacrylate Bonds After Outdoor Weathering

Material bonded	Percentage strength retention when stored at 73°F and 50% RH for (yr):						
	1	1.5	2	3	4	5	7
Neoprene to neoprene	100		80				60
Neoprene to acrylic	93		87			93	
Butyl rubber to aluminum	65[a]	92[a]			55[a]		55[a]
Polystyrene to polystyrene		80[a]		83[a]			
Rigid PVC to rigid PVC		93[a]		100[a]			

[a]Substrate failure.

may occur with the use of cyanoacrylate, depending on the bonding method, bonding conditions, atmospheric influences, and other factors. The usual reaction received from users when chlorosis does occur is concern over the possible effect on the strength of the bond and the unsightly external appearance. The phenomenon of chlorosis is nothing more than an adhesion of particles to the surface of the part and has no effect on the quality of the bond.

Chlorosis develops when a portion of liquid evaporates in the air without being solidified (polymerized). It becomes a powdery resin through a polymerization reaction with moisture in the air. The powdered resin falls back to the surface of the bonded parts. Chlorosis can develop under the following conditions:

1. Excessive amounts of adhesive.
2. Bonding under conditions of high room moisture content (relative humidity), as may be the case during a rainy period.
3. Subjecting the bonded portion to airtight or almost airtight conditions immediately after bonding.

Chlorosis can be avoided by taking the following actions:

1. Avoid the use of adhesive in excessive amounts.
2. Perform the bonding operation under appropriate moisture conditions (avoid excessive accumulation of water on the material surface); 50% relative humidity is ideal.
   a. Allow several minutes after removal of grease from the object being bonded.
   b. Thoroughly dry the object to be bonded before actual bonding.
   c. Wear polyethylene gloves during the bonding operation.
3. Attempt as much as possible to scatter vaporizing monomer.
   a. Avoid subjecting bonded material to airtight or nearly airtight conditions immediately after the bonding operation.
   b. Ventilate the bonded object immediately after the bonding operation by use of an electric hot-air blower or other appropriate means.
   c. Perform the bonding operation under conditions of free air circulation.
4. Promote the polymerization reaction (resinification).
   a. Assure appropriate heat conditions for the bonded object [up to about (70°C) 158°F].
   b. Blow heated air over the bonded area immediately after the bonding operation.

5. Form a protective shield on the surface where chlorosis is likely to develop by applying oil to the area surrounding the material surface immediately before or after the bonding operation.

Chlorosis can be eliminated by the following:

1. Wipe the particles carefully with a dry cloth.
2. Wipe off the particles with solvent.

The particles can readily be wiped off by the use of a cloth containing a solvent such as dimethylformamide, nitromethane, acetone, or chlorinated or fluorinated solvents. In this case, however, if the bonded object is plastic, caution must be exercised because plastics can be affected adversely by solvents.

Poor bonding arises from problems associated with the adhesive itself: for example, its decreased adhesive power. The simplest method of checking on the spot for any decrease in adhesive power is to test with Bakelite samples. If the adhesive is found to have decreased adhesive power, it has probably been stored improperly.

Material kept in the work environment up to 6 months will not suffer deterioration. For periods over 6 months, storage in 40°F or below can triple the useful shelf life.

IX. SUMMARY

Cyanoacrylates have shown themselves well in permanent outdoor assemblies as well as in temporary manufacturing aids. They are safe, convenient materials to incorporate in plant operations. New developments in technology have improved moisture resistance, setting times, gap filling, clarity, high-temperature resistance, and flexibility, and most recently, cyanoacrylates have become less surface sensitive.

As always, thorough testing of a specific application should precede specification. New research has led to improved commercially available models that are finding use in applications previously thought unsuitable for cyanoacrylates. Where appropriate, cyano-acrylates can increase productivity dramatically and reduce fundamental costs, to produce high-integrity parts for aircraft, industrial, agricultural equipment, electronic, automotive, and maintenance applications.

BIBLIOGRAPHY

Bittence, J. C., *Machine Des.* (June 1976).
Cagle, C. V., *Adhesives Bonding*, McGraw-Hill, New York, 1968.
Chastain, C. E., *Appliance Eng.* 8(4) (1974).
Coover, H. W., Jr., *Adhesives Mater.* (1958).
Graham, J. A., *Machine Des.* (1976).
Helmstetter, G., *Fundamentals of Cyanoacrylate Adhesives*, unpublished manuscript, Permabond International, 1988.
Lees, W. A., *Adhesives in Engineering Design*, Pitman, London, 1984.
Pagel, W. C., and A. G. McKown, *Industrial Adhesives for Metal and Plastic Symposium*, Madison, Wis., Feb. 1966.
Peace, R., *Adhesives Age* 22(9) (1979).
Petrie, E., *Assembly Eng.* (June 1976).

Repensek, W. G., *Technology and Use of Cyanoacrylate Adhesives in Industrial Assembly Applications*, American Fabricating Institute of Technology, Chicago, Sept. 1983.

Repensek, W. G., *Technological Advances in Cyanoacrylate Adhesives for Industrial Applications*, Society of Manufacturing Engineers, Dearborn, Mich., Nov. 1990.

Schneberger, G. L., *Adhesives Age* (Jan. 1980).

Weyher, D. F., Ch., *Adhesives in Modern Manufacturing*, Society of Manufacturing Engineers, Dearborn, Mich., 1970.

32
Silicone Adhesives and Sealants

Loren D. Lower and Jerome M. Klosowski *Dow Corning Corporation, Midland, Michigan*

I. INTRODUCTION

Silicone adhesives and sealants were introduced approximately 30 years ago and many of the silicones used in the early days are still performing. Products are available in a variety of forms, from pastelike materials to flowable adhesives. Both single- and multicomponent versions are available, with several different cure chemistries. Most of the silicones of commerce are based on polydimethylsiloxane (PDMS) polymers. Other siloxane polymers may be used when resistance to ultrahigh temperature, ultralow temperature, or solvents is required.

Applications are extremely broad. A partial list includes construction, highway, automotive, appliance assembly, original-equipment manufacture, maintenance, electronics, aerospace, and consumer uses. In some cases, silicones compete with other materials, such as polyurethanes, polysulfides, and acrylics, whereas in applications requiring long-term durability, silicones alone are specified. Silicones are often chosen for their excellent resistance to weathering and temperature extremes, their adhesion, and their ability to accommodate substrate movement. When silicone sealants and adhesives are mentioned, the thought of excellent durability comes to most readers' minds. Silicones [named for the similarity of the $(CH_3)_2SiO$ polymer repeat unit to the analogous organic ketones, $R_2C{=}O$] occupy a unique position between inorganic and organic materials. The saturated inorganic Si—O—Si polymer backbone provides flexibility and stability to sunlight, while the methyl groups ensure low intermolecular forces. Some of the key attributes of silicones, which are responsible for their unique properties and durability are [1]:

Low surface tension
High water repellence
Partially ionic backbone
Large free volume
Low apparent energy of activation for viscous flow
Low glass transition temperature
Freedom of rotation around bonds
Small temperature variations of physical constants
High gas permeability

521

High thermal and oxidative resistance
Low reactivity
Insolubility in water
High silicon–oxygen bond energy

Selected properties of PDMS are as follows:

Critical surface tension of wetting	24 mN/m
Water contact angle	110°
Glass transition temperature	150 K
Energy of rotation	0 kJ/mol
Activation energy for viscous flow	14.7 kJ/mol
Si-O Bond energy	445 kJ/mol
Percent polar contribution	41%

The saturated backbone and high Si—O bond energy result in products that perform very well in applications involving exposure to sunlight. Since the silicone polymer does not absorb energy in the ultraviolet (UV) region of the light spectrum, one must be cautious with the use of clear silicones. The silicones need no UV absorbers to be stable (and contain none); thus the UV light from the sun can pass through clear silicones to the surface below the sealant. If the surface is sensitive to UV light, deterioration of the substrate may occur. Except for light-protected areas and unsensitive substrates (such as glass), the most judicious choice is a pigmented silicone. The pigment acts as a UV blocker and protects the substrate beneath the silicone. Because of the unparalleled stability to UV radiation, silicones are the sealants of choice for wet glazing techniques and the only generic class of sealants allowed for structural glazing (the adhering of glass and other building materials to structures with no attachment other than the silicone). Structural glazing is used in all-glass buildings and skyscrapers.

Other types of sealants often contain large amounts of filler and UV stabilizers to afford some degree of longevity in sunlight. This makes the nonsilicones satisfactory for some applications, but not in applications in which the sun shines directly on the bond line. This application is reserved for silicones. A specialty application for silicones, which further illustrates their UV-light durability, is in the sealing of accelerated UV-weathering test machines. The excellent stability to UV light is true only for pure silicones and is not true of ''siliconized'' organics or ''modified silicones.'' These contain very little silicone and thus have durability characteristics determined primarily by their base polymer systems.

Silicones have low intermolecular forces that result in relatively flat physical property response with temperature change. An example of this flat response is shown in Fig. 1, in which the viscosity of silicone polymers and a hydrocarbon oil are plotted as a function of temperature [2,3]. The relatively low response of silicone properties to temperature is important during sealant application (e.g., no heating needed in cold weather and no flow in hot weather). Even more important, however, is the fact that the performance of the cured sealant or adhesive will be less temperature dependent than will most organic-based products. This has practical implications: in building joints, for example. In cold weather, the building components shrink, and joint sealants must maintain elasticity to accommodate this movement. This is also fundamental to their use as a structural glazing sealant/adhesive. The sides of all-glass buildings can get very warm in the summer sun, and the silicone must not lose strength at these temperatures. While this rather constant perfor-

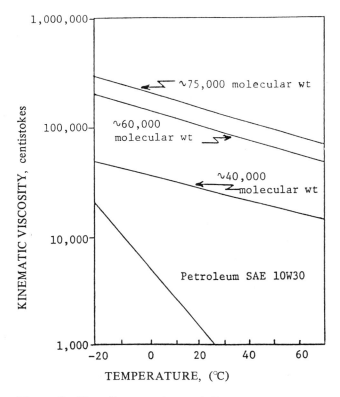

Figure 1 Viscosity versus temperature.

mance is critical in some construction applications, it is also important in many industrial and appliance applications, such as steam irons, where the sealant simultaneously prevents water leakage and acts as an assembly adhesive.

Silicone sealants are rated for their movement capability, with classes at ±12.5%, ±25%, ±50%, and even higher joint movement capability. This too is quite unique, since high-movement nonsilicone sealants rarely perform for long periods of time above ±25% joint movement.

II. CURE CHEMISTRY

Silicones are available in one- and multicomponent forms. The one-component types are commercially the most important and will be the focus of most of this discussion. These products, which generally cure by reaction with atmospheric moisture, are called RTV (room-temperature vulcanizing) sealants or adhesives. The surface cure rate of these products is a function of the cure system, but the rate of cure in depth depends on the ability to transmit water vapor through the mass of sealant. Silicones are highly permeable to moisture vapor, and generally the one-component types cure at a rate of about 0.3 cm/ day. Due to this high vapor permeability, the one-component silicones typically cure faster than do their nonsilicone counterparts.

The multicomponent products generally do not rely on moisture penetration for cure. Their chief attribute is fast cure in very deep sections. Thus many industrial production lines that demand fast cure use a two-component sealant (including the use of silicone encapsulants for electrical components). Cure of these two-part systems can be accelerated further by additional catalyst or exposure to elevated temperatures.

One of the more common two-part cure chemistries is based on the addition reaction of Si—H cross-links with vinyl functional polymers using platinum catalysts. This chemistry is shown below. One advantage of this addition chemistry is that it produces no cure by-products. Another common two-part chemistry involves condensation cure with alkoxysilane cross-linkers using Sn(IV) catalysts.

$$\text{—Si(Me)}_2\text{OSi(Me)}_2\text{CH=CH}_2 + \text{HSI—} \rightarrow \text{—Si(Me)}_2\text{OSi(Me)}_2\text{CH}_2\text{CH}_2\text{Si—}$$

A simplified cure mechanism for the one-component silicone RTV sealants or adhesives is shown below.

Reaction of cross-linker with polymer ends:

$$2\text{RSiX}_3 + \text{HOSi(Me)}_2\text{O[(Me)}_2\text{SiO]}_n\text{Si(Me)}_2\text{OH} \rightarrow$$
$$\text{X}_2\text{(R)SiO[(Me)}_2\text{SiO]}_n\text{Si(R)X}_2 + 2\text{HX}$$

Reaction of cross-linker-capped polymer end with moisture:

$$\text{X}_2\text{(R)SiO[(Me)}_2\text{SiO]}_n\text{Si(R)X}_2 + \text{H}_2\text{O} \rightarrow \text{OH(X)(R)SiO[(Me)}_2\text{SiO]}_n\text{Si(R)X}_2 + \text{HX}$$
$$\text{(A)} \qquad\qquad\qquad\qquad\qquad\qquad\qquad \text{(B)}$$

Reaction of resultant polymer end with another polymer:

$$\text{A} + \text{B} \rightarrow \text{X}_2\text{(R)SiO[(Me)}_2\text{SiO]}_n\text{Si(R)(X)-O-Si(X)(R)O[(Me)}_2\text{SiO]}_n\text{Si(R)X}_2 + \text{HX}$$

As indicated, the X groups above are hydrolyzable. Repeated hydrolysis and reaction of resultant polymer end groups leads to full cure, with elimination of HX as the leaving group. Examples of leaving groups, cross-linkers, and the common cure system names are given in Table 1.

Numerous other cross-linkers may be used. For the trifunctional cross-linkers, the R group may be methyl, ethyl, vinyl, and several other groups, with methyl the most common. In some cases tetrafunctional and higher-functionality cross-linkers or polymeric cross-linkers may also be employed. The acetic acid cure system should be avoided where substrates are subject to acid corrosion.

Two other classes of silicones deserve mention. These are the water-based silicones that are used in sealant and coating applications and the silicone pressure-sensitive adhesives. Water-based silicones can be prepared by anionic polymerization of siloxanes in water using a surface-active catalyst such as dodecylbenzenesulfonic acid [4]. The

Table 1 Examples of Leaving Groups, Cross-Linkers, and Cure Systems

Leaving group (HX)	Cross-linker	Cure system
HOC(O)CH_3	$\text{CH}_3\text{Si[OC(O)CH}_3\text{]}_3$	Acetic acid
HOCH_3	$\text{CH}_3\text{Si(OCH}_3\text{)}_3$	Alcohol
$\text{HONC(CH}_3\text{)(C}_2\text{H}_5\text{)}$	$\text{CH}_3\text{Si[ONC(CH}_3\text{)C}_2\text{H}_5\text{]}_3$	Oxime
$\text{CH}_3\text{C(O)CH}_3$	$\text{CH}_3\text{Si[OC(CH}_2\text{)CH}_3\text{]}_3$	Acetone
$\text{HN(CH}_3\text{)C(O)C}_6\text{H}_5$	$\text{CH}_3\text{Si[N(CH}_3\text{)C(O)C}_6\text{H}_5\text{]}_3$	Benzamide

resulting emulsion can then be cross-linked in several ways, including the use of alkoxy-silane copolymerization or tin catalysts in conjunction with colloidal silica. The result is essentially an emulsion of cured PDMS in water. Various fillers and other components are added, resulting in a sealant composition. Upon evaporation of water, a silicone elastomer results. These sealants have the advantages of low odor, ease of installation, and easy cleanup. Their properties are rather close to those of their conventional silicone counterparts.

The components of the silicone pressure-sensitive adhesives (PSAs) are analogous to their organic counterparts [5]. Generally, a silicate resin and a silicone polymer or gum are dissolved in solvent. Both the resin and the polymer typically contain silanol (Si—OH) groups that are reacted during processing of the PSA, leading to a cross-linked network. Additional reactions can be accomplished through the use of free-radical catalysts. The extent to which these cross-linking reactions occur, the resin/polymer ratio, as well as the respective molecular weights of these components, are important in setting the properties of the PSA.

Silicone PSA products are used in a number of medical and industrial applications, ranging from a variety of pressure-sensitive adhesive tapes and transfer films to automotive bonding. Advantages for the silicone PSA products include resistance to temperature extremes, chemical resistance, conformity to irregular surfaces, and electrical properties. They are also unique to most pressure-sensitive adhesives in their ability to adhere to difficult low-energy substrates, such as polytetrafluoroethylene and other silicones.

III. PROCESSING CONSIDERATIONS

Silicone adhesives and sealants typically contain polymer, fillers, cross-linker, catalyst, and other additives. The most common fillers are the reinforcing fumed silicas or the less reinforcing grades of calcium carbonate. Other fillers and pigments, such as carbon black and titanium dioxide, are also used. Silicones are typically made in high shear, vertical change can mixers, but continuous processing equipment may also be employed. Processing details are generally held proprietary, but some general guidelines are in order.

Since most silicones cure through reaction with water, it is important that the moisture content of fillers and other additives be controlled. The moisture content of fumed silica, for example, can vary from 0.2 to over 2%, depending on the humidity conditions during storage. It is also critical that introduction of moist air be kept to a minimum during mixing. Air incorporated during processing must generally be removed to reduce the tendency toward cured gels and related appearance problems in the final product.

The dispersion of the filler particles is also important to final sealant appearance. With increasing costs of fillers, it is also important to optimize dispersion to maximize the rheological and reinforcement benefits provided by the fillers. Manufacturers must balance the mixing time and energy required for complete dispersion with resultant product appearance and physical properties.

IV. PROPERTY DETERMINATIONS

Since most silicone adhesives and sealants are elastomeric in nature, their physical property testing often parallels classical rubber testing approaches. Common tests include durometer, tensile strength, elongation, and modulus. Several methods are available for the measurement of rubber properties, but the most commonly used are the American Society for Testing and Materials (ASTM) D-412, Test Method for Rubber Properties in Tension, and the ASTM C-661, Standard Test for Indentation Hardness of Elastomeric-Type Sealants by Means of a Durometer. These properties vary widely with the product and its intended application. Durometer can range from Shore A less than 20 to over 50. Tensile strength ranges from less than 0.2 to greater than 5 MPa, and elongation varies from about 100 to 2000%.

Cure time testing tends to be somewhat subjective, but again there are methods available, such as ASTM C-679, Tack-Free Time of Elastomeric Sealants. Tack-free time is the curing time required for the product to develop a skin that is not damaged when subjected to application and removal of a plastic film. It is important in all cases to determine the cure characteristics of the product in actual working conditions. Since most silicones cure by reaction with moisture in the air, the sensitivity of cure time to humidity should be determined. Surface cure rates can usually be tailored to meet application requirements.

The rheological properties of adhesives and sealants are important in many applications. When these products must be pumped or applied through automated equipment, the flow characteristics at pertinent shear rates are critical. Sophisticated rheological measurements can be performed to predict performance. The rheology of silicone adhesives and sealants can be tailored through adjustment of polymer viscosity, filler loading, and incorporation of various additives.

Often, only the extrusion rates of adhesives and sealants are measured, which is accomplished by subjecting the product to a given pressure and measuring its flow rate through a nozzle of known diameter (see ASTM C-1183, Extrusion Rate of Elastomeric Sealants). For many sealant applications, the sealant must not flow under its own weight in conditions of low shear rate. In this case, some measurement of "slump" is generally made. Several methods are available for measuring slump [see ASTM D-2202, Standard Test for Slump of Sealants, and ASTM C-639, Standard Test for Rheological (Flow) Properties of Elastomeric Sealants]. Again, it is important to determine the rheological performance of the product in the actual application.

Adhesion testing is a matter of some controversy. There is, however, a growing trend among manufacturers, specifiers, and standards organizations to move toward tests that better predict performance in application. The 180° peel adhesion test is often used as an internal quality control tool by manufacturers. This test allows for measurement of lot-to-lot consistency of products. The methodology for this test can be found in ASTM C-794, Test Method for Adhesion In-Peel of Elastomeric Joint Sealants. In its typical form this test involves placing a bead of the product onto the substrate, with a flexible member embedded in the product. The product is allowed to cure, and the member is then pulled away from the substrate. The force required for peel is noted, along with the mode of failure. The advantages of this test are that it is fast and lends itself well to subjecting the adhesive bond to environmental stresses such as hot-water exposure.

A series of tests that are better predictors of performance of sealants involves the preparation of tensile-adhesion joints or H pieces. In this case, the sealant bead is placed between blocks of the two substrates of interest and allowed to cure. This joint can then be pulled to destruction, with measurement of strength and failure mode. In addition, various environmental stresses may be applied, such as UV light exposure (weatherometry), water exposure, and cyclic movement. Testing in this way allows for improved prediction of movement capability and long-term performance. The methodology involved in this testing is included in the following methods: ASTM C-1135, Determining Tensile Adhesion Properties of Structural Sealants, and ASTM C-719, Test for Adhesion and Cohesion of Elastomeric Joint Sealants Under Cyclic Movement.

Any sealant or adhesive that is expected to perform in outdoor applications should be tested after exposure to light, heat, and water. For most forms of weatherometry, 500 to 1000 h is considered approximately the equivalent of 1 year outside in most climates (United States). If a minimum of 5 years of service is expected from the sealant, no less than 2500 h, and probably 5000 h in a UV fluorescent accelerated weathering machine should be used (as the conditioning cycle for rubber property testing). This is true for silicones and all other sealants that are expected to perform in such applications. This is a startling contrast to the 250 to 500 h used in most present standards (see Refs. 6 and 7).

V. BASIC FORMULATIONS

As mentioned in the processing section, silicone sealants and adhesives generally contain polydimethylsiloxane (PDMS) polymer, cross-linkers, fillers, catalysts, and other additives. These additives may be pigments, plasticizers (often unreactive polydimethylsiloxane polymers), and adhesion additives (such as silane coupling agents). Given below are simple formulations and properties (Table 2) for oxime-cured silicone sealants [8]. In these examples, the use of a nonreactive silicone plasticizer and a nonreinforcing carbonate filler results in substantial modulus reduction. This approach can also be used to modify the physical properties of silicones based on other cure chemistries. Low-modulus sealants are often used in sealant applications requiring high movement capability. High-modulus sealants are used more in structural and adhesive applications.

A. High-Modulus Oxime Sealant

	Percent by weight
Hydroxyl-ended PDMS polymer	80–85
Fumed silica	5–10
Oxime cross-linker	5–7
Sn(IV) catalyst	0.05–0.10

Table 2 Properties of Oxime-Cured Silicone Sealants

	Sealant	
Property	High modulus	Medium modulus
Tack-free time (min)	20 – 30	30 – 60
Durometer (Shore A)	25 – 35	20 – 30
Tensile strength (MPa)	1.2 – 2.1	0.9 – 1.4
Elongation (%)	200 – 400	400 – 700
100% Modulus (MPa)	0.5 – 0.9	0.35 – 0.5

B. Medium Modulus Oxime Sealant

	Percent by weight
Hydroxyl-ended PDMS polymer	60–80
Silicone plasticizer	5–20
Fumed silica	2–6
Calcium carbonate	20–30
Oxime cross-linker	5–7
Sn(IV) catalyst	0.05–0.10

VI. SUBSTRATE BONDING

Applications for silicones in bonding are numerous. Generally, one sealant will not bond to all substrates and it is common practice to develop new formulations to meet the ever-increasing list of requirements. In some instances primers are used for certain substrates, but silicones are usually self-priming. This self-priming feature is important from the standpoints of reducing installation costs and in reducing dependency on high-solvent primers, which are sometimes subject to environmental regulations. The surface characteristics for a given type of substrate can vary considerably between substrate manufacturers. For this reason it is always advisable to check adhesion before specifying a particular sealant. In addition, the importance of proper substrate cleaning and preparation should not be overlooked. Most adhesive and sealant producers will recommend the proper procedures for surface preparation. Some of the more common substrates and related applications for silicones are given in Table 3.

Table 3 Applications of Substrates

Substrate	Examples	Typical applications
Masonry	Concrete	Construction
	Mortar	Highway
	Brick	Consumer
Natural stone	Marble	Construction
	Granite	
	Sandstone	
Wood	Unpainted	Construction
	Painted	Glazing
		Consumer
		Maintenance
Glass	Float	Construction
	Reflective	Glazing
		Maintenance
		Original-equipment Manufacturing
		Consumer
Metals	Aluminum	Construction
	Steel	Glazing
	Copper	Electronics
	Stainless steel	Maintenance

Table 3 Continued

Substrate	Examples	Typical applications
Metals	Galvanized steel	Original-equipment Manufacturing Consumer
Coated metals	Paints Fluorocarbon Polyester	Construction Glazing Maintenance Original-equipment Manufacturing
Plastics	PVC PMMA Polyester Engineering plastics	Construction Glazing Maintenance Automotive Sanitary Original-equipment Manufacturing Consumer

REFERENCES

1. M. J. Owen and J. M. Klosowski, in *Adhesives, Sealants and Coatings for Space and Harsh Environments* (L. H. Lee, ed.), Plenum Press, New York, 1988, p. 283.
2. E. G. Rochow and H. G. LeClair, *J. Inorg. Nucl. Chem. 1*: 92 (1955).
3. J. M. Klosowski and G. A. L. Gant, *Plastic Mortars, Sealants, and Caulking Compounds*, ACS Series 113. (R. B. Seymour, ed.), American Chemical Society, Washington, D.C., 1979, p. 117.
4. D. T. Liles and N. E. Shephard, in *Science and Technology of Building Seals, Sealants, Glazing and Waterproofing*, Vol. 2, ASTM STP 1142 (J. M. Klosowski, ed.), American Society for Testing and Materials, Philadelphia, 1992.
5. L. A. Sobieski and T. J. Tangney, in *Handbook of Pressure Sensitive Adhesive Technology* (D. Satas, ed.), Van Nostrand Reinhold, New York, 1989, pp. 508–517.
6. L. B. Sandberg, *J. Mater. Civil Eng. 3*(4): 278–291 (1991).
7. G. R. Fedor, *2nd Symposium on Science and Technology of Building Seals, Sealants, Glazing and Waterproofing*, Ft. Lauderdale, Fla., ASTM C-24 FT., American Society for Testing and Materials, Philadelphia, 1992.
8. J. M. Klosowski, *Sealants in Construction*, Marcel Dekker, New York, 1989, pp. 269–270.

33

Epoxy Resin Adhesives

T. M. Goulding *Consultant, Johannesburg, South Africa*

I. INTRODUCTION

Epoxy or epoxide resins are a group of reactive compounds that are characterized by the presence of the oxirane group

$$CH_2 \overset{O}{\overset{\diagup \diagdown}{-}} CH -$$

They are capable of reacting with suitable hardeners to form cross-linked matrices of great strength and with excellent adhesion to a wide range of substrates. This makes them ideally suited to adhesive applications in which high strength under adverse conditions is a prerequisite. Their unique characteristics include negligible shrinkage during cure, an open time equal to the usable life, excellent chemical resistance, ability to bond non-porous substrates, and great versatility. Although they were hailed as wonder products when first introduced, it has now been accepted that they will not do everything. They have, however, clearly established niches, especially in high-technology applications, and have shown steady growth, generally ahead of the industry average. Sales of epoxy resins in Europe, for example, totaled 101,000 metric tons in 1980, 150,000 metric tons in 1985, and 205,000 metric tons in 1990.

Although work on epoxy resins started in the mid-1920s, the first commercially useful epoxy resins appeared during World War II. These were based on the diglycidyl ether of bisphenol A (usually referred to as DGEBA resins), and today these resins, in a range of molecular weights, constitute the majority of all epoxy resins used. By contrast, however, hardeners come in a variety of shapes and sizes, including amines and amides, mercaptans, anhydrides, and Lewis acids and bases. Choice of hardener depends on the application requirements, and the wide range of hardeners available increases the versatility of adhesives based on epoxy resins.

II. CHEMISTRY OF EPOXY RESINS

Epichlorhydrin is capable of reacting with hydroxyl groups, with the elimination of hydrochloric acid. The most widely used epoxy resins are the family of products produced by the reaction between epichlorhydrin and bisphenol A.

531

$$CH_2\text{—}CH\text{—}CH_2Cl \ + \ HO\text{—}\langle\bigcirc\rangle\text{—}\overset{\underset{\displaystyle CH_3}{|}}{\underset{|}{C}}\text{—}\langle\bigcirc\rangle\text{—}OH \ \longrightarrow$$

$$CH_2\text{—}CH\text{—}CH_2\text{—}O\text{—}\langle\bigcirc\rangle\text{—}\overset{\underset{\displaystyle CH_3}{|}}{\underset{|}{C}}\text{—}\langle\bigcirc\rangle\text{—}OH \ + \ HCl$$

This reacts with additional epichlorhydrin to produce a molecule of general structure

$$CH_2\text{—}CH\text{—}CH_2\text{—}\left[O\text{—}\langle\bigcirc\rangle\text{—}\overset{\underset{\displaystyle CH_3}{|}}{\underset{|}{C}}\text{—}\langle\bigcirc\rangle\text{—}O\text{—}CH_2\text{—}\overset{\overset{\displaystyle OH}{|}}{CH}\text{—}CH_2\text{—}\right]_n$$

$$O\text{—}\langle\bigcirc\rangle\text{—}\overset{\underset{\displaystyle CH_3}{|}}{\underset{|}{C}}\text{—}\langle\bigcirc\rangle\text{—}O\text{—}CH_2\text{—}CH\text{—}CH_2$$

Commercially useful grades are relatively low-molecular-weight products in which n ranges from 0 to about 4. When n is between 0 and 1, the product is a liquid, and this is the most useful product for adhesive applications. As n increases, the product moves toward a brittle solid. Solid grades find application principally in paints. Regardless of molecular weight, the resulting resin has two epoxy groups per molecule. Resins of greater functionality can be produced from polyols having more than two hydroxyl groups per molecule. Thus phenol novolac resins, having the general structure

$$\overset{\displaystyle OH}{\underset{\displaystyle }{\langle\bigcirc\rangle}}\text{—}CH_2\text{—}\left[\overset{\displaystyle OH}{\underset{\displaystyle }{\langle\bigcirc\rangle}}\text{—}CH_2\right]_n\overset{\displaystyle OH}{\underset{\displaystyle }{\langle\bigcirc\rangle}}$$

can be reacted with epichlorhydrin to produce epoxy novolac resins.

$$\begin{array}{ccc} CH_2\text{—}CH\text{—}CH_2 & CH_2\text{—}CH\text{—}CH_2 & CH_2\text{—}CH\text{—}CH_2 \\ | & | & | \\ O & O & O \\ \langle\bigcirc\rangle\text{—}CH_2\text{—} & \left[\langle\bigcirc\rangle\text{—}CH_2\right]_n & \langle\bigcirc\rangle \end{array}$$

These products may have much greater functionality, although stearic considerations limit the useful size of the molecule. Because of their higher functionality, epoxy novolacs have greater cross-link density, generally yielding better temperature resistance at the expense of increased brittleness. They are thus seldom used on their own, but make useful modifiers of the properties of DGEBA resins.

Other products that may be epoxidized in this way include dihydric and trihydric phenols, aliphatic polyols such as glycerol, and simple alcohols such as butanol or allyl alcohol. These products, especially the monofunctional glycidyl ethers, are used at relatively low percentages to modify properties of DGEBA resins, particularly to achieve lower viscosities.

Epoxy groups may also be produced by oxidation of olefinic unsaturation within animal and vegetable oils. The resulting products have too low a functionality for use as resins in their own right, but are added to DGEBA resins to introduce a measure of flexibility. Major manufacturers of epoxy resins include Shell, Dow and CIBA-GEIGY.

The resulting epoxy resin is capable of reacting with various products, or itself, to form a solid, infusible product of considerable strength. The fact that these reactions generally occur without the production of low molecular weight by-products means that shrinkage during cure is negligible. This reduces stresses in the cured structure, contributing to the strength of the cross-linked matrix and eliminating the need for sophisticated clamping techniques.

The two cross-linking reactions are external, by reaction of the oxirane group with active hydrogen, and internal, by homopolymerization through the oxirane oxygen. The former is typical of cross-linking by hardeners and the latter of catalyzed cross-linking. Both hardeners and catalysts are referred to as curing agents. The classic epoxy curing mechanism is illustrated by the reaction between a primary amine and an epoxy group:

$$RNH_2 + CH_2\text{--}CH\text{--}C\text{\Large\char`\~}\text{\Large\char`\~} \longrightarrow RN\text{\Large<}\begin{matrix}H\\CH_2\text{--}CH\text{--}C\end{matrix}\text{\Large\char`\~}$$

This product can react with an additional epoxy group to continue the cross-linking process.

$$RN\text{\Large<}\begin{matrix}H\\CH_2\text{--}CH\text{--}C\end{matrix} + CH_2\text{--}CH\text{--}C \longrightarrow RN\text{\Large<}\begin{matrix}CH_2\text{--}CH\text{--}C\\CH_2\text{--}CH\text{--}C\end{matrix}$$

This reaction is characteristic of hardeners having active hydrogens available, including amines, amides, and mercaptans. The reaction is catalyzed by hydroxyl groups, especially phenolic hydroxyls and tertiary amines. Because of the bulk of the substituent groups involved, steric factors have a major influence on the reaction rate. Thus low-molecular-weight hardeners tend to react more vigorously and produce more cross-linked structures, while hardeners of high molecular weight tend to react more sluggishly. Hardeners may thus be selected to produce highly exothermic reactions or reactions that take place only under the influence of external heat. Similarly, DGEBA resins having both epoxy groups at the ends of the molecule will react more readily with hardeners than will epoxy novolacs or other types of epoxies in which one or more of the functional groups may be hindered by the rest of the molecule.

Homopolymerization occurs readily in the presence of catalysts, especially at elevated temperatures.

$$n \left[\underset{CH_2-CH-C \sim\sim\sim}{\overset{O}{\triangle}} \right] \xrightarrow[R_3N]{BF_3} O \left[\underset{CH_2=C-C \sim\sim\sim}{\overset{O-}{\underset{|}{CH_2-CH-C \sim\sim\sim}}} \right]_{n-1}$$

Again this reaction is accelerated by hydroxyl groups or tertiary amines. This is also the predominant reaction with anhydrides. In fact, reactions with resin and hardener or catalyst are very much more complex than these idealized reactions, and both reactions as well as a number of side reactions probably occur to varying extents in any cross-linking mechanism. Major suppliers of curing agents include Anchor Chemicals, Dow, Shell, and Cray Valley Products.

III. PROPERTIES OF EPOXIES

A. Resins

Epoxy resins react with hardeners in stochiometric quantities. Thus a knowledge of the number of reactive sites is needed in order to calculate correct ratios. For the resin this is given by the epoxide equivalent weight (EEW), which is the quantity of resin required to yield one epoxy group. For a DGEBA type in which $n = 0$, the molecular weight is 340. Since there are two epoxy groups per molecule, the EEW is thus 170. Typically, the pure liquid DGEBA resins commercially available for adhesive applications have epoxide equivalent weight in the range 180 to 310, usually 190 to 210, while for paints or special applications, EEW may reach 2000 or more. The epoxy novolacs usually have EEW in the range 150 to 250, usually around 180.

The viscosity of a DGEBA resin is dependent primarily on molecular weight. Even at low molecular weight, viscosity is typically in excess of 6,000 cP, while at EEW 190 viscosity is usually around 12,000 cP. For applications requiring low viscosity it is thus necessary to include other types of epoxy resin or to use reactive or nonreactive diluents to achieve the desired viscosity.

B. Hardeners

Stochiometric ratios can be calculated similarly for hardeners. In principle, each active hydrogen will react with one epoxy group. Thus a low-molecular-weight aliphatic polyamine such as diethylene triamine (DETA) has a molecular weight of 103 and five active hydrogens. The hydrogen equivalent is thus 20.6. The stochiometrically correct ratio with an epoxy resin of EEW 200 would thus be 100 parts resin to 10.3 parts of DETA. In practice there is always a percentage of homopolymerization, especially at the temperature of reaction, and smaller amounts of DETA will still cause a complete cure, at the expense of increasing brittleness.

In general, suppliers of proprietary hardeners do not furnish detailed chemical descriptions. Instead, they supply data on recommended mix ratios, and from this the formulator can calculate the correct quantities. With catalysts, stochiometry is not critical, in theory. In practice, however, the quantity used will affect both the rate of cure and the cured properties. Thus with catalysts, in practice, the mix ratio is sometimes more critical than is the case with hardeners.

C. Mixed Product

During cure of epoxies, especially systems with a short pot life or large mixes, considerable heat is evolved. This accelerates the cure, leading to even greater heat evolution. Mixes larger than 5 kg can reach excessive temperatures even with systems that have relatively long pot lives in quantities of 100 g. In addition to shortening the pot life dramatically, exothermic reactions can push the peak temperature to the point where thermal degradation occurs, or at least to a level that creates excessive stresses in the curing matrix, causing it to crack on cooling. Except in certain circumstances, peak exotherm temperature should be limited by formulation to 150°C or preferably less in the mix quantities used. Cured epoxy resins may be formulated to be extremely hard, with Shore D hardeners of 80 or more, or soft, flexible products that barely produce a reading on the Shore A scale. When cured at approximately stochiometric ratios and unmodified with diluents or plasticizers, however, they are generally hard and tough to brittle, especially DGEBA and epoxy novolac types.

Heat distortion temperature (HDT) or deflection temperature (DT) is a measure of the tendency of cured product to soften when heated. It is a feature of the inherent thermoplasticity in cured epoxy compounds as a result of the relatively low cross-linking density, and may be any value from below 50°C to about 250°C, depending on formulation and cure cycle. Resins and hardeners of high functionality tend to have higher HDT. Postcuring at elevated temperature can increase HDT significantly.

IV. FORMULATING EPOXY ADHESIVES

Epoxy resins offer a unique combination of properties for adhesive applications. These include the ability to formulate liquid systems without solvents or carriers, the ability to convert these systems to cured products without the production of low-molecular-weight by-products, the ability to bond dissimilar or nonporous surfaces, and the ability to produce thick sections without subsequent stress cracking due to shrinkage.

Although epoxy resin and hardener may be used in unmodified form in adhesive systems, most systems will consist of components that have been modified by incorporation of various additives to achieve specific effects. Formulators will add catalysts or blend hardeners to obtain a specific usable life of the mix and to control the curing temperature. Reactive diluents may be added to modify viscosity or flexibility. Fillers impart improved compression strength and reduce shrinkage and cost. Solvents may be used to reduce viscosity or improve adhesion. Various additives may be added, usually at a low percentage, to reduce aeration, improve adhesion to difficult surfaces or minimize settlement of fillers. Depending on the application, particular properties such as flame retardency, electrical insulation or conductivity, or chemical resistance may be improved by formulating.

A. Resins

Although DGEBA resins provide the backbone of most epoxy formulations, they may be blended with other types to achieve modifications. Epoxy novolacs, having higher functionality, increase the cross-linking density, which improves heat resistance but decreases impact resistance. Incorporation of epoxidized oils increases flexibility at the expense of heat and chemical resistance. Low-viscosity polyfunctional epoxies based on polyols or polyhydric phenols reduce viscosity and can increase functionality without impairing cured properties. Monofunctional reactive diluents will also decrease viscosity and form part of the polymer backbone, to impart a measure of flexibility without the possibility of migration. Properties of commercially available epoxy resins and diluents from various suppliers are listed in Table 1.

B. Curing Agents

Use of mixed hardener systems is common. Hardeners may be blended to achieve properties intermediate to the individual components, to reduce exotherm or accelerate cure, to modify the cured properties, or simply to arrive at a more convenient mix ratio. Several classes of curing agents each having distinctive characteristics may be used.

1. *Aliphatic primary amines*. Common examples are diethylene triamine (DETA), tetraethylene pentamine (TEPA), n-aminoethyl piperazine, and isophorone diamine. They give good room-temperature cure at stochiometric ratios, but have poor HDT, inconvenient mix ratios, high peak exotherm, and are strongly irritant. Isophorone diamine produces very light colored mixes with good color stability.

2. *Aromatic primary amines*. These offer improved heat and chemical resistance and longer pot life with reduced exotherm, but poor color stability and sluggish cure. They are generally solids and require some formulating to produce easily handleable products. Reactions proceed best at elevated temperatures, where their irritancy can be a problem. For room-temperature cures they should be used with catalysts, of which phenols, BF_3 complexes, and anhydrides are the best. m-Phenylene diamine (MPDA) and methylene dianiline (MDA) are the best examples.

3. *Amine adducts*. Both aliphatic and aromatic amines can be adducted with small amounts of mono- or diglycidyl epoxies to produce amine adducts of medium to high viscosity that have decreased volatility and irritancy, more convenient mix ratios, and often, better reactivity.

4. *Tertiary amines*. Although their primary use is as catalysts with other hardeners, tris(dimethylaminomethyl) phenol (DMP 30) is an effective curing agent on its own, at both room and elevated temperatures.

5. *Amides*. Although amides on their own are too unreactive, reaction products of polyamines with fatty acids to produce amidopolyamines provide the largest group of commercial hardeners for adhesive applications. Reduced volatility and irritancy and a convenient mix ratio offer the compounder ease of handling. The mix ratio is relatively noncritical; increasing hardener levels yield increased flexibility and adhesion but reduced HDT and chemical resistance. Initially, amidoamines have poor compatibility and an induction period is necessary to allow the reaction to start. Pot lives are relatively long and exotherms low, but low-temperature cure is poor. Small quantities of imadazoline improve adhesion to metal. Dicyandiamide is a special example of an amide that can be used on its own. Its low reactivity yields a usable life in excess of 6 months, but at elevated temperatures it reacts quickly.

6. *Mercaptans*. Most mercaptans on their own are unreactive, but with catalysts produce flexible cures. Certain mercaptans in conjunction with DMP 30 provide extremely rapid cure, with low exotherm, making them ideally suited to retail applications.

7. *Acids and anhydrides*. This group of curing agents provides the best high-temperature performance. Boron trifluoride monoethylamine, oxalic acid, and maleic and phthalic anhydride are used for electrical or high-temperature applications, often with

Table 1 Epoxy Resins

Resin[a]	Chemical type	EEW	Viscosity (cP at 25°C)	Supplier
Low viscosity				
Araldite DY026	1:4 Butanediol diglycidyl ether	110–115	10	CIBA GEIGY
DER 732	Propylene glycol diglycidyl ether	305–335	55–100	Dow
DER 736	Propylene glycol diglycidyl ether	175–205	30–60	Dow
Epikote 812	Glycerol triglycidyl ether	140–160	100–170	Shell
Epikote 871	Linoleic acid diglycidyl ether	390–470	400–900	Shell
Medium viscosity				
Araldite GY260	DGEBA	185–196	12,000–16,000	CIBA GEIGY
DER 331	DGEBA	182–192	11,000–14,000	Dow
Epikote 880	DGEBA	185–192	10,000–16,000	Shell
Beckopox EP 140	DGEBA	180–192	9,000–12,000	Hoechst
High viscosity				
Epikote 834	DGEBA	230–280	(Softens at 35–40°C)	Shell
DER 337	DGEBA	230–250	Not given	Dow
Beckopox EP 151	DGEBA	400–500	20,000–30,000	Hoechst
Araldite EPN 1139	Epoxy novolac	170–180	50,000	CIBA GEIGY
Epikote 154	Epoxy novolac	176–181	3,500–7,000 at 52°C	Shell
DEN 438	Epoxy novolac	176–181	20,000–50,000 at 52°C	Dow
Monofunctional				
Allyl Glycidyl Ether	Allyl glycidyl ether	114	1	Shell
Dow BGE	Butyl glycidyl ether	143	3	Dow
Cardura E	Versatic acid glycidyl ether	240–250	7–8	Shell
Beckopox EP 080	2-Ethylhexyl glycidyl ether	190–205	2–3	Hoechst

[a]Note that trade names and grade designations may vary from country to country.

catalysts such as benzyldimethyl amine (BDMA) or DMP 30. Table 2 lists properties and characteristics of various curing agents.

C. Reactive Diluents

In addition to the monofunctional epoxies described under resins, products with active hydrogens, such as furfuryl alcohol, coal tars, or phenols, will react with the epoxy resins to form part of the cured structure. Triphenyl phosphite both reacts and accelerates. Lactams also react with the hardeners. Since all these products tend to degrade the performance of the cured product, it is preferable to use difunctional low-viscosity epoxies to reduce viscosity. Where applicable, functionality of reactive diluents must be allowed for when calculating ratios.

D. Plasticizers

Conventional plasticizers may be used in formulated products. Phthlate esters are the preferred plasticizers. They exhibit little tendency to migrate and have good compatibility with both resins and hardeners. Addition rates are typically 5 to 20%. Chlorinated plasticizers may be used to reduce flammability, especially in conjunction with antimony trioxide. The effect of plasticizer additions is generally to degrade most physical properties, although at low additions the effect is usually small. The effect of plasticizer additions on various important properties is as follows:

Pot Life: lengthened
Impact resistance: increased
Peak exotherm, tensile strength, chemical resistance, HDT: decreased

It should be noted that plasticizers do not introduce marked flexibility into epoxy resin systems. Nor is this generally a desirable attribute in adhesive applications, where epoxies are usually selected because of their great strength.

E. Fillers

Two types of fillers may be incorporated into formulated epoxy systems. Powder fillers are added to increase viscosity, improve abrasion resistance and gap-filling properties, impart specific electrical or mechanical properties, or reduce cost and shrinkage. Addition levels may be 50 to 300 parts by weight of resin (phr). Although most fillers will increase the density of the cured product, certain lightweight fillers will decrease density. Viscosity increases depend on surface area, oil absorption, and filler type. Chemical resistance may be improved or made worse, depending on fillers selected. Highly alkali fillers should be avoided, especially with acid-cured systems, as they may retard setting.

Fibrous fillers may be added to impart specific rheological properties or to reinforce the system. They will usually improve both tensile strength and impact resistance. Addition levels are much lower at 10 to 50 phr, as they usually cause much more rapid thickening. Table 3 lists common fillers.

Settlement of fillers during storage depends primarily on the particle size of the filler and its density, and the viscosity of the formulated product. Settlement can be reduced or eliminated by proper formulation. Fine particle fillers with relatively low specific gravity in high-viscosity products will settle much less, especially if the product is at all thixotropic. Where coarse fillers must be used, an approach toward a fully filled voidless

Table 2 Epoxy Curing Agents

Curing agent	Usage (phr)[a]	Uncatalyzed cure (°C)	HDT (°C)	Applications
Aliphatic primary amines				
Diethylene triamine	10–12	Ambient	80–100	Short pot life
Tetraethylene pentamine	13–15	Ambient		Ambient curing systems
Diethylamine propylamine	5–8	40–80		
n-Aminoethyl piperazine	22–25	Ambient		
Aromatic primary amines				
m-Phenylene diamine	12–15	60–100	150–180	Longer-pot-life general-purpose epoxies
Methylene dianiline	25–28	60–100		
Tertiary amines				
Benzyl dimethylamine	6–10	60–100	80–100	Catalysts, especially with polysulfide
Tris(dimethyl aminomethyl)phenol	3–6	20–60		
Amides				
Dicyandiamide	3–5	120–160	120	Latent catalysts for one-pack epoxies
Acids				
Boron trifluoride monoethylamine	2–4	120–150	175	Heat-resistant epoxies
Oxalic acid	5–10	120–160	60–120	Catalyst for anhydrides
Anhydrides				
Phthalic anhydride	60–80	120–140	120–150	Encapsulation
Maleic anhydride	50–80	80–120		

[a]DGEBA epoxy of EEW 200.

Table 3 Fillers and Extenders

Filler	Specific gravity	Oil absorption (%)	Relative cost	Usage level	Purpose
Silica	2.5–2.7	20–30	Low	50–500	Bulk, price
Quartz	2.5–2.7	15–30	Low	50–300	reduction,
Calcium carbonate	2.6–2.8	15–25	Low	50–300	stability, and
China clay	2.3–2.6	30–60	Low	20–100	reduced
					exotherm.
Carbon	2.0–2.2	—	Medium	5–50	Thermal and
Aluminum powder	2.5–2.7	—	High	20–100	electrical
Copper powder	8.8–9.0	—	High	20–100	conductivity
Silicon carbide	3.2	—	Medium	50–200	Abrasion resistance
Microballoons	—	—	High	10–50	Low density
Asbestos		—	Low	5–20	Reinforcement
Glass fiber	2.4–2.6	—	Medium	10–50	Reinforcement

system where the volume of liquid is such as just to fill the voids will solve the problem. Incorporation of fine fillers, use of a pigment-dispersing aid, and where application permits, use of a thixotroping agent will help to reduce or eliminate settlement.

Depending on addition level, fillers will generally increase the usable life and extend the cure time of the mix. Tensile and compressive strength usually increase maximally then decrease on further additions. Most fillers have relatively little effect on HDT. Chemical resistance will vary from filler to filler. Shrinkage is usually reduced.

F. Solvents

Although a major advantage of epoxy adhesives is their ability to be formulated without solvents, under certain circumstances solvents may be included. On porous substrates solvents may be added to reduce viscosity and assist penetration. On certain nonporous substrates, particularly some plastics, addition of a small percentage (1 to 3%) of a suitable solvent will improve adhesion. Common solvents are low-boiling aromatic solvents, ketones, or esters.

G. Additives

Additives are typically products added at levels of 0.1 to 0.5% to modify specific properties. Most commonly used additives are defoamers, antisettling or wetting agents, thixotropes, and adhesion promoters. Use of antioxidants or preservatives is rare. Because of their minimal shrinkage, compressive strength of cured epoxies is very high. Since aeration will reduce this substantially, use of defoamers, especially in heavily filled systems, is quite common. Many defoamers are suitable, but silicone-based defoamers should be avoided on surfaces where adhesion is critical. Addition levels of 0.05 to 0.2% usually suffice.

Antisettling agents, pigment dispersers, or wetting agents may be included in filled formulations. Depending on the formula, particularly the selection of fillers, such products may reduce or eliminate settlement. Usage will generally be at a level of 0.1 to 0.3% of formulation. These agents are best added prior to incorporation of the fillers. Various thixotropes are used in epoxy formulations to reduce or eliminate flow in products designed for use on vertical surfaces, to improve gap-filling properties or to reduce settlement of fillers. Fumed silica is widely used at levels of 0.1 to 3%. At low levels, the effect on viscosity is small except in high-viscosity systems, but settlement will be reduced. At higher addition levels, even low-viscosity products can be converted to firm pastes. To improve the efficiency of fumed silica, especially in the resin component, small quantities of polar liquids may be added.

Other thixotroping agents include Bentones and Tixogels, of which a number of grades are available, and China clay or kaolin, usually added as a filler, but which imparts thixotropy to the formulated product. Organofunctional silanes are extremely effective adhesion promoters. Added at levels of 0.05 to 0.2%, they can improve adhesion to certain nonporous substrates, such as glass, metals, and certain plastics. Formulators can select from a number of different functional groups, but generally epoxy functional types will be used in the resin component and amine functional grades in the hardener.

H. Elastomers

Occasionally, elastomers may be included in solvent-based formulations. Poly(vinyl butyral) improves adhesion to metal, as does nitrile rubber, while natural and synthetic rubbers may be incorporated to improve flexibility.

V. APPLICATIONS

Epoxy resin adhesives are used mainly in niche applications rather than as general-purpose adhesives. Due to the high strengths that can be achieved and the relatively high costs, they are generally used in structural applications in both concrete and metal bonding. Their good electrical properties allied to low shrinkage and durability suit them for potting and encapsulating. Low shrinkage and good gap filling make epoxies ideal for applications where clamping is difficult, while the fact that both components are generally liquid up to the moment of cure means that they can be used where application constraints require long open or assembly times. Conversely, systems with very short cure times are perfect for consumer applications. Good adhesion to nonporous surfaces allows them to be used in demanding situations. They find major outlets in the construction, automotive, and electronics industries.

A. Building and Construction

Water-based epoxy primers are ideal for damp porous substrates, as such primers will penetrate to an adequate depth to ensure good adhesion and produce a sound surface for bonding. Emulsifiable resins and hardeners are available, and the better systems deactivate the emulsifier system during cure to ensure that the cured system is not unduly water sensitive (Section VI.A). Solventless epoxy primers are used for bonding new concrete to existing concrete. Polyamide hardeners are preferred because of their ability to cure satisfactorily in the presence of water. Accelerators and diluents may be added, but fillers are generally omitted. The primer is applied to the existing concrete, and the fresh concrete cast before the resin has set (Section VI.B).

Epoxy adhesives are suitable for tiling, both for floor tiles in applications requiring acid or chemical resistance and in high-hygiene areas and also for tiling on vertical surfaces, where it is essential that tiles should not delaminate. Epoxy tiling systems are suitable for glazed tiles, clay and ceramic tiles, and decorative marble or granite tiles, where priming is recommended. Adhesives will usually be filled and thixotropic, especially for vertical tiling. Epoxies are also used for decorative paving in commercial and residential properties. Flooring made of small pebbles of different colors and textures bonded with relatively small proportions of epoxy are attractive and provide good drainage of water in areas such as swimming pool surrounds. Light-colored systems with good UV resistance are required. Hardeners should be based on Isophorone diamine because of its good color stability, and UV stabilizers may be included.

Self-leveling floors are produced from low-viscosity epoxy systems. Low-exotherm, unfilled systems are preferred. The entire floor should be cast in one operation, and thickness should preferably be at least 5 mm over the entire area. Because of their excellent chemical resistance to a wide range of chemicals, epoxies are often selected for flooring in chemical plant. Systems vary from trowelable to pourable or brushable and are usually filled. Choice of hardener and filler will depend on the specific chemicals encountered. Although tables of chemical resistance from suppliers will aid in selection of a suitable system, this system should always be tested using the chemicals that the floor is expected to withstand.

B. Metal Bonding

While construction applications usually require reactive hardener systems to give good room-temperature cure, many metal bonding applications require strength at elevated temperatures. Usually, however, they also permit heat curing and postcuring. Surface preparation is crucial to achieving high bond strength and will always involve at least degreasing and abrading. Because of the high strength of the substrates, joint design is also very important and should always aim to provide the largest practicable bonded areas. Since high-strength epoxies are generally hard, joint design should aim to produce bonds that are in tension or compression rather than shear or peel.

Solvent-free systems may include adhesion promoters, such as silicones, flexibilizers such as liquid polysulfide rubbers, and reinforcing fillers, either fibrous or micronized. Room-temperature to mildly elevated-temperature systems will be cured with amidopolyamine hardeners, often at ratios considerably in excess of stochiometric requirements. This increase in hardener quantity improves flexibility and adhesion at the expense of tensile strength and heat distortion temperature. For applications at higher temperatures, use is made of more reactive polyamine hardeners, often with metal powder as filler. These will have postcuring cycles of several hours at temperatures that are increased in steps up to 150 to 180°C. Alternatively, solvent-based hybrid systems can be formulated, incorporating phenolic resins, nitrile rubbers, and poly(vinyl acetals). These solvent-based systems are typically single-component, applied to both mating surfaces. After the solvent has flashed off, the assembly is clamped and cured at elevated temperature. This type of system is particularly suited to applications such as bonding of brake linings to their backing pads (Section VI.C).

C. Road Making

Epoxy adhesives find application in various aspects of road making. Two major areas are bullnosing and fixing of reflective road studs. Bullnosing requires a system with good impact and compressive strengths and a pot-life time of 1 to 4 h. Cast masses are on the order of 5 to 25 kg, so low-exotherm systems should be used. Fillers and flexibilizers may be included. Fixing of road studs is often performed on roads that are in use, so cure time should be as short as is practicable. Again, good impact and compressive strengths are required.

D. Wood Bonding

Although PVA adhesives for nonstructural applications and formaldehyde-based resins for structural applications have price advantages over epoxies and offer excellent performance, epoxies have advantages in certain applications. First, the open time may be as long as required. Second, with suitable primers they give more reliable bonds on difficult species, such as very dense or oily timbers. Third, they may be used to bond wood to other surfaces, such as metal or concrete, although in this case the formulator will concentrate more on the other substrate, as wood is not a difficult surface for epoxy adhesives. Primers may be emulsions in water or, particularly on oily surfaces, a conventional epoxy reduced to low viscosity with suitable solvents. The adhesive may also contain solvents to reduce viscosity and allow high filler loadings.

E. Engineering Applications

Epoxies find many applications in industry, especially on the engineering side. Grouting of bolts into concrete or rock surfaces, either to strengthen a rock face or for fixing of heavy machinery, is a common application. Such adhesives are usually filled systems, often using reactive hardeners to achieve rapid setting. Epoxy systems for horizontal grouting will usually be thixotropic to prevent the adhesive from slumping or flowing out (Section VI.D).

Crusher backing epoxies are used to fill the gap between the replaceable liners and the outer housing in industrial crushers. These products have conflicting requirements: fast setting but with low exotherm even in large volumes, low viscosity for easy pouring but with minimal settlement of fillers. In addition, they must have good impact resistance, negligible shrinkage, and be easy to remove when the liners are replaced.

F. Electrical Applications

Excellent electrical insulation makes epoxy systems suitable for potting and encapsulation of electrical and electronic components. Here use is made of one-pack epoxies employing latent curing agents such as dicyandiamide. At the curing temperature the system will have very low initial viscosity, ensuring good wetting and bubble release. The low residual stresses protect components from mechanical damage. Fillers or pigments may be added to render the cured article opaque.

The low shrinkage and good durability of epoxies also fits them for cable jointing compounds. Here a measure of flexibility is desirable. The cured article may need to be worked on from time to time, so systems that can be cut or peeled off may be required.

Usable lives of 15 min to 1 h are the norm. Components should be selected for good water resistance (Section VI.E).

G. Film Adhesives

Latent catalysts used in conjunction with either liquid or solid resins are cast in thin films on plastic or release paper for unsupported films or onto absorbent papers or cloths for supported films and then cured to B stage. These films can subsequently be cut to shape and placed between mating surfaces for subsequent heat curing (Section VI.F).

H. Miscellaneous Applications

Epoxy adhesives are popular for retail or consumer applications. They may be supplied as liquids, pastes, putties, or as one-pack systems in stick form. Considerable ingenuity has been employed in the packaging to ensure that stochiometric ratios are roughly correct. They are usually available as 5 min systems using mercaptan accelerators or standard setting systems where amidopolyamine hardeners are preferred because their mix ratios are relatively noncritical, and they lend themselves to formulating 1:1 mix ratios (Section VI.G).

Mercaptan accelerators are also used for systems with very short pot life, in automatic dispensing machines which meter the components, mix, and dispense in seconds. Systems with pot lives of 40 s can be handled in automated production lines. Epoxy resins are employed in a number of other niche applications, including adhesives to control static buildup in computer installations, carveable epoxies for pattern making and tooling, and acid-resistant adhesives for fastening tops to automotive batteries and for laminating and repair of glass-reinforced plastics. A selection of guide formulations for various applications follows.

VI. GUIDE FORMULATIONS

Parenthetical numbers that follow the components listed in the formulations below correspond to these suppliers:

1. Hoechst AG
2. Dow Chemicals
3. Anchor Chemicals
4. Degussa
5. Shell Chemicals
6. Union Carbide
7. Cray Valley Products
8. CIBA GEIGY
9. Thiokol Corp.
10. Sud Chemie AG
11. Diamond Henkel
12. SKW Trostberg

A. Water-Based Epoxy Primer

Beckopox EP 140 (1)	20.0
Versaduct 429 (7)	20.0
n-Butanol	5.0 – 10.0
Water	55.0 – 50.0

B. Epoxy Adhesive for Bonding New Concrete to Old

Base:			Hardener:		
	DER 331 (2)	80.0		Ancamine MCA (3)	30.0
	Epodil L (2)	15.0		Ancamide 500 (3)	24.0
	Aerosil 200 (4)	4.5		Anchor K54 (3)	6.0
	Water	0.5		Thiokol LP 3 (9)	40.0
		100.0			100.0

C. Metal-to-Metal Adhesives

1. Ambient Cure

Base:					
	Epikote 880 (5)	80.0		Beckopox EP 151 (1)	16.0
	Epikote 834 (5)	19.8		Beckopox EP 140 (1)	16.0
	Silane A 187 (6)	0.2		Quartz powder	50.4
		100.0		Beckopox EH 610 (1)	7.2
				Beckopox EH 652 (1)	10.4
					100.0

Hardener:	Versamid 125 (7)	100.0

2. Elevated Cure

Base:	Beckopox EP 140 (1)	100.0		DER 331 (2)	5.0
				DEN 438 (2)	20.0
Hardener:	Amicure CG 1200 (3)	7.0		Phenodur PR 263 (1)	25.0
	Ancamine 2014 AS (3)	2.0		Mowital B30H (1)	5.0
	Amicure UR (3)	3.0		Nitrile rubber	5.0
		12.0		Methyl ethyl ketone	10.0
				Toluene	30.0
					100.0

Cure schedule: 60 min/130°C   Cure schedule: 15 min/180°C

D. Grouting Adhesive

Base:			Hardener:		
	DER 331 (2)	45.0		MDA	15.0
	DER 732 (2)	5.0		Thiokol LP 3 (9)	15.0
	Silica 150 mesh	46.0		m-Cresol	15.0
	Tixogel VZ (10)	4.0		Silica 150 mesh	50.0
		100.0		Tixogel VZ (10)	5.0
					100.0

E. Cable Jointing Epoxy

Base: Araldite GY 260 (8)   100.0   Hardener: Ancamine LV (3)   60.0
      Epodil L (3)   20.0   Quartz flour   72.0
      Quartz flour   15.0   Magnesium silicate   3.0
                              135.0   135.0

F. Film Adhesives for Preimpregnation

Hot-melt type   Solvent type
  Epikote 880 (5)   100.0   DER 652 (2)   133.0
  Epiclon B570 (3)   86.0   Dyhard 100 S (12)   4.0
  Amicure DB/U (3)   3.0   Dimethyl formamide   15.0
                      189.0   Dowanol PM (2)   15.0
                                    BDMA   0.3
                                                         167.3

G. Fast-Setting Retail Epoxy Liquid

Base: DER 331 (2)   100.0   Hardener: Capcure 3-800 (11)   90.0
                                         Capcure EH-30 (11)   10.0
                                                              100.0

VII. SUMMARY

Advantages of epoxy resin adhesives may be summarized as follows:

1. Ability to bind a wide range of substrates.
2. Negligible shrinkage during cure, which minimizes stresses.
3. Elimination of galvanic corrosion when bonding dissimilar metals.
4. Solvent-free liquids with open times similar to pot life.
5. Minimal clamping requirements.
6. High strength, good durability, and resistance to a wide range of environments.
7. Flexible formulating, permitting a wide range of pot lives, application conditions, and cured properties.
8. Stochiometrically cured epoxies generally inert and physiologically harmless.

Disadvantages are:

1. Two-component systems require mixing in correct ratios, with attendant pot-life problems.
2. Many components toxic or irritants.
3. Relatively poor heat resistance of many cured systems.
4. Inherent brittleness, requiring careful joint design.
5. Poor cure at low temperatures.
6. Careful surface preparation required.
7. Need for skilled applicaters.
8. High cost.

ACKNOWLEDGMENTS

My thanks to Hoechst AG, Dow Chemicals, and Anchor Chemical Division of Air Products for the information and assistance provided.

BIBLIOGRAPHY

Dow Epoxy Resins in Adhesives, Dow Chemical Company, technical bulletin.
Formulating with Dow Epoxy Resins, Dow Chemical Company, technical bulletin.
Lee H., and K. Neville, *Handbook of Epoxy Resins*, McGraw-Hill, New York, 1967.
Reader, C. J., and N. T. Hunt, *Adhesives, Tooling, Electronic Insulation and Laminating*, Anchor Chemicals, Division of Air Products, technical bulletin.

34

Pressure-Sensitive Adhesives

T. M. Goulding *Consultant, Johannesburg, South Africa*

I. INTRODUCTION

Pressure-sensitive or permanent-tack adhesives are, as their name implies, adhesives that remain sticky even when dried or cured. This means that they are capable of bonding to surfaces simply by the application of light pressure. This makes them arguably the most convenient products available today from the end user's viewpoint and undoubtedly, accounts for the success they enjoy. Although figures are hard to come by, a survey by Business Trend Analysts quoted in the June 1990 issue of *Adhesives Age* shows that pressure-sensitive adhesives grew from 38% of total adhesive sales in the United States in 1980 to 44.6% in 1988, at an annual rate of 12%, to reach a sales value of $4.9 billion in 1989.

Tack is a word used to describe various phenomena, including *wet tack*, which is the ability of an adhesive to form a bond while still wet; *green tack*, which is the ability of certain polymers, specifically rubbers, to bond to themselves for several hours after drying, even though the surfaces do not feel sticky; and *pressure-sensitive tack*, which is the phenomenon of importance to this section. This relates to the ability of a dried film to bond tenaciously to most surfaces under light pressure. As pressure is increased, the bond improves. The classic theory of tack is that it arises from the presence of a two-phase system in which an elastic continuous phase provides the strength while a disperse phase acts as a viscous liquid that wets and adheres to the surface. Although this appears to be the dominant mechanism in the older rubber–resin systems, however, many modern systems do not rely on this apparently incompatible two-phase system. Acrylics, for example, can produce aggressive tack from a single component. Thus tack is also believed to stem from the viscoelasticity of many polymers, allowing them to conform to the substrate to be adhered and "wet" it even in the dry state. It follows that a fundamental requirement for tack is a glass transition temperature substantially below the application temperature to permit the necessary degree of flow.

Pressure-sensitive adhesives fall into three broad product categories: water based, solvent based, and hot melt. Application areas tend to overlap, and all three types can be used in most of the application areas. Despite this overlap, tapes tend to be produced from solvent-based adhesives, while water-based adhesives are preferred for label stock. Hot melts are used in both applications. Pressure-sensitive tapes for a variety of uses, such as masking, packaging, and insulation, are the largest application area, followed by self-adhesive labels. Although these applications appear outwardly similar, in fact there are fundamental differences. With tapes the adhesive fills the major role, ensuring adequate adhesion and requiring special properties, which may include high dielectric strength, heat resistance, or low toxicity. In labeling applications the major demands are on the backing, which needs the right lay-flat or curl properties and ease of cutting and printing, with relatively few demands on the adhesive. For certain applications, the adhesive may have to retain flexibility and tack at temperatures down to $-20°C$, or be easily removable. With tapes the adhesive is usually applied directly to the backing, while label adhesives are usually applied to the release paper and subsequently transferred to the backing. Other pressure-sensitive applications include self-adhesive floor tiles, adhesives for decor papers and flypapers, gloss lamination, disposable diapers and other personal hygiene products, and temporary assemblies.

II. PRODUCT TYPES

Traditional pressure-sensitive adhesives were solutions of rubber and resin in solvent, and these dominated the market until well after World War II. From that time, as an increasing array of elastomers became available, as the price of solvents soared and as environmental opposition to the use of solvents increased, water-based and hot-melt types made substantial inroads into the solvent-based market. This trend is likely to continue, although solvent-borne adhesives will probably always retain niches in areas where drying speed or ability to key into specific surfaces will outweigh environmental, handling, or price considerations.

A. Solvent-Based Adhesives

The three major components are an elastomer, which provides the elastic phase, the tackifier, and the carrier. The earliest pressure-sensitive adhesives used natural rubber tackified with wood rosins, or later, zinc oxide. With the advent of synthetic rubbers and other polymers, formulators have a very much larger range of elastomers at their disposal, including butyl rubber, styrene–butadiene rubber, polyisoprene, and the more recent thermoplastic rubbers, which are block copolymers of styrene with butadiene or isoprene, as well as acrylic polymers. Silicone elastomers are available for specialty applications, especially for use at elevated temperatures. Vinyl ethers and polyisobutylene can be used as both elastomer and tackifier, depending on the grade. Many of these types are not compatible, and where intermediate properties are required, they are generally achieved by blending homologs from the same or related families.

Tackifying resins fall mainly into two classes: wood rosin derivatives and hydrocarbon resins. Gum rosin is no longer widely used, as heat or aging lead to loss of tack through oxidation. Stable derivatives are produced by hydrogenation or esterification, and these are used extensively as tackifiers. Modern hydrocarbon resins are usually aliphatic, aromatic, or terpenes, although blends of these or certain specialty types may be suitable. There is no universal guide to selection, and a good deal of trial and error may be necessary to arrive at the ideal elastomer–tackifier combination and proportions. In addition, these may differ from one application to another, depending on whether the end use has tack, peel, or shear as the dominant criterion.

B. Hot-Melt Adhesives

The fundamentals of pressure-sensitive hot-melt adhesives are similar to those of solvent-based systems. Most elastomers and tackifiers are suitable, although ethylene–vinyl acetate copolymers are also used and the conventional rubber types are not. Pressure-sensitive hot melts are dominated by thermoplastic rubbers, which are ideal for use in these applications. Their unique properties arise from their essentially two-phase structure, in which thermoplastic regions of styrene end blocks lock the elastomeric mid-sections of butadiene or isoprene at room temperature but allow the elastomer to move freely at elevated temperatures or in solvent. This gives the polymer properties that are akin to those of vulcanized rubbers at room temperature, while allowing it to behave as a thermoplastic when heated or dissolved. This structure is illustrated in Fig. 1.

Early pressure-sensitive hot-melt adhesives used ethylene–vinyl acetate copolymers as elastomers, but they are seldom used now. Atactic polypropylene is sometimes used on its own or in admixtures. More recently, vinyl ethers and acrylic resins have become available and will probably play an increasingly important role as the technology is developed, especially on polar surfaces.

The major differences between solvent-based and hot-melt pressure-sensitive adhesives is that with hot melts the viscosity can no longer be controlled with solvents, and must, instead, be controlled either by temperature or by formulating. A further limitation is that waxes cannot, in general, be used for reducing viscosity as is the case with conventional hot melts, as waxes tend to reduce tack drastically. Hence the major influence

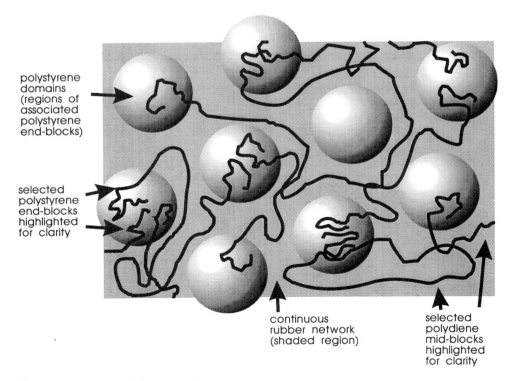

polystyrene
domains
(regions of
associated
polystyrene
end-blocks)

selected
polystyrene
end-blocks
highlighted
for clarity

continuous
rubber network
(shaded region)

selected
polydiene
mid-blocks
highlighted
for clarity

Figure 1   Structure of thermoplastic rubber.

Table 1 T_2 Values of Monomers in Common Use

Soft monomers	Tg (°C)	Hard monomers	Tg (°C)
Butyl acrylate	−54	Methyl methacrylate	105
Isobutyl acrylate	−40	Vinyl acetate	29
2 Ethyl hexyl acrylate	−85	Styrene	100
Ethyl acrylate	−22	Acrylonitrile	100

on viscosity in the formulation must come from the choice and quantity of tackifier resin. Low-melting-point resins or even liquid resins may be used to keep application temperatures as low as practicable.

Because pressure-sensitive hot melts will be applied typically at temperatures between 120 and 160°C, heat resistance is a critical factor. Double bonds accelerate oxidative degradation, leading to a loss of tack, while cleavage will usually result in increased tack, but a drop in viscosity. Use of antioxidants is essential, and trials should always be undertaken to ensure that enough of the right antioxidant has been incorporated to protect the hot melt adequately at the application temperature even if a machine stoppage leads to the molten adhesive being kept in a heated bath for much longer than usual.

C. Water-Based Adhesives

Various dispersions are available which even in unmodified form exhibit aggressive tack and good adhesion, especially to polar substrates. Produced by conventional emulsion polymerization techniques, the tack, peel, and shear properties of these dispersions can be varied within wide limits by the choice of monomers. A dispersion will usually consist of at least two monomers, one of high glass transition temperature (T_g) and the other with a low T_g value, and the ratio of the two will determine the final properties of the film. Table 1 lists the T_g values of monomers in common use.

Cross-linkable monomers may be included to make the formulated adhesive curable by catalysis, heat, or radiation, thereby improving the performance of the film, especially at higher temperatures. Since the dispersion has both toughness and tack built in, no further compounding is necessary, making pressure-sensitive acrylic dispersions the easiest products to work with. In most applications, however, the formulator will prefer to modify the properties to order, and use of tackifying resins added either in solution or as a dispersion is common. Vinyl ethers can again be used either as sole binders or as tackifiers to modify the properties of the base dispersion.

Water-based systems have good aging characteristics, resisting the effects of heat, ultraviolet light, and oxidation. Thus use of antioxidants is not normally necessary. Table 2 lists the major advantages and disadvantages of the various types of pressure-sensitive adhesives.

III. FORMULATING

As mentioned earlier, the critical characteristic is the correct tackifier and ratio. Although no rules exist for tackifier selection, there are certain shortcuts, based on chemical compatibility and melt point. Certain classes of tackifiers work well with specific types of

Table 2 Advantages and Disadvantages of Pressure-Sensitive Adhesives

Solvent-based	Water-based	Hot melt
	Advantages	
Quick drying	Easy cleaning	Very fast setting
Good adhesion to nonpolar substrates	Good adhesion to polar substrates	No solvent waste
Good key on certain plastics	Good heat and aging resistance	Environmentally acceptable
Versatile	Environmentally acceptable	100% active
	High solids	
	Ready to use	
	Disadvantages	
Flammability	Slow drying	High equipment cost
Toxicity	Requires heat to dry	Requires heat
Relatively low solids	Poor on nonpolar surfaces	Thermal degradation
Less easy to clean		Difficult to clean
		Can melt substrate
		Difficult to package

elastomers. For example, aliphatic hydrocarbons generally work better with natural rubbers, and aromatic types are preferred for SBR. With block copolymers, aliphatic resins of low melt point improve tack and low-temperature flexibility, while high-melting aromatic resins in small quantities stiffen the product, giving improved heat and shock resistance. Rosin derivatives and terpene resins offer good performance with most elastomers, generally at higher cost. In general, resins with solubility parameters close to those of the elastomer selected are most likely to offer good performance.

Tackifiers with melt points substantially above the T_g value of the elastomer can be expected to improve the strength of the adhesive at elevated temperatures but reduce the tack, while low-melting resins will impart greater tack and low-temperature flexibility at the expense of creep resistance and shear strength. The tackifier is responsible primarily for the balance of tack, peel, and shear properties in the finished adhesive. Usually, some of these properties must be traded off to optimize one property. For any given system, increasing tack is generally related to decreasing peel and shear strengths, and similarly, any modification intended to improve shear strength is likely to be at the expense of tack. High peel and shear strengths both require high cohesive strength within the film, but peel strength is dependent on adhesion to a much greater extent than is shear.

Thus most formulations are compromises which will favor the property that is most critical in the application intended. Figure 2 illustrates typical dependence of tack, peel, and shear performance in a given system as the resin/elastomer ratio is increased. The maxima occur at different tackifier percentages. Silicon elastomers for pressure-sensitive adhesives are invariably used in conjunction with silicon gums as tackifiers. Table 3 displays properties of various elastomers, and Table 4 contains information on tackifiers and plasticizers.

Solvents are selected primarily on the basis of solubility parameters and evaporation rate. Where mixed solvent systems are used to achieve the desired balance at the best cost,

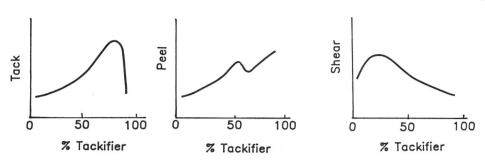

Figure 2 Dependence of tack, peel, and shear on resin/elastomer ratio.

the selection should be such that the slowest solvent remains an effective solvent for the system on its own. In addition, consideration should be given to the effect of the solvent on the substrate: too strong a solvent could degrade the substrate, but the right choice can assist in keying the adhesive to the surface.

The use of plasticizers is relatively uncommon in solvent-based pressure-sensitive adhesives, especially for use on tapes. Where plasticisers are included, their compatibility with the substrate should be considered, to ensure that plasticizer migration will not lead to transfer of the adhesive. Where plasticizing of block copolymers is intended, plasticizers should be selected that are compatible with the diene midblocks rather than with the polystyrene domains.

Additives used should include stabilizers or antioxidants, especially in products containing ethylenic unsaturation. Thickeners or thixotropes may be used to modify rheology. Fillers may be used in certain applications, in which case pigment-dispersing aids may be included to reduce settlement. Silane coupling agents may be used to improve adhesion to specific substrates. UV absorbers may be added to improve exterior durability, and pigments or dyes may be added to highlight the adhesive film.

Table 3 Elastomers in Common Use

Elastomer	Used in:
Rubbers	
Natural rubber	Solvent-based and water-based glues
Butyl rubber	Solvent-based glues
Styrene–butadiene rubber	Solvent-based and water-based glues
Block copolymers	
Styrene–butadiene–styrene	Solvent-based and hot-melt glues
Styrene–isoprene–styrene	Solvent-based and hot-melt glues
Other polymers	
Polybutene	Solvent-based and hot-melt glues
Poly(vinyl ether)	Solvent-based and water-based glues
Acrylic	Solvent-based and water-based glues
Ethylene–vinyl acetate	Hot-melt glues
Atactic polypropylene	Hot-melt glues
Silicon	Solvent-based glues

Table 4 Tackifying Resins and Plasticizers

Type and name	Melting point (°C)	Supplier
Aliphatic hydrocarbons		
Adtac LV	5	Hercules
Piccopale 100	100	Hercules
Quintone A100	100	Nippon Zeon
Mixed hydrocarbons		
Hercotac 1148	94	Hercules
Quintone N 180	80	Nippon Zeon
Aromatic hydrocarbons		
Picco 6100	100	Hercules
Piccodiene 2215	103	Hercules
Necires RF 85	85	Neville
Nevchem NL 100	100	Neville
Terpenes		
Piccolyte S25	25	Hercules
Zonarez B 115	100	Arizona
Alresen PT 191	70	Hoechst
Rosin esters		
Staybelite Ester 10	84	Hercules
Floral 105	105	Hercules
Zonester 65	65	Arizona
Oulupale XB 100	96	Veitsiluoto
Alresat KE 300	Liquid	Hoechst
Resin dispersions		
Dresinol 205	75	Hercules
Aquatac 6085	85	Arizona
Oulutac 80 D	80	Veitsiluoto
Plasticizers		
Shellflex 451	Liquid	Shell

Elastomers used in water-based systems include various rubber latices, especially natural rubber and SBR rubber, and occasionally, polychloroprene. The bulk of the market in water-based adhesives is now held by acrylic dispersions. Although these are designed for use without modification, it is normal to formulate, especially by addition of tackifiers. Commonly used tackifying resins include soft resins, or hard resins in solution, which may often be emulsified directly into the dispersion, and aqueous resin dispersions, which may be prepared separately and added or may be purchased from resin suppliers.

When using dispersions of tackifying resins, stability tests must be performed to ensure that there are no undesirable reactions between the emulsifier systems used in the resin dispersion and the elastomer dispersion. Resin dispersions often produce lower shear than resin solutions, thus necessitating reduced levels which result in lower tack. Additives used in water-based systems will include defoamers and preservatives as well as UV absorbers if necessary. Antioxidants are normally included only if service conditions require them. Catalysts may be added to cross-linkable grades to improve performance at elevated temperature, or self-cross-linking grades may be selected. Small quantities of solvents may be included to improve adhesion or penetration. Fillers are not generally

Table 5 Polymer Dispersions[a]

Name	Solid content (%)	Viscosity	T_g (°C)	Application area	Supplier
Acronol 81D	60	1400	−55	General	BASF
Arconal V205	69	1200	−40	Permanent labels	BASF
Lutanol I 65D	55	1700	N/S[b]	Tackifier	BASF
Mowiton DM758	60	5000	−65	Flooring adhesives	Hoechst
Revacryl A390	50	100	−55	Permanent labels	Harco
Revacryl 622	59	2000	−70	Deep-freeze labels	Harco
Vantac 300	58	1500	N/S[b]	Removable labels	Bevaloid

[a]Note that these are indicative values, not specifications.
[b]N/S, not stated in data sheets.

used, although in applications on vinyl or carpet floor tiles fillers may be included at levels up to 30% to reduce the glue-line shrinkage and the price. Rheology modifiers, including polylacrylates or inorganic thixotropes such as fumed silica, may be added. Table 5 shows the characteristics of some common dispersions.

IV. APPLICATIONS

The major application area for all pressure-sensitive adhesives is in tapes. Self-stick labels provide a second large area, while a range of miscellaneous applications make up the balance of pressure-sensitive adhesive use.

A. Tapes

Tapes may be classified according to application areas, such as electrical, packaging, or medical, or in terms of the type of tape, usually defined by the backing, which may be paper, fiber, film, foil, or foam. Figures drawn from various tables produced by the Fredonia Group Inc. and published in *Adhesives Age* in June 1991 and the Frost and Sullivan Report, ''The USA Market for Pressure Sensitive Adhesives,'' in *Adhesives Age* in August 1991 showed that of a total U.S. pressure-sensitive adhesives market in 1989 of $4.9 billion, $2.6 billion was for sales of pressure-sensitive tapes. Table 6 illustrates the relative importance of the various backings and application areas.

Table 6 Pressure-Sensitive Tape Market

Backing	Share (%)	Application area	Share (%)
Plastic and film	60.3	Packaging	38
Paper	23.1	Hospital and first aid	20
Cloth	11.9	Office and graphic art	17
Rubber	1.0	Construction	7.5
Other	3.7	Automotive	6.5
		Other	11

An apparently simple tape may comprise a number of elements, including a release coating, the backing, a primer, and the adhesive layer. The release coating ensures that the adhesive layer does not transfer partially or completely to the back of the tape from the coated side. With certain types of backing a release layer may not be necessary, and in some instances a separate release film may be necessary. The release coating should allow the tape to unwind easily but not spontaneously. The need for priming also depends on the nature of the backing and may take the form of an applied coating or layer, a chemical treatment such as corona treatment, or a physical treatment such as exposure to heat. Priming may sometimes be necessary to inhibit movement of plasticizer from backing to adhesive layer, or vice versa, but the usual purpose of priming is to obtain adequate adhesion to the substrate. The earliest backings used were cloth, mainly for first-aid dressing. Cloth backings allow the skin to breathe and offer good flexibility and tensile strength while permitting easy tear. Generally, no priming or release coating is needed. Fabrics used include cotton, nylon, and polyester. Paper backing is the cheapest type of backing available. Saturated paper backings have better physical properties than unsaturated paper, particularly for tensile and tear strengths, water resistance and permeability, and generally do not require release backings, while unsaturated papers usually require some aid to release. Saturated papers are predominant for general-purpose creped masking tapes.

Plastic films comprise more than 60% of all tape backings. Originally based on cellophane, a wide range of plastics is now available for various applications. Polyester, unplasticized poly(vinyl chloride) (PVC), and biaxially oriented polypropylene (BOPP) are used for packaging tapes, while PVC is still preferred for electrically insulating tapes. Cellulose acetate is used for "invisible" tapes, and poly(tetrafluoroethylene) (PTFE), particularly in conjunction with silicon-based pressure-sensitive adhesives, is used where resistance to elevated temperatures, chemical inertness, or low friction are the main requirements. Plastic films are impermeable, thin, uniform, and smooth and are generally inert with good dielectric properties. Reinforced with glass fibers or rayon embedded in the tape to distribute the load over greater areas, they are suitable for heavy-duty packaging applications.

Other backings used include foams, typically PVC or polyurethane, rubber, and metal foils. Foams are used for sealing and gasketing or thermal or acoustic insulation. They work particularly well on uneven surfaces. Adhesive may be applied to both surfaces, allowing their use as assembly adhesives. A double-sided, release-coated inter-layer is necessary for double-sided tapes. With foams, care must be taken to select an adhesive that does not cause the foam to collapse, either through solvent action or as a result of excessive heat. Rubber backings are used where flexibility is a primary requirement, although they also offer excellent electrical insulation. Aluminum foils are used mainly in the construction industry, to act as a moisture seal, to reflect heat or for insulation, or to offer a controlled leakage path for static electricity. Lead foils are occasionally used to screen harmful radiation.

Pressure-sensitive adhesives may also be coated directly onto release paper in order to produce transfer tapes, in which, as the name implies, the adhesive film will transfer from the release coating onto a substrate with which it is brought into contact. This permits exact placement of accurately controlled quantities of adhesive. Great care must be exercised in the choice of release paper to ensure successful application.

Packaging tapes represent the largest end use. Packaging, which includes closing of cartons, sealing containers, bundling, and protection of surfaces is based primarily on paper or plastic film backings, but other types may be used as well. Film-backed paper is replacing saturated paper for packaging applications, especially for masking and protection, while reinforced tapes are used for strapping large containers or bundling articles.

Hospital tapes and first-aid dressings still constitute an important area for pressure-sensitive tapes. Because of their ability to breathe, cloth tapes are still widely used, but other backings are now making their appearance for special applications, including invisible dressings and rigid or elastic support tapes. Choice of ingredients for adhesives is controlled by the need to produce a system that will not irritate the skin or inhibit healing. In addition, it must not lose adhesion as a result of perspiration, but must permit easy and clean removal. Traditional adhesives were based on natural rubber and zinc oxide, but water-based acrylic systems now dominate.

Office and graphic art applications require a diversity of tapes and backings. Block-out tapes are often multilayer constructions incorporating colored films. Printed films are used for graphics displays, while clear protective sheets may be used to protect printed graphics against smudging or erasure. Dye-cut lettering is available. A variety of clear-film backings is used for temporary fixing tapes.

Electrical tapes or insulating tapes require adhesives that will not corrode wiring, joints, or components. They should resist deterioration resulting from age and exposure to heat. Occasionally, chemical resistance may also be required. PVC is still the most widely used backing, but other films, especially polyester, are increasing in use. Rubber- or elastomer-based adhesives are suitable for service up to 130°C, acrylics to 155°C, while silicones, preferably on PTFE, can be used up to 180°C.

Other applications include thermal and accoustic insulation and tinted films for glass in the construction industry, wood-grained or other decorative vinyls for the automotive and furniture industries, double-sided tapes for mounting or splicing, and foams for gaskets and seals in the appliance industry.

B. Labels

Peel-and-stick labels provide a quick and easy way to apply labels to almost any surface. Die cut and supplied on release papers, they can be printed in computers for addresses, while preprinted labels for an enormous range of applications, particularly difficult substrates such as polyolefins or cold, moist containers make labeling a pleasure compared with traditional wet-applied systems such as dextrines or caseins. Pressure-sensitive label stock provides reliable bonding, is easy to use and virtually instantaneous, and offers a choice of properties, including permanence or easy removal, high-temperature resistance, or low-temperature flexibility. Large or small labels can be easily stored, handled, and applied: by hand, hand-held applicators, or semiautomatic or fully automatic industrial labelers.

Labels are generally regarded as falling into three classes: permanent label stock, removable labels, and labels for use at low temperatures. In addition, speciality applications include delayed-action labels, high-temperature applications, and decals. Permanent label stock is the mainstay of the label market. Label stock is invariably paper, and the labels, which are usually preprinted, are supplied on a release paper backing, leading to their popular name, "peel-and-stick" labels. Adhesives for permanent stock have high shear strength, and attempts to remove them will usually damage the label. Applications range from price stickers and address labels to inventory labels, shipping labels, warning signs, and labels for bottles, buckets, or drums.

By contrast, peelable or removable labels use adhesives with relatively low tack and shear strengths. On removal, no residue must remain on the surface from which the label was removed. Some removable labels use a water-soluble adhesive, permitting easy cleaning of the surface. These labels are used for temporary labeling or where they will frequently be replaced. Freezer labels use adhesives that have very good low-temperature flexibility to allow labels to be applied and remain adhered at temperatures down to −20°C or lower. They are characterized by very low glass temperatures, typically in the range −60 to −80°C.

Label adhesives are seldom solvent based. The majority are prepared using water-based adhesives, although hot melts are taking an increasing segment of this market. Important adhesive characteristics include the ability to be dye cut cleanly, low tendency to make the paper curl, coupled with ability to conform to the surface to which they are applied, very quick grab and good resistance to yellowing, and loss of tack with age. Acrylic dispersions, either with or without additional tackifiers, are the most widely used adhesives.

Production of label stock makes severe demands on the release paper. Since transfer coating is a common application method, the adhesive must wet the release paper adequately and yet transfer cleanly to the label when label and backing are united. The release paper must remove quickly and easily in use, but must remain firmly adhered during storage. Use of silicone releases is virtually universal, invariably requiring the use of wetting agents in the formulation to ensure that the adhesive wets the surface. There must be no transfer of the release coating to the adhesive surface, however, as this will destroy the tack of the adhesive.

C. Other Applications

The largest market outside tapes and labels is for adhesives for floor tiles. Peel-and-stick floor tiles are available from hardware stores for use by homeowners. These tiles usually have a hot-melt pressure-sensitive adhesive applied to the back so that laying to clean, prepared floors only needs the removal of a backing paper and pressing the tile into place. The adhesive layer in this application is substantially thicker than for tapes or labels, to ensure good contact over the entire area. Alternatively, vinyl flooring or carpets, either as tiles or in roll form, may be laid into water-based acrylic adhesives, although here the move is away from such adhesives in favor of more permanent adhesives typically based on vinyl acetate–ethylene copolymer dispersions. Acrylic dispersions are often used, however, usually in unmodified form, for application of vinyl tiles or sheeting over an existing impervious floor covering, where it is necessary to allow complete flash-off of the carrier prior to laying the new floor covering.

Personal hygiene products such as disposable diapers make use of self-adhesive strips, covered with a release tape that is removed at the point of use. Again unmodified acrylic dispersions as well as hot melts dominate this application. The same considerations apply here as in first-aid dressings, and in particular the adhesive must not cause skin irritation.

Gloss lamination, the application of thin films of polyester or polyolefin over printed paper to enhance gloss and protect the print, is an additional application area. Traditionally, this has been the preserve of solvent-based adhesives, which offer rapid drying, thus allowing high machine speeds, but water-based systems are increasing in popularity as the ability to formulate at very high solids reduces the drying time to acceptable limits.

V. COATING METHODS

Most of the popular coating methods are suitable for pressure-sensitive adhesives. Solvent-based adhesives are usually applied by roller coaters or occasionally by spray applicators. Water-based adhesives also use roller coaters predominantly, with nozzle feed machines the exception. Hot melts may be extruded, applied from slot orifice coaters, or calendared. Although virtually any type of roll coater may be used, reverse roll coating is the most common. This may incorporate a doctor roll, doctor blade, or Mayer bar to meter the spread rate. With tapes, adhesive is usually applied directly to the tape, which then passes through a drying station incorporating countercurrent air, usually warm. The dried tape is then rolled and slit. With label stock, adhesive is normally applied to the release paper, dried, and then transferred to the label when the label stock is united with the release paper in a nip roll.

Adhesives for reverse roll application will typically have viscosities in the range of 1000 to 10,000 cP. While coating speeds in excess of 200m/min are possible, machine speeds are normally limited by the speed at which the film can be dried. A simple reverse roll coater is shown in fig. 3.

For low application weights of hot-melt adhesives, slot-orifice coaters are preferred. Variation of slot width and temperature allow a wide range of viscosities and coating weights to be handled. Calendaring is used for high-viscosity adhesives and high coating weights. Extrusion is used for very high viscosity systems and permits both mixing and coating to be performed in a single operation.

VI. TESTING

Because of the unique properties of pressure-sensitive adhesives, special tests not applicable to other types have been developed. While standard physical tests such as nonvolatile content, viscosity, and specific gravity are performed to ensure consistency of application, these tests do not predict adhesive performance. For pressure-sensitive adhesives, three critical performance characteristics are usually measured: tack, peel, and shear strength.

A. Tack

The classic test for tack of a pressure-sensitive adhesive film is the rolling ball tack test. Here a ball is rolled down an inclined plane onto a film of the adhesive. The length the ball travels across the film before stopping is a measure of the tack of the film. This test gives a

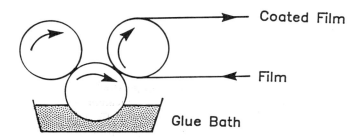

Figure 3 Reverse roll coater.

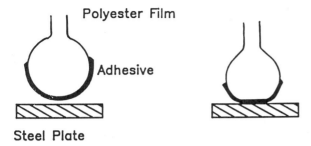

Polyester Film

Adhesive

Steel Plate

Figure 4 Loop tack test.

good indication of tack with elastomer adhesives but is unreliable with water-based systems.

A more universal test is the probe test, in which the end of a cylinder of standard diameter is brought lightly into contact with the film for a very short time and the force required to separate it from the surface is measured. Similar in principle is the loop tack test, in which a loop of coated film is lowered onto a steel plate, making contact under its own weight, and the force required to withdraw the plate is then measured. All of these tests are markedly affected by the cleanliness of the ball, probe, or plate. Figure 4 illustrates the loop tack test.

B. Peel

Peel strength is usually tested by laminating a coated film either to itself or to a specified substrate. The film is then peeled off the substrate at a steady speed at 90 or 180 degrees to the bond axis, and the force required for removal is measured. The result is always quoted as the force per unit width of film at a given rate of peel. Figure 5 shows the geometry of a 90- or 180-degree peel test.

C. Shear Strength

This test is a measure of the ability of a pressure-sensitive adhesive to withstand creep. A standard area of coated film is bonded to a steel plate and a weight suspended from it. The

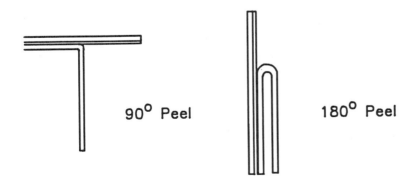

90° Peel   180° Peel

Figure 5 Geometry of 90- and 180-degree peel tests.

assembly is placed in an oven. In some shear tests the time for the assembly to delaminate at a fixed temperature is recorded, while in other tests the temperature at which failure occurs when the oven temperature is increased at a certain rate is the shear value quoted.

VII. GUIDE FORMULATIONS

Several guide formulations for various applications follow. It must be remembered that they should merely be used as a starting point to develop a satisfactory product. Not all materials will be available in all countries, and the nature and quality may vary from country to country. To simplify formulations, trade names have been used, and these may also differ. In most cases substitutes are available.

Parenthetical numbers that follow the components listed in the formulations below correspond to these suppliers:

1. Shell Chemicals
2. Hercules Inc.
3. Anchor Chemicals
4. Nippon Zeon
5. Hoechst
6. Firestone
7. Arizona
8. Harlow Chemicals
9. EKA Nobel
10. U.S. Cyanamid
11. Dow Chemicals
12. B.A.S.F.

A. Solvent-Based Adhesives

1. General-Purpose Adhesives for Tapes

	High shear	High tack
Cariflex TR 1101 (1)	100	—
Cariflex TR 1107 (1)	—	100
Piccolyte A 115 (2)	80	—
Abitol (2)	10	—
Foral 85 (2)	—	80
Piccolyte S 40 (2)	—	10
Ancazate BU (3)	1	1
SBP 62/82 (1)	250	300
Toluene	50	—

2. Paper-Splicing Adhesive

Natural rubber: pale crepe	7.5
Quintone D100 (4)	7.5
Antioxidant	0.2
Hexane	70.0
Toluene	14.8

3. Self-Stick Carpet Tiles

Synthacryl VSC 2291 (5)	20.0
Alresen PT 214 (5)	5.0
Methyl ethyl ketone	10.0
Toluene	30.0
Aerosil 200	0.0 – 5.0
Calcium carbonate, 10 μm	35.0 – 30.0

B. Hot-Melt Adhesives

1. Label and Tape Adhesives

	Soft, high tack	High strength
Cariflex TR 1102 (1)	—	100
Cariflex TR 1107 (1)	100	—
Floral 85 (2)	150	—
Shellflex 451 HP (1)	75	—
Piccolyte S 85 (2)	—	100
Abitol (2)	—	50
Ancazate BU (3)	5	5

2. Self-Adhesive Carpet Tiles

Cariflex TR 1102 (1)	75
Cariflex TR 1107 (1)	25
Piccolyte A 115 (2)	100
Abitol (2)	50
Calcium carbonate	100
Antioxidant	5

3. Disposable Industry Adhesive

Stereon 840 A (6)	25
Zonatac 105 Lite (7)	58
Shellflex 371 (1)	16
Antioxidant	1

C. Water-Based Adhesives

1. Adhesive for Permanent Label Stock

Revacryl 630 (8)	60.0
Snowtack 301 CF (9)	35.0
Aerosol OT (10)	0.3
Dowfax 2A1 (11)	0.1
Ammonia solution	0.3
Foamstopper 101 (8)	0.1
Water	4.2

2. Deep-Freeze Label Stock

Revacryl 622 (8)	94.5
Hercolyn D (2)	4.5
Aerosol OT (10)	0.5
Dowfax 2A1 (11)	0.1
Water	0.4

3. Removable Labels

Revacryl DP 3560 (8)	99.0
Aerosol OT (10)	0.5
Dowfax 2A1 (11)	0.1
Water	0.4

4. PVC Floor Tile Adhesive

Mowiton DM 758 (5)	40.0
Alresat KE 300 (5)	8.0
Calcium carbonate, 10 μ	32.0
Water	15.0
Collacryl D (12)	1.5–3.0
10% Caustic soda solution	2.0–3.0

ACKNOWLEDGMENTS

The author gratefully acknowledges the assistance of many companies, particularly the following, which supplied technical information and guide formulations: BASF, Carst & Walker, Chempro, Hoechst AG, Shell Chemicals, Taueber & Corssen, and, Harlow Chemical Co.

BIBLIOGRAPHY

Cariflex Polymers for Adhesives, Coatings and Sealants, Shell Chemicals, technical bulletin TR5.1.

Cariflex: Starting Formulations and Test Methods for Pressure Sensitive Adhesives, Shell Chemicals, technical bulletin TR5.2.

de Walt, C., *Factors in Tackification*, Hercules, Inc., bulletin R-218B.

Fredonia Group Study quoted in *Adhesives Age* (June 1991).

Frost, and Sullivan, quoted in *Adhesives Age* (Aug. 1991).

Hock, C. *J. Polymer Sci. C 3*: 139 (1963).

Satas, D., ed., *Handbook of Pressure-Sensitive Adhesive Technology*, Van Nostrand Reinhold, New York, 1982.

U.S. Department of Commerce, Business Trend Analysts Inc., quoted in *Adhesives Age* (June 1990).

Wotherspoon, T., *Technical Developments in Pressure Sensitive Adhesives*: Harlow Chemical Company, May, 1992.

35
Electrically Conductive Adhesives

Alan M. Lyons and D. W. Dahringer *AT&T Bell Laboratories, Murray Hill, New Jersey*

I. INTRODUCTION

Electrically conductive adhesives perform two primary functions. Like other types of adhesives, these materials provide a physical bond between two surfaces. In addition, an electrical interconnection between the two bonded surfaces is formed. This dual functionality is usually achieved by composite materials composed of metallic particles dispersed in an adhesive matrix. The electrical resistivity of conductive adhesives is compared to values for pure metals and polymers in Table 1.

There are two types of conductive adhesives: conventional materials that conduct electricity equally in all directions (isotropic conductors) and those materials that conduct in only one direction (anisotropic conductors). Isotropically conductive materials are typically formulated by adding silver particles to an adhesive matrix such that the percolation threshold is exceeded. Electrical currents are conducted throughout the composite via an extensive network of particle–particle contacts. Anisotropically conductive adhesives are prepared by randomly dispersing electrically conductive particles in an adhesive matrix at a concentration far below the percolation threshold. A schematic illustration of an anisotropically conductive adhesive interconnection is shown in Fig. 1. The concentration of particles is controlled such that enough particles are present to assure reliable electrical contacts between the substrate and the device (Z direction), while too few particles are present to achieve conduction in the X–Y plane. The materials become conductive in one direction only after they have been processed under pressure; they do not inherently conduct in a preferred direction. Applications, electrical conduction mechanisms, and formulation of both isotropic and anisotropic conductive adhesives are discussed in detail in this chapter.

II. APPLICATIONS

In this section we discuss three applications of electrically conductive adhesives: die attach adhesives, anisotropically conductive adhesives for liquid crystal display (LCD) assembly, and conductive adhesives for surface-mounted assembly of packaged components on printed wiring boards (PWBs). These applications were selected based on overall

Table 1 Bulk Electrical Resistivity Values for Selected
Metals, Polymers, and Composites

Material	Resistivity ($\Omega \cdot$ cm)
Silver	1.6×10^{-6}
Copper	1.7×10^{-6}
Aluminum	2.7×10^{-6}
Pb/Sn solder	15×10^{-6}
Conductive epoxy	ca. 500×10^{-6}
Carbon	3000×10^{-6}
Antistatic composites	ca. 10^4
Epoxy resin	ca. 10^{15}

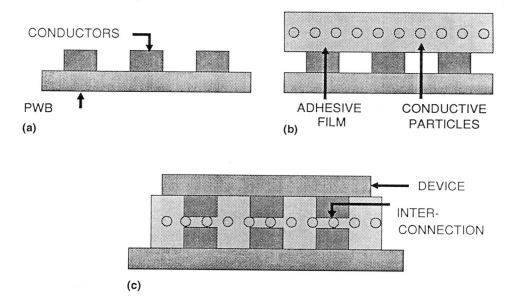

Figure 1 Anisotropically conductive adhesive assembly process: (a) bare substrate; (b) apply
anisotropically conductive adhesive; (c) align device, cure under pressure.

volume usage, importance to the industry, and future technological as well as political
impact.

Numerous other types of electrically conductive polymer composites are commercially available but are beyond the scope of this chapter. These materials are used in such
applications as conductive inks [1], thermoplastic molded monolithic objects for electrostatic dissipation (ESD) [2] and electromagnetic interference (EMI) shielding applications
[3] and a wide variety of other applications, including heating elements, switches, trans-
ducers, and batteries [2]. Similarly, the fabrication of conductive polymer materials via
metal vapor deposition or electrodeposition onto polymer surfaces will not be discussed
here.

A. Die Attach Adhesives

The most significant application for conductive adhesives in the manufacture of micro-electronics is the attachment of silicon chips to lead frames. Of the 40 billion integrated circuits (ICs) manufactured each year, approximately 90% are encapsulated in plastic molded packages, and most of these are assembled with conductive adhesives [4]. A schematic illustration of a plastic-molded IC package is shown in Fig. 2. The conductive adhesive forms the mechanical as well as electrical interconnection between the back side (ground plane) of the die and the plated copper lead frame. Initially, gold eutectic bonding was used for this die attach application. Several disadvantages to this process were encountered, however, including elevated assembly temperatures, cost of materials, and especially the stresses induced by coefficient of thermal expansion (CTE) mismatches [5]. Alternative interconnection materials were later developed, including silver glass materials for hermetic packages and conductive adhesives for nonhermetic applications. A review of these three die attach technologies has been published [6].

Adhesives gained acceptance rapidly despite the industry's reluctance to incorporate organic materials in microelectronic packages, due to the following advantages:

Low cost
Low assembly temperatures
Low-stress joints
High thermal conductivity
High purity
Excellent reliability
Easy integration into the manufacturing line

Conductive adhesives also form sufficiently robust joints that withstand the temperatures and pressures experienced during wirebonding and over molding. The formulation of die attach materials is discussed in Section IV.

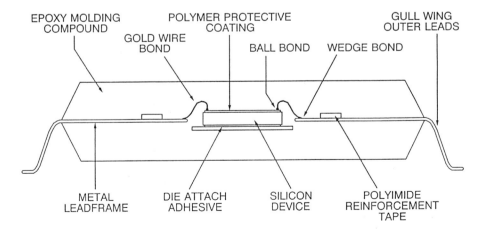

Figure 2 Cross section of a plastic-molded integrated circuit package.

B. Interconnections to Liquid Crystal Displays

A relatively new application for conductive adhesives, and a very important one in terms of technological advance and future manufacturing techniques, is the assembly of liquid crystal displays (LCDs). Interconnecting flexible circuit connectors, tape automated bonding (TAB) packages and bare IC chips to glass panels have been the principal technology drivers for the development of anisotropically conductive adhesives. A brief history of LCD interconnection technology [7] as well as an analysis of current manufacturing techniques for display-based consumer electronics equipment [8] have been reported that illuminate material requirements for this rapidly evolving type of product. The principal reasons for developing anisotropic adhesive systems for interconnections to glass panel displays are:

1. Low assembly temperatures to prevent glass fracture
2. Very fine pitch (repeat distance between adjacent conductors) of 2 to 20 mils
3. Low cost

Neither isotropic conductive adhesives nor conventional solder interconnections, can easily satisfy these requirements.

Early LCDs contained the electronic drivers on a rigid circuit board, whereas the glass panel contained only the active display area. The two substrates were connected by an adhesively bonded flex circuit or heat seal connector (HSC). HSCs are the most primitive type of anisotropically conductive adhesive material. They are in wide use in Japan and are commercially available in the United States.* HSCs are manufactured by screen printing conductive traces onto a flexible substrate (usually polyester) using either a graphite, silver, or silver–graphite ink. Alternatively, the conductor patterns may be formed by etching copper foils laminated to flex substrates. The conductive traces may be coated with a second layer of ink containing metalized spheres. A hot-melt adhesive, such as neoprene/ethylene–vinyl acetate [10], is applied over the entire top surface. The HSC is tacked into position at low temperatures onto one substrate, tested for accurate alignment, then fully cured at elevated temperatures (thermosets only). This process is repeated to join the second substrate. Thermoset hot-melts offer improved reliability over thermoplastic adhesives [10,11]; however, the thermoplastics offer processing advantages, as they can easily be removed and repositioned if the HSC is not properly aligned to either the glass panel or rigid PWB. Interconnection pitch as fine as 11 mils can be achieved.

Most current LCDs are assembled using tape automated bonding (TAB) to package the driver chips. The TAB packages can also act as connectors between the PWB and the glass display. Connections are made to the glass panel by use of anisotropically conducting polymer films and to the PWB by solder reflow [10,12]. Anisotropically conductive adhesives for attaching TAB packages to low-cost polymer thick film (PTF) flex circuits were first used for the assembly of low-cost calculators [9,13]. Principles of anisotropically conductive adhesive formulation are described in a separate section.

Flip-chip interconnection of silicon ICs to glass displays, as well as to multichip module substrates, is the focus of current research efforts. Although most of the flip-chip effort is directed toward solder bumping for multichip modules [14], adhesive interconnections dominate the emerging display assembly technologies. Technologies that employ

---

*For example, two US distributors of Japanese heat seal connectors are Nippon Graphite Industries, (702) 829-1905, and Nitto Denko America, Inc., (408) 432-5400.

anisotropically conductive films [15,16] isotropically conductive adhesive bumps [17,18] and nonconductive adhesives [19] have been reported. An especially novel system that ensures isolation in the X–Y plane, even at 2-mil pitch, has also been reported [7]. A UV-curable adhesive is applied to an LSI wafer and exposed through a mask such that only the material above the metallic contact pads on the die remains tacky. The surface is then coated with metallized polymer spheres, 10 μm in diameter; the spheres adhere only to the tacky regions above the 1-mil^2 pads. The chips are diced, assembled to glass substrates, and cured under pressure with an additional adhesive. Deformation of the elastomeric spheres, and flow of the adhesive during cure, ensures good electrical contact.

C. Surface-Mounted Assembly of Printed Wiring Boards

The future use of lead in the manufacture of electronic products is problematic. Lead is toxic, and although current worker exposure is low (and consumer exposure is negligible), regulatory and/or consumer forces may make the cost of using lead in electronics manufacturing prohibitive. A thorough assessment of lead use, as well as an evaluation of alternative joining technologies, has been published [20]. In addition, editorials have endorsed the elimination of lead from electronic products as a means of attracting consumers and increasing global market share in the next decade [21].

Conductive adhesives are one of the feasible alternatives to lead for electronics assembly. Isotropically conductive adhesives are suitable for standard pitch (50- to 100-mil) surface-mounted components and numerous commercial materials are available (see commercial supplier listing, Section VI.E). Anisotropically conductive adhesives are more suited to flex to rigid connections, fine pitch components (15- to 20-mil pitch), and flip-chip assembly (4- to 12-mil pitch) [22]. Adhesives are not ready to replace solder throughout the electronics industry, however, due to questions that remain concerning the reliability of electrical interconnections. Their implementation is currently limited to low-cost applications using polyester substrates and specialty applications where solder cannot be used. Additionally, the lack of equipment for large-volume assembly with anisotropically conductive adhesives, which require the simultaneous application of heat and pressure, impedes the acceptance of these promising materials.

III. CONDUCTION MECHANISMS IN METAL–POLYMER COMPOSITES

Increasing the concentration of metal particles in an insulating adhesive matrix changes the electrical properties of the composite in a discontinuous way. Assuming a random dispersion of the metal filler, as the concentration increases no significant change occurs until a critical concentration, p_c, is reached. This point, where the electrical resistivity decreases dramatically, called the *percolation threshold*, has been attributed to the formation of a network of chains of conductive particles that span the composite. A two-dimensional cartoon of a conductive adhesive below p_c and just above p_c is shown in Fig. 3. A typical plot showing the relationship between particle concentration and electrical resistivity is shown in Fig. 4.

Experimental [23] as well as theoretical [24–26] studies of percolation phenomena have been reported. In random and macroscopically homogeneous materials it has been demonstrated [27–29] that at concentrations of metal particles below the percolation

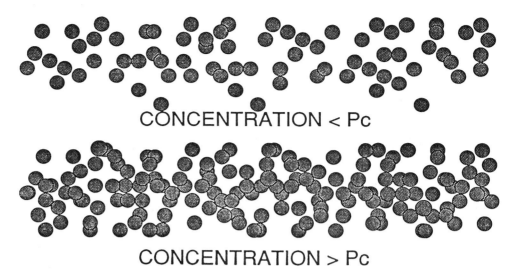

Figure 3   Electrical percolation threshold for conductive particles in an insulating matrix.

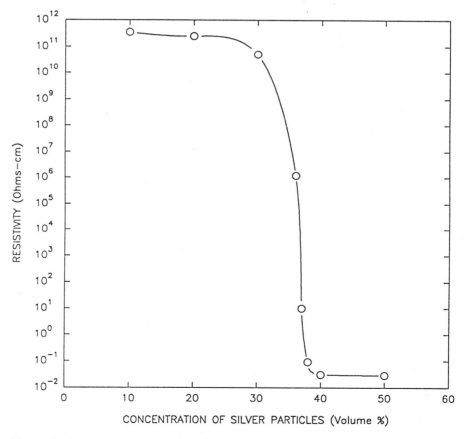

Figure 4   Resistivity of silver–phenolic composites. (From Ref. 23.)

threshold ($p < p_c$) a short-range percolation coherence length, ξ, exists. Electrical conductivity is probable for length scales less than ξ. Thus even if the metal-filled composite exhibits no bulk electrical conductivity, conduction can occur within domains that are smaller than ξ. As the concentration of metal particles approaches p_c, $\xi \rightarrow \infty$ and the composite becomes isotropically conductive.

The concentration of metal particles required to achieve p_c has been reported over a wide range, from less than 1 to more than 40 vol %. This range of values occurs due to several factors, including processing techniques [3,30,31], particle size in relatively monodisperse systems [32], particle size distribution [27], and particle aspect ratio. In many of the systems reported [23,25,33,34] random dispersions were assumed even when dense metal particles were employed. Recent work has demonstrated, however, that dense metal particles can settle, especially when the viscosity of the polymer matrix is low [27]. Particle settling is another factor that may influence the observed onset of percolation. Depending on how the electrical properties of the sample are measured, the observed value for p_c may be either higher or lower than the value of p_c in a truly random system.

The size of the metal particles relative to any structure present in the polymer matrix can also affect the value of p_c. Segregated composites have been prepared by compression molding a mixture of metal and polymer particles [35]. When the radius of the polymer particle (R_p) is significantly larger than the metal particle size (R_m), the metal is confined to the regions between polymer domains. Values of p_c as low as 6 vol % were achieved when $R_p/R_m = 16$. Metal-plated polymer spheres have been prepared where the effective $R_p/R_m \rightarrow \infty$ [37].

Another factor that influences the value of p_c is the aspect ratios of the metallic filler. Metal fibers, metal-plated glass fibers, and metal flakes can significantly lower the concentration required to achieve isotropic conduction as compared to spherical powders [3]. Values of p_c as low as 1 vol % have been reported with stainless steel fibers having an aspect ratio of 750 [37].

After mixing, some conductive epoxies do not always exhibit electrical conductivity. The electrical properties develop only after cure and the final resistance may be a function of the amount of time between mixing and cure [38–40]. This effect has been attributed to a fatty acid coating applied to the surface of the silver during manufacturer of the flake. The coating is removed at elevated temperatures during cure. Solvents such as polypropylene glycol may dissolve the coating before cure, rendering the pastes conductive [39]. Growth of insulating coatings about the silver flake particles is postulated as the cause for the increase in electrical resistivity of some conductive adhesives upon standing at room temperature before cure [40].

A. Anisotropic Conduction Mechanisms

Several processing techniques have been developed to achieve anisotropic conductivity. One method aligns the conductive metal particles in a magnetic field [41–43]. Alternatively, anisotropy can be achieved in materials with random dispersions of metal particles. Using an adhesive composite where the concentration of metal particles is below but close to p_c results in a material with a short-range percolation length ξ. If the separation distance between substrate and device is less than ξ, whereas the pitch is much greater than ξ, anisotropic conduction is achieved. This approach is difficult to implement, as shorts between adjacent conductors, as well as opens between the substrate and device, are statistically possible.

Most commercially available anisotropically conductive adhesives are formulated on the bridging concept, as illustrated in Fig. 1. A concentration of conductive particles far below the percolation threshold is dispersed in an adhesive. The composite is applied to the surface either by screen-printing a paste or laminating a film. When a device is attached to a PWB, the placement force displaces the adhesive composite such that a layer the thickness of a single particle remains. Individual particles span the gap between device and PWB and form an electrical interconnection. For successful implementation of anisotropically conductive adhesives, the concentration of metal particles must be carefully controlled such that a sufficient number of particles is present to assure reliable electrical conductivity between the PWB and the device (Z direction) while electrical isolation is maintained between adjacent pads (X-Y direction).

IV. FORMULATION OF ISOTROPICALLY CONDUCTIVE ADHESIVES

A. Requirements and Performance

A U.S. military hybrid specification (MIL-A-87172 of MIL-STD-883) established the selection and qualification requirements for polymeric adhesives used in military hybrid circuits and is used as a guide for die attach materials for nonmilitary applications as well. Table 2 lists the requirements for a type I (electrically conductive) adhesive and test results of a typical current generation isotropically conductive adhesive as reported by Estes [44]. These requirements specify test ranges for characteristics that will establish processing, performance, and reliability.

In the process category, viscosity and pot life determine the suitability of a material for a specific application technique (e.g., stencil and screen printing, syringe dispensing, and pin transfer printing) and the length of time the material can be used. Shelf life is important from an inventory control point of view as well as a cost factor. Cure schedule will govern the product flow during the manufacturing process and also the compatibility with temperature-sensitive components. The thermogravimetric analysis specification is an attempt to limit volatile evolution during cure. This can be very important in preventing voids in the conductive adhesive which will adversely affect electrical and thermal conductivity, joint strength, and die cracking [5]. In addition, the evolution of volatiles has been correlated to surface contamination of the die bonding pads and poor wire bonding yields [45].

The other test requirements shown in Table 2 deal with performance and reliability. The outgassed materials test in part is concerned with residual solvent and moisture in the starting adhesive, as well as by-products of thermal degradation. Ionic impurities and corrosivity have been associated with damage to the active die and thus poor reliability where found in excess. Bond-strength requirements are necessary to ensure processibility during subsequent manufacturing steps such as wirebonding and overmolding (plastic packages). The aged bond strength is intended to be another measure of thermal degradation resistance. Coefficient of linear thermal expansion and T_g relate to stresses in a bonded assembly; however, the significance of the value is not indicated by the requirements. Thermal conductivity is another check on void formation, while volume resistivity, especially after aging, is a measure of adequate particular content, sufficient particle to particle contact, and a stable particle–matrix–particle structure in the adhesive.

Table 2 Requirements for MIL-A-87172 and Typical Test Results

Test or condition	Requirement	Typical formulation
Material uniformity (3.4.1)	Inspect at $30\times$	Pass
Viscosity, cP (3.4.2)	Report for each batch	25,000–35,000
Pot life (3.4.3)	>1 h/25°C	3 months
Shelf life (3.4.4)	>12 months	>12 months
Cure schedule (3.5.1)	Per supplier data sheet	180°C/1 h
		150°C/1.5 h
Thermogravimetric analysis		
Wt. loss, 300°C, ASTM D-3850	$<1\%$	0.45%
Filler content (3.5.2.2)	$\pm2\%$ of reference lot	
Outgassed materials (3.5.3)	$H_2O < 3000$ ppm by RGA	User
Ionic impurities (3.5.4)	Total ionic (ppm)	
Hydrogen ion	$4.0 < \text{pH} < 9.0$	6.4
Extractable Cl$^-$	<300 ppm	<10
Na$^+$	<50 ppm	<10
K$^+$	<50 ppm	<10
NH$_4{}^+$	Report for information	<20
Other ions if present >5 ppm	Report for information	<20
Total ionic	<4.5 mS/m	3.3
Corrosivity (3.5.5)	No change, 48 h	Pass
Bond strength (3.5.6)		
Initial at 25°C	6.0 MN/m^2	12.4
Initial at 150°C	6.0 MN/m^2	7.8
At 25°C after 1000 h/150°C	6.0 MN/m^2	7.5
Coefficient of linear thermal	Below $T_g < 65 \times 10^{-6}$/°C	45×10^{-6}
expansion (3.5.7)	Above $T_g < 300 \times 10^{-6}$/°C	170×10^{-6}
T_g (glass transition temp.)	Information only	95°C
Second T_g		150°C
Thermal conductivity (3.5.8)	Type I > 1.5 W/m · K	>2.0
measured at 121°C		
Volume Resistivity (3.5.9)		
For type I adhesive at 25°C	$<5 \times 10^{-4}$ Ω · cm	0.0002
60°C	$<5 \times 10^{-4}$ Ω · cm	0.00025
125°C	$<5 \times 10^{-4}$ Ω · cm	0.00025
Measured after 1000 h/150°C	$<5 \times 10^{-4}$ Ω · cm	0.00027
Solvent		No

Source: After Ref. 44.

B. Formulations

1. Epoxy

Some of the first commercial conductive epoxy adhesives were simply based on silver powder dispersed in a liquid epoxy resin [e.g., diglycidyl ether of bisphenol A (DGEBA)] with an aliphatic amine [e.g., triethylene tetramine (TETA)] as a curing agent. Although capable of room-temperature cure, commercialization of this type of system was hampered by severe mix ratio disparity (typically, 50:1 by weight), problematic mixing due to

Table 3 Early (ca. 1960) Conductive Adhesive Formulation and Properties

	Parts by weight	
	Part A	Part B
Ingredient		
Epoxy resin (Epon 828)	20	
Silver flake (10–20 μm)	75–80	
Solvent (glycol ether)	0.5	
Curing agent Triethylene tetramine (TETA)		2.4
Total	100	2.4
Viscosity	50,000–100,000 cP	2 cP
Mixed properties		
Viscosity	10,000–100,000 cP	
Pot life	30–60 min.	
Cure cycles	5–7 days RT	
	4 h 50°C	
	2 h 75°C	
	1 h 100°C	
Cured properties		
Glass transition	50–100°C	
Volume resistivity	0.01–1 Ω · cm[a]	

[a]Value depends on amount of solvent and cure cycle.

viscosity differences, short pot life, and safety concerns. See Table 3 for typical properties of this type of a formulation.

Solvents were often used to help disperse the silver particles, to lower viscosity or to enhance conductivity. The presence of solvent can accomplish several objectives: first, by lowering the resin viscosity, additional silver can be added to the mix, increasing the likelihood of exceeding the percolation limit; and second, certain solvents can dissolve or displace lubricants on the surface of the silver flake (necessary for manufacturing the fine particles) that may interfere with the electrical contact between particles. The selection of a solvent involves consideration of its effect on both the epoxy cure reaction and the long-term performance of the adhesive. Most solvents reduce the cure reactivity of epoxy adhesives and some can prevent the development of a complete cure (lower alcohols are known to act in this fashion). The suitability of a particular solvent will also depend on its ability to leave or remain in the cured adhesive as intended. Low-boiling solvents may evaporate from a curing adhesive without void formation if the cure temperature is mild and the bond area (die) is small. Conversely, a high-boiling solvent may be retained in the cured adhesive to act as a plasticizer.

With improved formulations, organic-based conductive adhesives became feasible replacements for eutectic solders. Table 4 shows typical one- and two-part conductive adhesives from the 1970s. The one-part adhesive system employed a latent catalyst for long pot life but required a high-temperature cure cycle. The two-part system, formulated for a 1:1 by weight mix ratio could be cured at a lower temperature but had a shorter pot life. Formulations similar to the two-part system are still commercial today and are used

Table 4 Die Attach Conductive Adhesives

Ingredient	Parts by weight		
		Two-part adhesive	
	One-part adhesive	Part A	Part B
Liquid DGEBA	100	100	
Latent catalyst (DICY[a])	8		
Liquid anhydride (HHPA[b])			80
Tertiary amine (BDMA[c])			1
Silver flake	100–250	140	159
Properties			
Mix ratio W/W	—	1	1
Mixed pot life	3–6 months	16–24 h	
Cure temperature	175°C	100°C	
Cure time	1 h	2 h	

[a]Dicyandiamide.
[b]Hexahydophthalic anhydride.
[c]Benzyl dimethylamine.

where the low-temperature cure is critical. Some suppliers now offer this type of conductive adhesive in premixed, frozen syringes for added convenience.

The one-part formulation of Table 4 was improved over the years with respect to the curing temperature by the development of cure accelerators for DICY. However, concerns over residual outgassing of potentially corrosive cure by-products (e.g., ammonia) has resulted in the substitution of newer latent curing agents in modern formulations. An example of this type of curing agent is 1-cyanoethyl-2-ethyl-4-methylimidazole (sold as 2E4MZ-CN by Shikoku Chemicals Corporation), used at a level of 4 to 8 parts by weight of resin (phr), where the resin could be a liquid epoxy novolac [46].

As the die attach business became more competitive, adhesive suppliers refined their formulations to take advantage of small improvements in rheology, high-temperature performance, electrical conductivity, cure cycle, and ionic purity to satisfy customer's real or perceived needs. As a result, the current generation of conductive adhesive formulations may include new resins with lower viscosity, lower levels of ionic contamination, higher functionality, particulates with specific shapes, or combinations of shapes (e.g., flake and spheres); additives to control flow properties for better application techniques; conductivity enhancers to allow better particle-to-particle contact or higher concentrations of particles for improved conductivity; and adhesion promoters. Flow control additives are used in very low concentration and can be based on silicone oils, fluorocarbon or hydrocarbon surfactants, low-molecular-weight acrylic polymers, and so on. Some materials that provide conductivity enhancement include low-molecular-weight polypropylene glycols, select solvents, and some reactive diluents (low-molecular-weight mono- and difunctional epoxy resins). Organosilanes are the most commonly used adhesion promoters, but organotitinates can also be used. The use of additives with low boiling points that cannot be incorporated into the polymer structure should be minimized, however, to reduce void formation.

Residual stress in the die after assembly remains a concern when using epoxy die attach adhesives, especially with a large die. Various formulation techniques to reduce stress have been reported that include lowering the T_g of the matrix [47,48], formulating the matrix to exhibit two T_g values [49], reducing the modulus of the adhesive by minimizing silver content, and lowering cure temperature [47].

2. Polyimides

The development of polyimide-based conductive die attach adhesives resulted from the attractive properties exhibited by this generic class of materials. The high T_g value (typically, 100°C higher than epoxies) offers superior performance during high-temperature processing, such as wirebonding, overmolding, soldering, and lid sealing. Other properties of interest include low ionic contamination levels and a low CTE.

There are several disadvantages to using polyimides that limit their usefulness and increase manufacturing costs. In general, polyimide adhesives are formed by the thermally induced imidization of a polyamic acid precursor. The polyimide precursors are dissolved in a solvent such as n-methyl pyrrolidinone that must be removed before curing can begin. Most cure reactions generate water as a by-product which must also be removed carefully to prevent void formation. Finally, very high temperatures ($>250°C$) are required to cure the material fully. High residual stresses can result from a combination of the elevated process temperature, T_g, shrinkage during cure, and high modulus of the imidized resin.

Several major advances in polyimide chemistry have reduced some of the process difficulties. By end capping the polyimide precursor with acetylenic unsaturation, by-product free addition cures can be achieved. The development of a thermoplastic polyimide eliminates the high-temperature cure requirement [50]. Both of these modifications alter other characteristics, such as lowering of the high-temperature resistance over a traditional polyimide. The incorporation of dimethylsiloxane block segments into the polyimide backbone has also been reported [49]. This type of modification leads to an adhesive matrix with a lower modulus of elasticity and lower T_g value than those of unmodified polyimides and has been associated with lower-stress die attach assemblies.

3. Silicones

Silicone-based polymers have characteristics that make them desirable as the matrix of a conductive adhesive system. The excellent thermal stability, low ionic impurity content, and especially, the low modulus makes silicones desirable for bonding very large die. A typical formulation would include linear polydiorganosiloxane oligomers with both vinyl and silicon hydride functionality, a platinum catalyst, an electrically conductive particulate, adhesion promoter, reinforcing agents, and rheology modifiers [51]. The incorporation of sufficient silver flake to provide electrical conductivity will certainly increase the base polymer modulus and lower its elongation; however, from a mechanical property point of view, the silicones remain the primary choice for a low-stress die attach adhesive.

Reliability of the electrical properties of silicone-based isotropic adhesives has been the major difficulty to overcome and has essentially prevented commercialization. Another problem associated with silicones is that the addition polymerization reaction of silicones must be carefully controlled to prevent cure inhibition from various common chemical contaminants such as amines and sulfides. Other concerns include low-molecular-weight silicone polymer migration onto wirebond pads and very high CTE. There has been some activity in the development of hybrid resins that contain silicone blocks as commoner with epoxies such that the epoxy processing can be maintained with the added stress reduction property of the silicones [52].

V. FORMULATION OF ANISOTROPICALLY CONDUCTIVE ADHESIVES

A. Requirements and Performance

Materials for use as anisotropically conductive adhesives must satisfy requirements even more stringent than those defined previously for isotropically conductive adhesives. No specifications, however, have been defined specifically for these materials. When used for flip-chip applications, the adhesive serves not only as a physical and electrical interconnection between the device and the substrate, but also serves as the environmental protection and passivation layer. This fact, combined with high adhesive concentrations, makes the ionic contamination levels of these materials more critical than for isotropic conductive adhesives. In addition, the processing of these materials has a greater influence on joint reliability as the anisotropic electrical properties develop only after heat and pressure are applied to the joint.

Numerous geometrical factors of the specific interconnection will also influence anisotropic adhesive formulation and processing, including lead planarity, IC pad metallization, and IC test patterns. The planarity of the leads on the substrate and/or device and the compliance of the conductive particles will determine if anisotropically conductive adhesives can be used in a particular application. For systems with large disparities between lead height, no electrical interconnection will be formed, as shown in Fig. 5. Fine-pitch IC packages for surface-mounted applications, such as the plastic quad flat pack (PQFP), often use gullwing leads that offer much compliance to the joint. Even if the

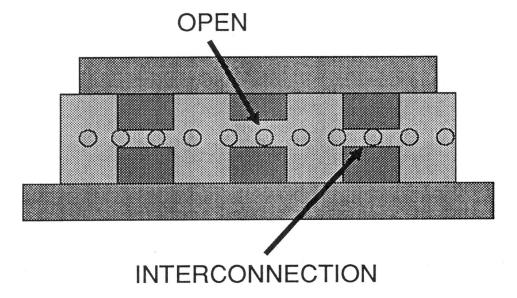

Figure 5 Effect of lead nonplanarity on anisotropically conductive adhesive interconnections.

leads are not initially coplanar, sufficient force can be exerted during assembly and cure to achieve electrical interconnection for all leads. Interconnecting flexible circuits and TAB packages to glass or rigid PWBs is relatively straightforward, whereas interconnecting bare silicon ICs to rigid substrates poses the greatest challenge.

Some of the requirements imposed on an anisotropically conductive interconnection used for flip-chip applications arise from the IC chip design. The metallization of the IC chip is usually aluminum or gold. Gold is preferred for the formation of reliable interconnections because an aluminum surface is coated with an insulating oxide that can be difficult for the conductive particles to penetrate (especially metal-plated polymer spheres). Most features on an IC, except for the bond pads, are passivated with an inorganic film such as Si_3N_4. However, some test patterns and grids located in the "streets" between chips are not electrically insulated. These metal features can cause electrical shorts between adjacent conductors via metal particle bridging. To prevent these problems, an additional insulating layer can be applied to the chip, or the bond pads can be bumped, raising the level of chip surface.

Data describing the reliability of joints assembled with anisotropically conductive adhesives are incomplete. Several papers have been published, but usually the sample size investigated is small, the accelerated stress tests are not standardized, and the results are highly dependent on device type (e.g., flexible circuit to rigid PWB, surface-mounted components, and flip-chip assembles). Further work is required in this area.

B. Formulation

1. Particle Specifications

The goal in formulating anisotropically conductive adhesives is to maximize particle concentration without compromising electrical isolation in the X–Y plane. Higher particle loadings increase the probability that an electrical interconnection will be made (especially for relatively small contact areas) and decrease contact resistance. Typical concentrations range from 5 to 15 vol % (30 to 60 wt % based on pure silver particles). The size of the particles usually range from 10 to 20 μm in diameter. Smaller particles offer the best results for very fine pitch applications.

To lower the probability of conduction in the X–Y plane (i.e., reduce the short-range percolation coherence length ξ), particles are used with an aspect ratio as close to 1 as possible. In contrast, isotropically conductive systems use flakes with high aspect ratios as fillers. Particle size distributions are minimized so that each particle can potentially serve as an electrical bridge between substrate and device.

Numerous types of particles are used in anisotropic adhesive formulations. Silver has high electrical conductivity and good resistance to corrosion; however, electromigration may cause problems in some applications. Nickel is a lower-cost alternative, but corrosion has been reported during accelerated aging tests. Solder particles offer the opportunity to form fusible linkages [53]. Gold offers the best properties, but the cost may be prohibitive for large-volume applications. Plated glass or polymer particles provide a lower-cost solution. Using a particle with a polymer core offers additional advantages, including low aspect ratio, good particle size uniformity, and compliance, that can help accommodate nonplanar surfaces.

The current-carrying capabilities of different particles will influence particle selection and concentration for a particular application. Measurements of several different materials have been reported [16] that exhibit a range of almost two orders of magnitude, as shown in Table 5.

Table 5 Current-Carrying Capability of Some
Anisotropically Conductive Adhesives

Particle type	Current (mA)
Solid metal	200
Metal-plated glass	20
Gold-plated polymer	5

2. Adhesive Matrix

Both thermoplastic and thermosetting adhesives are used to formulate anisotropically conductive adhesives. During assembly, thermoplastics must be heated above their glass transition temperature (T_g) to achieve good adhesion and electrical interconnection. The T_g value must be sufficiently high to avoid polymer flow during use but sufficiently low to prevent thermal degradation of the substrate and device. The ease with which joints can be assembled and repaired are the primary advantages of thermoplastic matrices. Typically, the electrical interconnections are characterized by moderate reliability and their used is restricted to consumer products.

Thermosetting matrices, such as epoxies and thermosetting hot-melt adhesives, are used where increased reliability is required. Repair of anisotropically conductive interconnections assembled with thermoset adhesives is problematic, however, as the adhesive matrix must be removed completely from the substrate and device prior to reassembly.

An additional consideration for the selection of an adhesive system is that robust bonds must be formed to all surfaces involved in the interconnection. Materials commonly found include metallizations on the substrate and components (e.g., gold, solder, copper, aluminum, and indium tin oxide), polymer substrates and coatings (e.g., polyimide, polyester, epoxy, and acrylic adhesives), and chip passivation layers (e.g., SiO_2 and SI_3N_4). Adhesion promoters may be required.

The thickness of the adhesive applied to the substrate should be considered in the early stages of formulation. The required thickness is a function of the geometry of the substrate and device to be interconnected. If a film adhesive is used, the film thickness must be sufficient to fill the gap between substrate and device to prevent void formation, as illustrated in Fig. 6a. The film thickness cannot be arbitrarily large, however, as the bonding temperature, pressure, and time must be sufficient to displace the excess adhesive. Loss of electrical contact can result when the final bond line is greater than the diameter of the filler particles, as shown in Fig. 6b.

Paste adhesives are less sensitive to height variations due to their lower viscosities. Particles in a low-viscosity matrix, however, are more susceptible to settling and agglomeration. This may lead to loss of isolation if the paste thickness applied is too great.

VI. MANUFACTURING WITH CONDUCTIVE ADHESIVES

A. Dispensing

Conductive adhesives are applied to the substrates using a wide range of standard techniques. For die attach applications, the isotropically conductive adhesive is applied using either a stamping tool or a syringe dispenser. To ensure complete void-free

VOID   INSUFFICIENT ADHESIVE

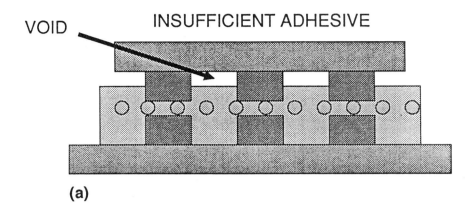

(a)

EXCESS ADHESIVE

LOSS OF
CONTACT

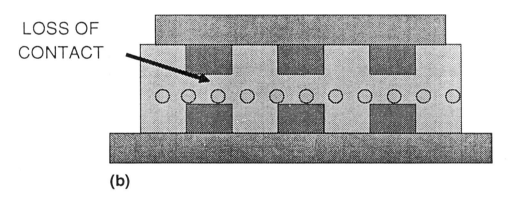

(b)

Figure 6 Effect of adhesive thickness on anisotropically conductive adhesive interconnections.

coverage, sophisticated multineedle syringe dispensers have been developed. Stencil and screen printing are used for the placement of isotropically and anisotropically conductive adhesive pastes on PWBs for the assembly of surface-mounted components. The development of ink-jet technology may prove to be another economical means of dispensing precise quantities of conductive adhesives paste [54]. Placement of anisotropically conductive films is accomplished by hand or by pick-and-place equipment.

B. Assembly

When using isotropically conductive adhesives, placement of components is performed by the same equipment as used for nonadhesive attachment techniques. Die bonders are similar to those used for eutectic bonding except for the type of adhesive dispenser. Surface-mounted placement machines developed for solder paste assembly can also be used with conductive adhesive pastes. After placement, the adhesives are cured in either batch or conveyorized ovens.

Assembly with anisotropically conductive adhesives is more complex, as the electrical interconnections require the simultaneous application of heat and pressure. For assembly of flex to rigid substrate interconnections, thermal compression bonders are used with either metal or ceramic hot bars. Pressures are typically in the range 25 to 100 kg/cm^2 [10]. Surface mounting of components with anisotropically conductive adhesives is usually accomplished by serial processes. Equipment similar to solder rework stations is used, where a device is aligned, placed into the adhesive, and cured under pressure with the application of heat by hot gas jets. This process is slow and difficult to control. Equipment costs are high due to low throughput resulting from curing each device in the placement machine. The pressure required to achieve interconnection will depend on the number and type of leads. A reliable batch processing technique, where all components on a PWB are cured simultaneously under pressure, is required before anisotropically conductive adhesives can be used for large-volume manufacturing of surface-mounted components.

Assembly of silicon chips onto substrates with anisotropically conductive adhesives uses specialized equipment, initially developed for flip-chip solder and TAB inner lead bonding. Heat and pressure are transmitted to the adhesive through a thermode attached to a robotic arm or a high-precision linear translator. Equipment requirements are more demanding than for solder assembly, as no self-alignment can occur. A minimum placement accuracy of ± 0.0005 in. is required. Coplanarity between the substrate and die is critical; one study reports maintaining coplanarity to within 0.00004 in. [19]. The pressure required to achieve interconnection depends on the size of the die, the type of conductive particle used, and the viscosity of the adhesive at the bonding temperature.

C. Reliability of Electrical Interconnections

The reliability of conductive adhesive electrical interconnections depends on the individual formulation and process employed [55]. In addition, the test vehicle configuration will strongly influence results. No comprehensive studies have been published, however, and no attempts to correlate chemical composition or a specific process variable to reliability performance have been reported.

Most conductive adhesive failures are accelerated by elevated temperature and humidity. In a study of 12 commercially available isotropically conductive adhesives, joint resistance increased between 160 and 35,000% when exposed to 65°C and 85% relative humidity (65/85) [56]. However, some adhesive manufacturers claim resistance changes of less than 10% after 1000 h at 60°C and 90% relative humidity [57] and less than 4% after 1000 h at 85/85 [58]. Anisotropically conductive adhesive joints are even more susceptible to early failures under accelerated test conditions due to process variations [16]. Reliability screening tests can be used effectively to iteratively optimize process parameters.

D. Repair

Isotropically conductive adhesive interconnections can be repaired using techniques similar to those used for solder rework. By application of heat locally at a temperature above the T_g value, a section of adhesive can be softened and the device can be removed mechanically. Adhesive residue on the substrate can be removed, if desired, by a mild scraping action. Clean components can be reassembled with the addition of fresh adhesive and the assembly can be cured either by additional application of local heat or by repeating the original cure cycle [59]. Anisotropically conductive adhesive interconnections are more difficult to repair, due to the need to remove previously cured adhesive from all affected surfaces prior to reassembly. Thermoplastic matrices greatly facilitate the repair process.

E. Suppliers of Commercial Conductive Adhesives

The following list includes several commercial manufacturers of conductive adhesives commonly used for die attach as well as other interconnection applications. Many manufacturers sell both isotropically as well as anisotropically conductive materials. The list is not intended to be inclusive but merely to provide initial guidance.

Ablestik Laboratories
AI Technology
Chromerics, Inc.
Emerson & Cumings Inc.
Epoxy Technology
Furane Products Company
Hitachi
Master Bond
Nippon Graphite
Polyflex Circuits
Quantum Materials
Tra-Con, Inc.
Zymet, Inc.

VII. CURRENT OUTLOOK

Conductive adhesive use should grow in the future. Die attach adhesive manufacturers are continually improving their formulations and developing new products in response to changing requirements (e.g., low stress and "snap cure" formulations). Anisotropically conductive adhesive use will become more widespread, as glass displays become an increasingly significant portion of the electronics market. Widespread replacement of solder paste as an interconnection material for surface-mounted assembly is more difficult to forecast. Isotropically conductive adhesives will probably prove inadequate for very fine pitch components. The use of anisotropics will be contingent on the development of a continuous process technology, as well as the demonstration of robust reliability.

REFERENCES

1. C. Wong, in *Handbook of Flexible Circuits* (K. Gilleo, ed.), Van Nostrand Reinhold, New York, 1992, p. 198.
2. R. P. Kusy, in *Metal-Filled Polymers* (S. K. Bhattacharya, ed.), Marcel Dekker, New York, 1986, p. 1.
3. D. M. Big, in *Metal-Filled Polymers* (S. K. Bhattacharya, ed.), Marcel Dekker, New York, 1986, p. 165.
4. L. T. Manzione, *Plastic Packaging of Microelectronic Devices*, Van Nostrand Reinhold, New York, 1990.
5. E. Suhir, in *Advances in Thermal Modeling of Electronic Components and Systems* (A. Bar-Cohen and A. D. Kraus, eds.), Hemisphere, New York, 1988,
6. R. K. Shukla and N. P. Mencinger, *Solid State Technol.*, 67 (July 1985).

7. T. Nukii, N. Kakimoto, H. Atarashi, H. Matsubara, K. Yamaura, and H. Matsui, *Proc. ISHM*, 1990.
8. B. Haskell and C. Lee, *Proc. NEPCON West*, 1992, p. 1601.
9. F. Juskey, *Adhesives Age*, 41 (Mar. 1988).
10. R. R. Reinke, *Proc. 41st Electronic Components Technical Conference (ECTC)*, 1991, p. 355.
11. P. B. Hogerton, K. E. Carlson, J. B. Hall, L. J. Krause, and J. M. Tingerthal, *Proc. International Electronics Packaging Conference*, 1990, p. 1026.
12. I. Tsukagoshi, A. Nakajima, Y. Goto, and K. Muto, *Hitachi Technical Report 16*, 1991, p. 123.
13. K. Gilleo, *Proc. 39th Electronic Components Technical Conference (ECTC)*, 1989, p. 37, and references therein.
14. R. R. Tummala and E. J. Rymaszewsik, eds., *Microelectronics Packaging Handbook*, Van Nostrand Reinhold, New York, 1989.
15. P. B. Hogerton, J. B. Hall, J. M. Pujol, and R. S. Reylek, *Mater. Res. Soc. Symp. Proc* 154:415 (1989).
16. D. D. Chang, J. A. Fulton, A. Lyons, and J. R. Nis, *Proc. NEPCON West*, 1992, p. 1381.
17. Y. Iinuma, T. Hirohara and K. Inoue, *Proc. ISHM*, 1987, p. 635.
18. F. W. Kulesza and R. H. Estes, *Hybrid Circuit Technol.* (Feb. 1992).
19. K. Hatada and H. Fujimoto, *Proc. 39th Electronic Components Technical Conference (ECTC)*, 1989, p. 45.
20. B. Allenby et al., *Proc. Surface Mount International*, 1992, p.1.
21. J. Murray, *Getting the Lead Out, Printed Circuit Fabrication 16*: 100 (1993).
22. J. Tuck, *Circuits Assembly*, 22 (Jan. 1992).
23. J. Gurland, *Trans. Met. Soc. AIME 236*: 642 (1966).
24. S. Kirkpatrick, *Rev. Mod. Phys. 45*: 574 (1973).
25. S. M. Aharoni, *J. Appl. Phys. 43*: 2463 (1972).
26. F. Bueche, *J. Appl. Phys. 43*: 4837 (1972).
27. A. Lyons, *Polymer. Eng. Sci. 31*: 445 (1991).
28. S. Etemad, X. Quan, and N. A. Sanders, *Appl. Phys. Lett. 48*: 607 (1986).
29. X. Quan, *J. Polymer. Chem. Polymer Phys. Ed. 25*: 1557 (1987).
30. D. M. Bigg, *J. Rheol. 28*: 501 (1984).
31. Z. V. Litvishko, Y. I. Khimchenko, N. K. Babenko, and A. L. Lobodyuk, *Sov. Powder Metall. Metal Ceram. 20*: 877 (1981).
32. G. R. Ruschau and R. E. Newnham, *J. Composite Mater. 26*: 2727 (1992).
33. L. Nicodemo, L. Nicolais, G. Romeo, and E. Scafora, *Polymer Eng. Sci. 18*: 293 (1978).
34. F. F. T. De Araujo and H. M. Rostenberg, *J. Phys. D Appl. Phys. 9*: 1025 (1976).
35. A. Malliaris and D. T. Turner, *J. Appl. Phys. 42*: 614 (1971).
36. F. Hochberg, U.S. patent 2,721,357 (1955).
37. D. M. Bigg and D. E. Stutz, *Polymer Composites 4*: 40 (1983).
38. B. Miller, *J. Appl. Polymer Sci. 10*: 217 (1966).
39. A. Lovinger, *J. Adhesion 10*: 1 (1979).
40. R. L. Opila and J. D. Sinclair, *Ann. Proc. Reliab. Phys. Symp. 23*: 164 (1985).
41. M. G. Golubeva, N. N. Turkova, L. Z. Shenfil, and V. E. Gul, *Colloid Journal 35*: 691 (1973).
42. I. O. Salyer, J. L. Schwendeman, and B. R. Hickman, U.S. patent 3,359,145 (1967).
43. S. Jin, R. C. Sherwood, T. H. Tiefel, J. J. Mottine, and S. G. Seger, Jr., U.S. patent 4,737,112, (1988).
44. R. H. Estes, *Proc. International Symposium on Microelectronics*, 1986, p. 642.
45. R. C. Benson, T. E. Phillips, and N. deHass, *Proc. 39th Electronic Components Technical Conference*, 1989, p. 301.

46. J. M. Pujol, C. Prud'homme, M. E. Quenneson, and R. Cassat, *J. Adhesion 27*: 213 (1989).
47. Y. Okabe, A. Kusuhara, M. Mizuno, and K. Horiuchi, *Proc. 38th Electronic Components Technical Conference*, 1988, p. 468.
48. L. M. Leung and K. K. T. Chung, *Hybrid Circuits 18*: 22 (Jan. 1989).
49. R. Pound, *Electron. Packaging Prod.*, 132 (Feb. 1989).
50. L. Ying, *Proc. International Symposium on Microelectronics*, 1986, p. 621.
51. M. A. Lutz, and R. L. Cole, *Hybrid Circuits 23*, 27–30 (Sept. 1990).
52. J. V. Crivello, *Polymer Eng. Sci. 32*(20) (1992).
53. K. Gilleo, *Electron. Packaging Prod.* 134 (Feb. 1989).
54. J. Tuck, *Circuits Assembly*, 20 (Apr. 1993).
55. A. O. Ogunjimi, O. Boyle, D. C. Whalley, and D. J. Williams, *J. Electron. Manuf. 2*: 109 (1992).
56. J. P. Honore, H. D. Rubin, and M. K. Zierold, *Proc. NEPCON West*, 1992, p. 1372.
57. S. L. Spitz, *Electron. Packaging Prod.* 64 (Feb. 1991).
58. G. P. Nguyen, J. R. Williams, and F. W. Gibson, *Circuits Assembly*, 36 (Jan. 1993).
59. J. C. Bolger, J. M. Sylva, and J. F. McGovern, *Surface Mount Technol.*, 66 (Feb. 1992).

IV
Applications of Adhesives

36
Adhesives in the Electronics Industry

Monika Bauer and Jürgen Schneider *Fraunhofer-Institute of Applied Materials Research, Teltow, Germany*

I. INTRODUCTION

Although the application of adhesives in the electronics industry is widespread, the production of printed circuit boards (PCBs) creates a demand for a wide spectrum of properties. Technical progress by research and development leads to more miniaturization of components and circuits, to a higher level of integration of printed circuits, and therefore to higher electrical power evolved per square unit. This trend continues. A lot of problems arising from this development could be solved by using special adhesives, sealants, or laminating resins only.

The aim of this short overview is to define the requirements of the electronics industry for adhesives, to derive their profile of properties, and to describe the state of the art. Additionally, encapsulating and sealing materials as well as binders for laminates are treated if there are analogous requirements regarding their properties. One well-known example of this situation is FR4-based material for PCBs, where the epoxy resin acts simultaneously as an adhesive for the copper foil and as a binder for the laminate. The primary fields for application of adhesives in the production of electronic circuits are adhesives for surface-mounted devices (SMDs), binders for laminates and/or adhesives for PCB base material, encapsulating and sealing materials for separate electronic components or complete circuits, and adhesive tapes used, for example, for flexible keyboards, feeding tapes, and covering materials. A variety of products are commercially available for all these applications.

II. REQUIREMENTS OF THE ELECTRONICS INDUSTRY ON ADHESIVES

As stated earlier, the main requirements for adhesives in the electronics industry result from the increasing miniaturization of the electronic devices, which is manifested in an increasing number of pins with decreasing distance between them. Therefore, both the number of leads per square unit of the circuit board increases and the conductor width decreases. Additionally, the ''concentration'' of devices on the board increases. Further requirements on the adhesives arise from the necessity of reaching high production rates using a simple technology with low energy consumption and of dealing with substances that are harmless for health and environment. From all these facts the following primary requirements result:

1.   Precise casting and dosing of the polymeric material is necessary to ensure electrical insulation between the conducting tracks on a PCB. Additionally, the adhesive should not shrink or creep irreversibly during manufacture of electric circuits (e.g., heating by soldering or curing processes, mechanical stresses by automatic insertion). For these reasons, the thermal expansion coefficients of the adhesives should be in the same range as those of the devices and the PCB.

2.   The increasing quantity of power dissipation per square unit as a result of the rising density of devices on PCBs leads to higher operating temperatures. Therefore, the thermal resistance of the applied adhesives has to correspond with the operating temperatures. Furthermore, a low dielectrical constant is essential for the adhesives and/or binders used for laminates because the heat dissipation is a result of dielectrical losses.

3.   Up to now the electrical connection of devices and conducting tracks is made by solder processes, and therefore the adhesives have to withstand temperatures of about 260°C for a short time without remarkable losses in their performance. Up to now, conducting adhesives for electrical connections as an alternative have been used in special applications only.

4.   The adhesives have to show adjusted rheological behavior. Depending on the application, materials with both high and low viscosity, with and without thixotropy, and with and without yield point are needed.

5.   Since electronic circuitries have a filigree structure, they are sensitive against corrosive media. For this reason they must be covered with coatings or sealants of high purity. Alkaline and chloride ions and residual solvents especially have to be avoided.

III. CHEMICAL BASE OF ADHESIVES FOR THE ELECTRONICS INDUSTRY

The chemistry of polymeric materials used for adhesives in the electronics industry does not differ from that of polymers for other applications. Taking into account the requirements mentioned in Section II, it becomes clear that reactive resins evolving no volatile substances during the curing process are of special interest. The following polymeric materials are of particular importance:

1.   *Epoxies* [unmodified types of bisphenol A, higher functionalized resins, novolac epoxies, cycloaliphatic resins, heterocyclic resins; one- and two-component systems; curing at room temperature, at elevated temperatures, or by ultraviolet (UV) irradiation], used as mounting adhesives, coverings, and binders for laminates [1,2]

2.   *Acrylics* (cyanoacrylates; anaerobic, by heating or UV-irradiation cross-linking acrylates) for mounting adhesives, sealants, and for adhesive tapes [2,3]

3.   *Silicone resins* (at room temperature or by heating and by UV-irradiation curing products) mainly for sealings and coverings [4]

4.   *Polyimides* as mounting adhesives [1]

Since epoxies and acrylics are the most important base resins for adhesives in electronic industry, they will be described in more detail.

A. Epoxy Resins

Examples of the resin components of epoxies are listed in Table 1. Several chemical substances can be used as hardeners for epoxies [5,6]. The two main groups of cross-linking agents used in industry are two or more functionalized amines and anhydrides of carboxylic acids. Cross-linking with amines is possible at room temperature as well as at elevated temperatures, depending on the chemical structure of the amines (Table 2). Anhydrides of carboxylic acids react at elevated temperatures only.

Table 1 Examples of Resin Components of Epoxies

Structural formula	Name
$H_2C-HC-CH_2-O-CH_2-CH_2-CH_2-CH_2-O-CH_2-CH-CH_2$	Butandiol–diglycidyl ether
	Bisphenol A–diglycidyl ether
	Bisphenol F–diglycidyl ether
	Novolac–epoxy–resin
	Cycloaliphatic epoxy resin
	Triglycidyl isocyanurate

Table 2 Reaction Temperatures of Amine-Based
Hardeners

Amine	Cross-linking temperature (°C)
Aliphatic	Room temperature
Cycloaliphatic	50–100
Aromatic	80–150

The cross-linking of epoxies by amines follows the addition of a primary amine with two epoxy groups, resulting in a tetrafunctional branching point, as one can see in Fig. 1. For the reaction of anhydrides with epoxy groups, the presence of a small fraction of hydroxyl groups (e.g., secondary hydroxyl groups of oligomeric epoxies) is necessary. At elevated temperatures a hydroxyl group reacts with an anhydride and then the carboxylic group of the resulting half-ester reacts with an epoxy group to get an adduct, as one can see from Fig. 2. Epoxies may be polymerized through cationic polymerization, initiated by UV radiation (Fig. 3).

B. Acrylic Resins

Acrylic resins used as adhesives are formed through radical or anionic polymerization [6]. Radical polymerization can be initiated by UV radiation as well as heat. The two reaction schemes are identical in principle (Fig. 4). Cyanoacrylates are of special interest for systems with very high reaction rates. Their reaction follows an anionic polymerization mechanism. Since the polarity of the cyanoacrylates is very high, water is able to act as an initiator (Fig. 5).

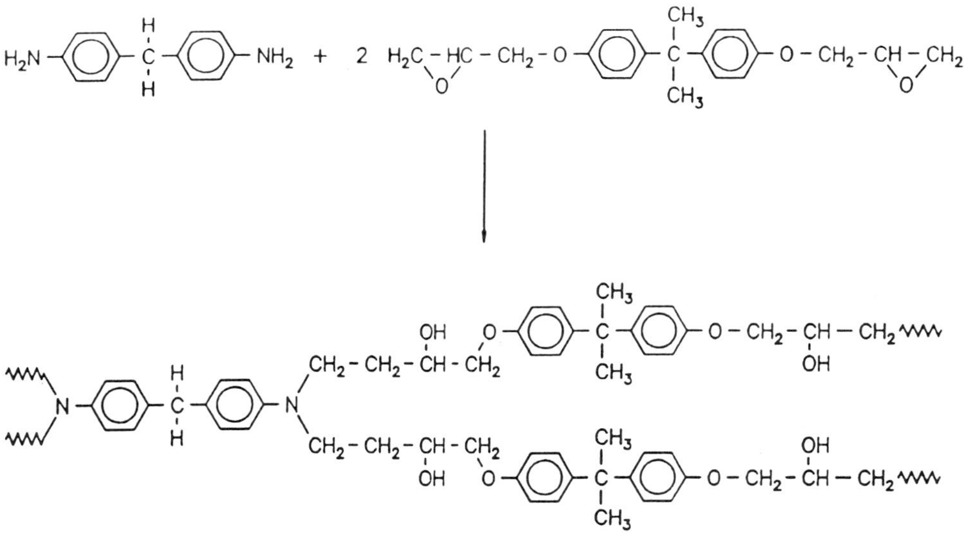

Figure 1 Cross-linking of epoxies by amines (simplified scheme).

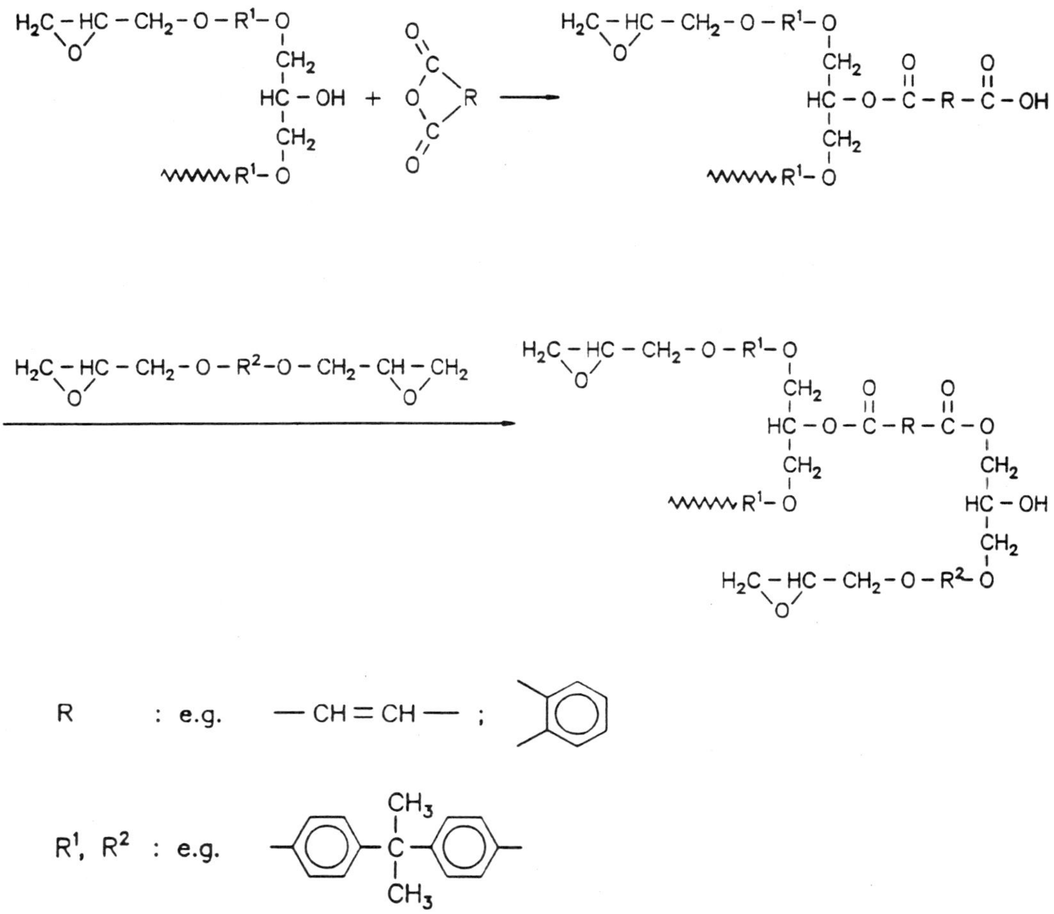

Figure 2 Cross-linking of epoxies by anhydrides (simplified scheme).

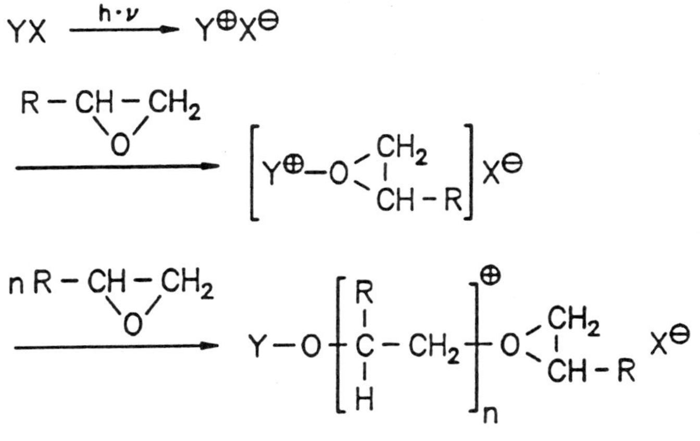

Figure 3 UV-initiated polymerization of epoxies.

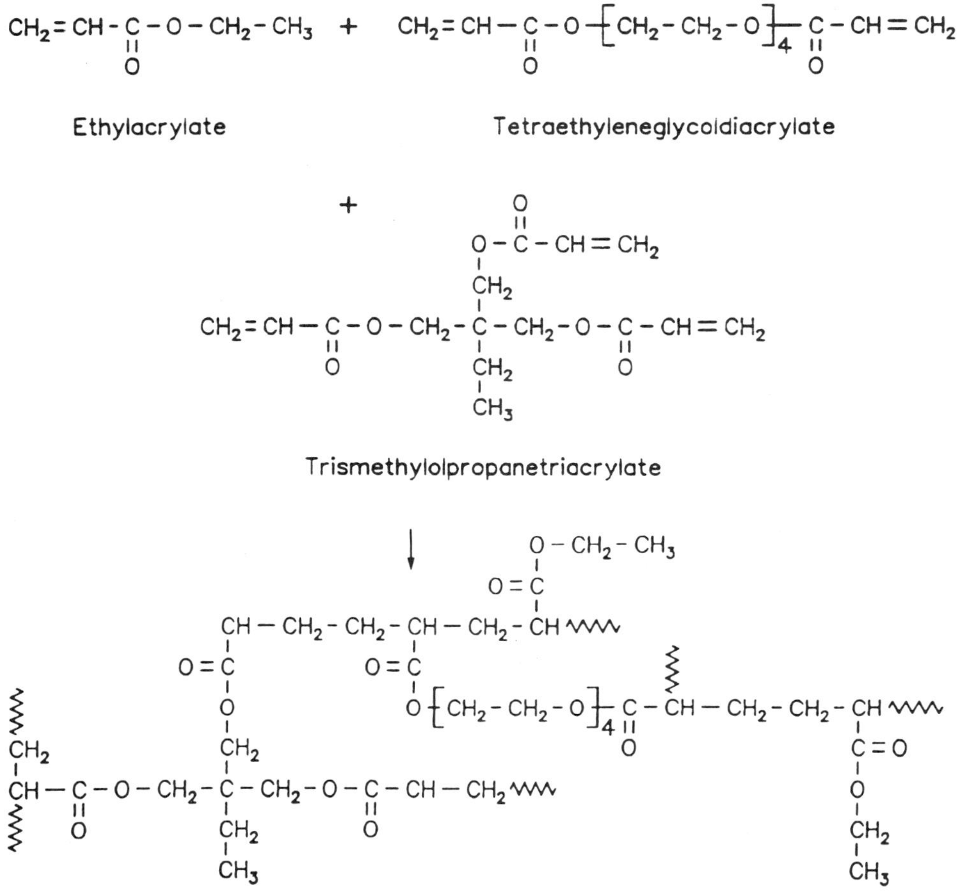

$$CH_2=CH-\underset{\underset{O}{\|}}{C}-O-CH_2-CH_3 \quad + \quad CH_2=CH-\underset{\underset{O}{\|}}{C}-O\left[CH_2-CH_2-O\right]_4\underset{\underset{O}{\|}}{C}-CH=CH_2$$

Ethylacrylate   Tetraethyleneglycoldiacrylate

$$+$$

$$CH_2=CH-\underset{\underset{O}{\|}}{C}-O-CH_2-\underset{\underset{CH_3}{\overset{CH_2}{|}}}{\overset{\overset{O-\underset{\|}{\overset{O}{C}}-CH=CH_2}{|}}{\underset{|}{C}}}-CH_2-O-\underset{\underset{O}{\|}}{C}-CH=CH_2$$

Trismethylolpropanetriacrylate

Figure 4   Cross-linking of acrylate resins through radical polymerization (simplified scheme).

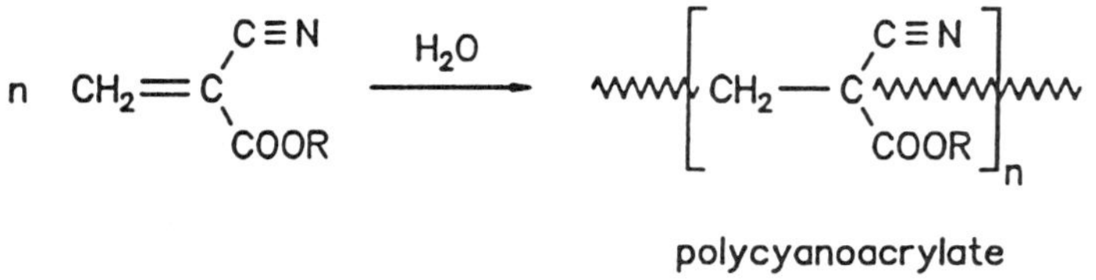

$$n \quad CH_2=\underset{\diagdown}{\overset{\diagup C\equiv N}{C}}\underset{COOR}{} \quad \xrightarrow{H_2O} \quad \text{www}\left[CH_2-\underset{\diagdown}{\overset{\diagup C\equiv N}{C}}\underset{COOR}{}\right]_n\text{www}$$

polycyanoacrylate

Figure 5   Polymerization of cyanoacrylates.

IV. MOUNTING ADHESIVES

Mounting adhesives are needed for the fixation of components in circuit boards during the loading process using surface-mounting technology (SMT) prior to soldering. As an alternative method, electrical-conducting adhesives can be used for the fixation of components. The advantage of this technique is that the soldering process can be avoided. Using SMD components for loading of PCBs, either an adhesive or a solder paste is needed to fix the devices. Which variant is selected for a particular application depends primarily on the board design. If a mixed loading of SMD components together with leaded devices must be used, fixation by an adhesive is the only applicable variant, but by using pure SMD loading, both adhesive and solder paste can be applied. It must be noted that the strength of fixation achievable by adhesives is higher than that achievable by solder pastes. On the other hand, an additional technological process, complete curing of the adhesive, is necessary. This disadvantage can be mitigated by using electrical-conducting adhesives, thus avoiding soldering altogether.

Requirements for SMD adhesives:

1. High adhesive strength of the uncured resins (wet adhesion) to get exact fixation of the components
2. Sufficient final strength after curing to guarantee the mechanical stability of the PCB
3. High curing rates at low temperatures to minimize the thermal stresses of thermally sensitive components
4. Low dielectric losses
5. Processing behavior that meets such technological requirements as:
   a. One-component systems to avoid mixing failures
   b. Homogeneous systems, especially if filled adhesives are used
   c. Sufficiently long pot life
   d. Rheological behavior adjusted to the dosing process used

A. State of the Art

1. Adhesives

The most commonly used adhesives in the electronics industry are thermosetting one-component epoxy resins and UV-irradiation cross-linking acrylates [7].

(a) *Thermosetting epoxy resins.* Generally, thermosetting epoxy resins are mixtures of the following:

Resins	50–80 wt %
Hardeners	20–50 wt %
Accelerators	0–3 wt %
Fillers, dyes, modifiers	0–20 wt %

At this time, thermosetting epoxy resins are the most important adhesive systems used for mounting [2]. One reason for their successful use for a long period is due to the great variability of their properties, which can be adjusted to a lot of requirements. Other advantages are simple processibility and good thermal stability. Furthermore, using modern efficient accelerator systems, the curing temperature of epoxies could be decreased to temperatures lower than 100°C and times shorter than 15 min. In this way, the thermal stressing of temperature-sensitive components can be minimized.

(b) *UV-cross-linking acrylate resins.* Generally, UV-curing acrylate resins contain [8]:

Reactive oligomers/prepolymers	50–80 wt %
Monomers/reactive thinners	10–40 wt %
Photoinitiators/stabilizers/accelerators	1–5 wt %
Inhibitors/dyes	1–5 wt %

The advantages of UV-curing acrylates are curing times under UV irradiation of only a few seconds and the low energy consumption of UV lamps compared to the thermal energy needed for thermosetting systems. On the other hand, these adhesives cannot be cured completely by UV irradiation. Therefore, postcuring is necessary. Commercially available adhesives (e.g., epoxy acrylates, urethane acrylates, polyester acrylates, silicon acrylates, and methacrylates) contain functional groups, which allow complete hardening. It must be noted that due to the lower glass transition temperature of these polymers than of thermosetting epoxy resins, their maximum operating temperature is limited.

(c)   *Electrical-and/or thermal-conducting adhesives* [1,9].   As described above, epoxies and acrylates are filled with metal powders to get electrical-conducting adhesives. For special applications polyimide and silicone adhesives are used also. Since the metallic particles must touch each other inside the resins to reach a sufficient level of conductivity, a metal content of 70 to 80 wt % is necessary. Silver is the metal generally used, since specific resistances of the filled adhesives down to about 10^{-4} $\Omega \cdot$cm can be achieved (metallic silver has a specific resistance of 1.6×10^{-6} $\Omega \cdot$cm). Using other metals, such as copper or nickel, the accessible electrical conductivity is too small. On the other side, copper-filled resins show good thermal conductivity and are therefore used for such applications, where heat dissipation is of importance. To reach high heat dissipation levels, ceramic fillers such as aluminum oxide or boron nitride in a quantity of about 60 to 75 wt % are used. The resulting adhesives reach thermal resistances of 5 to 7 K/W. The main advantages of electrical conducting adhesives are (1) better resistance against mechanical stresses, resulting from large temperature variations compared with solder connections, and (2) electrical connections obtained at low temperatures since soldering is not required.

2.   Deposition Processes

Adhesives of various rheological properties are available for the usual deposition processes: screen printing, pin transfer, and dispensing [7]. The adhesive droplet must have a definite height and size, which depend on the board design and the type of component, to bridge the distance between the surface of the circuit board and the electronical component. This can be achieved by a specific adjustment of the adhesive rheology, especially thixotropy and yield point. Usually, the adhesive is deposited on the board. If higher distances between board and device have to be bridged, an additional deposition of the adhesive on the underside of the component may be favorable to ensure wetting of both parts to bond and to counteract the drain-away of adhesive by gravitation. Some examples of distances between circuit board and various components are shown in Fig. 6. The usual

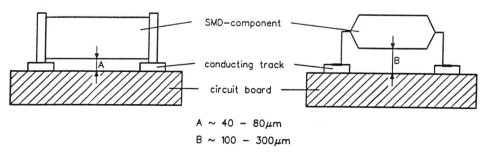

Figure 6   Droplet heights of SMD adhesives.

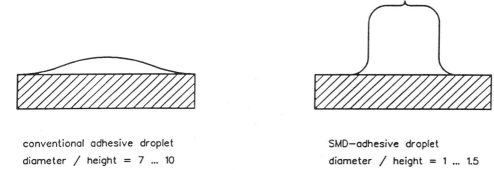

conventional adhesive droplet
diameter / height = 7 ... 10

SMD—adhesive droplet
diameter / height = 1 ... 1.5

Figure 7 Comparison of droplet heights for conventional and SMD adhesives.

diameter/height ratio of an adhesive droplet is about 10:1. Commercially available SMD adhesives may have ratios of up to 1:1 (Fig. 7).

V. ENCAPSULANTS AND SEALINGS

Protection of components or of the entire circuit board against environmental influences is necessary in many fields of application to ensure proper operation. For this purpose special sealants and encapsulants based on curable epoxies, acrylates, and silicons were developed. Polyurethanes and polyimides are also used as encapsulants and sealings but have only limited importance. Requirements for sealings and encapsulants are:

1. Good wetting of the substrate and sufficient adhesion
2. Resistance against high and low temperatures, moisture, and corrosive media
3. Good processibility
4. Adjusted rheological behavior
5. Good elasticity to withstand mechanical stresses

A. State of the Art

A lot of problems arising from the requirements for encapsulants and sealings can be solved using modified epoxy and acrylate resins [2,10]. Furthermore, silicones became important, which is due to their high heat resistance, good elasticity, maintenance of their electrical properties over a wide range of temperatures and frequencies, and nonflammability [4]. A special type of application of silicones are the silicone gels, which are weakly cross-linked silicone rubbers and behave like elastic liquids. Due to the good mechanical damping properties of silicon gels, they are used for encapsulating such components, which must be protected against vibrations. On the other hand, encapsulated components are easy to repair since the polymer molecules of these silicones have sufficient mobility for "self-healing."

VI. ADHESIVE TAPES

Two types of tapes are discussed: carrierless adhesive foils (also called transfer adhesives) used, for example, in mounting or laminating processes, and tapes consisting of an adhesive on a carrier tape, used for contact and distance films in foil keypads, covering

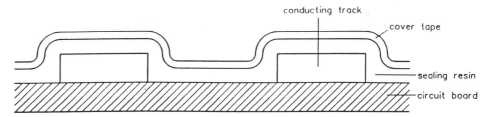

Figure 8   Covering tape with plastic sealing mass. (From Ref. 11.)

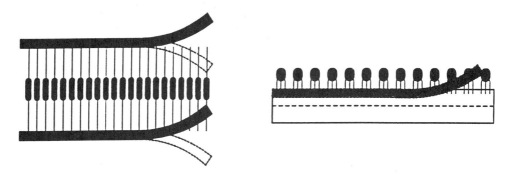

Figure 9   Two examples of adhesive tapes for component supply. (From Ref. 3.)

tapes in galvanizing and solder processes (Fig. 8), and as component supply in insertion installations (Fig. 9). The tape properties required depend, of course, on the type of application. Therefore, the following properties are important for component supply tapes [12]:

1.   Constant width of tapes to avoid malfunctions of the handler
2.   Sufficient adherence of the components on the tape
3.   Stability of the adhesive against aging
4.   Possible tape lengths up to 5000 m on a spool for effective production
5.   Well-defined strength–elongation behavior

It must be noted that the adhesion properties and aging behavior of the supply tapes are determined by the corresponding properties of the adhesives used, whereas the mechanical strength depends primarily on the carrier tape.

A.   State of the Art

Tapes for component supply in loading installations usually consist of an acrylate- or rubber-based adhesive (also thermosetting) on a paper or polyester foil carrier (dependent on the mass of the components). Depending on the type of application, different materials are used for the adhesive and the carrier for covering tapes. Most important carrier materials are foils (PVC, polypropylene, cellulose, polyester, polyimide), papers, and woven and nonwoven fabrics (cotton, glass). Adhesives frequently used are based on rubbers, silicones, and acrylates. Carrierless adhesive foils are used, for example, for the bonding of copper foils and polyimide films to get special base materials for circuit boards.

VII. ADHESIVES FOR BASE MATERIALS FOR CIRCUIT BOARDS

The majority of base materials for circuit boards are combinations of a copper foil with a laminate, where the laminate itself consists of a carrier material and a resin. Thus properties of the base material such as mechanical strength, dimensional stability, and processability are determined primarily by the carrier material. On the other hand, the resin materials are responsible for the thermomechanical and electrical properties as well as for its resistance against chemicals and moisture. Frequently used carrier materials are based on glass and carbon fibers, papers, and polyamide, whereas the majority of the laminating resins are thermosets such as epoxies, phenolics, cyanates, BT resins, maleimides, and various combinations of these [13].

Furthermore, the overall properties of the base materials for circuit boards are determined essentially by the joint between the copper foil and the laminate. This joint can be realized by both an added adhesive and by the laminating resin itself, which additionally, acts as adhesive. An additional adhesive is needed in the manufacture of flexible circuit boards (e.g., polyimide/copper) or of paper-based rigid circuit boards. During this process the adhesive is deposited on the bottom side of the copper foil after it is joined to the laminate by heating under pressure. The primary requirements for such adhesives are:

1. Adjustable processing behavior
2. Systems without or with low organic solvent contents
3. Sufficient bonding strength at elevated temperatures (250 to 320°C for short times)
4. Dielectrical stability, low dielectric constant
5. Resistance against etching and galvanic processes

For the production of base materials for circuit boards with higher performance (e.g., glass/epoxy or graphite/cyanate ester combinations) and of multilayer boards, the laminating resin acts as adhesive or special bonding prepregs must be used. The requirements for the resins, which act as adhesive, depend on both the processing conditions and the desired properties of the final circuit board and are similar to those described above.

A. State of the Art

Base materials for circuit boards based on phenolic resin/paper laminates are used primarily for low-performance materials, which are needed, for example, in consumer electronics. The actual properties standard of these materials meets the requirements of the market. The laminating adhesives used for the manufacture of base materials are organic solutions or aqueous dispersions of thermosetting resins such as poly(vinyl butyral)–phenolic resin or acrylate–phenolic resin.

Flexible circuit boards consist primarily of polyimide-based carriers. The problem of bonding the copper foil on the polyimide carrier has not yet been solved satisfactorily. Due especially to their low bonding strength at elevated temperatures, the production of such materials is very limited. Nevertheless, adhesives for copper–polyimide systems were developed, where one-component epoxy resins (e.g., epoxy–polyester mixtures) and reactive hot melts (e.g., phenolic resin–nitrile rubbers) reached importance.

A wide range of high-performance materials for circuit boards is available for such systems, where the laminating resin acts simultaneously as adhesive for the copper foil. The majority of the systems consist of resins, which are based on epoxies, bismaleimides, cyanates, BT resins, and carrier materials, made from glass and carbon fibers and polyamides. Since the properties of the resins are adjustable in a wide range using combinations and modifications of the resins, the bonding strengths desired for the resulting circuit board base materials can be achieved for almost all kinds of applications. On the other hand, further developments are needed to get base materials having, for example, high thermal and moisture resistance.

VIII. OUTLOOK

The future development of electronical circuitries will be characterized by miniaturizing of components and circuits as well as a more and more dense packing of components on the printed circuit board. For this reason, the surface-mounting technology of electronical components will become more and more important. Additionally, using SMD techniques, a higher level of production of electronical circuitries can be reached. The preparation of special adhesives is one important precondition for the success of SMD techniques. In this connection it is essential to look for new reaction principles for one-component adhesives having higher curing rates at lower temperatures to increase the production rates and reduce the energy consumption and to replace the solder process by electrical- and/or thermal-conducting adhesion bondings.

The next era in loading technology has already begun. This technology is characterized by conducting tracks made in thin-layer technology by metal-sputtering processes and by electronical functions realized in thick-layer technology by structurated deposition of inorganic or organic pastes with well-defined electrical properties. Concerning this matter, the development of organic pastes with well-defined and adjustable electrical properties and rheological behavior is necessary.

In the future, environmental protection will become more and more a matter of concern. Harmless technologies and recycling processes for both the adhesives and the entire circuits will have to be developed. In this connection, the following requirements must be met by the polymeric materials: nontoxicity, low, or no, volatile-product content during use, suited for repairing and disassembling processes of components and recycling.

ACKNOWLEDGMENT

M. B. is grateful to Fonds der Chemischen Industrie for financial support.

REFERENCES

1. A. Hof, *Adhaesion 33*: 21 (1989).
2. *AMICON*, product information, W. R. Grace & Co.,
3. *TESA Spezialklebebänder*, product information, Beiersdorf AG,
4. W. Brennenstuhl, *Swissbonding '89*, preprints, 1989, p. 93.
5. C. A. May and Y. Tanaka, *Epoxy Resins*, Marcel Dekker, New York, 1973.
6. A. J. Kinloch, *Structural Adhesives*, Elsevier, London, 1986.
7. V. Bejenke, *Adhaesion 33*: 13 (1989).
8. H.-J. Battermann, *Swissbonding '89*, preprints, 1989, p. 141.
9. E. Wipfelder and K. Orthmann, *Adhaesion 33*: 26 (1989).
10. G. Nosbuesch, *Swissbonding '89*, preprints, 1989, p. 61.
11. *Technicoll Lamibond*, product information, Beiersdorf AG.,
12. H. K. Engeldinger, *Elektro-Anzeiger 40*: 43 (1987).
13. L. R. Walling, *Galvanotech./Leiterplattentech 78*: 807 (1987).

37
Adhesives in the Wood Industry

Terry Sellers, Jr. *Mississippi State University, Mississippi State, Mississippi*

I. HISTORY OF WOOD ADHESIVES

The process of bonding wood materials with glue has been practiced since the early periods of civilization. Examples of glued wood materials exist which are thousands of years old, such as scarf-jointed wooden chariot wheels and decorative, thin-veneer-covered artifacts from the Pharaonic periods of Egypt. Ancient written history and paintings testify that early centers of civilization developed aesthetic, utilitarian, and warfare-related glued wood products. Adhesives were made from a wide range of natural materials, such as earthen substances (asphalt or bitumen and lime), animal parts (hides), fish skins, tree resins (pitch), plant gums, cereal grain starches, insect extracts (beeswax and lac), and other sticky materials (albumen from animal eggs and casein from milk). Gluing wood was an art form; these skills were passed down from generation to generation. The art of gluing advanced concurrently with the development of inks, paints, and papermaking. The gluing art form waxed and waned as the centers of civilization progressed or were interrupted by devastating wars and pestilence. As a result, the art of gluing wood made little overall progress until the eighteenth through twentieth centuries. During this time, the development of new machinery and discovery of new adhesive materials (synthetic resin-based adhesives from coal tar and petrochemicals) jointly led to very expansive production and use of old and new glued-wood products.

II. WOOD AS AN ADHEREND

Wood is a natural adhesive-bonded organic composite made up of approximately 65% polysaccharide (carbohydrate) fibers, 25% polyphenolic lignin binder, and 10% photo-plasmic-produced residues or cellular waste infiltrates (oleoresins, tannins, absorbed minerals, starches, etc.) which coat the porous cellular surfaces. The polysaccharides are composed of linear cellulose homopolymers (ca. 42%) interspersed with heterogeneous short-chain five- and six-carbon sugars called hemicellulose (ca. 23%), all of which can be metabolized by microorganisms [1]. Both cellulose and lignin have a density of about 1400 to 1500 kg/m^3 (1.4 to 1.5 specific gravity), similar to many synthetic resins and fibers. However, wood has a very porous, honeycomblike structure in which the longi-tudinally aligned intracellular-matrix cellulose fibers, or tracheids, have inner open spaces (lumen) with radial diameters of 4 to 25 μm (Fig. 1) [2,3]. Depending on the wood species, the fiber cells may have open or closed ends as well as interconnected pits (ca. 0.2-μm or 200-nm-diameter gaps) which may, or may not, allow air or liquids to penetrate from one fiber to another [2]. Thus wood is volumetrically lower in mass weight (density) than synthetic resin adhesives or fibers. When the cellulosic fibers are cut and exposed, they are polar (molecularly attractive) in nature. These cellulosic materials shrink and swell upon drying or wetting and are useful in wood adhesion.

In hardwoods, elements in longitudinal series comprise the segmented structure, which is termed a *vessel* or *pore*. Vessels, which are exposed in transverse section, constitute about 10 to 46% of the stem volume in deciduous hardwood trees and are relatively large-diameter cells (50 to 300 μm). Vessels form essentially open, vertical tubes within the wood because their end walls have partially dissolved. By comparison, the vessel diameter may be up to 10 times the diameter of a softwood fiber. Other porous cells in wood are resin canals and ray cells. Because the porous growth patterns of wood vary, the densities of various dry woods vary (200 to 1200 kg/m³). The porosity of wood

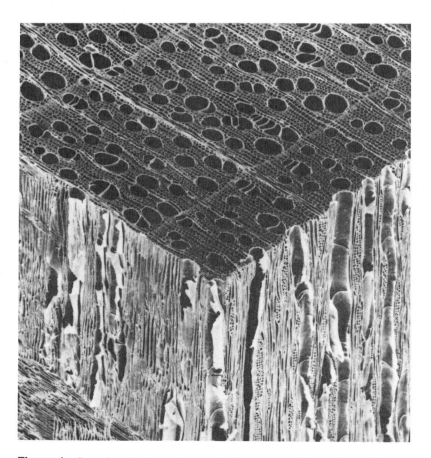

Figure 1   Scanning electron micrograph of red maple (*Acer rubrum* L.).

greatly influences the wood's utility as an adherend. Also, the wood's porosity influences the type and form of adhesive because porosity affects the wood's ability to absorb water and other solvents from the adhesives, and porosity allows some of the adhesive to be adsorbed over larger surface areas.

Polyphenolic lignin coats and bonds the intracellular cellulose fiber. When exposed, lignin is only moderately polar. Cavities of dried wood cells are lined with various nutrients, organic compounds generated by the trees, and minerals absorbed by the tree roots, all deposited in the cell structure. Collectively, these materials are called *extractives* because they are numerous and are relatively easy to extract with water and organic solvents. Depending on the chemical and physical characteristics of these substances, they may help or hinder the process of bonding wood with adhesives. The extractives dictate the pH of wood and the color of wood. Even though the extractives do not contribute to the wood's structural strength, they play an inordinately high role in the wood's permanence and gluability.

III. THE NATURE OF WOOD ADHESION

Adhesion is the state in which two surfaces are held together by chemical or physical forces, or both, with the aid of an adhesive [4]. When two pieces of wood are glued together for structural purposes, the basic objective is to hold the two pieces in a fixed position so that when applied stress exceeds the wood's strength, the failure occurs with the strength of the wood rather than with the adhesive. To achieve this goal, both the molecular attraction between the adhesive and the wood surface (adhesion forces) and the molecular attraction by the adhesive throughout its mass (cohesion forces) must be greater than the strength of the wood [3]. For nonstructural applications, the required joint strength is limited to that needed to hold the parts in place during handling, machining, and use. Often, a neat, tight appearance is the main criterion for acceptance. The total adhesion of adhesive (glue) to any wood is the sum of the varying magnitudes of the adhesive forces operating on the cut (excised and exposed cellulose and lignin) and uncut (unexcised and extractive coated) surfaces of the cell walls exposed during machining [5,6]. In addition to any mechanically related factors, the major contributors to adhesion are van der Waals forces, hydrogen bonding, and covalent bonds [7,8]. Since these forces operate over very short distances, 0.2 to 0.3 nm (2 to 3 Å), the adhesive must be in intimate contact with the wood for adhesion to take place. Stated differently, the adhesive must flow onto and wet the wood (principally the cellulose and to some degree the hemicellulose and lignin) before the adhesive reaches a rigid state. The adhesive may penetrate three to five cells deep into the wood and, depending on the wood density and type of adhesive, fill the open cells or coat their inner walls (Fig. 2). Once adhesion takes place, failure in joints is invariably a cohesive failure in either the adhesive or wood. Naturally, more area of cellulose and lignin is exposed by cutting wood along the grain (the edges and faces of lumber) than across the grain (Fig. 1). Hardwood fibers are 0.8 to 2.2 mm in length versus 3 to 6 μm in wall thickness. Softwood tree tracheids are up to 7 mm in length and about 35 μm in diameter. Thus in edge- and face-type wood glued with most industrial rigid adhesives, the adhesives are stronger than the wood when stressed in shear parallel or tension perpendicular to the grain. (For some hardwoods, this stress equals 13.8 MPa.) The glue-line thickness for these wood bonds is 50 to 250 μm (0.002 to 0.010 in.). It can be expected that failure of adequately bonded glue joints of this type will result in the cohesive failure from wood rather than from failure of the adhesive. On the

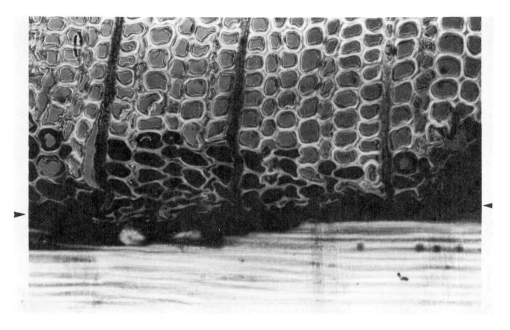

Figure 2 Fluorescent light micrograph of adhesive penetration in softwood. Arrows indicate bond-line region. (Courtesy of Robert L. Krahmer, Oregon State University, Corvallis, Oregon.)

other hand, it is practically impossible to make end grain-to-end grain butt joints sufficiently strong or permanent to meet load-carrying service requirements with present water-based adhesives and techniques. With the limited wood cellulose exposed to crosscut ends of wood (fiber or tracheid wall thicknesses), no more than 25% of the tensile strength of wood tested parallel with the grain can be developed, even with the best of techniques [9]. When the ends of wood are cut at slopes that longitudinally expose more cellulose, varying tensile strength parallel to grain compared to solid wood can be attained: 90% for end joints at 1:20 ratios, 80% for end-joint slopes of 1:16, and 60% for end-joint slopes of 1:6 [9,10]. The strengths attained in gluing individual wood fibers, fiber bundles, particles, flakes, and strands depend on the surface areas bonded. The surface area bonded depends on the adhesive type, the adhesive amount applied, and the adhesive distribution process. Bond formation for any wood consists of the following steps [3]:

1. Spread and distribute the adhesive evenly on the surface to be joined.
2. Bring the surfaces into intimate contact, often by applying pressure to mate the surfaces and facilitate adhesive flow and wetting.
3. Immobilize the adhesive by evaporation, absorption, and adsorption.
4. Apply heat to induce flow and to accelerate immobilization of the adhesive flow and wetting, and, in the case of hot melt, apply cooling to accelerate immobilization of adhesive.
5. Allow wetting and penetration of the adhesive into the cell wall (required for a fully waterproof bond).

6. Allow completion of the chemical–physical reactions that convert the polymer into a rigid solid.

IV. ADHESIVE BACKGROUND

The compositions of adhesives classified as resins and polymers (natural and synthetic) are myriad. To be a satisfactory adhesive, a material must be capable of both wetting the wood substrate (polar attraction) and being converted into a rigid solid through either evaporation of the solvent, physical cooling of a thermoplastic-type adhesive, or a chemical reaction that may require a catalyst and/or heat to complete. Synthetic resin and polymer adhesives need to have enough polymerization to produce adequate mechanical properties; as these adhesives increase to 500 molecular weight or above, useful cohesive strength, impact strength, and so on, are developed [11]. In the process of manufacturing synthetic adhesive resins, a range or distribution of molecular weights is produced. Similarly, all polymer molecules in adhesives are not the same size. The average molecular weight and molecular weight distribution of the adhesive resins and polymers significantly influence adhesive performance. The presence of low-molecular-weight components influences flow and wetting properties and enhances the bonding of higher-molecular-weight components. Some data suggest that the effective submicroscopic penetration in wood cellulose of polymers occurs at a molecular weight of 1000 or less [12,13], which would be about 10 cross-linked phenolic nuclei of a resol resin. A phenol–formaldehyde (PF) resol with a number-average molecular weight of 1700 has been estimated to be 5 nm (50 Å) in length [14]. Such penetration relates to the size, viscosity, mobility, and compounding of the resin–polymers as well as to the size and condition of the wood fiber pits, lumens, and vessel pores (Table 1). In structural applications, the cured adhesive must equal or exceed the substrate in strength. For all applications, the cured adhesive must be evaluated for four environmental degradation conditions: heat energy (pyrolysis) for thermal–chemical degradation; chemical energy (hydrolysis) for moisture content changes; physical energy for swelling and shrinkage stresses; and biological for bacteria and fungi degradation, especially for natural-based adhesives.

Table 1 Relative Comparison of Components Influencing Wood-Adhesive Bonding

Component	Distance		
	μm	nm	Å
Adhesive force	0.0002 – 0.0003	0.2 – 0.3	2 – 3
Fiber wall "pore" diameter	0.0017 – 0.002	1.7 – 2.0	17 – 20
Resin molecular length	0.0015 – 0.005	1.5 – 5.0[a]	15 – 50[a]
Bordered pit diameter	0.2	200	2000
Tracheid lumen diameter	4 – 25	—	—
Tracheid fiber diameter	35 ±	—	—
Vessel diameter	50 – 300	—	—
Proteinaceous extenders	<50	—	—
Lignocellulosic fillers	50 – 150	—	—
Glue-line thickness	50 – 250	—	—

[a]All resol molecules of a highly advanced plywood-type resin elute through a 25-nm (250-Å) pore-size Ultrahydrogel 250 GPC column.

Table 2   Some Mechanical and Physical Properties of Two Adhesives Compared to Wood

Adherend	Estimated surface energy[a] (dyn/cm)	Density (g/cm^3)	Thermal expansion coefficient[b] ($\Delta L/L/°C$)	Tensile strength (MPa)	Tensile stiffness, MOE (MPa)
Phenol–formaldehyde	51	1.4	45×10^{-6}	55	7600
Urea–formaldehyde	61	1.5	32×10^{-6}	48	9000
Wood	20–70	0.2–1.2	—	—	—
Parallel to grain[c]	—	—	4×10^{-6}	96	8300
Perpendicular to grain[c]	—	—	35×10^{-6}	5	—

[a]These energy values are estimated from critical surface tension of liquids on this solid.
[b]Coefficient of linear expansion (defined as increase in length per unit length for a temperature rise of 1°C) differs in the three directions of wood [15,16].
[c]These values are medium values for wood. Actual values for any species may be as much as 50% higher or lower.
Source: Ref. 6.

Fortunately, a number of materials compare favorably with wood in these requirements (Table 2), and they are available commercially at a reasonable cost.

V. DEVELOPMENT OF ADHESIVES

Early wood adhesives were natural-product based. Animal, casein, soybean, blood albumin, and starch glues were commonly used in quantity in the United States as late as the 1960s. Starches are still used extensively for paper and other industrial applications. Although some natural adhesives, such as casein and modified blood, formed a moderately water-resistant bond, a truly waterproof adhesive bond was not available to the wood industry until the advent of various coal tar or petroleum-based synthetic resins in the mid-1930s [17]. The versatile synthetic adhesives, known generically as urea–, melamine–, phenol–, and resorcinol–formaldehyde resins, dramatically increased the types and amounts of glued products that could be manufactured. In many respects, today's science of adhesion grew from World Wars I and II when evaluations were made of the adhesives and gluing processes that were used in constructing wooden aircraft and other war materials [17]. Poly(vinyl resin) emulsions and hot melts were later additions to the synthetic adhesives. The copolymer forms of these emulsions and melts are continually being upgraded in durability. Their quick tack and rapid setting (or cure) have made them widely used as adhesives for assembly gluing. Various combinations of emulsion polymers and hot melts with urethanes and phenolics (for nonreversible flow) are used increasingly for panel and specialty applications. During the 1980s, urethane (isocyanate) polymers were used in bonding boards made of particles, wafers, and strands.

Some natural adhesives have been of great historical importance to the adhesive industry. While they have been replaced in most applications by petroleum-based adhesives, which show improved performance or better economics, natural adhesives remain important in certain specialty areas due to their unique curing and bonding characteristics [18]. In 1991 in the United States, natural adhesive (casein, animal, and blood) consumption as wood binders was less than 5 kilotonnes (kt, metric), solids basis. However, as adhesives, natural renewable raw materials are receiving considerable attention and some use. The renewable materials include lignins, tannins, and proteins. Discussion of these very interesting renewable resources will of necessity be left to other work. Another area not covered is the growing worldwide use of cement (1800 kt/yr) and gypsum (1350 kt/yr) as a matrix for making cement-bonded particleboard and gypsum-bonded fiberboard and particleboard [19,20].

VI. TYPES OF ADHESIVES

A. Animal Adhesive

Animal adhesive is a natural high polymer of organic colloid derivation from collagen—a protein constituent of animal skins, connective tissue, and bones. The protein is derived from the simple hydrolysis of collagen covalent bonds and is essentially composed of polyamides of certain α-amino acids [21,22]. Both of the two types of animal adhesive—hide and bone—have a polydisperse system containing mixtures of similar molecules of widely differing molecular weights (10,000 to 250,000). As wood adhesives, their pH is fairly neutral to slightly acidic (>5), similar to wood. They are soluble only in water and are insoluble in oils, waxes, organic solvents, and absolute alcohol; however, they can be emulsified [21,22]. Animal adhesive can be produced in a number of forms, including the dry granulated products for use as hot adhesives, cold liquid adhesives for immediate use, and a variety of cake or jelly forms. Some liquid types are formulated between 35 and 65% dry solids, with an application viscosity of 3000 to 5000 mPa·s (cP) at room temperature. Many liquid forms contain clay or calcium carbonate for improved adhesive filming properties, as well as other modifiers as needed (fungicides, plastizers, etc.).

Animal adhesive acceptance stems from its unique ability to deposit a tacky viscous film from a warm water solution, which, upon cooling a few degrees passes to a firm gelled state, thereby providing an immediate, moderately strong initial bond. Gelation involves both intramolecular and intermolecular reorientation upon cooling the system, involving primary and secondary forces [22]. Subsequent natural or forced-air drying provides a permanent, strong resilient bond between polar materials, such as wood in a dry–ambient temperature interior state [17]. On the negative side, animal adhesives are subject to enzymatic and bacterial attack and are not durable under severe moisture exposure (>85% relative humidity for extended periods of time). Although animal adhesive consumption as a wood adhesive has essentially been replaced by petroleum-based adhesives, these tailored products have found applications in a wide range of items, including furniture, toys, sporting goods, musical instruments, and other household goods. There were only two producers in the United States in 1990 [22].

B. Casein Adhesives

Casein, the main protein obtained from milk (about 3% concentration), is produced from skimmed milk by precipitation with acid (e.g., lactic, sulfuric, or hydrochloric) [23]. Casein has been characterized as a globular polypeptide of 33,000 to 375,000 molecular weight [18,24]. The casein adhesive is produced by washing out the acid and dissolving the casein in calcium hydroxide/sodium hydroxide solutions with a resulting alkalinity varying between pH 9 and 13. For wood adhesives, the solutions are spray dried, and the adhesive is prepared in various particle sizes (180 to 500 μm). The powdered adhesive is rewetted (ca. 2 parts water to 1 part casein adhesive) and applied at a viscosity of about

4500 mPa·s (cP). To overcome mold and fungi attack, sodium pentachlorophenate, at about 3% dry weight, historically was added as a fungicide. However, adding toxic material to any product, including adhesives, is increasingly being regulated by governments, and these restrictions have had an unfavorable impact on such fungicides. Casein adhesives have been described as the first interior structural adhesive. They are cold setting (>15°C) and relatively successful at gap filling. These characteristics give them an advantage in situations where heat and pressure may be objectionable, as in door manufacturing. In the area of dry bond strength, most synthetic resins do not surpass casein adhesives. Caseins have been used in early airplane construction, furniture, plywood, flush doors, timber laminated beams (glulam), and numerous other applications. In the United States, they reached their greatest use during the 1960s–1970s (all imported) and are being almost totally replaced with synthetic adhesives. No casein has been produced in the United States since the 1950s.

C. Blood Adhesives

Historically, animal blood was used for an adhesive in fresh liquid form because it spoils in only a few hours [25]. About 1910, techniques were developed for drying whole animal blood in commercial quantities without denaturing its protein content, thus maintaining its water solubility. Its usefulness as a wood adhesive is classified by its solubility: 20 to 40% solubility for cold-press adhesives and 80% + solubility for hot-press adhesives. Virtually all the proteins in animal blood can be dispersed into useful adhesive form; thus dried blood is essentially 100% active protein [26]. All adhesive-grade blood from beef and hogs and to a lesser degree from sheep is spray dried. To develop the full adhesive potential, the dried blood is initially redissolved in water and subjected to alkaline dispersion at the mill site.

The blood protein molecules of animal blood have a molecular weight of 69,000 and are highly polar, complex, and heavily coiled [18,26]. Both calcium compounds and soluble silicates are frequently added to blood adhesives with a strong base to yield balanced dispersing systems of optimum effect on the blood protein [26]. The systems are buffered to prevent too rapid hydrolysis of the blood protein in alkaline solutions for processing. The blood content of the mixed adhesives in the mills will normally be less than 10%, with a mix viscosity of 5000 to 15,000 mPa·s (cP). When phenolic resins are added at about 10% of the dry adhesive weight, they impart semipermanent mold resistance to blood-based adhesives [26]. The hot-press varieties of blood adhesives played an important role in the early development of more durable adhesives, with particularly extensive use in plywood adhesives. The inherent variability of dried blood as a raw material and the improvements in performance and economics of synthetic adhesives have reduced the use of blood adhesives to very specialized applications in the United States [18]. One such application is the use of blood in PF adhesives that are foam extruded. The foam expands the adhesive volume four- to sixfold, resulting in a 20 to 30% adhesive reduction with equal bond quality [27]. In the United States, there is only one company manufacturing soluble-grade blood suitable for wood adhesives, but there are other manufacturing units available capable of major volumes if the economics warrant such conversion. In 1990 in the United States, less than 1 kt (solids basis) of soluble animal blood was used as a wood adhesive or adhesive additive.

D. Urea–Formaldehyde Resins

Urea–formaldehyde (UF) resins are the world's most widely used wood-product adhesive. The largest volumes of UF resins are used to bond particleboards and fiberboards. A smaller volume is used to bond interior-grade decorative and hardwood plywood. Since 1981, more particleboard has been produced worldwide than plywood [28]. UF resins were developed in the 1930s, and in 1990 their worldwide consumption was estimated to be >2500 kt, solids basis (<30% in North America).

As a thermoset adhesive, the resin is based primarily on formaldehyde (F) and urea (U) raw materials formulated in ratios of 1.1:1 to 2.0:1 [2]. Because of formaldehyde emission problems in the manufacture and use of UF bonded wood products, resins with the lowest molar ratio possible (1.1 to 1.2:1) are now commonly used. The prepolymer stage of the resin formulations (alkaline condensation) consists of methylol ureas which are further processed (acid condensation reactions where two polymer structures chemically combine, yielding water as a by-product) into long-chained molecules with methylene–ether bridges [8,29]. The resin formulations are neutralized to stop or slow the reaction, then they are cooled as aqueous solution adhesives (ca. 65% solids) or spray dried for powders. One advantage that UF resins have over phenolic resins is that UF resins continue to be substantially soluble in water during the resinification process [2]. There are conflicting thoughts on the exact nature of the resultant resin molecule configuration. Their molecular weights have been determined to be 1000 and 10,000, with a wide distribution of molecular size [8,29]. The higher values are sometimes considered to be due to association through hydroxyl groups. The reactive end groups of all these polymeric forms are predominately methylol groups [29]. Free methylol groups have a polar affinity toward cellulosic materials, like wood [2].

Resins with low F/U ratios (1.1 to 1.2:1) typically have lower water resistance (durability), lower strength, and a slower curing rate. To overcome these difficulties, improved prepolymers are formulated or various urea derivatives, melamine, or reactive scavengers (e.g., resorcinol-like tannins) are added [29]. UF resins are usually the cheapest of the synthetic resins and natural adhesives. Their price is dictated essentially by the price of fertilizer (as a nitrogen source) and to some degree by the international marketing of methanol, the raw material for formaldehyde.

The precondensed neutral UF resins are hardened or cured by reducing their pH. This reduction is accomplished by the action of an acid catalyst addition. Two typical catalysts are ammonium chloride and ammonium sulfate. Application of elevated temperatures and migration of water in the resin accelerate the curing, but curing of some formulations used for plywood can be accomplished by cold pressing (>15°C). For bonding particleboard, aqueous UF solutions, usually with catalysts, are suitable as manufactured (typically, 50 to 65% UF resin solids solutions). For plywood bonding, the aqueous UF resins are mixed with lignocellulosic or earthen fillers, proteinaceous extenders, catalysts, and water to a usable viscosity (1500 to 3000 mPa·s). The resultant plywood adhesive mix may have only about 28% UF resin solids, depending on the durability desired. UF resins are used strictly for interior applications because they are susceptible to hydrolytic degradation in the presence of moisture, especially at moderate to elevated temperatures. Degradation occurs slowly at 40°C and rapidly at 60°C [29].

UF resins in their cured state are nontoxic. Urea is nontoxic, but free formaldehyde is highly reactive and combines easily with proteins in the human body [8]. Regulations on the amount of free formaldehyde allowed over time are very restrictive in Europe and North America for UF-bonded wood products—0.1 to 0.3 part per million, depending on the European or U.S. test protocol [30]. Nevertheless, UF resins are some of the most versatile adhesives ever discovered, providing a multitude of products that would be hard to duplicate in cost, color, and variety. They are water resistant, light colored, and mold and fungi proof. Also, they blend with other resins and can be cured electronically.

E. Melamine–Formaldehyde Resins

Melamine–formaldehyde (MF) resins were first produced commercially in the United States in the 1930s, and they were used as particleboard binders in the 1950s [29,31,32]. These amino resins differ from UF resins in that the ratio of formaldehyde to melamine is 2.5 to 3.5:1 [2,32]. Both reactions use urea and formaldehyde as starting raw materials, and both reactions are carried out initially in alkaline pH. The intermediate products are soluble in water and condense into three-dimensional networks of insoluble and infusible adhesives. The melamine resin adhesives require acid catalysts and heat (>60°C) to cure. They are marketed principally as spray-dried powders. Once cured, MF resins are colorless, boiling-water resistant (up to 24 h), and resistant to molds and fungus [33]. MF resins as wood binders are often blended with UF resins to reduce MF costs and to upgrade UF durability [8,32]. They are applicable to radio-frequency curing as well.

The costs of MF resins are substantially greater than UF because of more technical processing, spray drying, and low demand [2,31]. Melamine, also called 2,4,6-tri-amino-1,3,5-triazine, is a cyclic compound containing nitrogen in the ring [31]. As a starting raw material, the molecular weight of melamine is 126 versus 60 for urea. The resultant resin can have a molecular weight of a few hundred to a few thousand. Like the UF resins, the methylol groups of MF resins may react with the primary hydroxyl groups of wood cellulose, yielding a very excellent bond [31,32]. In the United States, the MF resins are used more for laminating film and saturating paper than for binding wood. Worldwide, approximately 250 kt of MF resins are used as wood binders. Of that total, less than 2% is used in North America; the vast majority is consumed (solids basis) in Europe and Asia for bonding particleboard and plywood.

F. Phenol–Formaldehyde Resins

Phenol–formaldehyde (PF) resin is another synthetic condensate resin marketed commercially since the 1930s. In 1990, their volume as a wood adhesive (ca. 700 kt, solids basis) was second only to UF resins worldwide, with about 40% consumed in North America. They are the primary binder for exterior applications: construction-grade plywood, laminated veneer lumber, oriented strandboard (waferboard), and others. The PF resins are formulated into two types: one containing molar deficiency of formaldehyde to phenol in the presence of acid (novolak) and the other with molar excess of formaldehyde to phenol in the presence of alkali (resol). The novolak resin has phenol as an end group and does not have reactive methylol groups unless some hardening (cross-linking) agents are added. In other words, these resins are thermoplastic until a cross-linking formaldehyde donor is added, such as hexamethylenetetramine. The novolak molecular weight ranges between 200 and 1000 [8,34]. Because of their acidity and initial thermoplastic chemistry, novolak resins have experienced limited applications in wood bonding. One use is for molded wood-particle products and another is for waferboard face resin. Novolak resins are usually sold in powder form.

The resol resins, both powders and aqueous forms, have reactive methylol groups that condensate into larger molecules when heated without the addition of a hardening cross-linking agent. The heat-cured resol resin is three-dimensional, with chemical attachments to wood surfaces that are virtually indestructible. The molecular weight of resol resins may vary from a few hundred to 30,000 [34,35]. In the United States, the resol resins used to bond construction plywood are some of the highest-molecular-weight resoles and are water insoluble except for the presence of about 10 to 12% alkalinity on a solids basis. Resol resins, both powders and aqueous forms, have extensive markets for wood bonding. The typical molar ratio of F/P for plywood resol resins is 1.9 to 2.2:1, the F/P ratio for strandboard (waferboard) resol resins is near 2.5:1. In the United States the aqueous forms for plywood are between 42 and 45% nonvolatile solids and those for strandboard are 45 to 55% nonvolatile solids. For durability in bonding strength, the resol resins are most suitable in the pH range 10 to 11.5. The aqueous forms are shipped with viscosities ranging from 200 to 1200 mPa · s (cP). For strandboard manufacture, the resol resins are used as delivered to the mill with no other additive; thus the application viscosity is the same as the delivered viscosity and toward the lower end of the range (200 to 400 mPa · s). But for plywood, the resol resin is mixed with proteinaceous extenders (<50 μm size), lignocellulosic fillers (50 to 150 μm size), earthen minerals, and alkali to increase the viscosity, prevent overpenetration of the resin, and reduce the resin solids to about 28 to 31% in the adhesive mix [2,36]. The application viscosity of the plywood adhesive mixes is 3000 to 10,000 mPa · s, depending on the application method. The durability of cross-linked phenolic resins is unquestionable, being waterproof or boilproof in accelerated aging tests [17,33,36]. PF resins are not gap filling and are brittle unless altered by additives. Increasing attention is being given to additives (latent catalysts, faster reacting aldehydes, etc.) to resol resins, which allow bonding wood at moisture contents between 8 and 16%.

G. Resorcinol–Formaldehyde Resins

Resorcinol–formaldehyde (RF) resins began to be used in 1943 in the United States. RF resins cost 5 to 10 times more than PF resins because the RF resins are more complex to manufacture and have a smaller market share. Today, the majority of these adhesives manufactured for wood bonding are modified with phenol to make them cheaper. Such resins can be made by blending two separate resins or by starting the phenol–formaldehyde cook with an excess of formaldehyde and then adding resorcinol [37]. They are shipped in 45 to 55% nonvolatile solids form (aqueous-alcohol solutions, such as ethanol). Because the presence of two hydroxyl groups on its six-carbon nucleus (versus one on phenol), resorcinol reacts with formaldehyde, or with various phenol–formaldehyde resols, much faster than phenol [11,37]. With the introduction of a methylol group onto the ring, the activity of the other nuclear positions increases. Resorcinol–formaldehyde resins of between 0.5 and 0.7 mol of formaldehyde per mole of resorcinol are stable for several months [37].

The molecular weight of RF resins is less than 2000, and thus is good for wetting, penetrating, and bonding wood. The advantage of these resins is that they can be cured at ambient temperatures (>15°C) over a period of hours. With elevated temperatures, the cure times can be reduced drastically. The introduction of one-half or more of phenol with resorcinol on a mole basis will reduce the costs but still result in very durable exterior bonds with most woods [2]. However, the greater the phenol amount, the slower the curing rate. At the point of use, some additional formaldehyde is provided and the resin is converted to a highly cross-linked adhesive. The added formaldehyde donor is usually in the form of paraformaldehyde blended with a filler, such as wood flour or nutshell flours (50 to 150 μm size), and other additives to control thixotropic consistency of the two-component adhesive mix. The pH value of resorcinol–formaldehyde resins is similar to that of novolaks (5 to 7), but the pH of phenol–resorcinol–formaldehyde types is 8.5 to 9.0. When resorcinol reacts with formaldehyde, the reaction rate is affected by the R/P/F molar ratio, solution concentration, temperature, pH, presence of catalysts, and amount of certain alcohols [37]. Resorcinol resins for bonding wood are used for specialty applications such as bonding heavy timber laminated beams (glulam) and other special products that must meet military specifications. Resorcinol is an excellent additive to other adhesives and is used more extensively for nonwood applications, such as rubber tire cord bonding. The volume consumed for wood bonding in the world is estimated to be 6 kt, solids basis.

H. Diisocyanates (Urethanes)

Structural variations of urethane links (–O–CO–NH), formed by the reaction of an alcohol with an isocyanate (–N=C=O) group, are possible, depending on the properties desired [11]. The polarity and hydrogen-bonding capability of the urethane group give these polyurethane adhesives enhanced adhesion to a variety of surfaces, including wood and metal, by urethane bridges and polyurea bonding. The polymeric diisocyanates (PMDI) can continue to cross-link by reaction with themselves in the presence of absorbed moisture. The PMDI are sold as 100%-reactive adhesive liquids and are used as catalysts in vinyl emulsions. The reactivity of the isocyanate to atmospheric moisture requires the shielding of these reactive groups by using inert solvents or hydrophobic polymers as application vehicles. "Blocked" isocyanates have been developed which can be applied in aqueous suspension and which are capable of generating the free isocyanate group on heating. The chemical reaction of the wood's hydroxyl groups with isocyanate groups has led to increasing use of PMDI (diphenylmethane-4, 4′-diisocyanate) as an adhesive binder for structural waferboard and oriented strandboard products in North America and particleboard in Europe. The weight-average molecular weight of these PMDI is 365. The reaction of these PMDI with water is very slow under 50°C but is accelerated at a higher temperature and in the presence of alkalies, tertiary amines, or metal compounds [38]. PMDI use for wood binder in 1990 was estimated to be <50 kt, worldwide.

I. Poly(vinyl acetate) and Derivatives

Poly(vinyl acetate) (PVAc) is a synthetic, thermoplastic, fast-setting adhesive. Because of its bacteria resistancy, PVAc began during the 1940s to replace the natural adhesive made from hide and casein. PVAc is most widely used in the form of a dispersion of resin solids in water. In the emulsion form, these linear polymers (100,000 to 500,000 MW) can be controlled in particle size (0.7 to 1.5 μm) [2,39]. Most products used in the manufacture of adhesives are supplied in 55% solids content of varying viscosity grades (500 to 4000 mPa · s). They have excellent compatibility with plasticizers, wetting agents, solvents, thickeners, and other modifying agents, which allows a wide array of individual compounding. Often, poly(vinyl alcohol) (PVA) is added as a protective colloid to provide special properties to the homopolymer PVAc. Increasingly, copolymer forms are offered in the trade to decrease stress-related creep and to enhance exterior durability. On a molecular basis, these copolymer additives cross-link the polymer chains.

Poly(vinyl alcohol) (PVA) is produced by replacing the ester groups of PVAc and other polyvinyl esters with hydroxyl groups [39]. PVA can be produced at 2000 to 120,000 MW. The extent of replacement of the acetate groups can be controlled and determines the grade of PVA adhesive. Compared to PVAc, PVA has unique properties (hydroxyl grouping) and greater chemical reactivity (hydrogen-bonding potential). One useful reaction is condensation with aldehydes to form acetal resins. PVA is readily compounded with various materials to modify properties and may be solubilized with a wide array of chemicals in the presence of acid catalysts. In Japan, considerable quantities (ca. 7.5 kt, solids basis, in 1990) are consumed with isocyanate catalyst as a cross-linked polymer in bonding plywood and other wood products.

Poly(vinyl acetate) and poly(vinyl alcohol) are important raw materials for wood adhesives. They are frequently used in assembly adhesives and, increasingly, as binders for parallel- and cross-laminated wood products. They impart high bond strength, fast set, and colorless bond lines, and they lack the environmentally hazardous formaldehyde. Water and heat resistance are improved by special copolymer compounding by the adhesive manufacturer. Total worldwide consumption of PVAc and its derivatives as wood binders are estimated to be >150 kt, solids basis.

J. Hot-Melts

During the 1950s hot melts based on synthetic polymers began to appear commercially. Hot-melt adhesives for wood bonding are thermoplastic compounds, solid at room temperature, which become sufficiently fluid at elevated temperature to ''wet'' substrates. Hot-melt adhesives form bonds by resolidification upon cooling. They usually contain no water or volatile solvent. They do contain petroleum-based waxes (ca. 20 wt %) to reduce the melt viscosity of the hot-melt blends and a tackifer resin (ca. 25 to 50 wt %), such as rosin derivatives or a nonheat reactive substituted phenolic resin, and a semicrystalline polymer such as ethylene–vinyl acetate copolymer (EVAc) (ca. 25 to 35 wt %) or a polyamide [40–42]. Other materials (urethanes, etc.) are compounded for nonreversible flow or other specific purposes.

Rapid bond development has been the driving force behind the expanded use of hot melts, particularly in furniture assembly adhesives, where 11% of the adhesive products are hot melts [42]. Typical application temperatures are 160 to 220 °C. The EVAc is used more in Japan and Europe and the polyamide is used more in the United States [41]. The polyamide-based systems are faster setting, but they cost more. Polyamides used as wood adhesives are condensation thermoplastic polymers—long chain, vegetable oil-derived dibasic acids, and short-chain diamines—based on dimer acid [43,44]. Dimer acid is a 36-carbon-atom dibasic acid obtained by polymerization of unsaturated fatty acids, typically oleic or linoleic [43,44]. The purity of the dimer acid affects the linear nature of the polyamide and its MW as well as its cost. By control of constituent MW (5000 to 15,000), the type of application equipment and adhesive properties can be determined. The low-MW types are applied by hand-held equipment and the high-MW types by automated equipment. Hot melts bond well to porous substrates and polar surfaces, such as wood, and are increasing in use. Worldwide consumption of hot melts as a wood adhesive is estimated to be about 30 kt.

VII. SUMMARY

Adhesive technology has advanced to a sophisticated, semiquantitative science. In this chapter we review only a few of the major adhesive types used to bond wood. There are many more types and a multitude of combinations to satisfy an enormous array of needs. The use of adhesive-bonded wood products has made possible affordable housing and household goods for over 200 years. While the past contributions of wood adhesives to the well-being of humanity has been tremendous, the future contributions will be more extensive. The challenge of the future is to use wisely the renewable wood resources and the nonrenewable synthetic adhesives to enhance the quality of life of the world's peoples [45]. Research should be fostered to use and discover acceptable renewable adhesive binders. Combining thermoset resins with thermoplastic polymers resulting in "alloy" and copolymer adhesives is a science that should be encouraged. The resultant adhesives can offer improved toughness, elongation, elasticity, and resiliency and may well be superior to the individual resin–polymer adhesives. These properties are attributed to chemical reactions as well as to the morphology of the cured matrix adhesive system. Archaeologists and modern scientists have not done a thorough job of defining and documenting the history of past adhesives. Industry associations, companies, and individuals versed in wood adhesives should encourage development of centers and repositories on wood adhesives on each continent.

REFERENCES

1. B. H. River, C. B. Vick, and R. H. Gillespie, in *Treatise on Adhesion and Adhesives*, Vol. 7 (J. D. Minford, ed.), Marcel Dekker, New York, 1991, pp. 1–230.
2. F. F. P. Kollmann, E. W. Kuenzi, and A. J. Stamm, *Principles of Wood Science and Technology*, Vol. II, *Wood-Based Materials*, Springer-Verlag, New York, 1975, pp. 34–93, 100.
3. T. Sellers, Jr., J. R. McSween, and W. T. Nearn, Gluing of eastern hardwoods: a review, *General Technical Report SO-71*, U.S. Department of Agriculture Forest Service, Mississippi Forest Products Laboratory, Mississippi State, Mississippi, 1988.
4. International Organization for Standardization, Adhesives vocabulary, *Standard 6354-1982*, American National Standards Institute, New York, 1982.
5. G. Kitazawa, A study of adhesion in the glue lines of twenty-two woods of the United States, *Technical. Publication 66*, New York State College of Forestry at Syracuse University, Syracuse, N.Y., 1946.
6. J. D. Wellons, in *Adhesive Bonding of Wood and Other Structural Materials* (R. F. Blomquist, A. W. Christiansen, R. H. Gillespie, and G. E. Myers, eds.), Materials Research Laboratory, Pennsylvania State University, University Park, Pa., 1983, pp. 85–134.
7. J. E. Marian and D. A. Stumbo, *Holzforschung 16*(5/6): 134, 168 (1962).
8. A. Pizzi, ed., *Wood Adhesives: Chemistry and Technology*, Vol. I, Marcel Dekker, New York, 1983.
9. Forest Products Laboratory, *Wood Handbook: Wood as an Engineering Material*, Agriculture Handbook 72, U.S. Department of Agriculture Forest Service, Washington, D.C., 1987, pp. 9–17, 10–8, 10–9.
10. H. Sasaki, in *Concise Encyclopedia of Wood and Wood-Based Materials* (A. P. Schniewind, ed.), Pergamon Press, Oxford, 1989, pp. 163–165.
11. R. V. Subramanian, in *Adhesive Bonding of Wood and Other Structural Materials* (R. F. Blomquist, A. W. Christiansen, R. H. Gillespie, and B. E. Myers, eds.), Materials Research Laboratory, Pennsylvania State University, University Park, Pa., 1983, pp. 137–186.

12. H. Tarkow, W. C. Feist, and C. F. Southerland, *Forest Prod. J. 16*(10): 61 (1966).
13. A. J. Stamm, *Wood and Cellulose Science*, Ronald Press, New York, 1964, pp. 333–406.
14. M. Duval, B. Block, and S. Kohn, *J. Appl. Polymer Sci. 16*: 1585 (1972).
15. P. Koch, in *Utilization of the Southern Pines*, Vol. I, U.S. Department of Agriculture Forest Service, Washington, D.C., 1972, p. 382.
16. R. Summitt and A. Sliker, in *CRC Handbook of Materials Science*, Vol. IV (C. T. Lynch, ed.), *Wood*, CRC Press, Boca Raton, Fla., 1980, pp. 38–39.
17. R. A. G. Knight, *Adhesives for Wood*, Chemical Publishing Company, New York, 1952.
18. W. D. Detlefsen, in *Adhesives from Renewable Resources*, ACS Symposium Series 385 (R. W. Hemingway, A. H. Conner, and S. J. Branham, eds.), American Chemical Society, Washington, D.C., 1989, pp. 445–452.
19. A. A. Moslemi, *Chemtech 18*(8): 504 (1988).
20. A. A. Moslemi, *Panel World 32*(2): 34 (1991).
21. J. R. Hubbard, in *Handbook of Adhesives*, 2nd ed. (I. Skeist, ed.), Van Nostrand Reinhold, New York, 1977, pp. 139–151.
22. R. L. Brandis, in *Handbook of Adhesives*, 3rd ed. (I. Skeist, ed.), Van Nostrand Reinhold, New York, 1990, pp. 123–134.
23. C. N. Bye, in *Handbook of Adhesives*, 3rd ed. (I. Skeist, ed.), Van Nostrand Reinhold, New York, 1990, pp. 135–152.
24. H. K. Salzberg, in *Handbook of Adhesives*, 2nd ed. (I. Skeist, ed.), Van Nostrand Reinhold, New York, 1977, pp. 158–171.
25. A. L. Lambuth, in *Wood Adhesives: Chemistry and Technology*, Vol. II (A. Pizzi, ed.), Marcel Dekker, New York, 1989, pp. 1–29.
26. A. L. Lambuth, in *Handbook of Adhesives*, 2nd ed. (I. Skeist, ed.), Van Nostrand Reinhold, New York, 1977, pp. 181–191.
27. S. Nylund, in *Structural Wood Composites: New Technologies for Expanding Markets*, Proceedings 47359, Forest Products Research Society, Madison, Wis., 1988, pp. 120–126.
28. Staff report, *Wood-Based Panels: Proceedings of the Expert Consultation*, Food and Agriculture Organization of the United Nations, Rome, 1987.
29. J. M. Dinwoodie, in *Wood Adhesives: Chemistry and Technology*, Vol. I (A. Pizzi, ed.), Marcel Dekker, New York, 1983, pp. 1–57.
30. W. J. Groah, J. Bradfield, G. Gramp, R. Rudzinski, and G. Heroux, *Environ. Sci. Technol. 25*(1): 117 (1991).
31. M. Savla, in *Handbook of Adhesives*, 2nd ed. (I. Skeist, ed.), Van Nostrand Reinhold, New York, 1977, pp. 424–433.
32. I. H. Updegraff, in *Handbook of Adhesives*, 3rd ed. (I. Skeist, ed.), Van Nostrand Reinhold, New York, 1990, pp. 341–346.
33. R. H. Gillespie, *Forest Prod. J. 15*(9): 369 (1965).
34. F. L. Tobiason, in *Handbook of Adhesives*, 3rd ed. (I. Skeist, ed.), Van Nostrand Reinhold, New York, 1990, pp. 316–340.
35. T. Sellers, Jr., and M. L. Prewitt, *J. Chromatogr. 513*: 271 (1990).
36. T. Sellers, Jr., *Plywood and Adhesive Technology*, Marcel Dekker, New York, 1985.
37. R. H. Moult, in *Handbook of Adhesives*, 2nd ed. (I. Skeist, ed.), Van Nostrand Reinhold, New York, 1977, pp. 417–423.
38. Staff report, Rubinate M (polymeric MDI), *Material Safety Data Sheet 2285*, ICI Polyurethane Group, Mantua Grove, West Deptford, N.J., 1987.
39. A. E. Corey, P. M. Draghetti, and J. Fantl, in *Handbook of Adhesives*, 2nd ed. (I. Skeist, ed.), Van Nostrand Reinhold, New York, 1977, pp. 465–483.
40. M. Huntley, *Adhesives Age 34*(12): 28 (1991).
41. J. D. Domine and R. H. Schaufelberger, in *Handbook of Adhesives*, 2nd ed. (I. Skeist, ed.), Van Nostrand Reinhold, New York, 1977, pp. 495–506.

42. E. F. Eastman and L. Fullhart, Jr., in *Handbook of Adhesives*, 3rd ed. (I. Skeist, ed.), Van Nostrand Reinhold, New York, 1991, pp. 408–422.

43. R. D. Dexheimer and L. R. Vertnik, in *Handbook of Adhesives*, 2nd ed. (I. Skeist, ed.), Van Nostrand Reinhold, New York, 1977, pp. 581–591.

44. C. Rossitto, in *Handbook of Adhesives*, 3rd ed. (I. Skeist, ed.), Van Nostrand Reinhold, New York, 1990, pp. 478–498.

45. J. Perlin, *A Forest Journey: The Role of Wood in the Development of Civilization*, Harvard University Press, Cambridge, Mass., 1991.

38

Bioadhesives in Drug Delivery

Brian K. Irons*   *Columbia Research Laboratories, Madison, Wisconsin*

Joseph R. Robinson   *University of Wisconsin, Madison, Wisconsin*

I. INTRODUCTION

The use of bioadhesives in drug delivery systems is by no means new, although increased interest in its unique applications in therapy is evidenced by the recent spate of publications. Bioadhesives complement drug delivery systems through increased residence time in the various routes of administration. Prolonged contact time can offer very substantial improvements in local drug therapy as well as significant increases in bioavailability for some drugs. Indeed, for a number of drugs that can only be administered by injection, because of either poor membrane absorption or excessive drug degradation, prolonged residence time at a particular site can obviate the need for an injectable mode of drug administration.

A *bioadhesive* can be defined as any substance that can adhere to a biological substrate and is capable of being retained on that surface for an extended period of time [1,2]. Drug delivery systems using bioadhesives usually adhere to membrane surfaces or the mucin layer coating such surfaces. The majority of the targeted areas used in drug delivery have a coating of mucus, and bioadhesive polymers that attach to this mucus coating are generally called *mucoadhesives*. Their residence times on these surfaces are controlled by whether or not the bioadhesive is water soluble or insoluble. In the case of water-soluble bioadhesives, contact time is generally only a few hours, depending on the adhesive and flow of biological fluid at the site of drug administration. Water-insoluble polymers, in contrast, remain in place until the mucin or tissue replaces itself, typically a period of about 4 to 72 h.

Contact between the adhesive and the mucosal membrane or its coating can be seen as a two-step process, the initial contact between the bioadhesive and substrate and subsequent formation of bonds between the two surfaces. Success of the initial contact appears dependent on similarity of physicochemical properties between the adhesive and substrate and is often associated with "wetting" of the substrate surface. Formation of bonds, which can be electrostatic, hydrophobic, or hydrogen bonds, permit the bioadhesive (and drug delivery system) to attach to the substrate. To better understand the use of bioadhesives in drug delivery, it is necessary to consider the physicochemical characteristics of the bioadhesive, the substrate, and the drug. To optimize adhesion, physiological parameters of the targeted tissue must also be addressed.

**Current affiliation*: University of Wisconsin, Madison, Wisconsin.

II. BIOLOGICAL SUBSTRATE

A. Mucus Layer

All external cavities of the body are lined with a continuous, thick, gel-like structure called *mucin*. Although the thickness varies depending on the type of tissue [3,4], this layer serves as a protective barrier between the cell surface and its external surroundings. Mucin is secreted by goblet cells and special exocrine glands [5] and can be considered a natural bioadhesive capable of binding to the underlying epithelial tissue. This binding results in a continuous, unstirred gel layer over the mucosa and thus serves as a barrier between a drug delivery system and the underlying mucosal epithelium. Mucus is a mixture of mucin glycoproteins, water, electrolytes, enzymes, bacteria, and sloughed epithelial cells [5]. Most of the content of mucus is an aqueous fluid containing macromolecules [6], with the mucin glycoproteins making up less than 5% of the total weight [7].

Mucin glycoproteins are macromolecules linked together by cross-linking disulfide bonds, physical entanglement [6], and secondary bonds to form a continuous network. These glycoproteins have an abundance of oligosaccharide side chains [6], with their terminal ends usually being either sialic acid [8,9] or L-fucose [10,11]. The entire mucin network at physiological pH has a net negative charge due to these sialic acid residues ($pK_a = 2.6$) and additional sulfate residues [12]. Thus mucin can be viewed as an anionic polyelectrolyte consisting of hydrated, cross-linked, linear, flexible glycoprotein molecules with sufficient overlap and interpenetration to form a continuous network. Since mucus is continually being formed, secreted, and removed from these tissues, its turnover rate must be taken into consideration when designing a bioadhesive dosage form.

B. Epithelial Surface

Most animal cell membrane surfaces are covered with glycoproteins and glycolipids extending from the cell exterior [13]. Collectively, all the polysaccharide structures on the outer surface of the cell are referred to as the glycocalyx [14]. The glycocalyx is continually being synthesized by the underlying cells [14] and is thought to be partly responsible for the adhesive property of the cell. Like mucus, the surface of cell membranes has a net negative charge due to the presence of charged groups [8,9], and the binding of mucin to the cell layer then results primarily from interaction between two surfaces of the same charge with additional secondary forces providing stabilization. The primary adhesive force for most bioadhesives is thought to be hydrogen bonding.

Adherence of a drug delivery system directly to any mucosal membrane can occur if the mucus layer is disturbed or the bioadhesive penetrates the mucin. Disruption of the mucus layer can be by abrasion, cell sloughing, chemical alterations by mucolytic agents, or disease state of the tissue [15]. If such an interruption occurs, bioadhesives can serve (1) to maintain continuity of the mucus layer and minimize the exposed area, (2) replace the mucus layer and provide a protective covering for the underlying cell layers from physical and chemical injury, and (3) act as a platform for drug delivery to local tissues and facilitate recovery of the damaged or diseased cell layers.

III. BIOADHESIVES

A. Bioadhesive Examples

The majority of work using bioadhesives in drug delivery has been with a small number of water-soluble and water-insoluble polymers. Water-soluble polymers are not cross-

Table 1 Examples of Bioadhesives

Type	Example	Refs.
Water soluble		
Cationic	Polylysine	22
	Poly(vinylmethylimidazole)	22
	Polybrene	22
Anionic	Alginic acid	16, 17
	Carrageenan	16
	Carboxymethyl cellulose (sodium)	16, 17
Neutral	Polyethylene glycol	16, 22
	Poly(vinylpyrrolidone)	22
	Hydroxypropyl cellulose	68, 69, 70
Water insoluble		
Cationic	Gelatin	37
Anionic	Carbopol 934	70, 72
	Polycarbophil	23
	Cross-linked polymethacrylic acid	23
Neutral	Hydroxypropylmethyl cellulose	17
	Poly(methyl methacrylate)	91
	Ethyl cellulose	92

linked, whereas water-insoluble polymers are often swellable networks joined by cross-linking agents. Bioadhesives can be natural or synthetic in origin, but in drug delivery systems, nonbiological macromolecules or hydrocolloid materials are often used. Some examples of bioadhesives are given in Table 1.

B. Physicochemical Criteria for Bioadhesion

The physicochemical criteria for potential bioadhesion have been studied extensively for both natural and synthetic polymers. Past studies have shown that polyethylene glycols [16], sodium carboxymethyl cellulose [17], and potassium carrageenan [16] need a minimum molecular weight for bioadhesion. Further, molecular weight of a compound has been shown to be proportional to its bioadhesive strength. For most polymers, increasing the molecular weight means an increase in length of the molecule, which can have an effect on the physical penetration and subsequent entanglement of the polymer with the substrate. Interpenetration and entanglement of an adhesive polymer with a mucin substrate is partly responsible for its bioadhesive strength [17], and any parameter that alters this process will have an effect on bioadhesive–mucin interaction.

The chains of water-insoluble swellable polymers are connected to cross-linking agents. As the amount of cross-linking is increased, the diffusion coefficient of the polymer chains is decreased with a subsequent decrease in interdiffusion between the polymer and substrate and a decrease in the polymer's bioadhesive properties [18,19]. This increase in cross-linking also lowers chain-segment mobility and flexibility, which can reduce the amount of interpenetration and entanglement of the polymer with its substrate. It has been suggested that there is an optimal chain mobility because too little or too much flexibility of the side chains can lead to a decrease in interpenetration with the mucus [20].

The interactions between bioadhesives and their substrates occur through covalent bonds, electrostatic interactions, and hydrogen-bond formation. Due to the potential toxicity involved in covalent bonding of an adhesive to a biological substrate (e.g., cyanoacrylate ''superglue''), polymers that adhere via electrostatic interactions and hydrogen bonding are preferred. Anionic, cationic, and neutral polymers have been studied extensively for their bioadhesive properties [21–23]. When the bioadhesive strength of the hydrocolloids, acrylic acid and 2-hydroxyethyl methacrylate (containing carboxyl groups and neutral groups, respectively) were measured, the role of the negatively charged groups was clearly established [23]. It was also determined that both the charge sign and density are important [21,24]. When both toxicity and bioadhesive properties are considered, polyanionic polymers appear to be better bioadhesives than polycations. Also, polyanions with carboxyl groups appear to be better than those with sulfate groups when only toxicity is considered. Thus the pH of the media can play a significant role in a polymer's bioadhesive strength, depending on the pK_a of the adhesive. Since, as mentioned above, the mucus layer and the mucosal epithelium both carry a net negative charge, electrostatic interactions are likely to occur with polyanionic molecules leading to increased bioadhesion.

Many polymers show significant bioadhesive strength when they are not ionized. Bioadhesive polymers often have numerous hydrophilic functional groups such as carboxyl, hydroxyl, amide, and sulfate groups which can form hydrogen bonds with the biological substrates [16]. These bonds may play a larger role in bioadhesion than the electrostatic interactions mentioned above. Studies using cross-linked polyacrylic acid (pK_a = 4.75) show that the adhesion is greatest when the carboxylate groups are in the free acid form and show a significant drop in adhesive strength above pH 4.0 [18[, thus illustrating that hydrogen bonding is the dominant mechanism.

Sufficient hydration of a polymer is also of importance in bioadhesion. As bioadhesives hydrate in aqueous media, they swell and form gels with fixed charged groups inside the network. These fixed charged groups result in the development of a swelling force or a net osmotic pressure which drives the surrounding solvent from the more dilute external bulk solution into the polymer network [25]. It was found that the degree of hydration decreases as the number of charged acrylic acid groups decreases or the amount of uncharged groups on methyl methacrylate increases [26]. Thus the rate and extent of water uptake by a polymer is dependent on the type and number of hydrophilic functional groups present in the polymer and also on the ionic strength and pH of the surrounding media.

The degree of hydration of a polymer is pertinent to its adhesive properties because sufficient water is needed to properly hydrate and expand the adhesive. If insufficient amounts of solvent are available or hydration is slow, the polymer is not fully hydrated and this limits the flexibility and mobility of the polymer chains, which is crucial to their diffusion and penetration into a substrate.

Pores in the hydrated polymer are formed due to chain flexibility and chain movement [27] and are a characteristic of the expanded nature of the polymer network. Formation of pores is lowered with decreased hydration and this limits the active adhesive sites available on the polymer network. As the degree of hydration increases with an increase in the density of charged groups [26], so does the mesh size of the network. Indeed, it was determined that the tensile strength of a mucoadhesive is directly proportional to the mesh size of the polymer network [28]. Thus as with mucin, the expanded nature of the network of an adhesive is an important factor in controlling adhesive strength.

If a drug delivery system using bioadhesives is placed in an aqueous medium, the polymer will absorb water. This absorption of water leads to the formation of aqueous channels and subsequent desorption of water-soluble drug [29–31] (i.e., the hydration of the polymer allows the polymer chains to extend and form aqueous pores in the polymer matrices and allows diffusion of drug molecules out of the polymer matrix to the underlying absorbing epithelium). Controlling the swelling rate [32], the cross-linking density of the polymer network [33], tonicity and pH of the media [33,34], solubility of the active drug, and so on, of a drug delivery system containing a bioadhesive can all be manipulated to optimize release of drug from the delivery system to the targeted membrane.

C. Methods to Quantify Bioadhesion

Various techniques have been designed to study the strength of adhesion between biological substrates and water-soluble or water-insoluble polymers. Adhesive quantitation of water-insoluble polymers usually involves measurement of tensile and shear strength of adhesion. In an appropriate buffer solution, the polymer is sandwiched between either two biological membranes or a membrane and a nonbiological substrate, and the detachment force is measured. Using a modified tensiometer, the tensile strength can be measured from the vertical force of detachment [18,26] and the shear strength can be measured from the horizontal force of detachment [26]. When measured under controlled conditions of constant surface area, rate of removal, and applied force, these two parameters can give a comparative measure of adhesive performance. These methods allow the selection of suitable tissues for adherence as well as control of the bathing medium. Although the preferred substrate for bioadhesion, the use of tissue samples can be costly, and thus other less expensive methods are sometimes sought. One such approach is to use red colloidal gold particles which form a conjugate with mucin [35]. Upon interaction with a polymer, the intensity of the red color of the conjugate–polymer can be measured spectrophotometricly. Whereas the techniques discussed above measure adhesion strength, this method measures adhesion number.

Adhesive measurement of water-soluble polymers is more difficult, but a number of techniques have been reported to assess adhesive strength adequately. One method is to coat a plate of glass with a soluble polymer and, using a tensiometer, measure the force to move it through a mucus solution [17]. Fluorescent probes have also been used to measure bioadhesion of soluble polymers to cell membranes [22]. With this approach adhesive strength is measured as a function of membrane viscosity differences before and after polymer binding using the fluorescent probe pyrene, which is incorporated into the lipid bilayer. Others have described methods on the static and dynamic adhesiveness of polymers in mucin solutions based on fluid mechanics [36]. More recently, researchers have studied bioadhesion by measuring viscometric differences in a mixture of polymer and gastric mucin [37].

IV. APPLICATIONS

Along with the physicochemical characteristics of the bioadhesive and mucin–epithelium surface, physiological events in the area in which adhesion occurs must be addressed to optimize the drug delivery system. Most delivery systems utilizing bioadhesives are designed to be topically applied to a targeted tissue. Drug delivery systems using bioadhesives can be applied to many areas of the body, such as the oral cavity, gastric, intestinal, rectal, vaginal, ocular, and dermal areas. Each tissue type has its own unique properties which can be exploited for the delivery of drugs. Each biological membrane has its own permeability, enzymatic activity, and immunology, which have to be taken into consideration if both satisfactory bioadhesion and improved bioavailability of drug are to be achieved.

A. Gastrointestinal

Most drug delivery systems are taken orally with the absorption of drug occurring mainly in the proximal small intestine. Due to a harsh gastric environment, motility of the gastrointestinal (GI) tract, immunogenic responses, enzymatic degradation, and dynamic changes in localization of the drug, a bioadhesive drug delivery system must be able to overcome these parameters to be effective either locally or systematically. The intestinal route is a desirable one despite these conditions because of its high absorptive characteristics compared to other routes of administration, which often need permeability enhancement of the tissue to increase bioavailability of the drug.

For a bioadhesive to adhere to either the stomach or intestine for an extended period of time, it must overcome the shear force associated with the motility patterns (parastalsis) of the GI tract, which can physically dislodge the adhesive from the mucus surface. GI motility patterns differ whether the subject is in a fasted (interdigestive) or a fed (digestive) state. Fasted-state motility has distinct phases of varying contractile magnitude, with the largest force occurring during what is called the housekeeper wave [38,39]. This phase serves to clear the upper GI tract of indigestible materials, and any bioadhesive system must therefore bind strongly enough to withstand this physical force if it is to be localized for an extended period of time. The digestive state motility pattern differs from the fasted state in that there are continuous contractions of approximately equal magnitude but only half the magnitude of the housekeeper wave. These continual contractions, as well as the physical removal of the bioadhesive by food, also need to be considered if a bioadhesive is to adhere to the mucus or underlying mucosal layer for an extended period of time.

The gastric turnover of mucin in both the fasted and fed state is a significant issue for bioadhesion in the oral route. The relatively rapid and continual production and subsequent removal of older mucus by luminal peptic activity [40] makes long-term (i.e., 24 to 48 h) bioadhesion to the gastric mucin layer impractical. Some researchers have tried to deliver drugs to the intestine at a controlled rate using bioadhesives in the stomach [41,42] but because of the mucus exchange and the motility conditions discussed above, little can be expected in long-term gastric retention in humans.

Because of the high turnover rate of gastric mucin, for a bioadhesive to remain in the stomach for an extended period of time it would therefore need to adhere to the epithelial layer instead of the mucus. This has been exploited in the use of an antiulcer drug that can adhere to damaged gastric epithelial tissue. Ulcerations are formed in the gastric and intestinal regions, where the protective mucus layer has been altered, and the underlying tissue is thus subject to proteolytic degradation by pepsins and bacteria. The antiulcer drug Sucralfate is used for the treatment of peptic ulcers and has been shown to bind to damaged gastric mucosa [43]. Sucralfate, an aluminum salt of sulfated sucrose, has been shown to protect the gastric mucosa from noxious materials such as alcohol, aspirin, and nonsteroidal anti-inflammatory drugs (NSAIDS) [44–46]. Sucralfate polymerizes upon addition to acid and forms a viscous mass that binds to the gastric mucosa [47,48]. Its protective qualities against ulcerations are thought to be due to the enhancement of gel viscosity, hydrophobicity, and mucin content of the gastric mucus in the ulcer vicinity, as well as inhibition of prostaglandin synthetase [49].

Controlled intestinal release of drugs through the use of bioadhesives has certain advantages due to the high absorptivity and neutral pH of the intestinal lumen. Barring enzymatic and immunogenic responses, tissue absorption of drug from a bioadhesive platform can be high if retained in the intestine for extended periods of time. In situ experiments in rats [50] have shown increased residence time of certain cross-linked acrylic polymers in the intestine. This increased residence time in the lumen of the intestine increased bioavailability of poorly absorbed drugs.

Enzymatic and immunogenic degradation of both drug and bioadhesive must be addressed in any route of administration but seems to be very important in the GI tract. A detailed review of these parameters with regard to bioadhesion is beyond the scope of this chapter, but suffice it to say that any absorption of drug via the intestinal epithelium presupposes adequate protection against enzymatic degradation associated with the stomach and intestine as well as immune responses to antigens in the GI tract. Another drawback to the gastrointestinal route is that drugs which enter the general circulation are subject to first-pass metabolism as they pass through the hepatic–portal system leading to lower systemic availability. Most of the work to date associated with bioadhesives in drug delivery systems has focused on other routes of administration, which avoid such adverse conditions.

B. Rectal

Most rectally administered drugs for either local or systemic therapy are given in suppository form. Systemic availability of rectally administered drug is maximal when the dosage form is close to the anus [51]. Normally, after insertion, suppositories tend gradually to migrate and rest in the upper portion of the rectum. Drugs that are absorbed through this area into the blood stream enter the hepatic–portal system and are subject to first-pass metabolism, which in turn degrades many susceptible drugs and leaves them ineffective. The lower rectum's blood flow, however, drains directly into the general circulation, and first-pass metabolism of a drug can be avoided if the delivery system can be maintained in the lower region [52,53]. Suppositories containing bioadhesives can reduce this migration toward the upper rectum and hence improve bioavailability of drug.

Penetration enhancers that improve the uptake of compounds into the epithelium can also be incorporated into such a delivery system. These enhancers are often used for hydrophilic compounds (especially peptides and proteins), which show low permeability through the barrier membrane. Although penetration enhancers have obvious benefits in absorption of drugs through the epithelium, they may also cause adverse effects to the tissue as well as local or systemic side effects [54]. Yet because of the delivery system's localization, the concentration of enhancers can be minimized, thus reducing adverse effects. Indeed, promising results were shown in using enhancers in the rectum for compounds that normally show poor bioavailability. For example, insulin uptake into the bloodstream has been shown to increase when enhancers are used in rectal administration [55,56]. Controlled release of antipyrine and theophylline using cross-linked hydroxyethyl methacrylate (HEMA) as a bioadhesive was shown to sustain the availability of rectally applied drug in humans [57]. The combination of permeability enhancement and localization by bioadhesives has the potential to increase drug bioavailability significantly via the rectal route of administration.

C. Nasal

Nasal delivery systems are usually in the form of aqueous sprays in which the drug is distributed into the nasal cavity. This area provides an excellent route for drug absorption because of its large surface area and vascularity [58] as well as a thin layer of mucus secreted from local mucosal glands [59]. Absorption into the bloodstream via the nasal route also eliminates hepatic first-pass metabolism. This combination makes the nasal cavity an excellent route for localized treatment (e.g., nasal inflammation and allergic responses) as well as for systemic drug delivery. A suitable bioadhesive could then by hydrated by the nasal mucus and form a viscous gel covering the nasal cavity. The ciliary removal of mucus must be taken into consideration when using bioadhesives in the nasal cavity.

Many researchers have taken advantage of this potential route using bioadhesives as a delivery system. Hydrophillic compounds that are normally poorly absorbed in the nasal cavity [60] can still be utilized using penetration enhancers in conjunction with a retained delivery system [61]. Using degradable starch microspheres and a penetration enhancer, the nasal absorption of gentamicin [62] was shown to be improved. These microspheres form a gel when in contact with the moist nasal mucosa. Using these degradable starch microspheres, nasal administration of insulin has been shown to be improved when administered with penetration enhancers [63]. Insulin has also been administered in freeze-dried form with Carbopol 934 (a cross-linked polyacrylic acid polymer) to achieve a sustained release effect, which increased with increasing Carbopol concentration [64]. Using polyacrylic acid, other research has shown increased availability of both insulin and calcitonin by nasal administration in rats [65]. Using the nasal route of administration for insulin, as opposed to the daily subcutaneous injections commonly used, has obvious benefits with respect to patient compliance, although systemic levels of drug thus far are lower with the same dosage concentration, even with the addition of enhancers.

D. Vaginal

The vaginal and cervical route of administration is unique from other routes in that the tissue environment is subject to many changes throughout a women's life. Depending on whether the woman is pre- or postmenopausal, the tissue and mucus of the vaginal and cervical areas can be vastly different. Decreased endogenous levels of estrogen, cervical shrinkage, cell atrophy, and lower cervical mucus levels [66] are characteristic of postmenopausal women. Thus a vaginal bioadhesive delivery system geared to older women would need to address these conditions to optimize drug availability.

A women's menstrual cycle can also affect the vaginal environment. Vaginal mucus originates in the cervix, then migrates into the vaginal area. Monthly fluctuations in cervical mucin's properties have been documented [67], showing lower viscoelasticity when estrogen is dominant and thicker, more viscoelastic mucin when progesterone dominates. Again, a bioadhesive delivery system must take these considerations into account to optimize the bioavailability of drug to the tissue.

Most vaginally administered drugs are delivered via creams, foams, suppositories, gels, or tablets. The women's health care market is very large and profitable and hence a number of delivery systems utilizing bioadhesives have started to appear. The patented use of a soluble hydroxypropyl cellulose (HPC) cartridge for vaginal delivery of drug has been shown to release drug for an extended period of time [68]. The polymer forms a hydrated gel of sufficient viscosity in the vaginal cavity and releases drug directly to the vaginal area. The anticancer drugs bleomycin, carbazilquinone, and 5-fluorouracil have been administered directly to the cervix using disk- and rod-shaped dosage forms containing a combination of HPC and carbopol [69,70]. These dosage forms were shown to stay in the diseased area for a longer period of time than vaginal suppositories containing the same drug. Compared to suppositories, local side effects of these dosage forms were also reduced. Such a system has the possibility, then, to treat cancer of the cervix locally.

During a woman's reproductive years, the vaginal bacterial flora is capable of maintaining an acidic environment which can reduce vaginal infection by limiting the bacterial growth often associated with other disease states [71]. Maintenance of this slightly acidic pH is then crucial for vaginal health, and thus drug delivery systems that address this phenomena have obvious therapeutic benefits. A vaginal moisturizer containing the bioadhesive polycarbophil has been shown to alleviate postmenopausal vaginal dryness (Replens, Columbia Laboratories, Hollywood, Florida). The cream has the ability to remain in the vaginal cavity for 2 to 3 days after only one administration and maintains a healthier vaginal environment through its hydration of the mucosa. Because of polycarbophil's pK_a value of 4.75, the cream also can maintain a nearly normal acidic vaginal pH, which, as discussed above, has certain health-related advantages.

E. Oral Cavity

Drug administration to the oral cavity has many advantages from both a patient and a therapeutic point of view. Both local and systemic availability can be achieved using bioadhesives in the oral cavity. Anesthetic, anti-inflammatory, and antimicrobial agents can be administered locally for increased residence time using bioadhesives. Besides the use of adhesives for retention of dentures, the dental industry has taken advantage of using bioadhesives for other localized applications. The anesthetic lidocaine, used locally for toothaches, has been shown to have an increased duration of activity when administered in a mucoadhesive tablet containing a combination of freeze-dried hydroxypropyl cellulose and Carbopol 934 [72]. This has advantages over the usual forms of topical administrations, which show little precision in site specificity and can quickly be washed away by saliva. The analgesic lignocaine has also been studied when applied by a bioadhesive patch [73]. Various polymer systems have been employed to deliver fluorides to the oral cavity [74,75]. Others have reported therapeutic treatment of buccal lesions, such as aphthae and lichen planus using bioadhesives [76–78]. These dosage forms have the advantage over standard oral ointments of being applied directly to the lesion and achieving high drug levels because of increased duration at the site of inflammation.

Systemic delivery of drugs through the mouth has gained popularity in recent years. Drugs that are susceptible to degradation by the harshness of the gastrointestinal route can be administered via the mouth. This avoids first-pass metabolism of susceptible compounds by the hepatic system and offers the patient a more desirable route of administration than injection. Due to the limited area of the oral cavity, the delivery system itself is restricted in size, and hence potent compounds, such as proteins and peptides, are often more suited to such delivery systems.

The oral cavity can be divided into three distinct functional areas: the lining mucosa (buccal, sublingal, and soft palate), the masticatory mucosa (hard palate and gingiva), and the specialized mucosa (dorsal tongue). The thickness and keratinization of the tissue differs between these regions [79], and hence the permeability of each is unique [80]. The hard palate and gingiva are highly keratinized and subsequently offer limited permeability for drug delivery. The use of enhancers, however, has been shown to increase the permeability through keratinized tissues from bioadhesive platforms [81,82]. The majority of the work with systemic delivery systems using bioadhesives in the oral cavity has been concentrated on the buccal (cheek) route of administration because of its large surface area and nonkeratinization. Bioadhesive buccal tablets or patches has been utilized as delivery systems. They are usually designed to be unidirectional in their delivery (i.e., delivering the drug from the side of the patch attached to the buccal mucosa and not to the rest of the mouth). This is often accomplished by an impermeable backing facing the oral cavity. The bioadhesive of choice can then serve two purposes: as an adhesive keeping the delivery system in place and/or as a drug-containing matrix in which the compound diffuses from the matrix and permeates the mucosa into the general circulation.

When administered in a mucoadhesive tablet, similar to the tablet containing lidocaine above but with the addition of an oil base and the penetration enhancer glycocholate, insulin has shown increased absorption through the oral mucosa [83]. Insulin blood levels, however, were significantly lower in comparison to systemic levels achieved by intramuscular injection. The reason for the low bioavailability could be due to poor tissue permeability, even with the enhancer. Mucoadhesive dosage forms have also been used for the treatment of cardiovascular disorders such as angina and hypertension [84–86]. When administered by an adhesive delivery system, nifedipine showed plateau drug levels after 8 h and was sustained there until removal of the delivery system. Another delivery system using nitroglycerin in a bioadhesive buccal tablet has also been shown to have a sustained effect.

F. Ocular

Drug delivery to the eyes is made difficult by dilution of drug in the tears and the natural mechanisms of blinking and high tear turnover rate, which protect the eye from external contaminants. Traditional aqueous, ocular delivery systems are administered dropwise, and due to the foregoing conditions, bioavailability is severely limited for either local or systemic therapy. Although many attempts have been made to prolong drug release of ocular delivery systems, few have proven to be completely successful when patient acceptance, drug bioavailability, and cost are considered. For a drug to be sustained in the eye, it must be maintained in the precorneal area and deliver drug to this area for an extended period of time. Ocular bioadhesive delivery systems could therefore show a sustained effect if they penetrate the aqueous tear film and interact with the underlying mucin or cell layer. If firmly attached to the surface, the dosage form could remain in the preocular area longer than conventional ocular dosage forms, and if dissolution of drug release is controlled, utilization of water-soluble drug can be increased significantly.

Pilocarpine is a drug commonly used in glaucoma therapy to relieve intraocular pressure (IOP), which is a cause of great discomfort to the patient. Piloplex is a sustained-release product based on an emulsion system of pilocarpine bound to a polymeric carrier [87,88]. Piloplex was shown to prolong a reduce IOP as compared to standard pilocarpine hydrochloride drops. This is attributed to its bioadhesive properties, which keep the drug in the precorneal area longer than do conventional ocular dosage forms.

The release of progesterone used as a model drug in an ocular delivery system consisting of cross-linked acrylic acid has been shown to be sustained [89]. The delivery system showed increased bioavailability 4.2 times greater than a suspension without polymer and showed excellent bioadhesion to the conjunctival mucosa of the albino rabbit. Another system utilizing polycarbophil also showed increased bioavailability of a fluorometholone steroid suspension used for the treatment of inflammation [90]. Aqueous humor drug levels were maintained above the therapeutic minimum for 8 h in albino rabbits, and the mean residence time of fluorometholone was increased 1.7 times over an aqueous suspension.

REFERENCES

1. K. Park, S. L. Cooper, and J. R. Robinson, in *Hydrogels in Medicine and Pharmacy* (N. A. Peppas, ed.), CRC Press, Boca Raton, Fla., 1986.
2. H. Park and J. R. Robinson, *J. Controlled Release 2*: 47 (1985).
3. A. Alen, D. A. Mutton, J. O. Pearson, and L. A. Sellers, in *Mucus and Mucosa*, Ciba Foundation Symposium 109 (J. Nugent and M. O'Connor, eds.), Pitman, London, 1984.
4. B. A. Nichols, M. L. Chiappino, and C. R. Dawson, *Invest. Opthalomol. Vis. Sci. 26*: 464 (1985).
5. H. Schachter and D. Williams, *Advances in Experimental Medicine and Biology 144*: 3 (1982).
6. A. Silberberg and F. A. Meyer, *Advances in Experimental Medicine and Biology 144*: 53 (1982).
7. A. Allen and A. Garner, *Gut 21*: 249 (1980).
8. A. Gottschalk, *The Chemistry and Biology of Sialic Acid and Related Substances*, Cambridge University Press, London, 1960.
9. R. W. Jeanloz, in *Glycoproteins: Their Composition, Structure and Function* (A. Gottschalk, ed.), Elsevier, Amsterdam, 1972.
10. E. N. Chantler and P. R. Scudder, in *Mucus and Mucosa*, Ciba Foundation Symposium 109 (J. Nugent and M. O'Connor, eds.), Pitman, London, 1984.
11. T. A. Beyer, J. J. Rearick, J. C. Paulson, J. P. Prieels, J. E. Sadler, and R. L. Hill, *J. Biol. Chem. 254*: 12,532 (1979).
12. P. M. Johnson and K. D. Rainsford, *Biochim. Biophys. Acta 286*: 72 (1972).
13. H. Rauvala, *Trends in Biochemical Sciences 8*: 323 (1983).
14. G. W. Jones, in *Microbial Interaction* (J. L. Reissig, ed.), Chapman & Hall, London, 1977.
15. G. Forstner, J. Sturgess, and J. Forstner, *Advances in Experimental Medicine and Biology 89*: 349 (1977).
16. J. L. Chen and G. N. Cyr, in *Adhesion in Biological Systems* (R. S. Manly, ed.), Academic Press, New York, 1970.
17. J. D. Smart, I. W. Kellaway, and H. E. C. Worthington, *J. Pharm. Pharmacol. 36*: 295 (1984).
18. H. Park, Ph.D. thesis, University of Wisconsin–Madison, 1986.
19. R. M. Barrier, J. A. Barrie, and P. S. L. Wong, *Polymer 9*: 609 (1968).
20. R. Gurny, J.-M. Meyer, and N. A. Peppas, *Biomaterials 5*: 336 (1984).
21. K. Park, H. S. Ch'ng, and J. R. Robinson, in *Recent Advances in Drug Delivery Systems* (J. M. Anderson and S. W. Kim, eds.), Plenum Press, New York, 1984.
22. K. Park and J. R. Robinson, *Intern. J. Pharm. 19*: 107 (1984).
23. H. S. Ch'ng, H. Park, P. Kelly, and J. R. Robinson, *J. Pharm. Sci. 74*: 399 (1985).
24. N. A. Peppas and P. A. Buri, *J. Controlled Release 2*: 257 (1985).
25. P. J. Flory, *Principles of Polymer Chemistry*, Cornell University Press, Ithaca, N.Y., 1953.
26. S. H. S. Leung and J. R. Robinson, *J. Controlled Release 5*: 233 (1988).
27. R. P. Campion, *J. Adhesion 7*: 1 (1974).
28. S. H. S. Leung, Ph.D. Thesis, University of Wisconsin–Madison, 1987.
29. P. I. Lee, *J. Pharm. Sci. 73*: 1344 (1984).
30. P. I. Lee, *Polymer Commun. 24*: 45 (1983).

31. W. R. Good, in *Polymer Delivery Systems* (R. J. Kostelnik, ed.), Gordon and Breach, New York, 1976.
32. C. T. Reinhart, R. W. Korsmeyer, and N. A. Peppas, *Intern. J. Pharm. Technol. 2*: 9 (1981).
33. N. A. Peppas and J. Klier, *J. Controlled Release 16*: 203 (1991).
34. L. Brannon-Peppas and N. A. Peppas, *J. Controlled Release 8*: 267 (1989).
35. K. Park, *Intern. J. Pharm. 53*: 209 (1989).
36. A. G. Mikos and N. A. Peppas, *13th International Symposium on Controlled Release Bioactive Materials* (I. A. Chaudry and C. Thies, eds.), Controlled Release Society, Lincolnshire, Ill., 1986.
37. E. E. Hassan and J. M. Gallo, *Pharm. Res. 7*: 491 (1990).
38. J. H. Szurszewski, *Am. J. Physiol. 217*: 1757 (1969).
39. S. J. Konturek, in *Gastrointestinal Motility: Proc. 9th International Symposium on Gastrointestinal Motility* (C. Roman, ed.), MTP Press, Boston, 1984.
40. M. Feldman, in *Gastrointestinal Disease*, 4th ed., (M. H. Sleisenger and J. S. Fordtran, eds.), W. B. Saunders, Philadelphia, 1989.
41. M. A. Longer, H. S. Ch'ng, and J. R. Robinson, *J. Pharm. Sci. 74*: 406 (1985).
42. R. Khosla and S. S. Davis, *J. Pharm. Pharmacol. 39*: 47 (1987).
43. R. N. Brogden, R. C. Heel, T. M. Speight, and G. S. Avery, *Drugs 27*: 194 (1984).
44. W. C. Wu, E. L. Semble, and D. O. Castell, *Gastroenterology 88*: 1636 (1985).
45. D. Hollander and A. Tamawski, W. J. Krause, and H. Geregely, *Gastroenterology 88*: 366 (1985).
46. D. Hollander, A. Tamawski, H. Gergely, and R. D. Zipser, *Scand. J. Gastroenterol. 19*(suppl. 101): 97 (1984).
47. P. H. Ruth, *Dig. Dis. Sci. 32*: 647 (1987).
48. C. Tasman-Jones, G. Morrison, L. Thomsen, and M. Vanderwee, *Am. J. Med. 86*(suppl. 6A): 5 (1989).
49. B. L. Slomiany, J. Piotrowski, K. Okazaki, E. Grzelinska, and A. Slomiany, *Digestion 44*: 222 (1989).
50. C. Lehr, J. A. Bouwstra, J. J. Tukker, and H. E. Junginger, *J. Controlled Release 13*: 51 (1990).
51. L. G. J. de Leede, A. G. de Boer, C. P. J. M. Roozen, and D. P. Breimer, *J. Pharmacol. Exp. Ther. 225*: 181 (1983).
52. A. G. de Boer, J. M. Gubbens-Stibbe, and D. D. Breimer, *J. Pharm. Pharmacol. 33*: 50 (1981).
53. L. G. J. de Leede, A. G. de Boer, J. P. J. M. Havermans, and D. D. Breimer, *Pharm. Res. 1*: 164 (1984).
54. V. H. L. Lee, *J. Controlled Release 13*: 213 (1990).
55. A. Kamada, T. Nishibata, A. Kim, M. Yamamoto, and N. Yata, *Chem. Pharm. Bull. 29*: 2012 (1981).
56. T. Nishihata, J. H. Rutting, A. Kamada, T. Higuchi, M. Routh, and L. Caldwell, *J. Pharm. Pharmacol. 35*: 148 (1983).
57. L. G. J. de Leede, A. G. de Boer, E. Portzger, J. Feijen, and D. D. Breimer, *J. Controlled Release 4*: 17 (1986).
58. T. Nagai, *Med. Res. Rev. 6*: 227 (1986).
59. M. Taylor, *Laryngoscope 84*: 612 (1974).
60. G. S. M. J. E. Duchateau, J. Zuidema, W. M. Albers, and W. H. M. Merkus, *Intern. J. Pharm. 34*: 131 (1986).
61. S. Hirai, T. Yashiki, T. Matsuzawa, and H. Mima, *Intern. J. Pharm. 7*: 317 (1981).
62. L. Illum, N. F. Faraj, H. Critchley, and S. S. Davis, *Intern. J. Pharm. 46*: 261 (1988).
63. N. F. Farraj, B. R. Johansen, S. S. Davis, and L. Illum, *J. Controlled Release 13*: 253 (1990).

64. T. Nagai, Y. Nishimoto, N. Nambu, Y. Suzuki, and K. Sekine, *J. Controlled Release 1*: 15 (1984).

65. K. Morimoto, K. Morisaka, and A. Kamada, *J. Pharm. Pharmacol. 37*: 134 (1985).

66. A. C. Wentz, in *Novak's Textbook of Gynecology*, 11th ed. (H. W. Jones III, ed.), Williams & Wilkins, Baltimore, 1988.

67. E. Chantler, *Advances in Experimental Medicine and Biology 144*: 251 (1982).

68. B. L. Williams, Soluble medicated hydroxypropyl cellulose cartridge, U.S. patent 4,317,447 (Mar. 2, 1982).

69. Y. Machida, H. Masuda, N. Fujiyania, A. Ito, M. Iwater, and T. Nagai, *Chem. Pharm. Bull. 27*: 93 (1980).

70. Y. Machida, H. Masuda, N. Fujiyania, A. Ito, M. Iwater, and T. Nagai, *Chem. Pharm. Bull. 28*: 1125 (1980).

71. A. Bergman and P. F. Brenner, in *Menopause: Physiology and Pharmacology* (D. R. Mishell, Jr., ed.), Year Book Medical Publishers, Chicago, 1987.

72. M. Ishida, N. Nambu, and T. Nagai, *Chem. Pharm. Bull. 30*: 980 (1982).

73. I. M. Brook, G. T. Tucker, E. C. Tuckley, and R. N. Boyes, *J. Controlled Release 10*: 183 (1989).

74. D. B. Mirth, *Pharmacol. Ther. Dent. 5*: 59 (1980).

75. P. Bottenberg, J. Hermans, D. Coomans, C. de Muynck, J. P. Remon, D. Slop, and Y. Michotte, *STP Pharma 5*: 863 (1989).

76. T. Yotsuyanagi, K. Yamamura, and Y. Akao, *Lancet 14*: 613 (1985).

77. Attach: adhesive topical preparation for treatment of aphthous stomatitis, Teigin, Tokyo, 1982.

78. I. G. Tucker, H. A. M. Szylkarski, and K. Romaniuk, *J. Clin. Pharm. Ther. 14*: 153 (1989).

79. B. K. Berkovitz, G. R. Holland, and B. J. Mozham, eds., *A Colour Atlas and Textbook of Oral Anatomy*, Wolfe Medical Publications, London, England, 1978.

80. C. A. Squier and B. K. Hall, *J. Invest. Dermatol. 84*: 176 (1985).

81. Y. Kurosaki, T. Takatori, M. Kitayama, T. Nakayama, and T. Kimura, *J. Pharmacobio-Dyn. 11*: 824 (1988).

82. Y. Kurosaki, S., Hisaichi, T. Nakayama, and T. Kimura, *Intern. J. Pharm. 51*: 47 (1989).

83. M. Ishida, Y. Machida, N. Nambu, and T. Nagai, *Chem. Pharm. Bull. 29*: 810 (1981).

84. P. B. Deasy and C. T. O'Neill, *Pharm. Acta Helv. 64*: 231 (1989).

85. R. Konishi, *Proc. 9th Conference on Pharmaceutical Technology*, Shirakabako Nagano-ken, Japan, 1984.

86. J. M. Schor, S. S. Davis, A. Nigalaye, and S. Bolton, *Drug Dev. Ind. Pharm. 9*: 1359 (1983).

87. U. Ticho, M. Blumenthal, S. Zonis, A. Gal, I. Blank, and Z. W. Mazor, *Brit. J. Ophthalmol. 63*: 45 (1979).

88. J. R. Robinson and V. H. K. Li, *Recent Advances in Glaucoma* (U. Ticho and R. David, eds.), Elsevier, New York, 1984.

89. H. Hui and J. R. Robinson, *Intern. J. Pharm. 26*: 203 (1985).

90. D. L. Middleton and J. R. Robinson, *STP Pharma 1*: 200 (1991).

91. K. D. Bremecker, H. Strempel, and G. Klein, *J. Pharm. Sci. 73*: 548 (1984).

92. M. Marvola, M. Rajaniemi, E. Marttila, K. Vahervuo, and A. Sothmann, *J. Pharm. Sci. 72*: 1034 (1983).

39

Bonding Materials and Techniques in Dentistry

Eberhard W. Neuse and Eliakim Mizrahi *University of the Witwatersrand, Johannesburg, South Africa*

I. INTRODUCTION

The success of modern restorative dentistry in the repair, restoration, and replacement of tooth structure is critically dependent on the availability of specialized metallic and nonmetallic materials and on procedures for their proficient application. Most of the nonmetallics are polymeric in nature, and on a volume basis, the greatest share by far of these polymeric materials find use in the construction of dentures, and to a lesser extent also in the preparation of impression materials and prosthetic restorations such as crowns or bridges. Materials of this type do not display adhesive functions; at best, they may act as adherends. Yet there are different procedures, generally in operative dentistry, where certain polymers, sometimes in combination with inorganic compounds, have come to assume leading roles as active participants in adhesion processes in addition to other vital functions associated with their particular applications in restorative and preventive dentistry. In the classification of the FDI (Fédérale Dentaire Internationale), polymers of this type fall under the description of group M1 (dental filling and related materials) and include such items as luting agents, cavity liners, pit and fissure sealants, and finally, the important class of cavity-filling cements. Although some of these materials truly conform to the characteristics of an adhesive—namely, to bond two surfaces together—others, such as the cavity-filling cements, are in a sense half-sided adhesives insofar as they bond to one adherend surface only; yet they are included here because their one-sided bonding represents a realistic process of retention. The bonding reaction may involve mechanical interlocking, and this is indeed the mechanism utilized in the majority of adhesion processes encountered in dentistry. Alternatively, it may involve a chemical, generally ionic or covalent, bond-forming process, which one finds invariably coupled with smaller or larger contributions by the mechanical retention mode.

To facilitate presentation and digest of the subject, this chapter has been subdivided into four main sections in accordance with fields of application rather than composition of materials. Brief discussions of the applications and materials requirements are followed in each category by presentations of the working materials of choice, their mode of action, and where applicable, their strong and weak points in performance. In view of the abundance of publications in the field of dental materials, no attempt has been made to

629

provide a comprehensive compilation of references. Instead, preference has been given to the citation of selected recent publications by leading specialists, in which reference has been made to previous work on the subject. Fundamental facts and relationships presented as background information, which are considered textbook knowledge, have not been referenced. Exemplifying texts to be consulted for details are those of McCabe [1] and Wilson et al. [2].

II. LUTING

Cements are adhesive substances capable of bonding different bodies together; they are generally applied as liquids or viscous fluids, which set (harden, cure) in situ to solid materials. Dental cements used for luting or cavity lining are similar in composition and related in function; accordingly, certain data presented here are pertinent as well to Section IV. Luting cements are employed for the attachment of restorations and orthodontic bands to the tooth structure.* Luted restorations include crowns, inlays, and metal posts, as well as some of the older bridge constructions. In general, they are of a permanent nature. Temporary appliances, however (e.g., temporary crowns or splints), also require luting cements, and such appliances are similar in composition to those used for permanent luting except that they are designed for weaker bonding to facilitate future removal.

A. Requirements

A luting agent should have minimal solubility in the aqueous oral fluids while displaying good wetting properties. It should possess low initial viscosity to allow for proper seating of the restoration and for adequate narrowing of the margin between restoration and tooth. A wide margin, commensurate with a thick layer of poorly flowing, viscous cement in the luting space, will cause exposure of a proportionately large area of cement to the oral environment with consequent erosion effects, the development of microleakage, and potential secondary caries resulting from bacterial ingress. The cement should not be toxic or irritant to the pulp; it should provide thermal and, ideally, electrical insulation, features of particular importance for the luting of metallic restorations, such as gold crowns. In an ideal luting process the cement should bond chemically to the enamel and restoration adherends; with currently available materials, however, the bonding effect is largely or entirely one of micromechanical interlocking, as the material, utilizing existing surface roughness, flows into, and occupies, the microscopic interstices on the adherend surfaces. Once hardened in the assembly, the cement must assume sufficient strength for the microscopic protrusions into the interstices (tags) to withstand without fracture the occlusal masticatory (compressive and deformation) forces exerted onto the restored tooth.

---

*The hard tissue of the tooth substance consists of a protective outer coat of *enamel* and an underlying *dentin* phase. The latter, in turn, connects to the inner core of soft tissue (*pulp*), which is interpenetrated by nerve strands and blood vessels. The enamel, which covers essentially the visible part of the tooth and indeed represents the hardest tissue in the body, is composed almost entirely (97% by weight) of mineral-type hydroxyapatite (a crystalline calcium phosphate) in addition to a few percent of water and organic, mostly proteinaceous, matter. Dentin, constituting the major proportion of tooth substance, contains less mineralized phase (69% hydroxyapatite) but a comparatively large proportion of organic matter and water. Compositional and physical property data for enamel and dentin [3] are summarized in Tables 1 and 2.

Table 1 Composition of Tooth Structure

| | Content[a] | | | |
| | Enamel | | Dentin | |
Component	Percent by weight	Percent by volume	Percent by weight	Percent by volume
Mineral matter (mostly hydroxyapatite)	97	92	69	48
Organic matter (mostly proteinaceous)	1	2	20[b]	29[b]
Water	2	6	11	23

[a]From Ref. 3; data approximate.
[b]Essentially collagen.

B. Materials

The development of luting agents with ever more satisfactory bonding characteristics has been an ongoing objective in dental materials research for many decades. Although numerous adhesive compositions have been, and continue to be, commercialized, no perfect universal bonding systems have as yet reached the market. Important classes of luting materials include the zinc phosphate and silicophosphate cements, resin cements, chelating agents of the zinc oxide–eugenol and zinc oxide–ethoxybenzoic acid types, the polycarboxylates, the glass ionomer cements, and finally, a number of materials based on mono- and diacrylate resin systems, some of these possessing special adhesive properties.

1. Zinc Phosphates

Luting cements based on zinc phosphate have been known for more than a century and are still in major use today. The fundamental process leading to cementation is the formation of hydrated zinc phosphate from zinc oxide and phosphoric acid:

$$3ZnO + 2H_3PO_4 + H_2O \rightarrow Zn_3(PO_4)_2 \cdot 4H_2O$$

Table 2 Selected Physical Properties of Tooth Structure

Property	Enamel	Dentin
Compressive strength (MPa)	100 – 380[a]	250 – 350
Tensile strength (MPa)	10[b]	20 – 50
Modulus of elasticity (GPa)	10 – 80	11 – 15
Knoop hardness number	360 – 390	75
Vickers hardness number	350	60
Thermal conductivity ($W\ m^{-1}\ K^{-1}$)	0.88 – 0.92	0.59 – 0.63
Coefficient of thermal expansion ($10^{-6}\ °C^{-1}$)	11	8 – 9

[a]Depending on orientation of test sample and other factors.
[b]Measured in tension.
Source: Refs. 1 to 3.

Accordingly, the material is supplied as a two-part system: a powder component consisting predominantly of zinc oxide, and an aqueous solution of phosphoric acid (50 to 60 wt %) and other minor ingredients. The two components are mixed by the clinician in a controlled fashion immediately prior to use. The reaction is vigorous and strongly exothermic. For clinical application some retardation is called for; in most commercial products this is achieved by high-temperature (>1300°C) sintering of the zinc oxide reactant with some 10% of other, less reactive oxides, such as MgO, resulting in partial deactivation. To buffer the reaction further, up to 10% of aluminum and zinc phosphates are added to the phosphoric acid solution. With a powder/liquid ratio of 1.4 g/mL the mixture possesses adequate fluidity and working time to permit thin-film (ideally, 35 to 40 μm) application and allow for proper seating of the restoration. The lute so prepared sets within 4 to 7 min and undergoes further hardening thereafter; ultimate compressive strength after more than 24 h is typically in the vicinity of 80 MPa.*

To achieve satisfactory mechanical characteristics of the cement, the lute margin requires protection from moisture during the setting period (e.g., by varnish application); otherwise, phosphoric acid leaches out from the fresh cement, and the latter turns chalky and porous. The retention effects of the zinc phosphate cement rest on mechanical interlocking rather than chemical bond formation to the adherends. Surface roughness therefore contributes decisively to lute retention, although a limit is set by the inherent strength of the cured cement. The shear bond strengths of the phosphate and other

Table 3 Typical Compressive Strength of Luting, Cavity-Lining, and Endodontic Cements

Material	Compressive strength (MPa)
Calcium hydroxide	
Conventional	8
Resin-modified	20
Zinc phosphate	
Luting	80
Lining	140
Zinc oxide–eugenol (ZOE)	
Conventional	20
Poly(methyl methacrylate)–modified	40
Zinc oxide–ethoxybenzoic acid (EBA)	85
Methyl methacrylate resin, unfilled	85
Composite resin	180
Polycarboxylate	90
Glass ionomer	
Conventional	160
Light-cured	90
Calcium phosphate	35

Source: Ref. 1 and other literature.

---

*The retention of most crown restorations and the older types of bridge design is largely secured by the compressive forces exerted in vivo during mastication. Compressive strength data are therefore routinely specified for luting and cavity-lining cements. Representative data have been compiled in Table 3.

classical luting cements to the tooth structure are in a very low range, generally not exceeding 2 MPa. Cement solubility in the aqueous oral fluids is slight but noticeable and leads to slow lute erosion and loss of retention. The erosion is affected by lute margin width, zinc oxide load, and particle size. Taken together, such erosion effects render the zinc phosphates inferior in this respect to most resin and composite cements.

2. Silicophosphates

Closely related to the zinc phosphate luting cements, the silicophosphate materials are two-part bonding systems, which are mixed and applied essentially as described in the preceding section. Whereas the liquid component is a buffered aqueous phosphoric acid solution as before, the solid is a powdered mixture of a fluoride-containing, ion-leachable aluminosilicate glass and zinc oxide, and the hardening process yields a matrix of phosphates of zinc and aluminum, embedding zinc oxide, and glass particles. The fully set cement is less soluble than zinc phosphate in aqueous media, yet still prone to erosion, especially under acidic oral conditions. The fluoride content is beneficial in retarding secondary cavity formation, as fluoride ion gradually leaches from the lute. Because of their superior translucency, the silicophosphates are preferentially used for luting porcelain crowns, although their bonding mode is one of micromechanical interlocking, requiring specialized surface treatment of the porcelain restoration. The cements are weak under tensile and flexural loads but adequate in compression (Table 4).

3. Zinc Oxide–Eugenol Cements

Falling under the heading of metal chelate compounds, the zinc oxide–eugenol (ZOE) cements in the hardened state are essentially zinc phenolates formed by reaction of zinc oxide and eugenol (4-allyl-2-methoxyphenol) in the presence of moisture, embedding unreacted ZnO and eugenol. The products are stabilized through coordinative bonding of the metal center to two ether oxygen atoms with generation of five-membered chelate rings. The commercial products are two-part systems supplied either as powder–liquid or

Table 4 Selected Physical Properties of Restoratives

Material	Compressive strength (MPa)	Tensile strength (MPa)	Modulus of elasticity (GPa)	Thermal coefficient of expansion ($10^{-6}K^{-1}$)
Silicate	180 – 220	10 – 15	15 – 25	8 – 10
Acrylic				
Unfilled[a]	70 – 80	25 – 35	2	80 – 100
Glass ionomer				
Conventional and light-cured	180 – 250	11 – 13	1 – 6	13 – 16
Composite				
Conventional	200 – 260	35 – 55	9 – 15	20 – 35
Microfilled	250 – 260	30 – 40	6	50 – 75
Hybrid	300	50	14 – 16	
Amalgam				
Conventional	380 – 450	6	14 – 19	25

[a]No longer in use; data for comparison only.
Source: Ref. 1 and other literature.

paste–paste combinations. In the former case, the powder is composed of zinc oxide as the primary ingredient, usually containing 1 to 5% zinc acetate added to accelerate the reaction, and the liquid component is made up of eugenol mixed with small proportions (5 to 15%) of cottonseed or olive oils, added for viscosity control. The paste–paste combinations, supplied for greater ease of mixing, generally contain the ZnO–ZnOAc solids admixed to a plant oil, on the one hand, and the eugenol admixed to an inert mineral filler, on the other. In the absence of water, the two parts mixed together by the clinician can be handled conveniently without premature setting; once applied, the mixture, now exposed to the moist and slightly warmer oral environment, sets rapidly. The cement is weak, however, having a compressive strength of no more than 10 to 20 MPa (somewhat higher upon additional resin reinforcement). In addition, eugenol leaching from the lute and subsequent hydrolysis may lead to significant deterioration of the material. For this reason, the ZOE cements should be used only for temporary luting.

4. Zinc Oxide–Ethoxybenzoic Acid Cements

These two-part powder/liquid materials, close relatives of the ZOE cements, in the ultimate form are zinc chelates, resulting for the most part from reaction of ZnO with *ortho*-ethoxybenzoic acid:

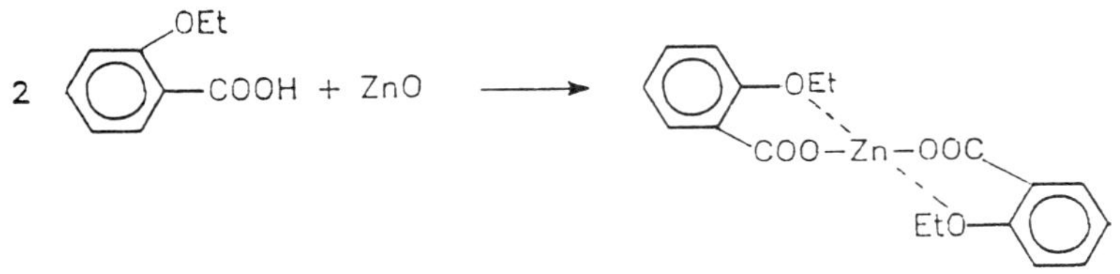

The zinc oxide–ethoxybenzoic acid (EBA) cements can thus be classified as zinc carboxylates in which the metal center is additionally bonded coordinatively to two ether oxygen atoms, forming electronically stabilized six-membered chelate ring structures. The powder component, again, contains zinc oxide, typically 60%, in addition to some 35% fused quartz filler and other resin ingredients, whereas the liquid part may typically consist of a 60:40 mixture of ethoxybenzoic acid and eugenol. Hence Zn eugenolate chelates are present as well in the hardened matrix. As with the ZOE materials, and much to the convenience of the clinician, the setting is accelerated by moisture in the oral environment. The hardened cement is comparatively strong under compressive load, ultimate compressive strength values reaching 80 to 85 MPa; in addition, it is less soluble than the ZOE cements in water. Both factors combine to render the EBA materials suitable for permanent luting.

5. Polycarboxylates

Developed some 25 years ago, the polycarboxylate materials are based on polycarboxylic acids, such as poly(acrylic acid), poly(maleic acid) and various acrylic acid copolymers, and their principal setting reaction involves carboxylate salt and chelate formation with polyvalent cations, mainly Zn^{2+}. Because of the polyvalent nature of the cations, the reaction leads to three-dimensional cross-linking. The polycarboxylates are generally supplied as two-part, solid-liquid systems. Finely powdered zinc oxide, sometimes admixed with magnesium oxide and other oxides, represents the solid component, whereas the liquid is a solution of poly(acrylic acid) (ca. 40%) and other polyacids in water. Rapid reaction occurs on mixing of the components; within 15 min some 75% of ultimate strength (compressive; approximately 80 MPa) is attained. Other commercial products are supplied as powders, which require mixing with water for cementation. These solids are composed of zinc oxide and anhydrous poly(acrylic acid) in the proper proportions, and reaction sets in upon admixture of water, which provides the vehicle for the ionic reaction sequence. Some materials contain fluoridation agents such as tin(II) fluoride in the powder component, thus providing protection of adjacent tooth structure against secondary caries without affecting the bonding characteristics. The polycarboxylate cements are more readily soluble than the phosphates in aqueous media. On the other hand, they offer an additional contribution to retention insofar as they form ionic bonds through salt formation of free carboxyl groups with the cationic calcium present in the hydroxyapatite tooth structure. The micromechanical bonding contribution in polycarboxylate cementation nevertheless is a major one, rendered highly efficacious as a result of surface porosity and wettability of the enamel adherend brought about by etching with the poly(acrylic acid). Bonding to dentinal tooth structure (invariably much weaker than to enamel) may also involve carboxyl group interaction with reactive groups (OH, NH_2) in the collagen constituent, which makes up a substantial proportion (Table 1). The polycarboxylates bond strongly to stainless steel, whereas there is little or no chemical adhesive bonding to noble metal alloys, porcelain, and resin restorations.

6. Glass Ionomers

For the development of cements of the glass ionomer (GI) type, features have been borrowed from both the polycarboxylates and the silicate cements. In the fundamental cementation process, polyacids, such as poly(acrylic acid) and acrylic–maleic or acrylic–itaconic acid copolymers, interact with inorganic cationic constituents of sodium aluminosilicate glass possessing a high Al content. The reaction involves a complex interplay between hydronium ions from the polyacids penetrating into the glass core and calcium and aluminum cations migrating out of the core into an outer gel phase for subsequent salt and complex formation with the polyacids. Tartaric acid is commonly added as a controlling agent of the setting characteristics. The application forms, powder–liquid and powder–water, are similar to those of the polycarboxylates. The cured cements are quite strong under compressive load, ultimate compressive strength values approximating 140 to 180 MPa, and sonication immediately after mixing of the components, resulting in void reduction, appears to have an enhancing effect on compressive strength. The adhesion characteristics resemble those of the polycarboxylates, provided that early moisture access is avoided, for example, by protecting the fresh lute margin with a coat of varnish. Typical shear bond strength data measure in vitro on conventional cements bonded to enamel and dentin, respectively, are 9 to 10 and 3 to 5 MPa [13,14]. Because of the rather low (ca. 12 MPa) inherent tensile strength of the glass ionomers, tensile failure is cohesive in the cement, setting a limit to overall expected bond strengths. For a novel stainless steel–reinforced GI cement, corresponding strength values of about 14 and 10 MPa, respectively, have been reported [14]. As the predominant use of the GI cements is in cavity lining, the reader is referred to Section IV.B.5 for additional details.

7. Resin Cements

Basically composed of poly(methyl methacrylate) upon setting, the acrylic resin cements in the unfilled state are simple, linear organic polymers.

$$nCH_2 = \overset{\overset{\displaystyle CH_3}{|}}{C} - COOCH_3 \longrightarrow \left[CH_2 - \overset{\overset{\displaystyle CH_3}{|}}{\underset{\underset{\displaystyle COOCH_3}{|}}{C}} \right]_n$$

Although known in dentistry for several decades, they have not as such enjoyed much acceptance because of considerable volume shrinkage on polymerization (21 to 22%) [27] and consequent microleakage, a high coefficient of thermal expansion (about 10 times greater than observed for tooth substance), high exothermicity of the polymerization reaction, and other shortcomings. Strong points, on the other hand, include low solubility in oral fluids, good thermal insulation, outstanding transparency, and ease of manipulation in the virgin (uncured) state. The commercial products generally are two-part systems made up, first, of a powder that contains fine (<50 μm) beads of poly(methyl methacrylate) and a peroxide-type initiator, and second, a liquid composed of methyl methacrylate monomer and a chemical activator, usually a tertiary amine, such as N,N-dimethyl-p-toluidine. More recent products use peroxide/alkylborane, peroxide/sulfinic acid, and other initiator/activator systems. The set cements bond mechanically to the tooth structure.

The development of filled resin cements (composites) has helped overcome some of the shortcomings of the unfilled resins. The two-part materials comprise a powder component, such as silanized silica of small (10 to 15 μm) particle size, combined with peroxide initiator, and a liquid component consisting of a bisacrylate monomer, such as 2,2-bis[4-(2-hydroxy-3-(methacryloyloxy)propoxy)phenyl]propane (bis-GMA), a diluent comonomer, usually triethylene glycol dimethacrylate (TEGDMA), and some 0.5% of a tertiary amine activator. Paste–paste and paste–liquid systems comprising the aforementioned reactants and activators are also on the market. The recent advent of light-activated composite materials for restorative applications has prompted the development of similarly composed, single-paste composite cements for luting purposes. Such light-cured cements contain certain diketones (0.03 to 0.09 wt %), the most popular being camphorquinone, which in the presence of amines generate free radicals upon irradiation with visible light. Halogen lamps (400 to 500 nm) are the standard light sources, although argon ion lasers (476.5 nm) are being used in current experimental studies with variable results. The free radicals so generated then initiate methacrylate polymerization. Also in use are *dual-cure composites*, in which a primary polymerization phase is photo-initiated, to be followed by chemically initiated secondary polymerization. Composite hardening proceeds with comparatively low polymerization shrinkage (1.2 to 2.7%) and low exothermicity, and the set cements feature low solubility and coefficients of thermal expansion significantly lower than observed with the unfilled resin materials. Both chemically and photochemically initiated low-viscosity resin systems of the bis-GMA-TEGDMA type as presented in the foregoing, yet containing little or no filler reinforcement, are now widely employed as pit and fissure sealing materials. Such sealants serve to protect natural enamel faults from becoming carious and thus represent a vital tool in preventive dentistry. Irrespective of the specifics of application, use of methacrylate-type resins requires brief (20 to 60 s) acid-etching pretreatment of enamel surfaces, generally with aqueous phosphoric acid as detailed in Section V.B.3. The surface roughening so achieved will allow resin tags to anchor to the enamel adherend; hence the mode of bonding to tooth structure is of the micromechanical kind. In view of the predominant part played by the composite resins as cavity-filling restoratives, the topic will be discussed in more detail in Section V.B.2.

8. Adhesive Resins

The luting cements based on silicates, phosphates, ZOE, EBA, and simple methacrylate resins provide little, if any, chemical adhesion, and as pointed out before, the existing bonding forces, for the most part involving micromechanical retention, are weak. However, special biphasic resin compositions are available which, on account of the presence of both hydrophilic and hydrophobic molecular constituents, experience enhanced retention to the enamel and dentinal domains of the tooth substance and to the restoration, although still largely by micromechanical interlocking. This feature is exploited for the luting of porcelain and composite inlays or onlays and for the bonding of porcelain veneers onto buccal tooth surfaces.* Resins of the biphasic type find their major use as dentin bonding agents and hence will be explored more thoroughly in Section V.B.3. Only two exemplifying luting agents for porcelain inlays are presented in this section. One of these is based on 4-(2-methacryloyloxyethoxycarbonyl)phthalic anhydride (4-META) as the key monomer. The material consists of a base, a mixture of 4-META and MMA, which is combined with the (preoxidized) tributylborane initiator prior to application. As the material also adheres strongly to etched base alloy, it is an efficacious bonding agent for the attachment of metallic bridges to acid-etched enamel of the tooth structure. Enamel–metal joints with this 4-META–based resin system have been reported to attain mean tensile bond strengths near 26 MPa [6]. Typical shear bond strengths to both dentin and base metal alloys are 20 MPa [7]. With similar values determined for the bond to porcelain, the resin is also useful for the repair of porcelain restorations [8]. The second bonding system consists of a weakly filled, modified phosphate ester of bis-GMA. The application procedure includes etching of the enamel with aqueous phosphoric acid for surface roughening, application of the bonding resin, and placement of the restoration. The resin hardens through chemically initiated polymerization; temporary superficial protection from oxygen is required for proper conversion [9]. In order to achieve acceptable retention, the porcelain surfaces are properly pretreated.* This affords enamel–porcelain joints with representative bond strengths of 14 to 18 MPa. Typical shear bond strength values for the adhesive itself, attached to silanized composite, are in the vicinity of 25 MPa [5]. Adhering strongly to base metal alloys [9–11], this bonding system also finds use in prosthodontics for the fastening of bridges as in the preceding case. Usefulness as an orthodontic bracket adhesive is also indicated [9], with shear bond strengths near 16 MPa and fracture in tension observed to occur at the metal surface [12]. For another highly efficacious adhesive-type luting agent, based on a biphenyldimethacrylate (BPDM) primer, the reader is referred to Section V.B.3.

---

*For optimal retention, the surfaces of the restoration require special pretreatment generally performed in the technician's workshop. For example, the bonding surfaces of porcelain restorations are commonly subjected to microsandblasting, followed by silanizing with a silane coupling agent, such as 3-methacryloyloxypropyl-(trimethoxy)silane, or simply by acid etching with hydrofluoric acid. Combined acid-etching and silanizing procedures are also popular, as are the more recently developed methods of silica coating by various techniques [4]. Similar silanizing treatments have been proposed for composite resin inlays [5]. The mechanism of adhesion to silica-coated and/or silanized adherend surfaces includes chemical bonding through Si—O links and major or minor contributions by micromechanical interlocking.

III. ENDODONTIC SEALANTS

Dental root treatment commonly comprises removal of the necrotic pulp or its remnants, cleaning, widening, and sterilization of the root canal, and filling of the prepared canal with core and sealing materials. Popular core materials are silver, gutta-percha, and silicon rubber points, and these are sealed in place by a cement sealer.

A. Requirements

The filling materials are in direct contact with the dentin of the canal walls and in a more indirect contact with the soft connective tissue in the apical area. Accordingly, various features of biocompatibility are a prime requirement of the endodontic sealant cements. These cements must also display acceptably low levels of solubility in aqueous media in addition to providing a good seal along the entire contact area for prevention of ingress of bacteria. Setting characteristics must be such that placement in the moist and warm endodontic environment can be accomplished without premature hardening; hence moisture activation, as observed with certain luting cements of the ZOE type, cannot be tolerated. Radiopaqueness is a property frequently called for whenever radiographic control of the filling geometry is indicated.

B. Materials

Although covered as a separate group (M2) in the FDI classification, the endodontic cements are to a large extent similar to the luting materials covered in Section II.B. One of the most frequently used cements is based on the ZOE system (see Section II.B.3) for reasons of simple and conventional application techniques and good setting properties. The irritating effects of the eugenol constituent are generally tolerated. Modified ZOE products containing various additives for consistency or setting time control dominate the market. Related sealer materials based on zinc oxide and ketone-type Zn-chelating agents, sometimes further modified with vinyl polymers, are also commercially available, as are certain retarder-modified calcium hydroxide–zinc oxide–salicylate combinations similar to the calcium hydroxide cavity-lining materials of Section IV.B.1. With a different class of sealing agents, which contain acrylate monomers resulting from epoxy–bisphenol A addition reactions, good dimensional stability and sealing capacity is attained, although irritation caused by acrylic monomers diffusing into soft connecting tissue may be problematic.

Among the newer materials advocated for root canal sealing and reviewed by Chow et al. [15] is an apatitic calcium phosphate cement formed under ambient conditions from calcium hydrogen phosphate and a tetracalcium phosphate formally composed of equimolar quantities of CaO and $Ca_3(PO_4)_2$:

$$CaHPO_4 \cdot 2H_2O + Ca_4(PO_4)_2O \rightarrow Ca_5(PO_4)_3OH + 2H_2O$$

Being hydroxyapatitic in structure, cements of this type are particularly tissue-compatible, and fluoride can be introduced (OH replaced by F) to provide protection against caries. Compressive strengths of these phosphates are in the relatively low range of 30 to 50 MPa, depending on formulation and application details. Increased strength and reduced setting time are achieved by compounding the phosphates with polyalkenoic acids, such as poly(acrylic acid).

For the most part, the sealer cements used at present exert bonding effects to adjacent surfaces of points or dentinal tooth structure by the micromechanical mode. Only the polyalkenoic acid–modified cements are likely to undergo additional weak retention through chemical bonding involving the carboxyl functions and apatitic hydroxyl groups of the dentin.

IV. CAVITY LINING

The dentinal tooth substance exposed by the clinician in the process of cavity preparation has direct access to the sensitive pulp via the dentinal tubules and so is highly responsive to irritating effects and attacks originating from, or transferred through, the cavity-filling material. Although such effects are minimal and clinically acceptable with many of the nonmetallic restoratives in current use, some of the more "classical" filling materials, notably amalgam, require pretreatment of the prepared cavity with a cement acting as a base or liner with the specific function of providing a protective barrier between restoration and dentin.

A. Requirements

In the process of hardening, the overlying filling material may exert considerable pressure on the liner. Additional forces, mostly compressive in nature, will be transmitted to the liner through the hardened restorative as the completed restoration is subjected to the stresses of mastication. To prevent liner deformation and flow under the packing load in the process of cavity filling, it is important for the lining cement to have undergone sufficient hardening before the filling step is initiated. Application of the cement requires a pastelike consistency, and this contrasts with the low-viscosity materials used for luting. As a rule, depending on type of cavity, type of restorative, filling technique, and other variables, a lining cement must attain a compressive strength of up to 26 MPa prior to filling, and a sufficiently long time interval between placement of the liner and that of the restorative is therefore indicated. Ideally, a liner should display good micromechanical and/or chemical bonding characteristics vis-à-vis both the tooth structure and the restorative so as to minimize microleakage. In terms of strength and mechanism, the bonding effects depend on the materials' composition; in practice, they are found to be rather weak.

The base or liner should provide a chemical barrier protecting the underlying dentin from attack by acids or acrylic monomers and other irritants that may diffuse out of the restoration. Needless to say, cements emitting irritants themselves may be restricted in use to the lining of shallow cavities where pulp irritation represents a less severe problem. Another important function of the cement base or lining, especially with amalgam fillings in deep cavities, is thermal and electrical insulation, so as to minimize heat transfer or transmission of electrical currents to the sensitive pulp area. Although electrical conduction is not always prevented by the common water-based ionic or metal-chelating cements, their thermally insulating properties generally are more than adequate for the purpose.

B. Materials

Not unexpectedly in view of the related functions of luting and lining materials, most of the materials discussed in Section II.B as luting agents are more or less equally useful as cavity base and liner cements. Cements classified as cavity liners include the calcium hydroxide materials, the zinc phosphates, zinc chelating agents, polycarboxylates, and glass ionomers.

1. Calcium Hydroxide Cements

One of the oldest lining materials in use, calcium hydroxide cement still enjoys some popularity in this field, although in a vastly modified form. It has antibacterial properties, is biocompatible, and promotes pulp recovery and regrowth of dentin. In the original form used—namely, as a suspension of calcium hydroxide in water—it yielded cements too weak for acceptable clinical performance. Coapplication of bonding agents such as methyl or carboxymethyl cellulose marginally raised the tensile and compressive strengths to 1 to 2 MPa and 7 to 8 MPa, respectively, and in order to apply the material as a liner under amalgam, an underlay by a different, stronger cement was required. More recent products are of the two-part type, with zinc oxide (typically, 10%) in combination with calcium hydroxide (50%) in one part, and salicylic esters exemplified by 1,3-butylene glycol disalicylate (40%) in the other. Cementation hence involves chelation, as in the more commonly used zinc oxide–eugenol cements discussed in Section II.B.3.

The latest development in the field of calcium hydroxide cements aims at light-activated compositions. The setting reaction in these products is quite different from the chelate formation mechanism of the calcium–zinc salicylate cements insofar as the materials harden through light-induced, chemically activated polymerization of dimethacrylate (bis-GMA) and monomethacrylate (HEMA) monomers as coingredients with calcium hydroxide. Although compressive strength values of present-day materials are still quite low (ca. 20 MPa), lining under amalgam is practicable under restricted conditions of cavity geometry. More often, these products are used as liners under silicate- or resin-based fillings. The bonding mode of the calcium hydroxide cements is largely micromechanical.

2. Zinc Phosphates

The zinc phosphate materials, discussed in Section II.B.1 as luting agents, are also in use for cavity lining. The major difference, compared to the luting cements, is the more puttylike consistency required for base or lining purposes and brought about by increasing the powder/liquid ratio to, maximally, 3.2 g/mL. A cement of this composition sets in 7 min or less to a hardness sufficient for amalgam filling without liner displacement, and the ultimate compressive strength may be as high as 140 MPa. Retention to the adherends, as in the luting application, utilizes micromechanical interlocking. The cements provide good thermal insulation, but the chemical barrier properties of these inherently acidic materials are poor. The former feature is beneficial, and the latter acceptable, for restoration with amalgam.

3. Zinc Oxide–Eugenol and Zinc Oxide–Ethoxybenzoate Acid Cements

The compositional and performance features of the zinc oxide–eugenol and zinc oxide–*ortho*-ethoxybenzoic acid cements were dealt with in Sections II.B.3 and II.B.4. While employed for temporary luting and filling, the ZOE materials find their major use in cavity lining. The fundamentally weak cement materials are usually reinforced for this purpose with poly(methyl methacrylate) filler in powder form, and compressive strengths of 40 to 50 MPa can thus be attained. The rather high degree of solubility and eugenol leaching, presenting a drawback in luting applications, can be tolerated for lining purposes even in deep cavities, where use of the phosphates is contraindicated. Occasionally, therefore, ZOE cements are placed as sublinings to be overlaid with zinc phosphate. They are not useful, however, as liners under resin-based restorations because free eugenol may interfere with the free-radical polymerization hardening of the resin filling materials.

Although employed predominantly as luting agents, the EBA cements, because of their high silica filler content (ca. 35%), are more resistant to flow in the uncured state, and possess higher compressive strength (80 to 85 MPa), than the ZOE cements. Therefore, they provide a useful lining function, notably under amalgam. As in luting applications, the bonding effects of both ZOE and EBA materials toward dentin and restorative are of the micromechanical mode and hence are quite weak.

4. Polycarboxylates

Once again, one is dealing here with a class of materials described in Section II.B.5 as luting agents, and indeed the only major difference compared to the latter is one of consistency, the polycarboxylates used as cavity base or lining materials having a higher viscosity as a consequence of a higher concentration of the aqueous poly(acrylic acid) solution. Because of the possibility of pulp irritation by the free polyacid present in the uncured material, use of the polycarboxylates is generally restricted to the lining of shallow cavities unless the linings are underlaid by protective calcium hydroxide or ZEO sublinings. As pointed out before, the polycarboxylates stand out against the phosphates and chelating cements because of their capability of bond formation between carboxyl groups and apatitic calcium cations of enamel and dentin. Compressive strength values of the hardened cement materials typically attain 90 MPa.

5. Glass Ionomers

Although originally used as direct filling materials, the glass ionomer (GI) cements have since proven their worth in a number of different dental applications, including the previously discussed luting of restorations. Their use as cavity base and lining materials has increased rapidly in recent years, and in this area the glass ionomers have established themselves as a major materials class for reasons of compatibility with resin restorations, biological acceptability, good thermal insulating properties, fluoride release,* and good strength and bonding characteristics, the latter accentuated by a low coefficient of thermal expansion (typically, $15 \times 10^{-6} \, °C^{-1}$), matching that of dentin (Table 1), thus minimizing microleakage caused by expansion differentials under thermocycling conditions. Similar in composition and hardening properties to the luting variety, the lining materials contain an ion-leachable (generally, calcium fluoride-modified) sodium fluoroaluminosilicate glass and an acrylic acid homo- or copolymer as the principal reactants. The powder–liquid products comprise the finely ground glass filler in the powder component, and an aqueous solution of poly(acrylic acid) or copolymer, sometimes in combination with tartaric acid. A high aluminum content in the glass serves to increase the reactivity with the polyacid. The tartaric acid additive undergoes early complex formation with Al^{3+} ions

---

*The release of fluoride ion to combat caries and encourage remineralization, although not directly pertinent to the adhesion problem, may affect bonding indirectly through creation of porous structures that would enhance leakage and ultimate weakening of the bond. The development of fluoride release mechanisms devoid of detrimental effects on existing bonds to restorations and restoratives or removable appliances is therefore a prime concern in dental materials research. See, for example, Cooley et al. [16,17].

liberated from the glass surface, thus facilitating calcium ion accessibility in the glass for acid attack; it remains an important participant in subsequent reaction steps leading to ultimate cross-linking.

Powder–water products differ from the powder–liquid products insofar as the solid component contains both the glass and the anhydrous polyacid, whereas the aqueous phase here is either plain water or a diluted aqueous tartaric acid solution. Both application forms produce the same type of end product, a cement comprising surface-gelled glass particle filler and polyacid matrix cross-linked through three-dimensional calcium and aluminum salt formation. Residual free carboxyl groups in the cement are left available for calcium salt formation involving adjacent dentin, and this represents an important, although weak chemical dentin-bonding mode utilized to advantage in GI applications. A further increase in the dentin–GI bond strength reportedly results from preconditioning the exposed dentin surface with aqueous poly(acrylic acid) solution; the conditioner etches the surface and serves to dissolve (and, perhaps, reprecipitate) the so-called smear layer, a thin (ca. 1 μm), mineral-rich zone of dentinal debris collecting on the freshly prepared dentin surface, which, if left untreated, is widely considered detrimental to the bonding process. Deep dentin surfaces, which possess a lower apatite content and show stronger resistance than upper dentin surfaces to bonding, can be activated for bonding by a mineralization treatment, which induces calcium phosphate crystallization and thus increases the Ca^{2+} ion concentration on the dentinal adherend.

GI adhesion to the enamel of the tooth structure is more efficacious than to dentin because, in addition to the calcium carboxylate bond formation with Ca^{2+} present in a higher concentration in the enamel adherend (Table 1), the free polyacid in the cement exerts an etching effect on the enamel surface, resulting in increased surface roughness and concomitantly improved micromechanical retention. Typical GI–enamel and GI–dentin bond strength data are provided in Section II.B.6. While GI liners are rarely used under amalgam, their beneficial application under composite fillings, where resin compatibility is important, has been widely accepted. Adhesion between liner and composite filling material can be improved further by etching the liner surface with phosphoric acid prior to packing of the restorative. Shear bond strength values of joints so prepared typically average 10 MPa (6.5 MPa without liner etching), and fracture occurs predominantly in the cement [18]. A weakened bond between liner and dentin, on the other hand, is often a consequence of this treatment. As the packed composite undergoes polymerization shrinkage, the firmly bonded GI liner, being subjected to tensile and/or shear stresses, tends to retreat from the dentin surface and, in the process, cause detrimental enhancement of leakage in the liner–dentin interface. With increasing success in research toward composite materials devoid of polymerization shrinkage (see Section V.B.2), one can expect this liner–dentin debonding problem to become less relevant.

The advent of light activation methods for resin composites has prompted research into light-activated GI cements. Representative products now on the market are powder/liquid combinations. The powder, again, constitutes an ion-leachable fluoroaluminosilicate glass containing a light-activated initiator. The liquid is an aqueous solution of a polycarboxylic acid modified with methacryloyloxy side groups, for example, a poly-(acrylic acid-co-methacryloyloxyethyl acrylate), hydroxyethyl methacrylate, and the light-sensitizer part of the light-activating system. Combinations of this type provide adequately long working times, as the purely chemical hardening process, utilizing the calcium cation–carboxylate interaction, proceeds at a conveniently slow rate. After completed placement, the material may be light activated, which initiates polymerization of the methacrylate side groups and entails rapid hardening. The presence of residual free carboxyl groups ensures chemical cement bonding to the enamel–dentin adherend as in the conventional products. Slow continuing reaction of polyacid and glass filler, following the light-curing step, leads to further maturation of the cement. Typical shear bond strength values for light-cured GI cements bonded to dentin range from about 3 to 5 MPa (occasionally even higher [20]), and similar values are obtained for bonds to amalgam.

Numerous other so-called ''light-curing'' GI cements have recently been commercialized that are related to the glass ionomers only insofar as they contain a powdery filler made up of GI powder and calcium phosphate as the principal ingredients. The matrix component of these materials is a light-curing mixture of mono- and diacrylate monomers. As a consequence, their setting shrinkage is considerably larger than that of the conventional GI cements [19]. Furthermore, containing no polyacids, these materials are unable to undergo the chemical bonding reaction to enamel–dentin characteristic of the glass ionomers proper, although other bonding mechanisms associated with the acrylate monomers may be quite efficacious. Procedural details for GI liner application have been described [21], and a good review of developments in this field is available [22].

V. CAVITY FILLING

While dental amalgam is still the most widely used cavity filling material for the direct restoration of defect posterior teeth, the retention of amalgam filling is due entirely to macromechanical containment in the undercut cavity. The same holds true for the silicate filling materials, which have for many decades been used for anterior restoration. These two classes of restoratives are therefore outside the scope of this chapter. Of interest as adhesion-active filling materials under the present heading are the glass ionomer cements, including their metal-reinforced varieties (cermets), and the composite resins.

A. Requirements

In addition to certain biological requirements, such as cariostatic properties and lack of pulp irritability or systemic toxicity, a filling material should possess low water absorption and should not dissolve in the oral fluids. The dimensional changes (generally involving contraction) on hardening of the material should be minimal so as to preclude tensile and/ or shear stress concentrations at the interface with tooth structure with resultant development of microleakage, and the thermal properties (e.g., coefficient of thermal expansion and thermal diffusivity) should resemble as far as possible those of the tooth substance so as to minimize the development of interfacial shear and tensile stresses. Ideally, the mechanical properties, notably strength and stiffness, should match those of enamel and dentin, and some bonding mechanisms, micromechanical and/or chemical, should be operative between cement and cavosurface. Additional requirements, of no major interest in the present context, are concerned with cosmetic considerations, radiopacity, and rheological behavior, the last-named two features being of importance in the clinical application.

B. Materials

1. Glass Ionomers

Although prevalently used as luting and cavity lining cements, the glass ionomers play a moderate part as cavity-filling materials, largely on the strength of their adhesion to the enamel and dentin of the tooth structure, the polyacid components participating in ionic bond formation with calcium cations of the hydroxyapatite in addition to undergoing weak ionic and/or covalent bonding with basic or nucleophilic sites in the dentinal collagen. The structural features and bonding mechanisms were discussed in Sections II.B.6 and IV.B.5. The compositions and properties of the GI filling materials are quite similar to those of the luting and lining varieties, the main difference being a more viscous consistency of the filling material, brought about by increased filler/liquid ratios and/or varied types and sizes of the glass–particulate fillers. As pointed out earlier, the weak link in GI–enamel bonding frequently is not so much the interface but the cement itself, which is quite brittle and posesses low flexural (15 to 20 MPa) and diametral tensile (8 to 12 MPa) strengths. It is largely for this reason that the GI cements are not routinely employed for restoration of permanent teeth, where premature failure would be expected under the load of masticating forces.

Metal-containing GI materials, known as cermets, are the latest in specialty development in the field of dental ionomer cements. The cermets contain a filler phase obtained by fusing silver and other metals or alloys together with aluminosilicate glass and pulverizing the molten mass. This is then combined with poly(acrylic acid) in one- or two-part fashion as described in Section II.B.6. The cermets display setting and bonding characteristics resembling those of the metal-free parent cements while displaying better fatigue limits, and in properly poly(acrylic acid)–conditioned cavities, cause significantly less marginal leakage. However, there appears to be no clear superiority with respect to other strength characteristics, both tensile and compressive strength values being in the same ranges as observed for representative GI cements, although for a silver–tin–zinc alloy as the metal component, encouraging compressive and diametral tensile strength data (187 and 18 MPa, respectively) have been reported [23]. An interesting potential application for reinforced-glass ionomers in restorative dentistry suggests itself for building up cores in severely destructed teeth prior to the placement of crowns. Perfect dimensional stability is required for a core to support a superimposed crown efficaciously. Conventional composites exposed to moisture are not sufficiently stable dimensionally for this kind of application as a consequence of unduly high water sorption, and there are indications that reinforced GI cements, on account of better dimensional stability, may more adequately fulfill that requirement [24].

2. Composites

The shortcomings of the unfilled acrylic resins as luting agents were emphasized in Section II.B.7, and for similar reasons, these clear acrylics have failed to establish themselves as restorative materials. The composite resins on the other hand, after a lengthy development period have come to be recognized as one of the most useful and versatile classes of dental materials now available to the clinician for both anterior and posterior restorations [25]. Composite resins are essentially ceramic-filled, polymerizable dimethacrylates, the curing (hardening) of which, as pointed out before, involves three-dimensional cross-linking through free-radical polymerization of the acrylic groups, initiated either chemically (i.e., through peroxide-amine redox initiation) or photo-lytically, (i.e., through a light-activated process commonly involving α-diketone photo-oxidants and amine-type photoreductants). Contrasted with the unfilled acrylic, the present-day composite resin systems feature low exotherms and comparatively low polymerization shrinkage (typically, 1.5%), low water absorption and solubility, yet improved thermal properties, esthetics, biocompatibility, and mechanical stiffness.

The dimethacrylate resins constituting the matrix generally contain aromatic ring structures to impart rigidity and high viscosity. The most common representative, bis-GMA, a bisphenol A derivative, was introduced in Section II.B.7. Other partly aromatic and highly viscous, yet less hydrophilic dimethacrylates as currently used matrix components, imparting enhanced dimensional stability, are 2,2-bis(4-methacryloyloxyphenyl) propane (bis-MA) and 2,2-bis[4-(3-methacryloyloxypropoxy)phenyl]propane (bis-PMA). To optimize clinical manipulation, the matrix contains low-viscosity comonomers, including the previously introduced TEGDMA and a large variety of aliphatic and aromatic urethanedimethacrylates. Although the degree of conversion and cross-linking increases with raised concentrations of the low-viscosity monomers, at the same time it causes increased polymerization shrinkage with obvious detrimental effects on adhesion to the tooth material. Although incremental placement of the composite, with intermittent partial curing of the individual layers, is being practiced in an effort to minimize contraction on curing, this technique tends to reduce the ultimate fracture toughness within the interface between the layers of the restorative. A recently described method of compensating for contraction during polymerization utilizes ammonia-treated montmorillonite as a low-percentage additive [26]. More promising pointers toward overcoming the polymerization shrinkage problem are found in the excellent work currently performed, inter alia, in the laboratories of Eick [27–29] and of Stansbury and Bailey [30] on cyclic monomers consisting of spiro-orthocarbonates, such as the cis-trans isomers of 2,3,8,9-di(tetramethylene)-1,5,7,11-tetraoxaspiro-[5.5]undecane or similar structures possessing exocyclic polymerizable double bonds. Monomers of this type undergo polymerization with volume expansion, and the reaction can be photoinitiated, for example, with (4-octyloxyphenyl)phenyliodonium hexafluoroantimonate. Structural design features have been discussed and methods for volume change measurement presented [31]. The presence of exocyclic double bonds may facilitate polymerization, and methacryloyloxy-substituted spiro-orthocarbonates, which also polymerize with volume expansion, offer the potential for copolymerization reactions with conventional resin systems. Further advancement in this field can be expected, and this should contribute significantly to the retention properties of composite materials.

The discontinuous, reinforcing phase of the composites, which on a mass basis constitutes some 50 to 85% of the total cement, consists of siliceous ceramic filler particles, generally crystalline quartz, barium or strontium aluminoborate silica, aluminosilicate glasses, prepolymerized composite material, and specialty biphasic glasses. Depending on filler particle size, one distinguishes the conventional composites, with a filler size of 1 to 50 μm, from an important intermediate class of composites featuring 1- to 5-μm filler size, a third class known as microfilled composites with a mean particle size of 0.04 μm, and finally, the so-called hybrid composites, which for most efficient packing and highest fracture toughness, typically incorporate some 70 to 75% of conventional filler and 8 to 10% of submicron-size silica filler. These variations of filler type, size, and concentration play a major part in affecting the physical and performance characteristics and thus the optimal clinical conditions for application of each one of the numerous types of compositions on the market.

The strength, fracture toughness, and general durability of the resin–filler combinations in the oral environment are all critically dependent on a strong bond between resin matrix and reinforcement particles. Weak interfacial bonding leads to marginal degradation, penetration of oral fluids, and premature wear under the masticatory forces. Untreated filler materials are anchored to the matrix essentially by the micromechanical mode, as the polymerizing resin locks into the surface voids and crevasses of the filler or penetrates into the pores of especially porous filler materials. Introduction of a chemical adhesion component in the form of coupling agents improves the bonding dramatically. The commonly utilized compounds are methacrylate-terminated alkoxysilanes [e.g., 3-methacryloyloxypropyl(trimethoxy)silane], occasionally in combination with zirconates and other co-coupling agents. The rationale behind this structural choice is the expectation that, upon treatment of the filler materials (glassy fillers requiring preetching) with coupling agents of this type, silyl ether bonds are formed with surface hydroxyl groups of the filler, while polymerizable vinyl groups protrude from the surface layer and, on compounding with the resin, should be available for copolymerization and cross-linking with the embedding matrix. In practice, however, most of the vinyl groups of the silanized filler surface appear to undergo homopolymerization, and the actual resin bonding involves formation of an interpenetrating, rather than cross-linking, network on the interface as the polymerizing matrix resin diffuses into the polymethacrylate surface layer. Irrespective of the actual bonding mechanism operative in the interface, silanizing of filler materials prior to compounding with the matrix is generally the accepted method of efficaciously enhancing resin–filler adhesion. Typical diametral tensile bond strength values reported for a light-cured, zirconate-treated bis-GMA resin composite containing a silanized glass filler are 55 to 56 MPa, as against 32 MPa for a composite containing untreated glass [32].

3. Bonding Agents

One of the most intensely pursued objectives in dental materials research over the past two decades is the achievement of clinically acceptable retention, by micromechanical and/or chemical bonding mechanisms, of the restorative to the prepared enamel and dentinal tooth structure. Perfect retention, in addition to providing a major contribution to the longevity of the restoration, would offer the best protection against microleakage of oral fluids along the tooth–restorative interface, with its detrimental consequences of bacterial ingress and secondary caries development. Optimally effective interfacial bonding re-quires complete wetting of the adherend surfaces by the adhesive and the attainment of durable bond strengths matching the inherent strength levels of the dental and restorative components of the joint. Although materials science is still a long way from reaching such perfection, much has been accomplished in recent years in pursuit of this goal. In view of the importance of dentinal and enamel bonding in restorative practice, the subject is being treated in this section under its own separate heading. Also covered here briefly are bonding methods used for prosthodontic and orthodontic attachments and repair.

The retention of restoratives and restorations to the tooth structure is customarily measured in terms of shear bond strength and, less commonly, tensile bond strength. Peel strength measurements, as routinely performed in other segments of adhesion technology, are not particularly predictive here and hence are seldom utilized in restorative dentistry. The bond strength data reported in the dental materials literature tend to show consider-able variability because of marked sensitivity to the materials and techniques employed. Type, age, and preconditioning of the tooth material, type and geometry of the prepared cavity (or other adhesion surface), and the application variables of primer and filling material all are of critical importance, and so are the details of postconditioning (e.g., storage in saline and thermocycling) of the prepared joints, and the techniques and devices used for bond strength testing.* The strength data given in the text should thus be accepted at best as representative, useful indicators of general bonding performance. It is equally important to realize that the data reported in the literature have been derived almost entirely from in vitro tests and thus cannot simply be correlated with in vivo results, although their value as predictors of clinical performance remains undisputed.

The composite materials presently on the market do not per se possess adhesive properties conducive to bonding to the hard tissue of tooth structure. Auxiliary techniques are available, however, which enable the clinician to overcome this inherent deficiency, and composite-type restorations are routinely placed nowadays under conditions leading to an acceptable, if not perfect degree of bond formation with the cavosurface. Thanks to these advances in dental material technology, cavity preparation with large undercuts, as with amalgam fillings, is no longer a necessity for successful restoration, and the beneficial consequences in terms of preservation of healthy tooth structure and minimization of secondary caries through reduction of microleakage are obvious. Because of differences in some of the bonding mechanisms between the resin–enamel and resin–dentin adherend pairs, the techniques required for resin bonding to enamel on the one hand, and to the dentinal tooth component on the other, differ in certain aspects. Enamel is a biomaterial of low free surface energy and thus will resist wetting by a potential adhesive. Moreover, as pointed out before, it consists of 97 wt% mineral constituent, essentially hydroxyapatite. Any adhesion process would therefore have to rely almost exclusively on reactions with the exposed apatitic hydroxyl groups, as has been established for the polycarboxylate and ionomer cements (Sections II.B.5 and II.B.6). Reactive partners of this type, however, are absent in the resin-based materials. For a mechanical joint, on the other hand, the cut enamel surface, having grooves substantially shallower than 100 nm, lacks the roughness required for retention of the intruding resin tags. The advent of the acid etch technique, developed by Buonocore in 1955, changed the situation dramatically. Acid etching, in essence an enamel-conditioning process, and by now a standard clinical procedure, involves a brief treatment of the clinically prepared enamel surface with acids, most commonly phosphoric acid, applied as an aqueous (30 to 50%) solution or, more conveniently, as an aqueous gel. The resultant increase in free surface energy enhances the wetting characteristics and so enlarges the interfacial contact area. In addition, the etching creates microporosity, which allows the subsequently placed resin to flow into the pores, forming resin tags with a typical length of 25 μm, thus efficaciously anchoring the composite to the enamel in a micromechanical fashion. The depth of hard-tissue penetration is not necessarily, however, the prime contributor to the bonding effect; tag density and inherent strength both are of at least equal importance. The placement of heavily filled

---

*The divergence of test methods currently employed in different laboratories has prompted numerous calls for international standardization, exemplified by recent proposals to standardize methods for dentinal bond strength determination and, herewith related, for the evaluation of microleakage and marginal gap dimensions [33]. On a more universal scale, several years ago, with the aim of developing standardized test methods, a working group was convened by D. R. Beech of the Australian Dental Standards Laboratory under the auspices of the International Standards Organisation (ISO) Technical Committee 106 (Dentistry). A draft report, completed in 1991, CD TR 11405, entitled *Dental Materials Guidance on Testing of Adhesion to Tooth Structure*, presents precise details of screening tests, bond strength measurements, gap and microleakage tests, and clinical usage tests. Hopefully, a final version will soon be universally available. A useful tool for assessment of the reliability of a bond is the Weibull analysis approach [34]. The method, utilized now in many laboratories, allows for determination of the probability of bond failure as a function of applied stress.

and viscous composites, including the hybrid types, which may find it difficult to penetrate into the pores, is frequently preceded by application of a layer of unfilled resin of low viscosity compatible with the composite, although the success of this method is questioned by others. Typical tensile bond strengths attained between composite resin and acid-etched enamel range from 16 to 23 MPa, highest bond strength values generally being associated with surfaces cut transversely to the enamel crystallites [35]. The topic of acid etching has been reviewed by Gwinnett [36] and by Retief [37]. In addition to the acid etching technique, methods of enamel etching by laser treatment have more recently been introduced and in general appear to be similarly effective, or even superior, although more cumbersome in clinical practice.

The development of chemical coupling or bonding agents for resin adhesion to hard tooth structure, pioneered by Bowen several decades ago [38] and recently reviewed by that author [39], represents a challenging chapter in contemporary dental materials research. Although applicable to resin–enamel bonding, the chemical adhesive materials currently available find their major use in resin–dentin bonding applications.

Contrasted with enamel, dentin contains only 69% hydroxyapatite matter in addition to an increased percentage of organic substance of low surface energy and aqueous fluids, with occupy the dental tubules (Table 1). On a volume basis, the overall organic–aqueous domain makes up more than one-half of the dentinal substance. The dentin surface is thus a strongly hydrophilic adherend. The bis-GMA and related resin components of the composite, on the other hand, represent hydrophobic constituents. A bonding agent intended to join dentinal and composite adherends durably must therefore be hydrophilic enough to displace the aqueous phase from the dentinal surface for subsequent bonding, by whatever mechanism, to the dentinal substrate. At the same time, however, it must comprise hydrophobic molecular entities compatible with, and capable of bonding to, the resinous restorative. Based on this rationale, early biphasic, surface-active dentin bonding agents, developed in Bowen's laboratory [38], were of the type N-[2-hydroxy-3-(meth-acryloyloxy)propyl]-N-phenylglycine (NPG-GMA), N-[2-hydroxy-3-(methacryloyloxy) propyl]-N-(4-tolyl)glycine (NTG-GMA), and related structures. These compounds are distinguished (1) by the presence of hydrophilic functional amino acid groups capable of chelating or ionic bonding to the apatitic surface calcium and other multivalent cations and to reactive amino groups in the organic (collagen) domains of dentin, and (2) by the presence of reactive vinyl groups capable of copolymerization with composite resin. Other first-generation bonding agents contained isocyanatoacrylates or diisocyanate-terminated oligourethanes designed so as to form cross-links between dentinal hydroxyl and amine functions and filler hydroxyl groups. Halogenated phosphate esters of bis-GMA, HEMA, and other methacrylate substrates, believed to function through calcium phosphate bonding to dentin and vinyl-type copolymerization with composite resin, were also developed at that time. The compounds were applied as thin layers to variously conditioned dentinal surfaces, followed by the placement of standard composites. Although initial results were by no means impressive, shear bond strengths at the very best attaining 10 MPa, these early pioneering investigations provided a powerful impetus to dental bonding research activities worldwide, and although many a development product fell by the wayside for reasons of poor long-term clinical performance, others were developed in the following years to a fairly high level of effectiveness and produced encouraging (although not necessarily clinically acceptable) results. Among the bonding systems that have reached the third-generation stage and compete for present-day clinical acceptance are those based on combinations of (1) glutaraldehyde with HEMA; (2) arylglycine-type surface-active monomers with PMDM, the adduct of HEMA to pyromellitic dianhydride; (3) hydrophilic HEMA with hydrophobic bis-GMA; and (4) methyl methacrylate with 4-META, the adduct of HEMA to trimellitic acid anhydride. A brief discussion of these examplifying bonding systems follows.

The original glutaraldehyde–HEMA system, developed in Asmussen's laboratory [40] and commonly known as GLUMA, contains as the critical component a primer consisting of an aqueous solution of glutaraldehyde (5%) and HEMA (35%), which was applied onto the dental surface precleansed with alkali-neutralized (pH 7.4) ethylenediaminetetraacetic acid (17% in water) for smear layer removal and superficial decalcification. This was overlaid with a sealer consisting of unfilled, light-cured resin of the bis-GMA type, onto which in turn the composite was placed. The primer mixture in this system interpenetrates and forms bonds with the top zone of the partly demineralized dentin matrix, to which it anchors the resinous overlays upon free-radical homo- and copolymerization. The bonding effects achieved with this early system were unsatisfactory; average shear bond strengths generally failed to exceed 10 MPa even after the implementation of further (minor) improvements. Bond failure occurred along the weakened decalcified dentin zone, as neither the primer nor the sealer diffused through that zone into the underlying calcified matrix. Adhesive failure at the sealer–composite interface was also observed [41]. Subsequent improvements and simplifications of the GLUMA system included changes in pretreatment and conversion of the primer into a self-contained bonding resin through inclusion of bis-GMA monomer and initiator. A typical present-day GLUMA bonding procedure [42] comprises the following steps:

1. Cavosurface cleansing by treatment with an aqueous solution of aluminum oxalate (ca. 5%) and glycine (2.5%) adjusted to pH 1.5. This results in both enamel and dentin etching and in amino acid infiltration into the etched dentin.
2. Brush application of bonding resin consisting of glutaraldehyde (5%), HEMA (33%), bis-GMA (2%), camphorquinone photoinitiator (0.1%), water (55%), and acetone (5%), followed by light curing.
3. Conventional placement of composite resin.

In this and similar systems (e.g., with pyruvic acid and glycine as cleanser components) [43,44] the amino acid infiltrated into the dentinal surface zone adds to the concentration of amino groups in that layer and thus contributes to glutaraldehyde bonding; in addition, it is believed to act as the reductant in conjunction with the comphorquinone photooxidant component in the interpenetrating resin, thus upon photoirradiation, initiating resin polymerization right along the contact surface with the cleanser. Shear bond strength values as high as 16 to 18 MPa to dentin, and up to 23 MPa to enamel, can be attained with this and similar third-generation GLUMA recipes.

In the field of bonding agents based on arylglycine–PMDM combinations, numerous advanced versions have originated from Bowen's early concept of biphasic monomers with both hydrophilic and hydrophobic functional sites as exemplified by the aforementioned NPG-GMA system. In our initial version, a second biphasic monomer, 2,5-bis [2-(methacryloyloxy)ethoxycarbonyl]terephthalic acid (PMDM), an addition product of HEMA to pyromellitic dianhydride, was added. The dentinal surface was first conditioned with an aqueous acidic solution of iron(III) oxalate, which removed the smear layer and deposited iron cations, contributing to the bonding effect through chelation. Next, an acetone solution of NPG-GMA or NTG-GMA was applied, followed by treatment with an acetone solution of PMDM and placement of the composite. The PMDM comonomer interacted synergistically with the precursor component, spontaneously inducing free-radical polymerization. Having passed through various stages of improvement, a current version, available commercially, comprises dentin conditioning with aluminum oxalate (6%) in dilute (2.5%) aqueous nitric acid, followed by application of a premixed acetone solution of NTG-GMA and PMDM. After solvent volatilization, this is overlaid with an unfilled, light-curing bis-GMA resin of low viscosity, to be followed by composite placement [39]. The micromechanical processes constituting the overall bonding effect have been studied by transmission and scanning electron microscopy* techniques [41,45]. Mean shear bond strengths of 17 to 18 MPa have been reported [46,47]; however, lower and quite variable values are also on record, once again stressing the need for standardization of bonding and testing techniques[48].

The recent finding in Bowen's laboratory that the oxalate conditioning and subsequent NPG-GMA coating steps can be replaced by a treatment with acidic NPG without loss of bonding strength has led to a related bonding system, also available commercially, in which the dentin is pretreated with a dilute (2.5%) aqueous nitric acid containing NPG (4%)[39]. This removes the smear layer, partially decalcifies the upper dentin layer, and permits interpenetration of the amino acid. Subsequent application of a 5% acetone solution of PMDM, with or without added HEMA, provides an overlay of resin, which penetrates into, and through, the decalcified zone and polymerizes spontaneously in contact with the amino acid, forming a resin-reinforced demineralized zone, which then bonds to the subsequently placed composite [45]. Tensile bond strengths are 12 to 16 MPa at best, and frequently much lower. On the other hand, and in contrast to the behavior of most other contemporary bonding agents, strength tends to increase slightly upon saline storage and thermocycling [49]. Failure typically occurs along the adhesive–tooth surface, and the adhesive resin itself is probably the weakest part of the joint.

Outstanding adhesion performance has recently been documented for a modified system in which the key ingredient is a combination of NTG-GMA and BPDM, a biphenyldimethacrylate derivative related to PMDM. The two components (called primers), dissolved in acetone, are premixed just prior to multiple brush application onto the dentinal surface preconditioned either by etching with 10% aqueous phosphoric acid or by treatment with a succinic anhydride-modified HEMA (SA-HEMA) (a hydrophilic/hydrophobic methacrylate possessing a propanoic acid terminal). The low-viscosity primer mixture displaces surface moisture on the dentin and interpenetrates the partly demineralized collagen layer exposed by the etching process and fills the dentinal tubule orifices. Subsequent application of an unfilled, photo-curing methacrylate bonding resin causes further resin reinforcement of the demineralized zone and subsequent copolymerization. This is followed by conventional composite application. Mean shear bond strengths range from about 27 to nearly 40 MPa, depending on details of the application technique, and failure is cohesive in dentin. The phosphoric acid–etching pretreatment and tolerance of a certain degree of surface moisture (by blotting or mild air drying) both combine to result in optimal bonding, whereas aggressively air-dried surfaces give

---

*Although not specifically indicated in the text, the techniques of transmission (TEM) and scanning (SEM) microscopy represent indispensable tools in the study of bonding processes and are widely used for the qualitative and quantitative evaluation of adherend surfaces, wetting and penetration, gap dimensions, and fracture mechanisms. Roulet et al. [61] have discussed the use of SEM in margin analysis, and publications dealing with preparatory methods for TEM and SEM investigations have been referenced by Eick et al. [45].

considerably weaker bonds [50]. The system described also lends itself exceedingly well to metal and porcelain bonding and has therefore found application in luting operations and prosthodontics [50,51]. For example, a Ni–Cr–Be base metal alloy is bonded to composite with a mean share bond strength in the vicinity of 25 MPa. Key aspects of the NTG-GMA-BPDM primer application have recently been discussed in some detail [51].

The development of HEMA-bis-GMA combinations as bonding agents has culminated in a number of recipes showing encouraging performance, and one major representative now on the market, defined as a dentin–enamel bonding system, has received wide attention. In a typical protocol, the enamel portions of the prepared cavity are conventionally acid etched, and the dentinal surfaces are primed with an aqueous solution of the hydrophilic HEMA and maleic acid as comonomers. This removes the smear layer and provides dentin interpenetration by the two monomers. Priming is followed by brush application, in a fairly thick layer (75 to 100 μm), of a resin adhesive composed of HEMA, bis-GMA, and a photoinitiator, with a few percent of a low-viscosity monomer added for viscosity reduction. After brief light curing of the adhesive coat, the composite is placed conventionally. Because of polymerization inhibition by oxygen, a reactive surface layer containing incompletely polymerized resin is left on the adhesive coat, and subsequent copolymerization with the composite resin overlay affords effective adhesive–composite bonding. Although earlier strength data reported were not particularly convincing, recent publications [41] cite mean shear bond strength values as high as 23 MPa, well on a par with enamel bonding data, with fracture for the major part cohesive in dentin or composite. Excellent performance with respect to minimal microleakage and marginal gap dimensions relative to competitive bonding systems tested are also on record [52]. On the other hand, this bonding system has been found to weaken on storage and thermocycling [41,49].

A combination of modified features of the last-named two bonding systems is realized in an adhesive application known as the Kanca technique, in which dentin and enamel pretreatment by phosphoric acid etching is followed by the consecutive layering of NTG-GMA, PMDM, and HEMA-bis-GMA adhesive resins, onto which the restorative is placed by conventional manipulation. Low microleakage, and composite shear bond strengths to enamel/dentin at the 18-MPa level, have been reported [53].

The last bonding system to be dealt with in this section, presented in Section II.B.8 as a luting agent, contains as the key monomer the addition product of HEMA to trimellitic acid anhydride, 4-(2-methacryloyloxyethoxycarbonyl)phthalic anhydride (4-META). Following early reports of excellent dentin–composite bonding results with 4-META–containing adhesives (tensile bond strengths typically 17 to 18 MPa), preeminently from Nakabayashi's group and reviewed by that researcher [54], the 4-META system has since been refined to the stage of commercialization and routine clinical use [7,55]. It typically comprises the following steps:

1. Short (10 to 30 s) pretreatment of prepared dentinal surface with the familiar citric acid–iron(III) chloride system (10% and 3%, respectively, in water)
2. Application of bonding resin, composed of 5% 4-META in MMA and premixed with the initiator, a partially oxidized tri-*n*-butylborane [56]
3. Overlaying of bonding resin coat with a thin layer of powdered poly(methyl methacrylate), followed by placement of composite

The acidic iron(III) chloride etchant, as pointed out before, removes the smear layer and acts as a decalcifying agent. In addition, just like 4-META itself, it appears to promote acrylate monomer penetration into the etched and partly demineralized dentinal surface. The interpenetrated bonding agent containing the hydrophilic–hydrophobic 4-META comonomer may be retained inside the demineralized zone by adsorption onto the hydrophilic and hydrophobic domains present in that zone so that, upon polymerization, a *hybrid* zone is generated, which consists of resin-reinforced dentinal matter capable of copolymerization with the adjacent overlay of composite restorative. Restricting the duration of the etching treatment to the short period indicated is a vital prerequisite for strong dentin–composite bond formation, as this will keep the depth of demineralization to less than 5 μm (ca. 2 μm in noncarious dentin) and maintain the collagen phase in a reactive (nondenatured) state, thus ensuring complete penetration of the demineralized stratum by the MMA/4-META agent down to the virgin (calcified) dentin matrix before polymerization sets in under the influence of the borane initiator. This, in turn, will ensure that no interlayer of decalcified and weakened dentinal material is left between virgin dentin and resin-impregnated stratum, as the exposed collagen, unprotected by infiltrated resin, is susceptible to degradation in an aqueous environment and thus would represent a weak link of the joint [55,57]. An outstanding advantage of the borane derivative as the initiator of this 4-META bonding system rests on its activation by water and oxygen as described by Nakabayashi et al. [57]. The moisture on the dental surfaces in combination with air triggers free-radical generation and thus the initiation of polymerization by the borane at the dentin interface rather than throughout the bulk of the resin layer as in other free-radical-initiated systems. This ensures that resin shrinkage proceeds toward the dentin adherend rather than away from it and so provides forceful counteraction against microleakage. In a further (commercialized) version, etching with citric acid–iron(III) chloride [containing poly(vinyl alcohol) for viscosity control] is followed by brush application of HEMA monomer (containing hydroquinone monomethyl ether), a subsequent application of the HEMA-4-META combination premixed with the tributylborane initiator, and the final placement of the restorative resin [58]. Excellent shear bond strength data, up to nearly 23 MPa, paired with a remarkably low degree of microleakage, have variously been reported [7,41,58,59], and fracture is cohesive in dentine and/or composite. The last-named adhesive system is also quite efficacious in prosthodontic and orthodontic bonding applications [62] and in the bonding of amalgam fillings, which in general practice, plugging into an undercut cavity, are retained solely by a micromechanical mode. Although dentin–amalgam shear bond strengths, just above 3 MPa, are weak in

Table 5   Representative Bond Strength Data for Present-Day Dentin Bonding Agents

Bonding agent[a]	Dentin conditioning	Composite–dentin shear bond strength (MPa)
Glutaraldehyde, HEMA	Al oxalate, glycine	10–18
NTG-GMA, PMDM	Al oxalate, diluted HNO_3	8–18
PMDM	NPG, diluted HNO_3	7–16
NTG-GMA, BPDM	SA-HEMA or diluted H_3PO_4	15–29
HEMA, BIS-GMA	HEMA, maleic acid	8–23
4-META, MMA	Fe(III) chloride, citric acid	10–23
4-META, HEMA	Fe(III) chloride, citric acid, then HEMA	9–22

[a]See Section V.B.3 and Table 6.
Source: Literature in period 1989–1993.

relation to corresponding dentin-composite strength data, the bond is effective in reducing microleakage appreciably in comparison to conventionally placed amalgam restorative.

Representative shear bond strength ranges for the bonding agents discussed in the foregoing are listed in Table 5, and the structural representations and universally used abbreviations for the principal methacrylate and dimethacrylate monomers are found in Tables 6 and 7. Detailed characterization techniques for methacrylates and derived polymers have been described by Ruyter and Øysaed [60].

Table 6 Structures and Abbreviations of Representative Monomethacrylate Monomers

Structure	Abbreviation
$CH_2=C(CH_3)-COOCH_3$	MMA
$CH_2=C(CH_3)-COO-CH_2CH_2-OH$	HEMA
$CH_2=C(CH_3)-COO-CH_2CH_2-OCO-CH_2CH_2-COOH$	SA-HEMA
$CH_2=C(CH_3)-COO-CH_2CH(OH)CH_2-N(C_6H_5)-CH_2-COOH$	NPG-GMA
$CH_2=C(CH_3)-COO-CH_2CH(OH)CH_2-N(C_6H_4CH_3)-CH_2-COOH$	NTG-GMA
$CH_2=C(CH_3)-COO-CH_2CH_2-OCO-C_6H_3(CO)_2O$	4-META

Table 7 Structures and Abbreviations of Representative Dimethacrylate Monomers

Structure	Abbreviation
$CH_2=C(CH_3)-COO-CH_2CH(OH)CH_2-O-C_6H_4-C(CH_3)_2-C_6H_4-O-CH_2CH(OH)CH_2-OCO-C(CH_3)=CH_2$	BIS-GMA
$CH_2=C(CH_3)-COO-C_6H_4-C(CH_3)_2-C_6H_4-OCO-C(CH_3)=CH_2$	BIS-MA
$CH_2=C(CH_3)-COO-CH_2CH_2CH_2-O-C_6H_4-C(CH_3)_2-C_6H_4-O-CH_2CH_2CH_2-OCO-C(CH_3)=CH_2$	BIS-PMA
$CH_2=C(CH_3)-COO-CH_2CH_2-O-CH_2CH_2-O-CH_2CH_2-OCO-C(CH_3)=CH_2$	TEGDMA
$CH_2=C(CH_3)-COO-CH_2CH_2-OCO-$ (benzene ring with COOH, HOCO, and $COO-CH_2CH_2-OCO-C(CH_3)=CH_2$ substituents)	PMDM
$CH_2=C(CH_3)-COO-CH_2CH_2-OCO-$ (biphenyl ring with HOCO, COOH, and $COO-CH_2CH_2-OCO-C(CH_3)=CH_2$ substituents)	BPDM

and positional isomer

VI. CONCLUSIONS

The foremost objective of operative dentistry is the durable placement of restoratives and the seating of restorations and prosthetic appliances with minimal loss of healthy tooth substance. With the realization that adhesion technology can be a powerful ally in this endeavor, advanced bonding techniques have in recent years been placed in ever-increasing numbers at the clinician's disposal in an effort to approach, and ultimately attain, this goal. Promising results are evident particularly in the design of bonding techniques permitting enhanced retention of composite restoratives to the enamel and dentinal phases of the tooth substance. Progress is also apparent in the development of adhesive systems allowing for the simplified and more efficacious attachment of bridges, inlays, onlays, and veneers to the tooth structure. Emphasis in future development work will focus less on the achievement of ever-greater bond strengths than on perfection of adhesion in terms of complete surface wetting, absence of interfacial microleakage with associated cariogenic factors, and enhanced durability of both the adhesive interface and the restorative' adherend.

ACKNOWLEDGMENTS

The authors are much indebted to Drs. W. W. Barkmeier, R. L. Bowen, R. L. Cooley, J. D. Eick, N. Nakabayashi, D. H. Retief, and B. I. Suh for helpful and informative correspondence. Thanks are also due to the numerous colleagues who provided reprints or preprints of their latest work, notably Drs. E. Asmussen, K. Hirota, K. Hotta, G. Øilo, J. F. Roulet, I. E. Ruyter, J. W. Stansbury, M. Suzuki, S. Takagi, and M. J. Tyas. Mrs. Mollie Pearmain is thanked for the proficient typing of the manuscript.

REFERENCES

1. J. F. McCabe, *Applied Dental Materials*, 7th ed., Blackwell, London, 1990.
2. H. J. Wilson, J. McLean, and D. Brown, *Dental Materials and Their Clinical Applications*, British Dental Association, London, 1988.
3. E. C. Combe, *Notes on Dental Materials*, 5th ed., Churchill Livingstone, Edinburgh, 1986.
4. J. G. Stannard and K. Kanchanatawewat, *J. Dent. Res. 69*: 209 (1990), Abstr. 804.
5. M. Nakayama, S. Utsumi, K. Inoue, and K. Suzuki, *J. Dent. Res. 69*: 127 (1990), Abstr. 150.
6. Y. Aboush and B. Jenkins, *J. Prosthet. Dent. 61*: 688 (1989).
7. R. L. Cooley, K. M. Burger, and M. C. Chain, *J. Esthet. Dent. 3*: 7 (1991).
8. R. L. Cooley, E. Y. Tseng, and J. G. Evans, *J. Esthet. Dent. 3*: 11 (1991).
9. W. Rux, R. L. Cooley and J. L. Hicks, *Quintessence Intern. 22*: 57 (1991).
10. W. W. Barkmeier, R. L. Looley, and C. J. Douville, *J. Dent. Res. 70*, 526, (1991), Abstr. 2076.
11. R. J. McConnell, D. R. Gratton, and T. Hafstede, *J. Dent. Res. 70*: 388 (1991), Abstr. 975.
12. L. Zardiackras, D. Givan, J. Fitchie, and L. Anderson, *J. Dent. Res. 70*: 391 (1991), Abstr. 1000.
13. T. E. Train and R. L. Cooley, *J. Dent. Res. 69*: 311 (1990), Abstr. 1617.
14. R. E. Kerby, *J. Dent. Res. 69*: 311 (1990), Abstr. 1624.
15. L. C. Chow, S. Takagi, P. D. Costatino, and C. D. Friedman, *Mater. Res. Soc. Symp. Proc. 179*: 3 (1991).
16. R. L. Cooley and J. W. Court, *J. Esthet. Dent. 2*: 114 (1990).
17. R. L. Cooley, V. A. Sandoval, and S. E. Barnwell, *Quintessence Intern. 19*: 899 (1988).
18. J. W. McLean, H. J. Prosser, and A. D. Wilson, *Brit. Dent. J. 158*: 410 (1985).
19. M. Irie, J. Tanaka, H. Nakai, K. Hirota, and K. Tomioka, *J. Dent. Res. 69*: 311 (1990), Abstr. 1620.
20. R. A. McCaghren, D. H. Retief, E. L. Bradley, and F. R. Denys, *J. Dent. Res. 69*: 40 (1990).
21. M. Suzuki and R. E. Jordan, *J. Am. Dent. Assoc. 120*: 55 (1990).
22. M. J. Tyas, *Current Opinion Dent. 2*: 137 (1992).
23. N. K. Sarkar, B. F. El Mallakh, and A. A. Kamar, *J. Dent. Res. 69*: 366 (1990), Abstr. 2061.
24. R. L. Cooley, J. W. Robbins, and S. Barnwell, *J. Prosthet. Dent. 64*: 651 (1990).
25. R. E. Jordan and M. Suzuki, *J. Am. Dent. Assoc. 122*: 31 (1991).
26. S. M. Collard, C. F. Liu, and C. D. Armeniades, *J. Dent. Res. 69*: 309 (1990), Abstr. 1603.

27. J. D. Eick, T. J. Byerley, R. P. Chappell, G. R. Chen, C. Q. Bowles, and C. C. Chappellow, *Dent. Mater. 9*: 123 (1993).

28. F. Millich, J. D. Eick, L. Jeang, and T. S. Byerley, *J. Polymer Sci. A: Polym. Chem. 31*: 1667 (1993).

29. J. D. Eick, S. J. Robinson, T. S. Byerley, and C. C. Chappellow, *Quintessence Intern. 24*: 632 (1993).

30. J. W. Stansbury, *J. Dent. Res. 70*: 527 (1991), Abstr. 2088. *J. Dent. Res. 71*: 239 (1992), Abstr. 1070.

31. H. W. Christie, C. C. Chappellow, T. J. Byerley, and J. D. Eick, *J. Dent. Res. 69*: 309 (1990). F. Millich, J. D. Eick, G. P. Chen, T. J. Byerley, and E. W. Hellmuth, *J. Polymer Sci. B: Polym. Phys. 31*: 729 (1993).

32. H. E. Strassler, J. M. Antonucci, and J. Marsh, *J. Dent. Res. 69*: 232 (1990), Abstr. 987.

33. D. H. Retief, *Am. J. Dent. 4*: 231 (1991).

34. S. M. Aasen, J. D. Oxman, and F. A. Ubel, *J. Dent. Res. 69*: 230 (1990), Abstr. 974.

35. T. Munechika, K. Suzuki, M. Nishiyama, M. Ohashi, and K. Horie, *J. Dent. Res. 63*: 1079 (1984).

36. A. J. Gwinnett, *Intern. Dent. J. 38*: 91 (1988).

37. D. H. Retief, *Operative Dent. 12*: 140 (1987).

38. R. L. Bowen, *J. Dent. Res. 44*: 895, 903, 906, 1369 (1965).

39. R. L. Bowen and W. A. Marjenhoff, *J. Esthet. Dent. 3*: 86 (1991).

40. E. C. Munksgaard and E. Asmussen, *J. Dent. Res. 63*: 1087 (1984).

41. R. P. Chappell, J. D. Eick, J. M. Mixson, and F. C. Theisen, *Quintessence Intern. 21*: 303 (1990). J. D. Eick, S. J. Robinson, R. P. Chappell, C. M. Cobb, and P. Spencer, *ibid. 24*: 571 (1993).

42. P. A. De Aranjo and E. Asmussen, *Intern. Dent. J. 39*: 253 (1989).

43. S. Uno and E. Asmussen, *Acta Odontol. Scand. 49*: 297 (1991).

44. S. E. Strickland, D. H. Retief, R. S. Mandras, and C. M. Russell, *J. Dent. Res. 70* (1991) 396, Abstr. No. 1043.

45. J. D. Eick, S. J. Robinson, C. M. Cobb, R. P. Chappell, and P. Spencer, *Quintessence Intern. 23*: 43 (1992).

46. W. W. Barkmeier and R. L. Cooley, *Am. J. Dent. 2*: 263 (1989).

47. W. W. Barkmeier, C.-T. Huang, P. D. Hammesfahr, and S. R. Jefferies, *J. Esthet. Dent. 2*: 134 (1990).

48. D. H. Retief, *J. Esthet. Dent. 3*: 106 (1991).

49. A. J. L. Carracho, R. P. Chappell, A. G. Glaros, J. H. Purk, and J. D. Eick, *Quintessence Intern. 22*: 745 (1991).

50. W. W. Barkmeier, B. I. Suh, and R. L. Cooley, *J. Esthet. Dent. 3*: 148 (1991).

51. B. I. Suh and F. A. Cincione, *Esthet. Dent. Update 3*: 61 (1992).

52. D. F. Rigsby, D. H. Retief, C. M. Russell, and F. R. Denys, *Am. J. Dent. 3*: 289 (1990).

53. J. Kanca, *J. Dent. Res. 69*: 231 (1990), Abstr. 984.

54. N. Nakabayashi, *CRC Crit. Rev. Biocompatibility 1*: 25 (1984); *Multiphase Biomedical Materials* (T. Tsuruta and A. Nakajima, eds.), Utrecht, The Netherlands, 1989, Chap. 6.

55. N. Nakabayashi, M. Ashizawa, and M. Nakamura, *Quintessence Intern. 23*: 135 (1992).

56. N. Nakabayashi and E. Masuhara, *J. Biomet. Mater. Res. 12*: 149 (1978).

57. N. Nakabayashi, M. Nakamura, and N. Yasuda, *J. Esthet. Dent. 3*: 33 (1991).

58. R. L. Cooley, E. Y. Tseng, and W. W. Barkmeier, *Quintessence Intern. 22*: 979 (1991).

59. R. P. Chappell, J. D. Eick, F. C. Theisen, and A. J. L. Carracho, *Quintessence Intern. 22*: 831 (1991).

60. I. E. Ruyter and H. Øysaed, *CRC Crit. Rev. Biocompatibility 4*: 247 (1988).

61. J. F. Roulet, T. Reich, U. Blunck, and M. Noack, *Scanning Microsc. 3*: 147 (1989); see also A. J. E. Qualtrough, A. Cramer, N. H. F. Wilson, J. F. Roulet, and M. Noack, *Intern. J. Prosthodont. 6*: 517 (1991).

62. K. Hotta, M. Mogi, F. Miura, and N. Nakabayashi, *Dent. Mater. 8*: 173 (1992). K. Hotta, J. Jpn. *Orthodont, Soc. 52*: 360 (1993).

40

Adhesives in the Automotive Industry

Eckhard H. Cordes *Mercedes-Benz AG, Bremen, Germany*

I. INTRODUCTION

Adhesive bonding and sealing are used for various applications in the modern automotive industry, ranging from flexible car body sealings to high-performance structural adhesives (Fig. 1). Adhesive types with specific properties are available for miscellaneous processing. The requirements for adhesive bonds are increased due to the extended life of the car. In adhesive processing, industrial health and environmental protection aspects are more and more important. Therefore, it is more difficult but nevertheless necessary to determine requirements for the adhesives to be used in the future. In addition, the demand for quality standards requiring better quality management is increasing.

II. ADHESIVE APPLICATIONS IN THE AUTOMOTIVE INDUSTRY

In this chapter, adhesive bonding and sealing in automobile production are subdivided schematically into five ranges of application: (1) mechanical parts production, (2) the body shop, (3) the paint shop, (4) the assembly shop, and (5) the manufacturing of components. Depending on the variety of applications, adhesives must satisfy a wide range of requirements. On principle, all body shop adhesives must be usable without risk to the paint shop and they must resist the high temperature of the paint bake ovens. Generally, the bond strength and/or sealing ability must perform under severe conditions for the life of the car. Further requirements depend on:

1. *Function of the material* (e.g., spot-weld sealants): good corrosion protection, weldability, no HCl or chlorine emitted to cause corrosion when overbaked, good adhesion on the substrates
2. *Processing technique:* manual or automatic application, bonding at the assembly line or at a separate working site
3. *Specific material characteristics* (e.g., moisture and/or hot-curing adhesive): curing time, stability in storage, flexibility at low temperatures, hydrolytic stability, aging resistance, adhesion properties

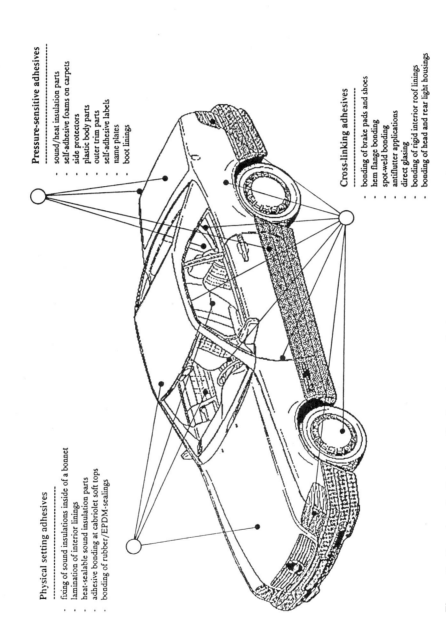

Pressure-sensitive adhesives
- sound/heat insulation parts
- self-adhesive foams on carpets
- side protectors
- plastic body parts
- outer trim parts
- self-adhesive labels
- name plates
- boot linings

Cross-linking adhesives
- bonding of brake pads and shoes
- hem flange bonding
- spot-weld bonding
- antiflutter applications
- direct glasing
- bonding of rigid interior roof linings
- bonding of head and rear light housings

Physical setting adhesives
- fixing of sound insulations inside of a bonnet
- lamination of interior linings
- heat-sealable sound insulation parts
- adhesive bonding at cabriolet soft tops
- bonding of rubber/EPDM-sealings

Figure 1 Samples of adhesive applications in car production.

For all ranges of application the adhesives must not only meet the functional requirements but also retain them under a wide variety of conditions during use: impacts, vibrations, climate conditions, extreme changes in temperature, corrosion, and so on.

A. Adhesives for Mechanical Applications

In this range of applications, adhesives are used for fastener locking, formed-in-place gaskets, and bonding of mechanical parts. Mainly cyanoacrylates, anaerobic and encapsulated adhesives, modified acrylates, and elastomer or resin-based compounds are applied. Examples of applications are listed in Table 1. To choose a suitable adhesive, the required strength and mechanical properties as well as the chemical conditions for the specific application should be well known. For example, for the curing of dimethacrylates, the catalytic effect of the glued surface, the absence of oxygen, the temperature, the mold of the bond line, and the type of material to be bonded are important. Depending on the type of bonding, the requirements differ. The bond strength, temperature and aging stability, and the resistance to chemical reagents must fit the application load. The specific conditions of the joint performance have to be taken into consideration.

Table 1 Adhesive and Sealant Applications in Mechanical Parts Production

Type of adhesive	Method of curing	Applications
Anaerobic adhesives	Absence of oxygen and metal contact	Gaskets Flat surface bonding Adhesive bonding of electric and electronical components Fastener locking Shaft/hub bonding
Cyanoacrylates	Moisture	Gaskets Thread sealing Flat surface bonding Bonding of caps in cylinder head covers, gearboxes, crankcases, axle housings Shaft/hub bonding
Epoxy resin adhesive foils	Heat treatment	Bonding of heat exchanger
Synthetic resin sealants	Solvent evaporation	Gaskets Common sealing
Encapsulated adhesives	Anaerobic or with hardener after bursting of the capsules	Fastener locking
Modified acrylates	Activator and absence of oxygen	Bonding of flat surfaces
Phenolic adhesive foils	Heat treatment	Bonding of brake straps Bonding of clutch and brake linings
Silicone rubbers	Moisture or hardener	Sealing of oil pans and housing covers

For adhesive bonding of plastics to plastics and plastics to metal the cyanoacrylates are usually better than anaerobic compounds, which are more suitable for metal-to-metal bonding because of their greater resistance to mechanical vibrations and impacts. Encapsulated adhesives can be used to coat on the fastener by the supplier. The curing takes place after fastening and locking are done. Figure 2 shows the great variety of available coated fasteners. Formed-in-place gaskets and adhesive sealants are used in various mechanical applications to seal and bond surfaces. There are cyanoacrylates, anaerobic adhesives, and modified acrylates and solvent-based rubber or resin compounds as well as silicones. The products are applied manually or automatically on the surfaces just prior to assembly. Figure 3 shows a sealant application extruded automatically onto an oil pan flange.

Figure 2 Examples of fasteners coated with encapsulated adhesives. (Courtesy of Loctite Deutschland GmbH.)

Figure 3   Application of a sealant automatically extruded onto an oil pan flange.

B.   Adhesive Applications in the Body Shop

There are adhesives and sealants in the body shop with basically four different functions (Table 2):

1.   Sealants for body joints
2.   Spot-welding sealants and tapes
3.   Antiflutter bonding
4.   Structural adhesive bonding and hem flange sealing

Examples of these applications are shown in Figs. 4 and 5. The sealants for body joints are applied after assembly. They are extruded over the welded joints and have to seal out dust

Table 2 Main Adhesive Bonding and Sealing Applications in the Body Shop

Shear strength range (MPa)	Application			
	Structural adhesive bonding and hem flange sealing	Antiflutter bonding	Body joint sealing	Spot-weld sealing
30 ⋮ 15	Epoxies			
⋮ 7	Polymer blends Polyurethanes	Polyurethanes	Polyurethanes	
⋮ 4	Acrylic plastisols	PVC plastisols Acrylic plastisols	PVC plastisols Acrylic plastisols	Acrylic plastisols
⋮ 2 ⋮ 0		Reactive butyls	Reactive butyls Butyls Nonsetting rubber compounds	Reactive butyls Butyls Nonsetting rubber compounds

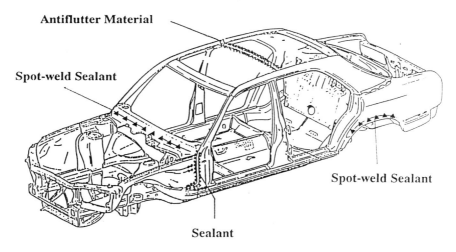

Antiflutter Material

Spot-weld Sealant

Spot-weld Sealant

Sealant

Figure 4 Examples of adhesive and sealant applications in the body shop.

and water and avoid corrosion. A typical application is shown in Fig. 6. The following materials are in use:

1. Moisture and/or heat curing one-component polyurethanes
2. PVC plastisols
3. Pregelling compounds based on synthetic rubber
4. Butyls

Rear Door Panels

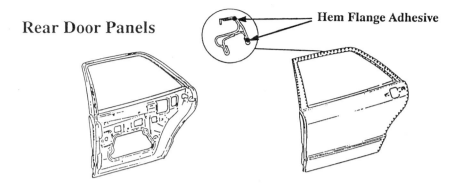

Hem Flange Adhesive

Figure 5 Application of a hem flange adhesive in the body shop.

Figure 6 Example of a sealant application for body joints in the body shop.

Figure 7 Extrusion of a spot-weld sealant on the inner side of a fender.

Spot-weld sealants and tapes are used in spot-welded flanges to protect against corrosion. They are applied to the flanges before joining, then the body parts are pressed together and spot-welded. Figure 4 shows the application points on a body shop car schematically. The application of spot-weld sealing on the flange between the front section and the lower windshield and on the inner flange of the rear fender can be seen as an example. The extrusion of a spot-weld sealant on the inner side of the fender is shown in Fig. 7. In use are:

1. PVC plastisols
2. Acrylic plastisols
3. Warm-applied butyls
4. Butyl tapes
5. Rubber-based pastes

The solvent-based compounds used formerly are no longer used.

Outer car body panels are stiffened with reinforcements to avoid flutter and the so-called "oil can" effect. In this way the strength of the body component is also improved. In Fig. 4 the use of an expandable butyl as antiflutter material is shown. Figure 8 gives an example of use of an intermediate layer for antiflutter bonding on a hood. The following materials are used:

1. Bituminous or acrylic-soaked foams
2. Warm-applied butyls
3. Vulcanizable expandable butyls
4. Hot-curing one-component polyurethanes

Metal-to-metal adhesives are used to bond and seal hem flanges as well as for structural bonding of body shop components. As an example, Fig. 5 shows the adhesive bonding of inner and outer door panels. In Fig. 9 the robotized application of an adhesive on a rotary table can be seen. The adhesive bonding of hem flanges enables a homogeneous stress distribution along the bond line, through which the stiffness of the bonded component is better than in a conventional spot-welded joint. Often, adhesive bonding is combined with spot welding, which provides some advantages: the adhesive can replace a large number of spot-welding points, which reduces expensive surface finishing at outer panels; the components can be handled immediately after joining, before the adhesive is cured; and improved strength is achieved. Moreover, the spot-welding points hinder the attack of peel forces, which is harmful to the bond line. Instead of spot welding, other joining techniques (rivet fastening, screw fastening, "clinchen," "toxen", etc.) can be used in combination with the adhesive bonding.

Corrosion protection is often mentioned as a principal advantage of adhesive flange bonding, but today, coated sheet metal and aluminum are used more and more in the body shop, so this advantage is no longer the primary one. With the increasing use of coated sheet metal, the adhesive choice becomes more important. The bond strength is poor if the

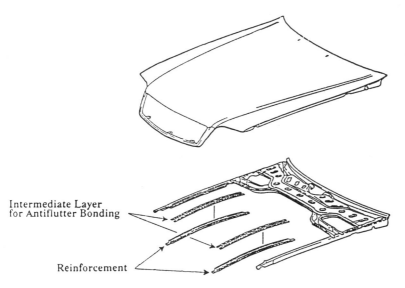

Intermediate Layer
for Antiflutter Bonding

Reinforcement

Figure 8 Sketch of inner and outer bonnet panels.

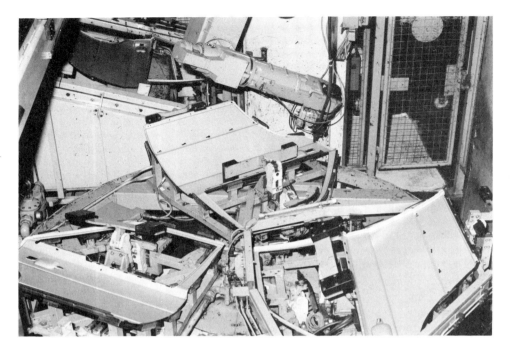

Figure 9 Robotized application of a hem flange adhesive on a rotary table.

adhesive used is not adapted to the particular properties of the coatings. For adhesive metal bonding the following compounds are in use:

1. PVC or acrylic plastisols if no higher strength is required
2. One-component hot-curing or two-component cold-setting polyurethanes with medium strength
3. Epoxy-based adhesives, hot or cold curing, as one- or two-component pastes or as adhesive films for structural bonding with significant loads
4. Polyurethane/epoxy-based polymer blends (so-called ''toughened'' epoxies)

The conventional epoxy–resin adhesives give good sheer strength results, but they show brittle fracture under impact loads, particularly in cold conditions. Polyurethane compounds are more flexible and tougher, but they give lower strength in adhesive bondings. Adhesive applications in the body shop had usually been confined to low-loaded components. Nowadays adhesive bonding is also used more and more for parts that have to transmit significant structural loads (e.g., chassis components, floor panels, and side rails). The deformation ability of the adhesive-bonded components must be high to absorb the impact energy to give the car body good crash behavior. That requires a toughened adhesive with as high a strength as possible and was the reason for the development of polymer blends.

Worthwhile mentioning also are the adhesive applications in car body manufacturing, where the traditional sheet steel construction is replaced by steel or light metal space frames with plastic exterior body panels. A growing demand exists for the use of exterior body components made of plastic or metal–plastic composites, as doors, tailgates, trunk lids, hoods, roofs, and so on. Using these components to build a car body has forced the manufacturing process to be altered from the conventional flow of manufacture. That has influenced adhesive processing and caused modifications in the adhesives. Polyurethane-based adhesives, which are one-component moisture-curing or two-component cold-setting compounds, are generally used in these applications.

C. Adhesives and Sealants in the Paint Shop

Apart from underbody coating (usually PVC based), which is not explained further, the majority of bonding and sealing products in the paint shop are also PVC compounds. Acrylat plastisols are not often used and polyurethane-based sealants are rarely found. The main applications are in seam sealing and antiflutter bonding. On a small scale there are adhesive applications to bond and seal caps and to fix sound deadeners. In addition, foams and butyls as well as bituminous or acrylic-based sealants are used to fill car body holes.

D. Adhesive and Sealant Applications in the Assembly Shop

There are lots of adhesive applications in the assembly shop and in the manufacture of components. A complete list is not given, but adhesive use is illustrated by examples. The larger quantities of adhesives used on the assembly line are for bonding of insulation pads, interior fittings, instrument panels, and roof modules, and for direct glazing. When the applications do not require a surface coating adhesive or when smaller assembly pads are to be bonded or a droplike or continuous extrusion of the adhesive is sufficient, hot melts can be used with advantage. When higher strength is not necessary, ethylene–vinyl acetate (EVA), polyamide (PA), or thermoplastic rubber compounds can be chosen. Otherwise, reactive hot-melt adhesives would be preferred. They can be applied as common hot melts, but they are cross-linking afterward and therefore provide bond strength like that of two-part urethane adhesives and good durability at higher temperatures. In some cases, adhesive bonding with cyanoacrylates or methacrylates, which cure in a few seconds, is used only as a temporary fixative to assist assembly.

When joining is required over the entire surface, pressure-sensitive adhesives, adhesive tape systems, or hot-melt adhesive foils can be used. For health reasons, conventional rubber-based solvent cements are used rarely today. There is a trend to replace them with hot melts or water-based adhesives. With new application techniques the hot melts can be coated on substrates like solvent-based adhesives. Water-based adhesive systems frequently have the disadvantages of a longer open time and insufficient tack. An additional mechanical fixing is often necessary. The adhesive bonding of interior roof linings is an example of the use of water-based instead of solvent-based adhesives.

The primary sealants in common use in the assembly shop (e.g., for sealing of assembled air-conditioning systems or air filters) are:

1. Polyisobutylene-based compounds
2. Butyls
3. Moisture-curing one-component polyurethane adhesives

The adhesive bonding of plastic assembly components such as instrument panels, spoilers, spare wheel boxes, roof parts, trim assembles, and fenders is generally with one- or two-component polyurethanes. Frequently, a primer is used as pretreatment to improve adhesion. Adhesive tape systems can provide good results for the bonding of dash panels, trim lines, insignia parts, and rear view mirrors, for example.

Today the use of polyurethane adhesives is a common practice for adhesive bonding of windows in a car body, called direct glazing. Compared to the former glazing technique using rubber seals or polysulfide materials, direct glazing has the following significant advantages:

1. Possibility of completely automatic application (see Fig. 10)
2. High-performance sealing, matching the safety standards
3. Higher body strength
4. Smoothly designed car bodies
5. Improved aerodynamics

In addition, with direct glazing windows can be used as design and engineering elements of a car body (flush glazing). One- or two-component adhesives are used, which can be applied either warm or at room temperature and which are moisture- or hot-curing or curing with hardener. One-component moisture-curing compounds are very common. Generally, the bonding process includes pretreatments using specific cleaner and primer for both the glass surface and the car body flange. To protect the adhesive joint against ultraviolet rays, ceramic silk-screen printing on the glass and a black glass primer are used. New developments utilize the requirement primerless direct glazing. As mentioned above, the complete process of direct glazing can be performed fully automatically. Figure 10 shows a sketch as an example of such manufacturing equipment. In Fig. 11 a robotized extrusion onto a windshield is shown. The robot is holding the windshield and leads it along a stationary swiveling nozzle, and after the adhesive is applied, puts it in the body opening (see Fig. 12). The accuracy in fitting is controlled by sensors.

E. Adhesive Applications in Component Manufacturing

The use of adhesives in component manufacturing ranges from automotive headlamps to plastic body components (e.g., hoods, tailgates) to interior fittings to cabriolet soft tops, including a wide range of adhesives employed. Table 3 lists applications without any claim to being complete. Looking at the plastic components it is obvious that there are many different types of polymers, but the adhesives selected are basically polyurethane and epoxy-based compounds. The latter are rarely used. The main difference among them is in the way they are formulated: one or two components, cold setting or curing at higher temperature, liquid or paste, and so on. Because of the easy processing, the newly developed two-part acrylic adhesives, which are applied in a no-mix formulation, are very interesting. The A component is applied to one side and the B component to the other side of the surfaces to be bonded. After being fixed together, the adhesive cross-links in a few minutes. Bondings manufactured using this type of adhesive show good shear and peel strength results and high durability at impact loads. Applications include the bonding of protection plates to the sill beam or the joining of exterior lighting housings.

In cabriolet-cover manufacturing, solvent adhesives, synthetic rubber- or polyurethane-based, are used for sealing the folding top seams and for bonding the soft cover to the hood linkage. Adhesives are often mixed with hardener to improve the heat resistance of the bonding. Hot-melt adhesive foils are also employed. Preformed butyls or

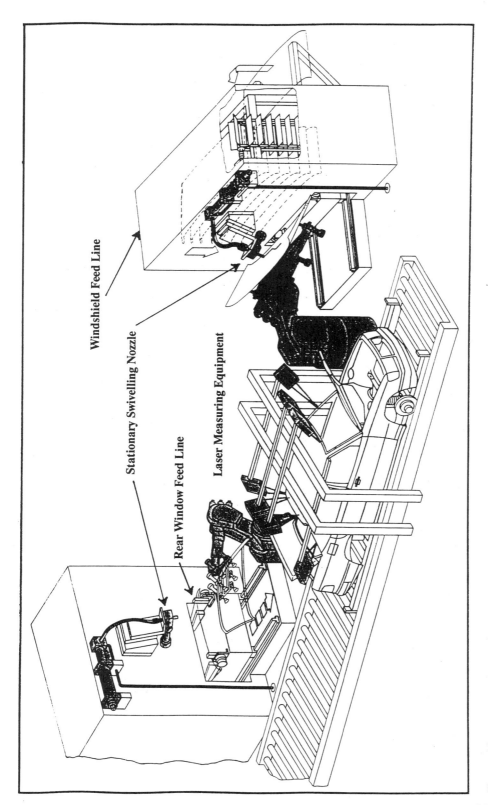

Windshield Feed Line

Stationary Swivelling Nozzle

Rear Window Feed Line

Laser Measuring Equipment

Figure 10 Sketch of automatic manufacturing equipment for direct glazing.

669

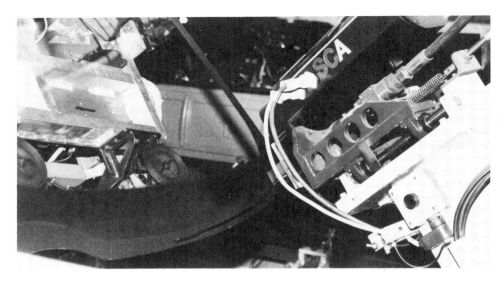

Figure 11 Robotized extrusion of a polyurethane adhesive onto a windshield.

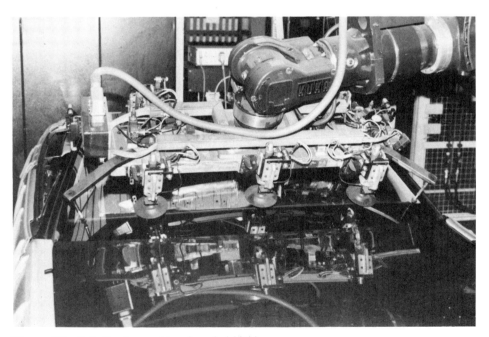

Figure 12 Robotized assembly of a windshield.

butyl sealant pastes or moisture-curing polyurethane adhesives are used for additional sealing.

Many interior fittings (e.g., instrument panel, door and sidewall panels, package trays, seat linings, boot carpetings, rear window shelves, etc.) are often manufactured by vacuum drawing and laminating the cover sheet material (leatherette, textile, leather, etc.)

Table 3 Adhesive Bonding and Sealing Applications in the Assembly Shop and in Components Manufacturing

Type of adhesive	Method of curing	Applications
Anaerobic adhesives	Absence of oxygen and metal contact	Fastener locking Flat surface bonding
Cyanoacrylates	Moisture	Gaskets Flat surface bonding Metal-to-metal bonding Metal-to-plastic bonding Plastic-to-plastic bonding Adhesive bonding or rubber and EPDM parts Adhesive bonding to assist assembly
Adhesive dispersion	Drying and/or heat supply, pressure	Laminating at interior fittings Adhesive joining of seat linings Adhesive bonding of headliners

Adhesive tape systems	No curing, joining under heat supply and pressure	Bonding of:
		  Moldings
		  Protective strips
		  Name plates
		  Pattern parts
		  Mirrors
		  Rubber seals
		  Wheelhouse covers
		  Flared wheel arches
		  Chassis beam panels
		  Insulating parts
		  Draught excludors
		  Reflectors

Epoxy resin adhesives	Heat supply or hardener	Radiators
Rubber sealant	Solvent evaporation	Common sealing in the assembly shop
Solvent-based cements	Solvent evaporation	Sundry (no further data, not state of the art)
Encapsulated adhesives	Anaerobic or with hardener after bursting of the capsules	Fastener locking
Modified acrylates	Activator	Bonding of flat surfaces Bonding of lighting housings Glass–metal bonding (e.g., quarter-window adjuster, rear view mirror)

Polyurethanes	Moisture or hardener	Assembly bonding of plastic components:
		  Spoiler
		  Side protectors
		  Spare wheel compartments
		  Fender
		  Instrument consoles

Table 3 *continued*

Type of adhesive	Method of curing	Applications

Hot-melt adhesives (also cross-linking)	Cool down (moisture)	Air duct systems
		Window guiderails
		Window lifter rails
		Roofs and sun roofs
		Rigid roof linings
		Bonding of plastic components:
		  Bonnets
		  Tailgates
		  Multipiece spoiler
		  Impact protection parts
		  Bumper
		  Heating and ventilation systems
		  Seat buckets
		  Backrest linings
		  Head and rear light housings
		Direct glazing
		Adhesive bonding and sealing at
		  cabriolet hard and soft tops
		Bonding of wiring harnesses
		Sealing of radiators
		Rear view mirrors
		Laminating at interior fittings
		Assembly bonding of moldings
		Bonding of headlight lenses
		Adhesive bonding of sound
		  systems
		Antiflutter bonding
		Bonding of insulation pads and
		  sound deadeners
		Adhesive bonding at filters and
		  filter housings, heating and
		  ventilation channels
		Bonding of insignia parts at
		  wheel caps
		Bonding of brackets at interior
		  door panels
		Adhesive bonding and sealing at
		  cabriolet hoods

onto the trim panel. The use of hot-melt (also cross-linking) adhesives and water-based polyurethane adhesives (with hardener) is state of the art for this type of component production.

III. Some Considerations Regarding Trends in Automotive Adhesive Bonding

For the future the evolution of adhesive bonding as a joining technique in automobile production points in two directions. On the one hand, well-known applications have to be optimized and improved to make them cost-effective but nevertheless reliable and trustful processes enjoying increasing acceptance for adhesive bonding. On the other hand, there will be new applications with different adhesive requirements, and adhesive suppliers must anticipate these changes and develop compatible adhesive compounds to satisfy the new requirements.

As to the first point mentioned above, cost-cutting steps have to be taken seriously. Increasing automation of the adhesive application is imperative. Adhesive bonding processes without extra pretreatment of joint surfaces and without using a primer but with reliable efficiency are required. Pregelling of body shop adhesives will be eliminated and oven temperatures for adhesive curing will be lowered to reduce energy costs. Increased use of reactive hot melts is conceivable. Multifunctional adhesives will be welcome: for example, hem flange adhesive bonding and seam sealing with only one material in one procedure. For ecological and personnel safety reasons, the use of harmful adhesives (e.g., solvent-based cements) will be reduced. Costs for toxic waste disposal, exhauster, reheat, or solvent recovery equipment will be reduced.

New applications of adhesive bonding can be expected where the specific advantages of this joining technique will be usable. Due to the lightweight construction, which will be more and more important, outside panels must be used as supporting parts of the body structure. Conventional sheet steel constructions often show welded joints at the visible outer skin of the car body, which should be avoided in a smooth aerodynamic body design. In hybrid constructions different materials must be bonded. For both techniques adhesive bonding is preferred to welding or soldering. Adhesive bonding can be also be combined with two new joining techniques, *clinchen* and *toxen*.

Structural adhesive bonding processes could be transferred from the body shop into the assembly shop to get clean and better defined glue surfaces. Temperature loadings to glue joints in paint bake ovens could be dropped, which would be an additional advantage. Components could be manufactured in a subsystem production process and adhesive bonded to the car body in the assembly shop. Adhesive bonding processes separate from the assembly line, performed at special working sites with specific adhesive equipment, would have advantages.

Recycling aspects will get more attention. Components should be recoverable and the adhesives applied must not disturb the reprocessing. New developments are being developed to manufacture laminated interior fittings, in which coverings and form substrates are made of the same or similar materials, so reprocessing can be done without prior delaminating of the layers. In this case the adhesives used had to fit with the substrate materials. The future number of adhesive bonding applications in the automotive industry will depend on the success of the adhesive bonding processes. The quality and the safety reproducibility, especially of high-performance structural adhesive bondings, will be more and more important for large-scale productions. A quality system including planning and surveillance should support these requirements.

BIBLIOGRAPHY

Achatz, D., and G. Kötting, *Sonderdokumentationsreihe 13* (1991).

Altenfeld, F., *Preprints, Euradh '92*, Dechema, Frankfurt am Main, Germany, 1992.

Altmann, O., *Adhaesion 3* (1991).

Becher, P., *Conference Proc., Eurobond '91*, Network Hagenburg, 1991.

Bistac, S., M. F. Vallat, and J. Schultz, *Preprints, Euradh '92*, Dechema, Frankfurt am Main, Germany, 1992.

Blank, N., and H. Schenkel, *Conference Proc., Eurobond '91*, Network Hagenburg, 1991.

Brockmann, W., and M. Bremont, *Preprints, Euradh '92*, Dechema, Frankfurt am Main, Germany, 1992

Burchardt, B., *Konstr. Elektron. 25* (1989).

Cordes, E. H., *Congress Proc. Swiss Bonding 1991*, Verlag IKD, Bietigheim-Bissingen, Germany, 1991.

Cordes, E. H., *Veranstaltung T-30-307-056-3*, Haus der Technik, Essen, Germany, 1993.

Cordes, E. H., and K. Bettenhausen, *Conference Proc. Eurobond '91*, Network Hagenburg, 1991.

Cordes, E. H., and K. Bettenhausen, *Kunststoffe 11* (1991).

Cordes, E. H., and B. Voigt, *Dechema Monogr. 119* (1990).

Cordes, E. H., and B. Voigt, *Preprints, Euradh '92*, Dechema, Frankfurt am Main, Germany, 1992.

Daniels, J., *Intern. J. Adhesion Adhesives 4* (1) (1984).

Dorn, L., *Conference Proc. Eurobond '91*, Network Hagenburg, 1991.

Engeldinger, H. K., *Adhaesion 4*, (1992).

Feinle, H., *Neue Klebetechniken im Fahrzeugbau*, PV-Weiterbildung Mercedes-Benz, Unter türkheim, Germany, 1990.

Gotthelf, H., *Veranstaltung T-30-228-056-2*, Haus der Technik, Essen, Germany, 1992.

Haberer, C., *Adhesive Bonding of Plastics to Metal in the Automotive Industry, Science and Technology*, Industry–University Short Course Program, CEI Europe, 1984, Chap. 11.

Hahn, O., and B. Motzko, *Preprints, Euradh '92*, Dechema, Frankfurt am Main, Germany, 1992.

Hälg, P., *Adhaesion 4* (1991).

Hennemann, O.-D., and Groß, A., *Adhaesion 4* (1992).

Hirthammer, M., and M. Schumann, *Preprints, Euradh '92*, Dechema, Frankfurt am Main, Germany, 1992.

Hussain, A., and C. Pflugbeil, *Conference Proc., Eurobond '91*, Network Hagenburg, 1991.

Jud, K., and U. Rempfler, *Preprints, Euradh '92*, Dechema, Frankfurt am Main, Germany, 1992.

Kleinert, H., P. Pochert, and C. Bär, *Conference Proc., Eurobond '91*, Network Hagenburg, 1991.

Kohl, M., and M. Krebs, *Preprints, Euradh '92*, Dechema, Frankfurt am Main, Germany, 1992.

Kötting, G., *Veranstaltung T-30-228-056-2*, Haus der Technik, Essen, Germany, 1991.

Kötting, G., *Sonderdokumentationsreihe 13* (1992).

Kötting, G., and M. Friederich, *Preprints, Euradh '92*, Dechema, Frankfurt am Main, Germany, 1992.

Krebs, M., *Sonderdokumentationsreihe 13* (1991).

Lehmann, H., R. Moser, and K. Mechera, *Conference Proc., Eurobond '91*, Network Hagenburg, 1991.

Lohse, H., and W. Meier, *Preprints, Euradh '92*, Dechema, Frankfurt am Main Germany, 1992.

Meier, W., and G. Cordes, *Sonderdokumentationsreihe 13* (1991).

Meyer, H. R., *Sonderdokumentationsreihe 13* (1991).

Müllenberg, L., *Conference Proc., Eurobond '91*, Network Hagenburg, 1991.

N. N., *Der Loctite, 1992/93*, Loctite Deutschland

Nobis, H., *Sonderdokumentationsreihe 13* (1991).

Pröbster, M., *Conference Proc., Eurobond '91*, Network Hagenburg, 1991.

Rademacher, D., *Sonderdokumentationsreihe 13* (1991).

Rohrer, P., *Conference Proc., Eurobond '91*, Network Hagenburg, 1991.

von Voithenberg, H., *Preprints Euradh '92*, Dechema, Frankfurt am Main, Germany, 1992.

Wesch, K., *Bonding, Coating and Sealing: PVC and Possible Alternatives*, Teroson, Heidelberg, Germany, 1992.

Yun Feng Chang, *Conference Proc., Eurobond '91*, Network Hagenburg, 1991.

Index